McGRAW-HILL SERIES IN CHEMICAL ENGINEERING
SIDNEY D. KIRKPATRICK, *Consulting Editor*

EDITORIAL ADVISORY COMMITTEE

MANSON BENEDICT · Professor of Chemical Engineering, Massachusetts Institute of Technology
CHARLES F. BONILLA · Professor of Chemical Engineering, Columbia University
JOHN R. CALLAHAM · Editor, *Chemical Engineering*
HARRY A. CURTIS · Commissioner, Tennessee Valley Authority
J. V. N. DORR · Chairman, The Dorr Company
A. W. HIXSON · Professor Emeritus of Chemical Engineering, Columbia University
H. FRASER JOHNSTONE · Chairman, Division of Chemical Engineering, University of Illinois
WEBSTER N. JONES · Vice President, Carnegie Institute of Technology
DONALD L. KATZ · Chairman, Department of Chemical and Metallurgical Engineering, University of Michigan
W. K. LEWIS · Professor Emeritus of Chemical Engineering, Massachusetts Institute of Technology

R. S. McBRIDE · Consulting Chemical Engineer
H. C. PARMELEE · Editor Emeritus, *Engineering and Mining Journal*
ROBERT L. PIGFORD · Chairman, Department of Chemical Engineering, University of Delaware
MOTT SOUDERS · Associate Director of Research, Shell Development Company
E. R. WEIDLEIN · President, Mellon Institute of Industrial Research
M. C. WHITAKER · Director, American Cyanamid Company
WALTER G. WHITMAN · Chairman, Department of Chemical Engineering, Massachusetts Institute of Technology
RICHARD H. WILHELM · Chairman, Department of Chemical Engineering, Princeton University

---

BUILDING FOR THE FUTURE OF A PROFESSION

Fifteen prominent chemical engineers first met in New York more than thirty years ago to plan a continuing literature for their rapidly growing profession. From industry came such pioneer practitioners as Leo H. Baekeland, Arthur D. Little, Charles L. Reese, John V. N. Dorr, M. C. Whitaker, and R. S. McBride. From the universities came such eminent educators as William H. Walker, Alfred H. White, D. D. Jackson, J. H. James, J. F. Norris, Warren K. Lewis, and Harry A. Curtis, H. C. Parmelee, then editor of *Chemical & Metallurgical Engineering*, served as chairman and was joined subsequently by S. D. Kirkpatrick as consulting editor.

After several meetings, this Editorial Advisory Committee submitted its report to the McGraw-Hill Book Company in September, 1925. In it were detailed specifications for a correlated series of more than a dozen text and reference books, including a chemical engineers' handbook and basic textbooks on the elements and principles of chemical engineering, on industrial applications of chemical synthesis, on materials of construction, on plant design, on chemical-engineering economics. Broadly outlined, too, were plans for monographs on unit operations and processes and on other industrial subjects to be developed as the need became apparent.

From this prophetic beginning has since come the McGraw-Hill Series in Chemical Engineering, which now numbers about thirty-five books. More are always in preparation to meet the ever-growing needs of chemical engineers in education and in industry. In the aggregate these books represent the work of literally hundreds of authors, editors, and collaborators. · But no small measure of credit is due the pioneering members of the original committee and those engineering educators and industrialists who have succeeded them in the task of building a permanent literature for the classical engineering profession.

THE SERIES

ARIES AND NEWTON—*Chemical Engineering Cost Estimation*
BADGER AND BANCHERO—*Introduction to Chemical Engineering*
CLARKE—*Manual for Process Engineering Calculations*
COMINGS—*High Pressure Technology*
COULSON AND RICHARDSON—*Chemical Engineering, Vols. 1 and 2*
DODGE—*Chemical Engineering Thermodynamics*
GRISWOLD—*Fuels, Combustion, and Furnaces*
GROGGINS—*Unit Processes in Organic Synthesis*
HUNTINGTON—*Natural Gas and Natural Gasoline*
JOHNSTONE AND THRING—*Pilot Plants, Models, and Scale-up Methods in Chemical Engineering*
KIRKBRIDE—*Chemical Engineering Fundamentals*
LEE—*Materials of Construction*
LEWIS, RADASCH, AND LEWIS—*Industrial Stoichiometry*
MANTELL—*Adsorption*
MANTELL—*Industrial Electrochemistry*
MCADAMS—*Heat Transmission*
MCCABE AND SMITH—*Unit Operations of Chemical Engineering*
MICKLEY, SHERWOOD, AND REED—*Applied Mathematics in Chemical Engineering*
NELSON—*Petroleum Refinery Engineering*
PERRY (EDITOR)—*Chemical Business Handbook*
PERRY (EDITOR)—*Chemical Engineers' Handbook*
PETERS—*Elementary Chemical Engineering*
PIERCE—*Chemical Engineering for Production Supervision*
RHODES, F. H.—*Technical Report Writing*
RHODES, T. J.—*Industrial Instruments for Measurement and Control*
ROBINSON AND GILLILAND—*Elements of Fractional Distillation*
SCHMIDT AND MARLIES—*Principles of High-polymer Theory and Practice*
SCHWEYER—*Process Engineering Economics*
SHERWOOD AND PIGFORD—*Absorption and Extraction*
SHREVE—*The Chemical Process Industries*
SMITH—*Chemical Engineering Kinetics*
SMITH—*Introduction to Chemical Engineering Thermodynamics*
STEPHENSON—*Introduction to Nuclear Engineering*
TREYBAL—*Liquid Extraction*
TREYBAL—*Mass-transfer Operations*
TYLER—*Chemical Engineering Economics*
VILBRANDT—*Chemical Engineering Plant Design*
WALKER, LEWIS, MCADAMS, AND GILLILAND—*Principles of Chemical Engineering*
WILSON AND RIES—*Principles of Chemical Engineering Thermodynamics*
WILSON AND WELLS—*Coal, Coke, and Coal Chemicals*
WINDING AND HASCHE—*Plastics, Theory and Practice*

HEAT TRANSMISSION

WILLIAM H. McADAMS

Professor of Chemical Engineering
Massachusetts Institute of Technology

THIRD EDITION

Sponsored by the
Committee on Heat Transmission
National Research Council

INTERNATIONAL STUDENT EDITION

McGRAW-HILL KOGAKUSHA, LTD.
Tokyo Auckland Beirut Bogota Düsseldorf Johannesburg
Lisbon London Lucerne Madrid Mexico New Delhi Panama
Paris San Juan São Paulo Singapore Sydney

HEAT TRANSMISSION

INTERNATIONAL STUDENT EDITION

Exclusive rights by McGraw-Hill Kogakusha, Ltd., for manufacture and export. This book cannot be re-exported from the country to which it is consigned by McGraw-Hill.

XIV

Copyright, 1933, 1942, 1954, by William H. McAdams. All rights reserved. No part of this publication may be reproduced, stored in a retrieval system, or transmitted, in any form or by any means, electronic, mechanical, photocopying, recording, or otherwise, without the prior written permission of the publisher.

Library of Congress Catalog Card Number: 53-12432

KOSAIDO PRINTING CO., LTD. TOKYO JAPAN

PREFACE TO THE THIRD EDITION

In the period of more than a decade since the writing of the second edition, the number of investigations in the field of heat transfer has increased at a tremendous rate. A considerable part of this research clearly reflects the military objectives for which it was performed. Some of the data have only recently been removed from the requirements of military secrecy and made available for general industrial application. The development of nuclear reactors and jet engines required the solution of novel heat-transfer problems which were beyond the range of previous knowledge. Government agencies such as the Atomic Energy Commission and the National Advisory Committee for Aeronautics have been responsible for important advances and for the exploration of many new fields.

The purpose of the third edition is to bring the previous text up to date by thorough discussion of the new developments in heat transfer. Any material which is not new has nonetheless been critically reviewed and modified where necessary in order to present a concise, unified picture of the present state of the art.

Through experimental investigation, the ranges of available correlations for heating and cooling of gases have been extended to cover many cases of supersonic compressible flow inside tubes and outside objects of various shapes. The friction and heat transfer of rarefied gases have been investigated into the region of free-molecule behavior, in which the Knudsen number becomes a significant parameter. An outstanding development with liquids has been the introduction of molten metals as practical heat-transfer media in high-performance applications. The extremely low Prandtl numbers of the liquid metals demand new correlations quite different from the conventional relations normally employed for water and oils.

New techniques have been developed. The use of fluidized beds for the contacting of fluids and solids has found widespread industrial acceptance. The exchange of heat between two fluids, as demanded in the regenerative gas turbine cycle, has been aided by the development of highly compact plate-and-fin and finned-tube exchangers. The phenomenon of local boiling of volatile liquids has been explored and is now used in a number of cases in which it is necessary to transfer heat at extremely high rates per unit of surface area.

Theoretical and mathematical approaches have yielded further insight into the mechanism of heat transfer between fluids and solids. The analogies between the transfer of heat, mass, and momentum have been further clarified. Analytical predictions of the local film coefficient of heat transfer for flow over a flat plate and for the entry region of a tube have met with considerable success. The combined effect of forced and natural convection with streamline flow in tubes has been treated mathematically for idealized conditions. The theory of extended surfaces has been applied to a number of new geometries. The mechanism by which heat is transferred in fixed beds of particles has received considerable study and is now subject to calculation in many cases.

Incremental methods for the computation of transient heat conduction have been further developed and explained. The enthalpy potential method has been expanded to include a new incremental graphical technique for calculation of the air temperature in cases of direct contact between water and air.

The need for further work in the next decade seems to be as great as in the past. The availability of improved correlations which may be applied over greatly extended ranges requires the determination of more accurate physical properties. For many common fluids the data are completely lacking.

The arrangement of the present edition is generally similar to its predecessor. Where recent data of greater accuracy are available, some of the earlier references have been deleted in order to keep the length of the volume within bounds. The references are grouped together near the Appendix, arranged in a separate alphabetical sequence for each chapter, thus serving as a subject bibliography. The Appendix has been enlarged to include new data. The nomenclature and abbreviations follow the American Standards Association as far as possible. Important design equations are marked with a boldface star.

Owing to the magnitude of the task, it is possible that some significant work has been overlooked; such accidental omissions are sincerely regretted. In general it has not been possible to mention data released later than 1952.

Many individuals in various fields of heat transfer have kindly furnished material for this book by means of personal contact or written communication with the author. The cooperation of every person who in this manner contributed his ideas or made available his data is hereby gratefully acknowledged. Thanks are due to Professor Hoyt C. Hottel who carefully revised and expanded the material on radiation, despite his many other professional commitments. Dr. Murray W. Rosenthal gave extensive assistance in the preparation of the manuscript and the illustrations. Mr. Robert L. Richards prepared the material for the chapter on applications to design. Mr. Thibaut Brian prepared many of the unsolved problems given at the end of each chapter. Mr. Roger Picciotto collected and arranged data for the Appendix. Professor James N. Addoms offered many valued suggestions and assumed responsibility for the final details before the volume went to press. The enthusiastic cooperation of my colleagues at the Massachusetts Institute of Technology is deeply appreciated. Special thanks are due to Mrs. Margaret B. Rose and Miss Jeanne Belisle for their assistance with the manuscript and the proofs.

William H. McAdams

PREFACE TO THE FIRST EDITION

This book is designed to serve both as a text for students and as a reference for practicing engineers. The problem of heat transmission is encountered in almost every industry, and because of the diversity in the fields of application there exist countless differences in detail. However, the principles underlying the problem are everywhere the same, and it is the purpose of this book to present fundamentals rather than to deal with the details of individual problems and special cases. This is done not merely because the discussion of individual cases would involve excessive space but because, in the first place, the solution of no individual problem can be adequate without a mastery of the principles underlying the whole field, and, in the second place, the solution of the specific problem becomes relatively easy once those fundamental principles are understood.

Because of the large number of fields in which heat transmission plays a part, the literature on the subject is extremely diffuse. Important data are often found in most unexpected places, and results are expressed in unfamiliar forms. Reliable data from the most diverse sources, including many unpublished results, have been reduced to a common basis for purposes of comparison and have been correlated in the light of the most helpful theoretical analyses. The results of these critical studies are presented as formulas and graphs carefully constructed to give due weight to all the data and at the same time to make the results readily available for use in engineering design. Because in many important cases the formulas represent not the results of a single investigation but concordant data from varied sources, it is believed that they afford the most dependable basis for engineering design.

The technique of employing the formulas is made clear by the solution of illustrative problems characteristic of the most important engineering uses to which the formulas are likely to be put. Where the general equations, based on the data for a number of fluids, contain a number of factors, calculations are simplified by the use of alignment charts, whereby the several physical properties are evaluated in terms of temperature of the fluid.

This book may be divided into three parts: conduction, radiation, and convection.

The section on conduction consists of two chapters, the first dealing with steady conduction of heat, thermal conductivities, the effect of the shape of bodies, and resistances in series and in parallel. In the second chapter, unsteady conduction, as in the heating and cooling of solids, is considered, and problems are solved by the use of charts involving four dimensionless ratios.

The second section treats radiation between solids for a number of important specific cases, radiation from nonluminous and luminous flames, and the general problem of furnace design, where heat is transferred simultaneously by several mechanisms.

The third section consists of seven chapters. The first of these deals with dimensional similarity. Because of the important relation between fluid motion and forced convection, a chapter is devoted to fluid dynamics. The third chapter serves as an introduction to convection and treats the relations between over-all and individual coefficients of heat transfer, the effect of deposits of scale, mean temperature difference in heat exchangers involving counterflow, multipass and cross flow, and the measurement of surface temperatures. The four remaining chapters deal with heat transfer by forced and free convection under the general headings of fluids inside pipes, fluids outside pipes, condensation, and evaporation. Photographs of convection currents and charts of distribution of velocity and temperature illustrate the mechanisms involved. The original data of a number of reliable investigators are plotted to develop the relations recommended for the various cases, and the experimental ranges of the various factors are tabulated. Optimum operating conditions are considered, and methods are given for determining the economic velocity of fluid in a heat exchanger and the optimum temperature difference in the recovery of waste heat.

The Appendix contains tables and charts of thermal conductivities, specific heats, latent heats of vaporization, and viscosities, and miscellaneous tables, such as steam tables, conversion factors, and dimensions of steel pipe.

In the text, reference numbers are used to designate the literature citations tabulated alphabetically in the Bibliography. Text page references are cited in the Bibliography, which thus serves as the Author Index.

This book has been written under the auspices of the National Research Council, Heat Transmission Committee; and the author wishes to express his appreciation of the active cooperation and assistance of the individual members of this Committee, particularly for the help of Paul Bancel of the Ingersoll Rand Company, Prof. C. H. Berry of Harvard University, John Blizard of the Foster Wheeler Corporation, W. H. Carrier of the Carrier Engineering Corporation, T. H. Chilton and A. P. Colburn of E. I. du Pont de Nemours & Company, Inc., H. N. Davis of Stevens

Institute of Technology, H. C. Dickinson of the Bureau of Standards, W. V. A. Kemp, formerly Secretary of the Committee on Heat Transmission, H. Harrison of the Carrier Engineering Corporation, C. F. Hirschfeld of the Detroit Edison Company, Prof. L. S. Marks of Harvard University, G. A. Orrok of New York, R. J. S. Pigott of the Gulf Companies, W. Spraragen of the National Research Council, T. S. Taylor of the Bakelite Corporation, M. S. VanDusen of the Bureau of Standards, and D. J. VanMarle of the Buffalo Foundry and Machine Company.

Help has also been received from so many individuals at such diverse times that it would be impossible to name them all, but this does not mean that the author fails to appreciate the value of such help and to be grateful for it. However, it is desired to mention the work of certain of my professional associates who have given freely of time and energy during the course of the work. In this group it is desired to mention T. B. Drew, J. J. Hogan, Prof. H. C. Hottel, C. R. Johnson, Prof. W. K. Lewis, W. M. Nagle, Prof. C. S. Robinson, Prof. W. P. Ryan, Prof. T. K. Sherwood, Prof. W. H. Walker, and Prof. G. B. Wilkes. Thanks are due to V. C. Cappello for the reading of proof.

William H. McAdams

CONTENTS

Preface to the Third Edition	v
Preface to the First Edition	ix
Index to Principal Relations for Convection	xiv
1. Introduction to Heat Transmission	1
2. Steady Conduction	7
3. Transient Conduction	31
4. Radiant-heat Transmission	55
5. Dimensional Analysis	126
6. Flow of Fluids	140
7. Natural Convection	165
8. Introduction to Forced Convection	184
9. Heating and Cooling Inside Tubes	202
10. Heating and Cooling Outside Tubes	252
11. Compact Exchangers, Packed and Fluidized Systems	282
12. High-velocity Flow; Rarefied Gases	309
13. Condensing Vapors	325
14. Boiling Liquids	368
15. Applications to Design	410
Appendix (Tables and Charts of Data)	443
Bibliography and Author Index	491
Subject Index	521

INDEX TO PRINCIPAL RELATIONS FOR CONVECTION

Relation between over-all and individual coefficients 187
Allowance for scale deposits; fouling factors 188–189, 376
Mean temperature difference 190–198

Forced Convection Parallel to Axis
Turbulent flow in tubes, usual case 219
   Simplified equations; gases, water 226–228
   Low Prandtl number; liquid metals 215–216, 218
Streamline flow in tubes with natural convection . . . 233–236
Streamline flow in tubes without natural convection . . . 237
Transition flow in tubes 239–241
Annular sections 242
Gravity flow in layer form 244–247
Rectangular sections 248
Flow parallel to plane 249
High velocities; supersonic flow 311–316

Forced Convection Normal to Axis
Air normal to single cylinders 260–261
Air normal to other shapes; spheres 265
Liquids normal to single cylinders 267
Finned surfaces 268–271
Gases normal to banks of tubes 271–275
Liquids normal to banks of tubes 276
Cross-baffled exchangers 277
Compact exchangers, comparison of types 287–289
High velocities; rarefied gases 311–316, 321

Natural Convection
Vertical surfaces 172–173
Horizontal cylinders 177
Horizontal plates 180
Enclosed spaces 181–182

Packed and Fluidized Systems
Coefficients between packed solids and flowing fluid . . . 295
Fluidized beds 303–307

Condensing Vapors
Film-type condensation on vertical tubes 331–334
Film-type condensation on horizontal tubes 338
Dropwise condensation 348
Mixtures of vapors 353–354
Mixture of vapor and non-condensable gas 355
Direct contact of air and water; cooling towers . . . 356–365

Boiling Liquids
Peak heat flux and critical temperature difference . . . 383–387
Film boiling outside tubes 387–388
Surface or local boiling 391–393
Vaporization inside tubes 408

CHAPTER 1

INTRODUCTION TO HEAT TRANSMISSION

The laws of heat transmission are of controlling importance in the design and operation of many diverse forms of steam generators, furnaces, preheaters, exchangers, coolers, evaporators, and condensers, in numerous different industries.

In some cases the primary objective is to obtain the maximum heat-transfer rate per unit surface compatible with economic factors. In other cases the object is the salvaging of heat as in exchangers, recuperators, and regenerators. In still other cases the purpose is to minimize the heat flux by the use of insulation. Frequently in a given problem, all three objectives are important.

MODES OF HEAT TRANSMISSION

1. Conduction. Conduction in a homogeneous opaque solid is the transfer of heat from one part to another, under the influence of a temperature gradient, without appreciable displacement of the particles. Conduction involves the transfer of kinetic energy from one molecule to an adjacent molecule; it is the only mechanism of heat flow in an opaque solid. With certain transparent solids such as glass and quartz, some energy is transmitted by radiation as well as by conduction. With gases and liquids, conduction may be supplemented by convection and radiation. Within a fluid flowing in streamline motion, heat is transferred by conduction at right angles to the direction of fluid flow.

2. Convection. Convection involves the transfer of heat by mixing one parcel of fluid with another. The motion of the fluid may be entirely the result of differences of density resulting from the temperature differences, as in natural convection, or the motion may be produced by mechanical means, as in forced convection. Energy is also transferred simultaneously by molecular conduction and, in transparent media, by radiation. In measuring thermal conductivity of gases and liquids, downward flow of heat is frequently employed to avoid transfer by convection; with transparent gases the radiation from the heat source to the heat sink must be deducted.

3. Radiation. A hot body emits radiant energy in all directions. When this energy strikes another body, part may be reflected. Part may be transmitted through the body, in which case the body is said to be diathermanous. The remainder is absorbed and quantitatively transformed into heat.† If two bodies, one hotter than the other, are placed within an enclosure, there is a continuous interchange of energy between them. The hotter body radiates more energy than it absorbs; the colder body absorbs more than it radiates. Even after equilibrium of temperature is established, the process continues, each body radiating and absorbing energy. Certain gases as well as solids are capable of radiating and absorbing thermal energy.

Since in most cases heat is transferred by more than one of the three modes, it is preferable to use the terms *transmission* to describe the overall process, reserving the use of the terms *radiation, convection*, and *conduction* for that portion of the heat transmission accomplished by the mechanism designated. The term *radiation* is probably the most generally misused of the three, such expressions as radiation through furnace walls, etc., being common.

ILLUSTRATIONS

Thus, in considering the brick walls of a dwelling, heat will be conducted through the solid from the warmer to the colder face. Also, when one end of an insulated iron bar is held in the fire, heat will be conducted through the metal toward the colder end, although a part will be conducted through the insulation and dissipated to the room by the mechanisms of conduction, natural convection, and radiation. When heat is transmitted through a cellular or porous solid, the heat may flow not only by conduction but also both by convection within the gas pockets and by radiation from surface to surface of the individual cells within the nonhomogeneous solid.

In the heating of a tank of liquid, if the steam coil is located near the top of the tank, the heated liquid, because of its reduced density, would not tend to mix readily with the colder, denser liquid at the bottom of the tank. In order to promote the thermosiphon circulation of the liquid, due to differences in density, the heating coil should be located near the bottom of the tank. For similar reasons, the heating element of the so-called "steam radiator" should be placed near the floor of the room rather than near the ceiling. Instead of depending wholly on natural convection, an increased rate of circulation can be obtained by mechanical means such as a pump or fan.

In a cold room heated only by a fireplace, an occupant near the fire may receive too much heat on one side, because of the radiant energy emitted by the fire, and yet be unpleasantly cold on the other side. Radiant energy from the sun travels through space, penetrates the earth's atmosphere, is partially absorbed, and the remainder reaches the surface of the

† Except in the relatively rare cases in which photochemical reactions are induced or energy is consumed in other special ways, as in nuclear fission.

earth, where part is reflected and the residue is transformed into heat. In the generation of steam, radiation from the fire to the tubes of the boiler plays an important part. The combustion gases pass over the boiler tubes, and thus additional heat is transferred to the tubes by conduction and convection. The carbon dioxide and water vapor of the flue gases radiate energy directly to the tubes. Hence the heat conducted through the metal walls of the tubes is supplied by radiation from warmer solids, by conduction and convection from the warm gases, and by radiation from certain constituents of the gases.

It is thus seen that in most cases of heat transmission more than one mechanism is involved. Therefore it will be necessary not only to present the basic laws governing the three mechanisms, together with auxiliary data, but also to develop sound methods for calculating the total heat flow due to the combined effects. Since the transfer of heat by conduction and convection between solids and fluids is determined by the velocity pattern in the fluid, a chapter on the mechanisms and laws of fluid flow is included.

It is interesting at this point to make a preliminary inspection of the basic equations of heat transfer.

LAWS OF HEAT TRANSFER

Conduction. Fourier's law[2],[†] for unidirectional conduction of heat states that the instantaneous rate of heat flow $dQ/d\theta$ is equal to the product of three factors: the area A of the section, taken at right angles to the direction of heat flow; the temperature gradient $-dt/dx$, which is the rate of change of temperature t with respect to the length of path x; and the thermal conductivity k, which is a physical property of the material. Mathematically expressed, Fourier's law is as follows:

$$\frac{dQ}{d\theta} = -kA\frac{dt}{dx} \qquad (1\text{-}1)$$

The thermal conductivities of various substances cover a range of ten-thousandfold.

The differential form given in Eq. (1-1) is general for unidirectional conduction and may be applied to cases in which the temperature gradient $-dt/dx$ varies with time as well as with the location of the point considered. In every case of heat flow by conduction, a temperature gradient must exist. If the temperature of a given point in the body varies with time, the rate of heat flow will also vary with time. The process of heat flow in a case in which temperature varies with both time and position is called *heat conduction* in the *unsteady state*, and problems

[†] Literature references are given in the Bibliography and Author Index at the back of the book.

involving the integration of the differential equations for this case are treated in Chap. 3.

Since, for conduction in the *steady* state, the temperature at each point does not vary with time, it follows that the temperature gradient $-dt/dx$ and, consequently, the rate of heat flow $dQ/d\theta$ are likewise independent of time. Hence, the rate $dQ/d\theta$ equals Q/θ and is designated by q; for a differential area Eq. (1-1) becomes

$$dq/dA = -k\, dt/dx \tag{1-1a}$$

Problems involving integration of this equation are treated in Chap. 2. Frequently it is possible to employ the potential form

$$q = \frac{\Delta t}{x_w/k_w A_m} = \frac{\Delta t}{R} \tag{1-1b}$$

where the mean area A_m is determined by the size and geometry and Δt is the temperature drop across the wall of thickness x_w. This equation is completely analogous with that for the flow of d-c electricity through a resistance

$$I = \frac{\Delta V}{X_E/k_E A_m} = \frac{\Delta V}{R_E} \tag{1-1c}$$

where I is the current (quantity of electricity per unit time), ΔV is the potential drop across the electrical resistance R_E, X_E is path length, k_E is the specific electrical conductivity (reciprocal of electrical resistivity), and A_m is the mean area of cross section of the electrical conductor. By analogy for steady heat flow through a series of resistances, one may write

$$q = \frac{\Delta t}{R} = \frac{\Sigma\, \Delta t}{\Sigma R} = \frac{\Sigma\, \Delta t}{\Sigma(x_w/k_a A_m)} \tag{1-1d}$$

The basic conduction equation of Fourier is the starting point in the theoretical treatment of numerous problems in heat transfer other than conduction in solids. Among these problems are heat transfer to fluids in streamline flow in tubes (Chap. 9), wetted-wall heaters (Chap. 9), and heat transfer to fluids moving by natural convection (Chap. 7). Additional applications of Eq. (1-1) to problems in steady flow are conduction along fins (Chap. 10) and heat transfer from condensing vapors (Chap. 13).

Thermal Radiation. The basic equation for total thermal radiation from the ideal radiator (the "black body") was discovered empirically by Stefan[5] in 1879 and was derived theoretically by Boltzmann[1] in 1884:

$$dq_r = \sigma\, dA\, T^4 \tag{1-2}$$

where dq_r is the rate of heat transfer by radiation from one side of the "black" element of area dA, T is the absolute temperature of the surface,

INTRODUCTION

and σ is the Stefan-Boltzmann dimensional constant, the magnitude of which depends only upon the units employed.

With gray surfaces (which have emissivities less than the black body) the net radiation between two surfaces in an enclosure containing a medium transparent to thermal radiation can be estimated by an equation developed by Hottel:[3]

$$q_{net,total} = \sigma A_1 \mathcal{F}_{12}(T_1^4 - T_2^4) \quad (1\text{-}3)$$

in which $\mathcal{F}_{12}$ is a function of the geometry and the emissivities of the two surfaces. Radiation is treated in detail in Chap. 4.

Convection. In 1701 Newton[4] defined the heat-transfer rate q_c from a surface of a solid to a fluid by the equation

$$q_c = h_m A(t_w - t) \quad (1\text{-}4)$$

where h_m is the coefficient of heat transfer from surface to fluid, excluding any radiation, A is the area of the surface, t_w is the surface temperature of the wall, and t is the bulk temperature of the fluid. For conduction in solids, and for radiation through transparent media, the mechanisms are known, and the heat-transfer equations are dependable. But with heat transfer from a solid to a fluid, the situation is far more complex, and Eq. (1-4) is *not* a law but merely a definition of the mean coefficient of heat transfer.

It will be found that the coefficient h_m depends not only on the units employed but on certain physical properties of the fluid, on dimensions of the apparatus, on velocity of the fluid past the surface, on whether or not the fluid is changing phase, and frequently on the temperature potential Δt. It is most helpful to employ point or local surface coefficients h_x defined by

$$\frac{dq}{dA} = h_x(t_w - t) \quad (1\text{-}4a)$$

although it is much more difficult to measure the local coefficient than the average coefficient. In a given set of units the values of h_m cover a range of one-hundred thousandfold, as shown in Table 1-1.

TABLE 1-1. APPROXIMATE RANGE OF VALUES OF h_m ORDINARILY ENCOUNTERED
Btu/(hr)(sq ft)(deg F)

Steam, dropwise condensation	5,000–20,000
Steam, film-type condensation	1,000– 3,000
Water boiling	300– 9,000
Organic vapors condensing	200– 400
Water, heating	50– 3,000
Oils, heating or cooling	10– 300
Steam, superheating	5– 20
Air, heating or cooling	0.2– 10

Local Over-all Coefficient of Heat Transfer. In the majority of industrial applications heat is transferred from one fluid to another by a three-step steady-state process: from the warmer fluid to the solid wall, through the solid wall, and thence to the colder fluid. It is customary to employ an over-all coefficient of heat transfer, U, based on the over-all difference between the temperatures of the two fluids, Δt_o:

$$dq = U \, dA \, \Delta t_o \tag{1-5}$$

based on some specified heat-transfer area dA.

Since the total temperature potential Δt_o between the warmer and colder fluids is the sum of the individual temperature potentials across each part of the path through which the same heat flows in series, Eqs. (1-4), (1-1b), (1-4), and (1-5) may be combined to give

$$\frac{1}{U' \, dA'} = \frac{1}{h' \, dA'} + \frac{x_w}{k_w \, dA_w} + \frac{1}{h'' \, dA''} = \frac{1}{U'' \, dA''} \tag{1-5a}$$

where the single prime refers to the warmer side and the double prime refers to the colder side.

Mean Δt. In a heat exchanger in which the warmer fluid is cooled by heating the colder fluid, it is clear that the over-all temperature potential usually varies from point to point. This type of problem may be solved (as shown in Chap. 8) by combining Eq. (1-5) with the total energy balance and integrating. The result may sometimes be expressed in terms of a mean over-all temperature potential Δt_m:

$$q = UA \, \Delta t_m = \frac{\Delta t_m}{(1/h'A') + (x_w/k_w A_w) + (1/h''A'')} \tag{1-6}$$

A large fraction of this text is devoted to a study of the factors which control the coefficient h and the correlation of these factors so that h may be predicted.

CHAPTER 2

STEADY CONDUCTION

Introduction. In the great majority of cases arising in engineering practice, heat flows from some medium into and through a solid retaining wall and out into some other medium. The flow through each medium is therefore but one step in a more complicated process, and the resistance to the conduction of heat through the retaining wall is only one of a series of resistances. A clear understanding of the mechanism of heat transfer by conduction through homogeneous solids is essential to the solution of the more complex problems.

This chapter deals with the basic differential equation for the steady conduction of heat and integrated forms for homogeneous bodies of various shapes. Equations are given for the conduction of heat through more than one resistance, for both series and parallel flow. The chapter contains several illustrative problems. The factors affecting the thermal conductivities of solids, liquids, and gases are discussed.

THERMAL CONDUCTION THROUGH HOMOGENEOUS SOLIDS

The mechanism of heat transfer by steady unidirectional conduction is given by the simple relation of Fourier[10] proposed in 1822:

$$q = -kA \frac{dt}{dx} \qquad (2\text{-}1) \star$$

where q is the rate of heat conduction along the x axis, A is the cross section of the path normal to the x axis, $-dt/dx$ is the temperature gradient along the path, and k is the thermal conductivity, a physical property of the substance. This equation is well established for conduction in homogeneous isotropic solids, and practical application of the equation to various problems requires (1) laboratory measurement of thermal conductivity of a representative specimen and (2) solution of the equation for the particular geometric and boundary conditions.

In the English system involving the units of feet, pounds, hours, degrees Fahrenheit, and British thermal units, k is expressed as British thermal units per hour per square foot taken at right angles to the direction of heat flow, per unit temperature gradient, degrees Fahrenheit per foot of length of path. This is often improperly abbreviated as "Btu per hr per

HEAT TRANSMISSION

Table 2-1. Nomenclature[a]

A Area through which heat flows at right angles, square feet; $A_a = (A_1 + A_2)/2$; A_m is true mean value
  a Temperature coefficient of thermal conductivity, reciprocal degrees Fahrenheit
  b Empirical constant
  c_p Specific heat at constant pressure, Btu/(lb)(deg F)
  c_v Specific heat at constant volume, Btu/(lb)(deg F)
  C Conductance, Btu/(hr)(deg F)
  C' Conductance per unit area, Btu/(hr)(sq ft)(deg F)
  d Prefix, indicating differential
  e Base of natural logarithms, 2.718 . . .
  h_j Contact coefficient at junction of two solids, Btu/(hr)(sq ft)(deg F)
  K_B Boltzmann constant, gas constant per molecule, in heat units; $K_B = 7.25 \times 10^{-27}$ Btu/(deg F)(molecule)
  K_L Lorenz constant, dimensional
  k Thermal conductivity at temperature t, Btu/(hr)(ft)(deg F); k_a at $t_a = (t_1 + t_2)/2$; k_0 at 0°F; k_m, true mean value
  k_e Electrical conductivity, $k_e = 1/\sigma$
  L Length, feet
  M Molecular weight
  M' Actual weight of one molecule, pounds; $M' = 3.65 M \times 10^{-27}$
  N Number of equal parts into which $-\int_1^2 k\, dt$ is divided
  N_L Number of lanes of equal heat flow
  N_{Mt} Maxwell number, $k/\mu c_v$, dimensionless
  q Steady rate of heat flow, Btu per hour
  q^* q/kz, degrees Fahrenheit
  R Thermal resistance $x/k_m A_m$, (deg F)(hr)/Btu
  R_T Total thermal resistance, $\Sigma(x/k_m A_m)$
  T Absolute temperature, degrees Fahrenheit absolute
  t Thermometric temperature, degrees Farhenheit
  V_a Acoustic velocity in a liquid, feet per hour
  x Length of conduction path, radius, feet
  y Width, distance, feet
  Z_c Mean distance between centers of molecules of a liquid, feet; $Z_c = (M'/\rho)^{1/3}$
  z Length, feet

Greek
  γ Ratio of specific heats, c_p/c_v, dimensionless
  Δ Prefix, indicating finite difference
  Δt Temperature difference along path, degrees Fahrenheit
  Δx Finite length of path, feet
  Δy Finite width of path, feet
  ϵ Elementary charge of an electron
  μ Viscosity, lb/(hr)(ft); $\mu = 2.42\mu'$
  μ' Viscosity, centipoises
  π 3.1416 . . . , a pure number
  Σ Prefix, indicating summation
  σ Electrical resistivity, ohms $\times$ centimeters
  ρ Density, pounds per cubic foot; ρ_a for apparent density
  ρ' Density, grams per cubic centimeter

[a] The symbols used in this book have, in general, been taken from the lists of the American Standards Association and the American Institute of Chemical Engineers. Literature references are given in the Bibliography and Author Index.

sq ft per deg F per ft," but it should be remembered that the units of k are

$$\frac{(\text{Btu})/(\text{hr})(\text{sq ft})}{\text{deg F}/\text{ft}} = \frac{\text{Btu}}{(\text{hr})(\text{ft})(\text{deg F})}$$

Throughout the text, unless otherwise specified, numerical values of k are expressed in these units, but factors are given in the Appendix for converting the numerical values of k from one system of units to another. Jakob[12] gives a comprehensive 50-page review of the theories regarding the thermal conductivities of various materials.

Thermal Conductivities of Solids. The thermal conductivities of various solids differ widely. Thus, k for copper is approximately 220, whereas k for corkboard is about 0.025. For a given solid, k is a function of temperature, and for most homogeneous solids hitherto investigated this temperature relationship is nearly linear over a moderate range in temperature: $k = k_0(1 + at)$, k being the value at $t°F$, k_0 at $0°F$, and a the temperature coefficient, which is equal to $dk/k_0\, dt$. The temperature coefficient is positive for many insulating materials, a notable exception being magnesite brick, as shown in the Appendix. For nonmetallic solids, as temperature is increased, k increases for amorphous substances and decreases for crystalline materials.

As temperature is increased, the values of k of industrial materials may pass through a maximum for some materials and a minimum for others. Both types of behavior can be accounted for by an equation given by Jakob:[12]

$$k = \frac{1}{b_1 T + b_2 + (b_3/T)}$$

The conditions for a maximum or minimum at T' are $T' = \sqrt{b_3/b_1}$ and the corresponding $k' = 1/(2b_1 T' + b_2)$; a maximum is possible if b_1 and b_3 are positive, and a minimum may occur if b_1 and b_3 are negative.

On the other hand, for good conductors, such as most metals, the temperature coefficient a is usually negative, aluminum and brass being exceptions. At a given temperature, the thermal conductivity of commercial samples of a metal may differ widely, because of the variation in the content of certain elements or compounds present in small amounts. The thermal conductivity of an alloy of two metals may be less than the value of k of either constituent. For example, constantan contains 60 per cent copper and 40 per cent nickel; the alloy has k of 13 vs. k of 222 for copper and 36 for nickel.

In the Lorenz-type equation[19] the ratio of the thermal and electrical conductivities of metals, divided by the absolute temperature, equals a dimensional constant K_L:

$$\frac{k}{k_e T} = \frac{k\sigma}{T} = K_L \qquad (2\text{-}2)$$

Originally Lorenz used K_L equal to $2(K_B/\epsilon)^2$, but revised theories lead to K_L equal to $3.3(K_B/\epsilon)^2$, where K_B represents the Boltzmann constant† and ϵ the elementary charge of an electron. Jakob[12] cites values of K_L for some metals and alloys at approximately 32°F; for 11 pure metals the values of $10^9 K_L$ ranged from 715 to 924, with k expressed in Btu/(hr)(ft)(deg F), T in degrees Fahrenheit absolute, and σ (the electrical resistivity) in ohms × centimeters; the theoretical value of $10^9 K_L$ is 783. But for 10 alloys $10^9 K_L$ ranged from 530 to 1420. As temperature increases to 1800°F, K_L for nickel increases by 26 per cent.[12]

Powell[25] summarizes values of k for various graphites. At 68°F the ranges of k values were as follows: 245 to 50 for Ceylon graphite, 180 to 32 for Cumberland graphite, and 95 to 69 for Acheson graphite. The corresponding Lorenz numbers are 150 to 450 times as large as the theoretical values based on electron conduction. Thus the Lorenz rule is inapplicable to amorphous materials.

For nonhomogeneous solids the apparent thermal conductivity at a given temperature is a function of the apparent or bulk density ρ_a. For 18 different kinds of wood at 85°F, MacLean[20] found that k was linear in apparent density: $k = 0.0137 + 0.00186 \rho_a$. ‡ Wilkes[36] notes that the apparent thermal conductivities of many fibrous materials and of certain powders reach minimum values at "critical" values of apparent density. With fibrous materials such as woods, the value of k for conduction parallel to the grain may be 1.7 times that for conduction normal to the grain. In determining the apparent thermal conductivity of granular solids, such as granulated cork or charcoal grains, Griffiths[11] found that air circulates within the mass of granular solid; under a certain set of conditions, the apparent thermal conductivity of charcoal grains was 9 per cent greater when the test section was vertical than when horizontal. For slag wool packed between vertical plates[21] the minimum k and the corresponding critical apparent density increased as the height of the plates was increased, but above the critical apparent density height was immaterial.

When the apparent conductivity of a cellular or porous nonhomogeneous solid is determined, the temperature coefficient may be much larger than for the homogeneous solid alone, since heat is transferred not only by the mechanism of conduction but also by convection in the gas pockets and by radiation from surface to surface of the individual cells. If internal radiation is an important factor, a plot of the apparent conductivity as ordinate vs. temperature should show a curve concave upward, since radiation increases with the fourth power of the absolute temperature. With certain fine powders, such as silica aerogel and diatomaceous earth, the apparent thermal conductivity decreases with decrease in the absolute pressure. At low pressures the mean free path x' of the molecules

---

† Gas constant per molecule, in heat units; see Table 2-1.
‡ The MacLean eq. is for conduction across the grain.

of air† in the pores exceeds the pore size, and the thermal conductivity of the air becomes proportional to pressure; in consequence the apparent conductivity of the powder decreases with decrease in pressure. With very small pores, the apparent conductivity of the powder at atmospheric pressure may be less than k of still air at 1 atm.

An example of low conductivity is a lightweight thermal insulation consisting of very thin layers of aluminum foil (0.0003 in. thick) spaced about ⅓ in. apart in either flat or crumpled sheets, the apparent density being only 0.19 lb/cu ft. For a given thickness, this material is said to have an apparent thermal conductivity at low temperatures equal to that of cork and less than magnesia at 400°F. The low apparent thermal conductivity is due to the small air spaces and to the low emissivity of the metal, approximately 5 per cent of that of a black body. Reflective insulation is treated in detail by Wilkes.[36] In cases in which heat is transferred through a porous solid by the combined mechanisms of conduction, convection, and radiation, it is preferable to express the results as conductances [Eq. (2-11a)] rather than as apparent thermal conductivities. In measuring the apparent thermal conductivity of diathermanous substances such as quartz, especially when exposed to radiation emitted at high temperatures, it should be remembered that a part of the heat is transmitted by radiation. The thermal conductivity of a solid, such as earth, is frequently measured by measuring the rate of rise of temperature of a heated probe (see Chap. 3).

Tables and plots of thermal conductivities of various solids, liquids, and gases are given in the Appendix.

In order to be of practical use, Eq. (2-1) must be integrated, and for this purpose the variation of k with t and of A with x must be known. As stated above, experimental data on the variation of k with temperature can often be expressed by the linear relationship

$$k = k_0(1 + at) \tag{2-3}$$

where k is the instantaneous value of the thermal conductivity at the temperature t and k_0 and a are constants. Combining Eqs. (2-1) and (2-3),

$$-k\,dt = -k_0(1 + at)\,dt = q\,dx/A$$

Integrating between the temperature limits t_1 and t_2,

$$-\int_{t_1}^{t_2} k\,dt = (t_1 - t_2)(k_0)\left(1 + a\frac{t_1 + t_2}{2}\right) = q\int_1^2 \frac{dx}{A}$$

Hence, where k is linear in t, one may use the equation

$$k_a(t_1 - t_2) = q\int_1^2 dx/A \tag{2-4}$$

† For air at 32°F, $x' = 3.17 \times 10^{-7}/P$, where x' is expressed in feet and P in atmospheres absolute.

where k_a is the arithmetic mean of k_1 at t_1 and k_2 at t_2, or k_a is evaluated at the arithmetic mean of t_1 and t_2.

If k is not linear in t, some mean value k_m will apply. Irrespective of the relation between A and x

$$k_m = \frac{1}{t_1 - t_2} \int_{t_1}^{t_2} k\, dt \qquad (2\text{-}5)$$

Equations for Common Shapes. In many practical cases the cross-sectional area A varies appreciably with the length of path x, so that the shape of the solid through which the heat is flowing must be known before Eq. (2-4) may be integrated. Four such cases will now be discussed, of which the first two include most of the problems arising in engineering practice.[34]

Case I. *Conduction of Heat through a Solid of Constant Cross Section.* For this case, where A is constant, integration of Eq. (2-4) between the limits of x_1 and x_2 gives

$$q = \frac{k_m A (t_1 - t_2)}{x_2 - x_1} = \frac{k_m A\, \Delta t}{x} \qquad (2\text{-}6)\ \star$$

For convenience, the temperature difference $t_1 - t_2$ has been replaced by Δt and the thickness $x_2 - x_1$ by x. Equation (2-6) states that the rate of heat flow in Btu per hour is directly proportional (1) to the mean thermal conductivity of the substance in question, (2) to the cross-sectional area of the path in square feet, and (3) to the temperature difference in degrees Fahrenheit between the points from and to which the heat is flowing and is inversely proportional to the length of the path in feet.

Illustration 1. Calculate the heat loss through a 9-in. brick and mortar wall ($k_m = 0.4$), 10 ft high and 6 ft wide, when the inner and outer surface temperatures are 330 and 130°F, respectively.

Solution. By Eq. (2-6),

$$q = \frac{k_m A\, \Delta t}{x} = \frac{(0.4)(60)(330 - 130)}{\tfrac{9}{12}} = 6400 \text{ Btu/hr}$$

On account of its simplicity, Eq. (2-6) is often employed even when the cross section of path is a variable quantity, average values A_m for the area being then employed. The next three cases resolve themselves into the problem of deriving rules for obtaining correct mean values of this area for various special shapes, employing the equation

$$q = k_m A_m\, \Delta t / x \qquad (2\text{-}7)\ \star$$

Case II. *Cross Section of Path Proportional to Radius (Long Hollow Cylinders).* Where heat is flowing through the sides of a cylindrical body of circular section, the direction of flow is at all points radial and perpendicular to the axis and the cross section of the path is proportional to the distance from the center of the cylinder. It will now be shown that in such cases the logarithmic-mean area is the proper average value.

Consider the flow through a section of thickness dx, at a distance x from the center (see Fig. 2-1).

By Eq. (2-4), $\int dx/A = k_m \Delta t/q$. The cross section of path is $A = 2\pi xL$, where L is the length of the cylinder. This gives

$$\int_1^2 \frac{dx}{x} = \frac{k_m \Delta t\, 2\pi L}{q}$$

On integration and substitution of limits, this becomes

$$q = \frac{k_m 2\pi L\, \Delta t}{\ln(x_2/x_1)} \quad (2\text{-}8)$$

As stated above, it is often desired to employ the simple form of Eq. (2-7), using a mean value of that area, A_m, which will give the correct heat flow. Equating Eqs.

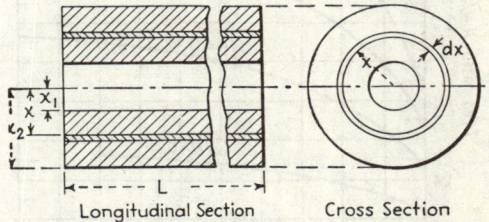

Fig. 2-1. Diagram for Case II: logarithmic-mean area.

(2-7) and (2-8), one obtains

$$q = \frac{k_m 2\pi L\, \Delta t}{\ln(x_2/x_1)} = k_m A_m \frac{\Delta t}{x_2 - x_1}$$

whence

$$A_m = \frac{2\pi x_2 L - 2\pi x_1 L}{\ln(2\pi x_2 L/2\pi x_1 L)} = \frac{A_2 - A_1}{\ln(A_2/A_1)}$$

$$= \frac{A_2 - A_1}{2.3 \log_{10}(A_2/A_1)} \quad (2\text{-}9)\,\star$$

Equation (2-9) requires that the area of cross section through which the heat is flowing in such a case be computed by dividing the difference of the external and internal areas by the natural logarithm of their ratio. For any two quantities, the average so obtained is called the *logarithmic mean* and is, as will later appear, a value frequently used in problems on flow of heat. For a given material and boundary temperatures, Eq. (2-8) indicates that the rate of heat conduction per unit length of hollow cylinder depends only on the ratio of outer and inner radii and is independent of thickness.

When the value of A_2/A_1 does not exceed 2, the arithmetic-mean area

$$A_a = (A_1 + A_2)/2 \quad (2\text{-}9a)$$

is within 4 per cent of the logarithmic-mean area; this accuracy is considered sufficient for many problems in heat conduction.

Optimum Thickness of Pipe Covering. The heat loss from an insulated steam pipe involves the coefficient of heat transfer from steam to metal, the conductivity and thickness of both the pipe wall and the insulation, and the coefficient of heat transfer from the surface of the insulation to the surroundings. This case will be treated in Chap. 15 after the surface coefficients have been studied in detail. As the thickness of the insulation is increased, the investment charges increase, but the heat loss

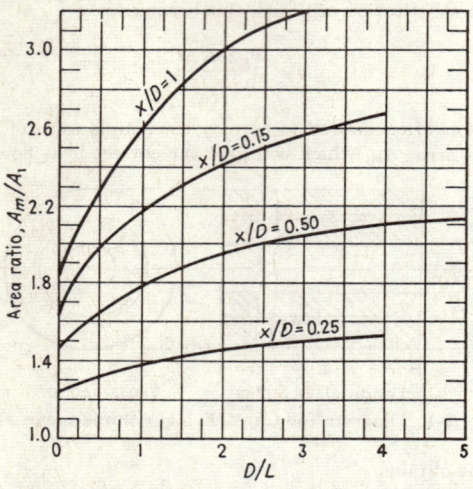

Fig. 2-2. A short cylinder of over-all length L and outside diameter D is insulated with a covering having uniform thickness x and total inside surface A_1 (including ends). The mean area A_m for use in Eq. (2-7) is found from the ordinate A_m/A_1. (*Courtesy Nickerson and Dusinberre,*[23] *Trans. ASME.*)

usually decreases, and the optimum thickness of insulation corresponds to a minimum total annual cost.

Case III. *Short Hollow Cylinders.* Consider a short cylinder of over-all length L and outside diameter D, covered on both ends and side with insulation of uniform thickness x, with inner and outer surfaces at uniform temperatures. Equation (2-9), derived for L/D of infinity, does not apply. Nickerson and Dusinberre[23] solved this problem for L/D from 0.25 to infinity by a relaxation procedure and found that the ratio of the mean area A_m to the *total* inside area A_1 is a function of the two ratios (D/L and x/D) as shown in Fig. 2-2.

Case IV. *Cross Section of Path Proportional to Square of Radius (Hollow Spheres).*

Consider a hollow sphere bounded by radii x_1 and x_2, through which heat is flowing radially in all directions at a uniform steady rate. At any radius x the cross-sectional area A equals $4\pi x^2$; upon integrating Eq. (2-4) between the limits x_1 and x_2, one obtains

$$\frac{q}{4\pi}\left(\frac{1}{x_1} - \frac{1}{x_2}\right) = k_m \, \Delta t \tag{2-10}$$

If it is desired to employ Eq. (2-7), involving a suitable mean area A_m, one may equate Eqs. (2-7) and (2-10) to obtain

$$A_m = 4\pi x_1 x_2 = \sqrt{A_1 A_2} \tag{2-10a} \star$$

In case, therefore, the cross-sectional area be proportional to the square of the linear dimension, the average area to be employed in calculating the heat flow is the geometric mean of the internal and external surfaces, as indicated by Eq. (2-10a). Were the outside diameter twice the inside, the arithmetic mean of the inside and outside areas would be 25 per cent too high. It is to be noted that both the logarithmic- and geometric-mean values are always lower than the arithmetic-mean value.

Conductance. Where heat is transferred by more than one mechanism through a structure having a mean cross-sectional area A_m, the conductance is defined as the gross rate of heat transfer Σq divided by the temperature drop Δt between its faces:

$$C = \Sigma q / \Delta t \tag{2-11}$$

The unit conductance C', or the conductance of a unit area, is defined by the equation

$$C' = \Sigma q / A \, \Delta t \tag{2-11a}$$

and equals C/A.† Where heat flows through a structure solely by conduction, $q = k_m A_m \, \Delta t / x$, the conductance would reduce to $k_m A_m / x$ and the resistance $x / k_m A_m$ would be equal to the reciprocal of the conductance. For cases in which heat is flowing through a hollow enclosure by conduction through one wall, thence by convection and radiation acting in parallel across the gas space to the other wall, and out by conduction, the concept of unit conductance is preferred [Eq. (2-11a)], although some writers report the results as apparent conductivities, based on (2-7). The conductance C of some structures is little related to their thickness, and in such cases the apparent conductivity k would be a function of thickness.

CONDUCTION THROUGH SEVERAL BODIES IN SERIES: RESISTANCE CONCEPT

The steady flow of heat through each of several bodies in series is often encountered.

† Although C' has the same dimensions as the coefficient h of heat transfer [Eq. (1-4)], the temperature difference $(t_s - t)$ used in defining h is that between the surface and the body of fluid flowing past the boundary, whereas the Δt employed in the definition of C' is that across the two faces of the structure.

With reference to Fig. 2-3, heat is being conducted at a steady rate through a wall composed of three different solids, a, b, and c. By applying Eq. (2-7) separately to each solid, it will now be shown that, for the steady state, the rate of heat flow may be calculated either by dividing the temperature drop through any individual resistance by its resistance $R = x/k_m A_m$, or by dividing the total temperature drop by the total resistance, which is the sum of the individual resistances. For the steady

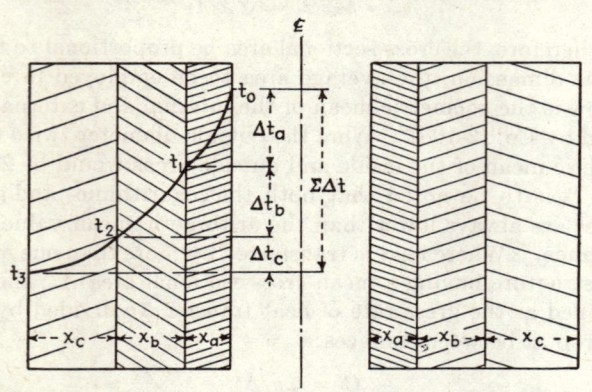

Fig. 2-3. Temperature distribution in walls of three concentric cylinders.

conduction of heat through a single homogeneous body, Eq. (2-7) may be written as follows:

$$q = \frac{\Delta t}{x/k_m A_m} = \frac{\Delta t}{R} \tag{2-12}$$

Applying Eq. (2-12) in turn to each solid,

$$q = \frac{t_0 - t_1}{R_a} = \frac{t_1 - t_2}{R_b} = \frac{t_2 - t_3}{R_c}$$

where R_a, R_b, and R_c are the individual resistances of the three solids, respectively. By adding the expressions $qR_a = t_0 - t_1$ and $qR_b = t_1 - t_2$, one obtains $q(R_a + R_b) = t_0 - t_2$. Hence

$$q = \frac{t_0 - t_2}{R_a + R_b} = \frac{t_0 - t_3}{R_a + R_b + R_c}$$

which may be written

$$q = \frac{\Delta t}{R} = \frac{\Sigma \, \Delta t}{R_T} = \frac{(\Delta t)_a + (\Delta t)_b + (\Delta t)_c}{R_a + R_b + R_c} \tag{2-13}$$

where R_T is the total series resistance, i.e., the sum of the individual resistances. Or

$$q = \frac{\Sigma \, \Delta t}{(x_a/k_a A_a) + (x_b/k_b A_b) + (x_c/k_c A_c)} \tag{2-13a} \star$$

It should be noted that k_a, k_b, and k_c represent the mean thermal conductivities of a, b, and c and A_a, A_b, and A_c the mean areas taken at right angles to the direction of heat flow. Thus, if the walls were concentric cylinders, each average area should be the logarithmic mean of the outer and inner surfaces of that hollow cylinder [Eq. (2-9)]. Equation (2-13a) applies not only to conduction through a series of concentric hollow cylinders but also to a series of flat walls or a series of concentric hollow spheres.

Illustration 2. A standard 2-in. steel pipe (see Appendix) carrying superheated steam is insulated with 1.25 inches of a molded high-temperature covering made of diatomaceous earth and asbestos ($k = 0.058$). This covering is insulated with 2.5 inches of laminated asbestos felt ($k = 0.042$). In a test, the temperature of the surroundings was 86°F, the average temperature of the steam side of the pipe was 900°F, and the temperature of the outer surface of the lagging was 122°F.

a. Calculate the heat loss, expressed as Btu/(hr)(ft of length of pipe).
b. Calculate the temperature at the interface between the two layers of insulation.
c. Calculate the surface coefficient of heat loss h, expressed as Btu/(hr) (sq ft of outside lagging surface) (deg F difference from surface to room).

Solution. The following diameters are needed: i.d. of pipe, 2.07 in.; o.d. of pipe, 2.37 in.; mean diameter of pipe, 2.22 in.; o.d. of first covering, 4.87 in.; logarithmic-mean diameter, 3.48 in.; o.d. of second covering, 9.87 in.; mean diameter 7.07 in.

a. The heat loss per foot is calculated from Eq. (2-13a), using k of 23.5 for wrought iron (see Appendix) and a wall thickness of 0.154 in.

$$q = \frac{900 - 122}{\frac{0.154/12}{(23.5)(2.22\pi/12)} + \frac{1.25/12}{(0.058)(3.48\pi/12)} + \frac{2.5/12}{(0.042)(7.07\pi/12)}}$$

$$= \frac{778}{0.00094 + 1.97 + 2.68} = \frac{778}{4.65} = 167 \text{ Btu/(hr)(ft)}$$

b. Since temperature drop is proportional to resistance, $900 - t_i = (778)\left(\frac{1.97}{4.65}\right)$ = 330, whence t_i equals 570°F.

c. $$h = \frac{q}{A_0 \Delta t} = \frac{167}{\frac{9.87\pi(1)(122 - 86)}{12}} = 1.8 \text{ Btu/(hr)(sq ft)(deg F)}$$

Contact Coefficients. In the preceding illustration, in which two solids were in contact, no allowance was made for a temperature drop at the boundary, which presupposes perfect contact. However, this requires the absence of gases or vacant spaces caused by those blowholes, bubbles, rough surfaces, etc., which are very likely to be present where two solids are brought together. Even traces of poorly conducting material between metals, such as oxide films on the surface, will cause abrupt drops in the temperature.[2,30] It is usually difficult to estimate accurately the thickness of such films, but their effect may be serious. Where the gap between two flat walls is so narrow that natural convection can be ignored, heat will be transferred by both radiation and conduction and a quantitative estimate of the temperature potential across the gap can be made.

Weills and Ryder[35] measured contact, or junction, coefficients h_j between metal blocks having surface temperatures t_1 and t_2:

$$h_j = \frac{q/A}{t_1 - t_2} \qquad (2\text{-}14)$$

which increased with increased pressure on the blocks. For example, with rough aluminum at 300°F, h_j ranged from 500 at 0 gauge pressure to 11,000 at 3,800 pounds per square inch. At 300°F and a given pressure, h_j increased as roughness decreased. Brunot and Buckland[5] report contact coefficients for machined joints.

Cetinkale and Fishenden[6] measured contact coefficients for metal blocks of steel, brass, and aluminum ground to various degrees of roughness, with air, spindle oil, or glycerol in the voids at the junction. Pressures ranged from 19 to 800 pounds per square inch and h_j ranged from 550 to 12,500. An interesting theory was developed for predicting contact coefficients from fundamental factors.

Heat Meter. Over-all resistances for structures in service may be determined by the use of the heat meter,[3a,36] which measures the temperature drop through the known resistance of the meter. By simultaneously measuring the temperature gradient through the wall itself, the thermal conductivity of the whole wall or of any layer may be measured, even though the use of the meter reduces the heat flow compared with that from the bare wall. Precautions should be taken to secure data under steady conditions. Van Dusen and Finck[32] report experimentally determined over-all thermal resistances of a number of walls and also individual resistances of the various components; in general, fairly satisfactory agreement was found between the predicted values and observed results.

UNIFORM INTERNAL GENERATION OF HEAT IN BODIES WITH HEAT DISSIPATION AT ONLY ONE SURFACE

Flat Plate. Consider a flat metal plate ideally insulated except on one surface. Heat is generated uniformly throughout the plate by steady flow of electricity and is dissipated at the colder surface by transfer to a boiling liquid. Let q_T represent the total generation in Btu per hour. Since the generation is uniform throughout the thickness, the local heat current q_x at distance x from the adiabatic surface is

$$q_x = q_T \frac{x}{x_T} = -kA \frac{dt}{dx}$$

Integration, from $q = 0$ at $x = 0$ to $q = q_T$ at $x = x_T$, gives

$$q_T = \frac{k_m A (t_1 - t_2)}{x_T/2} \qquad (2\text{-}14a) \;\star$$

STEADY CONDUCTION

If the same heat-transfer rate had been obtained by supplying q_T on one side and removing it on the other, Eq. (2-6) shows that twice as much Δt between the surfaces of the plate would be required as that with uniform generation [Eq. (2-14a)].

Other Shapes. Assuming uniform generation throughout the volume, similar derivations for other shapes give the following equations:

Rod:
$$\frac{q_2}{4\pi L} = k_m(t_0 - t_2) \qquad (2\text{-}14b) \star$$

Hollow Cylinder, Adiabatic Inner Surface:
$$\frac{q_2}{2\pi L}\left(\frac{1}{2} - \frac{r_1^2}{r_2^2 - r_1^2}\ln\frac{r_2}{r_1}\right) = k_m(t_1 - t_2) \qquad (2\text{-}14c) \star$$

Hollow Cylinder, Adiabatic Outer Surface:
$$\frac{q_1}{2\pi L}\left(\frac{1}{2} - \frac{r_2^2}{r_2^2 - r_1^2}\ln\frac{r_2}{r_1}\right) = k_m(t_2 - t_1) \qquad (2\text{-}14d) \star$$

Sphere:
$$\frac{q_2}{8\pi r_2} = k_m(t_0 - t_2) \qquad (2\text{-}14e) \star$$

COMPLEX PROBLEMS IN CONDUCTION

In both the long hollow cylinder and the hollow sphere, the two isothermal surfaces (the source and the sink) were parallel; consequently the heat flowed only along the radii, the cross-sectional areas were simple functions of the radii, and $\int_1^2 dx/A$ was readily evaluated. Sometimes, with two-dimensional conduction, the isothermal surfaces are *not* parallel.† An example would be a steam pipe in a rectangular box filled with insulation, the axes of the pipe and the box being parallel; a number of other examples could be cited. Such cases of steady conduction can be solved by a number of methods, such as (1) two-dimensional mapping, (2) two-dimensional relaxation, (3) computing machines, (4) one of several types of electrical models,[12] (5) the membrane analogy, or soap-film technique,[31] and (6) fluid-flow mappers;[21a] many of these methods are reviewed by Jakob.[12]

Two-dimensional Mapping. This graphical method of Lehmann[17,1] is based on the fact that the isotherms and adiabatics must intersect at right angles. In illustration, consider a hollow thick-walled flue L ft long, having a rectangular cross section. Figure 2-4 shows a typical portion, one-fourth of the total; the temperatures of the inner and outer surfaces

† Equations (2-18a) and (2-18b) for three-dimensional conduction can be employed to estimate q in two-dimensional problems, such as long hollow flues, by omitting the corner corrections and noting that there are now only 4 inside edges, instead of 12.

are uniform. The heat may be considered to flow steadily from the inner surface at t_0 to the outer surface at t_N through a number of lanes of *equal heat flow*. Let the integral $-\int_0^N k\, dt$ be divided into N equal parts, so that the $k_{\Delta x}\, \Delta t_{\Delta x}$ product is constant:

$$k_{\Delta x}\, \Delta t_{\Delta x} = -\frac{1}{N}\int_0^N k\, dt = \frac{1}{N} k_m(t_0 - t_N) \qquad (2\text{-}15)$$

The incremental form of the conduction equation is

$$q_L = k_{\Delta x} A\, \Delta t_{\Delta x}/\Delta x = k_{\Delta x}(L\, \Delta y)\, \Delta t_{\Delta x}/\Delta x = q_L = k_m(t_0 - t_N)L\, \Delta y/\Delta x N$$

Since the rate of heat flow throughout a lane is constant and N, k_m, $t_0 - t_N$, and L are constant, the ratio $\Delta y/\Delta x$ *must* be constant, although

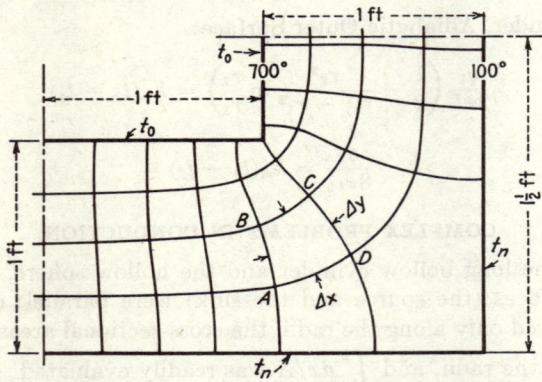

FIG. 2-4. Lanes of equal heat conduction in a representative portion of a hollow flue; see Illustration 3.

both Δy and Δx may vary. In Fig. 2-4, N has been arbitrarily taken as 4, and the adiabatics and isotherms have been drawn to intersect at right angles to form quadrilaterals with Δy equal to Δx, thus forming curvilinear squares. The total rate of heat flow through N_L lanes is

$$\Sigma_q = N_L k_m(t_0 - t_N)L/N \qquad (2\text{-}16) \star$$

In some cases it is more convenient to fix N_L, finding N from the map. Since the diagram is purely geometrical and is valid regardless of how k varies with t, it applies equally well when the direction of heat flow is reversed by interchanging the positions of the source and sink. Furthermore, if the locations of the adiabatic surfaces and those of the sources and sinks were interchanged, the same map applies but the new value of the ratio N_L/N would be the reciprocal of the former value. Veron[33] solves a variety of problems by the technique of two-dimensional mapping.

Illustration 3. In the example of Fig. 2-4, t_0 is 700°F, and t_N is 100°F. The map shows 8.2 lanes for one-fourth of the flue and N of 4. Consequently, using Eq. (2-16),

$$\frac{\Sigma q}{L} = \frac{N_L}{N} k_m (t_0 - t_N) = \frac{(4 \times 8.2)(k_m)(700 - 100)}{4} = 4920 k_m \quad \text{Btu}/(\text{hr})(\text{ft})$$

If $k = (0.43)(1 + 0.00038t)$, then $k_m = (0.43)(1 + 0.00038 \times 400) = 0.496$ and $\Sigma q = 2450$ Btu/(hr)(ft).

The spacing of the isotherms is independent of how k varies with t, but the temperatures corresponding to the intermediate isotherms, t_1, t_2, ..., t_{N-1}, depend on how k varies with t. If $k = k_0(1 + at)$, these intermediate temperatures may be obtained from equations of the form

$$\frac{t_0 - t_1}{t_0 - t_N} = \frac{1}{N} \frac{1 + a(t_0 + t_N)/2}{1 + a(t_0 + t_1)/2} \quad (2\text{-}16a)$$

If a equals zero, $t_0 - t_1 = t_1 - t_2 = (t_0 - t_N)/N$. In Illustration 3, t_1 is then 561°F; if k had been constant at 0.496, t_1 would have been 550°F.

Two-dimensional Relaxation. This procedure has been proposed in a number of publications.[7,8,29,37,†]

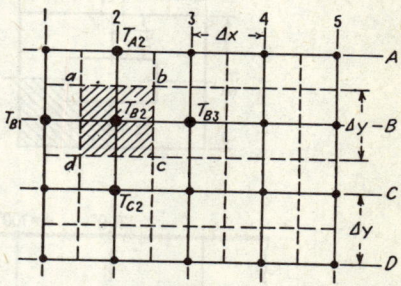

Fig. 2-5. Diagram[20a] for derivation of equation for two-dimensional conduction; method of relaxation.

Consider the case of a long rectangular flue with steady heat conduction along the x and y axes, but not along the z axis. Subscripts 1, 2, 3, etc., apply to planes parallel to the y axis, and letter subscripts A, B, C, etc., apply to planes parallel to the x axis. In the crosshatched portion $abcd$ of Fig. 2-5, it is pretended that there is an imaginary heat sink of strength q_{B2}. A heat balance on the volume (represented by the crosshatched zone $abcd$) gives

$$q_{B2} = k \frac{z \Delta y}{\Delta x}(t_{B1} - t_{B2}) + k \frac{z \Delta x}{\Delta y}(t_{A2} - t_{B2}) + k \frac{z \Delta y}{\Delta x}(t_{B3} - t_{B2}) + k \frac{z \Delta x}{\Delta y}(t_{C2} - t_{B2})$$

Using a square net ($\Delta y = \Delta x$), and designating q_{B2}/kz by $q_{B2}{}^*$, this equation becomes

$$\frac{q_{B2}}{kz} = q_{B2}{}^* = t_{B1} + t_{A2} + t_{B3} + t_{C2} - 4t_{B2} \quad (2\text{-}17) \star$$

When the steady state is reached, the imaginary sink $q_{B2}{}^*$ is inoperative and hence is equal to zero. The procedure is to guess the temperature

† The following treatment is due to Emmons.[8] The three-dimensional case has also been treated.[8]

pattern (using all knowledge available) and keep changing the guesses until q^* of Eq. (2-17) is zero at every point in the net, whereupon the correct solution for the distribution of temperature is obtained. Emmons,[8] Dusinberre,[7] and Jakob[12] give advice as to detailed procedure and solve illustrative problems.

Illustration 4. A hollow rectangular duct has inside dimensions of 2 by 1 ft, walls 1 ft thick, and outside dimensions of 4 by 3 ft. The inner walls are at a uniform temperature of 700°F; the outer walls are at a uniform temperature of 100°F. Find the steady rate of heat loss per unit length of duct (a) by the relaxation technique,

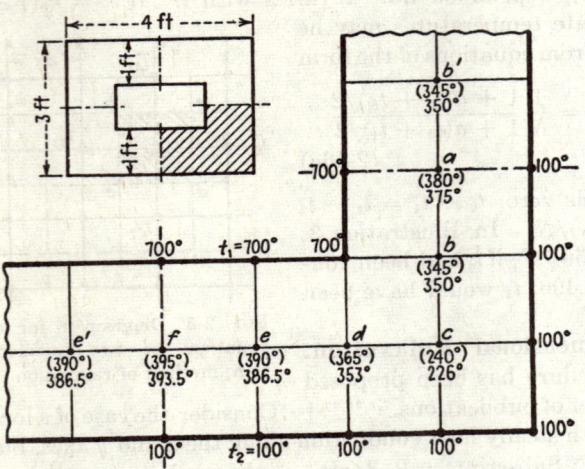

Fig. 2-6. Coarse net ($\Delta y = \Delta x = 0.5$ ft) for two-dimensional conduction in a representative portion of a hollow flue; temperatures in parentheses were estimated from a heat-flow map (Fig. 2-4); the other temperatures at points a to f were obtained by the relaxation procedure (Table 2-2).

using a square net ($\Delta x = \Delta y = 0.5$), and (b) using Eq. (2-18a), omitting the corner correction. This problem was solved in Illustration 3 and the map in Fig. 2-4.

Solution. (a) The duct is divided by adiabatic planes into N_s symmetrical sections; Fig. 2-6 shows a representative section of the flue, one-fourth in this case. At point c

$$q_c^* = \frac{q_c}{kz} = t_b + t_d + t_2 + t_2 - 4t_c \tag{2-17a}$$

Values of q_b^*, q_d^*, and q_e^* are calculated similarly. Points a and f lie on planes between symmetrical sections; by symmetry $t_b = t_b'$, and $t_e = t_e'$. Consequently

$$q_a^* = 2t_b + t_1 + t_2 - 4t_a \tag{2-17b}$$
$$q_f^* = 2t_e + t_1 + t_2 - 4t_f \tag{2-17c}$$

Estimates of temperature at points a through f are made from the map in Fig. 2-4 and are shown in parentheses in Fig. 2-6. Initial values of q^*, called the *residuals*, are calculated for points a through f and entered on the work sheet, Table 2-2. The largest residual (q_c^* of 50) is reduced to a value near 0 by changing the corresponding temperature t_c. Since Eq. (2-17a) contains the term $-4t_c$, a change of $-12°$ in t_c will change q_c* by $(-4)(-12) = 48°$ This change in t_c will also change the residuals at b and d by $-12°$. Consequently the new residuals are $q_b* = 40 - 12 = 28°$, $q_c* = -50 + 48 = -2$, and $q_d* = -30 - 12 = -42°$. The new value of t_c is $240 - 12 = 228°$; temperatures at other points are unchanged. The relaxation procedure is repeated until all values of $q*$ change by insignificant amounts. The final results are given in the last line of the work sheet. At step 6 the temperatures were within $2°$ of the final values in line 13. The value of q/L is found by considering

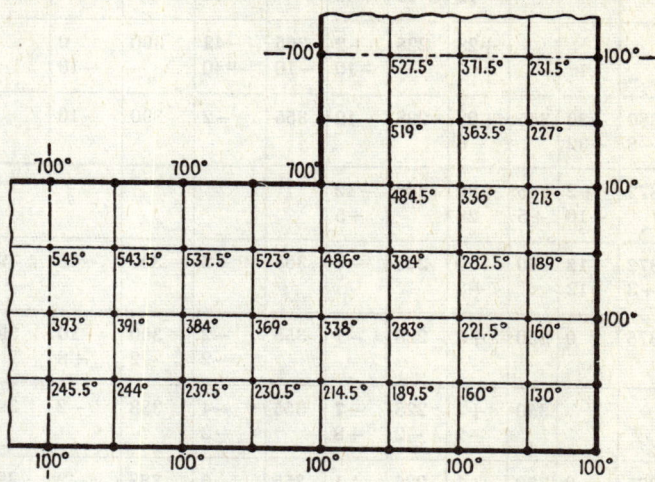

FIG. 2-7. Results of recalculation of Illustration 4, using finer net ($\Delta y = \Delta x = 0.25$ ft).

the points as junctions of "rods" which conduct the heat. The heat flow through the rods at planes of symmetry is halved since only half the flow is through each section. Considering the outer surface of the quarter section shown in Fig. 2-6,

$$\frac{q_2}{L} = k_m \Sigma \Delta t = k_m \left(\frac{293.5}{2} + 286.5 + 253 + 126 + 126 + 250 + \frac{275}{2}\right)$$
$$= 1326 k_m$$

or for the whole duct

$$\frac{q_2}{L} = 5304 k_m$$

(b) By Eq. (2-18a) the mean area per unit length neglecting the corner correction is

$$A_m = 6 + 0.54(1)(4)$$
$$= 8.16 \text{ sq ft per ft of length}$$

and

$$q/L = k_m(8.16)(600)/1$$
$$= 4900 \ k_m$$

which is close to the value of part (a). The solution of part (a) was repeated for a square net of $\Delta x = \Delta y = 0.25$ and the results are shown in Fig. 2-7. This new solution would indicate a slightly lower value than calculated in part (a).

TABLE 2-2. Work Sheet for Relaxation, Illustration 4

Point	a t	a q*	b t	b q*	c t	c q*	d t	d q*	e t	e q*	f t	f q*
Step 1	380	−30	345	+40 −12	240 −12	−50 +48	365	−30 −12	390	0	395	0
Step 2				+28	228	−2 −10	365 −10	−42 +40	390	0 −10		
Step 3	380 −8	−30 +32	345	+28 −8	228	−12	355	−2	390	−10		
Step 4	372	+2 +10	345 +5	+20 20	228	−12 +5						
Step 5	372 +3	+12 −12	350	0 +3	228	−7	355	−2	390	−10	395	0
Step 6	375	0	350	+3	228	−7	355	−2 −2	390 −2	−10 +8	395	0 −4
Step 7			350	+3 −2	228 −2	−7 +8	355	−4 −2	388	−2	395	−4
Step 8	375	0	350	+1	226	+1 −1	355 −1	−6 +4	388	−2 −1	395	−4
Step 9					226	0	354	−2	388	−3 −1	395 −1	−4 +4
Step 10							354	−2 −1	388 −1	−4 +4	394	0 −2
Step 11					226	0 −1	354 −1	−3 +4	387	0 −1	394	−2
Step 12					226	−1	353	+1	387	−1 −0.5	394 −0.5	−2 +2
Step 13							353	+1 −0.5	387 −0.5	−1.5 +2.0	393.5	0 −1.0
Result	375	0	350	+1	226	−1	353	+0.5	386.5	+0.5	393.5	−1

Steady Conduction from Buried Isothermal Heat Sources to the Air.
When the steady state is attained, all the heat from the surface of the isothermal source at t_1 is conducted through the medium of low thermal conductivity k to the surface of the ground at t_2 and thence to the air at t_a. If the dimensionless submergence (z/L) is relatively small, it is important to allow for the thermal resistance at the boundary between the surface of the ground at t_2 and the ambient air at t_a. The results[18a] are tabulated below for several cases in terms of the resistance factor R in the equation

$$q = (t_1 - t_2)/R$$

TABLE 2-3. RESISTANCE FACTORS R BETWEEN VARIOUS HEAT SOURCES AND A SEMI-INFINITE BODY OF LOW THERMAL CONDUCTIVITY k
(From A. L. London)[18a]

Description	Equation for $R = (t_1 - t_2)/q$
Sphere of diameter D with center at distance z below surface, z = positive	$R = \dfrac{1}{2\pi D k}\left(1 - \dfrac{D}{4z}\right)$
Horizontal cylinder of length L and diameter D with axis at distance z below surface[a]	$R = \dfrac{1}{2\pi L k}\ln\left(\dfrac{4z}{D}\right)$ $D \ll z,\ z \ll L$
Horizontal thin circular disk of diameter D well below surface, 10 per cent[b]	$R = \dfrac{1}{4Dk}$
Horizontal torus of mean diameter D_m and thickness y, with axis at distance z below surface	$R = \dfrac{1}{2\pi^2 D_m k}\ln\dfrac{4z}{y}$ $z > y,\ D_m \gtrsim 20y$
Thin horizontal rectangle with larger and smaller sides D_1 and D_2 well below surface	$R = \dfrac{1}{2\pi D_1 k}\ln\dfrac{2\pi z}{D_2}$ $D_1 \gg D_2,\ z > 2D_2$

[a] A more exact solution is
$$R = \frac{\cosh^{-1}(2z/D)}{2\pi k L} \quad \text{for } L \gg z$$
[b] A more exact solution is $(1 - D/5.66z)/4.44Dk$.

Three-dimensional Heat Conduction. Such problems arise in connection with furnaces of rectangular cross section having walls at least one-half as thick as the shortest inside diameter. Not only does heat flow at right angles to the entire inside area, but it also flows at various angles through the 12 edges and 8 corners of the outside walls. For such cases, the integration of the basic differential equation becomes difficult. Langmuir[16] solved the problem by experimentally determining the electrical conductance of a solution of copper sulfate in a container of the desired shape and, by comparing this with the electrical conductance of the same solution in a container of constant cross section, deduced the equations given below for an approximation to the average area A_m to be used in Eq. (2-7). This procedure is a good example of the theory of models; since the basic equations for the conduction of electricity and of heat are identical in form, the results are applicable to both phenomena. The empirical equations given below agree closely with those which Langmuir obtained from theoretical analyses based on isothermal surfaces at the boundaries. The results are also supported by subsequent data.[24]

Consider only bodies bounded by right rectangular parallelepipeds, corresponding inner and outer surfaces being parallel and in all cases the same distance x apart. All faces intersect at right angles. Four special cases are given, and it is believed that an understanding of these will enable the reader to handle any set of conditions of this type met in practice.

Case a. The Lengths y of All Inside Dimensions Exceed One-fifth of the Thickness x of the Walls. The sum total of the lengths of all the inside edges is Σy. To the actual inside area A_1 must be added $0.54x\Sigma y$ to correct for the 12 edges and $0.15x^2$ for each of the 8 corners, this sum being called the average area A_m:

$$A_m = A_1 + 0.54x\Sigma y + 1.2x^2 \qquad (2\text{-}18a) \star$$

Case b. Length y of One Inside Dimension Is Less than One-fifth the Thickness x of Walls. In this case, the lengths of the four inside edges, less than $x/5$, are neglected in determining Σy. Then

$$A_m = A_1 + 0.465x\Sigma y + 0.35x^2 \qquad (2\text{-}18b) \star$$

where Σy is the sum total of all the remaining 8 inside edges, each of which is greater than $x/5$.

Case c. Lengths of Two Inside Dimensions Are Each Less than One-fifth the Thickness x of the Walls.

$$A_m = \frac{2.78 y_{\max} x}{\log_{10}(A_2/A_1)} \qquad (2\text{-}18c) \star$$

where $y_{\max}$ is the longest dimension of the interior.

Case d. All Three Interior Dimensions Are Less than One-fifth the Thickness of the Walls. Using the same nomenclature,

$$A_m = 0.79\sqrt{A_1 A_2} \qquad (2\text{-}18d) \star$$

THERMAL CONDUCTION IN FLUIDS

The law for the conduction of heat is the same for liquids and gases as for solids, the values of k being, usually, very small. At room temperature, for example, k, expressed as Btu/(hr)(sq ft)(deg F/ft), is 0.34 for stationary water and 0.015 for stationary air, as compared with 220 for copper and 0.7 for firebrick. The thermal conductivities of molten metals[18] and fused salts are considerably higher than that of water. In most industrial heat exchangers involving forced convection of fluids, heat transfer by convection strongly supplements the heat transfer by conduction, and consequently such problems cannot be solved merely by reference to tables of thermal conductivities of the fluids; convection is treated in later chapters. With laminar flow of viscous oils in tubes under conditions where natural convection effects are minimized, conduction is controlling. With transparent gases such as air, heat may be transferred by conduction, convection, and radiation.

Thermal Conductivity of Liquids. In order to eliminate natural convection, Bridgman[4,†] employed a liquid layer $\frac{1}{64}$ inch in thickness and a temperature difference of about 1°F and obtained reliable values. Bates[3] developed an apparatus in which large temperature differences and thick layers of liquid are used; natural convection was eliminated by employing downward flow of heat and lateral guarding against heat loss.

The thermal conductivities of practically all nonmetallic liquids at normal temperatures lie between 0.050 and 0.150 Btu/(hr)(sq ft)(deg F per ft), although the value for water at 68°F is 0.344. The thermal conductivities of most liquids decrease with increased temperature, although water and certain aqueous solutions are exceptions, as shown in the Appendix. The effect of pressure on the thermal conductivities of liquids has been studied by Bridgman, who found that an increase from 1 to 2000 atm increased the thermal conductivities of 28 liquids from 11 to 15 per cent and that at 12,000 atm the conductivities were approximately doubled. Values of k are given in the Appendix.

In 1923, Bridgman[4] proposed the theoretical dimensionless equation for liquids:

$$k = 3K_B V_a / Z_c^2 \tag{2-19}$$

k being the thermal conductivity, K_B the Boltzmann gas constant per molecule in heat units, V_a the acoustic velocity in the liquid, and Z_c the mean distance of separation of centers of molecules, if there is assumed an arrangement cubical on the average and if Z_c in feet is calculated by the equation $Z_c = (M'/\rho)^{1/3}$, where M' represents the weight‡ of one molecule in pounds and ρ is density in pounds per cubic foot. The values of k calculated by means of Eq. (2-19) are shown to approximate Bridgman's experimental data for 11 liquids, the deviation being from -13 to $+38$ per cent. The velocity of sound in the liquids was not measured but was calculated from the known compressibility. The agreement is remarkable in view of the fact that the equation involves no empirical constant, and the equation gives the correct sign of the temperature coefficient at 1 atm for both water and the organic liquids.

---

† J. F. D. Smith[27] later used the same apparatus.
‡ $M' = 3.65M \times 10^{-27}$, where M is the molecular weight.

The Kardos modification,[14] which employs distance between *surfaces* of adjacent molecules rather than distance between centers, leads to the equation

$$k/\rho c_p V_a = 3.12 \times 10^{-10} \text{ ft} \qquad (2\text{-}20)$$

which agrees somewhat more closely with observed values than Eq. (2-19). The empirical equation of Smith,[28] based on data for water, 6 paraffin alcohols, 13 pure hydrocarbons, 6 petroleum fractions, and 20 miscellaneous liquids, is **recommended** for estimating the thermal conductivities of *nonmetallic* liquids at 86°F and 1 atm pressure, in those cases in which measured values are not available:

$$k = 0.00266 + 1.56(c_p - 0.45)^3 + 0.3(\rho'/M)^{1/3} + 0.0242(\mu'/\rho')^{1/6} \qquad (2\text{-}21) \star$$

In this equation, k is in Btu/(hr)(sq ft)(deg F per ft); c_p is the specific heat; ρ' is the specific gravity relative to water; M is the average molecular weight; μ' is the viscosity in centipoises. The thermal conductivities of binary mixtures of miscible liquids are a complex function of the concentration and the values of k of the components.

Riedel[26a] measured values of k at 68°F for a large number of liquids: 25 hydrocarbons; 86 organic compounds containing oxygen, nitrogen, phosphorus, or sulfur; 10 binary systems of organic compounds and 43 binary systems of inorganic salts in aqueous solution.

The thermal conductivities of molten metals are of interest in equations for forced convection for cooling of nuclear reactors or other high-temperature sources; a recent publication[18] tentatively selects values of k; where k values are not available, Powell[26] employed the Lorenz rule for estimating minimum values.

Thermal Conductivity of Gases. The relation of the thermal conductivity to the viscosity of a gas is of particular theoretical interest in that the data support to a remarkable degree the theoretical relation deduced from the kinetic theory.[13] Maxwell[21] used the dimensionless group

$$\frac{k}{\mu c_v} = N_{Ml} \qquad (2\text{-}22)$$

where μ is the absolute viscosity of the gas, c_v is the specific heat at constant volume, and N_{Ml} is the dimensionless Maxwell modulus. Eucken[9] has suggested the empirical relation

$$N_{Ml} = 0.25(9\gamma - 5) \qquad (2\text{-}22a)$$

where γ is the ratio of the specific heat at constant pressure to that at constant volume. Elimination of N_{Ml} from Eqs. (2-22) and (2-22a) gives the recommended relation for estimating the Prandtl modulus N_{Pr}, a dimensionless group $c_p\mu/k$ of importance in convection:

$$N_{Pr} = \frac{c_p\mu}{k} = \frac{4}{9 - (5/\gamma)} \tag{2-22b} \star$$

where experimental data are lacking. Values of the term $c_p\mu/k$ for gases at atmospheric pressure are shown in the Appendix.

According to the kinetic theory of gases, the mean free path of a gas molecule is *inversely* proportional to the absolute pressure, but the number of molecules per unit volume is *directly* proportional to the absolute pressure; consequently k should be independent of pressure. This simple rule fails at low pressures where the mean free path approaches or exceeds the width of the gas space and the thermal conductivity becomes proportional to the absolute pressure.[15]

For a gas above its critical temperature and pressure, the ratio of k to that at 1 atm increases with increase in reduced pressure P/P_c and with decrease in reduced temperature T/T_c (see Appendix).

PROBLEMS

1. A large flat furnace wall consists of three layers: 4 inches of kaolin insulating firebrick on the inside, 5 inches of kaolin insulating brick in the middle, and 5 inches of magnesite brick on the outside. If the inner surface is at 1375°F and the outer surface is at 215°F, what is the rate of heat loss per unit area through the wall? Use linear interpolation of the thermal-conductivity data in the Appendix.

2. A standard 2-inch steel pipe carrying steam is insulated with a 3-inch layer of glass wool. The outside surface of the pipe is at 400°F, and the outside surface of the covering is at 150°F. Estimate the rate of heat loss from the pipe, expressed as Btu per hour per 100 ft of pipe.

3. A large flat furnace wall is presently constructed of several layers of conventional insulating materials. The inner surface is maintained at 2400°F. Under normal operating conditions, the outer surface attains a temperature 350°F, and the heat loss through the wall is 700 Btu/(hr)(sq ft).

   a. In order to make an economic evaluation of a proposal for further insulation on the outside of the furnace, determine the percentage reduction in the heat loss as a function of the thickness and the mean thermal conductivity of the added insulating material. Assume that the rate of heat loss per unit area is directly proportional to the difference between the temperature of the outer surface of the wall and the temperature of the surroundings (60°F).

   b. What additional assumptions are necessary for such a calculation?

   c. In view of the general behavior of most insulating materials, are the assumptions of part *b* most likely to make the calculated percentage reduction greater or less than the percentage reduction determined if these assumptions were not made?

4. A long horizontal steam pipe 12 inches o.d. carries saturated steam at a temperature of 400°F. The outside of the pipe may be assumed to be essentially at the saturation temperature of the steam. The pipe is centered in a 2-ft-square sheet-metal duct. The space between the outside of the pipe and the duct is filled with powdered magnesia insulation, $k = 0.35$ Btu/(hr)(sq ft)(deg F per ft). It may be assumed that the sheet-metal duct is at a uniform temperature of 100°F. Calculate the rate of heat loss from the steam pipe per foot of pipe length by:

   a. Mapping.
   b. Relaxation.

5. Consider a rod having an o.d. of 1.0 inch, in which heat is generated internally according to the equation

$$w = w_1 \left[1 - \left(\frac{r}{R}\right)^2 \right]$$

where w is the local generation per unit volume at the local radius r, R is the total radius, and w_1 is the local generation per unit volume at the center line. The total amount of heat leaving the surface is uniform along the longitudinal axis at 500,000 Btu/(hr)(sq ft). Calculate the temperature drop from the center line of this rod to its surface. The thermal conductivity of the rod is 19 Btu/(hr)(sq ft) (deg F per ft).

6. A long steam pipe of o.d. D_1 is covered with thermal insulation having an o.d. D_2. The outer surface of the pipe remains at a constant temperature t_1 as more insulation is added (D_2 is increased). Is it ever possible that adding insulation will cause an increase in heat loss? If possible, under what conditions does this situation arise? The insulating material has a thermal conductivity of k.

The heat loss per unit area of outside surface of the insulation is assumed to be directly proportional to the difference between the temperature of the outer surface of the insulation and the temperature of the surroundings t_s (assumed to be constant). The proportionality constant is called the surface coefficient of heat transfer and is designated by the symbol h.

CHAPTER 3

TRANSIENT CONDUCTION

Abstract. The first part of this chapter presents the analytical solution to the transient heating or cooling of a slab, semi-infinite solid, sphere, long cylinder, and the buried long cylinder. The method for treating a brick-shaped solid and a short cylinder is given. Charts are also given for obtaining the midplane temperatures of a long square bar, a cube, and a cylinder, L/D of 1, with negligible surface resistances.

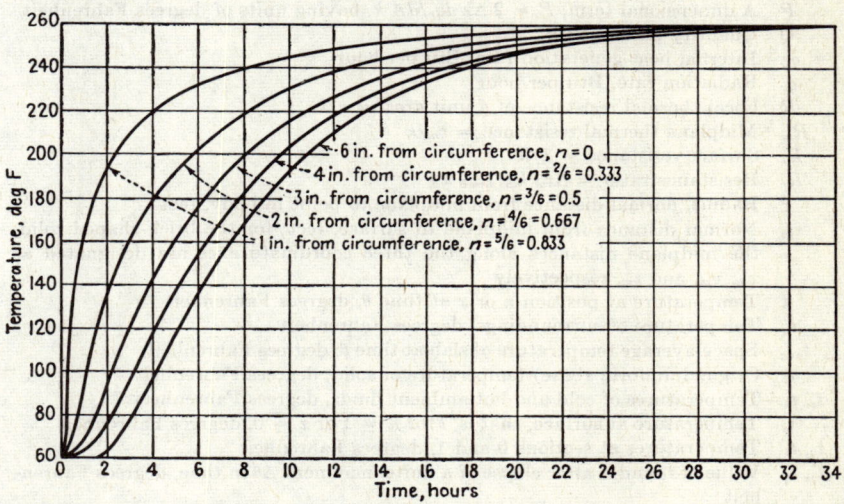

FIG. 3-1. Time-temperature relations for unsteady-state heating of round timbers in steam bath. (*Based on data of MacLean.*[28])

In many cases the actual operating conditions do not correspond to those for which the analytical relations were derived. Thus the ambient temperature, the heat-transfer coefficient, and even the physical properties of the solid may vary, and the initial temperature distribution may be of some nonuniform type not hitherto treated analytically. Heat may be supplied at a uniform rate at a surface or generated within the solid. Some of these complex cases are treated by numerical methods not requiring the calculus, and references are given for others. A graphical method is included for the slab.

The third part contains miscellaneous references.

Table 3-1. Nomenclature

A	Area through which heat flows at right angles, square feet
a_1	Constant, equal to $(\pi/2)^2$
c_p	Specific heat of solid, Btu/(lb)(deg F)
d	Prefix, indicating derivative
$G_{\Delta x, \Delta \theta}$	A dimensional term, equal to $q_G \, \Delta x / \Delta k$, having units of degrees Fahrenheit
h	Coefficient of heat transfer between surroundings at t_a and surface at t_s, Btu/(hr)(sq ft)(deg F)
k	Thermal conductivity of solid, Btu/(hr)(sq ft)(deg F per ft)
L	Length, feet
L'	Ratio of total volume to heat-transfer surface, feet
M	A dimensionless group, $M = (\Delta x)^2 / \alpha \, \Delta \theta$
m	Resistance ratio $R_s/R_m = k/r_m h$, dimensionless; for the semi-infinite solid, $m = k/hx$
N	A dimensionless group, $N = h \, \Delta x / k$
$N_{\Delta x}$	Number of imaginary slices, each Δx ft thick, in a slab
n	Position ratio r/r_m, dimensionless
P	A dimensional term, $P = 2 \, \Delta x \, dq_r / dA \, k$, having units of degrees Fahrenheit
Q	Quantity of heat, Btu
q_G	Internal heat-generation rate, Btu per hour
q_r	Radiation rate, Btu per hour
R	Local thermal resistance of a unit area $= x/k$
R_m	Midplane thermal resistance $= r_m/k$
R_s	Surface resistance $= 1/h$
R_r	Resistance ratio $= \Delta x_A \, k_B / \Delta x_B \, k_A$
r	Radius, normal distance from midplane to point in body, feet
r_m	Normal distance from midplane to surface, feet; for the brick-shaped solid, the midplane distances along the three coordinate axes are designated as x_m, y_m, and z_m, respectively
t	Temperature at position n or x at time θ, degrees Fahrenheit
t_a	Temperature of surroundings, degrees Fahrenheit
t_{avg}	Space-average temperature of slab at time θ, degrees Fahrenheit
t_b	Original uniform (base) temperature of solid, degrees Fahrenheit
t_c, t_h	Temperatures of cold and hot ambient fluids, degrees Fahrenheit
t_s	Temperature at surface, that is, t for $n = 1$ or $x = 0$, degrees Fahrenheit
t_0, t_1	Temperatures at sections 0 and 1, degrees Fahrenheit
t_0', t_1'	Value of t_0 and t_1 after elapse of a finite increment $\Delta \theta$ in time, degrees Fahrenheit
$t^\star$	Temperature of outer surface of fictive half slice, degrees Fahrenheit
V	Volume of solid, cubic feet
x	Normal distance from surface to point, feet
X	"Relative-time" ratio $= \alpha \theta / r_m^2$, Fourier number, dimensionless
Y	Unaccomplished temperature change $(t_a - t)/(t_a - t_b)$, dimensionless; Y_{avg} for space-mean value, Y_m at midplane, Y_s at surface
y	Coordinate at right angles to x and z axes, feet
Z	A dimensionless group $= x/(2 \sqrt{\alpha \theta})$
z	Coordinate at right angles to y and x axes, feet

Greek

α	$k/\rho c_p =$ thermal diffusivity, square feet per hour
Δ	Prefix, indicates finite increment
∂	Prefix, indicating partial derivative
θ	Time, from start of heating or cooling, **hours**
ρ	Density of solid, **pounds per cubic foot**
∞	Symbol for infinity

Introduction. In the transfer of heat by conduction, the case of the *steady* state, where the temperature at a given point is independent of time, has been treated in Chap. 2. The heat conduction is said to be in the *unsteady* state when the temperature at a given point varies with time. Unsteady-state conduction is involved in the quenching of billets, the annealing of solids, the manufacture of glass, the burning of bricks, the steaming of wood, and the vulcanization of rubber. Where chemical reactions occur during the heating period, as in the vulcanization of rubber, heat transfer plays a particularly important role, because the rate of the reaction increases rapidly with small increases in temperature. Figure 3-1 shows diagrammatically the types of curves obtained upon exposing a homogeneous solid to surroundings at a uniform temperature t_a. As the heating is continued, the temperature at a given point asymptotically approaches the temperature of the heating medium. Whereas points near the surface quickly approach the temperature of the surroundings, those in the interior lag far behind.

I. ANALYTICAL SOLUTIONS

The following section outlines the mathematics involved in calculating the relations between temperature and time for various points or sections of solids of several shapes and includes graphical solutions of the important equations. The relations involve the thermal conductivity, density, and specific heat of the body, its shape and size, and the external conditions, including the temperature of the surroundings and the coefficient of heat transfer between surroundings and the surface.

The following section on Theory outlines the procedure by which the integrated relations are obtained but may be omitted by readers interested only in applications of the theory.

Theory. As introduction, a simple limiting case of unsteady heat conduction is discussed. Consider a thin slab of metal of volume V, total surface area A, and thickness $2r_m$, at temperature t, in contact with warmer air at uniform temperature t_a. At any time θ from the start of the heating operation, the quantity of heat dQ transferred in the short time $d\theta$ depends upon the surface area of the slab, the difference in temperature between the air and the surface of the metal, and a factor h, called the coefficient of heat transfer from the surroundings to the surface, $dQ/d\theta = hA(t_a - t_s)$. Under such conditions the numerical value of h will be relatively small, and for a reasonable value of the temperature difference $t_a - t_s$ the corresponding rate of heat transfer per unit area $dQ/A\, d\theta$ will be small. Consequently, because of the high value of the thermal conductivity k of the metal and its small thickness, the metal temperature t will be substantially uniform and equal to t_s. Then, by a heat balance on the slab, having density ρ and specific heat c_p,

$$dQ = hA(t_a - t)\, d\theta = V\rho c_p\, dt$$

Assuming $hA/V\rho c_p$ constant,† integration from $t = t_b$ to $t = t$ and $\theta = 0$ to $\theta = \theta$

† If h is constant, a plot of $\ln(t_a - t)$ vs. θ would give a straight line having a negative slope equal to $hA/V\rho c_p$.

gives

$$\ln \frac{t_a - t_b}{t_a - t} = \frac{hA\theta}{V\rho c_p} \quad (3\text{-}1) \star$$

where ln represents a Napierian logarithm equal to 2.303 log$_{10}$. The same result could be obtained by using a *logarithmic*-mean value of the temperature difference $t_a - t$ in the expression

$$\frac{Q}{\theta} = hA(t_a - t)_m = \frac{V\rho c_p (t - t_b)}{\theta} \quad (3\text{-}1a)$$

The general case involving an appreciable temperature gradient through the solid is solved by application of an appropriate form of the conduction equation.

Derivation of Differential Equation for Transient Conduction. The general differential equation for transient conduction of heat is obtained from the familiar basic Fourier equation for the conduction of heat and a heat balance. Consider an element of volume with area $dy\,dz$ and thickness dx. The heat entering along the x axis is $-k_x(dy\,dz)(\partial t/\partial x)\,d\theta$, and that leaving is

$$-(dy\,dz)\left[k_x \frac{\partial t}{\partial x} + \frac{\partial}{\partial x}\left(k_x \frac{\partial t}{\partial x}\right)dx \right]d\theta$$

The difference between that entering and leaving along all three axes is equated to that stored in the element, giving the general differential equation for transient heat conduction, expressed in rectangular coordinates:

$$\frac{1}{\rho c_p}\left[\frac{\partial}{\partial x}\left(k_x \frac{\partial t}{\partial x}\right) + \frac{\partial}{\partial y}\left(k_y \frac{\partial t}{\partial y}\right) + \frac{\partial}{\partial z}\left(k_z \frac{\partial t}{\partial z}\right) \right] = \frac{\partial t}{\partial \theta} \quad (3\text{-}2)$$

Upon neglecting variations of k with temperature, and assuming the substance to be homogeneous and isotropic, k is taken outside the parentheses of Eq. (3-2), giving the term $k/\rho c_p$, called the *thermal diffusivity*:

$$\alpha \equiv k/\rho c_p$$

The desired temperature-time-position relations for the heating or cooling of various shapes are obtained by integration of Eq. (3-2), substituting the boundary conditions for the case in question. For example, in the case of the infinite slab, *i.e.*, one having a very large ratio of surface to thickness, the heat flow is unidirectional, and Eq. (3-2) reduces to

$$\frac{k}{\rho c_p} \frac{\partial^2 t}{\partial x^2} = \frac{\partial t}{\partial \theta} \quad (3\text{-}2a)$$

For the special case of a slab having a thickness $2r_m$ and a negligible surface resistance, corresponding to an infinite value of the surface coefficient h, the surface temperature changes to the temperature of the surroundings immediately at zero time. The boundary conditions are then $t = t_a$ at $x = 0$ and at $x = 2r_m$; $t = t_b$ at $\theta = 0$, and $t = t_a$ at $\theta = \infty$. A solution is given by the rapidly converging infinite series

$$\frac{t_a - t}{t_a - t_b} = \frac{4}{\pi}\left(e^{-a_1 X} \sin\frac{\pi x}{2r_m} + \frac{1}{3} e^{-9a_1 X} \sin\frac{3\pi x}{2r_m} + \frac{1}{5} e^{-25a_1 X} \sin\frac{5\pi x}{2r_m} + \cdots \right) \quad (3\text{-}3) \star$$

where a_1 equals $(\pi/2)^2$ and X represents the dimensionless ratio $\alpha\theta/r_m^2$. The total heat absorbed by the slab up to any time θ is obtained by evaluating the integral of $dQ = (t - t_b)\rho c A\,dx$, from 0 to $2r_m$, giving

$$\frac{Q}{2r_m A\rho c(t_a - t_b)} = 1 - \frac{8}{\pi^2}\left(e^{-a_1 X} + \frac{1}{9} e^{-9a_1 X} + \frac{1}{25} e^{-25a_1 X} + \cdots \right) \quad (3\text{-}4) \star$$

TRANSIENT CONDUCTION

For a slab having a thickness $2r_m$ and a *finite* surface resistance, corresponding to a definite and constant value of h, the boundary conditions become

$$k\left(\frac{\partial t}{\partial x}\right)_{x=0} = h(t_s - t_a) \qquad k\left(\frac{\partial t}{\partial x}\right)_{x=2r_m} = -h(t_s - t_a)$$

$t = t_b$ at $\theta = 0$; $t = t_a$ at $\theta = \infty$. Integration of Eq. (3-2a) for this case leads to a dimensionless relation among t, θ, x, and h. Solutions for solids of various shapes are available in the literature,[†] but computations directly from the equations are very tedious because of the large number of terms.

A number of writers[‡] have plotted the theoretical relations for a number of shapes, in terms of the dimensionless ratios involved.

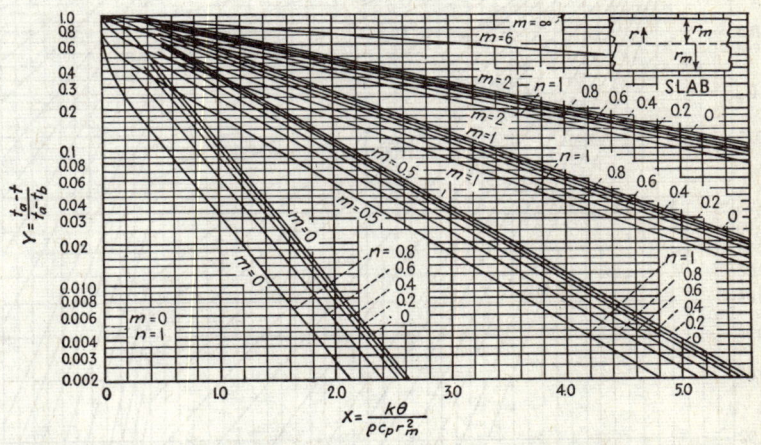

Fig. 3-2. Gurney-Lurie chart[19] for large slab; Figs. 3-3 to 3-5 give values of Y_m, Y_s, and Y_{avg} for a larger range of values of m and X. For small values of X, such that the change in midplane temperature is negligible, greater accuracy is obtained by evaluating Y from Fig. 3-6. Results for the brick-shaped solid may be obtained from Fig. 3-2 by use of the Newman method.[33]

The significance of all terms is given in Table 3-1, wherein any consistent units may be employed. For illustration, the symbols are defined in the system involving feet, pounds, hours, degrees Fahrenheit, and Btu.

The theoretical relations may be expressed in terms of four dimensionless groups:

1. An unaccomplished temperature change, $Y = (t_a - t)/(t_a - t_b)$.
2. A relative time, $X = \alpha\theta/r_m^2$.
3. A resistance ratio, $m = R_s/R_m = k/r_m h$.
4. A radius ratio, $n = r/r_m$.

[†] For example, see Fourier,[14] Carslaw,[5] Ingersoll, Zobel, and Ingersoll,[25] Gurney,[18] Byerly,[4] Newman,[33] Schack,[41] and Fischer.[12]

[‡] Williamson and Adams,[49] Gurney and Lurie,[19] Grober,[17] Schack,[41] Fishenden and Saunders,[13] Heisler,[21] and Goldschmidt and Partridge.[16]

In some charts Y or $1 - Y$ is plotted vs. X, for various values of m and n; alternatively, Y for a given n is plotted vs. m or $1/m$ for various values of X. These various procedures, or modifications of them, are allowable.

Slab. Figure 3-2 shows curves of Y vs. X for a slab, for various values of m and n, on semilogarithmic paper. Figures 3-3 and 3-4 show the

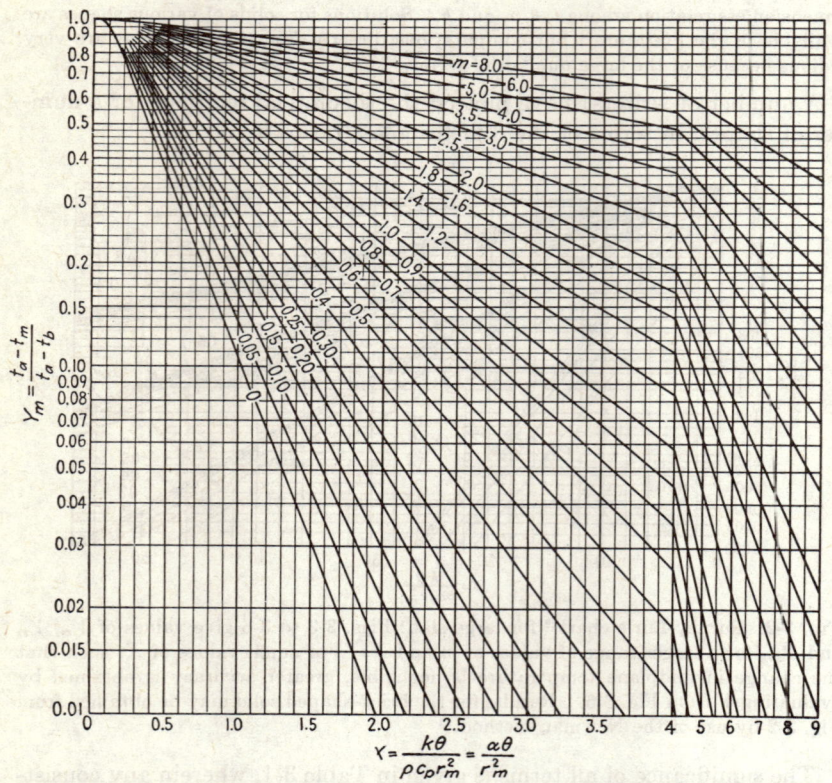

FIG. 3-3. Hottel chart[23] for large slab, for evaluation of midplane temperature t_m.

same predictions for a slab, plotted as Y_s at the surface, and as Y_m at the midplane, vs. X, for various values of m. Figure 3-5 shows the corresponding values of the space-mean values of Y_{avg} plotted vs. X for various values of m.†

Compare the heating of two slabs of the same material, the thickness in the first case being twice that in the second, and $m = 0$. For a given Y and a given position ratio n, it is seen from Fig. 3-2 that the value of

† Anthony[1] gives a number of analytical solutions for one-dimensional transients in single and composite slabs initially at uniform temperatures, for various boundary conditions.

X would be the same in both cases. However, $X = \alpha\theta/r_m^2$, and since the diffusivity α is the same in both cases, it is clear that the actual heating time θ will be directly proportional to the *square* of the thickness. In other words, it would take four times as long to heat the thick slab as the thinner one. If m were finite, the foregoing relation would not hold

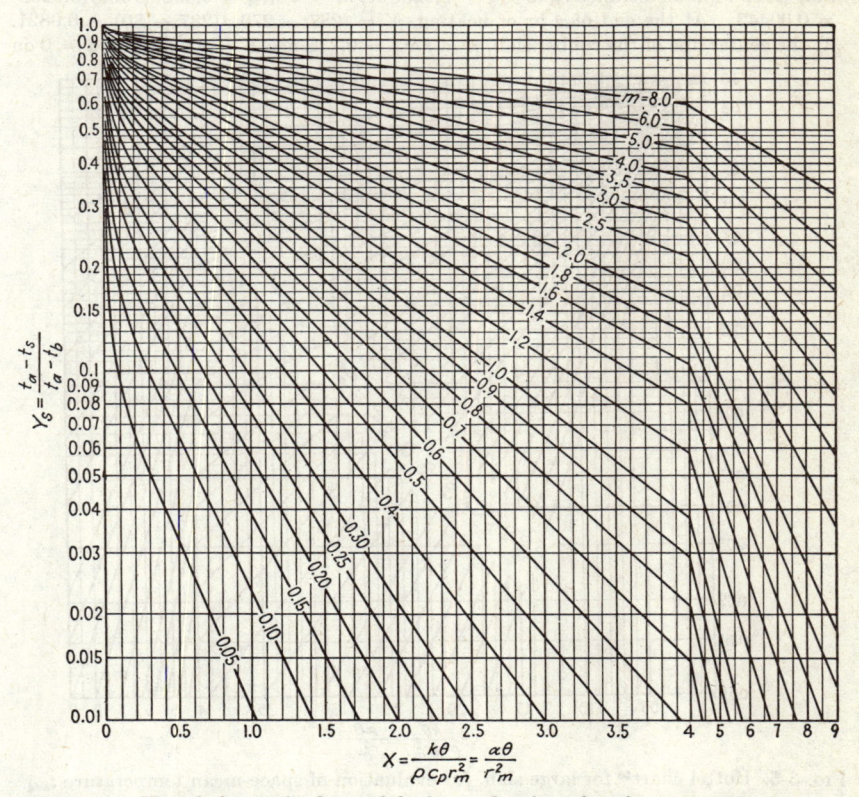

FIG. 3-4. Hottel chart[23] for large slab, for evaluation of surface temperature t_s.

unless m were the same in both cases, which would require that hr_m be constant.

The following examples illustrate the method of using the charts.

Illustration 1. A flat slab of rubber, 0.5 in. thick, initially at 80°F, is to be placed between two electrically heated steel plates maintained at 287°F. The heating is to be discontinued when the temperature at the center line of the rubber slab reaches 270°F.

 a. Calculate the length of the heating period.
 b. At the end of the run, what would be the temperature of the rubber in a plane 0.1 in. from the center line?
 c. How long, from the start of the heating, is required for the temperature to reach 270°F at the plane specified in part *b*?

HEAT TRANSMISSION

d. Repeat part *a* on the assumption that the rubber is heated from one face only, the other being perfectly insulated.

Data. Using the units of Table 3-1 (page 32) for the rubber, $k = 0.092$, and $k/\rho c_p = 0.0029$. Assume a constant coefficient h from metal to rubber of 1000.

Solution. *a.* All quantities will be expressed in the units mentioned above. Noting that the midplane distance r_m is $\frac{1}{48} = 0.0208$ ft, $m = k/hr_m = 0.092/(1000)(0.0208) = 0.00442$. At the end of θ hr of heating, $Y = (287 - 270)/(287 - 80) = 0.0821$. At the center line of the rubber slab, $n = r/r_m = 0$. Since $Y = 0.0821$ and $n = 0$ on

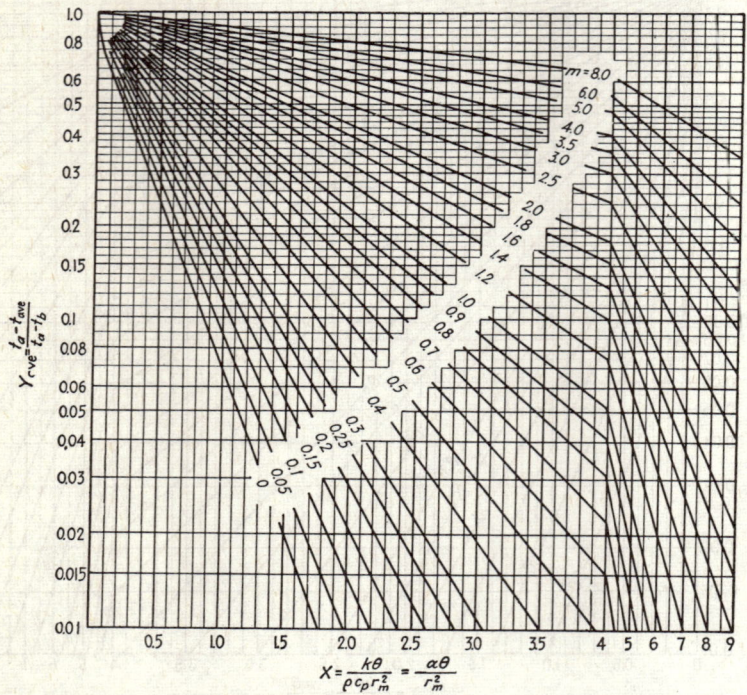

FIG. 3-5. Hottel chart[23] for large slab, for evaluation of space-mean temperature t_{avg}.

Fig. 3-2, interpolation to $m = 0.0044$ gives $X = 1.13 = k\theta/\rho c_p r_m^2 = 0.0029\theta/(0.0208)^2$, whence $\theta = 0.169$ hr, the answer to part *a*.

b. For the point 0.1 in. from the center line, $n = 0.1/0.25 = 0.4$, and, as before, $m = 0.0044$, and $X = 1.13$; from Fig. 3-2, Y is found to be 0.065. By definition, $Y = (287 - t)/(287 - 80)$, whence $t = 273.5°F$, the answer to part *b*.

In part *c*, where $Y = 0.0821$, $m = 0.0044$, and $n = 0.4$, Fig. 3-2 shows that $X = 1.03$. Since, by definition, $X = 0.0029\theta/(0.0208)^2$, $\theta = 0.153$ hr.

d. When heated from one side only, $r_m = 0.5/12 = 0.0417$ ft, $m = k/hr_m = 0.092/(1000)(0.0417) = 0.00221$. At the center line of the slab, n is 0.5, and $Y = (287 - 270)/(287 - 80) = 0.0821$. From Fig. 3-2, X is $0.955 = 0.0029\theta/(0.0417)^2$, whence $\theta = 0.573$ hr when heated from one side only, as compared with 0.169 hr when heated from both sides.

Olson and Schultz[34] give interpolation tables of Y for the slab, for values of X from 0 to 0.4.

TRANSIENT CONDUCTION 39

Brick-shaped Solid. For a brick-shaped solid, having total thicknesses $2x_m$, $2y_m$, and $2z_m$, the value of Y at a given time and position may be evaluated by the method of Newman[33] in which Y equals the product $Y_x Y_y Y_z$, where Y_x is evaluated from Fig. 3-2 at $X_x = k\theta/\rho c x_m^2$, at $n_x = x/x_m$, and at $m_x = k/hx_m$; similarly, Y_y and Y_z are read for the same θ at X_y, n_y, and m_y and at X_z, n_z, and m_z, corresponding to y_m and z_m.

Semi-infinite Solid. When heating a relatively thick body for a relatively short time, it is clear that the heat will penetrate only a short

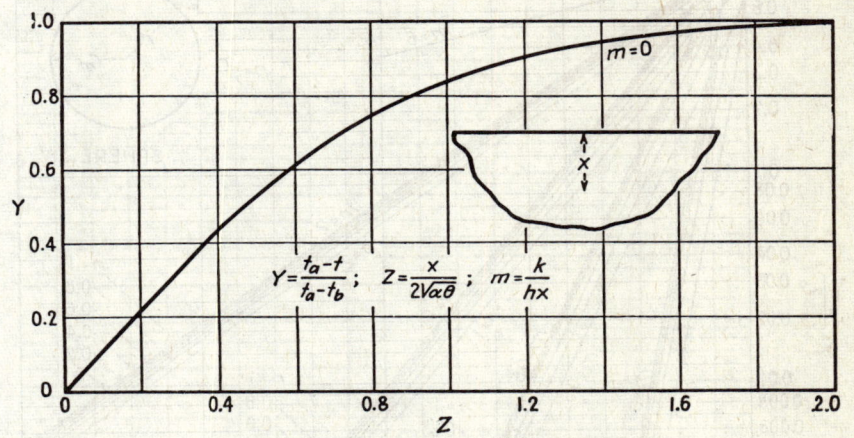

FIG. 3-6. Chart for semi-infinite solid.

distance in a zone near the surface and that the temperature at points farther below the surface would not be affected. This case corresponds to a semi-infinite solid. The unaccomplished temperature change Y, for a given value of $m(= k/hx)$, is plotted in Fig. 3-6 against the dimensionless term $Z[= x/(2\sqrt{\alpha\theta})]$, where x is the distance below the surface.†

† Where greater accuracy is desired than is obtainable from Fig. 3-6, for m of zero, one may use the equation of the curve:

$$Y = \frac{2}{\sqrt{\pi}} \int_0^Z e^{-Z^2} dZ$$

wherein $Z = x/(2\sqrt{\alpha\theta})$; values of this Gauss "error integral" Y are given in mathematical tables:

Z	0	0.1	0.2	0.3	0.4	0.5	0.6
Y	0	0.1125	0.2227	0.3286	0.4284	0.5205	0.6039

Z	0.8	1.0	1.2	1.4	2.0
Y	0.7421	0.8427	0.9103	0.9523	0.9953

Spheres and Long Cylinders.

Figures 3-7 and 3-8 show the values of Y for spheres and long cylinders plotted vs. X, for the various values of m and n. The curves given in Figs. 3-7 and 3-8 are based on the assumption of constancy of t_a, m, n, and $\alpha = k/\rho c_p$. Olson and Schultz[34] give interpolation tables of Y for the long cylinder, for values of X ranging from 0 to 0.4.

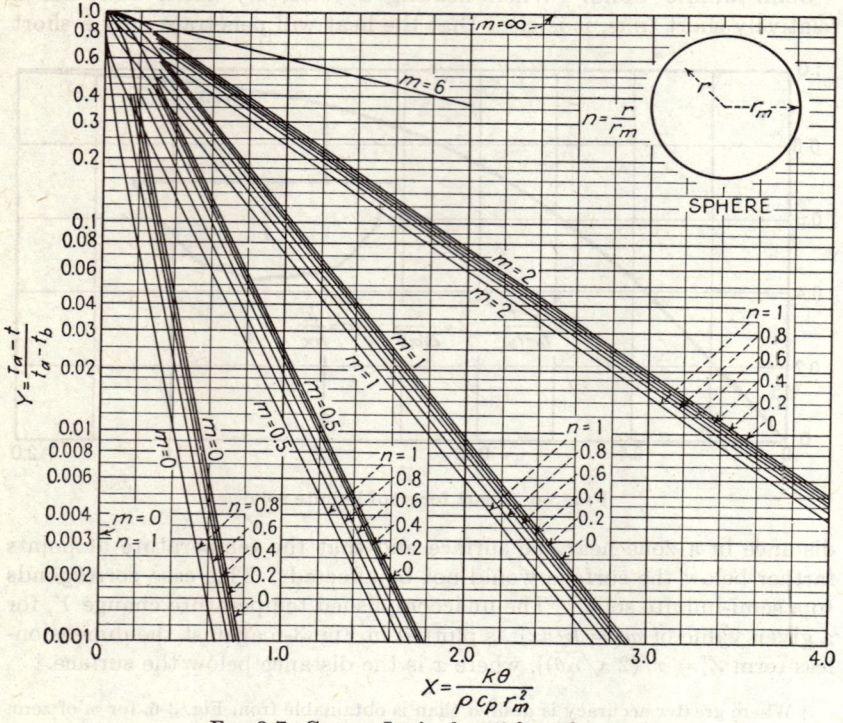

Fig. 3-7. Gurney-Lurie chart[19] for spheres.

Buried Long Cylinder.

Figures 3-8a and b are Gemant[15] graphs of the dimensionless accomplished temperatures and of the dimensionless instantaneous rates of heat transfer per unit length, respectively, plotted vs. the dimensionless time X after the surface temperature has suddenly been increased from t_b to t_s. Gemant used these relations to estimate the results of submerging multiple parallel steam-heated pipes near the surface of a concrete slab at the surface of the ground.

Hooper and Chang[22] measured the thermal conductivity of a sandy soil by means of a submerged heat probe and found that the thermal conductivity increased with increase in depth and with increase in ratio of water to soil.

Surface Resistance Controlling. In Figs. 3-2, 3-7, and 3-8, in order to facilitate extrapolation to values of m above 2, the approximate positions of curves for $m = 6$ are shown. The curves for $m = 6$ were based on the approximate equation (3-1), which was obtained by ignoring the temperature gradient in the solid. Substituting Y for the dimensionless

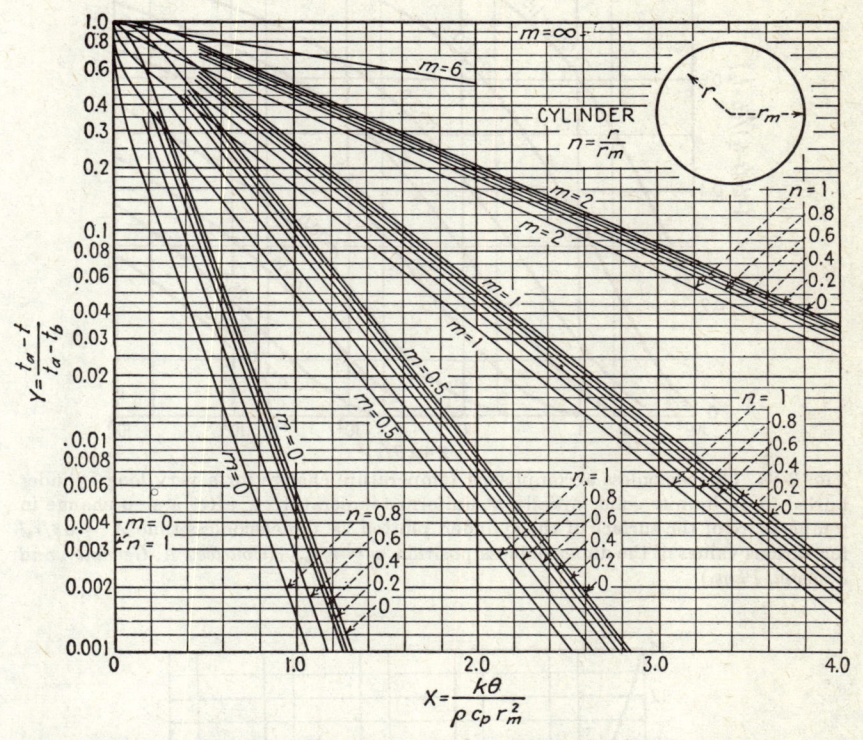

Fig. 3-8. Gurney-Lurie chart[19] for long cylinder; values of Y for short cylinder may be obtained from Figs. 3-2 and 3-8 by use of the Newman method.[33]

ratio of temperature differences $(t_a - t)/(t_a - t_b)$, X for the dimensionless term $k\theta/\rho c_p r_m^2$, and m for the dimensionless term k/hr_m, Eq. (3-1) becomes

$$-\ln Y = \frac{Ar_m}{V}\frac{1}{m}X \qquad (3\text{-}5) \star$$

For various shapes the dimensionless ratio Ar_m/V has the following values:

Ar_m/V
Slab having a large ratio A/r_m............ 1
Long cylinder........................ 2
Cube or sphere....................... 3

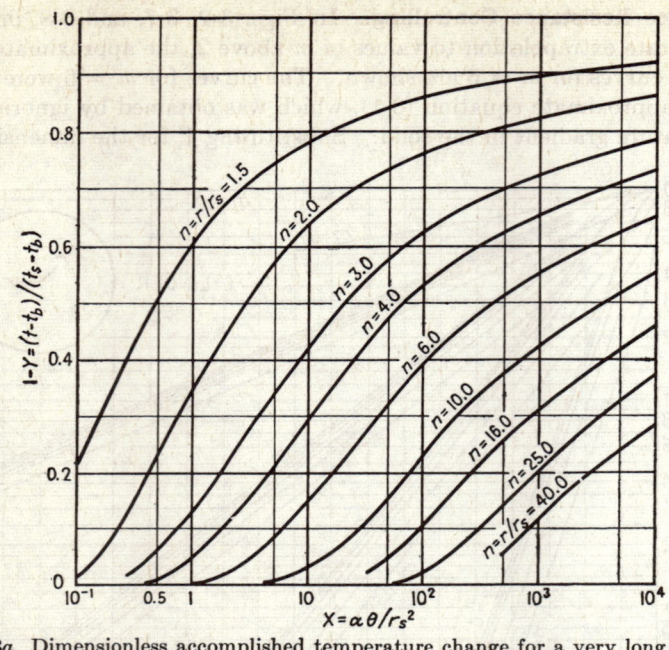

FIG. 3-8a. Dimensionless accomplished temperature change for a very long cylinder buried in an infinite solid initially at uniform temperature t_b, after a step change in temperature of the surface of the cylinder, plotted vs. dimensionless time $X = \alpha\theta/r_s^2$ for various values of the dimensionless position $n = r/r_s$. (*Courtesy A. Gemant*[15] *and J. Appl. Phys.*)

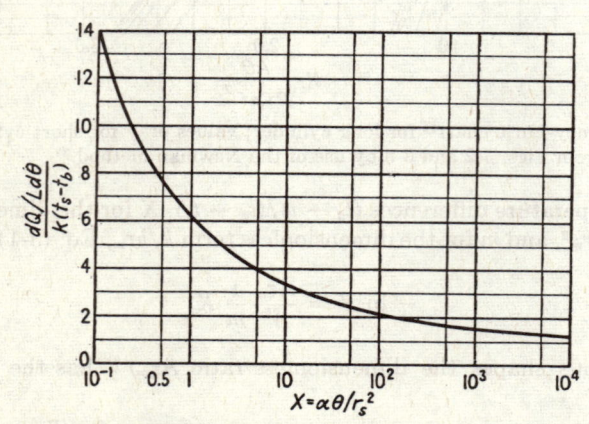

FIG. 3-8b. Dimensionless instantaneous rate of heat loss from a very long cylinder buried in a semi-infinite solid originally at uniform temperature t_b, after a step change in temperature of the surface of the cylinder. (*Courtesy A. Gemant*[15] *and J. Appl. Phys.*)

Negligible Surface Resistance. Figure 3-9 shows curves of unaccomplished temperature change Y_m vs. the dimensionless time $X = \alpha\theta/r_m^2$. For any fixed values of α and r_m, X is a direct measure of the time θ required to attain a given change in midplane temperature $(1 - Y_m)$; r_m is the *minimum* distance from surface to center.

Moving Heat Sources. Rosenthal[40] treats this subject, of importance in welding.

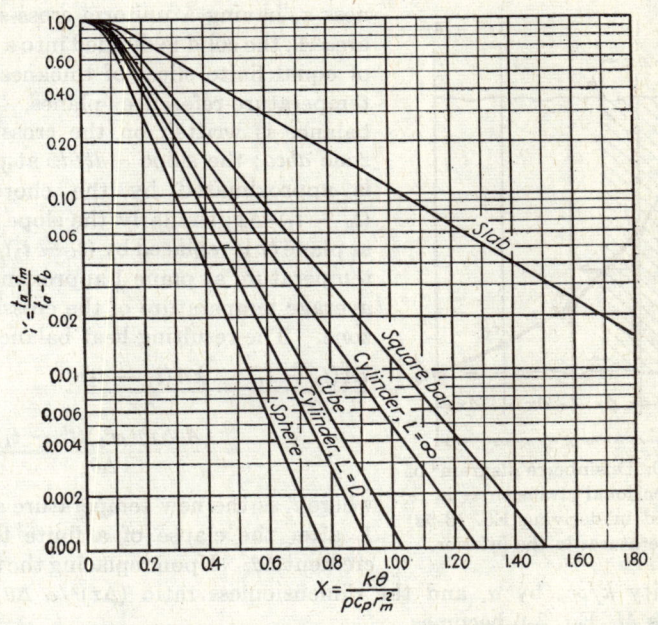

FIG. 3-9. Williamson-Adams chart[49] for midplane or mid-point of various solid shapes, having negligible surface resistance ($m = 0$).

II. NUMERICAL AND GRAPHICAL METHODS

In many cases the actual operating conditions do not correspond to those for which the analytical relations were derived. Thus the ambient temperature, the heat-transfer coefficient, and the physical properties of the solid may vary, and the initial temperature distribution may be of some nonuniform type not hitherto treated analytically. Heat might be supplied at a uniform rate at a surface or might be generated within the solid. Such complex cases can be readily handled without use of the calculus by employing numerical methods described below. Since the advent of the numerical method of Binder[3] or Schmidt[42] for unidirectional conduction, there have been many extensions of the method to problems involving one, two, and three dimensions.

1. One-dimensional Transient

In applying the numerical method the continuous process is replaced by a stepwise one. The following is based mainly on the Dusinberre[7,8] generalization of the increment method applied to one-dimensional transient conduction.

Conduction and Storage in the Interior. *a. General Numerical Method of Dusinberre.* Figure 3-10 shows a cross section of a large slab of thickness x, having a uniform cross-sectional area A; the solid is divided into a number of equal finite slices of thickness Δx by temperature-reference planes. A heat balance is written on the crosshatched zone $abcd$; the slope $-dt/dx$ at plane ad is approximated by the chord slope $(t_0 - t_1)/\Delta x$; similarly the slope $-dt/dx$ at plane bc is replaced by $(t_1 - t_2)/\Delta x$; the temperature at plane 1 approximates the average temperature of the crosshatched zone. The resulting heat balance is

$$\frac{kA(t_0 - t_1)}{\Delta x} - \frac{kA(t_1 - t_2)}{\Delta x} = \frac{(A\,\Delta x)(\rho c_p)(t_1' - t_1)}{\Delta \theta} \quad (3\text{-}6)$$

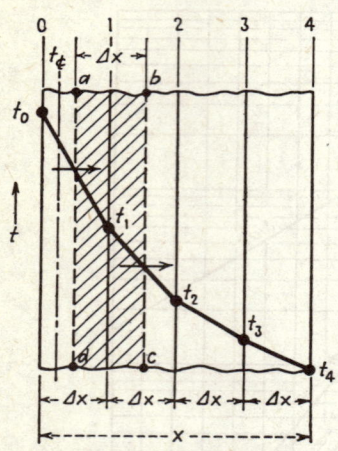

Fig. 3-10. Dusinberre diagram[8] of one-dimensional transients in a slab, used in deriving Eq. (3-6a) for temperatures in the interior.

where t_1' is the new temperature at plane 1, after the elapse of a finite time increment $\Delta \theta$. Upon replacing the thermal diffusivity $k/\rho c_p$ by α, and the dimensionless ratio $(\Delta x)^2/\alpha\,\Delta \theta$ by the modulus M, Eq. 3-6 becomes

$$t_1' = \frac{t_0 + (M - 2)(t_1) + t_2}{M} \quad (3\text{-}6a) \star$$

b. Schmidt Numerical Method. Schmidt[42] employed M of 2 and hence obtained the simple rules:

$$t_1' = \frac{t_0 + t_2}{2} \qquad t_2' = \frac{t_1 + t_3}{2} \qquad t_{n-1}' = \frac{t_{n-2} + t_n}{2} \quad (3\text{-}6b) \star$$

Thus, after elapse of one $\Delta \theta$, the *new* temperature at a given section is the arithmetic mean of the temperatures previously prevailing at adjacent sections. This method can be applied in tabular form, or graphically.

Illustration 2. A large slab of steel, 1 ft thick, initially has a uniform temperature of 700°F. Suddenly both faces are reduced to, and maintained at, 100°F. The steel has a density of 490 pounds per cubic foot, a specific heat of 0.13 Btu per pound per degree Fahrenheit, and a thermal conductivity of 25 Btu/(hr)(ft)(deg F). It is desired to predict the temperature of the midplane when 14.85 min have elapsed,

TRANSIENT CONDUCTION

(a) using the Schmidt tabular method based on M of 2, taking the initial surface temperatures as t_a ($= 100°F$), (b) the same as in (a), but employing an *initial* surface temperature of $(t_a + t_s)/2 = 400°F$, and (c) using the analytical solution, Eq. (3-3).

Solution. a. Schmidt Tabular Method ($M = 2$, $t_0 = t_s = t_a = 100°F$).

$$\Delta x = 2 \text{ inches} = \tfrac{1}{6} \text{ ft} \qquad \alpha = k/\rho c_p = 25/(490)(0.13) = 0.392 \text{ sq ft/hr}$$

$$\Delta\theta = \frac{(\Delta x)^2}{\alpha M} = \frac{(\tfrac{1}{6})^2}{(0.392)(2)} = 0.0354 \text{ hr}$$

$$\text{Number of time steps} = \frac{(14.85/60)}{0.0354} = 7 = N_{\Delta\theta}$$

$$t'_1 = \frac{t_0 + t_2}{2} \qquad t'_2 = \frac{t_1 + t_3}{2} \qquad t'_3 = \frac{t_2 + t_4}{2}$$

Since the system is symmetrical, $t_0 = t_6$, $t_1 = t_5$, and $t_2 = t_4$.

$\Delta\theta$	t_0	t_1	t_2	t_3	t_4	t_5	t_6
0	100	700	700	700	700	700	100
1	100	400	700	700	700	400	100
2	100	400	550	700	550	400	100
3	100	325	550	550	550	325	100
4	100	325	438	550	438	325	100
5	100	269	438	438	438	269	100
6	100	269	354	438	354	269	100
7	100	227	354	354	354	227	100

Once the temperatures of subsurface slices start to change, it is noted that they change only every other time increment, which is unrealistic. The final midplane temperature, 354°F, is 39°F below the analytical value. Even with 1-inch slices and the corresponding 28 time increments, t_3 is still 11°F below the analytical value.[46] These discrepancies are caused[8] by the ill-chosen initial value of t_s.

b. Initial $t_0 = t_6 = (t_a + t_s)/2 = (100 + 700)/2 = 400°F$, $M = 2$; thereafter $t_0 = 100°F$.

$\Delta\theta$	t_0	t_1	t_2	t_3	t_4	t_5	t_6
0	400	700	700	700	700	700	400
1	100	550	700	700	700	550	100
2	100	400	625	700	625	400	100
3	100	362	550	625	550	362	100
4	100	325	493	550	493	325	100
5	100	296	437	493	437	296	100
6	100	268	395	437	395	268	100
7	100	248	352	395	352	248	100

With this improved procedure,[8] the final midplane temperature of 395°F is only 2°F above the value from the analytical solution.

c. Equation (3-3), $t = t_3$, $x_3 = r_m$:

$$a_1 = \left(\frac{\pi}{2}\right)^2 \qquad X = \frac{\alpha\theta}{r_m^2} = \frac{(0.392)(14.85/60)}{(0.5)^2} = 0.388 \qquad a_1 X = 0.957$$

$$\frac{t_3 - 100}{700 - 100} = \frac{4}{\pi}\left(e^{-0.957}\sin\frac{\pi}{2} + \frac{1}{3}e^{-8.604}\sin\frac{3\pi}{2} + \frac{1}{5}e^{-23.9}\sin\frac{5\pi}{2} + \cdots\right)$$

$$= (4/\pi)(0.384 + \text{negligible}) \qquad t_3 = 393°F$$

In the previous example, M of 2 was quite satisfactory in part b, since the result agreed closely with the analytical solution. Dusinberre[8] considered a problem in which two very thick slabs of a given material, one originally at a high temperature and the other originally at a low temperature, were suddenly brought into perfect thermal contact. Upon employing M of 2, the calculated temperature of the first slice in the colder slab *oscillated* alternately above and below the analytical solution; the amplitude of the oscillations decreased as θ increased. When M of 3 (or 4)

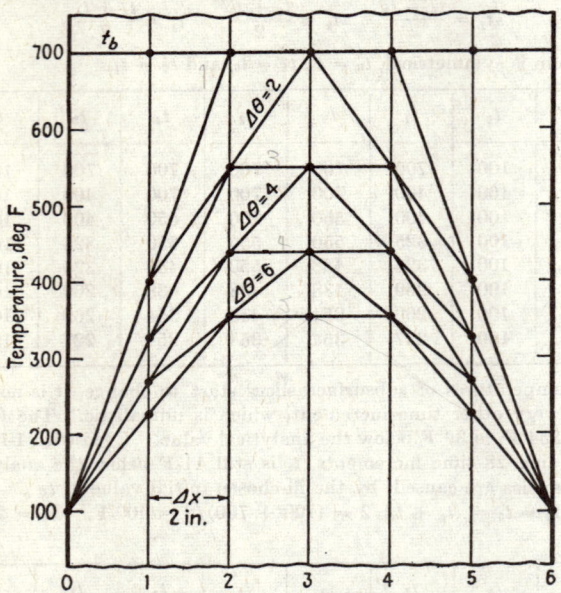

Fig. 3-11. Graphical method of Binder[3] and of Schmidt[42] for determining temperatures in the interior of a slab, subjected to a sudden decrease in surface temperature.

was used, the temperature of the specified slice changed at each time increment and agreed closely with the analytical solution.

c. Schmidt Graphical Method. Figure 3-11 shows the *graphical* solution to part a of this example. If the improvement of part b were utilized, the first alignment of t_2 and t_0 would be made with t_0 at 400°F, and all subsequent alignments of t_2 and t_0 would be made with t_0 at 100°F.

Heat Transfer at a Surface. *a. Dusinberre Numerical Method.* Let plane 00 be heated by a warmer fluid at instantaneous ambient temperature t_a. A heat balance on the adjacent half slice (Fig. 3-12) gives:

$$hA(t_a - t_0) - \frac{kA(t_0 - t_1)}{\Delta x} = \frac{(A\,\Delta x)(\rho c_p)(t'_{0.25} - t_{0.25})}{2\,\Delta \theta} \quad (3\text{-}7)$$

where $t_{0.25}$ and $t'_{0.25}$ are the old and new temperatures at the axis of this half slice. As an approximation[8] at plane 0.25 the temperature rise is replaced by that at the surface, $t'_0 - t_0$. Designating the dimensionless Nusselt or Biot modulus $h\,\Delta x/k$ by N, rearrangement of Eq. (3-7) gives

$$t'_0 = \frac{2Nt_a + [M - (2N+2)]t_0 + 2t_1}{M} \qquad (3\text{-}7a) \star$$

To retain a suitable influence of t_0 on t'_0, inspection of the bracketed term in Eq. (3-7a) shows that

$$M > 2N + 2 \qquad (3\text{-}7b) \star$$

In cases where N is large, this requires that M be large; and in view of the definition of M, the term $\Delta\theta/(\Delta x)^2$ may become inconveniently small. In such cases (large N) an alternate equation is employed. As an approximation the temperature gradient at the surface $(-dt'/dx)_0$ is assumed equal to the chord slope $(t'_0 - t'_1)/\Delta x$, and a heat balance at the surface gives $h(t'_a - t'_0) = k(t'_0 - t'_1)/\Delta x$, or

$$t'_0 = \frac{N}{N+1} t'_a + \frac{1}{N+1} t'_1 \qquad (3\text{-}7c) \star$$

When the heat-transfer coefficient at the surface is large, as with condensing steam or boiling water, the surface temperature changes substantially instantaneously from t_0 to t_a and consequently the initial surface temperature should be taken as the arithmetic mean of the original surface temperature and the ambient temperature.[8]

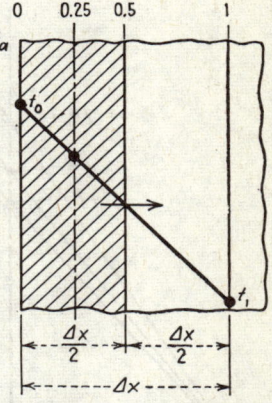

FIG. 3-12. Dusinberre diagram[8] used in deriving Eq. (3-7).

In applying the method one first tentatively fixes the number of slices n_s, which fixes Δx (since $n_s = x/\Delta x$); this, together with the highest value of k/h, fixes N. The modulus M is selected in the light of Eq. (3-7b), unless N is large, in which case Eq. (3-7c) is used. The modulus M need not be an integer, but it is quite convenient to employ integers (say 2, 3, or more). If the original distribution of temperature is seriously nonuniform, an M of 2 is undesirable, since the old value of t_1 would not influence the new value of t_1; with an M of 3 or more, t_1 is given weight along with t_0 and t_2. Nonintegral values of M are required (1) where it is desired to specify $\Delta\theta$, Δx, and α and (2) for a heterogeneous wall where the diffusivities and slice thicknesses do not permit use of a single value of M. A preliminary calculation, based on a relatively small number of slices, and correspondingly large time increments, gives the first estimate of the temperature distribution at the desired time θ. The calculation is then repeated, based on thinner slices and shorter time increments, until the change in the final temperature distribution is small.

b. Schmidt Revised Graphical Method. A fictive half slice having no heat capacity is added outside the surface, and the reference planes for temperatures are located at distances of 0.5 Δx, 1.5 Δx, 2.5 Δx, and so on; the outer boundary of the fictive slice is at a distance of $-0.5\ \Delta x$ from the surface. A heat balance at the surface gives

$$\frac{t_a - t_0}{k/h} = -\left(\frac{dt}{dx}\right)_0$$

The ambient temperature t_a is plotted as ordinate at a distance k/h from the surface. The term $(t_a - t_0)/(k/h)$ is the ratio of two *distances*. A

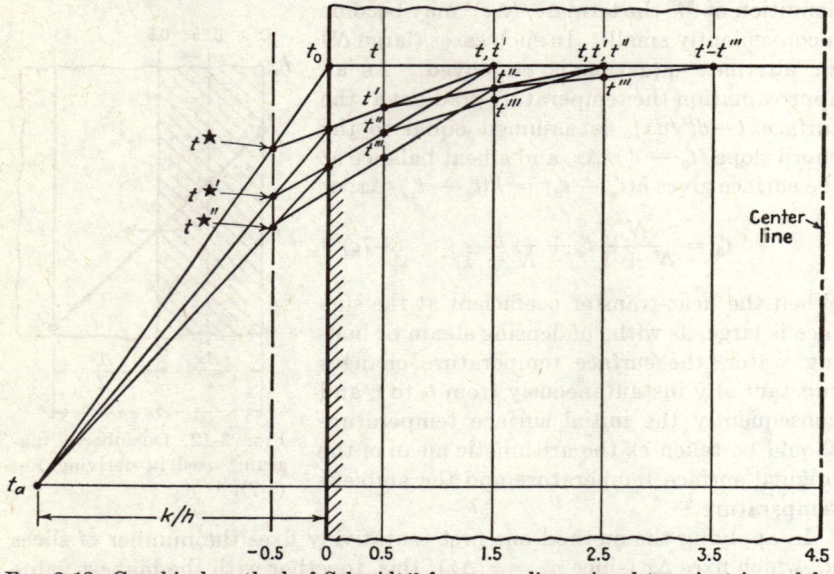

Fig. 3-13. Graphical method of Schmidt[43] for a one-dimensional transient in a slab, with allowance for heat transfer at one surface.

straight line from t_a to t_0 gives the slope $-(dt/dx)_0$ at surface, and the ordinate at the intersection with the -0.5 plane is designated as $t^\star$. As an approximation $(dt/dx)_0$ is replaced by the chord slope $(t^\star - t_{0.5})/\Delta x$, and the heat balance on the slice from plane 0 to plane 1 gives the Schmidt rule

$$t'_{0.5} = \frac{(t^\star + t_{1.5})}{2}$$

Heat balances on the interior give

$$t'_{1.5} = \frac{(t_{0.5} + t_{2.5})}{2} \qquad t'_{2.5} = \frac{(t_{1.5} + t_{3.5})}{2}$$

Figure 3-13 shows the graphical construction for a finite slab originally at uniform temperature, cooled at one face ($x = 0$) and adiabatic on the other face ($x = 4.5\,\Delta x$). Since there is no heat transfer at the adiabatic wall, the temperature gradient is always zero at this point.

This graphical method can be applied to problems with heat transfer at both surfaces, with allowance for changes in t_a and h, and to problems involving composite walls[29] including the effects of contact coefficients of heat transfer at the junctions of the two solids.

Schmidt[43] also developed graphical methods for cylinders and spheres. Veron[48] thoroughly treats graphical methods.

Internal Generation of Heat. If heat is being uniformly generated throughout a slice Δx at the *instantaneous* rate q_G, a derivation, similar to that used in obtaining Eq. (3-6a), gives

$$t_1' = \frac{t_0 + (M-2)t_1 + t_2}{M} + \frac{G_{\Delta x,\Delta\theta}}{M} \qquad (3\text{-}8)\;\star$$

where $G_{\Delta x,\Delta\theta} \equiv q_G\,\Delta x / Ak$.

For a half slice at the surface, Eq. (3-7a) becomes

$$t_0' = \frac{2Nt_a + [M - (2N+2)]t_0 + 2t_1}{M} + \frac{G_{\Delta x,\Delta\theta}}{M} \qquad (3\text{-}9)\;\star$$

Radiation and Convection at a Surface. Let one surface of a slab absorb a radiant flux of instantaneous value $\pm dq_r/dA$. The slab is in contact with an ambient fluid having an instantaneous temperature t_a. The heat balance on the half slice nearest the surface, and the approximation used in deriving Eq. (3-7a), gives

$$t_0' = \frac{2Nt_a + [M - (2N+2)]t_0 + 2t_1}{M} + \frac{P}{M} \qquad (3\text{-}10)\;\star$$

where

$$P \equiv 2\,\frac{dq_r}{dA}\,\frac{\Delta x}{k} \qquad (3\text{-}10a)$$

If heat were independently being generated within this half slice, term G should be added to term P.

Composite Slab. Consider two slabs of materials A and B, in good contact at the junction. If it is desired to employ the same M in both materials, since $\Delta\theta$ is the same for both,

$$M\,\Delta\theta = \frac{\Delta x_A{}^2}{\alpha_A} = \frac{\Delta x_B{}^2}{\alpha_B} \qquad (3\text{-}11)$$

and consequently the ratio of the thicknesses of the slices must be taken as equal to the square root of the thermal diffusivities:

$$\Delta x_A/\Delta x_B = \sqrt{\alpha_A/\alpha_B} \qquad (3\text{-}11a)$$

This procedure may lead to an inconveniently large number of slices in one material, in order to give integral numbers of slices in each material.

This difficulty can be avoided by using the same Δx in each material, and using
$$M_A/M_B = \alpha_B/\alpha_A \tag{3-11b}$$

The weighting equation at the interface may be obtained in one of two ways. (1) The temperature gradient in A, $k_A(dt'/dx_A)_i$, is replaced by $k_A(t'_{i-1} - t'_i)/\Delta x_A$, and similarly $-k_B(dt'/dx_B)_i$ is replaced by $k_B(t'_i - t'_{i+1})/\Delta x_B$. A heat balance at the interface gives

$$t'_i = \frac{1}{1+R_r} t'_{i-1} + \frac{R_r}{1+R_r} t'_{i+1} \tag{3-12} \star$$

where $R_r \equiv \Delta x_A\, k_B/\Delta x_B\, k_A$. (2) Alternatively, by considering half slices on each side of the interface, and assuming that the temperature rise at the midplanes of the half slices equals the rise at the interface, a heat balance on the two half slices gives

$$t'_i = \frac{t_{i-1} + \left(\dfrac{M_A}{2} + \dfrac{M_B R_r}{2} - 1 - R_r\right) t_i + R_r t_{i+1}}{(M_A/2) + (M_B R_r/2)} \tag{3-13} \star$$

If both M_A and M_B are taken as 2, this becomes

$$t'_i = \frac{1}{1+R_r} t_{i-1} + \frac{R_r}{1+R_r} t_{i+1} \tag{3-13a}$$

In Eq. (3-12), the *new* temperature at the interface depends on the *new* temperatures at the adjacent points, whereas, in Eqs. (3-13) and (3-13a), the *new* temperature at the interface depends on the *old* temperatures at the adjacent points.

The remainder of the points are calculated by use of Eq. (3-6a) and Eq. (3-7a) or Eq. (3-7c).

2. Two-dimensional Transient

This case is handled by the methods used for the one-dimensional transient. The nomenclature is shown in Fig. 3-14. The heat balance on the crosshatched zone $abcd$ having volume of $\Delta x\, \Delta y\, L$ is

$$\frac{k(L\,\Delta y)(t_{B0} - t_{B1})}{\Delta x} + \frac{k(L\,\Delta x)}{\Delta y}(t_{A1} - t_{B1}) + \frac{k(L\,\Delta y)}{\Delta x}(t_{B2} - t_{B1})$$
$$+ \frac{k(L\,\Delta x)(t_{C1} - t_{B1})}{\Delta y} = \frac{(L\,\Delta x\,\Delta y)(\rho c_p)(t'_{B1} - t_{B1})}{\Delta \theta}$$

For a square net ($\Delta x = \Delta y$) this reduces to[10]

$$t'_{B1} = \frac{t_{B0} + t_{A1} + t_{B2} + t_{C1} + (M-4)t_{B1}}{M} \tag{3-14} \star$$

which gives the new temperature at the central point in terms of the temperatures previously prevailing at the five points. To retain the influence

TRANSIENT CONDUCTION 51

of t_{B1} on t'_{B1}, equal to that of the four adjacent values, one employs an M of 5 in Eq. (3-14) and obtains the simple rule

$$t'_{B1} = \frac{t_{B0} + t_{A1} + t_{B2} + t_{C1} + t_{B1}}{5} \qquad (3\text{-}14a)$$

A heat balance on a half slice adjacent to the surface gives the relation[10]

$$t'_{B0} = \frac{2Nt_a + t_{A0} + t_{C0} + 2t_{B1} + (M - 4 - 2N)t_{B0}}{M} \qquad (3\text{-}15) \star$$

which is based on the assumption used in deriving Eq. (3-7a). To retain a suitable influence of the old surface temperature t_{B0} on the new value t'_{B0},

$$M > 4 + 2N \qquad (3\text{-}15a) \star$$

For large N a derivation, similar to that used in obtaining Eq. (3-7c), yields

$$t'_{B0} = \frac{Nt_a + 0.5t_{A0} + 0.5t_{C0} + t_{B1}}{N + 2}$$

$$(3\text{-}15b) \star$$

The numerical method can be adapted to variations of properties with temperature,[7] variations in boundary coefficients and ambient temperature,[7] composite walls,[29] and latent-heat effects due to changes in phase.[7] The same principles can be adapted to other fields, such as certain types of drying[47] or mass transfer, where the forms of the equations are the same as for heat conduction.

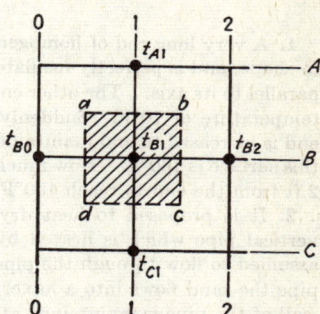

Fig. 3-14. Emmons diagram[10] for a two-dimensional transient, used in deriving Eq. (3-14).

III. MISCELLANEOUS

Analogue Methods. A mechanical integrator has been described by Nessi and Nisolle[32] and by Nessi.[31] Moore[30] devised a hydraulic analogue called the "hydrocal," and Coyle[6] devised a model employing air instead of water. Electrical analogues have been employed by Beuken,[2] Paschkis and Baker,[36] Heisler,[20] and others.

Wood. MacLean,[28] in a study of the data of Wirka[50] on the steaming of green southern pine timbers, found that the integrated relations (see Fig. 3-8 for the long cylinder) correlated the data on temperature gradients at various periods in the batch operation and obtained an average value of $k/\rho c_p = 0.00678$ sq ft/hr for the thermal diffusivity of the wood. The diameters of the 33 specimens ranged from 6.74 to 12 in., the average moisture content was 63 per cent by weight, and the average density was 65.4 lb/cu ft.

Rubber. Perks and Griffiths,[38] Shephard and Wiegand,[44] and Sherwood[45] give methods of computation of the rate of vulcanization of rubber as a function of temperature; Sherwood shows the substantial advantage, in the curing of rubber tires, of applying heat to both inner and outer surfaces.

Freezing and Melting. Ice formation on pipes is discussed by several writers,† and data[27] are available on the freezing and thawing of fruit juices. Fell[11] discusses theoretical approaches to problems in freezing molten metals flowing through externally chilled molds.

Heat Regenerators. These are treated in Chap. 11.

Periodic Variation in Temperature of Surroundings. Solutions are available in the literature.‡

PROBLEMS

1. A very long rod of homogeneous solid has a constant cross section, is hexagonal in shape, and is perfectly insulated from the surroundings on one end and on all faces parallel to its axis. The other end is not insulated. Originally the rod is at a uniform temperature of 70°F. Suddenly the temperature of the surface of the uninsulated end is increased to and maintained at 170°F. After 2.5 hr, the temperature 1 ft from this surface is 160°F. How much additional time will be required for the temperature 2 ft from the end to reach 160°F?

2. It is proposed to heat dry sand by passing it steadily downward through a vertical pipe which is heated by a vapor condensing on the outside. The sand is assumed to flow through the pipe with a uniform velocity profile. At the exit of the pipe the sand flows into a mixer, and its temperature becomes uniform. The inside wall of the pipe is maintained at 220°F, and the thermal resistance between the pipe wall and the sand is assumed to be negligible. The sand initially at a temperature of 120°F is fed to the pipe at a rate of 1.18 cu ft/hr. The pipe is 18 ft long and has an i.d. of 1.2 inches.

Estimate the temperature of the mixed sand leaving the heater.

Sand properties:

$$\text{Density} = 100 \text{ lb/cu ft}$$
$$\text{Thermal conductivity} = 0.20$$
$$\text{Specific heat} = 0.24$$

3. A large flat glass slab 6 in. thick ($k = 0.63$, $c = 0.3$, and $\rho = 139$) has been cooled very slowly in an annealing oven so that its temperature is substantially uniform at 900°F. It is proposed that the slab be further cooled from 900° by passing air parallel to the flat faces normal to the 6-in. dimension. To minimize thermal strains, the maximum temperature gradient allowable in the slab is 50°F/in. The air would be blown past the slab at a rate such that its temperature rise would be negligible and the coefficient from hot glass to air would be 5.0. Radiation may be neglected.

   a. When the slab is at 900°F, what is the lowest temperature of air that may be used for cooling?
   b. If air at the temperature in (*a*) is used for 3 hr, what is the lowest air temperature that could then be used?
   c. If the air temperature were so regulated that the slab was always being cooled at the maximum allowable rate, what would be the air temperature at the end of 3 hr?

† Elmer,[9] Planck,[39] and Pekeris and Slichter.[37]

‡ Carslaw,[5] Ingersoll, Zobel, and Ingersoll,[25] Grober,[17] Schack,[41] Goldschmidt and Partridge,[16] Houghten *et al.*,[24] and Jakob.[26]

TRANSIENT CONDUCTION

d. Sketch curves of temperature vs. thickness of the slab at the end of 3 hr for (*b*) and (*c*).

4. A composite slab consists of 9 inches of magnesite brick (*A*) and 8.70 inches of alumina brick (*B*), in good thermal contact. The initial distribution of temperature is specified below. Suddenly the exposed side of *A* is brought to, and maintained at, 100°F, while the exposed side of *B* is suddenly brought to, and maintained at, 0°F. Find the temperature distribution at the end of 3.05 hr, and when the steady state is reached.

Data and Notes. The initial distribution of temperature in *A* is given by the equation $t = 22.2x - 2.46x^2$, where x is distance (inches) from the bare face of *A*, toward *B*, and t is temperature in degrees Fahrenheit. The initial distribution of temperature in *B* is given by $t = 10x - 90$.

The following table shows t at various values of x:

x in A..	0	1	2	3	4	4.5	5	6	7	8	9
t.......	0	19.78	34.6	44.5	49.4	50.0	49.4	44.5	34.6	19.78	0

x in B..	9	10	12	14	16	17.7
t.......	0	10	30	50	70	87

It is desired to use at least three slices in *A* and in *B*.

	A	*B*
k, Btu/(hr)(sq ft)(deg F per ft)...........	2.15	0.58
ρ, lb/cu ft..............................	159	90.0
c_p, Btu/(lb)(deg F)......................	0.22	0.20
α, sq ft/hr.............................	0.0615	0.0322

5. A large slab of dry wood, 0.75 inch thick, initially at 88°F, is suspended horizontally in a tunnel. Suddenly the upper face is subjected to a uniform radiant flux of 393 Btu/(hr)(sq ft), from a number of infrared lamps suitably arranged above the upper face. At this same zero time, dry air at a mean temperature of 114°F is blown past the upper and lower faces, giving surface coefficients of heat transfer h equal to 4.48 Btu/(hr)(sq ft)(deg F) on each face. It is desired to predict the temperature distribution within the slab as a function of time.

a. Derive the necessary equations for an approximate numerical solution of this problem.
b. Taking $\Delta x = \frac{1}{8}$ inch, proceed with the calculation of the temperature profile for three time increments, using an *M* of 3.
c. Compute the temperature profile after an infinite time interval.

Wood properties:
$$k = 0.101$$
$$\alpha = k/\rho c_p = 0.00318 \text{ sq ft/hr}$$

6. A long hollow flue of square cross section is initially at a uniform temperature of 60°F. Suddenly hot gases are blown through the flue at a high velocity, maintaining the inner surface of the walls at 500°F. The inside cross section is a 2-ft square, and the total cross section is a 5-ft square, the walls being 1½ ft thick. The wall has the following three properties: k of 2.2, ρ of 115, and c_p of 0.18. It is assumed that the outside surface coefficient of heat transfer is given by $h = 0.27(\Delta t)^{1/4}$, where h is the local surface coefficient in Btu/(hr)(sq ft)(deg F) and Δt is the difference between the local surface temperature and the ambient temperature of the surroundings, 60°F.

Derive the necessary equations for a numerical solution to this problem. Outline in detail the procedure for calculating the temperature distribution within the walls of the flue at the end of any time interval and the total heat transferred from the hot gases to the wall since time zero.

CHAPTER 4

RADIANT-HEAT TRANSMISSION

By Hoyt C. Hottel

The relative importance of the several mechanisms of the transfer of heat from one body to another differs greatly with temperature. The phenomena of conduction and convection are affected primarily by temperature difference and very little by temperature level, whereas radiation interchange increases rapidly with increase in temperature level. It follows that, at very low temperatures, conduction and convection are the major contributors to the total heat transfer; at very high temperatures, radiation is the controlling factor. The temperature at which radiation accounts for roughly one-half of the total heat transmission depends on such factors as the emissivity of the surface or the magnitude of the convection coefficient. For large pipes losing heat by natural convection, this is room temperature; for fine wires of low emissivity, it is above a red heat.

Subject matter will be divided into (1) the nature of thermal radiation (pages 55 to 63), (2) radiant-heat interchange between the surfaces of solids separated by a nonabsorbing medium (pages 63 to 82), (3) radiation from nonluminous gases (pages 82 to 98), (4) radiation from clouds of particles (pages 99 to 105), and (5) the general problem of radiation in an enclosure (pages 105 to 124). Nomenclature is summarized in Table 4-1.

THE NATURE OF THERMAL RADIATION

When a body is heated, radiant energy is emitted at a rate, and of a quality, dependent primarily on the temperature of the body. Thus, when the filament of an incandescent lamp is heated electrically, both the quantity of energy emitted per unit time and the proportion of visible radiation (light) emitted are found to increase rapidly with increase in temperature of the filament. At temperatures below approximately 1000°F, the radiation is not perceived by the human eye but may be recognized by the sense of warmth experienced when the hand is held near a slightly warmer body. Certain materials, when suitably excited by electric discharge, bombarded by electrons, exposed to radiation of

Table 4-1. Nomenclature

- A Area of a surface, square feet; $A_{1,2,3}$, area of a source-sink-type surface; $A_{R,S,T}$, area of no-flux surface; A', area exclusive of that swept by gases leaving combustion chamber (on page 66, plane area replacing a dimple)
- a $\partial \ln \epsilon_G / \partial \ln P_G L$, dimensionless; $a_c(a_w)$, due to $CO_2(H_2O)$
- a_i Weighting factor of ith term in evaluating ϵ_G, dimensionless
- B Length of radiant-heat exchanger, feet
- b $\partial \ln \epsilon_G / \partial \ln T_G$, dimensionless; $b_c(b_w)$, due to $CO_2(H_2O)$
- C Pressure correction for gas emissivity, dimensionless; $C_c(C_w)$, due to $CO_2(H_2O)$
- c Exponent on temperature ratio in obtaining α_G from ϵ_G; $c_c(c_w)$, due to $CO_2(H_2O)$. (In Planck equation, c = velocity of light.)
- c_1, c_2 Dimensional constants in Planck equation
- c_p Specific heat at constant pressure, Btu/(lb)(deg F)
- D Basic determinant for a system of simultaneous linear equations
- $_m D_n$ Determinant formed by replacing nth column of D by a column of m-functions
- E Radiation-emission rate from a gas per unit of volume, energy/(volume)(time), in all directions from a small volume element
- F View factor. F_{12} = fraction of radiation emitted by black area A_1 in solid angle 2π from every element of its surface, which is intercepted directly by A_2
- $\bar{F}$ Black-surface over-all interchange factor. $\bar{F}_{12}(\bar{F}_{1G})$ = fraction of radiation initially emitted by black surface A_1, which reaches and is absorbed by A_2 (by gas) directly and by aid of reflection and/or reradiation from no-flux zones, but involving no other source-sink zones
- $\mathcal{F}$ Over-all interchange factor. $\mathcal{F}_{1G}(\mathcal{F}_{12})$ = radiation reaching gas (surface A_2) due to original emission from A_1 only, but including assistance given by reflection at other source-sinks and from the no-flux zones, expressed as a ratio to W_{B1}
- g Dimensional proportionality constant in relation between α_{12} and T_1 and T_2
- h Convection heat-transfer coefficient, Btu/(sq ft)(hr)(deg F); h_1, at surface A_1; h_r, equivalent value due to radiation. (In Planck equation, Planck's constant of action.)
- I Intensity of emission from a black surface in direction ϕ with normal, energy/(area)(time)(solid angle)(cos ϕ); I_λ = above, per unit of wavelength interval
- i Hourly enthalpy of fuel, air, and any recirculated flue gas entering a furnace, Btu per hour, above base temperature T_0
- KL Absorption strength of a luminous flame, dimensionless
- k Absorption coefficient of a gas, (ft-atm)$^{-1}$; k_x, value applicable to spectral-energy fraction x; k_λ, monochromatic value at wavelength λ. [In Planck equation, k = molecular gas constant (Boltzmann constant).]
- L Mean beam length in evaluating gas emissivity, feet; see Tables 4-2 and 4-3
- L^0 Mean beam length when $P_G X = 0$
- m Exponent on absorber temperature in relation of surface absorptivity to temperature, dimensionless; as a subscript on A, F, $\bar{F}$, or $\mathcal{F}$, represents a source-sink zone
- n Exponent on emitter temperature in relation of surface absorptivity to temperature; as a subscript on A, F, $\bar{F}$, or $\mathcal{F}$, represents a source-sink zone
- P Pressure, atmospheres; P_G, partial pressure of radiating gas; P_T, total pressure; $P_c(P_w)$, partial pressure of $CO_2(H_2O)$

RADIANT-HEAT TRANSMISSION

TABLE. 4-1. NOMENCLATURE.—(*Continued*)

- p Number of zones in an enclosure, both source-sink and no-flux
- q Heat-transfer rate, Btu per hour; q_L = external heat-loss rate from a combustion chamber
- R Relative flux density, dimensionless; $_1R_2$, flux density away from A_2, due solely to radiation originating at A_1, and expressed as a ratio to W_{B1}; $_GR_1$, same, but due to radiation originating in the gas
- R, S, T As subscripts, these designate no-flux zones
- $\overline{RS}$ Abbreviated designation of $A_R F_{RS} \tau_{RS}$, square feet
- r Distance between two surface elements, feet
- r_{af} Weight ratio of air to fuel
- r_f Ratio of average billet-pushing rate over a period of several hours to pushing rate during periods steady operation, dimensionless
- S Stock heat-transfer surface per unit length, feet
- T Temperature, degrees absolute (usually Rankine); T_{avg}, arithmetic mean temperature; T_G, gas temperature; T_G', mean gas temperature along an exchanger; T_1, temperature of surface A_1; T_1', mean surface temperature along an exchanger; T_0, ambient-air temperature; T_F, true flame temperature; T_r, red brightness temperature [in Eqs. (4-67) and (4-68)]
- U Over-all coefficient of heat transfer through a wall, based on inside-surface and outside-air temperatures
- U_R Over-all coefficient of heat transfer through area A_R, based on inside-gas and outside-air temperature
- V Volume of a radiating gas mass, cubic feet
- V/A Mean hydraulic radius of a radiating gas mass, feet
- W Emissive power, Btu/(sq ft)(hr) in solid angle 2π, from plane element; W_{B1}, emissive power of black surface A_1; W_G, emissive power of a gas
- $W_{B\lambda}$ Monochromatic emissive power of a black surface, ergs/(sq cm)(sec)(cm) in solid angle 2π, from plane element
- w Mass flow rate, lb per hour; w_s, mass flow rate of stock; w_A, firing rate in pounds of equivalent coal per hour; w_G, mass rate of flow of combustion products
- X Characteristic dimension of a gas shape, feet
- x Distance, feet
- x, y, z Spectral-energy fractions, adding to 1
- $\overline{12}$ Abbreviated designation of $A_1 F_{12} \tau_{12}$ or $A_2 F_{21} \tau_{21}$

Greek

- α Absorptivity for radiation, dimensionless; α_1, of surface 1; α_{12}, of surface 1 for radiation from surface 2; α_{G1}, of a gas for radiation from surface 1
- ϵ Emissivity, dimensionless; ϵ_1, of surface 1; ϵ_G, of a gas; ϵ_c, gas emissivity due to CO_2; ϵ_w, gas emissivity due to H_2O; ϵ_{G1}, emissivity of gas radiating to A_1; ϵ_{avg}, evaluated at arithmetic-mean temperature; ϵ_{12}, gas emissivity averaged over all paths with ends on A_1 and A_2; ϵ_C', in Eq. (4-44b) only, = pseudo emissivity due to convection
- $\Delta\epsilon$ Correction on emissivity, for superimposed radiation from CO_2 and H_2O, dimensionless
- η Combustion-chamber heat-transfer efficiency, dimensionless
- λ Wavelength, microns; $\lambda_{0.5Q}$, wavelength below which half the energy lies, in the spectrum of a black body
- μ Microns (1 μ = 10^{-4} cm); on page 64, refractive index
- σ Stefan-Boltzmann constant, energy/(area)(time)(deg abs)^4, in solid angle 2π above a plane element

TABLE 4-1. NOMENCLATURE.—(*Continued*)

- τ Transmittance of gas for radiation from a surface, dimensionless; τ_{12}, transmittance of gas for radiation leaving one surface, 1 or 2, for the other; τ_1, transmittance of gas for radiation from all surface zones, properly weighted, toward A_1; τ_x, transmittance of a gas for the spectral-energy fraction x of black radiation; $\tau_{n \cdot G}$, transmittance of a gas layer n times as thick as that having transmittance τ_G corresponding to emissivity ϵ_G; τ_λ, transmittance at wavelength λ
- ϕ Angle made by a radiant beam with normal to surface element
- ω Solid angle

suitable wavelength, or allowed to react chemically, emit a *characteristic* radiation which shows a discontinuous spectrum, with energy concentrated in certain wavelengths characteristic of the emitting substance. Examples of characteristic radiation include the mercury-arc and neon lamps. Certain solids and liquids, when illuminated by light of suitable wavelength without rising appreciably in temperature, emit a characteristic radiation described as *fluorescence* if emission ceases with the illumination and as *phosphorescence* if emission continues an appreciable time after illumination ceases. The term *thermal radiation* is used broadly to describe radiant energy emitted in consequence of the temperature of a body, more narrowly to describe radiation the quality and quantity of which depend solely on temperature and not on the nature of the emitting body. This chapter will deal only with radiation resulting directly from thermal excitation, referred to hereafter merely as *radiation*.

Kirchhoff's Law. If two small bodies of areas A_1 and A_2 are placed in a large evacuated enclosure perfectly insulated externally, then, when the system has come to thermal equilibrium, the bodies will emit radiation at the rates $A_1 W_1$ and $A_2 W_2$, respectively, where W is the total emissive power,† energy per unit time per unit area of the surface [Btu/(sq ft)(hr)] emitted throughout the hemisphere above each element of surface. Let the energy impinging on unit area of any small body in the enclosure, due to radiation from the walls of the latter, be W_B. If the bodies have *absorptivities* (fraction of incident radiation that is absorbed) of α_1 and α_2, then energy balances on the bodies will have the form

$$W_B A_1 \alpha_1 = A_1 W_1 \quad \text{and} \quad W_B A_2 \alpha_2 = A_2 W_2$$

from which $W_1/\alpha_1 = W_2/\alpha_2 (= W_x/\alpha_x$, where x is *any* body$) = W_B$. This generalization, that *at thermal equilibrium* the ratio of the emissive power of a surface to its absorptivity is the same for all bodies, is known as *Kirchhoff's law*. Since α cannot exceed unity, Kirchhoff's law places an upper limit on W, called W_B, dependent on temperature alone; and any surface having this upper limiting emissive power is called a *perfect radiator*. Since such a surface must have an absorptivity of unity and

† Sometimes called *emittance, total hemispherical intensity,* or *radiant-flux density.*

therefore a reflectivity of zero, the perfect radiator is more commonly referred to as a *black body*. The ratio of the emissive power of an actual surface to that of a black body is called the *emissivity* ϵ of the surface. Kirchhoff's law restated is as follows: *At thermal equilibrium the emissivity and absorptivity of a body are the same.*

Black-body Radiation Laws. The emissive power of a black body depends on its temperature only, and the second law of thermodynamics may be used to prove a proportionality between emissive power and the fourth power of the absolute temperature. The resulting relation

$$W_B = \sigma T^4 \qquad (4\text{-}1)$$

is known as the *Stefan-Boltzmann law*: and the proportionality constant σ is known as the Stefan-Boltzmann constant [0.1713 × 10^{-8} Btu/(sq ft)(hr)(deg R)^4; 5.67 × 10^{-5} erg/(sq cm)(sec)(deg K)^4; 4.88 × 10^{-8} kg-cal/(sq m)(hr)(deg K)^4]; 1.00 × 10^{-8} Chu/(sq ft)(hr)(deg K)^4].[†]

Other properties of black-body radiation of interest in heat transmission are its distribution in the spectrum and the shift of that distribution with temperature. If $W_{B\lambda}$ is the *monochromatic emissive power* at wavelength λ such that $W_{B\lambda} \, d\lambda$ is the energy emitted from a surface throughout a hemispherical angle per unit area per unit time in the wavelength interval λ to $\lambda + d\lambda$, the relation among $W_{B\lambda}$, λ, and T is given by *Planck's law*,

$$W_{B\lambda} = \frac{2\pi h c^2 \lambda^{-5}}{e^{ch/k\lambda T} - 1} = \frac{c_1 \lambda^{-5}}{e^{c_2/\lambda T} - 1} \qquad (4\text{-}2)$$

where c = velocity of light, 2.9979 × 10^{10} cm/sec; h = Planck's constant, 6.6236 × 10^{-27} erg-sec; k = Boltzmann's constant, 1.3802 × 10^{-16} erg/deg K; c_1 = 3.7403 × 10^{-5} erg cm^2/sec; c_2 = 1.4387 cm deg K. Planck's law is perhaps better visualized by converting Eq. (4-2) to the form

$$\frac{W_{B\lambda}}{T^5} = f(\lambda T) \qquad (4\text{-}3)$$

presented graphically in Fig. 4-1. This may be visualized as the intensity-wavelength relation at 1° absolute. The monochromatic emissive power at any temperature varies from 0 at $\lambda = 0$ through a maximum and back to 0 at $\lambda = \infty$; at any wavelength it increases with temperature, but values at shorter wavelengths increase faster so that the maximum value shifts to shorter wavelengths as the temperature rises. The wavelength of maximum intensity is seen to be inversely proportional to the absolute temperature (*Wien's displacement law*). The relation is $\lambda_{\max} T = 0.2898$ cm deg K. This can be misleading, however, since the wavelength of maximum intensity depends on whether wave-

[†] Deg R (K) designates degrees Fahrenheit (centigrade) absolute. Radiation constants are 1951 values.[2]

length λ or frequency ν is used in defining intensity. It may readily be shown† that the displacement law giving $W_{B\nu,\text{max}}$ is $\lambda_{\text{max}} T = 0.5099$ cm deg K. This possible ambiguity is eliminated if one uses the top scale of Fig. 4-1, which gives the fraction of the total energy at wavelengths

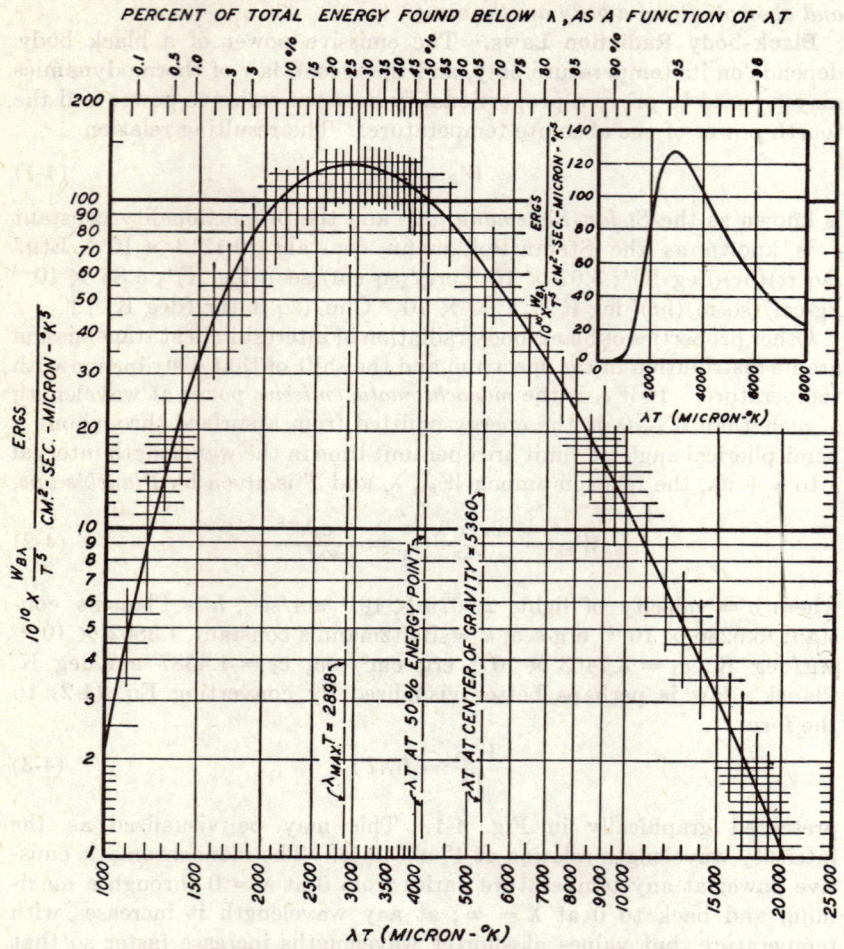

FIG. 4-1. Distribution of energy in the spectrum of a black body.

below λ as a function of λT. From this, the displacement law for *equal-energy division* is $\lambda_{0.5Q} T = 0.411$ cm deg K. The area under the curve of Fig. 4-1 (arithmetic coordinates) is the Stefan-Boltzmann constant.

The Emissivity and Absorptivity of Surfaces. In evaluating radiant-heat transfer between surfaces one could consider monochromatic radi-

† By use of the two relations: $-W_{B\nu} \, d\nu = W_{B\lambda} \, d\lambda$ and $\nu = c/\lambda$ or $d\lambda/d\nu = -\lambda^2/c$.

ation exchange and integrate throughout the spectrum; and certain advantages would appear. For most engineering purposes, however, a simplification is achieved by handling total radiation, expressing it in terms of the fourth-power temperature law applicable strictly only to the black body or perfect radiator, and letting the more or less weak residual temperature function be taken care of by the emissivity, absorptivity, or transmissivity of the pertinent bodies.

Since the radiation leaving the surface of a nonblack body originates within the volume of the body, some ambiguity exists in the term emissivity when a strong temperature gradient exists at and normal to the surface. However, metallic conductors are so opaque to radiation that a negligible portion of the radiation leaving their surfaces originates more than 0.0001 in. within the interior; and most nonconductors emit negligibly from more than a few hundredths of an inch below their surface. When the few exceptions are involved in problems of radiant-heat exchange, *e.g.*, molten sodium chloride or high-silica glass, they should be treated as problems of radiation from a volume rather than from a surface, and knowledge of their absorption coefficient as a function of wavelength is necessary.

The emissivity ϵ of a surface (more properly the total hemispherical emissivity, to differentiate it from monochromatic emissivity ϵ_λ, the ratio of radiating powers at the wavelength λ, and from directional emissivity ϵ_θ, the ratio of radiating powers in a direction making the angle θ with the normal to the surface) varies with its temperature, its degree of roughness, and, if a metal, its degree of oxidation. Table A-23 (page 472) gives the emissivities of various surfaces and emphasizes the large variation possible in a single material. Although the values in the table apply strictly to normal radiation from the surface (with few exceptions), they may be used with negligible error for hemispherical emissivity except in the case of well-polished metal surfaces, for which the hemispherical emissivity is 15 to 20 per cent higher than the normal value.[3]

A few generalizations may be made concerning the emissivity of surfaces: (1) The emissivities of metallic conductors have been shown to be very low and substantially proportional to the absolute temperature; and the proportionality constant for different metals varies as the square root of the electrical resistance at a standard base temperature.[21,64] Unless extraordinary pains are taken to prevent any possibility of oxidation or imperfection of polish, however, a specimen may exhibit several times this theoretical minimum emissivity. (2) The emissivities of nonconductors are much higher and, in contrast to metals, generally decrease with increase in temperature. Refractory materials may be expected to decrease in emissivity one-fourth to one-third as the temperature increases from 1850 to 2850°F; their grain structure and color are more important than chemical composition.[49] (3) The emissivities of most nonmetals are above 0.8 at low temperatures, in the range 0.3 to 0.8 at furnace refractory temperatures. (4) Iron and steel vary widely with the degree of oxidation and roughness, clean metallic surfaces having an emissivity of 0.05 to 0.45 at low temperatures to 0.4 to 0.7 at high temperatures; oxidized and/or rough surfaces, 0.6 to 0.95 at low temperatures to 0.9 to 0.95 at high temperatures.

Key

1, Slate composition roofing
2, Linoleum, red-brown
3, Asbestos slate
4, Soft rubber, gray
5, Concrete
6, Porcelain
7, Vitreous enamel, white
8, Red brick
9, Cork
10, White dutch tile
11, White chamotte
12, MgO, evaporated
13, Anodized aluminum
14, Aluminum paint
15, Polished aluminum
16, Graphite

The two dotted lines bound the limits of data on gray paving brick, asbestos paper, wood, various cloths, plaster of Paris, lithopone, and paper.

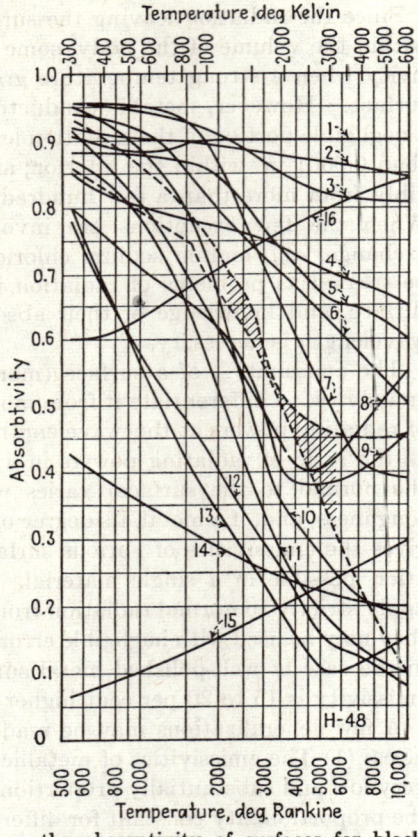

Fig. 4-2. Effect of source temperature on the absorptivity of surfaces for black radiation.

The absorptivity α of a surface depends on the factors affecting emissivity and, in addition, on the quality of the incident radiation, measured by its distribution in the spectrum. One may assign two subscripts to α, the first to indicate the temperature of the receiver and the second that of the incident radiation. It has already been seen that, according to Kirchhoff's law, the emissivity of a surface at temperature T_1 is equal to the absorptivity $\alpha_{1,1}$ which the surface exhibits for black radiation from a source at the same temperature; *i.e.*, a surface of low radiating power is also a poor absorber (or good reflector or transmitter) of radiation from a source at its own temperature. If the monochromatic absorptivity α_λ varies considerably with wavelength and much less with temperature (which is generally the case for nonmetals), it follows that the total absorptivity $\alpha_{1,2}$ will vary more with T_2 than with T_1. Data of Sieber[68a] on $\alpha_{1,2}$ at $t_1 = 70°F$ for a large group of nonmetals (see Fig. 4-2) indicate a decrease, with increase in T_2, from 0.8 to 0.95 at 500°R to 0.1 to 0.9 at 5000°R. The absorptivity of metallic conductors, on the other hand, increases approximately linearly with $\sqrt{T_1 \cdot T_2}$.

If α_λ is a constant independent of λ, the surface is called *gray* and its total absorptivity α will be independent of the spectral-energy distribution of the incident radiation; then $\alpha_{1,2} = \alpha_{1,1} = \epsilon_1$; that is, emissivity ϵ may be used in substitution for α even though the temperatures of the incident radiation and the receiver are not the same.

RADIATION BETWEEN THE SURFACES OF SOLIDS SEPARATED BY A NONABSORBING MEDIUM

Radiant interchange between any two surfaces forming part of an enclosure involves considerations of two kinds, (1) the view the surfaces have of each other and (2) their emitting and absorbing characteristics. The only case in which the first of these need not be considered—because each surface has an unobstructed view of the other alone—is the case of infinite parallel planes. Consider gray plane 1 of area A_1 and emissivity and absorptivity ϵ_1, opposite gray plane A_2 of emissivity and absorptivity ϵ_2. In unit time, unit area of plane 1 emits $\epsilon_1 \sigma T_1^4$ of which the fraction ϵ_2 is absorbed and $(1 - \epsilon_2)\epsilon_1$ is reflected back toward A_1 and absorbed, etc. The resulting infinite geometric series expressing absorption at A_2 is

$$q_{1 \to 2} = A_1 \sigma T_1^4 [\epsilon_1 \epsilon_2 + \epsilon_1(1-\epsilon_2)(1-\epsilon_1)\epsilon_2 + \epsilon_1(1-\epsilon_2)^2(1-\epsilon_1)^2 \epsilon_2 + \cdots]$$

$$= A_1 \sigma T_1^4 \frac{\epsilon_1 \epsilon_2}{1 - (1-\epsilon_1)(1-\epsilon_2)} = A_1 \sigma T_1^4 \frac{1}{1/\epsilon_1 + 1/\epsilon_2 - 1} \quad (4\text{-}4)$$

Since the emissivity term is symmetrical, the net radiation interchange between the two surfaces is given by

$$q_{1 \rightleftarrows 2} = A_1 \sigma (T_1^4 - T_2^4) \frac{1}{1/\epsilon_1 + 1/\epsilon_2 - 1} \quad (4\text{-}5)$$

The View Factor F. The more complicated but important case of radiation interchange in a system of several surfaces at different temperatures and emissivities involves the concept of a geometrical view factor F. *F_{12} is defined as the fraction of the radiation leaving black surface A_1 in all directions which is intercepted by surface A_2.* Its evaluation necessitates considering radiation exchange between two small surface elements, dA_1

and dA_2, located on A_1 and A_2. Figure 4-3 shows black element dA_1, of total emissive power W_{B1}, radiating in all directions from one side, with some of its radiation being intercepted by black element dA_2 at distance r. Connecting line r makes angles ϕ_1 and ϕ_2 with the normals to dA_1 and dA_2, respectively. The rate of radiation from dA_1 to dA_2, called $dq_{1\to 2}$, will be proportional to the apparent area of emitter dA_1 viewed from dA_2, or $dA_1 \cos \phi_1$; to the apparent area of interceptor dA_2 viewed from dA_1, or $dA_2 \cos \phi_2$; and inversely proportional to the square of the distance separating the elements. Calling the proportionality constant I_1, one may write

$$dq_{1\to 2} = I_1 \, dA_1 \cos \phi_1 \, dA_2 \cos \phi_2 / r^2 \qquad (4\text{-}6a)$$

This equation is sometimes expressed in a different form. Let the small *solid angle* subtended by dA_2 at dA_1 be called $d\omega_1$ (see Fig. 4-4). By definition a solid angle is numerically the area subtended on a sphere of unit radius, or, for a sphere of radius r, the intercepted area divided by r^2. Hence one may write $d\omega_1 = dA_2 \cos \phi_2 / r^2$, and Eq. (4-6a) becomes

$$dq_{1\to 2} = I_1 \, dA_1 \cos \phi_1 \, d\omega_1 \qquad (4\text{-}6b)$$

Since Eq. (4-6a) is symmetrical with respect to dA_1 and dA_2, and to $\cos \phi_1$ and $\cos \phi_2$, a third way of writing Eq. (4-6a) is

$$dq_{1\to 2} = I_1 \, dA_2 \cos \phi_2 \, d\omega_2 \qquad (4\text{-}6c)$$

The various forms of Eq. (4-6) are equivalent; the choice among them in subsequent use will depend on the particular problem. The proportionality factor I_1 of Eq. (4-6) is known as the *intensity of radiation* from the surface dA_1.†

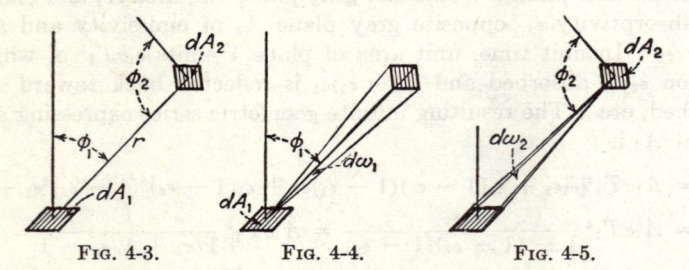

FIG. 4-3.   FIG. 4-4.   FIG. 4-5.

The rate of radiation dq_1 in *all* directions from one side of dA_1 is given by integration of Eq. (4-6b) over the complete hemispherical angle 2π above dA_1; $dq_1 = dA_1 I_1 \int d\omega_1 \cos \phi_1$. Since by definition W_{B1} equals dq_1/dA_1, the relation between W_B and I is established. The integral, over the solid angle 2π, may readily be shown to have the value π, from

---

† Equations (4-6a) to (4-6c) are restricted in application, in this chapter, to evaluation of interchange between surfaces both of which are immersed in a common medium of refractive index 1 (adequately approximated by all gases). More generally if $I_{0,1}$ is the intensity of emission of a black surface at T_1 when immersed in a vacuum (refractive index = 1), its intensity when immersed in a medium of refractive index μ is $I_{0,1} \mu^2$.

which *the emissive power of a black surface is π times its intensity of radiation* as used in the cosine law.

By integration of Eq. (4-6a) over finite areas A_1 and A_2 to obtain the rate of radiation from A_1 to A_2 and division of the result by $A_1 W_{B1}$ one obtains F_{12}, the desired fraction of the radiation leaving surface A_1 in all directions which is intercepted by surface A_2. Although the discussion has been restricted to black surfaces, it is apparent that, for a nonblack surface A_1 the emissivity of which is independent of angle of emission, F_{12} calculated by the method above will continue to represent the fractional radiation from A_1 intercepted by A_2 (though not necessarily absorbed unless A_2 is black). Plainly, from the meaning of F it follows that

$$F_{11} + F_{12} + F_{13} + \cdots = 1 \tag{4-7}$$

and that if A_1 cannot "see" itself, $F_{11} = 0$.

The symmetrical character of Eq. (4-6a) indicates that $dq_{2\to 1}$ is identical to $dq_{1\to 2}$ except for the replacement of I_1 by I_2 and that therefore the net exchange is given by replacing I_1 by $I_1 - I_2$ in Eqs. (4-6). It further follows that in the integrated relation

$$q_{1 \rightleftarrows 2} = A_1 F_{12} \cdot \sigma T_1^4 - A_2 F_{21} \sigma T_2^4 \tag{4-8}$$

the terms $A_1 F_{12}$ and $A_2 F_{21}$ *must be identical*. That this must be so is independently obvious from the fact that when T_1 equals T_2 the net heat flux must equal zero. The general equation of *direct* radiant-heat interchange between two black surfaces, exclusive of the effects of any other partially reflecting surfaces that augment the interchange between the black surfaces, is consequently

$$q_{1 \rightleftarrows 2} = (W_{B1} - W_{B2}) AF = 0.171 \left[\left(\frac{T_1}{100}\right)^4 - \left(\frac{T_2}{100}\right)^4\right] AF \tag{4-9}$$

in which A is the area of one of the surfaces and F is a geometrical factor dependent only on the shape and relative orientation of the two surfaces and on which one of the two surfaces is used in evaluating A. The importance of this conclusion needs emphasis by repetition; the net interchange between any two black surfaces (and as we shall see later, gray as well) can be evaluated by formulating the one-way radiation from either surface to the other, whichever is more convenient, and then replacing the emissive power by the difference of emissive powers of the two surfaces.

In an *enclosure of black surfaces* the net heat flux from A_1 is, by application of Eq. (4-9),

$$q_{1,\text{net}} = (A_1 F_{12} \sigma T_1^4 - A_2 F_{21} \sigma T_2^4) + (A_1 F_{13} \sigma T_1^4 - A_3 F_{31} \sigma T_3^4) + \cdots \tag{4-10}$$

$$\equiv A_1 F_{12} \sigma (T_1^4 - T_2^4) + A_1 F_{13} \sigma (T_1^4 - T_3^4) + \cdots \tag{4-10a}$$

$$\equiv A_1 \sigma T_1^4 - (A_1 F_{11} \sigma T_1^4 + A_2 F_{21} \sigma T_2^4 + A_3 F_{31} \sigma T_3^4 + \cdots) \tag{4-10b}$$

Although the formulation of the product term AF in general necessitates integration of a relation such as (4-6), there exists an important class of surface-interchange problems capable of simple direct evaluation. Consider areas of infinite extent in one direction, generated by a straight line moving always parallel to itself; all cross sections normal to the infinite dimension are identical. On the upper or radiating face of surface A_1 (Fig. 4-6) draw all possible straight lines representing tangents to pairs of points, to form the new surface A_1' containing no positive curvature and therefore unable to see any of itself. To complete an enclosure,

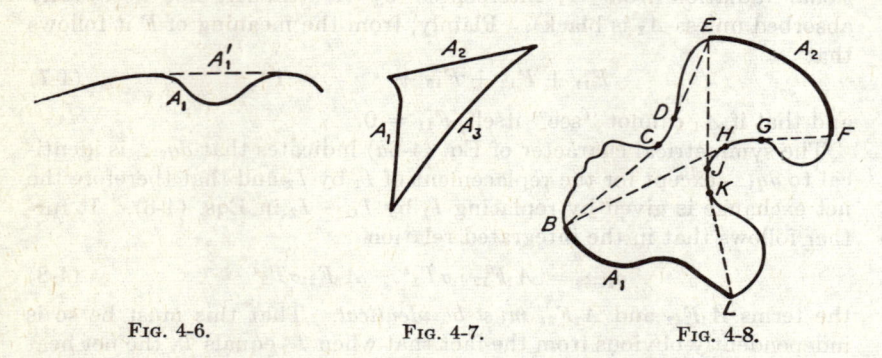

FIG. 4-6.   FIG. 4-7.   FIG. 4-8.

imagine a surface A_2 equal in shape and size to A_1' and displaced upward from it an infinitesimal amount. $A_1F_{12} = A_2F_{21}$, but surface A_2 "sees" only A_1; so $F_{21} = 1$. Consequently,

$$A_1F_{12}[\equiv A_1(1 - F_{11})] = A_1' \qquad (4\text{-}11)$$

In words, the net radiation streaming away from a black surface (measured by A_1F_{12}) is the same as the total radiation streaming away from the minimum surface formed by replacing all dimples or areas of positive curvature by plane areas. In the class of systems about to be discussed, a string tightly stretched from edge to edge over the radiating face represents the effective area. Now consider an enclosure formed by the three surfaces A_1, A_2, A_3 of Fig. 4-7, and let each surface represent the effective area as previously described; *i.e.*, all surfaces are without positive curvature in the direction of their mutual irradiation and can therefore "see" only the other two. Consequently,

$$\begin{aligned} A_1F_{12} + A_1F_{13} &= A_1 \\ A_2F_{21} + A_2F_{23} &= A_2 \\ A_3F_{31} + A_3F_{32} &= A_3 \end{aligned} \qquad (4\text{-}12)$$

From the reciprocal relation between the AF products for two mutually radiating surfaces, one may reduce the number of unknown F's in

(4-12) from six to three, giving

$$A_1F_{12} + A_1F_{13} = A_1$$
$$A_1F_{12} + A_2F_{23} = A_2 \qquad (4\text{-}13)$$
$$A_1F_{13} + A_2F_{23} = A_3$$

Solution of these for $A_1F_{12} (\equiv A_2F_{21})$ gives

$$A_1F_{12} = \frac{A_1 + A_2 - A_3}{2} \qquad (4\text{-}14)$$

Consider now the more complex enclosure of cross section represented by the solid lines in Fig. 4-8. It is desired to formulate radiant-heat interchange between the heavy-lined surfaces A_1 and A_2, or to determine A_1F_{12}. Between the edges B and L of A_1 stretch a tight string representing the effective area A_1'. Stretch a minimum-length line over the surface connecting edge B of A_1 to edge E of A_2, dotted line $BCDE$; also a minimum-length line from edge L of A_1 to edge F of A_2, line $LKJHGF$. It is plain that the direct radiant interchange between A_1 and A_2 is the same regardless of whether they are connected by the solid-line surfaces BE and LF or the corresponding dotted-line surfaces, since no part of the field of view either A_1 or A_2 has of the other is affected by the substitution. Now stretch a minimum-length line from B to F, line $BHGF$, and a second one from L to E, line $LKJE$. Consider the three-sided enclosure formed by the surfaces A_1, $BCDE$, and $EJKL$. By analogy to Eq. (4-14),

$$A_1F_{1 \to \overline{BCDE}} = \frac{A_1' + \overline{BCDE} - \overline{LKJE}}{2} \qquad (4\text{-}15)$$

where $\overline{BCDE}$ is a shorthand representation of the area given by the product of the line length $BCDE$ and the dimension of the system normal to the sketch. Similarly, consider the three-sided enclosure formed by surfaces A_1', BH, and $HJKL$, from which

$$A_1F_{1 \to \overline{HJKL}} = \frac{A_1' + \overline{HJKL} - \overline{BH}}{2} \qquad (4\text{-}16)$$

Inspection of the figure indicates that, in addition to $\overline{BCDE}$ and $\overline{HJKL}$, the only surface A_1' can see is A_2. Consequently,

$$A_1F_{1 \to \overline{BCDE}} + A_1F_{1 \to \overline{HJKL}} + A_1F_{12} = A_1'$$

Substitution from (4-15) and (4-16) then gives

$$A_1F_{12}(\equiv A_2F_{21}) = \frac{(\overline{LKJE} + \overline{BH}) - (\overline{BCDE} + \overline{HJKL})}{2}$$
$$= \frac{(\overline{LKJE} + \overline{BHGF}) - (\overline{BCDE} + \overline{LKJHGF})}{2} \qquad (4\text{-}17)$$

68   HEAT TRANSMISSION

In words, the AF product for interchange between two surfaces *in this class*, per unit of length normal to the sketch, is the sum of the lengths of crossed strings stretched between the ends of the lines representing the two surfaces, less the sum of the lengths of uncrossed strings similarly

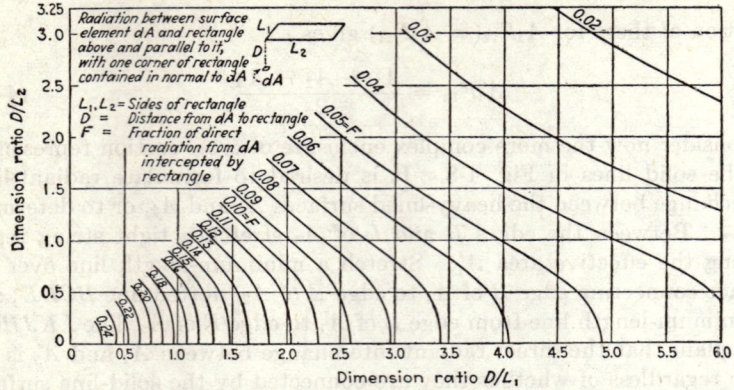

FIG. 4-9. View factor F for direct radiation between an element dA and a parallel rectangle with corner opposite dA.

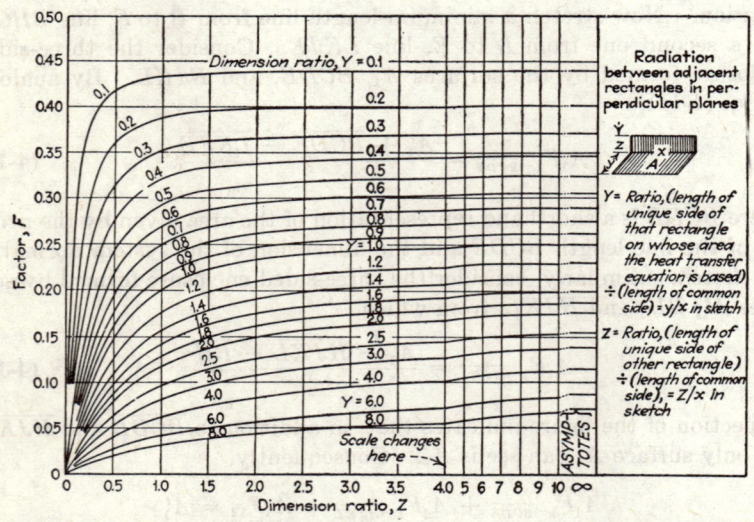

FIG. 4-10. View factor F for direct radiation between adjacent rectangles in perpendicular planes.

stretched between the surfaces, all divided by 2. This general case will be seen to include the three-sided enclosure covered by Eq. (4-14).

Values of F have been calculated for various surface arrangements on the assumption that emissivity ϵ_θ is constant, independent of θ (exact for

RADIANT-HEAT TRANSMISSION

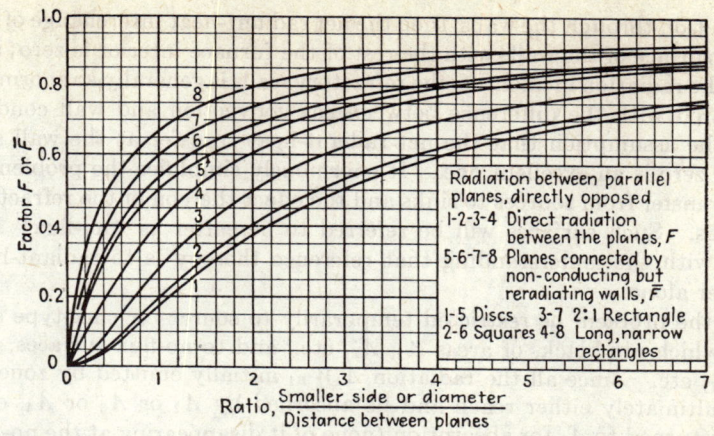

Fig. 4-11. View factor F and interchange factor $\bar{F}$ for opposed parallel disks, squares, and rectangles.

black surfaces, quite good for most nonmetallic or tarnished or rough metal surfaces). These values of F for a surface element dA and a rectangle in a parallel plane appear in Fig. 4-9; for adjacent rectangles in perpendicular planes in Fig. 4-10; for opposed parallel rectangles and disks of equal size as line 1 to 4 of Fig. 4-11; for an infinite plane parallel to a system of parallel tubes as lines 1 and 3 of Fig. 4-12. Other cases are treated in the literature.[27,28,29,65,25,23a]

Allowance for Refractory Surfaces. The Factor $\bar{F}$. One of the commonest problems of radiant-heat transfer in industrial-furnace design is that in which a portion of the enclosure constitutes a heat source or heat sources (such as a fuel bed, a carborundum muffle, a row of electric resistors), another portion a heat sink or heat sinks (such as the surface of a row of billets, the tubes of a tube still or boiler furnace, etc.), and another portion an intermediate refractory connecting-wall system which is a heat sink only to the extent that it loses heat by conduction through its walls to the furnace exterior. If the convection from gases on the inside of such refractory walls is approximately equal to the loss by

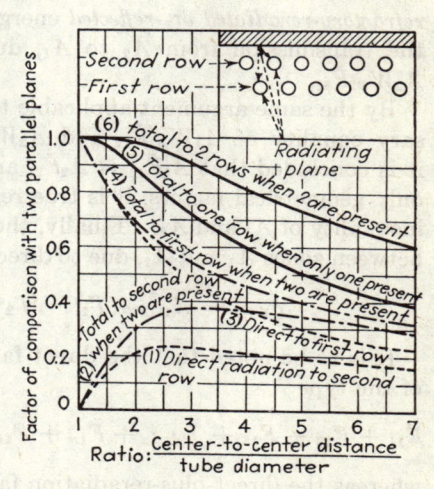

Fig. 4-12. View factor F and interchange factor $\bar{F}$ for radiation between a plane and one or two rows of tubes parallel to it.

conduction through the walls, then the net radiant-heat interchange of the inside surface of the walls with the rest of the furnace interior is zero; and since the radiation incident on the refractory walls is generally so enormous compared with the difference between gas convection and wall conduction, the assumption that the net radiant-heat transfer at the wall surface is zero is an excellent one. It enormously simplifies the problem of heat transfer from sources to sinks and the effect thereon of the refractory surfaces. Such surfaces will be referred to hereafter as "no-flux" surfaces, with the understanding that reference thereby is to radiant-heat transfer alone.

Let the problem be restricted temporarily to source- or sink-type surfaces which are black, of areas A_1, A_2, etc., and to no-flux surfaces, A_R, A_S, A_T, etc. Since all the radiation $A_1 W_{B1}$ initially emitted by zone A_1 must ultimately either reach and be absorbed by A_2 or A_3 or A_4, etc., or be returned to A_1 for absorption (none of it disappearing at the no-flux surfaces unless an equal quantity is emitted), it becomes desirable to define a new kind of factor $\bar{F}$, $\bar{F}_{12}$ being the fraction of the beam $A_1 W_{B1}$ streaming away from A_1 which reaches A_2 directly *and* by the assistance of the no-flux surfaces. Then, just as the *direct* radiant transfer from A_1 to A_2 due to initial radiation from A_1 was $A_1 W_{B1} F_{12}$, so the *direct plus refractory-reradiated or -reflected* energy transfer is $A_1 W_{B1} \bar{F}_{12}$. Similarly, the transmission from A_2 to A_1, due to initial radiation from A_2, is $A_2 W_{B2} \bar{F}_{21}$.

By the same argument applicable to the factor AF, namely, the necessary equality of $A_1 W_{B1} \bar{F}_{12}$ and $A_2 W_{B2} \bar{F}_{21}$ when $T_1 = T_2 (W_{B1} = W_{B2})$, it is concluded that $A_1 \bar{F}_{12} = A_2 \bar{F}_{21}$ and that, since this relation contains only geometrical factors, it is true regardless of temperature equality or inequality of A_1 and A_2. Finally, then, the net radiant-heat interchange between zones A_1 and A_2, due to direct-plus-refractory action, is given by

$$q_{1 \rightleftarrows 2} = A_1 \bar{F}_{12} \sigma (T_1^4 - T_2^4) \equiv A_2 \bar{F}_{21} \sigma (T_1^4 - T_2^4) \qquad (4\text{-}18)$$

It is to be noted that the direct factors F by definition obey relations of the type

$$F_{11} + F_{12} + F_{13} + \cdots + F_{1n} + F_{1R} + F_{1S} + F_{1T} + \cdots = 1 \qquad (4\text{-}19)$$

whereas the direct-plus-reradiation factors $\bar{F}$ obey relations of the type

$$\bar{F}_{11} + \bar{F}_{12} + \bar{F}_{13} + \cdots + \bar{F}_{1n} = 1 \qquad (4\text{-}20)$$

The factor $\bar{F}$ has been determined exactly for a few geometrically simple cases[33] and may be approximated for others. If A_1 and A_2 are equal parallel disks, squares, or rectangles connected by nonconducting but reradiating refractory walls on which the only restriction is that the angular distribution of their emission is like that of a black surface, and any reflection is diffuse,† then $\bar{F}$ is given by Fig. 4-11, lines 5 to 8. If A_2 represents an infinite plane and A_1 is one or two rows of infinite parallel tubes in a parallel plane and if the only other surface is a no-flux surface far enough behind the tubes to be substantially uniform in temperature (one diameter is sufficient), $\bar{F}_{21}$ is given by line 5 or 6 of Fig. 4-12.‡

If the no-flux surface $A_R + A_S + A_T \cdots$ is so disposed in the enclosure that the "view" of A_1 from all elements of it is the same, *i.e.*, all elements are equally irradiated by A_1; if similarly the view factors to A_2 from all elements of the no-flux surface are the same; and if the no-flux surface is either black or diffuse-reflecting, then a general expression for $\bar{F}_{12}$ is readily obtainable. All of the no-flux surface may be grouped together in a single zone of area designated by A_R. Black surface A_1 emits $A_1\sigma T_1^4$, sending fraction F_{12} directly to A_2, where complete absorption occurs, and fraction F_{1R} directly to A_R and uniformly distributed over it. A_R, being a no-flux surface, must get rid of all the radiation incident on it, either by reflection or by reradiation, sending by either mechanism fractions F_{R1}, F_{R2}, etc. to A_1, A_2, etc., and fraction F_{RR} toward itself for another reflection or reradiation which distributes between A_1 and the other sinks as before. Thus A_2 ultimately receives the fraction $F_{R2}/(1 - F_{RR})$ of the radiation initially incident on A_R. One may therefore write

$$\bar{F}_{12} = F_{12} + F_{1R}\frac{F_{R2}}{1 - F_{RR}} \tag{4-21}$$

If the system is further restricted to one containing but two source-sink surfaces, A_1 and A_2, Eq. (4-21) readily simplifies to

$$A_1\bar{F}_{12} = A_1F_{12} + \frac{1}{(1/A_1F_{1R}) + (1/A_2F_{2R})} \tag{4-21a}$$

Equation (4-21) covers a large fraction of problems of radiant-heat interchange between source and sink in a furnace enclosure and is in error only to the extent to which the assumption of uniform temperature of the no-flux surface is not permissible.

The development of expressions permitting any desired degree of approach to the exact answer, depending on the number of zones into which the no-flux surface is divided, will be made later (see pages 75 to 76).

It is to be noted that so long as interest is restricted to black source-sink surfaces, all of such surfaces at any one temperature may be grouped

---

† A diffuse-reflecting surface is one which reflects with the same energy distribution in the solid angle above the surface as that corresponding to black-body emission; *i.e.*, its reflection follows the cosine law. Nonmetallic surfaces do not depart greatly from this characteristic.

‡ See page 117 for limitations on use of Fig. 4-12.

If the single source and the single sink cannot see themselves, Eq. (4-21a) becomes:

$$\bar{F}_{12} = \frac{A_2 - A_1 F_{12}^2}{A_1 + A_2 - 2A_1 F_{12}} \tag{4-21b}$$

together into a single zone of total area A_1 without introducing any approximation into the formulation of $\bar{F}_{1n}$. Of interest also is the fact that the evaluation of $\bar{F}_{12}$ necessitates but a partial description of the enclosure, including A_1, A_2, and the no-flux surfaces, but no other source-sink surfaces.

Gray Enclosures. The Factor $\mathfrak{F}$.[30b] Consider an enclosure composed of gray source-sink surfaces $A_1, A_2, \ldots$ having emissivities and absorptivities $\epsilon_1, \epsilon_2, \ldots$, and of no-flux surfaces A_R, A_S, A_T which are partially diffuse-reflecting and partially absorbing, but not necessarily gray. In the steady state the net flux from A_1 is the sum of its net interchanges with $A_2, A_3 \ldots$ because there is by definition no net interchange at $A_R, A_S, \ldots$. The net flux between A_1 and A_2 occurs by a complex process involving multiple reflection from all surfaces including A_3, A_4, $\ldots$ and $A_R, A_S, \ldots$, as well as reradiation from surfaces $A_R, A_S, \ldots$; and one might at first consider the contribution of $A_R, A_S, \ldots$ to the net flux between A_1 and A_2 impossible to disentangle, because the equilibrium temperature of A_R, for example, depends on contributions from $A_3, A_4 \ldots$ as well as A_1 and A_2. The new concept necessary here is that the refractory zone A_R can be thought of as having a partial emissive power due to the presence of each of the source-sink zones, and a total emissive power equal to their sum. Thus, the term $q_{1\rightleftarrows 2}$ represents net flux between A_1 and A_2 due to their respective emission rates and including, in addition to direct interchange $A_1 F_{12}\epsilon_1\epsilon_2\sigma(T_1^4 - T_2^4)$, the contributions due to multiple reflection at all surfaces as well as such contributions by reradiation from the no-flux surfaces as are consequent on their partial emissive powers due to the existence of A_1 and A_2 alone as net radiators in the system. This is the *necessary* meaning of $q_{1\rightleftarrows 2}$ if it is to become zero when $T_1 = T_2$. It is apparent that $q_{1\rightleftarrows 2}$ must take a form equal to $\sigma(T_1^4 - T_2^4)$ multiplied by some factor which depends on the geometry of the whole enclosure and the emissivity of its source-sink surfaces, and which can be expressed in the form

$$q_{1\rightleftarrows 2} = A_1 \mathfrak{F}_{12} \sigma(T_1^4 - T_2^4) \equiv A_2 \mathfrak{F}_{21} \sigma(T_1^4 - T_2^4) \qquad (4\text{-}22)$$

The problem is to evaluate $A\mathfrak{F}$. Plainly, it cannot depend on any system temperatures. Consequently, if all source-sink surfaces except A_1 are kept at absolute zero, and $q_{1\rightleftarrows 2}$ (which now becomes $q_{1\rightarrow 2}$) is evaluated and used to determine $\mathfrak{F}$ in Eq. (4-22), that value of $\mathfrak{F}$ will be generally applicable regardless of the particular combination of temperatures of the source-sink surfaces. In addition to the assignment of zero temperature to all source-sink surfaces except A_1, one more simplification can be introduced. Let the temperature of A_1 be such that if black it would have an emissive power of 1. Surfaces $A_R, A_S, A_T, \ldots$ will assume equilibrium emissive powers (partial values due to the radiation of A_1

only) between zero and 1. At each surface there will be radiant flux toward and away from the surface due to reflection of radiation initially emitted from A_1 and involving multiple reflections from all the source-sink surfaces and reflections and/or reradiation from the no-flux surfaces. For surface A_m call this radiant flux per unit area $_1R_m$, the presubscript as a reminder of the original source of the flux, and R instead of W to indicate that the quantity is a relative flux density scaled down in the ratio $1:\sigma T_1^4$ because of the assumed value of the emissive power of A_1. The flux densities streaming away from A_2, A_3, ... will be, $_1R_2$, $_1R_3$, ..., geometrically analogous to emissive powers in their fractional interception by various other surfaces, but due exclusively to reflection; they would be zero in a black system. The flux density from A_1 will be $_1R_1 + \epsilon_1$, to include its own original emission ϵ_1 as well as its contribution due to mutual reflection within the system. Flux densities away from (and toward) the no-flux surfaces will be $_1R_R$, $_1R_S$, ...

The radiation absorbed at surface A_2 will be due to beams from the various surfaces "seen" by it, and each incident beam will be partially absorbed and partially reflected, in the ratio $\epsilon_2/(1 - \epsilon_2)$. Then, since the flux away from A_2 is $A_2 \cdot {_1R_2}$, the total rate of absorption at it is $A_2 \cdot {_1R_2} \cdot \epsilon_2/(1 - \epsilon_2)$. Since this absorption is the result solely of emission originating at A_1 when σT_1^4 is 1, it follows that

$$A_1 \mathcal{F}_{12} = {_1R_2} \cdot A_2 \frac{\epsilon_2}{1 - \epsilon_2} \tag{4-23}$$

or

$$A_1 \mathcal{F}_{1n} = {_1R_n} \cdot A_n \frac{\epsilon_n}{1 - \epsilon_n} \tag{4-23a}$$

The problem is to find the values of the R's for use in (4-23a), by setting up energy balances on all the surfaces. The incidence of radiation on A_1 includes that from itself, from A_2, A_3, A_R, etc., and their sum is represented by the bracketed term below. The fraction $1 - \epsilon_1$ of all this is reflected and therefore equals the flux away from A_1 exclusive of its original emission, or $A_1 \cdot {_1R_1}$. Equating these two gives

$$[A_1 F_{11}(\epsilon_1 + {_1R_1}) + A_2 F_{21} \cdot {_1R_2} + \cdots A_R F_{R1} \cdot {_1R_R} + \cdots](1 - \epsilon_1)$$
$$= A_1 \cdot {_1R_1} \tag{4-24}$$

Similar relations may be formulated for A_2, A_3, ..., A_R, ... One thus obtains as many equations as there are unknown flux densities R, permitting a solution for the latter and correspondingly an evaluation of any interchange factor $A_1 \mathcal{F}_{1n}$ by Eq. (4-23a).

In assembling the equations like (4-24) for solution, a shorthand nomenclature is desirable. Let $A_1 F_{1R}$ or $A_R F_{R1}$ each be designated by $\overline{1R}$. Replace reflectivity $1 - \epsilon$ by the symbol ρ, and divide both sides of each equation by it. The system of energy balances is then

$$
\begin{aligned}
&\left(\overline{11}-\frac{A_1}{\rho_1}\right)\cdot {}_1R_1 + \overline{12}\cdot {}_1R_2 + \overline{13}\cdot {}_1R_3 + \cdots + \overline{1R}\cdot {}_1R_R + \overline{1S}\cdot {}_1R_S + \cdots = -\overline{11}\cdot \epsilon_1 \\
&\overline{12}\cdot {}_1R_1 + \left(\overline{22}-\frac{A_2}{\rho_2}\right){}_1R_2 + \overline{23}\cdot {}_1R_3 + \cdots + \overline{2R}\cdot {}_1R_R + \overline{2S}\cdot {}_1R_S + \cdots = -\overline{12}\cdot \epsilon_1 \\
&\qquad \vdots \\
&\overline{1R}\cdot {}_1R_1 + \overline{2R}\cdot {}_1R_2 + \overline{3R}\cdot {}_1R_3 + \cdots + (RR-A_R){}_1R_R + \overline{RS}\cdot {}_1R_S + \cdots = -\overline{1R}\cdot \epsilon_1 \\
&\overline{1S}\cdot {}_1R_1 + \overline{2S}\cdot {}_1R_2 + \overline{3S}\cdot {}_1R_3 + \cdots + (RS\cdot {}_1R_R) + (\overline{SS}-A_S){}_1R_S + \cdots = -\overline{1S}\cdot \epsilon_1
\end{aligned}
\tag{4-25}
$$

To express the solution, define the determinant D:

$$
D \equiv \begin{vmatrix}
\overline{11}-\dfrac{A_1}{\rho_1} & \overline{12} & \overline{13} & \cdots & \overline{1R} & \overline{1S} & \cdots \\
\overline{12} & \overline{22}-\dfrac{A_2}{\rho_2} & \overline{23} & \cdots & \overline{2R} & \overline{2S} & \cdots \\
\overline{13} & \overline{23} & \overline{33}-\dfrac{A_3}{\rho_3} & \cdots & \overline{3R} & \overline{3S} & \cdots \\
\vdots & & & & & & \\
\overline{1R} & \overline{2R} & \overline{3R} & \cdots & \overline{RR}-A_R & \overline{RS} & \cdots \\
\overline{1S} & \overline{2S} & \overline{3S} & \cdots & \overline{RS} & \overline{SS}-A_S & \cdots \\
\vdots & & & & & &
\end{vmatrix}
\tag{4-26}
$$

Solution of (4-25) for ${}_1R_n$ gives

$$
{}_1R_n = {}_1D_n/D \tag{4-27}
$$

in which ${}_1D_n$ is the determinant formed by replacing the nth column of D by the quantities on the right side of (4-25). Then. from Eq. (4-23),

$$
A_1 \mathfrak{F}_{1n} = \frac{\epsilon_n A_n}{\rho_n}\frac{{}_1D_n}{D} \tag{4-28}
$$

More generally,

$$
A_m \mathfrak{F}_{mn} = \frac{\epsilon_n A_n}{\rho_n}\frac{{}_mD_n}{D} \tag{4-29}
$$

where ${}_mD_n$ is formed by replacing the nth column of D by

$$
-\overline{m1}\cdot \epsilon_m,\; -\overline{m2}\cdot \epsilon_m,\; -\overline{m3}\cdot \epsilon_m,\; \ldots,\; -\overline{mR}\cdot \epsilon_m,\; -\overline{mS}\cdot \epsilon_m,\; \ldots
$$

A statement previously made about $\mathfrak{F}$ now needs qualification. $\mathfrak{F}_{12}$ is seen to depend on the geometry of the *whole* system and on the emissivities of *all* the source-sink surfaces, but not on the grayness of the no-flux surfaces.

To indicate method of use, $A_1\mathfrak{F}_{12}$ will be partially evaluated. On formulation of $_1D_2$ for insertion into Eq. (4-28), the first and second columns, after factoring $-\epsilon_1$ out of the second column, will be found alike except for the top members, $\overline{11} - A_1/\rho_1$ and $\overline{11}$, respectively. Since the value of a determinant is unchanged by replacing any column (or row) by a new one the members of which are formed by subtracting any other column from the one in question, column 1 may be replaced by the members $-A_1/\rho_1, 0, 0, 0, \ldots$; and the determinant then equals $-A_1/\rho_1$ times the minor determinant formed by crossing out the first row and column. Then

$$A_1\mathfrak{F}_{12} = \frac{\epsilon_1\epsilon_2}{\rho_1\rho_2} A_1 A_2 \frac{\begin{vmatrix} \overline{12} & \overline{23} & \cdots & \overline{2R} & \overline{2S} & \cdots \\ \overline{13} & \overline{33} - A_3/\rho_3 & \cdots & \overline{3R} & \overline{3S} & \cdots \\ \overline{1R} & \overline{3R} & \cdots & \overline{RR} - A_R & \overline{RS} & \cdots \\ \overline{1S} & \overline{3S} & \cdots & \overline{RS} & \overline{SS} - A_S & \cdots \end{vmatrix}}{D}$$

(4-30)

The number of unique view factors F necessary for evaluation of $\mathfrak{F}$ by (4-30) may be determined. By noting that in a p-zone system there are p^2 F's in the determinant D but that (1) D is skew-symmetrical and (2) any row or column of F's adds to 1, it is seen that the number of unique F's necessary is $p(p-1)/2$. If in addition each of the zones except one cannot see itself, the number is further reduced to $(p-1)(p-2)/2$.

Equation (4-29) can be used to make allowance for any degree of complexity of an enclosure and to approach the true solution to any degree of approximation dependent on the number of zones into which a surface is divided. The guiding principle in deciding upon the number of zones necessary is that any reradiation or reflection must come from a zone small enough so that different parts of its surface do not have a significantly different view of the various other surfaces. As already seen, black source-sinks need be zoned only according to temperature; but light gray ones may require further subdivision. The problem left unsolved on page 71—the formulation of $\bar{F}_{12}$ with proper allowance for variation in refractory temperature—may now be completed. Using Eq. (4-30) and setting in the condition that $\epsilon_1, \epsilon_2, \epsilon_3, \ldots = 1$ or that $\rho_1, \rho_2, \rho_3 = 0$, one can eliminate all rows and columns containing reference to any source-sink surfaces except 1 and 2 as follows:

Cancel A_1/ρ_1 out of the numerator of (4-30), and multiply the denominator first column by ρ_1/A_1, which, being zero, makes the first column $-1, 0, 0, 0, \ldots$ and reduces the order of D by one. Similarly, multiply the second column of the numerator and the second column of the denominator by ρ_3/A_3, making them become $0, -1, 0, 0, \ldots$ each.

One thus obtains

$$A_1\bar{F}_{12} = \begin{vmatrix} \overline{12} & \overline{2R} & \overline{2S} & \cdots \\ \overline{1R} & \overline{RR} - A_R & \overline{RS} & \cdots \\ \overline{1S} & \overline{RS} & \overline{SS} - A_S & \cdots \\ \cdot & \cdot & \cdot & \\ \cdot & \cdot & \cdot & \\ \cdot & \cdot & \cdot & \end{vmatrix} \div \begin{vmatrix} \overline{RR} - A_R & \overline{RS} & \cdots \\ \overline{RS} & \overline{SS} - A_S & \cdots \end{vmatrix}$$

(4-31)

If there is but one refractory zone, all rows and columns mentioning others may be crossed out and (4-31) yields

$$A_1\bar{F}_{12} = \overline{12} - \frac{\overline{1R} \cdot \overline{2R}}{\overline{RR} - A_R} \tag{4-32}$$

which is readily shown to be the same as Eq. (4-21).

Many practical problems are covered by formulation of $\mathfrak{F}$ for the case of an enclosure divided into any number of no-flux zones A_R, A_S, ... but only two source-sink zones A_1 and A_2—an especially justifiable assumption if the emissivities of A_1 and A_2 are so high as to make reflections from their surfaces relatively unimportant in the over-all heat transfer. From (4-30) it may readily be shown that for this case

$$\frac{1}{A_1\mathfrak{F}_{12}} = \frac{1}{A_1}\left(\frac{1}{\epsilon_1} - 1\right) + \frac{1}{A_2}\left(\frac{1}{\epsilon_2} - 1\right) + \frac{1}{A_1\bar{F}_{12}} \tag{4-33}$$

If the two-zone assumption for source-sink surfaces is accepted, the numerical value of $\mathfrak{F}$ for this case is as good as the method used for evaluating $\bar{F}$.

Allowance for Space-varying Temperature of Source-sink Surfaces. In a continuous furnace in which the temperature of the heat sink varies from its point of entry to exit, it is seldom justifiable from a practical standpoint to make rigorous allowance for the temperature variation. Some kind of engineering approximation must be made, of which two possibilities will be discussed. Consider first an electric stock-heating furnace so long in the direction of stock movement relative to its other two dimensions that the stock locally at distance x from one end can "see" only those parts of the heat source, itself, and associated no-flux surfaces which characterize the distance x. If the stock heat-transfer surface per unit length is designated by S, the local sink and source temperatures by $T_{1,x}$ and $T_{2,x}$, and the mass flow rate and heat capacity of the stock by w_s and c_p, then

$$w_s c_p \, dT_{1,x} = q/A \, \big]_{\text{local}} \cdot S \, dx \tag{4-34}$$

in which $q/A\,]_{\text{local}}$ is the local heat-transfer rate to the stock per unit of its area, due to whatever mechanisms of heat transfer are operative. If $q/A\,]_{\text{local}}$ is expressible as a function of $T_{1,x}$ or quantities dependent on it, (4-34) can be integrated.

$$q/A\,]_{\text{local}} = \mathfrak{F}_{12}\sigma(T_{2,x}{}^4 - T_{1,x}{}^4) + q/A\,]_{\text{convection}} \qquad (4\text{-}35)$$

in which $\mathfrak{F}_{12}$ is the local value of the radiant-exchange factor taking into account emissivities and the relative disposition of stock, electric resistors, and refractory surfaces. Along the middle of the furnace, $\mathfrak{F}$ would be evaluated as though A_1 and A_2 were of infinite extent in the x direction, the "view" that dA_1 has of cooler surfaces to one side of dA_2 being assumed to be compensated by the view it has of warmer surfaces on the other side; near the furnace ends the presence of a no-flux end surface should be included for rigor. Considering $T_{2,x}$ to be known and constant or related to $T_{1,x}$ by an energy balance and $q/A\,]_{\text{conv}}$ to be expressible as a function of $T_{1,x}$ and $T_{2,x}$, integration of (4-34) over the full furnace length B is possible. It yields

$$\left(\frac{S}{w_s c_p}\right) B = \int_{T_{1,0}}^{T_{1,B}} \frac{dT_{1,x}}{q/A\,]_{\text{local}}} \qquad (4\text{-}36)$$

Therefore, plotting the reciprocal of $q/A\,]_{\text{local}}$ against $T_{1,x}$ will give a curve the area under which equals $(S/w_s c_p)B$.

The above method is in increasing error as the surface elements in length dx see with a larger view factor those parts of the system lying too far from x. For the other case to be discussed, that of an enclosure in which all three dimensions are of the same order of magnitude, it is suggested that if there is a space variation in sink temperature the sink area be divided into several zones (seldom more than three for practicality). The general interchange factor $\mathfrak{F}$ can then be evaluated for interchange between each sink zone and every other source-and-sink zone. Together with knowledge of sink enthalpy change from zone to zone, the heat-transfer equations will yield enough information for a solution; but simultaneous equations will be encountered, more the more zones used, and some problems will involve trial and error as well.

Nongray Enclosures. Although the analysis of an enclosure which leads to the formulation of a radiant-interchange factor $\mathfrak{F}$ for gray systems could be modified to allow for real (nongray) surfaces, the complications, superimposed on an already elaborate procedure, would be prohibitive from an engineering standpoint. Moreover, the necessary data are not available for most surfaces; and in addition, as indicated on page

61 and Table A-23 (page 472), there is a wide variation in surfaces of similar description. It is illuminating, however, to consider one case which is geometrically simple enough to permit rigorous treatment—the case of a small nongray body of area A_1 in black surroundings at T_2. The interchange is plainly given by

$$q_{1 \rightleftarrows 2} = A_1 \sigma(\epsilon_1 \cdot T_1^4 - \alpha_{12} \cdot T_2^4) \tag{4-37}$$

in which α_{12} is the absorptivity of surface 1 at T_1 for black radiation at T_2. From page 62 it was seen that over a modest temperature range α_{12} can be represented by $gT_1^m T_2^n$, with n considerably greater numerically than m for nonmetals, and both negative; and n about the same as m for bright metallic surfaces, and both positive. Then

$$\epsilon_1 = \alpha_{11} = gT_1^{m+n} \tag{4-38}$$

and

$$q_{1 \rightleftarrows 2} = A_1 \sigma g(T_1^{4+m+n} - T_1^m T_2^{4+n}) \tag{4-39}$$

This may be put arbitrarily into the form of a fourth-power difference by the following procedure: Take the derivative of $q_{1 \rightleftarrows 2}$ with respect to T_2,

$$\frac{-dq}{dT_2} = A_1 \sigma g(4 + n)T_1^m T_2^{3+n} \tag{4-40}$$

As T_1 and T_2 approach T, this becomes

$$\frac{-dq}{dT} = A_1 \sigma g(4 + n)T^{3+m+n} \tag{4-41}$$

or

$$\frac{-dq}{4T^3 \, dT} = \frac{-dq}{dT^4} = A_1 \sigma g \left(1 + \frac{n}{4}\right) T^{m+n} = A_1 \left(1 + \frac{n}{4}\right) \sigma \epsilon \tag{4-42}$$

Over a moderate range of temperature, then,

$$q_{1 \rightleftarrows 2} = A_1 \sigma \left[\epsilon_{\text{avg}} \left(1 + \frac{n}{4}\right)\right] (T_1^4 - T_2^4) \tag{4-43}$$

with ϵ_{avg} evaluated from (4-38) at the arithmetic-mean temperature. Thus, although ϵ approaches α as T_2 and T_1 approach one another, the emissivity factor by which $\sigma(T_1^4 - T_2^4)$ should be multiplied is not ϵ, but $\epsilon(1 + n/4)$. Equation (4-43) can be used with small error over an absolute temperature ratio up to 2.

A corollary of the above derivation is the expression of radiation in the form of a first-power difference relation for use in combining with convection. From Eq. (4-42) it is apparent that if $q_{\text{rad}} = h_r(T_1 - T_2)(A)$, the transfer coefficient h_r is given by

$$h_r = (4 + n)\sigma \epsilon_{\text{avg}} T_{\text{avg}}^3 \tag{4-44a}$$

In the field of interest of the present chapter the reverse procedure is of greater interest, *viz.*, the expression of a convection transfer coefficient in the form of an equivalent "emissivity due to convection," ϵ_C, for adding to the true emissivity. Plainly,

$$\epsilon_C = \frac{h_c}{(4 + n)\sigma T_{avg}^3} \qquad (4\text{-}44b)$$

The more general case of net interchange in an enclosure consisting of two nongray source-sink zones and some no-flux zones will conform to the relation

$$q_{1 \rightleftarrows 2} = \sigma A_1 \mathcal{F}_{1 \to 2} T_1^4 - \sigma A_1 \mathcal{F}_{1 \leftarrow 2} T_2^4 \qquad (4\text{-}45)$$

The arrows between the subscripts are to indicate that, contrary to previous derivations for black or gray systems, the value of $\mathcal{F}_{12}$ *does* depend on which surface is the source of the radiation and which the sink. In fact, $\mathcal{F}$ now depends on temperature because it depends on ϵ and α, and only in the limit as T_1 approaches T_2 do the two $\mathcal{F}$'s become the same. As stated above, however, the evaluation of these rigorous $\mathcal{F}$'s is seldom simple enough for practical use. If $T_1 \gg T_2$, the first term on the right of Eq. (4-45) is plainly controlling and $\mathcal{F}_{1 \to 2}$, evaluated approximately by use of ϵ_1 and α_{21}, may be used for both terms. If the temperatures are close together, $\mathcal{F}_{1 \leftarrow 2}$ may also require evaluation, using ϵ_2 and α_{12}. Either this procedure or that of evaluating a single $\mathcal{F}_{12}$ by use of the same kind of effective emissivity to replace ϵ and α as appears in Eq. (4-43) is believed not to lead to significant error for most engineering purposes. The limitations on this statement, however, have not been adequately examined.

Recommended Procedure. The use of the preceding principles is best illustrated by some examples.

Illustration 1. What per cent of black-body radiation from a source at (*a*) 2500°F, (*b*) 1000°F will a quartz window transmit? Additional data: The transmittance of quartz changes rapidly with wavelength; it may be assumed transparent in the range 0.16μ to 3.7μ, otherwise opaque (except in far infrared, of no consequence here).

a. 2500°F = 1642°K. Below $\lambda_{min}T = 0.16 \times 1642 = 263~\mu$ deg K, according to Fig. 4-1, a negligible fraction of the energy lies. Below $\lambda_{max}T = 3.7 \times 1642 = 6100$, 73.5 per cent of the energy is found. Therefore, transmittance = 0.735.

b. 1000°F = 810°K. Lower limit still less significant. Below $\lambda_{max}T = 3.7 \times 810 = 3010$, 26 per cent of the energy is found. Transmittance = 0.26. Plainly, a total radiation pyrometer with quartz lens, which is generally calibrated for black-body sources, cannot be used to measure the radiation from sources of unknown spectral energy distribution.

Illustration 2. Show that, if parallel planes are not gray [see Eq. (4-5)], their net interchange is given approximately by $A_1 \sigma T_1^4/(1/\epsilon_1 + 1/\alpha_{21} - 1) - A_1 \sigma T_2^4/(1/\epsilon_2 + 1/\alpha_{12} - 1)$. Discuss the nature of the approximation.

Illustration 3. Long parallel opposed flat strips A_1 and A_2 of width W each, separated by spacing S, and of length L which is great compared with W or S, constitute a radiating system of interest in electric-furnace design. What is the value of the view factor F_{12}? *Ans.*: $\sqrt{1 + (S/W)^2} - S/W$.

Illustration 4. Long parallel circular tubes of equal diameter D and on center-to-center distance C are of interest in furnaces heating fluid carried in tubes. What is the value of the view factor from a tube to one of the two it can see? *Ans.*: $\dfrac{1}{2} - \dfrac{C/D}{\pi} - \dfrac{1}{\pi}\left(\tan^{-1}\sqrt{\left(\dfrac{C}{D}\right)^2 - 1} - \sqrt{\left(\dfrac{C}{D}\right)^2 - 1}\right)$ What fraction of the radiation from a round-rod resistor is intercepted by another rod in contact along their full length? *Ans.*: 0.182. By six such rods?

Illustration 5. What is the heat transfer by radiation between an oxidized nickel tube 4 in. o.d., at a temperature of 800°F and an enclosing chamber of silica brick at 1800°F, the brick chamber being (*a*) very large relative to the tube diameter and (*b*) 8 in. square inside, gray, and having an emissivity = 0.8?

a. Since the surroundings are large compared with the enclosed tube, it is unnecessary to allow for the emissivity of the silica brick, because the surroundings, viewed from the position of the small enclosed body, appear black; hence Eq. (4-37) is applicable. The emissivity of oxidized nickel at 800°F is, by interpolation from Table A-23, about 0.43; its absorptivity for radiation from a source at 1800° is approximately its emissivity at 1800°F, which by extrapolation is about 0.58. The tube area per foot is $\pi 4/12 = 1.05$ sq ft/ft. From Eq. (4-37)

$$q(\text{per foot length}) = 1.05 \times 0.171 \left[0.43\left(\dfrac{800 + 460}{100}\right)^4 - 0.58\left(\dfrac{1800 + 460}{100}\right)^4\right]$$
$$= -25{,}300 \text{ Btu}/(\text{hr})(\text{ft of tube})$$

The more usual procedure of using a single value for α and ϵ would give, for $\epsilon = 0.58$, q per foot $= -24{,}500$.

b. Since the enclosure is not large compared with the tube, it is necessary to allow for the emissivity of the silica brick, using Eq. (4-33). Since T_2 is considerably higher than T_1, $\mathfrak{F}_{1 \leftarrow 2}$ is more important than $\mathfrak{F}_{1 \rightarrow 2}$, and α_{12} will therefore be used instead of ϵ_1.

As before, $\bar{F}_{12} = 1$. When $\alpha_{12} = 0.58$ and $\epsilon_2 = 0.8$, Eq. (4-33) gives

$$\mathfrak{F}_{12} = \dfrac{1}{1 + \left(\dfrac{1}{0.58} - 1\right) + \dfrac{1.05}{2.67}\left(\dfrac{1}{0.8} - 1\right)} = 0.549$$

Therefore $q = -24{,}500 \times 0.549/0.58 = -23{,}200$.

Illustration 6. A muffle-type furnace in which the carborundum muffle forms a continuous floor of dimensions 15 by 20 ft has its ultimate heat-receiving surface in the form of a row of 4-in. tubes on 9-in. centers above and parallel to the muffle and backed by a well-insulated refractory roof; the distance from the muffle top to the row of tubes is 10 ft. The tubes fill the furnace top, of area equal to that of the carborundum floor. The average muffle-surface temperature is 2100°F; the tubes are at 600°F. The side walls of the chamber are assumed substantially nonconducting but reradiating and are at some equilibrium temperature between 600 and 2100°F, such that they radiate just as much heat as they receive. The tubes are oxidized steel of emissivity 0.8; the carborundum has an emissivity of 0.7; both are gray. Find the radiant-heat transmission between the carborundum floor and the tubes above, taking into account the reradiation from the side walls.

Call the area of the roof tubes A_1, that of the carborundum floor A_3, that of the refractory side walls of the furnace A_R. The problem must be broken up into two parts, first considering the roof with its refractory-backed tubes. To an imaginary plane A_2 of area 15 by 20 ft located just below the tubes, the tubes emit radiation

$A_1\mathfrak{F}_{12}T_1^4$, equal to $A_2\mathfrak{F}_{21}T_1^4$. To obtain $\mathfrak{F}_{21}$, one must first evaluate $\bar{F}_{21}$, which comes from Fig. 4-12, line 5, from which $\bar{F}_{21} = 0.84$. From Eq. (4-33)

$$\mathfrak{F}_{21} = \cfrac{1}{\cfrac{1}{0.84} + \left(\cfrac{1}{1} - 1\right) + \cfrac{9}{4\pi}\left(\cfrac{1}{0.8} - 1\right)} = 0.73\dagger$$

This amounts to saying that the system of refractory-backed tubes is equal in radiating power to a continuous plane A_2 replacing the tubes and refractory above them, having a temperature equal to the tubes and an equivalent or effective emissivity of 0.73.

The new simplified furnace now consists of an enclosure formed by a 15- by 20-ft rectangle A_3 of emissivity 0.7, above and parallel to it a 15- by 20-ft rectangle A_2 of temperature T_1 and emissivity 0.73, and refractory walls A_R to complete the enclosure. The desired heat transfer is $q_{2\rightleftarrows3}$.

$$q_{2\rightleftarrows3} = \sigma(T_1^4 - T_3^4)A_2\mathfrak{F}_{23}$$

Normally to evaluate $\mathfrak{F}_{23}$, one would use Eq. (4-29), with the refractory walls divided into a number of zones. For the present case, however, Fig. 4-11, line 6, presents an exact allowance for the continuous variation in side-wall temperature from top to bottom. The interchange factor between parallel 15- by 20-ft rectangles separated by 10 ft may be taken as the geometric mean of the factors for 15-ft squares separated by 10 ft and 20-ft squares separated by 10 ft. Then, from Fig. 4-11, line 6, $\bar{F}_{23} = \sqrt{0.63 \times 0.69} = 0.66$. From Eq. (4-33),

$$\mathfrak{F}_{23} = \cfrac{1}{\cfrac{1}{0.66} + \left(\cfrac{1}{0.73} - 1\right) + 1\left(\cfrac{1}{0.7} - 1\right)} = 0.433 = \mathfrak{F}_{32}$$

i.e., the floor and tubes interchange 43.3 per cent as much radiation as parallel black planes close together, each of area equal to the floor. The net interchange is

$$q_{\text{net}} = 0.171 \times (15 \times 20)(25.6^4 - 10.6^4)0.433 = 9{,}270{,}000 \text{ Btu/hr}$$

Illustration 7. The distribution of radiant heat to the different rows of tubes in a tube nest irradiated from one side is desired when the tubes are 4.0 in. o.d. on 8-in. triangular centers. Let the area of the continuous plane below the tube nest be A_1 and the area of the tubes, A_2. According to Fig. 4-12, curve 3, the first row of tubes will intercept directly 0.66 of the total. According to curve 1, the second row will intercept 0.21 of the total, leaving $1 - 0.66 - 0.21 = 0.13$ to be intercepted by the remaining rows.

Suppose the tube nest replaced by a single row of tubes A_2 with refractory back wall A_R. Equation 4-32 gives an approximation for $\bar{F}_{12}$. For the present case $F_{RR} = 0$ and $F_{1R} = 1 - F_{12}$ and $F_{R2} = F_{12}$; so Eq. (4-32) becomes

$$\bar{F}_{12} = F_{12} + (1 - F_{12})F_{12} = 0.66 + 0.34 \times 0.66 = 0.88$$

a value that could have been read from Fig. 4-12, curve 5. A single tube and back wall will therefore be 88 per cent as effective a heat receiver as an infinite number of rows, so far as radiant-heat transmission is concerned.

Suppose the one plane had been replaced by two rows of tubes with refractory back wall, instead of by a single row. According to Fig. 4-12, curves 4 and 2, the total

† The use of Eq. (4-33) was hardly justifiable here, since the "views" from spots on the top and the bottom of the tubes comprising the area A_1 are so different; but when A_1 is divided into two zones, the value of $\mathfrak{F}_{21}$ is raised to only 0.74.

radiation to the first row is 0.69, to the second 0.29, to both 0.69 + 0.29, or 0.98 as much as to an infinite number of rows (or to a continuous plane).

From Fig. 4-12, it is seen that only when the tubes are of small diameter relative to their distance apart is there any considerable quantity of radiant-heat penetration beyond the second row. The solution of a three- or four-row problem may be made readily by application of the principles discussed on page 66.

RADIATION FROM NONLUMINOUS GASES

If black-body radiation passes through a gas mass containing, for example, carbon dioxide, absorption occurs in certain regions of the infrared spectrum. Conversely, if the gas mass is heated, it radiates in those same wavelength regions. This infrared spectrum of gases has its origin in simultaneous quantum changes in the energy levels of rotation and of interatomic vibration of the molecules and, at the temperature levels reached in industrial furnaces, is of importance only in the case of the heteropolar gases. *Of the gases encountered in heat-transfer equipment, carbon monoxide, the hydrocarbons, water vapor, carbon dioxide, sulfur dioxide, ammonia, hydrogen chloride, and the alcohols are among those with emission bands of sufficient magnitude to merit consideration.* Gases with symmetrical molecules, hydrogen, oxygen, nitrogen, etc., have been found not to show absorption bands in those wavelength regions of importance in radiant-heat transmission at temperatures met in industrial practice.

Carbon Dioxide and Water Vapor. Consider a hemispherical gas mass of radius L containing carbon dioxide of partial pressure P_c, and let the problem be the evaluation of radiant-heat interchange between the gas at temperature T_G and a black element of surface at temperature T_s, located on the base of the hemisphere at its center. Per unit of surface the emission of the gas to the surface is $\sigma T_G^4 \epsilon_G$, where ϵ_G denotes gas emissivity, the ratio of radiation from gas to surface to the radiation from a black body at the same temperature. For carbon dioxide ϵ_G depends on T_G, the product term PL, and the total pressure P_T. Figure 4-13 is a smoothed plot of carbon dioxide emissivity ϵ_c at a total pressure of 1 atm, based on direct measurements of total emission by Hottel and Mangelsdorf[34] and Hottel and Smith,[35] and indicating by dotted lines those regions of the plot unsupported by data. As the total gas pressure is increased, the lines of the CO_2 spectrum are broadened; Fig. 4-14 gives an approximate correction factor C_c for a total pressure differing from 1 atm.[32] The absorption by the gas of radiation from the surface is $\sigma T_1^4 \alpha_c$, where α_{G1} is the gas absorptivity for black-body radiation from a source at T_1. Although the gas absorptivity must equal its emissivity when $T_1 = T_G$, as T_G increases above T_1 the absorptivity is affected by temperature in two ways: the molecular-absorption coefficient of the gas increases somewhat, but the number of absorbing gas molecules in a fixed path length at a fixed partial pressure decreases. Approximate

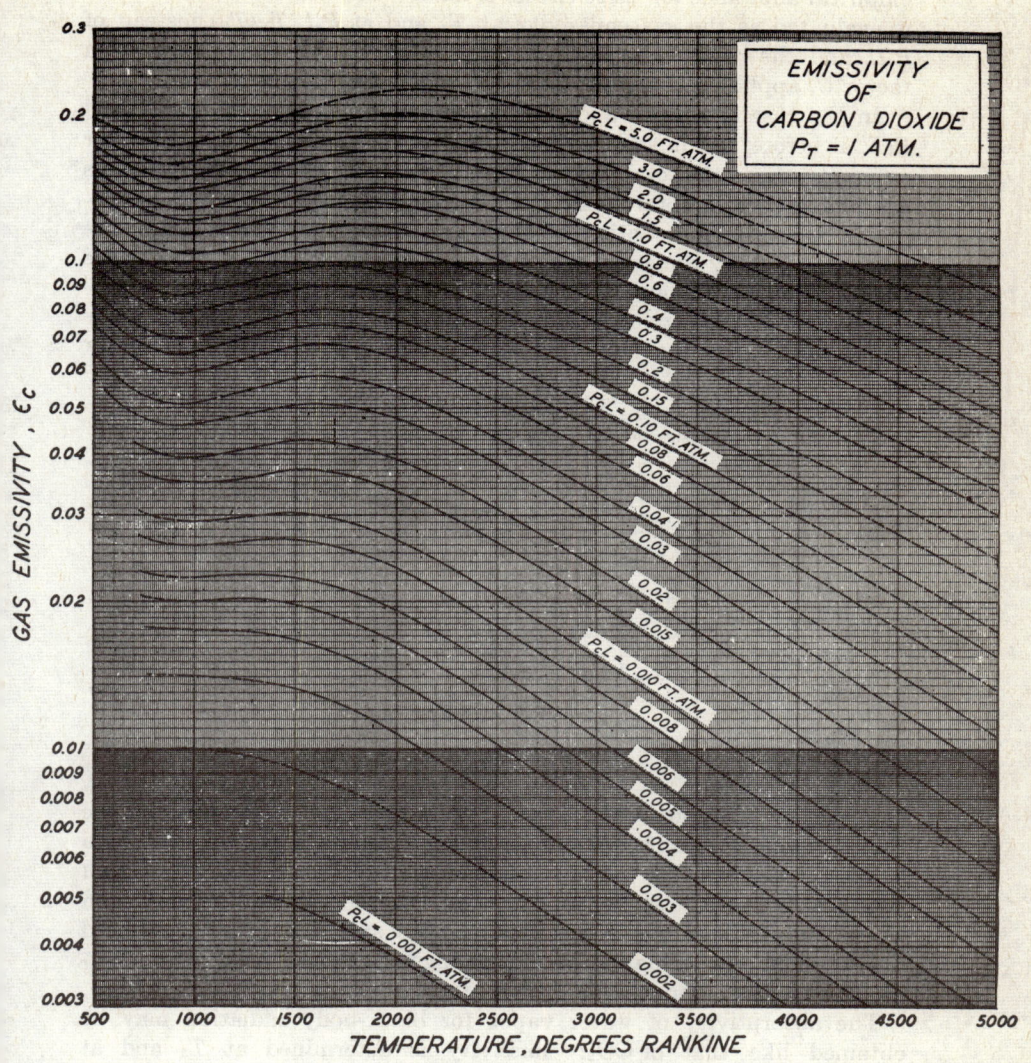

FIG. 4-13. Emissivity of carbon dioxide.

empirical allowance for these effects is made when α_{G_1} for carbon dioxide is evaluated as the gas emissivity at T_1 and at $P_cL(T_1/T_G)$ instead of P_cL, and the result is multiplied by $(T_G/T_1)^{0.65}$. The same correction factor C_c applies to absorptivity if the total pressure is not 1 atm.

In the case of water vapor the gas emissivity ϵ_G depends on T_G and P_wL, as before, and in addition somewhat on the partial pressure of water vapor P_w and on the total pressure P_T. Correlation of the data of various experimenters at a total pressure of 1 atm was found possible [32] by reducing all measured emissivities to values corresponding to an idealized case where $P_w = 0$, by the use of a factor depending on P_w and on P_wL. The smoothed curves through the resulting corrected data of various investigators[61,34,35,18,32,5,19] appear in Fig. 4-15 as a plot of ϵ_G vs. T_G for the various values of P_wL, for the "ideal" system at zero partial pressure of water

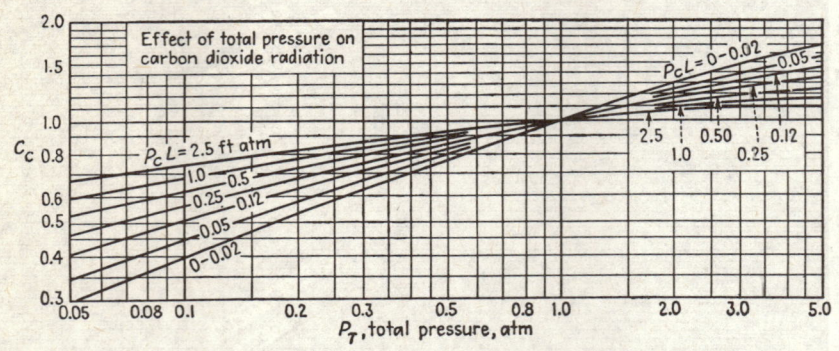

FIG. 4-14. Correction factor C_c for converting emissivity of CO_2 at 1 atm total pressure to emissivity at P_T atm.

vapor and a total pressure of 1 atm. Allowance for departure from the "ideal" state is then made by multiplying ϵ_G as read from Fig. 4-15 by a factor C_w read from Fig. 4-16 as a function of $P_w + P_T$ and P_wL. Although Fig. 4-16 is satisfactory for showing the separate effects of P_w and L on water-vapor radiation at a total pressure of 1 atm, it must be considered a very rough approximation for estimating the effect of total pressure P_T when that differs appreciably from 1 atm.

The absorptivity of water vapor for black-body radiation may be obtained like that of CO_2: emissivity is determined at T_1 and at $P_wL(T_1/T_G)$ and the result is multiplied by $(T_G/T_1)^{0.45}$. The correction factor C_w still applies.

When carbon dioxide and water vapor are present together, the total radiation due to both is somewhat less than the sum of the separately calculated effects, because each gas is somewhat opaque to the other. The correction for this effect may be read from Fig. 4-17, which gives the amount $\Delta\epsilon$ by which to reduce the sum of ϵ_G for CO_2 and ϵ_G for H_2O

RADIANT-HEAT TRANSMISSION 85

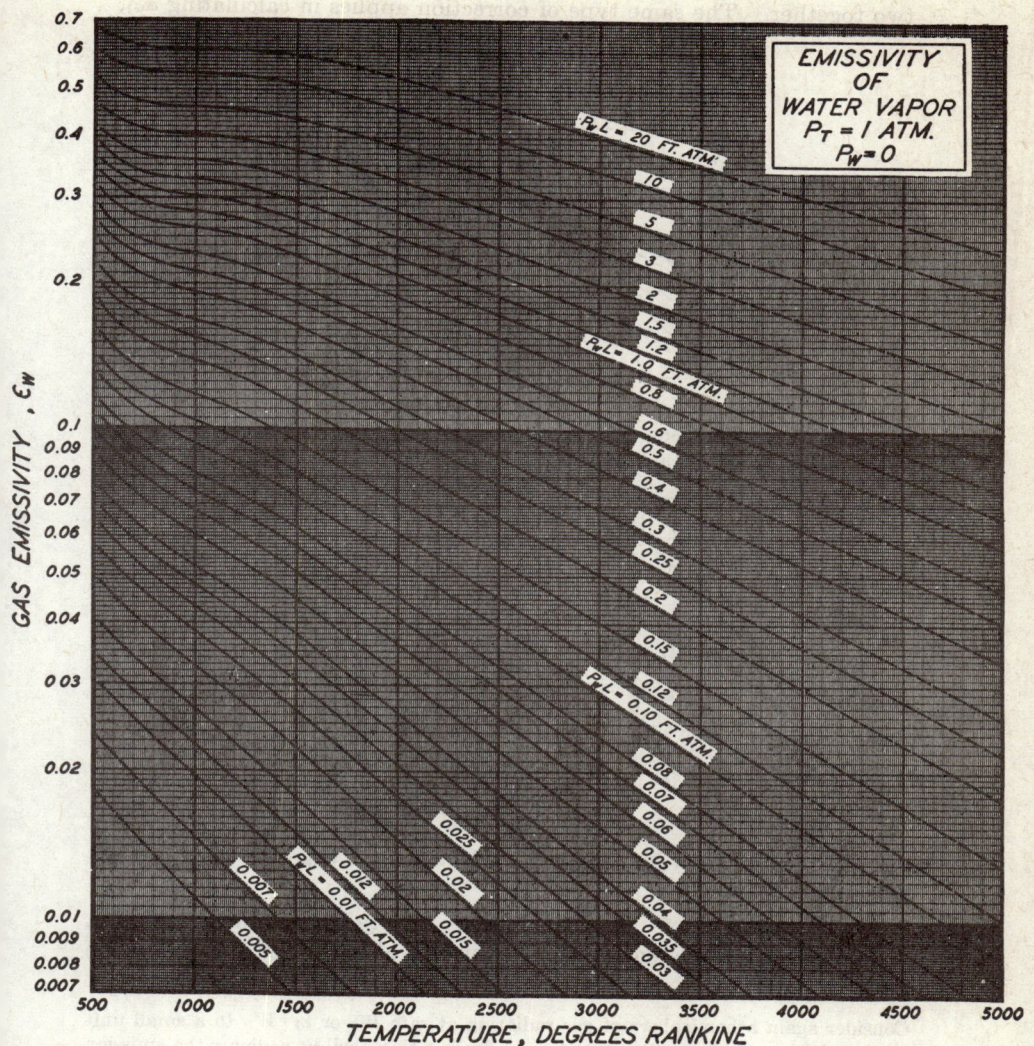

FIG. 4-15. Emissivity of water vapor.

(each evaluated as if the other gas were absent) to obtain the ϵ_G due to the two together. The same type of correction applies in calculating α_{G1}.

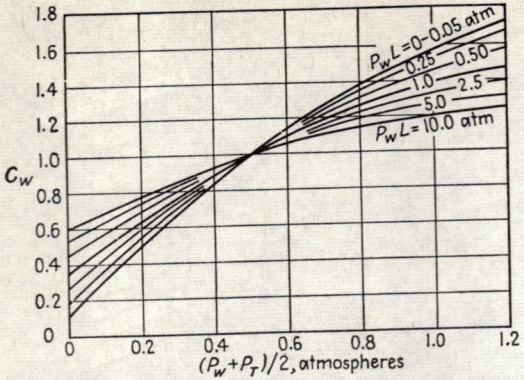

FIG. 4-16. Correction factor C_w for converting emissivity of H_2O to values of P_w and P_T other than 0 and 1 atm, respectively.

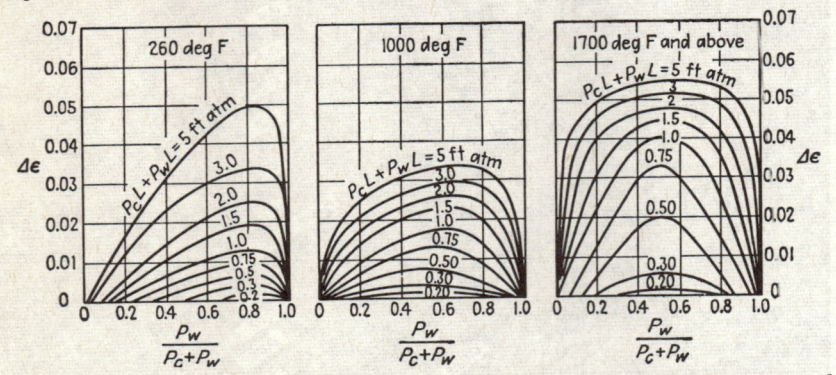

FIG. 4-17. Correction to gas emissivity due to spectral overlap of water vapor and carbon dioxide.

The Geometry of Gas Radiation. Before gas-surface radiation exchange can be evaluated, the method of allowing for the gas shape must be presented.

Consider again a hemisphere of gas radiating at rate W_G or $\epsilon_G \cdot W_B$ to a small unit surface element at the center of its base. From a thin shell at radius x the emission to the surface element, per unit of its area and per unit of thickness of the shell, is dW_G/dx, or $W_B \, d\epsilon_G/dx$, or $W_1 P_G \, d\epsilon_G/d(P_G x)$. Since emissive power is π times intensity, the intensity of irradiation at the base center due to a hemispherical shell of thickness dx is then $[W_B \, d\epsilon_G/dx] \, dx/\pi$. By application of Eq. (4-6b), the emission rate toward surface element dA from a gas volume bounded by the solid angle $d\omega$ and the radii x and $x + dx$ extending from dA, when the radius makes an angle ϕ with the normal to dA, is given by

$$\frac{W_B}{\pi} d\omega \cos \phi \, dA \, \frac{d\epsilon_G}{dx} dx \qquad (4\text{-}46)$$

Correspondingly, the emission rate from a finite isothermal gas mass to a finite portion of its confining surface is given by integration of the above,

$$\frac{W_B}{\pi} \int_A \int_0^{x_{max}} \int_\omega \cos\phi \, d\omega \, \frac{d\epsilon_G}{dx} \, dx \, dA \tag{4-47}$$

The order of integration depends on the problem. Generally, the best order is that indicated above, since the tedious and usually graphical allowance for shape is thereby made once and for all in the first step and need not be repeated as the partial pressure of radiating gas P_G or the absolute system size X changes. $\int \cos\phi \, d\omega$ is evaluated for spot dA on the surface and is obtained as a function of x, the variable distance to the opposite bounding surface. If this is represented by $f_1(x)$ and $d\epsilon_G/dx$ by $f_2(x)$, the emission becomes

$$\frac{W_B}{\pi} \int_A \int_0^{x_{max}} f_1(x) \cdot f_2(x) \, dx \, dA \tag{4-48}$$

A few representative spots dA on a surface will generally suffice for an integration.

A procedure similar to the above has been carried out for various shapes to find what value of mean beam path length L will give the true emission when the latter is evaluated as $A \cdot W_G \cdot \epsilon_{P_G L}$. The value of L so obtained is found to depend on the emitting gas and on $P_G X$, where X is a characteristic dimension of the gas shape, but is fortunately representable for practical purposes as an almost constant fraction of X.[26,62] The mean beam length L may be thought of as the radius of a gas hemisphere which will radiate to unit area at the center of its base the same as the average radiation, over A, from the actual gas mass.

For the special case of a gas radiating to the whole of its confining surface, with $P_G X$ sufficiently small to make the gas substantially non-self-absorbing, L is readily obtained.

The volume of the emitting element producing the emission represented by Eq. (4-46) is $x^2 \, d\omega \, dx$, from which the emission per unit volume, to dA, is

$$\frac{W_B}{\pi} \frac{d\epsilon_G}{dx} \frac{dA \cos\theta}{x^2} \tag{4-49}$$

Since the area dA subtends the solid angle $dA \cos\theta/x^2$ at the emitting volume, the emission per unit volume per unit solid angle of emission is $(W_B/\pi)(d\epsilon_G/dx)$. This, times 4π, is the reception on a sphere of radius x from the unit volume at its center. As x approaches zero the attenuation of the emission by absorption in the intervening gas is eliminated. Then

$$4W_B \frac{d\epsilon_G}{dx} \bigg]_{x \doteq 0}, \quad \text{or} \quad 4W_B P_G \frac{d\epsilon_G}{dP_G L} \bigg]_{P_G L \doteq 0} \tag{4-50}$$

represents total gas emission E per unit volume, without attenuation; or

$$E/W_B = P_G(4 \, d\epsilon_G/dP_G L) \bigg]_{P_G L \doteq 0} \tag{4-51}$$

The gas-radiation plots thus yield, by differentiation at small $P_G L$, the emission per unit volume of gas expressed as a ratio to the emissive power of a black surface. Returning now to the problem of evaluating the radiation from a gas of volume V to its whole bounding surface A when PX is small, one may equate the two ways of expressing the rate of gas emission:

$$AW_B \cdot \epsilon_{P_GL} = V \cdot W_B \cdot 4P_G(d\epsilon_G/dP_GL)_{P_GL \doteq 0} \quad (4\text{-}52)$$

At $P_GL = 0$, $d\epsilon_G/dP_GL = \epsilon_{P_GL}/P_GL$, and this, combined with (4-52), yields

$$L^0 = 4V/A \quad (4\text{-}53)$$

the superscript being used as a reminder that $P_GL = 0$.

Thus at low values of P_GL *the mean beam length L^0 for radiation to all the surfaces forming an enclosure is four times the mean hydraulic radius of the enclosure.* Values of L^0 for various gas shapes, including radiation to *single* sides of two rectangular parallelepipeds,[56] are given in Table 4-2, column 3. For finite values of P_GX the mean beam length is less than L^0.

TABLE 4-2. BEAM LENGTHS FOR GAS RADIATION

Shape	Characterizing dimension X	When $P_GL = 0$	For average values of P_GL
Sphere	Diameter	⅔	0.60
Infinite cylinder	Diameter	1	0.90
Semi-infinite cylinder, radiating to center of base	Diameter	....	0.90
Right-circular cylinder, height = diameter, radiating to center of base	Diameter	....	0.77
Same, radiating to whole surface	Diameter	⅔	0.60
Infinite cylinder of half-circular cross section. Radiating to spot on middle of flat side	Radius	....	1.26
Rectangular parallelepipeds:			
1:1:1 (cube)	Edge	⅔	
1:1:4, radiating to 1 × 4 face		0.90	
radiating to 1 × 1 face	Shortest edge	0.86	
radiating to all faces		0.89	See Table 4-3
1:2:6, radiating to 2 × 6 face		1.18	
radiating to 1 × 6 face	Shortest edge	1.24	
radiating to 1 × 2 face		1.18	
radiating to all faces		1.20	
1:∞:∞ (infinite parallel planes)	Distance between planes	2	
Space outside infinite bank of tubes with centers on equilateral triangles; tube diameter = clearance	Clearance	3.4	2.8
Same as preceding, except tube diameter = one-half clearance	Clearance	4.45	3.8
Same, except tube centers on squares; diameter = clearance	Clearance	4.1	3.5

Factor by which X is multiplied to obtain mean beam length L

Average values of L from the results of tedious graphical or analytical treatment of various shapes appear in Table 4-2, last column. The ratio L/L^0 may for most practical purposes be taken at about 0.85, though Port's study of rectangular parallelepipeds[56] indicates the variation of the ratio both with $P_G X$ and with shape. Table 4-3, adapted from Port's results, indicates the effect of $P_G X$ on the ratio. At high $P_G X$'s the exact determination of the ratio is plainly unimportant because gas emissivity is then not sensitive to $P_G L$.

TABLE 4-3. PARALLELEPIPEDS. RATIO OF MEAN BEAM LENGTH L TO L^0

$P_c L$ or $P_w L$	0.01	0.1	1
Ratio for CO_2	0.85	0.80	0.77
Ratio for H_2O	0.97	0.93	0.85

A corollary of the above discussion is that the magnitude of gas radiation shifts from proportionality to gas volume at very low PX's to proportionality to bounding-surface area at very high PX's; but to simplify its combination with other heat-transfer mechanisms, it is formally expressed in terms of bounding surface over the whole range of PX.

Identification of some of the terms above with another approach to gas radiation is desirable. The attenuation of monochromatic radiation in length dx obeys the relation

$$-dI_\lambda = k'_\lambda I_\lambda \, dx \tag{4-54}$$

the proportionality constant k'_λ—the absorption coefficient—being proportional to P_G and a function of temperature. Integration over the whole path length L and summation over the various spectral ranges of interest yields

$$\epsilon_G = \sum a_i (1 - e^{-k_i P L}) \tag{4-55}$$

From Eqs. (4-51) and (4-55) it may be shown that the emission per unit volume, expressed as a ratio to the emissive power of a black surface, is given by

$$\frac{E}{W_B} = P_G \sum 4 a_i k_i \tag{4-56}$$

Interchange between Gas and Black Surroundings. Mean Temperature. With beam length L defined it is now possible to formulate the net radiant-heat interchange between a gas containing carbon dioxide and water vapor and its bounding surface. With a restriction temporarily to a black bounding surface at uniform temperature, the radiation interchange is

$$q/A = \sigma(\epsilon_G T_G^4 - \alpha_{G1} T_1^4) \tag{4-57}$$

To keep straight on nomenclature, a series of subscripts will be appended to the value of ϵ_G read from Fig. 4-13 or 4-15, the first representing the gas (c for CO_2; w for H_2O), the second the temperature on the plot (whether T_G or T_1), the third the value of $P_G L$ at which ϵ_G is read. In this nomenclature, the terms in Eq. (4-57) are defined as follows:

$$\epsilon_G = \epsilon_{c,T_G,P_cL} \cdot C_c + \epsilon_{w,T_G;P_wL} \cdot C_w - \Delta\epsilon_{T_G}$$

$$\alpha_{G1} = \alpha_c + \alpha_w - \Delta\alpha$$

$$\alpha_c = \epsilon_{c,T_1,P_cLT_1/T_G} \left(\frac{T_G}{T_1}\right)^{0.65} C_c \qquad (4\text{-}58)$$

$$\alpha_w = \epsilon_{w,T_1,P_wLT_1/T_G} \left(\frac{T_G}{T_1}\right)^{0.45} C_w$$

$$\Delta\alpha = \Delta\epsilon_{T_1}$$

Although α_{G1} approaches ϵ_G as T_1 approaches T_G, an argument such as appears on page 78 indicates that ϵ_G cannot then be factored out of

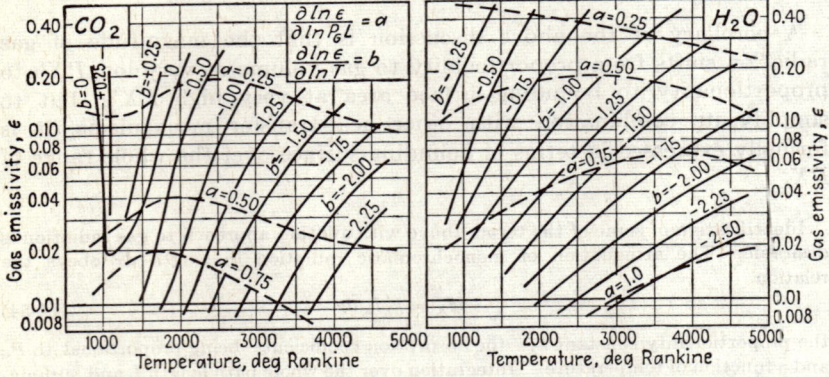

Fig. 4-18. Rate of change of emissivity with T and $P_G L$, for CO_2 and H_2O.

Eq. (4-57). If over a restricted range of variables ϵ_G is assumed proportional to $(P_G L)^a T_G^b$, then α_{G1} will be proportional to $(P_G L)^a (T_1/T_G)^a T_1^b (T_G/T_1)^c$ or to $(P_G L)^a T_1^{a+b-c} T_G^{c-a}$, where c is the previously encountered exponent used in evaluating absorptivity—0.65 for CO_2 and 0.45 for H_2O. By the same process that led to Eq. (4-43), it may be shown that when T_G and T_1 are not too far apart, q/A may be expressed in either of two ways:

$$\frac{q}{A} = \sigma\left(\epsilon_{G,\text{avg}} \frac{4+a+b-c}{4}\right)(T_G^4 - T_1^4) \qquad (4\text{-}59)$$

or

$$q/A = [(4+a+b-c)\epsilon_{G,\text{avg}}\sigma T_{\text{avg}}^3](T_G - T_1) \qquad (4\text{-}60)$$

in which $\epsilon_{G,\text{avg}}$ is the gas emissivity evaluated at T_{avg}, the arithmetic mean of T_G and T_1. The bracketed term of (4-60) will be recognized as an equivalent heat-transfer coefficient due to gas radiation. Values of a and b, which represent $\partial \ln \epsilon_G/\partial \ln P_G L$ and $\partial \ln \epsilon_G/\partial \ln T_G$, respectively, are given in Fig. 4-18 for CO_2 and H_2O vapor. When a mixture of the two gases is present, it is recommended that mean values of a, b, and c be used in Eq. (4-59), the factors for the different gases being weighted in proportion to their emissivities. (A roughly approximate value of $a + b - c$ suffices, however, since an error of 0.1 in it produces an error of only 2.5 per cent in q/A.)

The use of Eq. (4-59) when T_G and T_1 do not differ greatly leads to results of generally higher accuracy than (4-58) and will usually save time as well by eliminating the necessity for the fairly tedious evaluation of absorptivity. Even when T_G and T_1 differ by a factor as great as 2, the error due to use of (4-59) seldom exceeds 5 per cent.

Discussion has so far been restricted to a bounding surface which was black. Now consider gray walls of emissivity ϵ_1, and focus attention

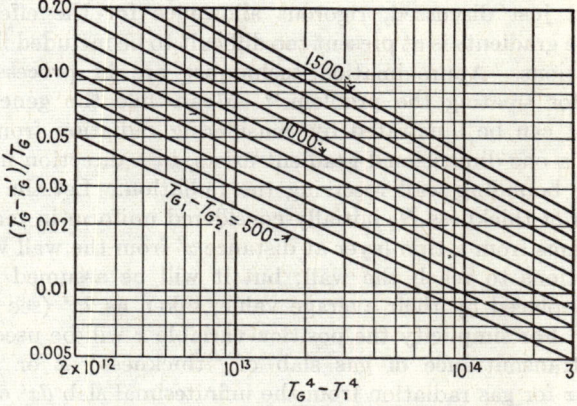

Fig. 4-19. Plot for calculation of mean gas temperature in countercurrent gas radiant-heat exchangers, in terms of arithmetic-mean gas and surface temperatures T_G and T_1 and gas-temperature change, $T_{G_1} - T_{G_2}$. Temperatures degrees Rankine. (Cohen.[9])

on the radiation initially emitted from a gas column toward a bounding-surface element. Partial absorption occurs at A_1, partial reflection, partial transmission of the reflected radiation through the gas, partial absorption on second incidence, etc., giving rise to an infinite series capable of evaluation from the gas-radiation plots (4-13) and (4-15). The resulting sum is the interchange for a black-walled system [Eq. (4-57)], reduced by a factor which must lie between ϵ_1 and 1 and is nearer the former the higher P_GL. Since ϵ_1 is for many surfaces above 0.8, the use of the factor $(\epsilon_1 + 1)/2$ as a multiplier on (4-58) cannot lead to much error. If ϵ_1 is small, the method of a later section may be used (see pages 104 to 117).

If gas radiation occurs in a system in which there is a continuous change in temperature of the gas and the surface from one end to the other of the exchanger, and if the dimension of the exchanger in the flow direction is so great relative to the other two as to conform to the case discussed on page 76, then the graphical method suggested there is applicable, with $q/A]_{\text{local}}$ given by Eq. (4-57) or its equivalent. However, the tedious graphical integration can be eliminated by evaluation of suitable mean gas and surface temperatures T'_G and T'_1 to be used in Eq. (4-57). By examining a wide range of the variables involved, Cohen[9] was able to construct Fig. 4-19, which satisfactorily presents the true mean gas temperature T'_G as a function of the four terminal temperatures of gas and surface. The mean surface temperature T'_1 to be used is the surface temperature at the point in the exchanger where the gas temperature is T'_G; it may be obtained by an enthalpy balance.

Temperature Nonuniformity in a Gas. When gas temperature varies rapidly with distance rather than so slowly as to justify application of the method just discussed, rigorous allowance for the effect of gas-temperature gradients is at present too difficult to be included in engineering calculations. Approximation techniques are in process of being developed for treating the problem.[10] Meanwhile the general nature of the effect can be illustrated by considering radiation from a gas in which only a one-dimensional gradient exists, in a direction normal to a plane black bounding wall receiving the radiation. Let the gas be an infinite slab of thickness H, initially considered uniform in temperature. Radiant beams from a thin layer at distance x' from the wall will radiate in all directions to reach the wall; but it will be assumed that these can all be replaced by their average value, taken as $2x'$ (see Table 4-2, column 3). For simplicity the position-variable x will be used, equal to $2x'$. The transmittance of gas slab of "thickness" x or absorption strength $P_G x$ for gas radiation from the infinitesimal slab dx' can now be determined. The emission to one side of a slab of area A and thickness $\delta x' (= \delta x/2)$ is (volume) [emission per unit volume, from Eq. (4-51)], or

$$\left(A \frac{\delta x}{2}\right)\left[W_B \cdot 2P_G \left(\frac{d\epsilon_G}{dP_G L}\right)_{L=0}\right] = A W_B \left(\frac{d\epsilon_G}{dP_G L}\right)_{L=0} \delta(P_G x) \quad (4\text{-}61)$$

This, multiplied by the transmittance τ of the layer $P_G x$, is equal to the increase in emission to the wall due to increasing a slab thickness from x to $x + \delta x$, or

$$A \cdot W_B \left(\frac{d\epsilon_G}{dP_G L}\right)_{L=0} \cdot \tau \, \delta(P_G x) = A W_B \left(\frac{d\epsilon_G}{dP_G L}\right)_{L=x} \delta(P_G x) \quad (4\text{-}62)$$

Then

$$\tau = \frac{(d\epsilon_G/dP_G L)_{L=x}}{(d\epsilon_G/dP_G L)_{L=0}} \quad (4\text{-}63)$$

In words, the transmittance of a gas layer $P_G x$ for radiation from a thin slab is the ratio of slopes on an ϵ_G-$P_G L$ curve at $P_G x$ and at 0. When

τ-$P_G L$ curves are calculated from the CO_2 or H_2O plots, Fig. 4-13 or 4-15, the curves for different temperatures are found to differ little. If now the assumption is made that transmittance is a function of the temperature of the emitting layer only, the original objective can be accomplished. It will be illustrated by a specific example. Let the gas layer contain 5 per cent CO_2 and be 1 ft thick, and let the gas tem-

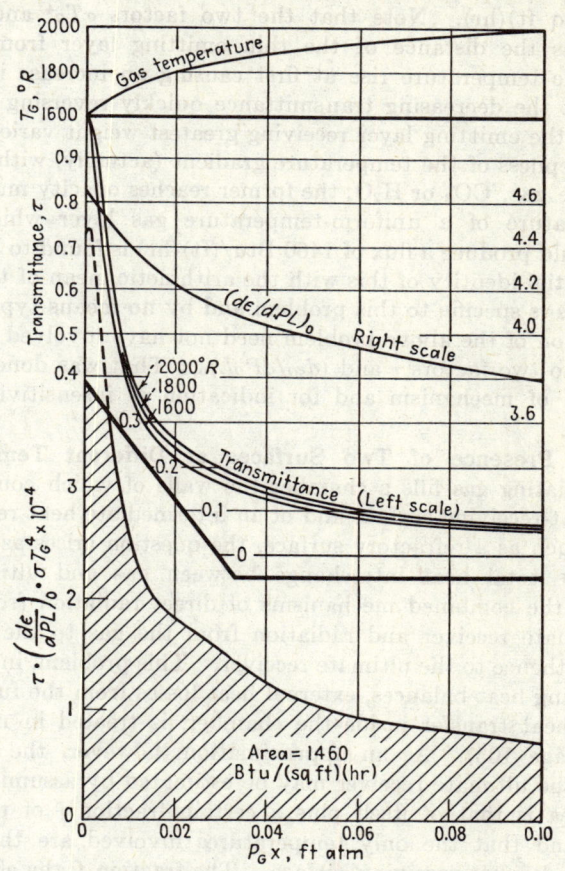

FIG. 4-20. Effect of temperature nonuniformity on gas radiation.

perature vary from 1600°R at the surface to 2000°R at 1 ft, as indicated in the top curve of Fig. 4-20. The first step is the calculation of transmittance curves for the temperature range of interest. Three of these, for 1600, 1800, and 2000°R, are presented. Then the thin-layer gas-emission values $(d\epsilon_G/dP_G L)_0$ are obtained as a function of temperature and, from the top curve, converted and plotted as a function of $P_G x$ (second curve, scale on right). The total energy reaching the wall per unit area is now given by

$$\frac{q}{A} = \int_{P_x=0}^{P_x=2PH} \tau(d\epsilon_G/dP_GL)_0 \cdot \sigma T_G^4 \, d(P_Gx) \qquad (4\text{-}64)$$

At the bottom of Fig. 4-20 the product $\tau(d\epsilon_G/dP_GL)_0 \cdot \sigma T_G^4$ is plotted against P_Gx; and the area out to $P_Gx = 2P_GH = 0.1$ is found to be 1460 Btu/(sq ft)(hr). Note that the two factors σT_G^4 and τ are in opposition as the distance of the thin emitting layer from the wall increases, the temperature rise at first causing an increase in received emission but the decreasing transmittance quickly reversing the trend. The part of the emitting layer receiving greatest weight varies of course with the steepness of the temperature gradient (actually, with dT/dP_Gx) and with the gas, CO_2 or H_2O; the former reaches opacity much sooner. The temperature of a uniform-temperature gas layer which in this problem would produce a flux of 1460 Btu/(ft)(hr) is found to be 1811°R. The substantial identity of this with the arithmetic mean of the limiting temperatures is specific to this problem and by no means typical.

The solution of the above problem need not have involved separating $d\epsilon_G/dP_GL$ into two factors τ and $(d\epsilon_G/dP_GL)_0$. That was done solely for visualization of mechanism and for indication of insensitivity of τ to temperature.

Effect of Presence of Two Surfaces at Different Temperatures. When a radiating gas fills a chamber the walls of which consist of the ultimate heat-receiving surface and of an intermediate heat receiver and reradiator such as a refractory surface, the question arises as to how to evaluate the total heat interchange between gas and ultimate heat receiver by the combined mechanisms of direct radiation from the gas to the ultimate receiver and radiation from the gas to the refractory surface and thence to the ultimate receiver. This problem, in its general form involving heat balances, external heat losses from the furnace, and convection heat transfer inside the chamber, is treated in more detail later (see page 105). As an approximation, however, the total heat transfer to the ultimate receiver may be estimated by assuming that its effective area is that of itself plus a certain fraction f of that of the refractory and that the only temperatures involved are those of the gas and the ultimate receiving surface. The fraction f, the effectiveness of the refractory surface, varies from zero when the ratio of refractory surface to ultimate receiving surface is very high, to unity when the ratio is very low and the value of ϵ_F is low. When the refractory-surface area and ultimate heat-receiving surface area are of the same order of magnitude, a value of 0.7 may be used for f, although for more exact calculations the method of page 105 *et seq.* should be used.

Radiation from Other Gases. In the design of sulfur burners and of sulfur dioxide coolers, the radiation from the gas may be a major factor in the evaluation of the total heat transferred. The data of Coblentz[8] on the infrared absorption spectrum of sulfur dioxide, although hardly adequate as a basis for quantitative calculations, have been used for want of something better. The results are given in Fig. 4-21, based on calculations by Guerrieri.[23]

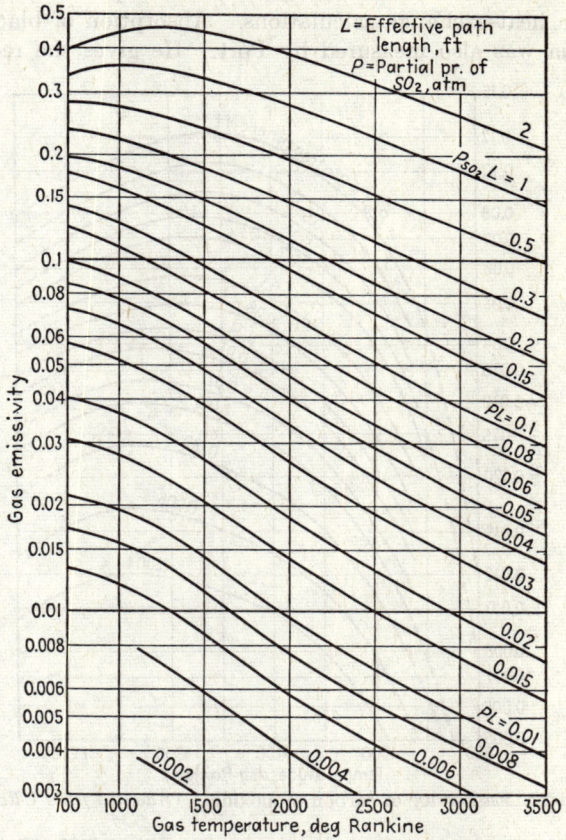

FIG. 4-21. Emissivity of sulfur dioxide. (*Adapted from Guerrieri.*[23])

Total radiation from carbon monoxide in nitrogen at a total pressure of 1 atm has been measured by Ullrich,[72] using a total path length L of 1.68 ft. Figure 4-22 has been prepared from the data. Although absorption measurements indicated a strong effect of gas temperature, their accuracy was low. It is tentatively recommended that absorption be calculated in the same way as for CO_2 and H_2O, using a value of c of 0.7. The spectra of CO and CO_2 overlap sufficiently to make the incremental emissivity due to CO equal 0.8 (0.7) of its value if present alone in the amount $P_{Co}L = 0.2$ (1.0), when carbon dioxide is present to the extent $P_{CO_2}L = 1$.

Total radiation from ammonia vapor–nitrogen mixtures at a total pressure of 1 atm has been measured by Port,[56] using a path length L of 1.68 ft. His recommended values of emissivity are presented in Fig. 4-23, which shows the large effect to be expected from ammonia radiation in heat-exchange calculations. Absorption of black radiation by ammonia was also measured by Port. He gives the results in the

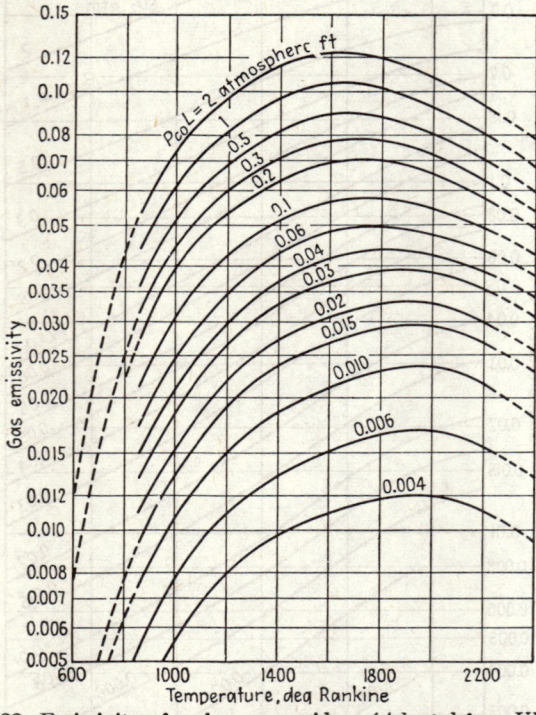

FIG. 4-22. Emissivity of carbon monoxide. (*Adapted from Ullrich.*[72])

form of a series of correction charts to gas emissivity as read at the surface temperature and at $P_G L$. When the surface temperature is 500° below that of the gas, the correction factor is about 0.75, 0.85, and 1.0 for gas temperatures of 2500, 2000, and 1500°R. For larger differences in temperature, absorptivity becomes unimportant in a calculation of net interchange.

For other gases of interest one must rely on evaluations based on the infrared absorption spectra of the gases in question. For the methods of such calculation and for a more complete story on gas radiation the reader is referred to the literature.[58,26,59,60,54,55]

Illustration 8. Fuel oil of composition $CH_{1.8}$ is to be burned with 20 per cent excess air and fed to one end of a 6 in. i.d. pipe losing heat to air blown across the outside of the pipe. Conditions are such that the gas enters at 2000°F where the surface temperature is 800°F and leaves the other end, of surface temperature 600°F, at 1000°F. Fuel is fed at such a rate that the hourly heat capacity of the combustion products is equal to 90 Btu/(hr.)(deg F) and may be assumed constant. What is the necessary length of the pipe (a) if it is black, (b) if it is gray, with emissivity 0.8?

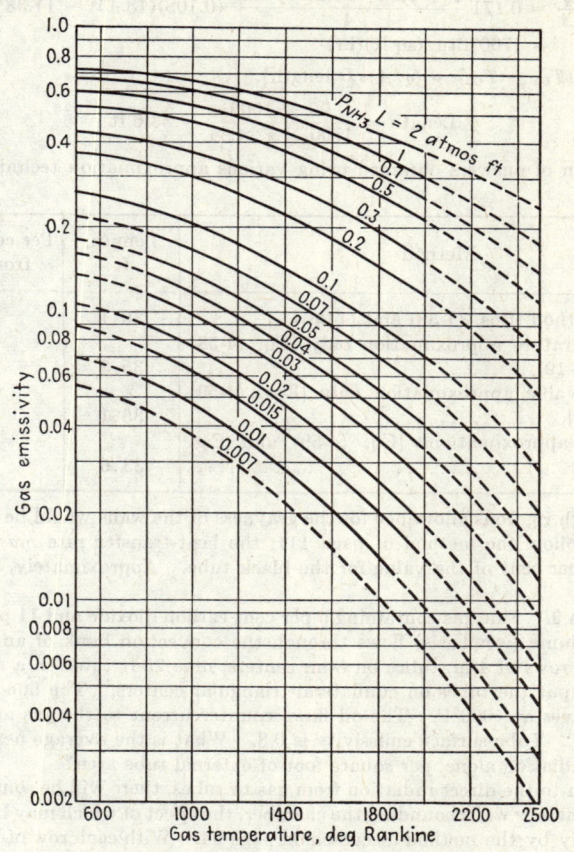

FIG. 4-23. Emissivity of ammonia. (*Port.*[56])

By stoichiometry, $P_w = 0.103$, and $P_c = 0.114$. Since L for a long tube equals $0.9 \times$ diameter (see Table 4-2), $P_c L = 0.052$, and $P_w L = 0.047$. Rigorous integration along the tube will be avoided by using mean values for gas and surface temperature. Referring to Fig. 4-19, $T_G = (1000 + 2000)/2 + 460 = 1960$; $T_1 = (600 + 800)/2 + 460 = 1160$; $T_G^4 - T_1^4 = 1.29 \times 10^{13}$; $T_{G1} - T_{G2} = 1000$; reading from the figure, $(T_G - T'_G)/T_G = 0.057$; so $T'_G = 1960 - 0.057 \times 1960 = 1849°R$. The corresponding surface temperature $T'_1 = (800 + 460) - \dfrac{2460 - 1849}{1000}(800 - 600) = 1138°R$. The approximate value of average heat-transfer rate will be obtained by use of Eq. (4-59). At $T_{\text{avg}} = (1849 + 1138)/2 = 1493°R$, ϵ_c from Fig. 4-13 = 0.0625; from Fig. 4-15, $\epsilon_w = 0.042$; from Fig. 4-16, $C_w = 1.08$; $\Delta\epsilon = 0$. Therefore, $\epsilon_{\text{avg}} = 0.0625 + 1.08 \times 0.042 = 0.108$. Using Fig. 4-18, $a_c = +0.37$, $b_c = +0.2$, $a_w = +0.7$, $b_w = -0.9$; also $c_c = +0.65$, and $c_w = +0.45$. Weighting these in proportion to the emissivities, using $0.42 \times 1.08 = 0.045$ for emissivity of H_2O, gives $a = +0.50$, $b = -0.26$, $c = +0.57$. Equation (4-59) gives

$$\frac{q}{A} = 0.171 \frac{4 + 0.50 - 0.26 - 0.57}{4} (0.108)(18.49^4 - 11.38^4)$$
$$= 1700 \text{ Btu}/(\text{sq ft})(\text{hr})$$

But $q = wc_p(T_{G1} - T_{G2}) = (q/A)\pi D(\text{length})$.

a. $$\text{Length} = \frac{90 \times 1000}{1700 \times \pi \times \frac{1}{2}} = 33.6 \text{ ft}$$

A comparison of answers obtained using various approximation techniques is given below:

Method	Length, ft	Per cent deviation from rigorous
Rigorous method [Eqs. (4-58) and (4-36)]............	36.1	
Using temperature approximation only [Eq. (4-58) and Fig. 4-19]................................	35.3	−2.2
Using local-value approximation only [Eqs. (4-59) and (4-36)].................................	35.1	−2.8
Using both approximations [Eq. (4-59) and Fig. 4-19]..	33.6	−6.9

b. Although rigorous allowance for the grayness of the walls would be quite tedious and would follow the method of page 111, the heat-transfer rate *must* lie between 80 and 100 per cent of the value for the black tube. Approximately, $L = 33.6/0.9 = 37.4$ ft.

Illustration 9. Flue gas containing 6 per cent carbon dioxide and 11 per cent water vapor by volume (wet basis) flows through the convection bank of an oil tube still consisting of rows of 4-in. tubes on 8-in. centers, nine 25-ft tubes in a row, the rows staggered to put the tubes on equilateral triangular centers. The flue gas enters at 1600 and leaves at 1000°F. The oil flows countercurrent to the gas and rises from 600 to 800°F. Tube-surface emissivity is 0.8. What is the average heat input rate, due to gas radiation alone, per square foot of external tube area?

In addition to the direct radiation from gas to tubes, there will be some reradiation from the refractory walls bounding the chamber, the effect of which may be determined approximately by the method discussed on page 94. With each row of tubes there is associated $\frac{8}{12} \times \sqrt{3}/2$ or 0.577 ft of wall height, of area $(\frac{8}{12} \times 9 \times 2 + 25 \times 2) \times 0.577 = 35.8$ sq ft. One row of tubes has an area of $\pi \times \frac{4}{12} \times 25 \times 9 = 235$ sq ft. If the recommended factor of 0.7 on the refractory area is used, the effective area of the tubes is $\frac{235 + 0.7 \times 35.8}{235} = 1.11$ sq ft/sq ft of actual area. The exact evaluation of outside tube temperature from the known oil temperature would involve a knowledge of oil-film coefficient, tube-wall resistance, and rate of heat flow into the tube, the evaluation usually involving trial and error. However, for the present purpose the temperature drop through the tube wall and oil film will be assumed 75°F, making the tube surface temperatures 675 and 875°F; average 775°F The remainder of the problem is handled substantially like Illustration 8 above.

RADIATION FROM CLOUDS OF PARTICLES

The treatment of radiation from powdered-coal or atomized-oil flames, from dust particles in flames, and from flames made luminous by the thermal decomposition of hydrocarbons to soot involves the evaluation of radiation from clouds of particles. Powdered-coal flames contain particles varying in size from 0.01 in. down, with an average size in the neighborhood of 0.001 in. (25 μ), and a composition varying from a high percentage of carbon to nearly pure ash. The suspended matter in luminous gas flames has its origin in the thermal decomposition of hydrocarbons in the flame due to incomplete mixing with air before being heated, consists of carbon and of very heavy hydrocarbons, and has an initial particle size, before agglomeration, in the range 0.006 to 0.06 μ.[53,52,39,40] Flames of heavy residual oils, in addition to their soot luminosity due to the cracking of evolved gaseous hydrocarbons, contain solid particles produced by the coking of the heavy bituminous material present in each droplet.[7,22,69] These particles are of a size comparable with the original oil-drop size which, in industrial-furnace practice, may have a mass-median diameter of from 0.008 in. (200 μ) to 0.002 in. (50 μ), or even less. Powdered-coal particles and the coke particles in oil flames are sufficiently large to be substantially opaque to radiation incident on them, whereas the soot particles in a luminous flame are so small as to interact with thermal radiation like semitransparent or scattering bodies. Consequently the two kinds of luminosity follow different optical laws.

Luminous Flames. Reference here is to flames the luminosity of which is due primarily to soot rather than to suspended macroscopic particles. Evidence of the dominant role of luminosity in radiation from many industrial-furnace flames is long-standing. Lent[38] has made a blast-furnace gas flame practically black by addition of benzol; Haslam and Boyer[23b] found that a small luminous acetylene flame radiated roughly four times as intensely as when nonluminous; Sherman[67] in experiments on a large model furnace found the flame emissivity to increase from about 0.2 along the whole furnace length with a nonluminous flame to 0.6 with a luminous natural-gas flame; Trinks and Keller[71] were able to produce an emissivity up to 0.95 in a 2-ft thickness of a natural-gas flame, and Mayorcas[45] an emissivity of 0.8 in a 3-ft depth near the burner end of an open hearth.

The combustion process by which soot luminosity is produced is a complex one, far from completely understood. If before a hydrocarbon vapor is heated it is mixed *homogeneously* with enough oxygen to form CO and H_2 on reaction, the tendency to form soot is substantially eliminated. Consequently, although the production of luminosity is dependent on the type of hydrocarbon, with aromatic compounds most prone to form soot, it is even more dependent on the air-fuel mixing process. This in turn makes luminosity depend on the primary air-fuel ratio, the fuel firing rate, the momentum of the jet of fuel and its atomizer to the extent that momentum transfer is responsible for induction of air into the jet, the mode of admission of the air, and the scale of the system. Other factors are the mode of atomization of oil (whether pressure- or air- or steam-atomizing) and the ratio of heat-sink to no-flux area as that affects the rate of cooling of the flame. To aid in progress toward solution of this complex problem an International Committee on Flame Radiation has been organized, to conduct cooperative research on the problem primarily at Ijmuiden, Holland.[4,44,49a] The broad objective of developing a recommended procedure for predicting luminous-flame radiation will require some years to accomplish. Up to the present a number of flame types have been studied in great detail on a model furnace large enough to yield results applicable to full-scale furnaces.[46,13,14,16,57,6,36,47,15]

The luminosity of a cloud of soot particles will conform to Eq. (4-55), the monochromatic emissivity taking the form

$$\epsilon_\lambda = 1 - e^{-cL \cdot f(\lambda)} \qquad (4\text{-}65)$$

where c is the soot concentration. Evaluation of the total emissivity thus depends on knowing the function $f(\lambda)$ representing the absorption coefficient, the soot concentration c, and, if $f(\lambda)$ depends on particle size, the particle size and size-distribution function for the cloud.

Hottel and Broughton,[31] reviewing the experimental results of Ångstrom, Ladenburg, and Rubens and Wood on the variation, with λ, of the absorption coefficient of soot deposited on glass, concluded that $f(\lambda)$ was representable by $k/\lambda^{0.95}$ in the infrared down to 0.8 μ. In the visible spectrum near 0.6 μ the following data were presented:

	Exponent on λ
Stark, avg of 6 gas-soot deposits	0.45
Becker, thinnest layer of amyl acetate soot	1.53
Avg of 12 amyl acetate or turpentine soots	1.42
Amyl acetate flame	1.48
Hottel and Broughton, avg of 4 acetylene soots	1.23
Avg of 4 amyl acetate soots	1.30
Avg of 4 city-gas soots	1.75
Avg of 22 measurements on amyl acetate flames	1.39

The possible effect of particle size was mentioned, but an average exponent of λ of 1.39 was recommended. Mie's theory of fine-particle scatter[50] has been applied to soot absorption by Senftleben and Benedict[66] and by Yagi.[78] The latter concluded that $f(\lambda)$ is representable by $[1 - 15(r/\lambda)^2]/\lambda$, where r is the particle radius. In contrast with the experimental data presented above, this relation calls for an equivalent exponent on λ which decreases with decreasing λ, is always less than 1, and for 0.02-μ-diameter particles is above 0.99 throughout the spectral range of interest. The evidence against the general validity of such a relation appears strong, but it does suggest the need for further clarification of the problem (see reference 77).

Measurements on Luminous Flames. Although the prediction from first principles of flame luminosity in a proposed furnace is not now possible, a simpler approach of engineering value is often feasible.

If a furnace for a given operation exists for study, measurements can be made which are of assistance in improving performance or in designing a new furnace of different capacity or shape. The significant property of a flame, for the present interest, is its emission rate, separable into its temperature and its emissivity. Accordingly, the experimental determination of flame emissivity and temperature will be considered briefly, with restriction to the use of two commonly available instruments, the optical pyrometer and the total-radiation pyrometer.

The optical pyrometer is an instrument for determining the "brightness temperature" of an object, defined as the temperature of a black body having the same intensity of emission, at the mean wavelength of the red screen in the instrument, as the object viewed. The high precision of which a good optical pyrometer is capable is not generally appreciated; it will be examined briefly. An intensity match to within 1 per cent is possible. From Eq. (4-2) converted to its Wien's-law form (1 in the denominator dropped because of the high value of $c_2/\lambda T$), differentiation gives

$$\frac{d \ln W_{B\lambda}}{d \ln T} = \frac{c_2}{\lambda T} \tag{4-66}$$

At 1800°K and with a red screen of 0.65 μ effective wavelength, this gives

$$-\Delta \ln T = (0.01)(0.65)(1800)/14{,}387 = 0.00081$$

from which the error in temperature is 0.08 per cent, or 1.5°C.

Let an optical pyrometer be sighted through a flame onto a mirror and back through the flame, thereby permitting two readings to be taken, the red brightness temperatures T_{r1} and T_{r2} for one and two flame thicknesses. If the red emissivity and true temperature of the flame are $\epsilon_{\lambda,F}$ and T_F and the mirror reflectivity is ρ_M, one may write from Wien's law

$$c_1 \lambda^{-5} e^{-c_2/\lambda T_{r1}} = \epsilon_{\lambda,F} c_1 \lambda^{-5} e^{-c_2/\lambda T_F} \tag{4-67a}$$

and

$$c_1 \lambda^{-5} e^{-c_2/\lambda T_{r2}} = \epsilon_{\lambda,F} c_1 \lambda^{-5} e^{-c_2/\lambda T_F}(1 + \tau_{\lambda,F} \cdot \rho_M) \tag{4-67b}$$

where $\tau_{\lambda,F}$ is the red transmittance of the flame for radiation from one thickness of itself. Since monochromatic absorptivity is independent of the intensity of the source within wide limits, $\tau_{\lambda,F}$ may be taken as $1 - \epsilon_{\lambda,F}$. With this substitution, (4-67a) and (4-67b) may be solved for flame temperature and emissivity to give

$$\epsilon_{\lambda,F} = 1 - \frac{e^{-\frac{c_2}{\lambda}\left(\frac{1}{T_{r2}} - \frac{1}{T_{r1}}\right)} - 1}{\rho_M} \tag{4-68}$$

and

$$T_F = 1 \Big/ \left[\frac{1}{T_{r1}} + \frac{\lambda \ln \epsilon_{\lambda,F}}{c_2}\right] \tag{4-69}$$

This gives, however, only the monochromatic emissivity. The total emissivity can be calculated if the variation of ϵ_λ with λ is known, *i.e.*, if the function $cLf(\lambda)$ of Eq. (4-65) is known. If the recommendations of Hottel and Broughton[31] are accepted [$cLf(\lambda)$ equals $KL/\lambda^{1.39}$ in the

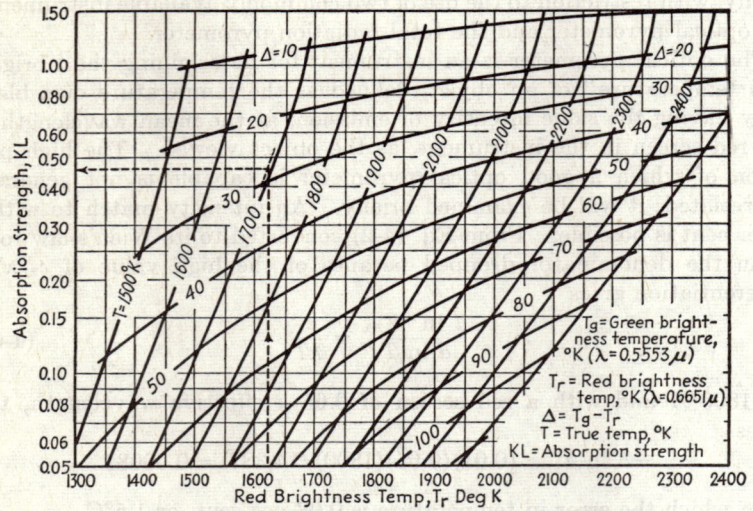

Fig. 4-24. Absorption strength KL of luminous flames.

visible, $K'L/\lambda^{0.95}$ in the infrared], it may be shown that the true temperature and red brightness temperature are sufficient to determine the *absorption strength KL* of the flame (proportional to soot concentration and to mean beam length); and KL and T_F suffice to determine the total flame emissivity ϵ_F. Figures 4-24 and 4-25[31] provide the necessary relations in graphical form.

Instead of using a mirror behind the flame for doubling its thickness (sometimes difficult to arrange) one may make use of relation (4-65) between emissivity and wavelength. If the optical pyrometer is provided with absorption screens of two different effective wavelengths, say red and green, then the red and green brightness temperatures of the flame suffice to determine both KL and its true temperature (see Fig. 4-24).

RADIANT-HEAT TRANSMISSION 103

The above two methods necessitate careful measurements and, if the flame is unsteady, comparable averaged pairs of measurements. Although both methods have been used successfully on luminous flames of less than ½ in. diameter, they are strictly research methods. A method suggested by H. Schmidt[63] involves use of a total-radiation pyrometer—one, however, which is nonselective and therefore contains no quartz windows or lenses (see Illustration 1, page 79). If the pyrometer is sighted through the flame onto a cold background to give reading W_F, through the flame onto a black hot background wall to give reading W_{F+W}, and directly onto the wall to give reading W_W, one

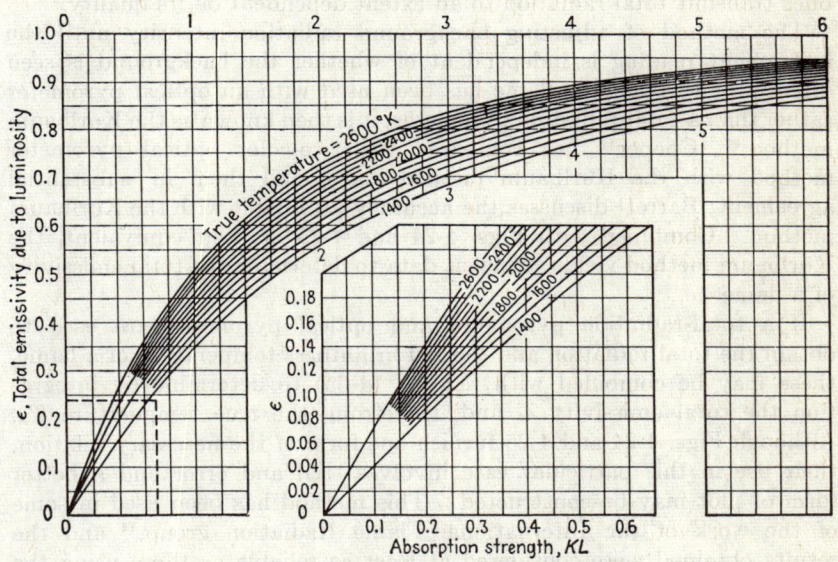

FIG. 4-25. Emissivity of luminous flames.

may estimate therefrom the flame temperature and total emissivity. Let the emissive power of a black body at flame temperature be $W_{B,F}$. Then if the flame transmittance for radiation from the back wall is designated by τ_F, one has

$$W_{F+W} = W_F + W_W \cdot \tau_F \tag{4-70}$$

and

$$W_F = \epsilon_F \cdot W_{B,F} \tag{4-71}$$

If the background temperature is changed until readings W_W and W_{F+W} are the same, Kirchhoff's law of the equality of ϵ_F and α_F or $1 - \tau_F$ applies and the above relations reduce to $W_W = W_{B,F}$; that is, the wall temperature, at match, equals the true flame temperature. If, however, the assumption is made that with wall and flame temperatures not too far apart $1 - \tau_F$ and ϵ_F are still equal, then Eqs. (4-70) and (4-71) yield

$$W_{B,F} = \frac{W_F \cdot W_W}{W_F + W_W - W_{F+W}} \tag{4-72}$$

and

$$\epsilon_F = \frac{W_F + W_W - W_{F+W}}{W_W} \tag{4-73}$$

This modified Schmidt method has been used by the International Flame Radiation group.[47,14] Although it is apparently satisfactory, it cannot be considered rigorous since not all luminous flames are gray and nongray ones transmit total radiation to an extent dependent on its quality.

The method of adjusting background radiation intensity until the instrument reading is independent of whether the background is seen directly or through the flame has been used with an optical pyrometer rather than a total-radiation pyrometer; it is then known as the Kurlbaum method.[37] Cooper[12] has compared the two-color optical pyrometer method with the Kurlbaum method and found them in substantial agreement; Barret[1] discusses the accuracy attainable with the Kurlbaum method. Combined with Figs. 4-24 and 4-25 or their equivalent, the Kurlbaum method yields sufficient data to determine the total emissivity of a flame.

If a total-radiation pyrometer and optical pyrometer are used to obtain the total radiation and the red brightness temperature of a flame, these may be combined with relation (4-65) to determine by integration the total emissivity ϵ_F and, therefrom, the true temperature T_F. Although Figs. 4-24 and 4-25 furnish one form of the necessary relation, their use in this particular case involves trial and error and a better form of plot may be constructed. This method has been used in some of the work of the International Flame Radiation group,[14] and the results obtained were considered at least as reliable as those using the modified Schmidt method. No background measurement is involved.

When temperature measurements have been made on an industrial flame of one size to determine its absorption strength KL for the purpose of estimating the emissivity ϵ_F of a similar but larger flame, the absorption strength KL determined from Fig. 4-24 should be multiplied by the dimension ratio L_2/L_1 before Fig. 4-25 is used. In addition, the value of KL should correspond to the particular shape of flame under consideration, in accordance with the principles discussed in connection with the use of Table 4-2. An example will be found on page 105.

Powdered-coal Flames. The radiation from powdered-coal flames has been treated analytically by Wohlenberg and his associates,[75,74] by Haslam and Hottel,[24] and by Lindmark and Edenholm.[41] Experimental studies have been conducted by Lindmark and Kignell[42] and by Sherman.[68]

It may be shown that the emissivity of a cloud of opaque particles, based on the area of the envelope of the whole cloud, is of the form $1 - e^{-\gamma}$, in which γ is the product term, (concentration of particles) (time-average cross section of a particle) (length of radiant beam through cloud), the last term being defined as in Table 4-2. By making suitable assumptions as to the laws of particle-size distribution in pulverized coal and the rate of combustion of individual particles, one may use the foregoing exponential relation to calculate the emissivity of a pulverized-coal flame. Values so obtained, however, are almost invariably considerably lower than measured flame emissivities. The discrepancy is probably due to the contribution of cracked hydrocarbons producing luminosity as well as to residual ash particles not allowed for in the theoretical derivation. Fortunately, modern pulverized-coal installations usually involve such large flames that their emissivity is not far from unity.

Illustration 10. It is desired to determine approximately the radiation from a proposed luminous-flame burner installation, from measurements made on a similar combustion chamber all the dimensions of which are one-half those of the proposed installation. It is intended to keep aeration and mixing conditions as similar as possible in the two chambers. The flame in each case is roughly spherical in form. An optical pyrometer with red and green screens is sighted through the smaller flame at its diameter, the apparent temperatures obtained being $T_r = 2699°F$ ($= 1755°K$) and $T_g = 2740°F$ ($= 1778°K$). (a) What are the emissivity and true temperature of the smaller flame? (b) What will be the probable rate of heat transfer per square foot of flame envelope of the proposed larger installation if the flame temperature is the same and the surrounding walls are at 2600°F and black?

a. From Fig. 4-24, when the red brightness temperature T_r of the flame is 1755°K and the difference between T_g and T_r is $1778 - 1755 = 23°K$, the true-flame temperature is found to be 1811°K and the absorption strength KL is 0.7. This value of KL, however, corresponds to length of radiant beam L equal to the diameter of the flame sphere. According to Table 4-2, the average value of L is 0.6 times the diameter when the radiating shape is spherical. Then the average absorption strength is $0.7 \times 0.6 = 0.42$. From Fig. 4-25, when $KL = 0.42$ and true-flame temperature $= 1811°K$, the flame emissivity is 0.20.

b. If the flame dimensions are doubled, other things being equal, the absorption strength KL will double. When $KL = 2 \times 0.42 = 0.84$, Fig. 4-25 indicates that the flame emissivity will be 0.365 (not double the value 0.20). The net radiation per square foot of flame envelope will be

$$\frac{q}{A} = 0.171 \times 0.365 \times \left[\left(\frac{1811 \times 1.8}{100} \right)^4 - \left(\frac{2600 + 460}{100} \right)^4 \right] = 15{,}900 \text{ Btu/(sq ft)(hr)}$$

THE GENERAL PROBLEM OF RADIATION IN AN ENCLOSURE CONTAINING GAS[30c]

The multizoned gray enclosure containing nonabsorbing gas and the uniform-temperature black enclosure containing absorbing and emitting gas have been considered quantitatively on pages 72 and 89. A marked increase in complexity of the solution may be expected when one considers the general case of a multisurface system of source-sink and no-flux surfaces enclosing an absorbing and emitting gas. Plainly, this is the description of a fuel-fired furnace cavity. No claim can be made that the present knowledge of this problem permits rigorous allowance for all factors, but the solution to be presented is sufficiently close to reality to be useful when judged by engineering needs.

Enclosure of Gray Surfaces Containing a Gray Gas of Uniform Temperature. Consider first the gray-gas case. A gray gas is here defined as one which exhibits, for radiation from whatever source, an absorptivity α equal to its emissivity ϵ. Emissivity and absorptivity will vary, however, with path length. If the gas transmittance τ, equal to $1 - \alpha$, is established for radiation from one zone to another, the transmittance for twice the mean path length between zones will be τ^2, a conclusion which can be true only for a gray gas, which does not produce a change in quality of the radiation transmitted by it. The problem is considered solved when the net radiation interchange between any two source-sink zones (such as $q_{1\rightleftarrows 2}$) or between any source-sink zone and the gas (such as $q_{1\rightleftarrows G}$) can be evaluated. The same remarks concerning the disentangling of radiation by identification of its source and sink apply here as were made on page 72 relative to an enclosure containing no gas; and this section should not be read until the derivation leading to Eq. (4-29) is understood. As before, one expects that $q_{1\rightleftarrows 2}$ will be expressible in the form $A_1 \mathfrak{F}_{12} \sigma (T_1^4 - T_2^4)$, but with a new meaning for $\mathfrak{F}_{12}$ since it must now allow for the presence of the gas. The net radiation loss by the gas will be

$$q_{G,\text{net}} = q_{G\rightleftarrows 1} + q_{G\rightleftarrows 2} + \cdots \tag{4-74}$$

and $q_{G\rightleftarrows 1}$ will be given by

$$q_{G\rightleftarrows 1} = A_1 \mathfrak{F}_{1G} \cdot \sigma (T_G^4 - T_1^4) \tag{4-75}$$

If as in the earlier derivation $\mathfrak{F}_{12}$ or $\mathfrak{F}_{1G}$ is evaluated on the assumption that A_1 is the only original radiator present in the system—all other source-sink zones and the gas being at absolute zero—the $\mathfrak{F}$ factor obtained will have general validity. As before, let A_1 have such a temperature that, if black, its emissive power would be 1. A little consideration will show that, if an energy balance on A_1 analogous to Eq. (4-24) for the diathermanous-gas system is to be made, the *only change necessary is the insertion, into every term containing a direct view factor such as F_{12} or F_{1R}, of an additional multiplying term* τ_{12} or τ_{1R} to allow for the transmittance of the radiation leaving A_1 for A_2 or A_R. And since $\tau_{12} \equiv \tau_{21}$, the change in the derivation can be expressed as a redefinition of the meaning of the shorthand symbols typified by $\overline{12}$ or $\overline{1R}$ from

$$\overline{12} = A_1 F_{12} \equiv A_2 F_{21} \tag{4-76}$$

to

$$\overline{12} = A_1 F_{12} \tau_{12} \equiv A_2 F_{21} \tau_{21} \tag{4-77}$$

With this modification made on *all* bar terms, Eqs. (4-22) to (4-32)† all become valid for systems containing a gray gas.

Now consider the use of the above [relations (4-26) and (4-29), with bar terms redefined by Eq. (4-77)] to evaluate radiation interchange between A_1 and the gas, *i.e.*, to determine $\mathfrak{F}_{1G}$ for use in Eq. (4-75). Let A_1 continue to be the only original emitter, radiating at rate $A_1\epsilon_1$ per unit value of black emissive power at temperature T_1. Radiation streaming away from $A_2, A_3, \ldots, A_R, A_S, \ldots$ is due solely to reflection at $A_2, A_3, \ldots$, to reflection and/or reradiation at $A_R, A_S, \ldots$. Of the total original emission from A_1, the amount $A_1\mathfrak{F}_{11}$ (evaluated using the new meaning of bar terms) returns to and is absorbed by A_1, $A_1\mathfrak{F}_{12}$ goes to and is absorbed by $A_2, \ldots, A_1\mathfrak{F}_{1n}$ is absorbed by A_n. The residue must have been absorbed by the gas since all other surfaces are nonretaining. Then

$$A_1\mathfrak{F}_{1G} = A_1(\epsilon_1 - \mathfrak{F}_{11} - \mathfrak{F}_{12} - \cdots - \mathfrak{F}_{1n}) \qquad (4\text{-}78)$$

For completeness as well as to introduce a useful new concept, $\mathfrak{F}_{1G}$ will be derived independently of $\mathfrak{F}_{11} \cdots \mathfrak{F}_{1n}$. Let the *gas* in the system be the only original emitter, with all source-sink zones $A_1 \cdots A_n$ kept at absolute zero; and let the flux density away from the various surfaces be designated by $_GR_1, _GR_2$, etc. Then the incidence on A_1 from A_2 and the intervening gas is $A_2 F_{21}\tau_{21} \cdot {}_GR_2$ from the surface and $A_2 F_{21}(1 - \tau_{21})$ from the gas. Similar terms for incidence on A_1 from zones A_1, A_3, etc., appear in the bracket below which, multiplied by reflectance ρ_1, must equal the flux $A_1 \cdot {}_GR_1$ away from surface 1. Using the new bar nomenclature, the equating of the two ways of expressing flux away from A_1 yields

$$\overline{11} \cdot {}_GR_1 + A_1 F_{11}(1 - \tau_{11}) + \overline{12} \cdot {}_GR_2 + A_1 F_{12}(1 - \tau_{12}) + \cdots$$
$$+ \overline{1R} \cdot {}_GR_1 + A_1 F_{1R}(1 - \tau_{1R})\rho_1 = A_1 \cdot {}_GR_1 \qquad (4\text{-}79)$$

When the terms $A_1 F_{11} + A_1 F_{12} + \cdots + A_1 F_{1R} + \cdots$ are replaced by A_1 and the unknown R's are grouped on the left, the above becomes

$$(\overline{11} - A_1/\rho_1) \cdot {}_GR_1 + \overline{12} \cdot {}_GR_2 + \cdots + \overline{1R} \cdot {}_GR_R + \cdots$$
$$= -A_1 + (\overline{11} + \overline{12} + \cdots + \overline{1R} + \cdots) \qquad (4\text{-}80)$$

Similar energy balances on $A_2 \cdots A_R \cdots$ yield a system of equations with coefficients on R's identical to those previously encountered in Eqs. (4-25). Solution for $_GR_n$, the relative partial flux density from

---

† The new generalization stops short of including Eq. (4-33), the derivation of which involved use of such relations as

$$\overline{11} + \overline{12} + \overline{13} + \cdots + \overline{1R} + \overline{1S} + \cdots = A_1$$

This is true when the bar terms denote AF, not when they denote $AF\tau$.

surface n due to the existence of gas G as the sole emitter, is then

$$_GR_n = {_GD_n}/D \tag{4-81}$$

in which $_GD_n$ is obtained by inserting, into the nth column of D [relation (4-26), with redefined bar terms], the terms

$$-A_1 + (\overline{11} + \overline{12} + \cdots + \overline{1R} + \cdots),$$
$$-A_2 + (\overline{21} + \overline{22} + \overline{23} + \cdots + \overline{2R} + \cdots), \ldots,$$
$$-A_R + (\overline{R1} + \overline{R2} + \cdots + \overline{RR} + \cdots)$$

Any one of these expressions may be written

$$-A_n[1 - (F_{n1}\tau_{n1} + F_{n2}\tau_{n2} + \cdots + F_{nR}\tau_{nR} + \cdots)]$$

the parenthetical term of which equals the weighted-mean transmittance, or transmittance for the total radiation arriving at or leaving A_n (and therefore identifiable with a single subscript n). The complement of τ_n is the gas absorptivity and, because it is gray, the gas emissivity ϵ_{Gn}. Note that ϵ_{G1} and ϵ_{G2} differ only because the mean path lengths through the gas to A_1 and A_2 differ. Recapitulating, $_GR_n$ is obtained by inserting into the nth column of D the terms $-A_1\epsilon_{G1}, -A_2\epsilon_{G2}, \ldots, -A_R\epsilon_{GR}$, ... With reflected flux density $_GR_1$ known, $\mathfrak{F}_{1G}$ may be evaluated. Since the ratio of absorbed to reflected energy at A_1 is ϵ_1/ρ_1, one may write

$$A_1\mathfrak{F}_{1G} = \frac{A_1\epsilon_1}{\rho_1} {_GR_1} = \frac{A_1\epsilon_1}{\rho_1} \frac{{_GD_1}}{D} \tag{4-82}$$

This expression may be shown to be the equivalent of (4-78). Although (4-82) appears to be simpler because it involves fewer terms to evaluate, (4-78) is as easy to use because an interest in $\mathfrak{F}_{1G}$ is usually accompanied by an interest in $\mathfrak{F}_{12}, \mathfrak{F}_{13}, \ldots$ as well; and the only new term to evaluate in (4-78) is $\mathfrak{F}_{11}$.

As an example of the application of the above principles, consider the simplest possible system capable of representing the case under discussion, viz., a gas at T_G interchanging radiation with a heat sink represented by a single zone of area A_1, the enclosure being completed by refractory surfaces grouped together into a single no-flux zone A_R. With net interchange between the gas and the sink being given by Eq. (4-75), the problem is to evaluate $\mathfrak{F}_{1G}$.

From (4-78),

$$A_1\mathfrak{F}_{1G} = A_1\epsilon_1 - A_1\mathfrak{F}_{11}$$

$$= A_1\epsilon_1 \left(1 + \frac{\epsilon_1}{\rho_1} \frac{\begin{vmatrix} \overline{11} & \overline{1R} \\ \overline{1R} & \overline{RR} - A_R \end{vmatrix}}{\begin{vmatrix} \overline{11} - A_1/\rho_1 & \overline{1R} \\ \overline{1R} & \overline{RR} - A_R \end{vmatrix}} \right) \tag{4-83}$$

Designating the determinant in the numerator by B, partially expanding the determinant in the denominator into $B - (\overline{RR} - A_R)(A_1/\rho_1)$, taking reciprocals, and replacing ρ_1 by $1 - \epsilon_1$, one obtains

$$\frac{1}{A_1 \mathcal{F}_{1G}} = \frac{1}{A_1 \epsilon_1}\left(1 - \epsilon_1 - \frac{\epsilon_1 A_1(\overline{RR} - A_R)}{B - A_1(\overline{RR} - A_R)}\right)$$

$$= \frac{1}{A_1}\left(\frac{1}{\epsilon_1} - 1\right) + \frac{1}{A_1 - \overline{11} + \overline{1R}^2/(\overline{RR} - A_R)} \tag{4-84}$$

When ϵ_1 becomes 1, $\mathcal{F}_{1G}$ by definition becomes $\bar{F}_{1G}$ and, from (4-84),

$$A_1 \bar{F}_{1G} = A_1 - \overline{11} + \overline{1R}^2/(\overline{RR} - A_R) \tag{4-85}$$

Eq. (4-84) may therefore be written

$$\frac{1}{A_1 \mathcal{F}_{1G}} = \frac{1}{A_1}\left(\frac{1}{\epsilon_1} - 1\right) + \frac{1}{A_1 \bar{F}_{1G}} \tag{4-86}$$

For this system all direct view factors can be expressed in terms of the single one F_{R1} as follows:

$$\overline{1R} = A_1 F_{1R} \tau_{1R} = A_R F_{R1} \tau_{1R}$$
$$\overline{11} = A_1 F_{11} \tau_{11} = A_1(1 - F_{1R})\tau_{11} = (A_1 - A_R F_{R1})\tau_{11}$$
$$\overline{RR} = A_R F_{RR} \tau_{RR} = A_R(1 - F_{R1})\tau_{RR}$$

Substitution of these values in (4-85) and replacement of τ by $1 - \epsilon_G$ gives

$$\bar{F}_{1G} = \epsilon_{11}\left(1 + \frac{A_R/A_1}{\frac{\epsilon_{11}/\epsilon_{RR} - \epsilon_{11}}{1 - \epsilon_{11}} + \frac{\epsilon_{11}}{1 - \epsilon_{11}}\frac{1}{F_{R1}}}\right)$$
$$- \frac{(A_R/A_1)F_{R1}^2}{F_{R1} + \epsilon_{RR}(1 - F_{R1})}(\epsilon_{11} + \epsilon_{RR} - 2\epsilon_{R1} + \epsilon_{R1}^2 - \epsilon_{11}\epsilon_{RR}) \tag{4-87}$$

in which double-subscript ϵ's refer to an evaluation of gas emissivity based on a path length specific to the two surface elements mentioned. The problem, simple as it appeared to be, has a solution rather formidable for engineering use. If the various gas emissivities are assumed to be alike or if each one is replaced by their average value, called ϵ_G, it will be noted that the second term on the right of (4-87) will vanish, giving

$$\bar{F}_{1G} = \epsilon_G\left(1 + \frac{A_R/A_1}{1 + \frac{\epsilon_G}{1 - \epsilon_G}\frac{1}{F_{R1}}}\right) \tag{4-88}$$

Some practical consequences of (4-86) and (4-88) are these: increasing the flame emissivity increases the heat transmission, but not proportionately; decreasing surface emissivity ϵ_1 (and absorptivity) from unity when the flame is very transparent produces almost no effect on the heat transmission; but decreasing ϵ_1 from unity when the flame is substantially opaque ($\epsilon_G = 1$) produces a proportional decrease in heat transmission.

The limitations on the validity of (4-88) and (4-86) must be borne in mind. They are restricted to a one-zone sink, a one-zone refractory or no-flux surface, and a gray gas. The first two assumptions are rigorously justifiable only when each element of A_1 (or A_R) shares its "view" of its own zone and of A_R (or A_1) in the same ratio as every other element; and this in turn is true only when the two kinds of surfaces are intimately mixed in the same ratio on all parts of the enclosure, forming what one might call a "speckled" enclosure. Under those circumstances, the assumption made in going from (4-87) to (4-88), that the various ϵ_G's are representable by an average value, becomes valid; and F_{R1} becomes $A_1/(A_1 + A_R)$. If (4-88) is used as an approximation for a system which does not have a speckled enclosure, however, use of the true value of F_{R1} is preferable.

Equations (4-86) and (4-88) are well-known solutions, available and in use for many years[30a,43a] before the determinant method of derivation was available; and their derivation from first principles was perhaps as simple as the one here presented, but only because of restriction to a two-zone system. With the new method available, summarized in Eq. (4-29), the decision as to the number of zones of heat-sink or refractory area into which the enclosure should be divided can be made to depend, as it should, on the importance of the particular problem and the time available for handling it, rather than on whether the engineer can see his way through a multizone solution.

Equilibrium Temperature of No-flux Surfaces (Gray Gas Present). The determination of the general exchange factor $\mathfrak{F}$ involved determination of relative partial flux densities R of some particular surface due to the presence of the various original emitters in the system. Each relative partial value can be changed to an absolute partial value by multiplying by the σT^4 value of the emitter responsible, and the sum of such partial values equals the true total value. For no-flux surface A_R this flux density is identifiable with σT_R^4. Then

$$T_R^4 = (T_1^4 \cdot {}_1R_R + T_2^4 \cdot {}_2R_R + \cdots + T_n^4 \cdot {}_nR_R) + T_G^4 \cdot {}_GR_R \quad (4\text{-}89)$$

or

$$T_R = \sqrt[4]{({}_1D_R \cdot T_1^4 + {}_2D_R \cdot T_2^4 + \cdots + {}_nD_R \cdot T_n^4 + {}_GD_R \cdot T_G^4)/D} \quad (4\text{-}90)$$

Since the sum of the D_R's in the numerator can be shown to equal D, Eq. (4-90) indicates that T_R^4 is a weighted mean of the fourth powers of the various original emitters present, including all source-sink surfaces and the gas—as it must of course be.

Application of Eq. (4-90) will be made to determine the equilibrium refractory temperature for two cases of possible interest. For the previously discussed case of a gray gas enclosed by A_1 and A_R, with all values of ϵ_G taken to be the same, Eq. (4-90) yields, for the equilibrium refractory temperature,

$$T_R^4 = T_G^4 - \frac{(T_G^4 - T_1^4)}{1 + \dfrac{1}{F_{R1}} \dfrac{\epsilon_G}{1-\epsilon_G} + \epsilon_G \dfrac{1-\epsilon_1}{\epsilon_1}\left(\dfrac{1}{F_{R1}} \dfrac{\epsilon_G}{1-\epsilon_G} + \dfrac{A_R}{A_1} + 1\right)} \quad (4\text{-}91)$$

When A_1 is black, the last term in the denominator drops out. For the case of an enclosure formed by two source-sink zones A_1 and A_2 and a single no-flux zone A_R and containing no radiating gas, Eq. (4-90) yields

$$T_R^4 = \frac{aT_1^4 + bT_2^4}{a+b} \qquad (4\text{-}92)$$

where

$$a = \frac{A_1\epsilon_1}{1-\epsilon_1}\left(\frac{A_2\epsilon_2}{1-\epsilon_2}\,\overline{1R} + \overline{12}(\overline{1R}+\overline{2R}) + \overline{1R}\cdot\overline{2R}\right)$$

$$b = \frac{A_2\epsilon_2}{1-\epsilon_2}\left(\frac{A_1\epsilon_1}{1-\epsilon_1}\,\overline{2R} + \overline{12}(\overline{1R}+\overline{2R}) + \overline{1R}\cdot\overline{2R}\right)$$

This simplifies greatly if A_1 and A_2 are black.

It should be noted that the solutions of various problems involving gray gas in a gray enclosure have never involved the emissivity of the no-flux surfaces. It is immaterial whether the no-flux surfaces get rid of incident radiation by reflection or by absorption and reradiation or by one of these mechanisms in one spectral region and the other in another, since the gas exhibits the same transmittance for all kinds of radiation and the source-sink absorptivities are likewise unaffected by quality of radiation incident on them.

Enclosure of Gray Surfaces Containing a Real (Nongray) Gas. In the derivation of interchange factors for gray-gas systems, the single value of transmittance τ_{12} applied to *all* radiation leaving A_1 for A_2, whether originally emitted by A_1 or reflected from it after any number of trips through the gas. Only for gray gas is this true. For a *real* gas, with its characteristic absorption in certain spectral regions (see Fig. 4-26), the absorbable wavelengths are filtered out after a number of passages through the gas, and the transmittance of the gas for the remainder of the radiation approaches 1. The gray-gas assumption thus leads to prediction of too large an interchange between gas and sinks and too small an interchange between the source-sink surfaces. In obtaining a value of $A_1\mathfrak{F}_{1G}$ applicable to a real gas, it is desirable to retain the mechanics of gray-gas formulation. Fortunately, this is possible.

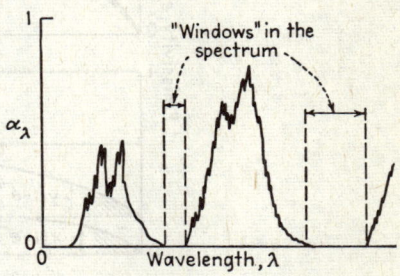

Fig. 4-26. Variation of gas absorptivity with wavelength.

For a gray gas the transmittance for the absorption path length represented by P_GL is e^{-kP_GL}, where k is the absorption coefficient of the gas, a constant independent of wavelength and therefore applicable to the integrated spectrum; and the absorptivity and emissivity equal $1 - e^{-kP_GL}$ (see Fig. 4-27, top). The relation of transmittance to P_GL for a real gas can be represented to any desired degree of accuracy by

$$\tau = x \cdot e^{-k_xPL} + y \cdot e^{-k_yPL} + z \cdot e^{-k_zPL} + \cdots \quad (4\text{-}93)$$

and the emissivity relation by

$$\epsilon = x(1 - e^{-k_xPL}) + y(1 - e^{-k_yPL}) + z(1 - e^{-k_zPL}) + \cdots \quad (4\text{-}94)$$

with k_x representing the absorption coefficient applicable to fraction x of the total energy spectrum of the gas, and with the condition

$$x + y + z + \cdots = 1 \quad (4\text{-}95)$$

The lower part of Fig. 4-27 shows the real-gas emissivity of the upper part approximated by the sum of two e functions giving an asymptotic

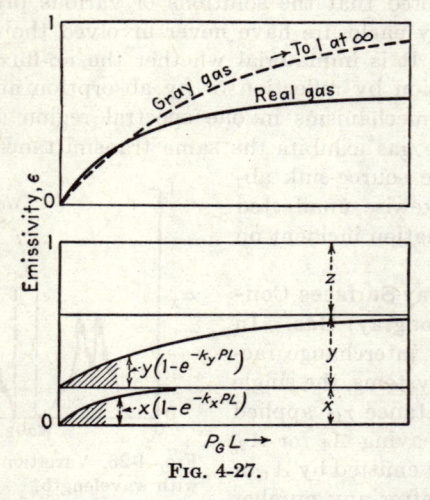

FIG. 4-27.

emissivity of $x + y$, or $1 - z$. Representing e^{-k_zPL} by τ_z, Eq. (4-93) yields for the transmittance $\tau_{n\cdot G}$ of n layers of gas, each of absorption path length P_GL:

$$\tau_{n\cdot G} = x\tau_x^n + y\tau_y^n + z\tau_z^n + \cdots \quad (4\text{-}96)$$

The components of which the total real-gas transmittance is composed are thus a series of gray-body transmittances, each used with a weighting factor, $x, y, \ldots$.

Consider now an enclosure all the surfaces of which aid in transmission of radiation from the enclosed gas to A_1 only by the process of reflection, *i.e.*, a system of gray source-sink surfaces and completely reflecting or white no-flux surface. A little consideration will show that the real-gas solution for $A_1\mathcal{F}_{1G}$ is obtainable as the weighted sum of a number of gray-gas solutions, using successively τ_x, τ_y, τ_z for the gas transmissivity and weighting each solution by the factor x, y, z, etc., or

$$\mathfrak{F}_{iG} = x \cdot \mathfrak{F}_{iG}\Big]_{\substack{\text{based on}\\ \text{use of } \tau_x}} + y \cdot \mathfrak{F}_{iG}\Big]_{\substack{\text{based on}\\ \text{use of } \tau_y}} + \cdots \qquad (4\text{-}97)$$

and similarly,

$$\mathfrak{F}_{12} = x\mathfrak{F}_{12}\Big]_{\substack{\text{based on}\\ \text{use of } \tau_x}} + y\mathfrak{F}_{12}\Big]_{\substack{\text{based on}\\ \text{use of } \tau_y}} + \cdots \qquad (4\text{-}98)$$

The reason for the restriction on the validity of (4-97) or (4-98), that any no-flux surfaces if present must be white rather than gray, needs consideration. A white refractory surface reflects all incident radiation without changing its quality, *i.e.*, without changing the fractions of it for which the gas will exhibit absorptivity $1 - \tau_x$, $1 - \tau_y$, etc. But a gray refractory surface, to the extent that it absorbs and reemits, changes the quality of the radiation. If, for example, a beam of radiation incident on A_R from the gas has an emissivity of $\frac{1}{2}$ in one-half of the energy spectrum, or a total emissivity and absorptivity of $\frac{1}{4}$, the resulting radiation leaving A_R would be half absorbed by the gas on next passage through it if it left A_R by reflection without change in character, and only one-fourth absorbed if it left A_R by emission as black radiation. Since the derivation of $A_1\mathfrak{F}_{1G}$ when A_R, A_S, ... are present is based on attenuation by the gas in an amount independent of the history of a beam of radiation, the nongray-gas solution represented by (4-97) applies rigorously only to systems in which any no-flux surfaces present are *perfect diffuse reflectors*. If allowance must be made for the grayness of any no-flux surface, it must be reclassified as a source-sink surface, say A_3, of unknown temperature the value of which is obtained by introducing the condition that the sum of the interchanges of A_3 with the other surfaces and with the gas must be zero.

The abandonment of the gray-gas restriction introduces the complication that gas absorptivity and gas emissivity are no longer equal; and $\mathfrak{F}_{1 \leftarrow G}$ and $\mathfrak{F}_{1 \rightarrow G}$ are alike only in the limit as T_1 approaches T_G [see the discussion concerning Eq. (4-45)]. If $x, y, z, \ldots$ and $\tau_x, \tau_y, \tau_z, \ldots$ are based on gas emissivity evaluated at T_G, their use in (4-97) yields the quantity $\mathfrak{F}_{1 \leftarrow G}$; if they are based on the absorptivity of a gas at T_G for radiation from a surface at T_1 (see page 90), their use in (4-97) yields the quantity $\mathfrak{F}_{1 \rightarrow G}$. If $\mathfrak{F}_{1 \rightarrow 2}$ is to be evaluated, the absorptivity of a gas at T_G for radiation from T_1 is used; if $\mathfrak{F}_{1 \leftarrow 2}$, the absorptivity of a gas at T_G for radiation from T_2. To save the time of evaluating two complete sets of $\mathfrak{F}$'s, however, it is suggested that if the total spread in absolute temperatures of the various zones and gas is not too great, $\mathfrak{F}_{1G}$ evaluated by use of an effective emissivity given by the term in parentheses in Eq. (4-59) will suffice for both $\mathfrak{F}_{1 \leftarrow G}$ and $\mathfrak{F}_{1 \rightarrow G}$.

Although Eq. (4-97) or (4-98) can in principle be used to handle an enclosure of any degree of complexity as to zoning and filled with gas of radiating characteristics producing any shape of curve of ϵ_G vs. P_GL, a little consideration shows what an enormous amount of effort is involved if these expressions for $\mathcal{F}$ contain many terms. A simplification is mandatory and, fortunately, feasible. If wall reflectivities are not very large, a beam of radiation from the gas is rapidly attenuated in its succession of reflections and transmissions, and the fitting of the ϵ_G-P_GL curve is important only for a few units of P_GL. The transmittance τ given by Eq. (4-96) can be made to equal true transmittance at 0 and at two integral multiples of P_GL by assuming the gas gray throughout the energy fraction x and clear throughout the fraction $y + z + \cdots = 1 - x$, that is, by assuming that all the τ's but τ_x are zero and that an asymptotic transmissivity of $1 - x$ is approached as $P_GL = \infty$. An examination of Eq. (4-97) now indicates that, since all values of $\mathcal{F}_{1G}$ on the right-hand side except the first are for nonabsorbing gas and therefore are zero,

$$\mathcal{F}_{1G} = x \cdot \mathcal{F}_{1G} \Big]_{\substack{\text{based on} \\ \text{use of } \tau_x}} \qquad (4\text{-}99)$$

For source-sink surface interchange, Eq. (4-98) yields

$$\mathcal{F}_{12} = x \cdot \mathcal{F}_{12} \Big]_{\substack{\text{based on} \\ \text{use of } \tau_x}} + (1 - x) \cdot \mathcal{F}_{12} \Big]_{\substack{\text{based on} \\ \text{clear gas,} \\ \tau = 1}} \qquad (4\text{-}100)$$

There remains only the evaluation of x and τ_x from a gas-radiation plot such as Fig. 4-13. Let the objective be to fit the ϵ_G-P_GL curve at one and two units of P_GL, and call the corresponding ϵ's read from the plot, ϵ_G and $\epsilon_{2.G}$. From Eq. (4-96),

$$1 - \epsilon_G = \tau_G = x\tau_x + (1 - x)$$

and $\qquad (4\text{-}101)$

$$1 - \epsilon_{2.G} = \tau_{2.G} = x\tau_x^2 + (1 - x)$$

Solution of these gives

$$x = \frac{\epsilon_G^2}{2\epsilon_G - \epsilon_{2.G}}$$

and $\qquad (4\text{-}102)$

$$\tau_x = 1 - \frac{\epsilon_G}{x} = 1 - \frac{2\epsilon_G - \epsilon_{2.G}}{\epsilon_G}$$

Recapitulating, $\mathcal{F}_{1G}$ equals x times a value of $\mathcal{F}_{1G}$ from Eq. (4-82) [or (4-86) and (4-88)] using a transmittance of $1 - \epsilon_G/x$ or an emissivity of ϵ_G/x, with x defined by (4-102).

The determination of x and τ_x from values of ϵ_G at $1P_GL$ and $2P_GL$ is recommended when A_1 has a low reflectivity and/or when P_GL is large;

but a small enclosure with heat-sink zones of high reflectivity may make the many-times reflected radiation relatively more important. In that case some other pair of ϵ_G values may be used, such as ϵ_G and $\epsilon_{3 \cdot G}$, or $\epsilon_{2 \cdot G}$ and $\epsilon_{4 \cdot G}$.

Application of Principles. The procedures presented in the previous ten pages permit allowance for the effects of factors somewhat casually handled in the past. The relations presented are not as easy to use as the relations of convective-heat transmission; but this is because the mathematics of radiation in an enclosure, where every part of the system affects every other part, is intrinsically more complicated than the mathematics of heat-transfer processes capable of expression in the form of a differential equation. With a little practice in manipulation of determinants[†] the reader should be able to evaluate $\mathfrak{F}$ factors for systems of a considerable degree of complexity in a reasonable time. If the higher-order determinants encountered are evaluated numerically for the specific example of interest rather than algebraically to obtain results like (4-87), the time required for a solution is not prohibitive. Some of the special cases encountered are used so frequently, however, that algebraic formulation of their general solution is desirable. A few such cases will be presented here.

Nongray Gas, Gray Sink A_1, and Reflecting No-flux Surface A_R. The gray-gas solution for this case, simplified by the use of a single path length and therefore a single τ_G for all zone pairs, was given in Eqs. (4-86) and (4-88). Modification to allow for nongray gas gives

$$A_1 \mathfrak{F}_{1 \leftarrow G} = \cfrac{x}{\cfrac{1}{A_1}\left(\cfrac{1}{\epsilon_1} - 1\right) + \cfrac{1}{\cfrac{\epsilon_G}{x}\left(A_1 + \cfrac{A_R}{1 + \cfrac{\epsilon_G/x}{1 - \epsilon_G/x}\cfrac{1}{F_{R1}}}\right)}} \qquad (4\text{-}103)$$

with x equal to $\epsilon_G{}^2/(2\epsilon_G - \epsilon_{2 \cdot G})$, and with ϵ_G evaluated for a path length given by Table 4-2. If $A_1 \mathfrak{F}_{1 \to G}$ is wanted, α_{G1} replaces ϵ_G. If $q_{G \rightleftharpoons 1}$ is wanted and T_G and T_1 do not differ by more than a factor of 2, then ϵ_G evaluated at the arithmetic-mean temperature and multiplied by $(4 + a + b - c)/4$ can be used to obtain a value for $A_1 \mathfrak{F}_{1G}$ which will serve as an approximation for both the emission and the absorption term. In evaluating the net flux from the gas, convection from gas to A_1 and from gas to A_R must of course be included. It is to be remembered that the latter was assumed equal to the external loss through the refractory.

Nongray Gas, Enclosed by Two Gray Sinks A_1 and A_2, and No A_R. This variation on the previous case has interest for at least two reasons. Suppose a gas, a primary heat sink A_1, and a refractory surface the external loss from which is so large that it cannot be treated as a no-flux surface (the term no-flux still refers to radiant-heat

---

[†] Evaluation of higher-order determinants by mechanical computers is of course feasible. The labor of evaluation with pencil and slide rule has been so greatly reduced by the method of Crout,[12a] however, that fifth- or sixth-order determinants need no longer be considered formidable by the engineer not equipped with the newer devices.

transmission only), because the internal gain by convection is so much less than the loss through the wall. Then the refractory surface becomes a secondary heat sink and is A_2 rather than A_R. Or consider a gas, a heat sink A_1, and a no-flux surface which is not justifiably classed as completely reflecting. As the discussion on page 113 indicated, the no-flux surface must be treated as a source-sink-type surface A_2. The three $\mathcal{F}$'s necessary for a complete solution will be obtained first. With the various τ's between zone pairs considered alike, Eqs. (4-82) and (4-99) give

$$A_1\mathcal{F}_{1G} = x \frac{A_1 \epsilon_1}{\rho_1} \frac{\begin{vmatrix} \overline{11} + \overline{12} - A_1 & \overline{12} \\ \overline{21} + \overline{22} - A_2 & \overline{22} - A_2/\rho_2 \end{vmatrix}}{\begin{vmatrix} \overline{11} - A_1/\rho_1 & \overline{12} \\ \overline{12} & \overline{22} - A_2/\rho_2 \end{vmatrix}}$$

$$= \frac{xA_1\epsilon_1}{\rho_1} \frac{\begin{vmatrix} A_1(\tau - 1) & A_1 F_{12}\tau \\ A_2(\tau - 1) & A_2\left(\tau - \dfrac{1}{\rho_2}\right) - A_1 F_{12}\tau \end{vmatrix}}{\begin{vmatrix} A_1\left(\tau - \dfrac{1}{\rho_1}\right) - A_1 F_{12}\tau & A_1 F_{12}\tau \\ A_1 F_{12}\tau & A_2\left(\tau - \dfrac{1}{\rho_2}\right) - A_1 F_{12}\tau \end{vmatrix}}$$

$$= x(\tau - 1)\frac{\epsilon_1}{\rho_1} \frac{\begin{vmatrix} A_1 & 1 \\ 1 & \dfrac{1 - 1/\rho_2\tau}{A_1 F_{12}} - \dfrac{1}{A_2} \end{vmatrix}}{\begin{vmatrix} \tau - 1/\rho_1 & 1/A_1 \\ \tau - 1/\rho_2 & \dfrac{1 - 1/\rho_2\tau}{A_1 F_{12}} - \dfrac{1}{A_2} \end{vmatrix}}$$

$$A_1\mathcal{F}_{1G} = x(1 - \tau_x)\frac{\epsilon_1}{\rho_1}$$
$$\frac{1 + A_1/A_2 + (1/\rho_2\tau_x - 1)/F_{12}}{(1/\rho_1 - \tau_x)/A_2 + (1/\rho_1 - \tau_x)(1/\rho_2 - \tau_x)/A_1 F_{12}\tau_x + (1/\rho_2 - \tau_x)/A_1} \quad (4\text{-}104)$$

$A_2\mathcal{F}_{2G}$ is obtained from the above by interchange of subscripts 1 and 2. $A_1\mathcal{F}_{12}$ is obtained from Eqs. (4-28) (with redefined bar terms) and (4-100), which yield on simplification

$$A_1\mathcal{F}_{12} = \frac{x(\epsilon_1/\rho_1)(\epsilon_2/\rho_2)}{\text{same denominator as}}$$
$$\text{(4-104)}$$
$$+ \frac{(1-x)(\epsilon_1/\rho_1)(\epsilon_2/\rho_2)}{(1/\rho_1 - 1)/A_2 + (1/\rho_1 - 1)(1/\rho_2 - 1)/A_1 F_{12} + (1/\rho_2 - 1)/A_1} \quad (4\text{-}105)$$

These three $\mathcal{F}$'s are for use in the heat-transfer equations:

$$q_{G \rightleftarrows 1} = A_1\mathcal{F}_{1G}\sigma(T_G{}^4 - T_1{}^4) + h_1 A_1(T_G - T_1) \quad (4\text{-}106a)$$
$$q_{G \rightleftarrows 2} = A_2\mathcal{F}_{2G}\sigma(T_G{}^4 - T_2{}^4) + h_2 A_2(T_G - T_2) \quad (4\text{-}106b)$$
$$q_{2 \rightleftarrows 1} = A_1\mathcal{F}_{12}\sigma(T_2{}^4 - T_1{}^4) \quad (4\text{-}106c)$$

(The difference between $\mathcal{F}_{x \leftarrow y}$ and $\mathcal{F}_{x \rightarrow y}$ is here ignored for simplicity of treatment) If surface A_2 is losing heat to the outside at a rate $A_2 U(T_2 - T_0)$, a heat balance on A_2 yields

$$q_{G \rightleftarrows 2} = q_{2 \rightleftarrows 1} + A_2 U(T_2 - T_0)$$

or

$$A_2\mathcal{F}_{2G}\sigma(T_G{}^4 - T_2{}^4) = A_1\mathcal{F}_{12}\sigma(T_2{}^4 - T_1{}^4) + A_2 U(T_2 - T_0) - h_2 A_2(T_G - T_2) \quad (4\text{-}107)$$

Assuming the source temperature T_G and the primary sink temperature T_1 to be known, (4-107) permits a solution for the unknown refractory temperature T_2 (but trial and error because mixed in first and fourth powers).†

The other application mentioned for this system of equations was to make allowance for grayness of a refractory surface. In this case A_2 is truly a no-flux surface, with $A_2U(T_2 - T_0) = h_2A_2(T_G - T_2)$. Then (4-107) can be readily solved for $T_2{}^4$. If this is put into (4-106a) and (4-106b) and the radiation terms of those two equations are added, one obtains

$$q_{G,\text{ net loss by radiation}} = q_{1,\text{ net gain by radiation}} = \left[A_1\mathfrak{F}_{1G} + \cfrac{1}{\cfrac{1}{A_1\mathfrak{F}_{12}} + \cfrac{1}{A_2\mathfrak{F}_{2G}}}\right]\sigma(T_G{}^4 - T_1{}^4)$$

(4-108)

The term in the brackets, allowing as it does for the radiation from gas to A_1 with the aid of A_2, is like the term $A_1\mathfrak{F}_{1G}$ for the system gas-A_1-A_R (with A_2 representing A_R) except that it now allows for the grayness of A_2. If in (4-108) A_2 is assumed to be a white surface ($\rho_2 = 1$), it may be shown that the term in the brackets reduces to the $A_1\mathfrak{F}_{1G}$ of (4-103).

Refractory-backed Tubes in an Enclosure. When the walls of a furnace chamber include gray tubes mounted on a refractory backing as well as other refractory surfaces, two alternatives are possible. The system may be considered to consist of several no-flux zones, one of them the wall behind the tubes, and of source-sink zones such as the tubes and any other. However, the performance of the tubes and their refractory backing can instead be treated by itself, by momentarily visualizing a new enclosure consisting of the gray tubes of area A_2, the no-flux refractory backing of area A_R, and an imaginary black plane of area A_1 ($= A_R$) located a small distance on the other side of the tubes from the no-flux surface. Let the factor $A_2\mathfrak{F}_{21}$ ($\equiv A_1\mathfrak{F}_{12}$) now be evaluated [use of Fig. 4-12 and Eq. (4-33)]. Then, since A_1 represents the area of the plane of the tubes, it is apparent that the tubes and their refractory backing act *in all respects in the original furnace enclosure like a gray plane of emissivity* $\mathfrak{F}_{12}$. It is to be understood that, although this procedure may be used regardless of whether the main furnace chamber contains radiating gas or not, it is not valid for the gas-containing case if the radiation or absorption of any gas between the tubes and their refractory backing is very significant. If the tubes are mounted parallel to a refractory wall with sufficient clearance to make gas circulation behind the tubes significant, the problem is far more difficult to handle.

Estimation of Heat Transfer in a Combustion Chamber. Although relations have been presented for evaluating radiant-heat transmission in chambers filled with the combustion products of fuels, these relations have been restricted to idealized cases in which the gas temperature was uniform or was changing in one dimension along a flow path long compared with the transverse dimensions. Plainly, the average combustion chamber, in which combustion and mixing are occurring simultaneously and in a complicated flow path which involves recirculation as well, is far from typical of the idealized systems discussed. Those systems can nevertheless provide an indication of the performance to be expected and can in many cases be used for quantitative prediction. The simplest

† Trial and error may be shortened by use of the procedure discussed on page 79 and Eq. (4-44b), that of formulating an "emissivity" due to convection, for use in fourth-power relations.

case to discuss is the limiting one in which all dimensions of the chamber are of the same order of magnitude and in which the mixing energy provided in the incoming fuel and air produces a turbulent gas mixture uniform in temperature throughout and equal to the temperature of the gas leaving the chamber. Assume the problem to be the determination of heat transfer in the chamber, given the mean radiating temperature T_1 of the stock or heat sink, the chamber dimensions, and the fuel and air rates. Let the unknown mean gas temperature (and exit temperature) be T_G. Then the net heat-transfer rate from the gas is given by

$$q_{G,\text{net}} = A_1 \mathfrak{F}_{1G} \sigma (T_G^4 - T_1^4) + h_1 A_1'(T_G - T_1) + U_R A_R (T_G - T_0) \quad (4\text{-}109)$$

The sink area A_1' at which convection heat transfer occurs is indicated as possibly different from the area A_1 at which gas radiation occurs, because heat-sink surfaces such as a row of tubes covering the gas outlet from the chamber, and therefore receiving by convection no heat which affects the mean gas temperature in the chamber, should be included in A_1 but not in A_1'. Convection from gas to A_R has been assumed equal to the loss through A_R, which replaces it in the equation. With the outside-air temperature T_0 known, Eq. (4-109) expresses a relation between two unknowns, $q_{G,\text{net}}$ and T_G. The other relation is an energy balance, such as

$$q_{G,\text{net}} = i - w_G c_p (T_G - T_0) \quad (4\text{-}110)$$

where i represents the hourly enthalpy of the entering fuel, air, and recirculated flue gas if any, above a base temperature T_0 (water as vapor); and c_p represents the heat capacity (mean value between T_G and T_0) of the gas leaving the chamber, at hourly mass rate w_G. Equations (4-109) and (4-110) may be solved, usually by trial and error, to give $q_{G,\text{net}}$ and T_G. The limitation on q_G, that it does not include gas convection at area $A - A'$, must be borne in mind.

The pair of equations just discussed applies strictly to one of two limiting furnace types—that one in which the assignment of a mean flame temperature equal to the temperature of the gases leaving is justifiable. The method consequently predicts the minimum performance of which the system is capable. Better agreement between predicted and experimental results is obtained on some furnaces when the assumption is made that flame temperature and exit-gas temperature are not the same but differ by a constant amount. In a number of furnace tests used to determine what value of this difference produces agreement between experiment and the equations recommended, the difference was found to be about 300°F.

The other extreme in furnace types is that one in which combustion occurs substantially instantaneously at the burners (through complete premixing of fuel and air); the temperature attained is that generally known as theoretical flame temperature or adiabatic-combustion temperature; and the temperature falls continuously as the gases flow from burner to outlet. When such a furnace is long compared with its cross section normal to the direction of gas flow, Eq. (4-109) may be considered as applying to a differential length of furnace, and the remarks on pages 76 to 77 or 91 are applicable. One must, however, be prepared to examine the validity of the assumption that radiant flux in the gas-flow direction is of secondary significance.

Simplified Treatment of Combustion-chamber Heat Transmission. Equation (4-109) or its equivalent has been used as a basis for deriving various simplified relations, easier to use but restricted in applicability in proportion to the degree of simplification. Several of these will be presented.

Billet-reheating Furnaces. For continuous billet-reheating furnaces Eq. (4-109) has been modified as follows: (1) q is heat transferred to the stock, not from the flame; (2) convection terms have been omitted; (3) to compensate therefor and to allow for steadiness of furnace operation, $\mathfrak{F}_{1G}$ is evaluated using $1.2 r_f \cdot \epsilon_G$ instead of ϵ_G, where r_f is the ratio of average billet-pushing rate over a period of several hours to pushing rate during periods of steady operation; (4) ϵ_G is flame emissivity due to CO_2 and H_2O only; (5) $F_{R1} = A_1 F_{1R}/A_R = A_1/A_R$; (6) an average value of $(T_G^4 - T_1^4)$ is used, equal to the geometric mean of its value at the two ends of the furnace, and at the hot end T_G is taken as the calculated "theoretical" flame temperature, or adiabatic-combustion temperature. Although this equation has been tested on a group of reheating furnaces of various types and found satisfactory by Eberhardt and Hottel,[17] the furnaces were in most cases equipped with gas burners using premixed fuel and air. In consequence, there was a closer approach to the adiabatic flame temperature than in many furnaces of this type. Heiligenstädt[24a] and Trinks[70a] recommend, for estimating the gas temperature at the hot end of furnaces of this type, the use of a so-called *pyrometric efficiency* by which the adiabatic flame temperature is multiplied. This efficiency, obviously a function of burner design, has been found for many furnaces to be of the order of 0.75 to 0.8. Smith,[69a] in measurements on a model reheating furnace burning unpremixed coke-oven gas, measured a peak gas temperature which was about 75 per cent of the adiabatic-combustion temperature. For further discussion of furnaces of this class, the reader is referred to Heiligenstädt[24b] and Trinks.[70a]

Petroleum Heaters. For cracking-coil and tube-still furnaces Eq. (4-109) has been modified[30a] as follows: (1) by omitting the last term, q becomes heat transferred to oil instead of heat lost by flame; (2) $h_1 A_1'$ has for simplification been assigned an average value equal to $7 A_1 \mathfrak{F}_{1G}$ (the term is unimportant relative to the radiation term). The relation

is then
$$q_c = [\sigma(T_G^4 - T_1^4) + 7(T_G - T_1)]A_1\mathfrak{F}_{1G} \qquad (4\text{-}111)$$

Comments on page 117 concerning the proper value of A_1 and A_R for the case of tubes mounted on a wall apply. In evaluating $\mathfrak{F}_{1G}$, ϵ_G is calculated allowing for gas radiation only; $\epsilon_1 = 0.9$. In applying Eq. (4-111) to data for 19 furnaces, Lobo and Evans[43] found that F_{R1} was represented closely by $A_1/(A_1 + A_R)$ for values of A_R/A_1 from 0 to 1, by A_1/A_R for values of A_R/A_1 from 3 to 6.5. Since Eq. (4-111) involves heat received by oil rather than heat lost by the flame, when it is combined with the energy balance represented by Eq. (4-110) the latter must be modified. The term i is replaced by $i - q_L$, where q_L is the external heat loss from the combustion chamber. A simplified graphical treatment of the solution of Eqs. (4-110) and (4-111) is available (Lobo and Evans[43]) together with a comparison of results with 85 tests on 19 furnaces of widely different types and excess air, burning fuel oil, or refinery gas; the average deviation was 5.3 per cent; excluding tests almost certainly faulty, the average deviation was less than 4 per cent (q_{exp} vs. q_{calc}).

A relation for petroleum heaters, somewhat easier to use than Eq. (4-111) but not so safe, is obtainable by assuming certain terms in Eq. (4-109) constant, combining with Eq. (4-110) to eliminate T_G, and finding an expression different in form but numerically similar over the range of interest. The relation is

$$\eta = \cfrac{1}{1 + \cfrac{\sqrt{i/A_1\mathfrak{F}_{1G}}}{1.4\left[\cfrac{i/w_G(c_p)_m}{100}\right]^{1.6}}} \qquad (4\text{-}112)$$

where η is the ratio of heat transferred to oil to the enthalpy of the entering air and fuel (net value). Other equations applicable in this field are those of Wilson, Lobo, and Hottel,[73] and of Mekler.[48]

Steam-boiler Furnaces. For calculating heat transmission in the radiant sections of steam-boiler furnace settings, many empirical relations are available. One of the simplest is the Orrok-Hudson equation

$$\eta = \frac{1}{1 + (r_{af}\sqrt{w_A}/27)} \qquad (4\text{-}113)$$

in which r_{af} is the weight ratio of air to fuel; w_A is the firing rate expressed as pounds of equivalent good bituminous coal per hour per square foot of exposed tube area (complete circumference if not buried in wall).

Mullikin[51] assumes that the flame emissivity ϵ_G is unity for large pulverized coal-, oil-, or gas-fired furnaces and that compensation for this somewhat too high value comes from use of the same value for gas temperature in Eqs. (4-109) and (4-110). When ϵ_G is unity, the term

$A_1\mathcal{F}_{1G}$ of Eq. (4-109) becomes simply $A_1\epsilon_1$ if A_1 is a plane surface (if tubes on a wall, see discussion on page 117). Mullikin introduces additional multiplying factors on A_1 to allow for resistance of overlying slag or refractory facing on metal-block walls. These are 0.7 for bare-faced metal blocks on tubes and 0.35 for refractory-faced metal blocks on tubes. The simplification suggested is unsafe to use on small furnaces, where ϵ_G is certainly not unity.

Wohlenberg and Mullikin[76] have presented a somewhat more rigorous analysis of the same data.

Wohlenberg[75,76] uses a relation intrinsically similar to Eqs. (4-88) and (4-109) together with a heat balance involving the assumption of equality of flame and exit-gas temperatures; he presents the relation for η in the form of the product of a number of quantities each making separate allowance for one of the variables under control.

Illustration 11. The effect of grayness of refractory walls in the calculation of heat transfer from flame to heat sink in an enclosure has been discussed; the relation presented is rather tedious to use. To see whether or not some simplifying generalization can be reached, it is desirable to examine the effect of refractory grayness in the simple case of a "speckled" furnace (see page 110) in which the ratio of refractory surface to heat sink is 1 and the heat sink area A_1 is gray, with emissivity 0.9. It is also desired to examine the effect of minor departure of refractory surfaces from the "no-flux" condition assumed in most of the relations presented.

The gas will be assumed nonluminous, uniform in temperature at 2140°F, and containing 10 per cent each of CO_2 and H_2O; and the mean beam length will be taken as 10 ft. The heat sink A_1 will be assumed at 1440°F.

By use of the gas-radiation charts the values ϵ_G and ϵ_{2G} are found to be 0.253 and 0.340. From Eq. (4-102), $x = 0.253^2/(0.506 - 0.340) = 0.386$; $\epsilon_G/x = 0.253/0.386 = 0.656$; $\tau_x = 0.344$.

a. First consider the refractory a no-flux surface, but gray, necessitating classification of it as A_2, with Eq. (4-108) applicable. Then $A_1\mathcal{F}_{1G}$, $A_2\mathcal{F}_{2G}$, and $A_1\mathcal{F}_{12}$ must be obtained, from (4-104) and (4-105). Substitution in (4-108) gives the over-all transfer rate from the gas by radiation

$$q_{G,\text{ loss by radiation}} = [\quad]\sigma(T_G^4 - T_1^4)$$

Numerical values are:

ϵ_2	0	0.5	1.0
[ ]	0.280	0.342	0.380

The first value, corresponding to a white refractory, is obtained much more simply than the others, directly as $A_1\mathcal{F}_{1G}$ of Eq. (4-103). It is apparent that increasing the grayness of the refractory increases the net flux to the heat sink materially and that the assumption of white refractory is hardly permissible if it is known to have a moderately high emissivity. Recalling that the assumption of a *gray gas* eliminates any effect of refractory grayness but tends to predict too high a heat transfer, let $\mathcal{F}_{1G}$ be calculated for the gray-gas case, Eqs. (4-88) and (4-86). This is found to give the value for $\mathcal{F}$ of 0.379. The closeness of this to the black-refractory real-gas case is probably accidental, but in one other case tested the same close agreement occurred. If the properties of the refractory are well enough known to assign an emissivity to it, a reasonably good estimate of the heat-exchange factor should be obtained by interpolation between the gray-gas case and the white-refractory real-gas case, both fairly readily calculated.

b. The effects of convection at the refractory and loss through it can be allowed for, following the discussion associated with Eq. (4-106). When the refractory is white, there is no effect on the net radiant flux to sink A_1 (except through the effect on gas temperature, assumed here constant) because the refractory's contribution to that flux is all by reflection and its temperature is immaterial. With black refractory the interchange factor to A_1 changes from its value above for the no-flux case (0.380) to the following, all based on an inside gas-convection coefficient, flame to refractory, of 2:

Wall type	9 in. good insulation $U = 0.186$	9 in. firebrick $U = 1$	3 in. silica brick $U = 1.5$
[ ]	0.382	0.370	0.360
Ratio to no-flux case	1.005	0.972	0.947

These small effects justify the recommendation that refractory surfaces can be treated as surfaces of zero net radiant flux except for special cases, for example, like silica roofs of open-hearth furnaces.

This problem was studied and the results evaluated by Smith.[70]

Illustration 12. Natural gas is being burned for steam generation in a combustion chamber of which the back wall and floor are water-cooled. The gas passes through a tube nest directly above and covering the top of the combustion chamber. The chamber is 16 ft wide by 16 ft long by 20 ft high. The gas, fired at the rate of 130,000 cu ft/hr (measured and fired at 60°F, 30 in. Hg, saturated) with 15 per cent excess air (saturated), has the equivalent composition $C_{1.25}H_{4.5}$ and a net heating value of 1070 Btu/cu ft. The "cold" surfaces of the chamber have an average temperature of 350°F. What is the rate of heat input to the water-cooled walls, floor, and tubes above, exclusive of any convection to the roof tubes as the gas passes up through them? What percentage of the enthalpy of the entering fuel does this represent?

Derived Data. By stoichiometry, the products of combustion contain 8.60 per cent CO_2, 16.36 per cent H_2O, 2.44 per cent O_2, and 72.60 per cent N_2, wet basis; their total is 4910 lb-moles. From specific-heat charts the average molal heat capacity of the products between 2000 and 60°F is 8.25; between 2500 and 60°F it is 8.45.

Assumptions. The external loss from the refractory walls will be assumed equal to the convection to them on the inside. Convection coefficients inside the chamber $= 2.0$. Refractory wall conductance $k_w/x_w = 0.9$. The flame completely fills the chamber. To the emissivity of the flame due to nonluminous gases will be added 0.1 to allow for the luminosity due to cracked hydrocarbons in the flame (this varies enormously with burner type). The emissivity of the "cold" surfaces $= 0.8$, and absorptivity equals emissivity. The mean flame temperature is 100°F above the exit-gas temperature (these approach one another as firing rate increases). The problem will be simplified by allowance for but one source-sink and one no-flux zone.

Solution. Equations (4-109) and (4-110) are to be solved for q_G and T_G. $A_1 = 16 \times 20 + 16 \times 16 \times 2 = 832$ sq ft. (The effective area of the tube nest, for radiation reception, is that of a plane replacing the tubes.) $A_R = 16 \times 20 \times 3 = 960$ sq ft. $A'_1 = 16 \times 20 + 16 \times 16 = 576$ (plane of tube nest is excluded here). Evaluation of $\mathfrak{F}$ involves F_{R1} (or F_{1R}) and ϵ_G. In this problem F_{R1} and F_{1R} are equally tedious to evaluate; F_{R1} shall be chosen. Since the three refractory rectangles do not all "see" the same arrangement of surfaces above them, it is necessary to determine the product $A_R F_{R1}$ for each and to add them, then to divide by the total A_R

Consider first the front wall, 16 by 20 ft, which "sees" three cold faces, one directly opposite, one above, and one below. The fraction of its radiation intercepted by the wall opposite comes from Fig. 4-11, line 2. By the method of Illustration 6, $F = \sqrt{0.196 \times 0.26} = 0.225$, the fraction of the radiation from the front refractory wall intercepted by the rear water-cooled wall. To find the fraction intercepted by the water-cooled floor, reference is made to Fig. 4-10. From that figure, when $Y = 20/16$ and $Z = 16/16$, $F = 0.17$. Since the imaginary top plane replacing the tubes intercepts the same fraction as the water-cooled floor, the total fraction intercepted by cold surfaces is $0.17 \times 2 + 0.225 = 0.565$; and $A_R F_{R1}$ for the front refractory wall is $16 \times 20 \times 0.565 = 181$ sq ft. A similar procedure leads to the value $0.17 \times 2 + 0.213$, or 0.553, as the fraction of the radiation from either refractory side wall which is intercepted by the three cold faces. Then the final value of F_{R1} is

$$F_{R1} = \frac{(16)(20)(0.565) + (16)(20)(0.553)(2)}{(16)(20) + (16)(20)(2)} = 0.56$$

Flame emissivity ϵ_G is next to be evaluated. The gas temperature is so much higher than sink temperature T_1 that $\mathcal{F}_{1G}$ will be based exclusively on gas emissivity.

One must first make a provisional guess as to the value of t_G and adjust later if necessary. Temporarily assume 2500°F. The effective beam length for gas radiation would be 0.6 times one side if the chamber were cubical (see Table 4-2); 0.6 times an average side of 18 ft, or 10.8 ft, may be used (a considerable error in this assumption will not materially affect the result). Then $P_C L = (0.086)(10.8) = 0.93$, and $P_W L = (0.1636)(10.8) = 1.77$. At $t_G = 2500$°F, using Figs. 4-13 to 4-17, one obtains

$$\epsilon_G = 0.115 + 0.191 \times 1.09 - 0.05 = 0.27$$

due to gas radiation. Similarly for double the path length, $\epsilon_{2G} = 0.36$. Adding 0.1 for soot luminosity to the first and $1 - 0.9^2 = 0.19$ to the second, one obtains: $\epsilon_G = 0.37$; $\epsilon_{2G} = 0.55$. From Eq. (4-102), $x = 0.37^2/(0.74 - 0.55) = 0.72$, and $\epsilon_G/x = 0.514$. In allowing for the effect of receiver-surface emissivity, one should note that the radiation-receiving surfaces are of two kinds, plane surfaces in floor and back wall and a nest of tubes in the roof. The former will have an emissivity (or absorptivity) of 0.8. The tube nest will exhibit an effective absorptivity much higher, because any beams penetrating up between tubes will have many chances for absorption after reflection. In the present example a mean value of 0.9 will be used on the whole of A_1. Calculating first the value of $A_1 \mathcal{F}_{1G}$ for the white-refractory real-gas case, one obtains from Eq. (4-103)

$$A_1 \mathcal{F}_{1G} = \frac{0.72}{\dfrac{1}{832}\left(\dfrac{1}{0.9} - 1\right) + \dfrac{1}{0.514\left(832 + \dfrac{960}{1 + \dfrac{0.514}{0.486}\dfrac{1}{0.56}}\right)}} = 399 \text{ sq ft}$$

Calculating for orientation the value of $A_1 \mathcal{F}_{1G}$ for the gray-gas case, one obtains from (4-86) and (4-88)

$$A_1 \mathcal{F}_{1G} = \frac{1}{\dfrac{1}{832}\left(\dfrac{1}{0.9} - 1\right) + \dfrac{1}{0.37\left(832 + \dfrac{960}{1 + \dfrac{0.37}{0.63}\dfrac{1}{0.56}}\right)}} = 447 \text{ sq ft}$$

The preceding illustration indicated that a medium-gray refractory would make the answer come about halfway between these values; or $A_1 \mathcal{F}_{1G} = 423$ sq ft. This amounts to saying that the flame-wall system interchanges 423/832, or 51 per cent, as much heat as a system of parallel black planes close together, having an area A_1 and temperatures T_G and T_1. The over-all refractory-wall coefficient $= U = 1/(\frac{1}{2} + 1/0.9 + \frac{1}{2}) = 0.47$. Substitution into Eq. (4-109) now gives

$$q_{G,\,net} = 423 \times 0.171 \left[\left(\frac{T_G}{100}\right) - \left(\frac{810}{100}\right)^4 \right] + (2)(576)(T_G - 810) + (0.47)(960)(T_G - 520)$$

An energy balance, Eq. (4-110) (with the gas-exit temperature assumed 100°F below T_G), gives

$$q_{G,\,net} = (130,000)(1070) - (4911)(8.45)(T_G - 100 - 520)$$

Solution by trial and error of these two simultaneous equations gives $T_G = 2810$ (2350°F) and $q_G = 48{,}200{,}000$ Btu/hr. If the flame emissivity and heat capacity are adjusted to 2300° instead of 2500° and the solution of equations repeated, one obtains $t_G = 2320°$ and $q_G = 49{,}400{,}000$ Btu/hr, indicating that the final result is insensitive to the temperature at which ϵ_G and Mc_p are evaluated. Not all the heat q_G goes to the water-cooled surfaces; the third term in the heat-transfer equation represents loss through refractory walls. This is $(0.47)(960)(2260)$, or $1{,}000{,}000$ Btu/hr. Then, finally, the heat received by the water-cooled surfaces, exclusive of convection to the first tube row, is $49{,}400{,}000 - 1{,}000{,}000 = 48{,}400{,}000$ Btu/hr, or $48{,}400{,}000/(130{,}000)(1070) = 34.8$ per cent of the enthalpy of the entering fuel. The chief function of this type of calculation, with its arbitrary allowance for luminosity, is to find what value of luminosity makes the calculation fit good furnace data and then use the same value in the design of other furnaces of similar character.

PROBLEMS

1. A large plane, perfectly insulated on one face and maintained at a fixed temperature T_1 on the bare face, which has an emissivity of 0.90, loses 200 Btu/(hr)(sq ft) when exposed to surroundings at absolute zero. A second plane having the same size as the first is also perfectly insulated on one face, but its bare face has an emissivity of 0.45. When the bare face of the second plane is maintained at a fixed temperature T_2 and exposed to surroundings at absolute zero, it loses 100 Btu/(hr)(sq ft).

Let these two planes be brought close together, so that the parallel bare faces are only 1 in. apart, and let the heat supply to each be so adjusted that their respective temperatures remain unchanged. What will be the net heat flux between the planes, expressed in Btu/(hr)(sq ft)?

2. A furnace having walls 28.25 in. thick contains a peephole 7 by 7 in. in cross section. If the temperature of the inner walls of the furnace is 2200°F, what would be the heat loss through the peephole to surroundings at 70°F?

3. An electric furnace of rectangular cross section is to be designed for batch heating of a stock from 70 to 1400°F. The hearth, covered with stock, is 6 by 12 ft in area. The refractory side walls are well insulated. Parallel to the plane of the roof, in a plane several inches below it, is a system of round-rod resistors, each 10 ft long and 0.5 in. in diameter, spaced on 2-in. centers. The plane of the resistors is 4 ft above the top of the stock.

What is the heating time for a 6-ton batch of stock having a mean specific heat of 0.16, when the resistor temperature is maintained at 2000°F?

Notes. Assume that the emissivities of the resistors and stock are 0.6 and 0.9, respectively, and neglect heat losses and heat storage in the walls of the furnace.

4. An annealing furnace 10 ft long has a cross section normal to length, as shown in Fig. 4-28. The firebox a is at a uniform temperature of 2200°F, and the 10- by 6-ft hearth b is covered with stock at a temperature of 1400°F. So far as radiant-heat transfer is concerned, assume that the firebox acts like a uniformly black plane c,

2 ft high over the bridge wall. Neglect the contribution of the combustion products to the radiant-heat interchange in the system, and neglect convection. Assume no external losses from the furnace, and assume that the **refractory surfaces have emissivities of 0.65**.

   a. Calculate the direct interchange of heat by radiation between the firebox and the stock, if the latter is a black surface.
   b. Calculate the total net interchange between the two if the stock is a black surface.
   c. Repeat *b*, assuming that the stock has an emissivity of 0.75.
   d. Calculate the average temperature of the working chamber walls and roof for condition *c*.

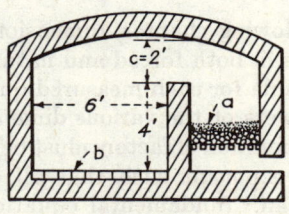

Fig. 4-28.

5. On a clear night, when the effective black-body temperature of space is *minus* 100°F, the air is at 60°F and contains water vapor at a partial pressure equal to that of ice or water at 32°F. A very thin film of water, initially at 60°F, is placed in a very shallow well-insulated pan, placed in a spot sheltered from the wind ($h = 0.46$), with a full view of the sky.

State whether ice will form, and support this with suitable calculations.

6. The convection section of an oil pipe still on which performance data are available is composed of a bank of tubes 24 ft long, 3.5 in. i.d., and 4.0 in. o.d. There are six tubes in each horizontal row; center-to-center spacing of the tubes, arranged on equilateral triangular centers, is 8 in. The minimum free area for gas flow is 53.8 sq ft. Oil enters the bottom row of tubes at 420°F and flows upward through each row in series, leaving the convection section at 730°F. Flue gas at atmospheric pressure from the combustion chamber, flowing transverse to the tubes, enters the top of the bank at 1550°F, with a mass velocity of 810 lb/(hr)(sq ft of minimum free area) and leaves at the bottom of the section at 590°F.

The flue gas contains 7.1 per cent CO_2 (dry basis); the ratio of $H_2O:CO_2$ in the flue gas is 1.38; the molecular weight of the flue gas is 27.5; the mean molal heat capacity between 1550 and 590°F is 7.77.

Calculate the number of rows of tubes required.

Additional Data and Assumptions.

1. External heat losses in the section are negligible.
2. The emissivities of all surfaces are 0.9. Because of the high emissivities, radiation received on a surface by reflection may be ignored.
3. Preliminary calculations indicate that the temperature drop through the oil film and tube wall is 30°F.
4. Average length of radiant beams through the gas will differ for the tube surface and for the refractory surface, but the latter will be assumed equal to the former. Likewise, the convection coefficient of heat transfer from gas to refractory will be assumed the same as from gas to tubes.
5. The effect of the refractory end walls, through which the tubes pass, will be neglected.
6. For convection heat transfer from flue gases, simplified Eq. (10-11*b*) is satisfactory.

CHAPTER 5

DIMENSIONAL ANALYSIS

Abstract. In the following chapter, dimensional analysis is discussed and applied to problems in both forced and natural convection.

The dimensional formula for each measured quantity or factor has the form of products of powers of the various dimensions involved. Hence each dimensionless combination of factors must be a product of the factors, each entering with an exponent such that all dimensions cancel. As pointed out by Bridgman,[1] fundamental equations can be so arranged that the quantities enter the equations through certain combinations that are dimensionless, and the *form of such equations is independent of the size of the units involved* in the various terms in the equation.

Granting that all the factors controlling a physical situation are known, dimensional analysis is a method by which this knowledge may be capitalized and put into a form useful for planning experiments and in interpreting the data obtained.[7] The method is particularly valuable where the mathematical relations are unknown or complex and will indicate the logical grouping of the factors into dimensionless combinations. The latter feature is helpful in interpreting data where two or more factors have been varied in different experiments. Wherever possible, correlations should be supported by data where each factor is varied separately.

It is shown that dimensionless groups may be found by three methods:

1. Algebraic (the classical method of Rayleigh, and the "improved" Π theorem).
2. Rearrangement of differential equations.
3. Consideration of the requirements for similarity (geometric, kinematic, and dynamic).

The theory of models is outlined, and the utility of self-consistent units in dimensionless ratios is illustrated. A table of dimensionless groups is included.

DIMENSIONS

In applying the method, the first step is to write the dimensions of each of the quantities or factors entering the physical situation. The dimensions of a given factor or quantity are determined by its definition in terms of dimensions that are arbitrarily selected as fundamental. The choice of dimensions is largely a matter of convenience, and hence there is nothing absolute about the dimensions of any quantity. In problems in

DIMENSIONAL ANALYSIS

fluid dynamics, the ordinary mechanical quantities mass† M, length L, and time θ may conveniently be selected as fundamental dimensions. One can treat force F as a dimension. The law of Newton, that force is proportional to the product of mass and acceleration, is

$$F = ma/g_c = ML/\theta^2 g_c \tag{5-1}$$

where g_c is a conversion factor or a **dimensional constant,** which has a numerical value dependent on the units chosen. If FML and θ are treated as having separate dimensions, the dimensions of g_c are those of ma/F or $ML/F\theta^2$. Alternatively, by calling g_c dimensionless in Eq. (5-1), the dimension F may be replaced by ML/θ^2 or the dimension M by $F\theta^2/L$; if g_c is thus made dimensionless, its numerical value is unity. All these

TABLE 5-1. SOME MECHANICAL QUANTITIES AND THEIR DIMENSIONS

Symbol	Quantity	Net dimensions $FML\theta$	$ML\theta$	$FL\theta$
D, L	Diameter, length	L	L	L
V	Velocity, average	L/θ	L/θ	L/θ
g	Acceleration due to gravity	L/θ^2	L/θ^2	L/θ^2
m	Mass	M	M	$F\theta^2/L$
F	Force	F	ML/θ^2	F
g_c	Conversion factor, ma/F	$ML/F\theta^2$	None	None
w	Mass rate of flow	M/θ	M/θ	$F\theta/L$
ρ	Density, mass per unit volume	M/L^3	M/L^3	$F\theta^2/L^4$
γ	Specific weight, $\rho g/g_c$	F/L^3	$M/L^2\theta^2$	F/L^3
p	Pressure	F/L^2	$M/L\theta^2$	F/L^2
μ_F	Viscosity, $F\theta/L^2$	$F\theta/L^2$	$M/L\theta$	$F\theta/L^2$
μ	Viscosity, $\mu_F g_c = \mu$	$M/L\theta$	$M/L\theta$	$F\theta/L^2$
N_{Re}	$DV\rho/\mu = DV\rho/\mu_F g_c$	None	None	None

practices are allowable. From these arbitrarily selected fundamental dimensions other quantities may be derived; for example, velocity is expressed as L/θ. Table 5-1 lists certain factors or quantities, and the corresponding dimensions in each of three systems $FML\theta$, $ML\theta$, and $FL\theta$.

Note: Throughout this book the symbol μ designates the product $\mu_F g_c$, hence the dimensionless Reynolds number is written $DV\rho/\mu$.

In making a dimensional analysis, one may use any of the three sets of dimensions listed in Table 5-1 or other appropriate sets, but fundamentally the same result will be obtained regardless of the choice of dimensions. If a system containing more than the minimum necessary number of dimensions, such as the $FML\theta$ system, is selected, the corresponding conversion factor, g_c in this case, should be included. The

† In this book, mass **always** denotes the absolute quantity of matter, regardless of the units such as pounds, grams, or tons in which M might be expressed numerically.

TABLE 5-2. SOME HEAT-TRANSFER QUANTITIES AND THEIR DIMENSIONS

Symbol	Quantity	Net dimensions $ML\theta TFH$	Net dimensions $ML\theta T$
A, S	Area of surface, cross section	L^2	L^2
b	Breadth, perimeter	L	L
c	Specific heat, on mass basis	H/MT	$L^2/\theta^2 T$
d	Prefix indicating differential	None	None
D, r	Diameter, radius	L	L
D_v	Diffusivity, volumetric	L^2/θ	L^2/θ
$-dt/dL$	Temperature gradient	T/L	T/L
E	Elastic modulus	F/L^2	$M/L\theta^2$
F	Force	F	ML/θ^2
g, a	Acceleration	L/θ^2	L/θ^2
g_c	Conversion factor	$ML/F\theta^2$	None
G	Mass velocity, w/S	$M/\theta L^2$	$M/\theta L^2$
h, U	Coefficients of heat transfer	$H/\theta L^2 T$	$M/\theta^3 T$
J	Mechanical equivalent of heat	LF/H	None
k	Thermal conductivity	$H/\theta LT$	$ML/\theta^3 T$
$k/\rho c$	Thermal diffusivity	L^2/θ	L^2/θ
K_H	Kinetic-energy equivalent of heat	$ML^2/\theta^2 H$	None
L	Length	L	L
p	Pressure per unit area	F/L^2	$M/L\theta^2$
Q	Quantity of heat	H	ML^2/θ^2
q	Steady rate of heat flow	H/θ	ML^2/θ^3
R	Thermal resistance	$T\theta/H$	$T\theta^3/ML^2$
t, T	Temperature	T	T
V	Velocity	L/θ	L/θ
w	Mass flow rate	M/θ	M/θ
x_m	Mean free path of molecule	L	L
Greek			
α	Proportionality factor	None	None
β	Coefficient of expansion	$1/T$	$1/T$
Γ	w/b	$M/\theta L$	$M/\theta L$
∂	Prefix, indicating partial derivative	None	None
Δ	Prefix, indicating finite difference	None	None
ϵ	Emissivity	None	None
κ, γ	Specific-heat ratio, c_p/c_v	None	None
λ	Enthalpy change	H/M	L^2/θ^2
μ	Viscosity, $M/L\theta$	$M/L\theta$	$M/L\theta$
μ_F	Viscosity, $F\theta/L^2$	$F\theta/L^2$	$M/L\theta$
Π	A dimensionless group	None	None
ρ	Density, mass per unit volume	M/L^3	M/L^3
σ	Surface tension	F/L	M/θ^2
τ	Tractive force per unit area	F/L^2	$M/L\theta^2$
ϕ, ψ	Prefixes, indicating functions	None	None

introduction of g_c, through choice of the $FML\theta$ system, throws no light on whether or not the acceleration due to gravity affects the problem.

In applying dimensional analysis to problems in heat transfer, it is customary to employ *at least* four dimensions: mass M, length L, time θ, and temperature or temperature difference T. It is allowable to add force F as a dimension, by including the conversion factor g_c in the dimensional analysis; similarly the dimension of heat H may be added by including either the mechanical equivalent of heat J (having dimensions LF/H) or the product Jg_c or K_H (the kinetic-energy equivalent of heat, which has the dimensions $ML^2/\theta^2 H$).

The symbols and dimensions of a number of quantities are shown in Table 5-2. The net dimensions in the $FML\theta TH$ system are readily obtained by inspection, since they correspond to the technical units in which many engineers evaluate these factors. Table 5-2 also shows the net dimensions in the $ML\theta T$ system, in which neither force nor heat is assigned a separate dimension; the dimensions of force are obtained by calling g_c dimensionless, which gives force the dimensions of mass times acceleration, ML/θ^2; the dimensions of heat are obtained by calling Jg_c dimensionless, which gives heat the dimensions of kinetic energy, ML^2/θ^2. The use of either set of dimensions leads to fundamentally the same result; the advantage of using the first set is that the result of the dimensional analysis will automatically contain[2] the necessary conversion factors for use in any system of units, including technical units. The results of the dimensional analysis, using the $ML\theta T$ system, are sound and can be made applicable to all systems of units by later insertion of the proper conversion factors in the result.

THREE METHODS OF FINDING DIMENSIONLESS GROUPS

These may be classified under several headings:

1. Algebraic.
   a. The classical method of Rayleigh.
   b. The "improved" Π theorem.
2. Use of differential equations.
3. Geometric, kinematic, and dynamical similitude, which indicates the significance of dimensionless groups.

1. Algebraic. *a. Rayleigh Algebraic Method.*[10] The following problem illustrates this procedure:

Illustration 1. For heating or cooling a fluid flowing without phase change in turbulent motion through heated or cooled tubes, it is desired to determine the logical grouping of the factors affecting the coefficient of heat transfer h. The heat is conducted through the liquid film; hence k should be a factor. Because the film thickness depends on the mass velocity G of the fluid, tube diameter D, and viscosity μ, these factors should affect h; and since, for a given q, the specific heat affects the bulk temperature of the stream, c_p also should enter. These same factors appear in the basic energy and hydromechanical equations. The $ML\theta TFH$ system is to be used; since both M and H appear in the factors, the dimensional constant $K_H (= ML^2/\theta^2 H)$ is included. Letting ϕ represent any function, one writes

$$h = \phi(G, D, c_p, \mu, k, K_H)$$

For convenience, the foregoing is replaced by an infinite series:

$$h = \alpha G^a D^b c_p^e \mu^f k^i K_H^m + \alpha_1 G^{a_1} D^{b_1} c_p^{e_1} \mu^{f_1} k^{i_1} K_H^{m_1} + \cdots$$

where the dimensionless factor α and the dimensionless exponents may have any values required. Since all terms of the series are alike in form, one may deal with only the first term:

$$h = \alpha G^a D^b c_p^e \mu^f k^i K_H^m \tag{5-2}$$

Substitution of the dimensions gives

$$\frac{H}{\theta L^2 T} = \left(\frac{M}{\theta L^2}\right)^a (L)^b \left(\frac{H}{MT}\right)^e \left(\frac{M}{L\theta}\right)^f \left(\frac{H}{\theta L T}\right)^i \left(\frac{ML^2}{H\theta^2}\right)^m$$

Summation of the exponents of like dimensions gives the condition equations

$$\begin{aligned}
\Sigma H: & \quad 1 = e + i - m \\
\Sigma M: & \quad 0 = a - e + f + m \\
\Sigma L: & \quad -2 = -2a + b - f - i + 2m \\
\Sigma \theta: & \quad -1 = -a - f - i - 2m \\
\Sigma T: & \quad -1 = -e - i
\end{aligned}$$

Simultaneous solution† gives $b = a - 1$, $f = e - a$, $i = 1 - e$, and $m = 0$; substitution in Eq. (5-2) gives

$$\frac{hD}{k} = \alpha \left(\frac{DG}{\mu}\right)^a \left(\frac{c_p \mu}{k}\right)^e \tag{5-2a}$$

It is noted that the exponents arbitrarily retained, a and e, were those appearing on G and c_p, respectively, and hence G and c_p each appeared only once in Eq. (5-2a). If other pairs of exponents had been retained, the corresponding factors would have appeared only once but all the various results would be mutually convertible. For the case under discussion the dimensionless ratios usually retained are DG/μ, $c_p\mu/k$, and either h/c_pG or hD/k, the last equaling the product of the other three. The final result is

$$hD/k = \phi[(DG/\mu), (c_p\mu/k)] \tag{5-2b}$$

in which the unknown functions may be of any kind and must be experimentally determined. After the function has been experimentally determined, a simplified (*dimensional*) equation of limited application can be obtained by substituting the physical properties of some *one* fluid as functions of parameters, such as temperature and pressure. The size of the proportionality constant in such a dimensional equation will depend on the size of units in the factors remaining in the equation.

b. Alternate Algebraic Method, "Improved" Π Theorem.[4,7] Consideration of the physical situation indicates that there are n_f physical factors of importance, all of which may be defined by some minimum number of dimensions, n_d, in the dimension-system employed. The next step is to find the *maximum* number (n_i) of dimensionally incompatible factors, *i.e.*, a set of physical factors which cannot be combined to form

† If it is assumed that h depends on both V and ρ, rather than upon the product $V\rho = G$, Eq. (5-2a) is again obtained.

a dimensionless group. The numerical value of n_i cannot exceed the minimum number n_d of dimensions necessary to describe the factors: $n_i \leq n_d$. The maximum number of independent dimensionless groups n_{Π} will equal the number of physical factors less the maximum number of incompatible factors: $n_{\Pi} = n_f - n_i$.[†] This method is frequently called the "Π theorem," with n_{Π} dimensionless groups designated as $\Pi_1, \Pi_2, \ldots, \Pi_{n-1}, \Pi_n$.

Illustration 2. Consider heat transfer by natural convection between a vertical heated (or cooled) plate of height L at uniform temperature t_w and the surrounding cooler (or warmer) fluid at uniform temperature t_a. For convenience the local heat-transfer coefficient h_{1x} at height x is based on the local difference in temperature between the plate and that of the ambient fluid at t_a outside the thermal boundary layer: $h_{1x} = dq/dA(t_w - t_a)$. Superficial consideration of the physical situation indicates that the following physical factors are involved, but the functional relations are unknown:

$$\phi(h_{1x}, x, k, c_p, \mu, \rho, \beta, \Delta t, g) = 0$$

In this example the $ML\theta T$ system will be employed in which the conversion factors J, g_c, and $Jg_c = K_H$ are considered dimensionless. Table 5-3 shows the net dimensions of the factors involved. Usually more than one set of dimensionally incompatible factors may be found, but the maximum number cannot exceed the number of

TABLE 5-3. DIMENSIONS OF FACTORS, IN $ML\theta T$ SYSTEM

Dimensions	Factors								
	h_{1x}	x	k	c_p	μ	ρ	β	Δt	g
M	1	..	1	...	1	1			
L	...	1	1	2	-1	-3	...	..	1
θ	-3	..	-3	-2	-1	...	...	..	-2
T	-1	..	-1	-1	...	...	-1	1	

dimensions (four in this case). In order that the factor of greatest interest (h_{1x} in this case) appear in only one dimensionless group, it will not be included in the dimensionally incompatible factors. All the necessary dimensions ($ML\theta T$ in this case) must be included in the dimensionally incompatible factors; one of the possible sets contains $x(=L)$, $\rho(=M/L^3)$, $\beta(=1/T)$, and $\mu(=M/L\theta)$. No combination whatever of these dimensionally incompatible factors will yield a dimensionless group. As suggested by Hunsaker and Rightmire,[4] when a large number of factors are involved (as in the present example), the time required to manipulate the equations is reduced considerably by solving for the dimensions in terms of the dimensionally incompatible factors:

$$L = x \qquad M = \rho L^3 = \rho x^3 \qquad T = 1/\beta$$
$$\theta = M/L\mu = \rho x^3/x\mu = \rho x^2/\mu$$

Each of the remaining factors (h, k, c_p, Δt, and g) contains one or more of the dimensions, and the incompatible factors contain all the dimensions. Consequently, each

[†] If one fails to find the maximum number of dimensionally incompatible factors, the procedure will eventually reveal the missing factor.

of the remaining factors *must* produce a dimensionless group when its dimensions are eliminated by use of one or more of the above four equations:

$$\Delta t = T = 1/\beta \qquad \beta \, \Delta t = \Pi_1$$

$$g = \frac{L}{\theta^2} = \frac{x}{\dfrac{\rho^2 x^4}{\mu^2}} = \frac{\mu^2}{\rho^2 x^3} \qquad \frac{x^3 \rho^2 g}{\mu^2} = \Pi_2$$

$$c_p = \frac{L^2}{\theta^2 T} = \frac{x^2}{\dfrac{\rho^2 x^4}{\mu^2} \cdot \dfrac{1}{\beta}} = \frac{\mu^2 \beta}{\rho^2 x^2} \qquad \frac{c_p \rho^2 x^2}{\mu^2 \beta} = \Pi_3$$

$$h = \frac{M}{\theta^3 T} = \frac{\rho x^3}{\dfrac{\rho^3 x^6}{\mu^3} \cdot \dfrac{1}{\beta}} = \frac{\mu^3 \beta}{\rho^2 x^3} \qquad \frac{h_{1x} \rho^2 x^3}{\mu^3 \beta} = \Pi_4$$

$$k = \frac{ML}{\theta^3 T} = \frac{\rho x^3 \cdot x}{\dfrac{\rho^3 x^6}{\mu^3} \cdot \dfrac{1}{\beta}} = \frac{\mu^3 \beta}{\rho^2 x^2} \qquad \frac{k \rho^2 x^2}{\mu^3 \beta} = \Pi_5$$

It is concluded that five dimensionless groups fully replace the nine physical factors. For convenience these may be rearranged as desired:

$\Pi_1 = \beta \, \Delta t$ the thermal expansion modulus

$\Pi_2 \Pi_1 = \dfrac{x^3 \rho^2 g}{\mu^2} \cdot \beta \, \Delta t$ the local Grashof (buoyancy) modulus

$\dfrac{\Pi_3}{\Pi_5} = \dfrac{c_p \mu}{k}$ the Prandtl modulus

$\dfrac{\Pi_4}{\Pi_5} = \dfrac{h_{1x} x}{k}$ the local Nusselt number

$\dfrac{\Pi_3}{\Pi_2} = \dfrac{c_p}{x \beta g}$ or $\dfrac{\Pi_5}{\Pi_2} = \dfrac{k}{x \beta g \mu}$

Each of these five groups is dimensionless in the $ML\theta T$ system. In the technical $ML\theta TFH$ system, the dimensional constants g_c and K_H are not equal to unity:

$$g_c = 32.2 \, \frac{\text{(lb matter)(ft)}}{\text{(pounds force)(sec)}^2}$$

or

$$g_c = 4.17 \times 10^8 \, \frac{\text{(lb matter)(ft)}}{\text{(pounds force)(hr)}^2}$$

$$K_H = J g_c = (778)(32.2) = 25{,}000 \, \frac{\text{(lb matter)(ft)}^2}{\text{(Btu)(sec)}^2}$$

or

$$K_H = (778)(4.17 \times 10^8) = 3.24 \times 10^{11} \, \frac{\text{(lb matter)(ft)}^2}{\text{(Btu)(hr)}^2}$$

These five Π's will now be checked to determine whether or not dimensional constants are required for the technical $ML\theta TFH$ system:

$$\beta \, \Delta t = \left(\frac{1}{T}\right)(T) = \text{dimensionless}$$

$$\frac{x^3 \rho^2 g}{\mu^2} \cdot \beta \, \Delta t = \frac{L^3 \cdot \dfrac{M^2}{L^6} \cdot \dfrac{L}{\theta^2}}{\dfrac{M^2}{L^2 \theta^2}} = \text{dimensionless}$$

$$\frac{c_p \mu}{k} = \frac{\dfrac{H}{MT} \cdot \dfrac{M}{L\theta}}{\dfrac{H}{\theta LT}} = \text{dimensionless}$$

$$\frac{h_{1x}x}{k} = \frac{\dfrac{H}{\theta L^2 T} \cdot L}{\dfrac{H}{\theta L T}} = \text{dimensionless}$$

$$\frac{c_p}{x\beta g} = \frac{\dfrac{H}{MT}}{L \cdot \dfrac{1}{T} \cdot \dfrac{L}{\theta^2}} = \frac{H\theta^2}{ML^2} = \frac{1}{K_H}$$

$$K_H \frac{\Pi_3}{\Pi_2} = \frac{K_H c_p}{x\beta g} = \text{dimensionless}$$

With K_H included in the last group, all five of these groups are dimensionless in both the $ML\theta T$ system and the $ML\theta TFH$ system.† These same groups would apply to vertical cylinders of large diameter; for small diameters, x/D might be required.

If the wall temperature had been linear in height, rather than uniform, another dimensionless group would be required:

$$\pm (dt/dx)_w \cdot x / (t_{w1} - t_{w2})$$

2. Finding Dimensionless Ratios from Differential Equations. If one can write sound and adequate differential equations describing the laws of nature involved in a given problem, these laws can be put into dimensionless form, with proper consideration of the boundary conditions. Although the resulting equations may be complex, dimensionless groups can be obtained and used to correlate experimental data. When problems in dimensional analysis fall within the analytical power of the worker, this is an excellent method of finding dimensionless ratios, since unsound assumptions cannot enter.

Stokes[11] utilized this method for problems in the flow of fluids, and later Nusselt[8,9] applied it to heat-transfer problems.

Illustration 3. Consider the evaporation of a spherical drop of any volatile liquid of any initial radius r_0 at any constant equilibrium temperature t_s, into a current of a gas having any uniform average temperature t_G. The equation for the instantaneous rate of heat transfer is $dQ = hA(t_G - t_s)\,d\theta$. The boundary conditions are as follows: at $\theta = 0, r = r_0$; at $\theta = \theta, r = r$; t_G and t_s are constant. Since dQ equals $-\lambda\rho d(4\pi r^3/3)$ and A equals $4\pi r^2$, one obtains

$$-dr = \frac{h(t_G - t_s)}{\lambda \rho}\,d\theta \tag{5-3}$$

Clearly the instantaneous value of h will depend on the mass velocity G of the gas, the thermal conductivity k_G of the gas, the viscosity μ_G of the gas, and the instantaneous radius r of the drop. For the moment, considering only gases of the same Prandtl

---

† The last group, or some rearranged form, appears in the literature,[5] but its importance has not been determined. Upon further consideration of the physical situation, it is noted that the thermal buoyant force involves the product $g\rho\beta\,\Delta t$, and consequently this product and ρ should have been used, instead of g, ρ, β, and Δt. Had this been done, the number of physical factors would have been seven instead of nine, and only three dimensionless groups would have been obtained: $h_{1x}x/k$, $x^3\rho^2 g \cdot \beta\,\Delta t/\mu^2$, and $c_p\mu/k$.

modulus, $\phi(c_p\mu/k)_G$ may be omitted from the analysis. By inspection, two dimensionless groups are required to describe these five variables: $hr/k_G = \phi(rG/\mu)$. Eliminating h, Eq. (5-3) becomes

$$-dr = \left(\frac{k_G}{r}\right) \phi \left(\frac{rG}{\mu}\right) \left(\frac{t_G - t_s}{\lambda \rho}\right) d\theta \tag{5-3a}$$

Having utilized the basic relations involved, the variables (r and θ) should be made dimensionless. As can be seen from the boundary conditions, the radius can be put into dimensionless form by employing a radius ratio, r/r_0, which equals 1 at the start and 0 at the end if the drop is evaporated to dryness. Equation (5-3a) now becomes

$$-\frac{(r/r_0)d(r/r_0)}{\phi[(r/r_0)(r_0G/\mu)]} = d\left[\frac{k_G(t_G - t_s)\theta}{r_0^2 \lambda \rho}\right] \tag{5-3b}$$

On the right-hand side all terms except θ are constants fixed by the initial conditions and consequently have been placed under the derivative sign. It is seen that three dimensionless parameters are required: the dimensionless radius (r/r_0), the initial-drop Reynolds number (r_0G/μ), and the dimensionless time: $k_G(t_G - t_s)(\theta)/r_0^2\lambda\rho$. Since the mass of the drop is proportional to the cube of the radius, r/r_0 could be replaced by the dimensionless mass (M/M_0). The dimensionless radius could be obtained by high-speed photography, or the dimensionless mass could be obtained by use of a sensitive balance.

3. Geometric, Kinematic, and Dynamic Similarity. Consider steady, well-developed turbulent flows in two round pipes 1 and 2 having uniform diameter D_1 and D_2. In order that these two systems be similar, three criteria must be satisfied:[12]

1. Geometric similarity, which requires that

$$dx_1/dx_2 = dy_1/dy_2 = dz_1/dz_2 = D_1/D_2 \tag{5-4}$$

2. Kinematic similarity, which requires that velocities u and v and velocity gradients be in the same proportions in the two systems at the same reduced position y/D:

$$\frac{u_1}{v_1} = \frac{u_2}{v_2} \quad \text{and} \quad \frac{\partial u_1/\partial x_1}{\partial u_1/\partial y_1} = \frac{\partial u_2/\partial x_2}{\partial u_2/\partial y_2} \tag{5-5}$$

3. Dynamic similarity, which requires that the ratios of the several types of forces in each system be equal. The ratio of the viscous and acceleration forces is

$$F_{v1}/F_{a1} = F_{v2}/F_{a2} \tag{5-6}$$

In each system the viscous force is given by $F_v = dx\,dz\,\mu(\partial u/\partial y)/g_c$, and the acceleration force is given by $F_a = ma/g_c = \rho\,dx\,dy\,dz(\partial u/\partial \theta)/g_c$. Also $\partial u/\partial \theta = (\partial x/\partial \theta)(\partial u/\partial x) = u(\partial u/\partial x)$. Substitution in Eq. (5-6) gives

$$\frac{\rho_1\,dx_1\,dy_1\,dz_1\,u_1(\partial u_1/\partial x_1)/g_c}{dx_1\,dz_1\,\mu_1(\partial u_1/\partial y_1)/g_c} = \frac{\rho_2\,dx_2\,dy_2\,dz_2\,u_2(\partial u_2/\partial x_2)/g_c}{dx_2\,dz_2\,\mu_2(\partial u_2/\partial y_2)/g_c}$$

By use of Eqs. (5-4) and (5-5), this becomes

$$D_1 V_1 \rho_1/\mu_1 = D_2 V_2 \rho_2/\mu_2$$

Thus in the two geometrically and kinematically similar systems, equal Reynolds numbers assure dynamic similarity.

SOME DIMENSIONLESS GROUPS

In the literature certain dimensionless groups have been named as shown in Table 5-4. The physical significance of various dimensionless ratios, and their interrelations, are discussed by Klinkenberg and Mooy.[6]

TABLE 5-4. SOME DIMENSIONLESS GROUPS

Group	Symbol	Name
hr/k	N_{Bi}	Biot number
$\rho V^2/E$	N_{Ca}	Cauchy number
$(h/k)(\mu^2/\rho^2 g)^{1/3}$	N_{Co}	Condensation number
$\lambda L^3 \rho^2 g/k\mu \, \Delta t$	N_{Cv}	Used in equation for condensing vapor
$g_c \, \Delta p/\rho V^2$	N_{Eu}	Euler number
$2g_c \, \Delta p_f \, D/4\rho V^2 L$	f	Fanning friction factor
$k\theta/\rho c_p r^2$	N_{Fo}	Fourier number
V^2/Lg	N_{Fr}	Froude number
$(L^3 \rho^2 g/\mu^2)(\beta \, \Delta t)$	N_{Gr}	Grashof number
wc_p/kL	N_{Gz}	Graetz number
x_m/L	N_{Kn}	Knudsen number
V/V_a	N_{Ma}	Mach number
hL/k	N_{Nu}	Nusselt number
$DV\rho c_p/k$	N_{Pe}	Peclet number
$c_p \mu/k$	N_{Pr}	Prandtl number
$LV\rho/\mu$	N_{Re}	Reynolds number
$\mu/\rho D_v$	N_{Sc}	Schmidt number
$h/c_p V \rho$	N_{St}	Stanton number
$L/\theta V$	N_{Sl}	Strouhal number
$L\rho V^2/\sigma g_c$	N_{We}	Weber number

LIMITATIONS TO RESULTS OBTAINED BY DIMENSIONAL ANALYSIS

As stated at the beginning of this discussion, the result obtained by applying dimensional analysis is limited by the validity and completeness of the assumptions made prior to the analysis. Thus, if one assumes that h depends upon a number of factors that happen to be dimensionally compatible, the analysis will show this to be the case but will throw no light on the validity or completeness of the assumptions. Experiment is the only safe basis for determining the correctness and adequacy of the assumptions. If the assumptions are known to be correct and complete, then the result of dimensional analysis, *i.e.*, the logical grouping of the factors or variables into dimensionless groups, can be accepted without hesitation.

THEORY OF MODELS

Dimensionless ratios, found by any sound procedure, can be used in the theory of models. Thus if n dimensionless ratios fully describe a physical situation, fixing $n - 1$ ratios fixes the remaining one. Resistance to flow of air around airplanes or of liquids through a new type of heat exchanger can be determined by experiments with scale models. The data, determined from experiments with scale models, can be plotted in dimensionless groups and used for full-sized equipment. If the term $g_c \Delta p_F D/L\rho V^2$ is some unknown function of $DV\rho/\mu$, experiments on the scale model should be run at the *same* values of $DV\rho/\mu$ that will be used with the full-sized equipment. If a scale model one-tenth full size is used and the same fluid and same temperatures are used for both the scale model and the full-sized apparatus, since D will be one-tenth as large, in order to make $DV\rho/\mu$ the same in both cases, V should be ten times as large as on the full scale. If this velocity is excessive, a different fluid or a different temperature can be used, giving a sufficiently large value of μ/ρ for the experiments with the model.

Thus, based on dimensional analysis, this "theory of models" has a firm basis[3,10a] and is of great value in planning experiments. Nevertheless, the method demands care in the analysis of all controlling factors and thorough familiarity with the technique of allowing for their variations. For example, the effect of roughness, scale deposits, and wettability should not be overlooked.

Illustration 4. Assume that one wishes to design a full-scale apparatus to heat hydrogen gas at 1 atm flowing at a steady mass velocity of 4210 pounds per hour per square foot, inside vapor-heated horizontal pipes having uniform i.d. of 2.07 inches (0.1727 ft) and heated lengths of 18.5 ft. The Reynolds number DG/μ will be $(0.1727)(4210)/(0.0242) = 30{,}000$. Since the flow is turbulent, consideration of the physical situation indicates that

$$\frac{h_m}{c_p G} = \phi\left(\frac{c_p \mu}{k}, \frac{DG}{\mu}, \frac{L}{D}\right)$$

where the functions relating these four dimensionless groups are conceded to be unknown. Because of the inconveniently high mass rate of flow and the inflammable nature of the gas, it is proposed to plan a small-scale experiment with air (c_p of 0.24 and μ of 0.047) flowing inside a tube of small diameter (0.0358 ft). In order that measurement of h_m in the single small-scale experiment on air be adequate to predict h_m for the large-scale apparatus for hydrogen, what values of L and G should be employed with air, and what will be the ratio of h_m for hydrogen to that for air?

Fortunately, as a close approximation, both diatomic gases have the same value of $c_p\mu/k$. The design value of L/D may be obtained by employing a tube length of $(18.5)(0.0358/0.1727) = 3.86$ ft. The design value of DG/μ may be obtained by employing G of $(30{,}000)(0.047)/(0.0358) = 39{,}400$. Since $c_p\mu/k$, L/D, and DG/μ are the same in both cases, the values of $h_m/c_p G$ must be identical. Consequently

$$\frac{h_m \text{ for hydrogen}}{h_m \text{ for air}} = \frac{c_p G \text{ for hydrogen}}{c_p G \text{ for air}} = \frac{(3.4)(4210)}{(0.24)(39{,}400)} = \frac{1.51}{1}$$

At somewhat smaller Reynolds numbers natural convection would be important, in which case it would also be necessary to make the Grashof number

$$\frac{D^3 \rho^2 g}{\mu} \beta \, \Delta t$$

the same in both cases, which could be done by adjusting the temperatures in the experiment with air.

CONSISTENT UNITS

It should be clear that the numerical value of any **dimensionless** group is independent of the units employed, provided a consistent set is employed. Thus if the ratio of length to diameter of a given pipe is 48:1, the same result will be obtained regardless of whether the units employed are inches per inch, feet per foot, or meters per meter. The same principle applies to the more complex dimensionless groups. Thus, consider the isothermal flow of a liquid, having a density of 56.0 lb/cu ft and a viscosity of 0.0056 lb/(sec)(ft), at an average velocity of 5.0 ft/sec through a pipe having an inside diameter of 0.0833 ft. Evaluating all terms on the basis of the fps system of units, the Reynolds number would be $DV\rho/\mu = (0.0833)(5.0)(56.0)/0.0056 = 4165$ in this set of consistent units. If each of the terms be converted to the cgs system of units, $DV\rho/\mu = (2.54)(152.5)(0.898)/0.0833 = 4165$ in this set of consistent units. **In other words, in any one set of self-consistent units the numerical value of a dimensionless group is the same as in any other set of self-consistent units.** Hence, if convenient, one may use one set of self-consistent units in one dimensionless group and a different set of self-consistent units in a second dimensionless group appearing in the same equation.

Throughout the book, the **preferred** primary units are those shown in Table 5-5.

TABLE 5-5. PREFERRED PRIMARY UNITS (TECHNICAL SYSTEM)

Quantity	Dimensions	Units
Mass	M	Mass pound of commerce, 454 grams
Force	F	Pound force = 444,000 dynes = 32.2 poundals
Length	L	Feet
Time	θ	Hours (**seconds used in flow of fluids**)
Temperature	T	Degrees Fahrenheit for t, degrees Fahrenheit absolute for T
Heat	H	Btu
g_c	$ML/F\theta^2$	$\dfrac{32.2 \text{ pounds matter} \times \text{feet}}{\text{pounds force} \times \text{seconds}^2} = \dfrac{4.17 \times 10^8 \text{ pounds matter} \times \text{feet}}{\text{pounds force} \times \text{hours}^2}$
J	LF/H	778 ft × pounds force per Btu

The corresponding units of various secondary quantities are given in the various nomenclature tables in later chapters and in the Appendix.

If it is desired to employ an English *absolute* system, using force in poundals and mass in pounds matter, all dimensionless equations in the book will give the correct results by taking g_c as unity. Similarly, one can employ a *gravitational* English system, involving force pounds and slugs (pounds matter/32.2); again g_c equals unity. A similar situation exists with respect to the corresponding systems of metric units.

PROBLEMS

1. A gas having a molecular weight of 86 is flowing at a rate of 1.38 kg/min through a tube having an actual i.d. of 1.20 in. At a cross section where the absolute pressure is 30 atm and the temperature is 340°F, the true density is 1.2 times that predicted by the perfect-gas law, and the viscosity is 2.0×10^{-6} force-pound $\times$ seconds per square foot. Calculate the numerical value of the *dimensionless* Reynolds number.

2. It is desired to investigate the drag force on geometrically similar solid objects placed in a free stream of flowing fluid. In the general case, the force F exerted on the body is thought to depend on the characteristic length of the object x, the approach velocity V, the acceleration due to gravity g, and the following properties of the fluid: the density ρ, the viscosity μ, the surface tension σ, and the velocity V_a at which sound waves are propagated in the fluid.

   a. What is the number n_π of algebraically independent dimensionless groups which may be used to correlate the variables?
   b. Determine a set of these n_π dimensionless groups, using the force F in only one group.
   c. Calculate the numerical value of each of these groups, except the one in which the force F appears, for the condition in which $x = 7$ inches, $g = 32.2$ ft/(sec)(sec), $V = 11$ ft/sec, and the fluid is water at 68°F (assume $V_a = 4900$ ft/sec).

3. Below are some data from an early article on the performance of a packed regenerator. In heating runs, the packed spheres were initially at an elevated uniform temperature, and air at room temperature entered at constant mass rate. In cooling runs, the spheres were initially at room temperature, and air at approximately 200°F entered at constant mass rate. In the data given below the relation between the temperature of entering air (t_1), the temperature of exit air (t_2), and initial uniform temperature (t_0) of the bed of spheres is given by the dimensionless term $y = (t_2 - t_0)/(t_1 - t_0)$. The heat capacity of the vertical cylindrical column was small compared with that of the spheres used for packing, and the column was well insulated.

DIMENSIONAL ANALYSIS

Correlate all the values of y given below, plotted as ordinates against some suitable abscissa:

Material	Steel	Steel	Steel	Steel	Lead	Glass
Bed depth, inches.......	5.125	10.25	5.125	2.56	9.55	9.71
Bed diam, in............	4	8	4	2	8	8
Particle diam, in........	0.125	0.25	0.125	0.0625	0.233	0.237
Sp ht..................	0.111	0.111	0.111	0.111	0.0347	0.168
Sp gr..................	7.83	7.83	7.83	7.83	11.34	2.60
k, Btu/(hr)(ft)(deg F)...	26	26	26	26	19.5	0.4
Air superficial mass vel., lb/(sec)(sq ft)........	0.15	0.15	0.30	0.30	0.3	0.3

y	θ, sec	θ, sec	θ, sec	θ, sec	θ, sec	θ, sec
0.1	300	600	140	...	113	127
0.2	360	680	160	...	133	149
0.3	380	740	190	...	147	166
0.4	410	810	205	100	162	182
0.5	450	890	220	...	176	199
0.6	480	970	240	...	191	215
0.7	510	1050	250	...	206	232
0.8	570	1140	300	140	222	250
0.9	650	1300	330	170	248	280

The quantity θ is the elapsed time at which the indicated y was measured.

4. Two metals, A and B, are being considered for use in constructing tuning forks which are to produce vibrations of a given frequency. What would be the ratio of the costs of the metal required for the two tuning forks, expressed as a function of the frequency, the physical properties of the metals, and the cost per unit mass of each metal (assumed independent of the quantity purchased)? The tuning forks are to be geometrically similar.

5. It is agreed to assume that the thermal conductivities of gases and vapors depend on the following factors: c_p, specific heat at constant pressure; c_v, specific heat at constant volume; μ, viscosity; T, absolute temperature; $\alpha = \partial \rho / \rho (\partial p)_T$, coefficient of compressibility; $\beta = \partial \rho / \rho (\partial T)_p$, coefficient of thermal expansion.

List all conclusions which may be drawn from dimensional considerations. If one additional physical variable were to be included, what should it be?

CHAPTER 6

FLOW OF FLUIDS

Abstract. The introduction to this chapter compares the mechanisms of isothermal streamline and turbulent flow in tubes and discusses the intermediate region of transition. The first section covers the basic relations: material balance, force balance, and total energy balance. The second section treats streamline flow in circular and noncircular sections. The third section deals with turbulent flow along plates and inside tubes and presents data on velocity profiles and friction factors for smooth and rough tubes and methods of calculating pressure drop. The fourth section deals with pressure changes owing to sudden contraction and enlargement in cross section, the resistance of fittings, and the pressure drop in packed tubes. The fifth section treats pressure drop for flow outside tubes. Several illustrative problems are included.

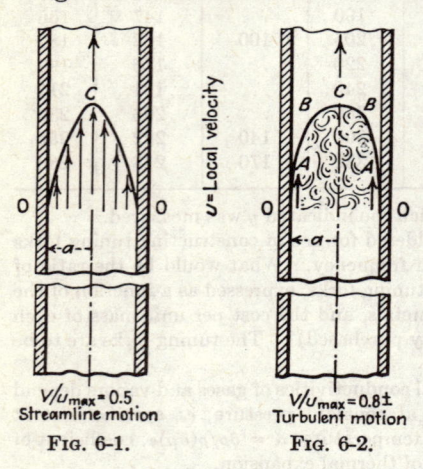

Streamline vs. Turbulent Motion. With steady isothermal flow of a fluid in a long straight pipe, at a sufficient distance (calming length) downstream from the inlet, the velocity pattern becomes fixed, giving what is termed *well-developed* flow. As has long been known from the classical researches of Osborne Reynolds[55] and others, the steady isothermal flow of a fluid through a long straight pipe may occur by one of several mechanisms. When the dimensionless Reynolds number $N_{Re} = DV\rho/\mu$ is sufficiently small, the individual particles of fluid flow in straight lines parallel to the axis of the pipe, without appreciable radial component,† as shown in Fig. 6-1. This type of motion is variously described as *streamline, viscous,* and *laminar*. At sufficiently high values of N_{Re},

$V/u_{max} = 0.5$
Streamline motion
FIG. 6-1.

$V/u_{max} = 0.8\pm$
Turbulent motion
FIG. 6-2.

---

† Of course, the minute molecules are moving rapidly over the almost infinitesimal mean free paths, but the motion of large groups of molecules (parcels of fluid) is substantially axial.

the motion is said to be *turbulent*, because of the presence of innumerable eddies or vortexes present in the major portion of the pipe, as indicated in Fig. 6-2. The shapes of the velocity-distribution curves are different for the two types of motion. Thus, for streamline flow the curve is of parabolic shape and $V/u_{max} = 0.50$, whereas, for turbulent flow, the curve rises more sharply near the wall and is flatter in the central section, and V/u_{max} is approximately 0.8.† Mathematical relations between pressure drop, rate of flow, and other factors are discussed later, and it is shown that the equation for streamline motion is different from that for turbulent motion. As shown below, there is considerable evidence that, in the zone OA of Fig. 6-2, streamline flow exists, in spite of the turbulence in the range BCB. The critical Reynolds number $N_{Re,c}$ is that value of $DV\rho/\mu$ below which streamline motion exists over the entire cross section. Table 6-2 shows values determined by several methods. The lower limit of the critical value of $DV\rho/\mu$ is normally taken as 2100, although streamline flow may persist at much larger values of N_{Re}. At some higher value of N_{Re} the flow is turbulent; between these lower and upper limits lies the zone of **transition** from streamline to turbulent flow. Both these limits are affected by type of entry, initial disturbances, roughness, etc. In the inlet region of the pipe, where the velocity profiles are still changing, the situation is more complex.

I. BASIC RELATIONS

The basic relations for flow of fluids are the material balance, the force balance, and the total energy balance.

Material Balance. For steady mass rate of flow, the law of conservation of matter may be written

$$w = \frac{V_1 S_1}{v_1} = \frac{V_2 S_2}{v_2} \qquad (6\text{-}1)$$

This is sometimes called the *equation of continuity*. The ratio w/S appears so often in heat-transfer calculations that it is designated by a single symbol G and is called *mass velocity*:

$$G = w/S = V/v = V\rho \qquad (6\text{-}1a)$$

It is noted that G is independent of temperature, pressure, and state of aggregation.

Force Balance. Consider steady mass rate of flow of a compressible fluid in well-developed motion through a conduit having uniform cross section S and uniform perimeter b. Consider the volume $S\,dL$ between two planes normal to the stream, a differential distance dL apart, as

† Figure 6-10 shows a plot of V/u_{max} vs. N_{Re}.

TABLE 6-1. NOMENCLATURE FOR FLUID FLOW

In the dimensionless equations in this book, any consistent units may be employed; the units shown in the table are from the technical system: linear dimensions in feet, time in **seconds**,[a] mass of fluid in pounds (1 lb equals 454 gm), forces in pounds force (as for pressure gauges and calculation of work in foot-pounds), heat in Btu, and temperatures in degrees Fahrenheit.

a_p Ratio of surface to volume for a particle, 1/feet
B Ratio of c_p to R'_G, dimensionless; $B = \gamma/(\gamma - 1)$
b Breadth, wetted perimeter, feet
C Constant
c_p Specific heat at constant pressure, Btu/(lb of fluid)(deg F)
c_v Specific heat at constant volume, Btu/(lb of fluid)(deg F)
d Prefix, indicating differential, dimensionless
D Diameter (inside) of conduit, feet
D_H Diameter of helical coil of pipe, feet
D_o Diameter (outside) of tube, feet
D_p Diameter of packing, average, feet = $6/a_p$
D_v Volumetric hydraulic diameter, $4 \times free$ volume/surface area of tubes, feet
E Internal energy, Btu per pound of fluid
f Friction factor, dimensionless, in Fanning equation; f' and f'' are factors in Eqs. (6-12) to (6-13a)
g Gravitational acceleration, local, usually taken as standard value, 32.17 ft/(sec)(sec)
g_c Conversion factor in Newton's law of motion, equals 32.2 ft × pounds matter/(sec)(sec)(pounds force)
G Mass velocity, equals w/S, lb of fluid/(sec) (sq ft of cross section); G equals $V\rho = V/v$
G_{max} Maximum mass velocity through minimum cross section, lb of fluid/(sec)(sq ft)
i Enthalpy, $E + (pv/J)$, Btu per pound of fluid
J Mechanical equivalent of heat, 778 ft × force pounds per Btu
K_c Dimensionless factor in contraction-loss equation
K Dimensionless factor in Eqs. (6-7) and (6-7a)
L Length of straight pipe, feet
L_e Equivalent length, feet; fictive length of straight pipe having same resistance as fittings and valves (Table 6-4)
M Average molecular weight of a gas. For a perfect gas, M equals pounds matter in 359 cu ft at 32°F and normal barometric pressure
m Exponent, dimensionless; also mass having dimension $[M]$, Chap. 5
N Number of rows of tubes over which fluid flows
N_{Re} Reynolds number, dimensionless; $N_{Re} = 4r_h G/\mu$; $N_{Re,c}$ is critical value
n Exponent, dimensionless
p Absolute pressure (intensity of), pounds force per square foot
Q Heat added (net) from surroundings, Btu per pound of fluid
q Volumetric rate of flow, cubic feet per second
R Universal gas constant equals pv/MT, 1544 ft × force pounds/(pound mole of gas)(deg F abs)

[a] In chapters dealing with heat-transfer rates and heat-transfer coefficients, time is expressed in *hours* instead of *seconds*, but in all dimensionless equations in all chapters, the time unit is immaterial provided that the dimensional factors g_c and g are assigned appropriate values (Chap. 5).

TABLE 6-1. NOMENCLATURE FOR FLUID FLOW.—(*Continued*)

R_G Gas constant in *mechanical* units equals pv/T, equals $1544/M$, ft × force pounds per pound mass per degree Fahrenheit absolute

R'_G Gas constant in *heat* units, Btu/(lb of gas)(deg F abs), $R'_G = R_G/J$

r Local radius, feet

r_w Total radius, feet

r_h Hydraulic radius, feet; r_h equals cross section of stream, divided by wetted perimeter $r_h = S/b$

S Cross section normal to fluid flow, square feet

s_L Longitudinal pitch, center to center, feet

s_T Transverse pitch, center to center, feet

s_{min} Minimum pitch of adjacent tubes, feet

t Bulk temperature, degrees Fahrenheit; t' equals $t + 0.25(t_s - t)$

t_f Equals "film" temperature, degrees Fahrenheit; $t_f = t + 0.5(t_s - t)$

t_s Surface temperature, degrees Fahrenheit

T Absolute temperature, degrees Fahrenheit absolute (degrees Fahrenheit + 460); T_m for mean of terminal temperatures

u Local velocity, feet per second; u_{max} at axis

$u^\star$ Friction velocity, feet per second; $u^\star = V\sqrt{f/2}$

u^+ A dimensionless ratio; $u^+ = u/u^\star$

V Average velocity, feet per second, equals cu ft of fluid/(sec)(sq ft of cross section)

v Specific volume, cubic feet per pound of fluid, equals $1/\rho$

w Mass rate of flow, pounds of fluid per second

W_s Work (net) added from surroundings, feet × force pounds per pound of fluid

X Abbreviation for the dimensional term $\sqrt{2g_c D\rho(-dp_f/dL)}$, pounds matter/(sec)(sq ft)

x Ratio of pitch to outside diameter of tubes, dimensionless; x_L is longitudinal value in direction of fluid flow, and x_T is transverse to direction of fluid flow

Y Abbreviation for the dimensionless term fT_f/T_b

y Distance from wall, feet; clearance, feet

y^+ A dimensionless ratio; $y^+ = yu^\star \rho/\mu$

z Vertical distance above any arbitrarily chosen datum plane, feet

z_1, z_2 Smaller and larger dimensions of a rectangle, feet; see Fig. 6-5

Subscripts

b Bulk
f Film
i Isothermal
m Mean
s Surface
w Wall
0, 1, 2, 3   References to stations

Greek

α Dimensionless ratio in Eq. (6-2)
α' Dimensionless ratio in Eq. (6-4)
Γ Mass rate of flow per unit perimeter, lb of fluid/(sec)(ft), equals w/b
γ Ratio of specific heats of gas, $\gamma = c_p/c_v$

TABLE 6-1. NOMENCLATURE FOR FLUID FLOW.—(*Continued*)

- ϵ Fraction voids, dimensionless
- θ Time, seconds
- μ Absolute viscosity of fluid, lb/(sec)(ft), equals $\mu_F g_c$, where μ_F is expressed in pounds force $\times$ seconds per square foot; μ equals $0.000672 \times$ viscosity in centipoises
- μ/ρ Kinematic viscosity, square feet per second
- π 3.1416 . . .
- ρ Density, pounds fluid per cubic foot
- τ Local tractive force at *local* radius r, pounds force per square foot of cylindrical surface
- τ_w Tractive force at wall due to fluid friction, pounds force per square foot of wall
- ϕ Function of aspect ratio (z_1/z_2); see Fig. 6-5

TABLE 6-2. CRITICAL REYNOLDS NUMBER FOR CIRCULAR PIPES

Reference	Method	$N_{Re,c}$
a	Pipe friction	Between 2100 and 2300
b	Color band	Above 2000
c	Stethoscope	Between 1890 and 2130
d	Motion of colloidal particles	Between 2000 and 3700

[a,b] Schiller[56] gives a comprehensive review of data obtained by these two methods.
[c] Bond.[6]
[d] Fage and Townsend.[28]

shown in Fig. 6-3. The forces acting on the upstream and downstream planes are Sp and $S(p + dp)$, and the drag force at the wall is the shear stress τ_w at the wall, multiplied by the wall surface $b\,dL$. The net difference of these forces, overlooking the effect of gravity, is used to accelerate the fluid, according to the Newtonian equation $F = ma/g_c$, where the acceleration a is taken as $dV/\alpha\,d\theta$:

$$-S\,dp - \tau_w b\,dL = \frac{1}{g_c} \frac{S\,dL}{v} \frac{dV}{\alpha\,d\theta} \quad (6\text{-}2)$$

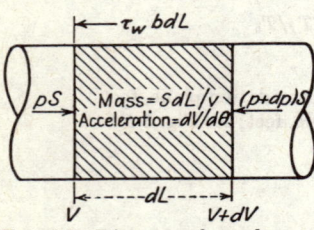

FIG. 6-3. Diagram of steady one-dimensional flow in a horizontal pipe.

where the dimensionless velocity-distribution factor α allows for variations in local velocity with the radius. Upon multiplying by v/S, replacing $dL/d\theta$ by V and the ratio S/b by the hydraulic radius r_h, this equation becomes

$$-\bar{v}\,dp = \frac{V\,dV}{\alpha g_c} + \frac{\tau_w v\,dL}{r_h} \quad (6\text{-}2a)$$

If the pipe had been inclined or vertical, another term would have been present because of the work required to lift the mass of fluid a distance dz

FLOW OF FLUIDS

against the force of gravity, giving

$$-v\,dp = \frac{V\,dV}{\alpha g_c} + \frac{\tau_w v\,dL}{r_h} + dz\frac{g}{g_c} \qquad (6\text{-}2b)$$

This equation also applies to a uniformly tapered duct. A dimensionless friction factor or drag coefficient f is defined by the relation†

$$f \equiv \frac{\tau_w}{\rho V^2/2g_c} \qquad (6\text{-}2c)$$

It is not implied that the shear stress at the wall is at the expense of the term $\rho V^2/2g_c$; this definition is *purely arbitrary*. As shown later, f is a function of a dimensionless Reynolds number $4r_h V\rho/\mu$, and of the roughness of the wall. Eliminating the shear stress at the wall from Eqs. (6-2b) and (6-2c), dividing by v, and replacing V/v by G yields

$$-dp = \frac{G\,dV}{\alpha g_c} + \frac{fG^2 v\,dL}{2g_c r_h} + \frac{dz}{v}\frac{g}{g_c} \qquad (6\text{-}2d)$$

Occasionally this equation is written as

$$-dp = -dp_a - dp_f - dp_z$$

where the subscripts a, f, and z refer to acceleration, friction, and lift. Thus

$$-dp_f = \frac{fG^2 v\,dL}{2g_c r_h} \qquad (6\text{-}2e)$$

which is called the *Fanning equation*.

Consider *incompressible* flow in a pipe of uniform diameter, with uniform pressure across any plane normal to the stream. The force balance on a central cylinder of radius r is $-\pi r^2\,dp = \tau 2\pi r\,dL$, and that for the entire cross section is $-\pi r_w^2\,dp = \tau_w 2\pi r_w\,dL$; the ratio gives

$$\tau/\tau_w = r/r_w \qquad (6\text{-}3)$$

which shows that local shear stress varies linearly from zero at the axis to τ_w at the wall.

Total Energy Balance. For steady mass flow the law of conservation of energy requires that the total energy entering at the first section plus all energy added between sections must equal the total energy leaving at the second section (see Fig. 6-4). At each section the fluid possesses *three* types of mechanical energy. Owing to its elevation of z ft above an arbitrarily selected horizontal datum plane, the fluid possesses a *potential energy* zg/g_c, recoverable by permitting unit mass of fluid to

---

† Some writers multiply the left-hand side of Eq. (6-2c) by factors of 2 or 4 and thus change the definition and consequently the numerical value of the friction factor; throughout this book the symbol f is used consistently as the factor defined in Eqs. (6-2c) and (6-2d).

fall to the datum plane. Owing to its average velocity V, the fluid possesses a *kinetic energy* $V^2/2g_c\alpha$, recoverable by reversibly bringing the stream to rest. Owing to its pressure and volume, the fluid possesses a mechanical energy pv. This *flow work* is the energy required to maintain the flow; the force acting is the intensity of pressure p times the cross section S, that is, pS, the distance is the specific volume v divided by S, namely, v/S, and the work (force times distance) is consequently $(pS)(v/S)$ or pv, foot-pounds per pound of fluid.

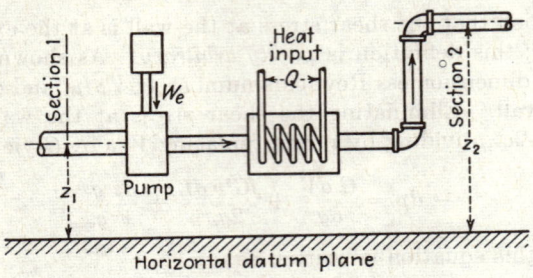

FIG. 6-4.

The fluid also possesses *internal energy* (thermal, chemical, and otherwise) E, Btu per pound of fluid. The net heat added between sections is designated by Q, Btu per pound of fluid, and the energy added by all other means, such as a pump between sections, is designated by W_e, foot-pounds per pound of fluid. The mechanical equivalent of heat is designated by J, equal to 778 ft-pounds/Btu. Based on a time interval such that unit masses of fluid enter at section 1 and leave at section 2, and expressing all terms in Btu per pound of fluid, the total energy balance is

$$\frac{1}{J}\left(z_1\frac{g}{g_c} + \frac{V_1^2}{2\alpha'g_c} + p_1v_1\right) + E_1 + \frac{W_e}{J} + Q$$
$$= \frac{1}{J}\left(z_2\frac{g}{g_c} + \frac{V_2^2}{2\alpha'g_c} + p_2v_2\right) + E_2 \quad (6\text{-}4)$$

The *increase* in internal energy, $E_2 - E_1$, is in general obtained experimentally, by measuring the other terms in Eq. (6-4), employing a calorimeter with no pump between sections. It is sometimes convenient to combine two or more terms of Eq. (6-4). For example, the sum of the internal energy and the flow work is designated as the enthalpy, i:

$$E + pv/J = i \quad \text{and} \quad dE + d(pv)/J = di$$

The total energy balance for a differential length, containing no pump, is then

$$dQ = di + \frac{1}{J}\left(\frac{g}{g_c}dz + \frac{V\,dV}{\alpha'g_c}\right) \quad (6\text{-}4a)$$

where di is the differential change in enthalpy. Numerical values of the enthalpy difference above a stated "base" temperature are given by tables for various fluids, such as water, air, and various refrigerants.

Strictly, the kinetic-energy term $V^2/2g_c\alpha$ should be obtained by evaluating the integral

$$\frac{V^2}{2\alpha' g_c} = \frac{1}{w} \int \frac{(u\rho 2\pi r\, dr)u^2}{2g_c} \tag{6-4b}$$

where u is the local velocity at a given radius. Thus, for isothermal streamline flow in a long straight circular pipe, where the velocity distribution is parabolic $[u/V = 2(r_w^2 - r^2)/r_w^2]$, the kinetic term can be shown to be V^2/g_c; that is, α equals $\frac{1}{2}$. For flow in the turbulent range, the velocity gradient is less steep than for streamline flow, and as an approximation α is taken as unity, which ordinarily introduces little error. Where kinetic-energy terms are important relative to other terms in Eq. (6-4a) and the flow is streamline at one section and turbulent at the other, it is worth while to compute kinetic energy from Eq. (6-4b), employing empirical equations for the velocity field.

Using the data of a number of observers for isothermal flow of liquid mercury, air, or water, Isakoff and Drew[33a] correlated the ratio V/u_{max} in terms of N_{Re} as shown by the upper curves of Fig. 6-10; the lower curve shows the exponent n in the power law $u/u_{max} = (y/r_w)^n$, which was preferred to the conventional graph of u^+ vs. y^+. If the power law applied at all values of y/r_w, then V/u_{max} would be equal to $2/(2+n)(1+n)$.

Since for perfect gases $pv = RT/M$, E equals $c_v T$, and $c_p - c_v$ equals R/JM, or $1.985/M$, the enthalpy change for perfect gases becomes

$$di = c_p\, dT \tag{6-4c}$$

Illustration 1. Air is flowing at constant mass rate through a horizontally arranged air heater and is heated from a temperature of 70°F at section 1 to 170°F at section 2, the cross section being equal at both sections. The absolute pressure is normal barometric at the first section and less by 1 inch of water at the second section. The average air velocity at section 1 is 20 ft/sec. Calculate the net heat input Q, taking $(c_p)_m$ as 0.24.

Solution. The absolute pressure at the second section $29.92 - (1/13.6) = 29.85$ in. mercury, and the velocity, assuming the perfect-gas law, is

$$V_2 = 20 \left(\frac{460 + 170}{460 + 70}\right)\left(\frac{29.92}{29.85}\right) = 23.8 \text{ ft/sec}$$

Expressing all terms in Btu per pound of air [Eq. (6-4a)],

$$Q = (0.24)(630 - 530) + \frac{(23.8)^2 - (20)^2}{(64.3)(778)} = 24 + 0.0033$$

Illustration 2. If in the preceding illustration the heater had been vertical, the air leaving the apparatus at a level 10 ft above the air inlet, would the net heat input have been materially different from that for the horizontal heater?

Solution. Per pound of air, the term $(z_2 - z_1)/J = 10/778 = 0.0129$ Btu, which is negligible compared with the 24 Btu necessary to increase the enthalpy.

For the conditions of Illustrations 1 and 2, the change in kinetic energies of the air amounted to only 0.014 per cent of the increase in enthalpy, and the change in potential energies was only 0.054 per cent of the change in enthalpy. Under ordinary conditions, such as those of Illustrations 1 and 2, the changes in both potential and kinetic energies are usually neglected, giving what might be termed the "usual heat balance": $Q = i_2 - i_1$. Under conditions involving the heating or cooling of gases flowing at very high velocities, the change in kinetic energies may be important compared with changes in enthalpy (Chap. 12).

II. STREAMLINE FLOW

Many, but not all,[†] fluids follow the Newtonian definition of viscosity:

$$g_c \tau \equiv \mu \frac{du}{dy} \tag{6-5}$$

Tubes and Pipes. Consider steady *incompressible* flow of a Newtonian fluid in a horizontal tube or pipe of uniform diameter, and neglect gravitational effects. Elimination of the *local* shear stress τ from Eqs. (6-3) and (6-5) and replacement of dy by $-dr$ gives $du = -g_c \tau_w r\, dr/\mu r_w$. For isothermal flow μ is constant. *Assuming* no slip at the wall ($u = 0$ at $r = r_w$), integration from r_w to r gives

$$u = \frac{g_c \tau_w}{\mu r_w} \frac{r_w^2 - r^2}{2} \tag{6-5a}$$

The volumetric rate of flow dq in a small annulus is $(u)(2\pi r\, dr)$. Elimination of u by means of Eq. (6-5a), and integration from the wall to the axis, gives

$$q = \frac{\pi g_c \tau_w}{\mu r_w} \frac{r_w^4}{4} = \pi r_w^2 V \tag{6-5b}$$

Comparison of Eqs. (6-5a) and (6-5b) shows that the velocity distribution is parabolic,

$$u/V = 2[1 - (r^2/r_w^2)] \tag{6-5c}$$

and that the maximum velocity (at the axis) is twice the average velocity. Upon eliminating q by $\pi r_w^2 V$, τ_w by Eq. (6-2a) (with dV of zero), replacing $r_h (= S/b)$ by $D/4$ and r_w by $D/2$, one obtains the **Hagen-Poiseuille** law of isothermal streamline motion in well-developed incompressible flow in round tubes or pipes:

$$\frac{-dp_f}{dL} = \frac{32 \mu V}{g_c D^2} = \frac{40.75 \mu w}{g_c D^4 \rho} \tag{6-5d}$$

This equation has been verified for isothermal flow of hydrocarbon oils in sizes ranging from capillary tubes up to 12-inch steel pipe.[66] This equation can be applied to compressible flow by utilizing Eq. (6-2a), noting

[†] Alves et al.[1] describe the various types of non-Newtonian suspensions and solutions and give a method for predicting the pressure drop in pipes.

FLOW OF FLUIDS

that α is 0.5 for laminar flow. The equation does not apply to gases at very low pressures, where one may encounter either slip flow or free molecule flow, instead of continuum flow (see Chap. 12). The effect of pressure on the viscosity of liquids and the consequent modification of Eq. (6-5d) are discussed by Hersey and Snyder.[31]

Instead of using Eq. (6-5d) it is generally customary to equate the pressure-drop terms of Eqs. (6-5d) and (6-2e), thus determining the value of f in Eq. (6-2e) to be used for streamline flow in circular tubes or pipes.

$$f = 16/(DG/\mu) = 16/N_{Re} \tag{6-5e}$$

which is plotted on Fig. 6-11, discussed later.

Nonisothermal Flow. Few data are available on the friction factors for nonisothermal laminar flow. Data[†] for heating or cooling petroleum oils inside horizontal pipes may be correlated[39,46] with those for isothermal flow by employing Eq. (6-5d) with μ evaluated at a special temperature: $t' = t + (t_s - t)/4$.

Upon heating a petroleum oil flowing upward inside a steam-heated pipe, it was found[46a] that the local pressure gradient owing to friction decreased as the oil flowed upward; the over-all pressure drops were two to five times those for oil entering at the temperature of the steam-heated wall. A tentative procedure was given for predicting the pressure drop, utilizing the following type of equation for viscosity: $\mu = c/t^n$.

Noncircular Cross Sections. For *isothermal streamline* flow inside straight ducts of various shapes, theoretical equations have been derived by many writers.[‡] The pressure drop for isothermal streamline flow through a *rectangular* duct of sides z_1 and z_2 is given by the theoretical equation

$$\frac{-dp_f}{dL} = \frac{4\mu V}{z_1 z_2 g_c \phi} = \frac{4\mu G}{z_1 z_2 g_c \rho \phi} = \frac{4\mu w}{z_1^2 z_2^2 \rho g_c \phi} \tag{6-6}$$

where ϕ is a function of the aspect ratio z_2/z_1, given by Fig. 6-5. For broad parallel plates, having a clearance z_1,

$$-dp_f/dL = 12\mu G/z_1^2 g_c \rho \tag{6-6a}$$

The pressure drop through an *annular space* of inner diameter D_1 and outer diameter D_2 is given by the equation

$$\frac{-dp_f}{dL} = \frac{32\mu G}{\rho g_c \left(D_2^2 + D_1^2 - \dfrac{D_2^2 - D_1^2}{2.3 \log_{10}(D_2/D_1)} \right)} \tag{6-6b}$$

which can be approximated by Eq. (6-6a).

The equations for streamline flow in straight ducts of various shapes do not coincide with that for straight circular pipes, even when expressed in

[†] References 9, 38, 39, and 65.
[‡] Boussinesq,[7] Graetz,[27] Greenhill,[28] and Lamb.[44]

terms of the hydraulic radius. Data[†] on streamline flow in rectangular sections (z_1/z_2 from 0.05 to 1.0) and annular spaces agree[16] with the theoretical equations. The available data for flow in long rectangular ducts and annular spaces indicate that the transition from streamline to turbulent flow starts at a Reynolds number $4r_hG/\mu$ of about the same values as for straight circular pipes, that is, 2100, where r_h, the hydraulic radius, is the cross-sectional area divided by the wetted perimeter. Flow in short ducts is treated elsewhere.[52]

Dukler and Bergelin photographed films of water flowing under the influence of gravity down the vertical wall of a broad plate. Wave motion in the outer layer was evident at Reynolds numbers as low as 760.

Effect of Curvature of Pipe upon Friction. For the isothermal flow of fluids in curved pipes, the friction loss may be considerably more than in

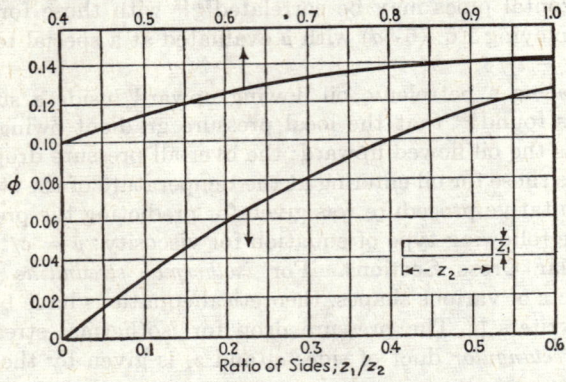

Fig. 6-5. Values of ϕ as a function of the aspect ratio for streamline flow in rectangular ducts.

straight pipes, conditions being otherwise the same. For isothermal *streamline* flow in curved pipes, sufficient data are available[‡] to show the mechanism of the flow and to allow calculation of the pressure drop. Color-band experiments showed that the particles of fluid follow tortuous paths, traveling from the center of the pipe toward the outside wall and then crossing back toward the inside wall. Near the wall, therefore, the particles of fluid travel faster in a curved section of pipe than in a straight section because of their spiral path. Because of this increased velocity near the wall and the longer path traveled per foot of pipe length by the individual fluid particles, a higher friction drop in curved pipes is to be expected. Color-band and pressure-drop experiments also show that streamline flow can exist at much higher Reynolds number in curved pipe

[†] Cornish,[10] Davies and White,[12] Drew,[16] Lea,[45] Nikuradse,[50] and Trahey and Smith.[62] According to Lea, Eq. (6-6b) gives pressure drops much higher than the experimental values when D_2/D_1 is very great.

[‡] Eustice,[22] White,[64] and Taylor.[61]

FLOW OF FLUIDS

than in straight pipe. Table 6-3, prepared by Drew,[15] shows some experimental values of the critical Reynolds number for flow in pipe coils.

TABLE 6-3. EFFECT OF CURVATURE ON CRITICAL VALUES OF $N_{Re,c}$

Diameter of pipe to diameter of coil, D/D_H	$N_{Re,c}$ White (friction)	$N_{Re,c}$ Taylor (color band)
1/15.15	7590	....
1/18.7	....	7100
1/31.9	....	6350
1/50	6020	....
1/2050	2270	....

To evaluate the friction factor for fluids flowing in *streamline motion* through pipe coils, use Fig. 6-6, prepared by Drew.[15] If the friction factor for curved pipes, so calculated, is less than 0.009, the flow is turbulent, not viscous, and the foregoing rule fails.

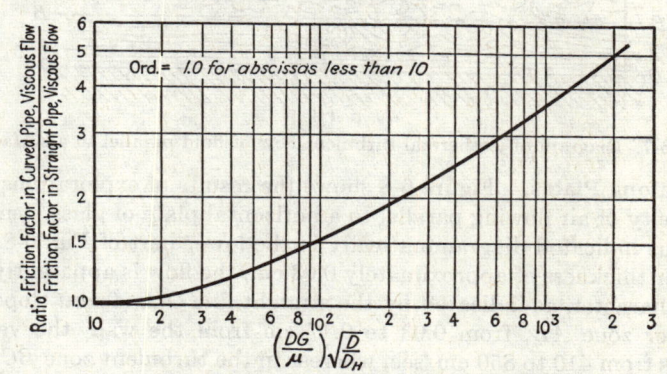

FIG. 6-6. Drew chart for streamline flow in helical coils.

Miscellaneous Shapes. Pressure drop for flow across banks of tubes is treated in Sec. VI, and pressure drop for flow through packed beds is discussed in Chap. 11.

III. TURBULENT FLOW

For the usual case, in which the main stream flows in well-developed turbulent motion, the work of a number of investigators indicates that the flow conditions resemble those shown diagrammatically in Fig. 6-7. At the wall OO the velocity is zero; and very near the wall OA the fluid is moving in laminar motion. In the buffer layer AB the motion may be either streamline or turbulent at a given instant. In the zone BC the motion is always turbulent.

Couch and Herrstrom[11] introduced a color band at the axis of a glass pipe through which water was flowing in turbulent motion and found that the color was quickly dissipated by the turbulence, as was the case in the classical color-band experiments of Reynolds. A second color band, laid down simultaneously near the wall, was not disturbed, thus indicating either that near the wall the flow was streamline in character or that if eddies existed, they were too feeble to dissipate the color band. This experiment showed substantial turbulence at the axis and no noticeable turbulence near the wall.

It is generally believed that the thickness AB of the transition, or "buffer," layer varies with time, because of the more or less periodic formation of vortexes, but that, nevertheless, an *average* thickness may be assigned to the buffer zone.

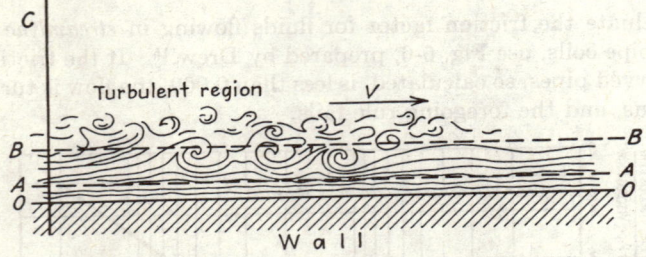

Fig. 6-7. Diagram of isothermal turbulent flow of fluid parallel to a surface.

Flow along Plates. Figure 6-8 shows the results of explorations[63,†] of the velocity of air flowing parallel to a horizontal plate of glass, using the apparatus indicated diagrammatically in the lower part of Fig. 6-8. Up to a layer thickness of approximately 0.03 cm, the flow is apparently laminar in character, as indicated by the straight line $O'A$ of *unit* slope. In the buffer zone AB, from 0.03 to 0.17 cm from the wall, the velocity increases from 410 to 850 cm/sec, whereas in the turbulent zone BC, from 0.17 to 1.7 cm from the wall, the velocity increases from 850 to 1200 cm/sec. The turbulent zone BC is represented by the straight line having a slope of $1/7$; in the buffer zone AB the slope of the curve gradually changes from unity to $1/7$. Similar results for the isothermal flow of air over smooth plates have been obtained by a number of workers.‡ Although air velocities were obtained by pitot tubes in some cases and by hot-wire anemometers in others, the results support those of Fig. 6-8. Data are available to show that wide variations in $DV\rho/\mu$ cause some variation in the slope of the curves of u vs. y in the turbulent region.

† Air velocities were measured by a calibrated hot-wire anemometer, made of wire 0.005 cm in diameter. For locations very close to the wall, the anemometer readings were corrected for abnormal behavior when near solids.

‡ Jürges,[35] Hansen,[30] Dryden and Kuethe,[19] and Elias.[20]

FLOW OF FLUIDS 153

Effect of Air Velocity. Figure 6-8 also shows velocity explorations for five different air velocities. It is seen that increasing the velocity of the main stream of air from 13.1 to 78.7 ft/sec decreases the thickness of the laminar sublayer (zone where the slope is unity) from 0.087 to 0.018 cm.

Effect of Roughness. Velocity explorations[63] in a stream of air flowing over an artifically roughened surface (Fromm waffle plate) gave curves similar to those for the smooth plate (Fig. 6-8), but in the turbulent zone BC the slope $d \ln u / d \ln y$ was 0.25 for the rough plate, compared with 0.14 for the smooth plate. Thus, the marked roughness of the plate affected the velocity gradient in the turbulent zone.

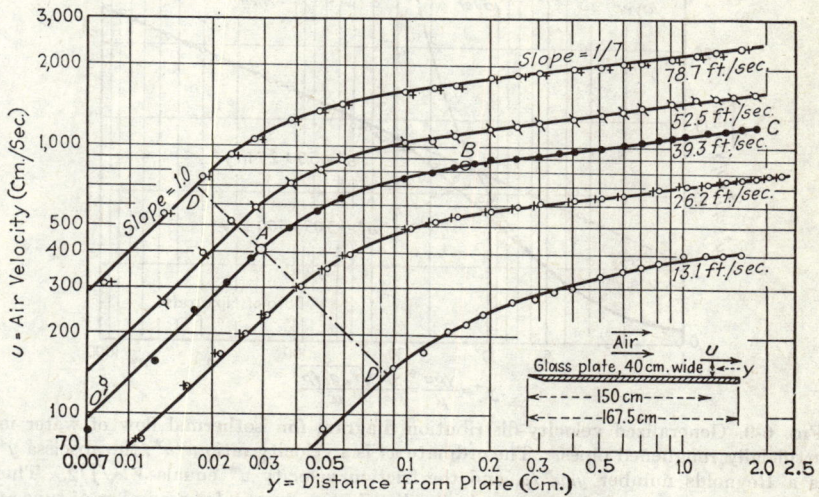

FIG. 6-8. Effect of air velocity upon thickness of laminar sublayer, 150 cm from upstream edge. (*Data of Van der Hegge Zijnen.*[63])

Flow inside Pipes. Available graphs[59] of u/u_{max} vs. y/r_w show flatter curves for smooth pipes than for artificially roughened pipes. In comparing his results for rough and smooth pipes, Stanton[59] made the important discovery, later confirmed and extended by Nikuradse,[51] that the velocity-distribution curves could be correlated by plotting the "velocity-deficiency ratio" $(u_{max} - u)/u^\star$ vs. the position ratio r/r_w. The term $u^\star$ is defined as $\sqrt{\tau_w g_c/\rho}$, and since it involves the shear stress at the wall and has the dimensions of velocity, it is called the *friction velocity*. In obtaining equations for velocity-distribution curves in the turbulent core, Prandtl[54] used the form

$$\frac{u_{max} - u}{u^\star} = \frac{1}{K} \ln \frac{r_w}{y} \qquad (6\text{-}7)$$

$$\frac{du}{dy} = \frac{u^\star}{Ky} \qquad (6\text{-}7a)$$

Elimination of $u^\star/K$ and integration gives[36] the equation of the velocity field in the turbulent core, for *both* smooth and artificially roughened pipes:

$$\frac{u}{u^\star} = 5.5 + 2.5 \ln \frac{yu^\star\rho}{\mu} \tag{6-7b}$$

Figure 6-9[36,5] shows velocity-distribution data of Nikuradse[51],† for water leaving pipes having diameters ranging from 0.394 to 3.94 inches, with various degrees of wall roughness. The ratio $u/u^\star$ is designated as

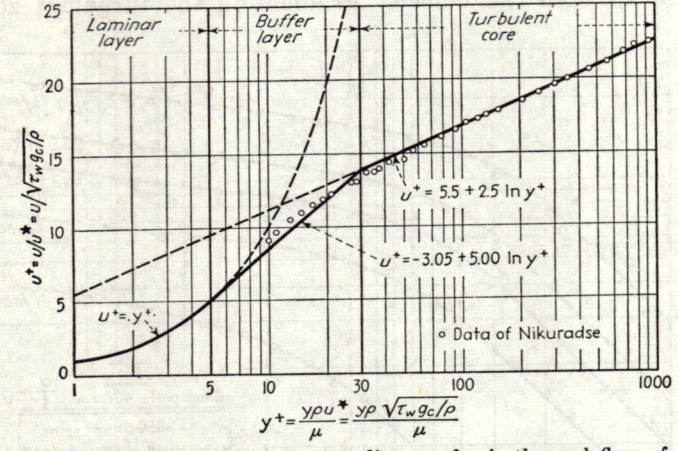

Fig. 6-9. Generalized velocity-distribution diagram for isothermal flow of water in artificially roughened pipes. The ordinate u^+ is a velocity ratio $u/u^\star$, the abscissa y^+ is a Reynolds number $y\rho u^\star/\mu$, and the friction velocity $u^\star$ equals $V\sqrt{f/2}$. This method of plotting correlates the velocity distribution u vs. y for a number of runs at various average velocities in pipes having various diameters and various degrees of roughness at the wall. (*von Kármán.*[36])

u^+, and the ratio $yu^\star\rho/\mu$ is designated by y^+. As for flow of air over flat plates (Fig. 6-8), the data fall into three zones:

A *laminar sublayer* extending from the wall to y^+ of 5, where $u^+ = y^+$.

A *buffer layer* extending from y^+ of 5 to y^+ of 30, for which von Kármán[36] writes $u^+ = -3.05 + 5.0 \ln y^+$.

A *turbulent core*, for which various writers give $u^+ = 5.5 + 2.5 \ln y^+$.

This "universal" velocity-distribution diagram (Fig. 6-9) has been used successfully[3] in predicting shear stress τ_w at the wall, despite the fact

† As pointed out by Miller,[48] these equations were fitted to the data by Nikuradse after deducting a correction of 7 from the experimental value of y^+; if this had not been done, a single equation similar to Eq. (6-7b) would have fitted the uncorrected data. Also Miller pointed out that the velocities were measured at a short distance beyond the end of the pipe. Nevertheless, subsequent data of Deissler[14] on air agree closely with the corrected data of Nikuradse.

FLOW OF FLUIDS 155

that Eq. (6-7a) calls for a finite velocity gradient at the axis where du/dy is actually zero. This diagram has also been used in Chap. 9 in deriving analogies between the transfer of heat and momentum.

In ordinary work, *average* velocities V over the entire pipe are used, rather than local velocities. Figure 6-10 shows a logarithmic graph[60] of the ratio of average to maximum velocity, V/u_{max}, vs. Reynolds numbers $DV\rho/\mu$ and $Du_{max}\rho/\mu$. Up to $DV\rho/\mu$ of 2100, V/u_{max} is 0.5 as predicted by theory for isothermal streamline flow in a long straight pipe; in the range from 2100 to 3000, V/u_{max} rises sharply from 0.5 to 0.726 and thereafter increases more slowly as the Reynolds number is increased.

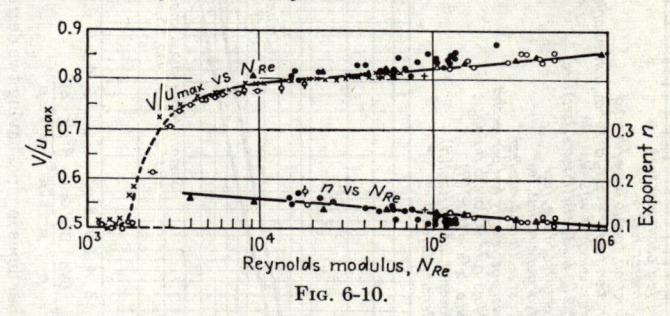

Fig. 6-10.

Friction Factors for Smooth Pipes. The friction factor f of Eq. (6-2c) is based on test data and, as shown by Stanton and Pannell,[60] Blasius, and others,[17] is found to depend upon two dimensionless groups: the Reynolds number $N_{Re} = DV\rho/\mu = DG/\mu$ and "relative roughness," *i.e.*, the actual average roughness divided by the diameter. Based on the data of a number of investigators on the flow of air, water, and oils in clean, smooth pipes of brass, copper, lead, and glass, with diameters ranging from 0.5 to 5.0 in., it is found that the values of f, when plotted vs. the Reynolds number, lie in a relatively narrow band, giving the recommended curve[17,18] for smooth pipes, Curve AB of Fig. 6-11. Based on the correlation of 1380 experimental points, lying in the range of $N_{Re} = 3000$ to 3,000,000, the empirical equation of Koo[41] for long smooth pipes is

$$f = 0.00140 + 0.125/(N_{Re})^{0.32} \qquad (6\text{-}8)$$

Over the limited range of N_{Re} from 5000 to 200,000

$$f = 0.046/(N_{Re})^{0.2} \qquad (6\text{-}8a)$$

von Kármán gives the following equation

$$\frac{1}{\sqrt{f}} = 4.0 \log_{10}(N_{Re}\sqrt{f}) - 0.40 \qquad (6\text{-}8b)$$

which fits the data for friction in smooth pipes and is recommended for extrapolation to high values of N_{Re}. This equation can be rearranged to

156 HEAT TRANSMISSION

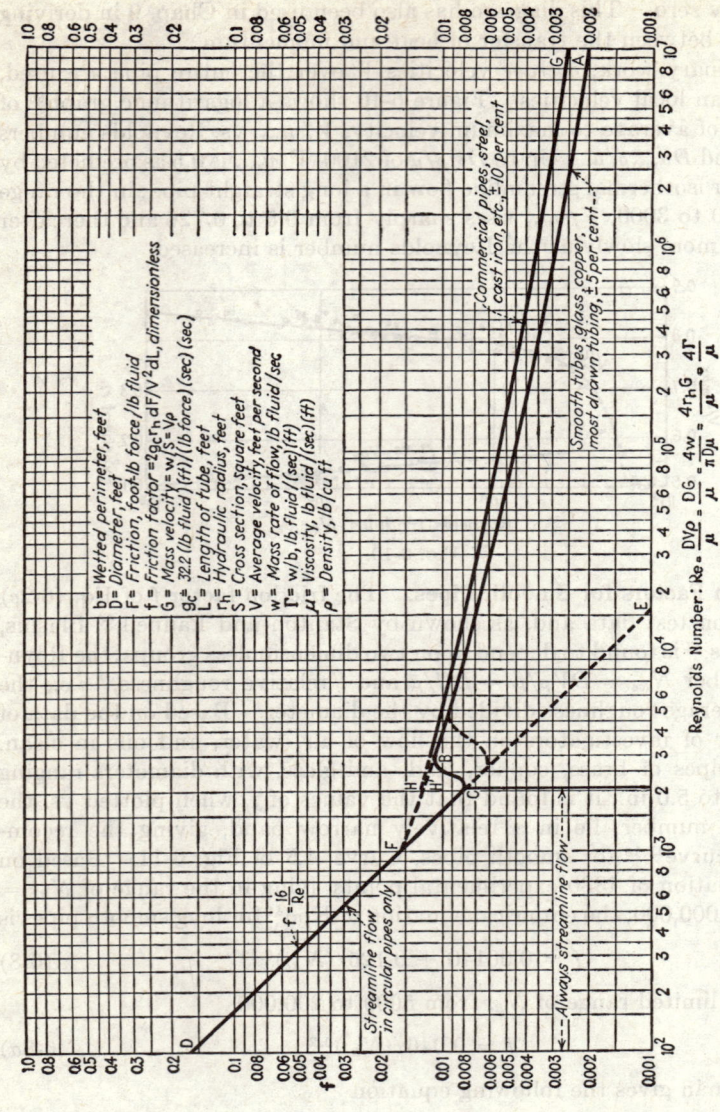

FIG. 6-11. Friction factors for isothermal flow of Newtonian fluids in pipes having two degrees of roughness. For very large ratios of wall roughness to pipe diameter, use f of 0.020 for turbulent flow.

give
$$\frac{G}{X} = 2 \log \frac{DX}{2.52\mu} \qquad (6\text{-}8c)$$

which is useful in calculating G for a given pressure gradient and diameter; $X \equiv \sqrt{2g_c D\rho(-dp_f/dL)}$.

Liquids. The friction factor depends on DG/μ; if the temperature varies, μ and consequently f will vary; f_m designates the length-mean value $f_m = (1/L)\int_0^L f\,dL$. Friction factors for heating or cooling petroleum oils flowing in turbulent motion in tubes may be correlated either[46] by use of Fig. 6-11, with μ evaluated at the "film temperature," $t_f = (t_s + t)/2$, or[58] by considering the ordinate of Fig. 6-11 to be the product of the nonisothermal f and the term $(\mu/\mu_s)^{0.14}$.

Gases. Humble et al.[33] measured friction factors for turbulent flow of air inside horizontal heated tubes, having ratios of T_s/T_b as high as 2.5:1. As Δt was increased, the values of f fell progressively lower than conventional values of f without heat transfer. All data could be correlated by the equation

$$\frac{1}{\sqrt{Y}} = 4 \log X' \sqrt{Y} - 0.4 \qquad (6\text{-}8d)$$

where

$$Y = f\frac{T_f}{T_b} \qquad \text{and} \qquad X' = \frac{DG}{\mu_b}\frac{(\mu/\rho)_b}{(\mu/\rho)_f}$$

where $T_f \equiv (T_s + T_b)/2$. It is noted that Eq. (6-8d) reduces to Eq. (6-8b) for turbulent flow without heat transfer in smooth tubes.

Friction Factors for Rough Pipes. Data of a number of observers† for clean pipes of steel and cast iron are represented (within a deviation of ± 10 per cent in f) by Curve HG[17,18] or Fig. 6-11. Curve HG may be represented by the equations

$$\frac{1}{\sqrt{f}} = 3.2 \log_{10} N_{Re} \sqrt{f} + 1.2 \qquad (6\text{-}8e)$$

$$\frac{G}{X} = 1.6 \log_{10} \frac{1.19\,DX}{\mu} \qquad (6\text{-}8f)$$

For badly corroded or tuberculated pipes, due especially to a decrease in diameter, f, based on the observed friction and the original diameter, may rise to values as high as 0.020, depending primarily on the average actual diameter of the opening in the tuberculated pipe and the ratio of wall roughness to pipe diameter.[48a,40,53]

Flow inside Rectangular and Annular Sections. For turbulent flow in rectangular ducts and annular spaces, f may be evaluated from data for circular pipes by using an "equivalent diameter" equal to $4r_h$. The

† Data for flow of fluids at acoustic velocities[24,37] are in reasonably good agreement with the data of Fig. 6-11; see also Chap. 12.

available data† on oil, water, and air support this procedure. If the core of an annulus is not concentric with the larger pipe, substantial deviations may occur.

Calculation of Pressure Drop in Tubes. The basic relation is a combination of the force balance, definition of f, and the equation of continuity, derived in Eq. (6-2d):

$$-dp = \frac{G^2\,dv}{\alpha g_c} + \frac{fG^2 v\,dL}{2g_c r_h} + \frac{g}{g_c}\frac{dz}{v} \qquad (6\text{-}9)$$

Incompressible Flow. For incompressible fluids, such as liquids at moderate pressure or gases with small fractional change in pressure, the specific volumes, and consequently the velocities in a pipe of uniform diameter, are substantially constant, and integration gives

$$p_1 - p_2 = \frac{f_m G^2 v_m L}{2g_c r_h} + \frac{g}{g_c}\frac{z_2 - z_1}{v_m} \qquad (6\text{-}9a)$$

Gases and Vapors. For compressible fluids, changes in specific volume may be such that allowance for this variation may be necessary. For example, consider steady mass flow of a compressible fluid through a horizontal duct of constant cross section. Where v_2/v_1 is less than 2:1, little error is introduced by assuming v constant at $v_m = (v_1 + v_2)/2$, giving

$$p_1 - p_2 = \frac{G^2(v_2 - v_1)}{\alpha g_c} + \frac{f_m G^2 v_m L}{2g_c r_h} \qquad (6\text{-}9b)$$

Over the range in pressures involved, the term R_G in the gas law ($v = R_G T/p$) will be taken as constant. Upon dividing Eq. (6-9) by v, one obtains

$$-\frac{1}{R_G}\int \frac{p\,dp}{T} = \frac{G^2}{\alpha g_c}\int \frac{dv}{v} + \frac{G^2}{2g_c r_h}\int f\,dL \qquad (6\text{-}9c)$$

Where the absolute temperature of the fluid does not vary greatly, a mean value T_m is used as an approximation. As explained above, f varies but little with temperature, especially for turbulent flow, and integration gives

$$\frac{p_1^2 - p_2^2}{2R_G T_m} = \frac{G^2}{\alpha g_c}\ln\frac{v_2}{v_1} + \frac{f_m L G^2}{2g_c r_h} \qquad (6\text{-}9d)$$

Compressible flow is further treated in Chap. 12.

The equation for *adiabatic* flow of a perfect gas in a horizontal pipe may be derived as follows[37]: The sum of enthalpy and kinetic energy is constant, $i + G^2 v^2/2g_c = i_1$. For a perfect gas, $i = c_p T = c_p pv/R'_G = Bpv$, where $B = c_p/R'_G = \gamma/(\gamma - 1)$. These relations give the equation of the Fanno line:

$$\frac{G^2 v^2}{2g_c} + Bpv = i_1$$

† Atherton,[2] Cornish,[10] Drew,[16] Fromm,[25] Huebscher,[31a] Kratz, Macintire, and Gould,[42] Lea,[45] Mikrjukov,[47] and Nikuradse.[50]

which is a pure pressure-volume relation. Solving this equation for p, differentiating and dividing by v, one obtains

$$\frac{dp}{v} = -\frac{G^2}{2g_cB}\frac{dv}{v} - \frac{i_1}{B}\frac{dv}{v^3}$$

which is used to eliminate the term dp/v in Eq. (6-9). Integration gives

$$\frac{G^2}{g_c}\left(1 - \frac{1}{2B}\right)\ln\frac{v_2}{v_1} + \frac{i_1}{2B}\left(\frac{1}{v_2^2} - \frac{1}{v_1^2}\right) + \frac{f_{\text{avg}}G^2L}{2g_c r_h} = 0 \quad (6\text{-}9e)$$

The equation of the Fanno line may be used to evaluate v_1 and v_2 at the terminal pressures.

IV. END EFFECTS

Sudden Contraction. The pressure drop caused by a sudden contraction of the cross-sectional area of a pipe (Fig. 6-12) may be calculated from the relation for turbulent flow

$$v_{01}(p_0 - p_1) = \frac{V_1^2 - V_0^2}{2g_c} + \frac{K_c V_1^2}{2g_c} \quad (6\text{-}10)$$

where the dimensionless coefficient K_c is a function of the ratio of the smaller cross section to the larger, as shown[32] in Fig. 6-13.† Equation (6-10) does not apply for streamline flow, which has been treated by Kays.[36a]

For rounded or conical entrances, there is little entrance loss ($K_c = 0.05$) if the flow in the smaller pipe be turbulent. Other cases are treated on page 388 of reference 52.

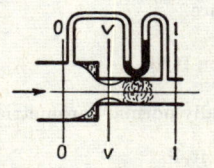

FIG. 6-12. Diagram of sudden contraction.   FIG. 6-13. K_c vs. S_1/S_0, for Eq. (6-10).

Sudden Enlargement. The pressure change on sudden enlargement of a cross section is calculated from the equation

$$v_{23}(p_2 - p_3) = \frac{V_3^2 - V_2^2}{2g_c} + \frac{(V_2 - V_3)^2}{2g_c} \quad (6\text{-}11)$$

where subscript 2 refers to the smaller cross section and 3 to the larger. This equation has been shown by Schutt[57] to be exact for liquid flowing in turbulent motion and may be used safely for gases flowing at moderate

† From Hughes and Safford,[32] "Hydraulics," p. 330, Macmillan, New York, 1911.

velocities. Where the enlargement is gradual (total angle of divergence not greater than 7 deg), the pressure drop may be calculated by integrating a relation based on Eqs. (6-1), (6-2b), (6-2c), and (6-8a). Where the angle of divergence is greater than 7 deg, no simple treatment will apply.†

The following example illustrates the application of Eqs. (6-9b), (6-10), and (6-11):

Illustration 3. Air is flowing at constant mass rate inside the straight horizontal tubes of a cooler at a mass velocity of 2 lb/(sec)(sq ft of cross section). The air enters the tubes at 500°F and normal barometric pressure and leaves at 180°F. The tubes have an actual i.d. of 2.00 inch and are 19 ft long. Calculate (a) the pressure drop in the tubes, expressed as inches of water, and (b) the over-all pressure drop between upstream and downstream chambers, if these have cross sections twice those of the steel tubes.

Solution. a. From the Appendix, the viscosity of air, expressed as lb/(ft)(sec), is 0.0000195 at 500° and 0.0000141 at 180°F. Since DG is $(\frac{2}{12})(2)$, the values of DG/μ are 17,100 and 23,600, and, from Fig. 6-11, the corresponding values of f are 0.0077

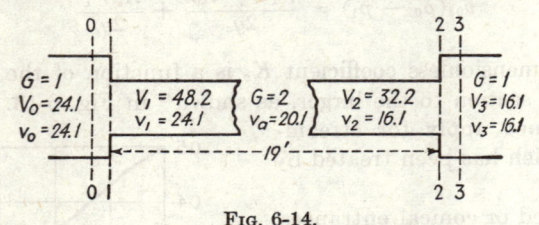

```
0 1                                          2 3
| |                                          | |
| | G=1      | V₁=48.2 | G=2   | V₂=32.2 |   | | G=1
| | V₀=24.1  | v₁=24.1 | V₀=20.1| v₂=16.1|   | | V₃=16.1
| | v₀=24.1  |         |       |         |   | | v₃=16.1
| |<---------------- 19' ------------------->| |
0 |                                          | 2 3
```

Fig. 6-14.

and 0.0072, with $f_m = (0.0077 + 0.0072)/2 = 0.0075$. The hydraulic radius $r_h = D/4 = \frac{1}{24}$ ft. The average molecular weight of air is 29.0, that is, 29 lb of air occupies 359 cu ft at 32°F and normal atmospheric pressure.

$$v_1 = \frac{359}{29} \frac{500 + 460}{492} = 24.1 \text{ cu ft/lb}$$

Assuming that the final absolute pressure is substantially normal barometric pressure,

$$v_2 = \tfrac{359}{29} \tfrac{640}{492} = 16.1 \text{ cu ft/lb}$$
$$v_a = (16.1 + 24.1)/2 = 20.1 \text{ cu ft/lb}$$

The values of v, V, and G are shown in Fig. 6-14. By Eq. (6-9b)

$$p_1 - p_2 = \frac{(2)^2}{32.2}(16.1 - 24.1) + \frac{(0.0075)(19)(2)^2(20.1)}{(64.4)(\frac{1}{24})}$$
$$= -0.993 + 4.27 = 3.28 \text{ pounds force/sq ft}$$

$$\left(3.28 \frac{\text{pounds force}}{\text{ft}^2}\right)\left(\frac{32.2 \text{ lb} \times \text{ft}}{\text{pounds force} \times \text{sec}^2}\right)\left(\frac{1}{32.2 \text{ ft/sec}^2}\right)\left(\frac{1}{62.3 \text{ lb/ft}^3}\right)\left(\frac{12 \text{ in.}}{1 \text{ ft}}\right)$$
$$= 0.630 \text{ in. water}$$

The change in kinetic energy is only 0.033 per cent of the net heat input [Eqs. (6-4a) and (6-4c)].

† Dönch,[13] Gibson,[26] Kröner,[43] and Nikuradse.[49]

FLOW OF FLUIDS

$$Q = (0.24)(500 - 180) + \frac{(32.2)^2 - (48.2)^2}{(778)(64.4)} = 76.8 - 0.0257 \text{ Btu/lb}$$

b. From Fig. 6-13, $K_c = 0.3$, whence by Eq. (6-10)

$$p_0 - p_1 = \frac{V_1{}^2 - V_0{}^2}{2g_c v_{\text{avg}}} + \frac{K_c V_1{}^2}{2g_c v_{\text{avg}}} = \frac{(48.2)^2 - (24.1)^2}{(64.4)(24.1)} + \frac{(0.3)(48.2)^2}{(64.4)(24.1)} = 1.12 + 0.45$$
$$= 1.57 \text{ pounds force/sq ft}$$

By Eq. (6-11)

$$p_2 - p_3 = \frac{V_3{}^2 - V_2{}^2}{2g_c v_{\text{avg}}} + \frac{(V_2 - V_3)^2}{2g_c v_{\text{avg}}} = \frac{(16.1)^2 - (32.2)^2}{(64.4)(16.1)} + \frac{(32.2 - 16.1)^2}{(64.4)(16.1)}$$
$$= -0.75 + 0.25 = -0.5 \text{ pound force/sq ft}$$

Summary

Differences	Pounds force / Square foot	Inches of water
$p_0 - p_1 = 1.12 + 0.45$	+1.57	+0.301
$p_1 - p_2 = -0.99 + 4.27$	+3.28	+0.630
$p_2 - p_3 = -0.75 + 0.25$	-0.50	-0.096
$p_0 - p_3$	4.35	0.835

Since 1 atm is equivalent to a pressure of $(14.69)(144) = 2115$ pounds force/sq ft, little error was made in neglecting the change in p in calculating v_2. Assuming 500 air tubes, the volumetric rate of flow at the exit would be $(500)(3.14)(1)/(144)(32.2) = 351$ cu ft/sec at normal pressure. With the fan at the exit, the power theoretically required would be $(351)(4.35/550) = 2.77$ hp.

Fittings. These are allowed for by assigning a fictitious or equivalent length L_e of straight pipe to the existing straight pipe and using the sum $L_e + L$, in Eq. (6-2d) for turbulent flow. For DG/μ above 2100 to 3000,

Table 6-4. L_e/D Ratios for Screwed Fittings, Valves, Etc., Turbulent Flow Only[a]

Fitting	L_e/D	Fitting	L_e/D
45-deg elbows	15	Tee (used as elbow entering branch)	90
90-deg elbows, standard radius	32	Couplings, unions	Negligible
90-deg elbows, medium radius	26	Gate valves, open	7[b]
90-deg elbows, long sweep	20	Gate valves, one-fourth closed	40[b]
90-deg square elbows	60	Gate valves, one-half closed	200[b]
180-deg close return bends	75	Gate valves, three-fourths closed	800[b]
180-deg medium-radius return bends	50	Globe valves, open	300[b]
Tee (used as elbow, entering run)	60	Angle valves, open	170[b]

[a] Perry,[52] p. 390.
[b] Rough estimate; values depend on construction of valve.

the dimensionless ratio L_e/D for the fitting apparently varies but little with DG/μ, and values are taken from Table 6-4.

V. PACKED TUBES

A modified friction factor for flow of gases through packed tubes is defined by the dimensionless relation[21]

$$f' = \frac{\Delta p_f g_c D_p \rho_m}{L G_0^2} \frac{\epsilon^3}{1 - \epsilon} \tag{6-12}$$

where Δp_f is the pressure drop owing to friction, L is the depth of the packing, G_0 is the **superficial** mass velocity based on the cross section of the empty tube, ρ_m is the mean density of the fluid, ϵ is the fraction voids, and D_p is a surface-mean particle diameter, defined by $6/a_p$, where a_p is the ratio of surface to volume of a particle. Ergun[21] correlates data of a number of observers for solid packings of various shapes, by use of the dimensionless relation

$$f' = 1.75 + \frac{150(1 - \epsilon)}{D_p G_0/\mu} \tag{6-12a}$$

for values of $D_p G_0/(\mu)(1 - \epsilon)$ ranging from 1 to 2500.

In a recent report Coppage (reference 9, Chap. 11) presents and correlates data for Fanning friction factors and heat-transfer coefficients for flow of air through cylindrical containers packed with wire screens or spheres, for Reynolds numbers $4G_0/a\mu_f$ from 5 to 1000 (see also Chap. 11).

VI. FLOW OUTSIDE TUBES

Flow Parallel to Axis. For turbulent flow outside and parallel to the axis of the tubes, as in certain types of heat exchangers, data are available for both gases and liquids, and the recommended treatment is given in Fig. 6-11.

Flow Normal to Axis. The friction for turbulent flow of fluid normal to a bank of tubes, N rows deep, may be considered as due to N contractions and enlargements, which leads to the dimensionless equation[8]

$$(p_1 - p_2)_f = 4f'' N G_{max}^2/2g_c \rho \tag{6-13}$$

For **turbulent** flow of air across banks of staggered pipes, of various diameters, with a wide range of transverse and longitudinal spacings, Andreas[1a] and Grimson[29] find that in Eq. (6-13) f'' depends on arrangement and a Reynolds number. Grimson gives the functions for a large number of arrangements both for staggered and in-line positions of the tubes, for $D_o G_{max}/\mu_f$ from 2000 to 40,000; the data are approximated by the dimensionless equations of Jakob:[34]

FLOW OF FLUIDS

For tubes in line and x_T from 1.5 to 4.0†

$$f'' = \left(0.044 + \frac{0.08x_L}{(x_T - 1)^n}\right)\left(\frac{D_o G_{max}}{\mu_f}\right)^{-0.15} \qquad (6\text{-}13a)$$

in which $n = 0.43 + (1.13/x_L)$.

For staggered tubes, with x_T from 1.5 to 4.0:

$$f'' = \left(0.25 + \frac{0.1175}{(x_T - 1)^{1.08}}\right)\left(\frac{D_o G_{max}}{\mu_f}\right)^{-0.16} \qquad (6\text{-}13b)$$

in which x_L is the ratio of the longitudinal pitch to tube diameter, x_T is the ratio of the transverse pitch to tube diameter, and pitch is center-to-center distance. Thus for tubes in line, with x_L of 2 and x_T of 2, Eq. (6-13a) gives $f'' = 0.20(D_o G_{max}/\mu_f)^{-0.15}$; for staggered tubes, with x_T of 2, Eq. (6-13b) reduces to $f'' = 0.34(D_o G_{max}/\mu_f)^{-0.15}$. These equations are based on the data of Grimison,[29] which cover values of $D_o G_{max}/\mu_f$ ranging from 2000 to 40,000. In all cases G_{max} is based on the *minimum* free area, either in the transverse or diagonal openings.

As later pointed out by Boucher and Lapple,[6a] for values of x_T below 1.5 and above 4, Eqs. (6-13a) and (6-13b) predict unduly high values of f''; for x_T of 1.25, the deviation is roughly 50 per cent. These equations are applied quantitatively in Chap. 15, where diagrams of tube layouts are also shown.

Bergelin et al.[4] find that values of f'' for nonisothermal flow of oils normal to staggered or in-line banks of tubes are roughly correlated by

$$f'' = \frac{f_i''}{(\mu_b/\mu_s)^{0.14}} \qquad (6\text{-}14)$$

where the values of f_i'' for isothermal flow were based on the Grimison data;[29] the values of $D_o G_{max}/\mu$ ranged from 2000 to 10,000.

In the region of **streamline** flow, for modified Reynolds numbers $D_v G_{max}/\mu$ ranging from 1 to 70, their results were correlated by

$$f'' = \frac{70}{D_v G_{max}/\mu_b}\left(\frac{D_o}{s_{min}}\right)^{1.6}\left(\frac{\mu_s}{\mu_b}\right)^m \qquad (6\text{-}14a)$$

where $m = 0.57/(D_v G_{max}/\mu)^{0.25}$. At very high ratios of pitch to diameter, Eq. (6-14a) predicts results which are too small. An earlier correlation is that of Chilton and Genereaux:[8]

$$f'' = \frac{26.5}{D_v G_{max}/\mu_b}\frac{s_L}{D_v} \qquad (6\text{-}14b)$$

In the *transition range*, $D_o G_{max}/\mu_b$ from 70 to 6000, the values of $f''(\mu_b/\mu_s)^{0.14}$ are plotted vs. $D_o G_{max}/\mu_b$ on logarithmic paper as shown in

† Below $D_o G_{max}/\mu_f$ of 10,000, f'' should be taken constant at the value given by Eq. (6-13a) at $D_o G_{max}/\mu_f$ of 10,000.

Fig. 10-21. It is noted that the in-line arrangements gave minimum values of $f(\mu_b/\mu_s)^{0.14}$, while the staggered arrangements did not.

PROBLEMS

1. Water at 68°F flows from a lake through 500 ft of 4-inch i.d. cast-iron pipe to a water turbine located 250 ft below the surface of the lake. After flowing through the turbine, the water is discharged into the atmosphere through a horizontal 50-ft section of the same pipe. The turbine power output is 10 hp when the water in the discharge pipe is flowing at 5 ft/sec. What is the turbine efficiency, defined as the actual power output of the turbine divided by the power output that would be withdrawn if there were no friction within the turbine?

2. If the turbine in Prob. 1 were by-passed, what would be the mass flow rate of water through the 550 ft of 4-inch pipe?

3. Air is to be delivered from a compressor to a distribution line by means of standard 1½-in. steel pipe, 300 ft long. The maximum rate of consumption by the equipment connected to this distribution system is 600 cu ft of free air (measured at 68°F) per minute. In order for the pressure in the distribution line to be always 70 lb/sq in. gauge or higher, what must be the pressure rating of the compressor? Assume that the air flow is isothermal.

4. Glycerol at 68°F, specific gravity 1.26, is pumped at a rate of 28,000 cu ft/hr by a single pump through two horizontal pipes connected in parallel. Both pipes have a length of 100 ft. One is a standard 4-in. steel pipe, the other a standard 9-in. steel pipe. It is agreed to neglect pressure drop due to fittings. What is the velocity in the smaller pipe? What is the pressure drop through the lines? If the lines discharge into an open tank, what head must be developed by the pump?

5. It is desired to heat 27 lb/min of dry air from 40 to 1040°F (moving-stream temperatures) by passing it through a heated horizontal section of smooth pipe, having an i.d. of 2 inches. The heat is to be supplied by electrical wires wrapped around the outside of the pipe, and these wires will supply 4.0 kw of electrical power/ft of heated pipe, distributed uniformly over the heated length. The air is to leave the heated section at a pressure of 1 atm. abs. What must be the length of the heated section and the pressure of the air entering the section?

6. Water is pumped over a high pass in mountain country from a lake in the valley. The pump is located 10 ft above the lake surface. From the pump, the pipe line extends 2000 ft to the top of the pass, which has an altitude of 1010 ft above the lake surface. The pipe line then runs 1000 ft down the other side of the pass, losing 500 ft in altitude. The pipe line next travels another 1000 ft horizontally and empties into a reservoir.

Neglecting end losses and assuming that the lengths of pipe are reported as "equivalent lengths" (thus containing allowance for bends, etc.), compute the shaft horsepower which must be delivered to the pump to move water through this system at a rate of 120 gal/sec. The pipe line has an i.d. of 1.5 ft and is made of steel. The over-all pump efficiency is 75 per cent. The discharge end of the pipe is several feet above the surface of the reservoir.

CHAPTER 7

NATURAL CONVECTION

Abstract. This chapter deals with heat transfer by conduction and natural convection between surfaces and fluids under conditions where changes in phase do not occur. The four sections treat vertical surfaces, horizontal cylinders, horizontal plates, and enclosed spaces (between vertical plates, between horizontal plates, and in annuli).

In the first section the mechanism and theory are discussed and data for *vertical* surfaces are correlated by dimensionless function involving the Nusselt, Grashof, and Prandtl numbers (Fig. 7-7). Simplified relations [Eqs. (7-5a) and (7-5b)] for air and an alignment chart (Fig. 7-9) for various fluids are included.

Data for *horizontal cylinders* and various fluids are correlated in Fig. 7-10. Simplified relations [Eqs. (7-7) and (7-7a)] for air and an alignment chart (Fig. 7-11) for various fluids are given. Table 7-2 gives values of $h_c + h_r$ for various values of diameter and Δt.

Recommended general relations [Eqs. (7-8) to (7-8b)] for *horizontal plates* and simplified relations [Eqs. (7-8c) to (7-8e)] for air are included.

Recommended relations for *enclosed spaces* are given in the last section.

Introduction. Visualize a heated surface exposed to colder air in a room. Since the density of the air near the heated surface is less than that of the main body of the air, buoyant forces cause an upward flow of air near the surface. If the surface were colder than the air, because of greater density near the surface, the air would flow downward. In either case heat is conducted through the gas layers and is carried away by bulk motion or convection. Although both conduction and convection are involved, the process is called "natural" or "free" convection. Measurements of heat-transfer coefficients from surfaces date back many years, but most of the experimental work on the mechanism has been done in the twentieth century.

The heat transferred by radiation is usually substantial compared with that by convection, and the total is calculated by the following equation:

$$q = h_c A_s(t_s - t_a) + \epsilon h_{rb} A_s(t_s - t_e)$$

where t_s, t_a, and t_e are the temperatures of the surface of the body, the air, and the enclosing walls, respectively, and h_{rb} is obtained from Eq. (4-44a) or Eq. (10-4). Where t_e and t_a are the same, the equation becomes

$$q = (h_c + h_r)(A_s)(t_s - t_a) \qquad (7\text{-}1) \star$$

HEAT TRANSMISSION

Table 7-1. Nomenclature

A_s Area of heat-transfer surface, square feet; A_1 for smaller area in concentric annulus

C Constant in equation: $Y = CX^n$, dimensionless

C' Constant in equation: $h_c = C'(\Delta t)^n / L^{1-3n}$, dimensional

c_p Specific heat of fluid, Btu/(lb)(deg F)

D Diameter, feet; D_o for outside diameter of cylinder, D_1 and D_2 for smaller and larger diameters of concentric cylinders

g Acceleration owing to gravity, 4.17×10^8 ft/hr^2

h Surface coefficient of heat transfer, Btu/(hr)(sq ft)(deg F); h_c for natural convection from a surface to an ambient fluid; h_c' for an enclosed space, based on $t_{s1} - t_{s2}$, h_{c1}', based on A_1; $h_c + h_r$, combined coefficient for natural convection plus radiation, based on $t_s - t_a$; h_r, for radiation only; h_{rb}, for radiation between black surfaces

k Thermal conductivity of fluid at bulk temperature, Btu/(hr)(sq ft)(deg F per ft); k_f at t_f, k_s at t_s

L Geometrical factor, feet; in general L denotes height of vertical surface; L_{horiz} for length of horizontal surface

N Dimensionless term

$N_{Gr,f}$ Grashof number, dimensionless: $L^3 \rho_f^2 g \beta_f \Delta t / \mu_f^2$ for a vertical surface, $D_o^3 \rho_f^2 g \beta_f \Delta t / \mu_f^2$ for a horizontal cylinder, and $x^3 \rho_f^2 g \beta_f \Delta t / \mu_f^2$ for an enclosed gas space

$N_{Nu,f}$ Nusselt number, dimensionless: $h_c L / k_f$ for a vertical surface, $h_c D_o / k_f$ for a horizontal cylinder, and $h_c' x / k_f$ for an enclosed gas space

$N_{Pr,f}$ Prandtl number, dimensionless, $c_p \mu / k$ evaluated at t_f

n Exponent, dimensionless

P Absolute pressure, atmospheres

q Rate of heat transfer, Btu per hour; q_c for natural convection, $q_c + q_r$ for natural convection plus radiation, q_k for conduction only, and q_r for radiation only

T Absolute temperature, degrees Rankine

t Thermometric temperature, degrees Fahrenheit; t_a, ambient temperature; t_e, for enclosing walls; t_f, for "film" temperature $= t_f = (t_s + t_a)/2$; t_s for surface

u Local velocity, feet per hour

X Product of Grashof and Prandtl numbers, dimensionless

x Clearance in enclosed space, feet

x' Mean free path of gas molecules, feet

Y Nusselt number, dimensionless, $h_c L / k_f$, $h_c x / k_f$, and $h_c' x / k_f$

Greek

α Thermal diffusivity of fluid $= k / \rho c_p$, square feet per hour

β Coefficient of volumetric expansion, reciprocal degrees Fahrenheit

Δt Temperature potential, degrees Fahrenheit

ϵ Emissivity of surface for radiation, dimensionless; see Appendix

μ Viscosity of fluid, pounds/(hr)(ft) $= 2.42$ times viscosity in centipoises

π Constant, equal to $3.1416 \ldots$

ρ Density of fluid, pounds per cubic foot

$\phi(t_f)$ Dimensional term, equal to $k^3 \rho^2 g / \mu^2$, evaluated at t_f, [Btu/(hr)(sq ft)(deg F)]^3

$\psi(t_f)$ Dimensional term equal to $\rho^2 g \beta c_p / \mu k$, evaluated at t_f, 1/(cu ft)(deg F)

I. VERTICAL SURFACES

Mechanism. The mechanism of heat loss by conduction and convection from vertical surfaces has been studied by Griffiths and Davis,[11] Nusselt and Jurges,[29] Schmidt,[33] and Schmidt and Beckmann.[34]

Figure 7-1 shows the velocity and temperature fields near a vertical heated plate, 1 ft high, measured by Schmidt.[33] At a given distance z from the bottom, the local upward air velocity increases with increase in distance y from the wall, goes through a maximum at a distance of from 2 to 3 mm from the plate, and then decreases with further increase in y. The air streams approach the plate laterally at the bottom and diagonally further up the plate. The temperature gradients are steepest near the

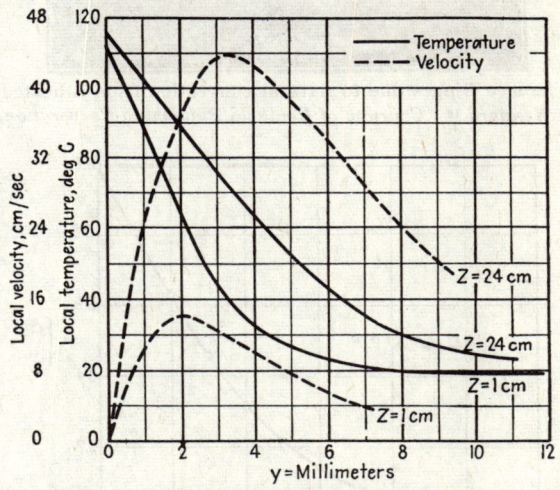

Fig. 7-1. Velocity and temperature explorations by Schmidt[33] at various distances from a heated vertical plate exposed to air.

bottom of the plate, where the air is the coldest, and decrease as the air flows up the plate, approaching an asymptotic value for tall plates. Results for tall plates show that the local air velocity ultimately approaches an asymptotic value, depending on the height of the plate and the Δt employed.

The *average* coefficient for the plate is an inverse function of height for small heights and is substantially independent of height for tall plates. Similar results for plates of various heights are given in the other references cited above. The results of Koch[21] for tall vertical cylinders are similar to those for vertical plates.

Kennard[19] describes a method for measuring temperature distribution near heated bodies exposed to colder air, which depends on the change of the index of refraction of air with change in temperature. By the use of a suitable interferometer, the interference fringes are measured, and from the resulting photographs it is possible to compute the temperature gradients. Figure 7-2 shows a photograph of the fringes in front of a vertical plate, 53°F warmer than the ambient air. Figure 7-3 shows the corresponding calculated dimensionless distribution of temperature, which is similar to that obtained by use of thermocouples.

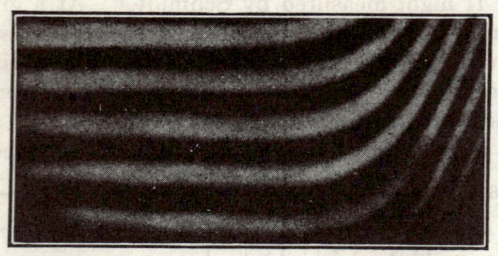

Fig. 7-2. Interference fringes due to natural convection from a heated plate to air. (*Photograph by Kennard.*[19] *Courtesy of Reinhold Publishing Corporation, New York.*)

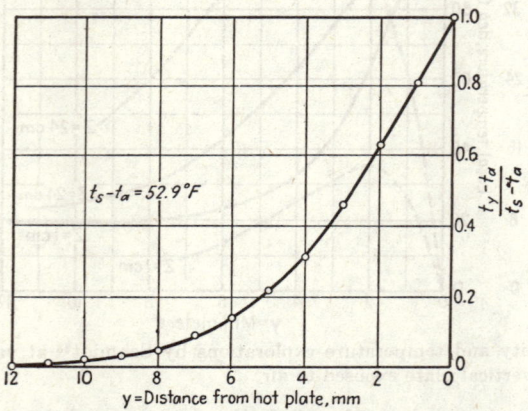

Fig. 7-3. Temperature gradients calculated by Kennard[19] from results shown in Fig. 7-2.

Eckert and Soehnghen[8,9] used an interferometer to study the mechanism of natural convection from a heated vertical aluminum plate to air; the plate was 3 ft high and 1.5 ft wide. Figure 7-4 shows the results of one experiment. The flow is laminar for the first 20 inches from the bottom; then the transition from laminar to turbulent flow starts at 21 inches at a critical value of the Grashof number, $N_{Gr} = x^3 \rho^2 g \beta \, \Delta t / \mu^2$ of 4×10^8. Near the top of the plate the transition has proceeded further, and the laminar sublayer is thinner than at 21 inches. By employing a larger temperature potential, the transition starts nearer the bottom of the plate, and a fully developed turbulence occurs near the top of the plate at a

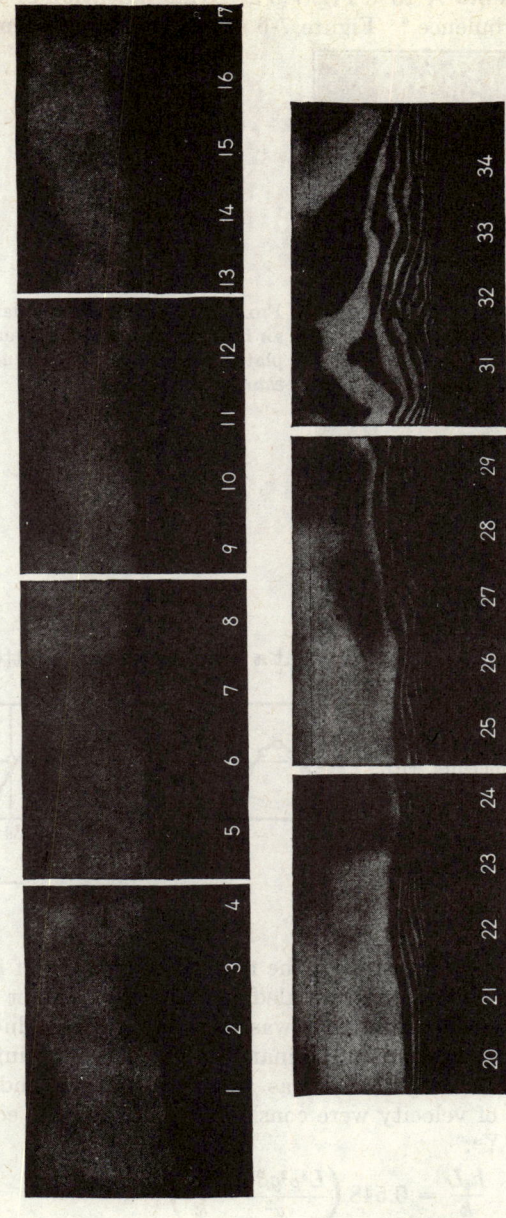

FIG. 7-4. Laminar and turbulent flow of air along a heated vertical plate, interference photographs by Eckert and Soehnghen. (*Courtesy of E. Eckert and McGraw-Hill Book Company, Inc., New York.*)

170   HEAT TRANSMISSION

Grashof number of 3.8×10^{11}; Fig. 7-5 gives an indication of scale of the fluctuations in turbulence.[9] Figure 7-6 shows the nonuniformity in the

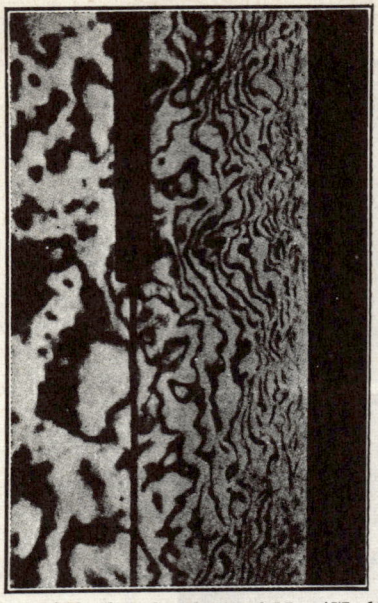

FIG. 7-5. Turbulent boundary layer for air flowing upward along a vertical heated plate.[9] (*Courtesy of Institution of Mechanical Engineers.*)

value of the local values of $N_{Nu}/N_{Gr}^{1/4}$ at a certain moment, plotted against the local height.[9]

FIG. 7-6. Instantaneous values of local heat-transfer coefficients, with turbulent boundary layer for air flowing upward on a vertical heated plate. (*Courtesy of Institution of Mechanical Engineers; data of Eckert and Soehnghen.*[9])

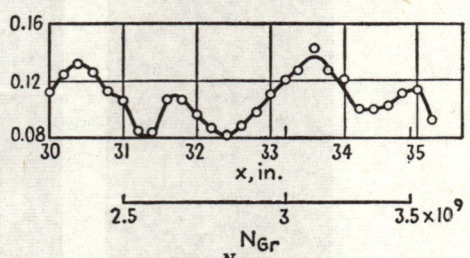

Theory. The factors involved in the natural convection of heat from a vertical plane to a fluid were revealed in 1881 in a pioneer paper by Lorenz.[23] It was assumed that heat was transferred by conduction to a layer of fluid flowing upward in laminar motion under the influence of buoyancy. The wall temperature was assumed uniform, and no horizontal components of velocity were considered. The Lorenz equation is of the form $Y = CX^n$:

$$\frac{h_c L}{k} = 0.548 \left(\frac{L^3 \rho^2 g \beta \, \Delta t}{\mu^2} \frac{c_p \mu}{k} \right)^{0.25} \tag{7-2}$$

As shown by Tribus,[37] specification of geometrical similarity, and of equal ratios of buoyant and viscous forces, gives

$$\frac{L_1^2 \rho_1 g \beta_1 \, \Delta t_1}{\mu_1 u_1} = \frac{L_2^2 \rho_2 g \beta_2 \, \Delta t_2}{\mu_2 u_2}$$

This dimensionless group, together with the Reynolds modulus, defines similarity in a natural convection system, but since it is difficult to evaluate the velocity in the Reynolds modulus in a natural-convection system, it is more convenient to eliminate the velocity by multiplying the above relation by the Reynolds modulus, giving the Grashof modulus:

$$\frac{L_1^3 \rho_1^2 g \beta_1 \, \Delta t_1}{\mu_1^2} = \frac{L_2^3 \rho_2^2 g \beta_2 \, \Delta t_2}{\mu_2^2}$$

For geometrically similar systems of different sizes, equal Grashof numbers assure equal ratios of inertia forces, viscous forces, and buoyant forces.

As pointed out by Boussinesq[3] in 1902, if the influence of viscous forces is negligible (as might be nearly true for highly turbulent flow), then the effect of viscosity would disappear, requiring that the Nusselt number be a function of the product $N_{Gr} N_{Pr}^2$:

$$N_{Gr} \cdot N_{Pr}^2 = \frac{L^3 \rho^2 g \beta \, \Delta t}{\mu^2} \left(\frac{c_p \mu}{k}\right)^2 = \frac{L^3 g}{\alpha^2} \cdot \beta \, \Delta t$$

where α is the molecular thermal diffusivity of heat, $k/\rho c_p$. In 1936, Saunders[32] showed that data for water were above those for air, when the Nusselt numbers were plotted vs. $N_{Gr} N_{Pr}$, at high values of X, and that better agreement would be obtained by plotting N_{Nu} vs. $N_{Gr} N_{Pr}^{1.5}$, instead of employing the conventional product $N_{Gr} N_{Pr}$. Further data are needed to clarify the situation.

Based on measurements of the temperatures of the air at various locations, Schmidt and Beckmann[34] derived the following equation for natural convection from short vertical plates, not more than 2 ft high:

$$\frac{h_c L}{k_f} = 0.52 \left[\left(\frac{L^3 \rho_f^2 g \beta_f \, \Delta t}{\mu_f^2}\right)\left(\frac{c_p \mu}{k}\right)_f\right]^{0.25} \qquad (7\text{-}3)$$

In a recent paper dealing with the theory of natural convection from a vertical plate to ambient fluids having Prandtl numbers from 0.01 to 1000, Ostrach[30] derives equations for both the velocity and temperature fields and finds reasonable agreement with data for air. It was concluded that the ratio of the Nusselt number to the fourth root of the Grashof number was a unique function of the Prandtl number; at a Prandtl number of 0.01, this ratio was predicted as 0.08, compared with 0.07 from Eq. (7-3a) and 0.17 from an empirical relation [Eq. (7-3)] for Prandtl numbers of 0.7 to 500.

Low Prandtl Numbers. Eckert[8] derived a theoretical relation for natural convection for vertical planes:

$$\frac{h_m L}{k_f} = 0.68 \left(\frac{N_{Pr,f}}{0.952 + N_{Pr,f}}\right)^{\frac{1}{4}} (N_{Pr,f} N_{Gr,f})^{\frac{1}{4}} \qquad (7\text{-}3a) \ *$$

This equation† successfully correlated the data of Schmidt and Beckmann[34] for air and short vertical planes. For molten metals, which have low Prandtl numbers, this equation predicts much lower results than conventional relations, such as Eq. (7-3).

Correlation of Data for Vertical Planes and Cylinders. In 1932, King[20] correlated his own data for short vertical planes (½ to 12 inches) with those of other observers for vertical plates and vertical tubes. Data of Colburn[5] for water flowing at low velocities in vertical tubes were included. Using height L in both the Nusselt and Grashof numbers, the data were closely correlated by a logarithmic plot of $N_{Nu,f}$ vs. $N_{Gr,f} N_{Pr,f}$. The data could be represented by equations of the form $Y = CX^n$:

$$\frac{h_c L}{k_f} = C \left[\left(\frac{L^3 \rho_f^2 g \beta_f \, \Delta t}{\mu_f^2} \right) \left(\frac{c_p \mu}{k} \right)_f \right]^n \tag{7-4}$$

At values of X from 3.5×10^7 to 10^{12}, C is 0.13, and n is ⅓; at values of X from 3.5×10^7 to 10^4, C is 0.55, and n is ¼, and h_c is inversely proportional to the fourth root of height. Data for horizontal cylinders (with L replaced by diameter), for a sphere with L replaced by radius and for blocks with L given by

$$\frac{1}{L} = \frac{1}{L_{vert}} + \frac{1}{L_{horiz}}$$

were correlated by the same curve as for vertical surfaces; a similar finding was made by Jakob and Linke.[16] It was conceded that there might be some small differences in curves for the various shapes. In the turbulent range where n is ⅓, the geometrical factor cancels from the equation in relating the Nusselt and Grashof numbers; consequently erroneous specification of the geometrical factor is immaterial. In the laminar range, where n is 0.25, h_c is inversely proportional only to the fourth root of the geometrical factor. The data for spheres and blocks fall in a range where the geometrical factor has substantially no influence.

The subsequent extensive data for short vertical plates, of Weise[39] and of Saunders,[32] are shown in Fig. 7-7; the solid curve shows the recommended correlation. In the turbulent range, X from 10^9 to 10^{12}, the **recommended** equation for vertical planes and vertical cylinders is

$$\frac{h_c L}{k_f} = 0.13 \left[\frac{L^3 \rho_f^2 g \beta_f \, \Delta t}{\mu_f^2} \left(\frac{c_p \mu}{k} \right)_f \right]^{1/3} \tag{7-4a} \star$$

In the laminar range, X from 10^9 to 10^4, the **recommended** equation is

$$\frac{h_c L}{k_f} = 0.59 \left[\frac{L^3 \rho_f^2 g \beta_f \, \Delta t}{\mu_f^2} \left(\frac{c_p \mu}{k} \right)_f \right]^{0.25} \tag{7-4b} \star$$

For X smaller than 10^4, the recommended curve should be used.

† An analogous calculation was made for gases by Squire.[35]

NATURAL CONVECTION

Simplified Equations for Air. For air at ordinary temperature and at atmospheric pressure Eq. (7-4) reduces to simplified *dimensional* equations for vertical surfaces:

$$h_c = C' \Delta t^n / L^{1-3n} \tag{7-5} \star$$

For X from 10^9 to 10^{12}

$$h_c = 0.19 \, \Delta t^{1/3} \tag{7-5a} \star$$

For X from 10^3 to 10^4

$$h_c = 0.29(\Delta t/L)^{0.25} \tag{7-5b} \star$$

To facilitate evaluation of X, Fig. 7-8 shows a graph of $\psi(t_f) = \rho^2 g \beta c_p / \mu k$ vs. film temperature, for air at atmospheric pressure, based on the recent physical properties from Keenan and Kaye.[18]

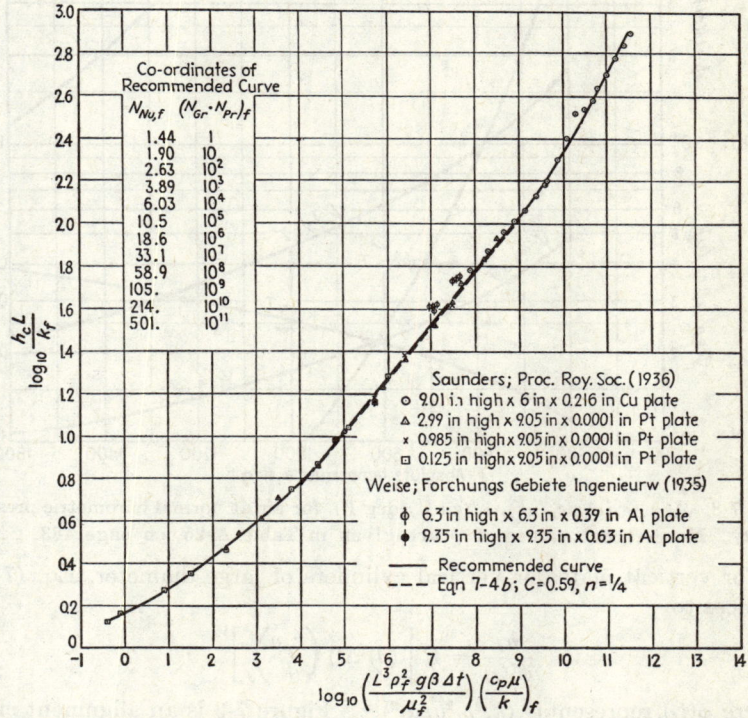

FIG. 7-7. Natural convection from short vertical plates to air; see also Fig. 7-8.

Illustration 1. For example, assume that the surface is at 210°F and the air is at 70°F; the film temperature would be 140°F, and, from Fig. 7-8, $\rho_f^2 \beta_f g c_p / \mu_f k_f$ is 0.88 $\times 10^6$. The corresponding value of X is

$$X = \frac{L^3 \rho_f^2 g \beta_f \, \Delta t}{\mu_f^2} \left(\frac{c_p \mu}{k}\right)_f = 0.88 \times 10^6 L^3 \, \Delta t$$

Since Δt is 140°F, $X = 1.23 \times 10^8 L^3$. If L were, say, 2 ft, X would be 1.0×10^9 and

Eq. (7-5a) applies: $h_c = (0.19)(140)^{1/3} = 0.99$. If the surface had an emissivity of 0.9 and the walls of the room were at 70°F,

$$h_{rb} = \frac{(0.171)[(6.70)^4 - (5.30)^4]}{670 - 530} = 1.49$$

Consequently h_r would be $(0.9)(1.49)$, or 1.34, $h_c + h_r$ would be 2.33, and the total flux q/A would be $(2.33)(140)$, or 326 Btu/(hr)(sq ft).

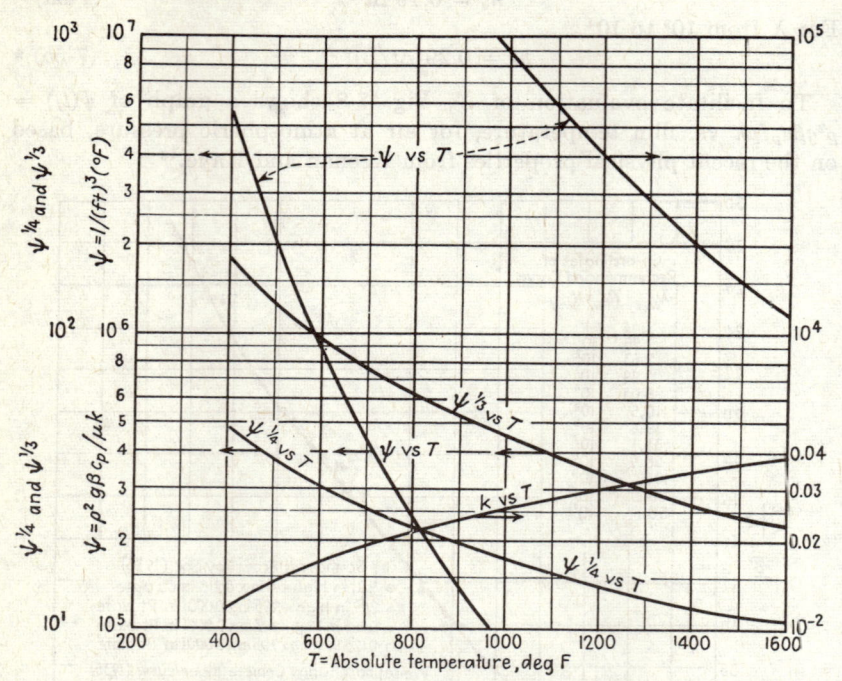

Fig. 7-8. $\psi(t_f) = (\rho^2 g \beta c_p/\mu k)$, 1/(cu ft)(deg F), for air at normal barometric pressure vs. t_f. More recent values of ψ are given in Table A-25, on page 483.

For vertical plates or vertical cylinders of large diameter, Eq. (7-4a) reduces to

$$\frac{h_c}{\phi(t_f)} = C\left[(\beta_f \Delta t)\left(\frac{c_p \mu}{k}\right)_f\right]^{1/3}$$

where $\phi(t_f)$ represents† $(k_f^3 \rho_f^2 g/\mu_f^2)^{1/3}$. Figure 7-9 is an alignment chart based on Eq. (7-4a) for *vertical* surfaces for X from 8×10^7 to 10^{12}. In the case of liquids, only the film temperature, *i.e.*, the arithmetic mean of the temperatures of the tube and the fluid, and the Δt are needed to determine h_c. For gases, a pressure term must be included, expressed as $P^2 \Delta t$. In the laminar range this chart would give conservative results, and Fig. 7-7 should be used.

† Values of $\phi(t_f)$ are given for water in the Appendix.

Touloukian et al.[36] reported data for a heated vertical cylinder inside an unheated cylinder containing liquid; the diameter ratio was 4.4. The heater had unheated cylinders at both ends.

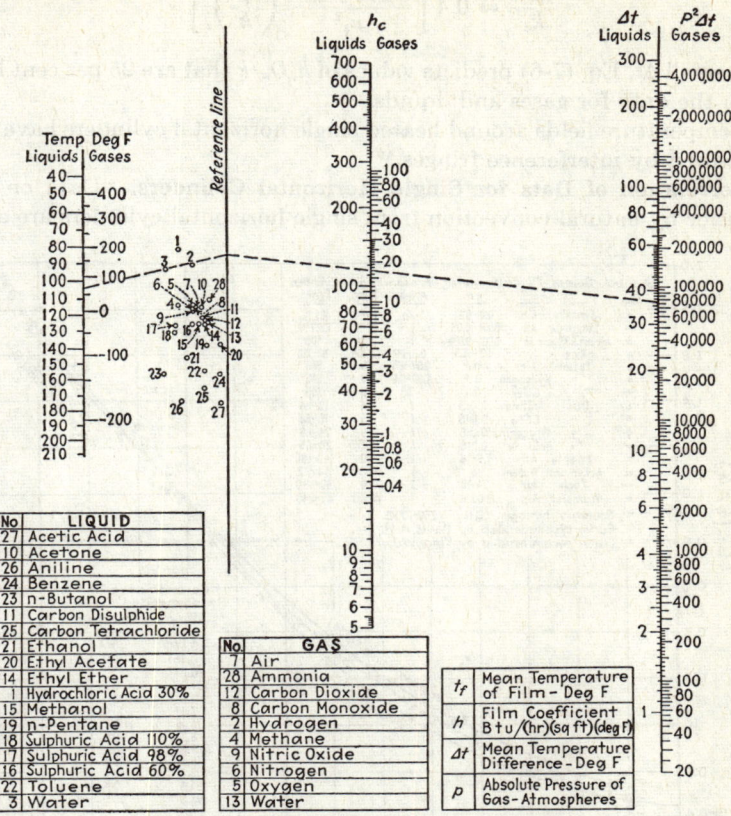

Fig. 7-9. Alignment chart[4] for h_c with a *turbulent* boundary layer of air or liquid on *vertical* surfaces, based on Eq. (7-4a), for $X_L = N_{Gr,f} \cdot N_{Pr,f}$ from 10^8 to 10^{12}.

Note that Fig. 7-11 is for h_c with a *laminar* boundary layer on a *horizontal* cylinder, based on Eq. (7-6b), for X_D from 10^3 to 10^9. For vertical surfaces with X_L from 10^8 to 10^4, use Fig. 7-11, taking D' equal to $2L'/\pi$, expressed in *inches*.

For X_L below 10^4, use recommended curve on Fig. 7-7. With molten metals and fused salts, use Eq. (7-3a).

II. HORIZONTAL CYLINDERS

Theory. Temperature and velocity fields near a heated horizontal cylinder have been measured by Jodlbauer.[17] These results have been roughly predicted from theoretical considerations by Hermann;[13] for Grashof numbers exceeding 10^4, Hermann predicts $h_c D_o / k_s = 0.37 N_{Gr,s}^{0.25}$ for natural convection from single horizontal cylinders to diatomic gases

($c_p\mu/k = 0.74$). If this equation for gases were generalized by adding $(c_p\mu/k)^{0.25}$, one would obtain

$$\frac{h_c D_o}{k_s} = 0.4 \left[\frac{D_o^3 \rho_s^2 g \beta_s \, \Delta t}{\mu_s^2} \left(\frac{c_p \mu}{k} \right)_s \right]^{0.25} \tag{7-6}$$

For small Δt, Eq. (7-6) predicts values of $h_c D_o/k$ that are 25 per cent lower than the data for gases and liquids.

Temperature fields around heated single horizontal cylinders have been measured by interference fringes.[19]

Correlation of Data for Single Horizontal Cylinders. Data on heat transfer by natural convection from single horizontal cylinders are avail-

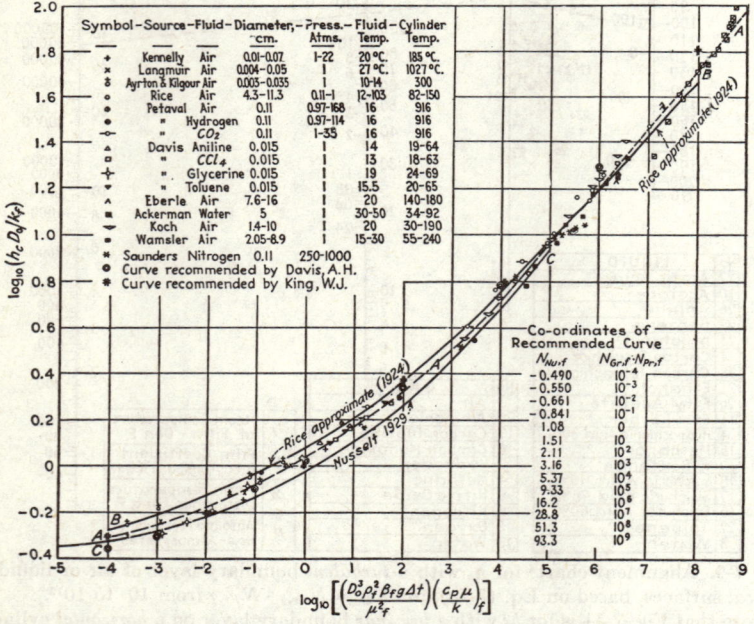

FIG. 7-10. Natural convection from horizontal cylinders to gases and liquids; see also Fig. 7-8.

able on air, hydrogen, carbon dioxide, oxygen, water, aniline, carbon tetrachloride, glycerin, toluene, and olive oil, covering a 7300-fold range of diameters, a 10,000-fold range of pressures, an 11,000-fold range in the value of $c_p\mu/k$, and a range of Δt from a few degrees to 2850°F. Figure 7-10 shows the dimensionless correlation.[24] The physical properties c_p, k, ρ, μ, and β were evaluated at a film temperature halfway between the temperature of the surface of the cylinder and the temperature of the ambient fluid.

NATURAL CONVECTION

For estimation of heat-transfer coefficients from single horizontal wires or pipes to any fluid, the use of Curve AA[†] (Fig. 7-10), is **recommended**. The data on heat transfer from horizontal *pipes* to air (Eberle,[7] Wamsler,[38] Rice,[31] and Koch[21]) and to water (Ackermann[1]), for X from 10^3 to 10^9, are represented by the dimensionless equation

$$\frac{h_c D_o}{k_f} = 0.53 \left[\frac{D_o^3 \rho_f^2 g \beta_f \, \Delta t}{\mu_f^2} \left(\frac{c_p \mu}{k}\right)_f \right]^{0.25} \qquad (7\text{-}6a)\ddagger$$

At zero Δt, $h_c D_o / k_f$ approaches 0.45. If the characteristic linear dimension in Eq. (7-6a) had been taken as the maximum distance traveled by the fluid $L = \pi D_o/2$, instead of D_o, the constant would have been 0.59, the same as in Eq. (7-4b) for vertical surfaces.

The heat-transfer coefficient from an internally heated 0.75-inch horizontal tube 1 ft long, immersed in a water bath was increased[27] up to fivefold by vibrating the pipe sinusoidally in a vertical direction, but when the velocity of vibration was less than a certain critical value, no increase was obtained.

Simplified Equations for Air. For air at ordinary temperature and at atmospheric pressure, the following simplified *dimensional* equations may be employed:

For X from 10^9 to 10^{12}

$$h_c = 0.18 \, \Delta t^{1/3} \qquad (7\text{-}7) \star$$

For X from 10^3 to 10^9, the *usual* case for pipes

$$h_c = 0.27 (\Delta t / D_o)^{0.25} \qquad (7\text{-}7a) \star$$

Figure 7-11 is an alignment chart,§ based on Eq. (7-6b), for the rapid estimation of heat-transfer coefficients for natural convection from a single horizontal pipe to various gases and liquids. This chart gives conservative values for h_c, about 10 per cent lower than Eq. (7-6a). In the case of liquids, only the film temperature, *i.e.*, the arithmetic mean of the temperature of fluid and pipe, the pipe diameter, and the temperature difference are needed for the estimation of h_c. In the case of gases, a pressure term is included.

Illustration 2. It is desired to estimate the value of h_c between ammonia gas at 10 atm pressure and 392°F and a 2-in. cylinder at 212°F.
Solution. Using Fig. 7-11, align $t_f = 302°F$ on the left-hand scale with point 9. Since $P^2 \Delta t / D' = (10)^2(180)/2 = 9000$, locate this point on the right-hand scale

[†] This curve differs approximately 10 per cent from that developed by Davis[6] from a portion of the data used in Fig. 7-10.

[‡] Rice[31] employed a constant of 0.47 instead of 0.53:

$$\frac{h_c D_o}{k_f} = 0.47 \left[\frac{D_o^3 \rho_f^2 \beta g \, \Delta t}{\mu_f^2} \left(\frac{c_p \mu}{k}\right)_f \right]^{0.25} \qquad (7\text{-}6b)$$

§ Prepared by Chilton, Colburn, Genereaux, and Vernon.[4]

marked $P^2 \Delta t/D'$, join this point by a straight line with the previously determined point on the reference line, and read $h_c = 4.3$ Btu/(hr)(sq ft)(deg F) on the scale for gases. This value corresponds to Eq. (7-6b). Eq. (7-6a) would give a value 13 per cent higher and $h_c = 1.13(4.3) = 4.8$.

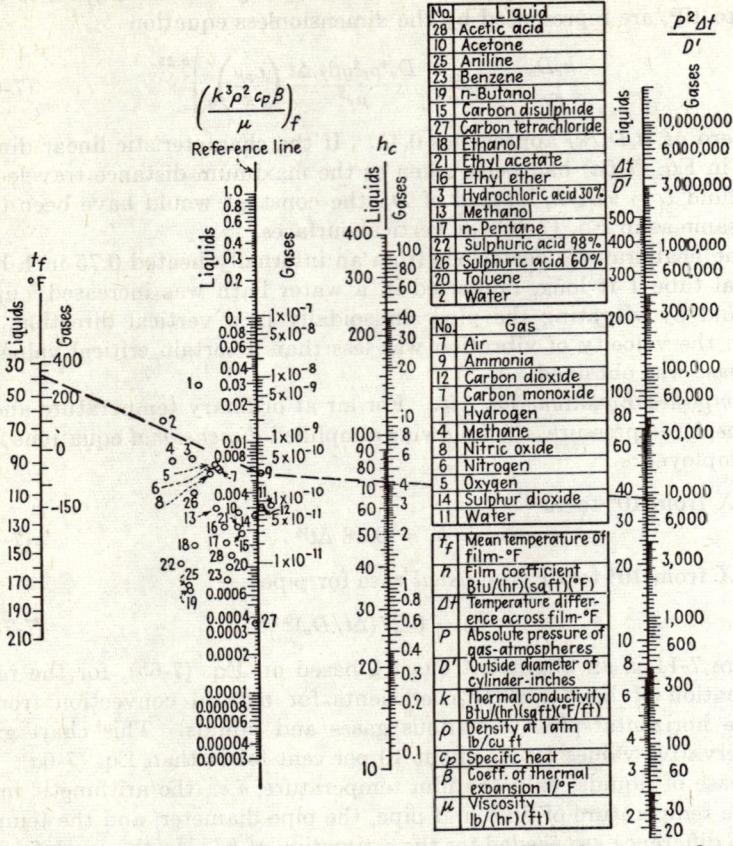

FIG. 7-11. Alignment chart[4] for h_c with a *laminar* boundary layer on a *horizontal* cylinder, based on Eq. (7-6b), $Y = 0.47X^{0.25}$. Since the recommended relation, Eq. (7-6a), $Y = 0.53X^{0.25}$, has a higher constant, multiply h_c from Fig. 7-11 by 1.13. The chart is based on values of X_D from 10^3 to 10^9. For higher values of X, use Fig. 7-9 for vertical surfaces.

Simultaneous Loss by Radiation. Radiation is an important factor in the loss of heat from surfaces where the gas movement is due to natural convection, except for small wires, where the values of h_c for the wires are normally so very large that substantially the entire heat loss is due to natural convection, even with large differences in temperature between the wires and the surrounding solids. The heat radiated from the warmer solid to the surrounding walls may be expressed in terms of the coefficient of heat transfer by radiation h_r obtained from Eq. (4-44a). Table 7-2 shows values of $h_c + h_r$ from single horizontal pipes of steel, with oxidized surfaces, based on data of Heilman[12] and McMillan.[25]

TABLE 7-2. VALUES OF $h_c + h_r$[a]

(For horizontal bare or insulated standard steel pipe of various sizes in a room at 80°F)

Nominal pipe diam, in.	\multicolumn{14}{c}{$(\Delta t)_s$, temperature difference, deg F, from surface to room}
	50	100	150	200	250	300	400	500	600	700	800	900	1000	1100	1200
½	2.12	2.48	2.76	3.10	3.41	3.75	4.47	5.30	6.21	7.25	8.40	9.73	11.20	12.81	14.65
1	2.03	2.38	2.65	2.98	3.29	3.62	4.33	5.16	6.07	7.11	8.25	9.57	11.04	12.65	14.48
2	1.93	2.27	2.52	2.85	3.14	3.47	4.18	4.99	5.89	6.92	8.07	9.38	10.85	12.46	14.28
4	1.84	2.16	2.41	2.72	3.01	3.33	4.02	4.83	5.72	6.75	7.89	9.21	10.66	12.27	14.09
8	1.76	2.06	2.29	2.60	2.89	3.20	3.88	4.68	5.57	6.60	7.73	9.05	10.50	12.10	13.93
12	1.71	2.01	2.24	2.54	2.82	3.13	3.83	4.61	5.50	6.52	7.65	8.96	10.42	12.03	13.84
24	1.64	1.93	2.15	2.45	2.72	3.03	3.70	4.48	5.37	6.39	7.52	8.83	10.28	11.90	13.70

[a] Values of $h_c + h_r$ for Δt up to 1000°F are given by Bailey and Lyell.[2]

The data of Table 7-2 for bare horizontal pipes may be used in predicting the heat loss from an insulated horizontal steam pipe by employing Eq. (7-1), as shown in the following calculation.

Illustration 3. A standard horizontal 2-in. steel steam pipe (o.d. = 2.38 in.) is insulated with a 2-in. layer of pipe covering, $k = 0.04$ Btu/(hr)(sq ft)(deg F per ft). The steam is condensing at 370°F, and the room is at 80°F. Required, the heat loss in Btu/(hr)(100 ft length), insulated as described above, and if bare.

For the bare pipes the controlling resistance is that on the outer surface. From Table 7-2, by interpolation for standard 2-in. pipe and Δt of 290°F, $h_c + h_r$ is 3.4. Since q equals $(h_c + h_r)A_o \Delta t$,

$$q = (3.4)\left(\pi \times \frac{2.38}{12} \times 100\right)(290) = 61{,}300 \text{ Btu/hr}$$

For the insulated pipe, x/kA_m and $1/(h_c + h_r)A_o$ are the important resistances. By Eq. (2-9), the logarithmic-mean area A_m is

$$\frac{100\pi}{12} \frac{6.38 - 2.38}{2.3 \log_{10}(6.38/2.38)} = 105.5 \text{ sq ft}$$

and the outer surface is $100\pi 6.38/12 = 166.8$ sq ft. Assuming $h_c + h_r$ of 1.8, and neglecting the thermal resistances on the steam side and that of the pipe wall, one obtains

$$q = \frac{370 - 80}{\frac{2/12}{(0.04)(105.5)} + \frac{1}{(1.8)(166.8)}} = \frac{290}{0.0395 + 0.00333} = 6780 \text{ Btu/hr.}$$

Assuming $h_c + h_r = 1.8$ for an o.d. of 6.38 in. gives a surface resistance 0.00333, which is 7.8 per cent of the total resistance, 0.0428; hence the corresponding difference from outer surface to air would be $(0.078)(290) = 23°F$, and from Table 7-2 it is seen that $h_c + h_r$ would be approximately 1.8 for this condition. Since the surface resistance was such a small fraction of the total resistance, no closer approximation is warranted. Methods for calculating the most economical thickness of insulation are given in Chap. 15.

Liquid Metals and Salts. Based on data for horizontal pipes with a variety of liquids, including liquid metals, Hyman et al.[14] found that the constant of Eq. (7-3a) was 0.54, which was obtained by substituting $2.5D$ for L in Eckert's equation.

Rarefied Gases. Madden and Piret[26] measured natural-convection coefficients from horizontal wires to rarefied gases (air, helium, and argon) at products of the Grashof and Prandtl numbers from 10^{-1} to 10^{-8}, as discussed in Chap. 12.

III. HORIZONTAL PLATES

General Correlation. Fishenden and Saunders[10] correlate data from horizontal square plates exposed to air by employing the dimension of a side as the characteristic geometrical factor.

For heated plates facing upward, or cooled plates facing downward, in the *turbulent* range, X from 2×10^7 to 3×10^{10}:

$$\frac{hL}{k_f} = 0.14 \left[\frac{L^3 \rho_f^2 g \beta_f \, \Delta t}{\mu_f^2} \left(\frac{c_p \mu}{k} \right)_f \right]^{1/3} \qquad (7\text{-}8) \star$$

and in the *laminar* range, X from 10^5 to 2×10^7:

$$\frac{hL}{k_f} = 0.54 \left[\frac{L^3 \rho_f^2 g \beta_f \, \Delta t}{\mu_f^2} \left(\frac{c_p \mu}{k} \right)_f \right]^{0.25} \qquad (7\text{-}8a) \star$$

For heated plates facing downward, or cooled plates facing upward in the *laminar* range, X from 3×10^5 to 3×10^{10}:†

$$\frac{h_c L}{k_f} = 0.27 \left[\frac{L^3 \rho_f^2 g \beta_f \, \Delta t}{\mu_f^2} \left(\frac{c_p \mu}{k} \right)_f \right]^{0.25} \qquad (7\text{-}8b) \star$$

Simplified Equations for Air. The corresponding simplified equations for heated plates facing up, or cooled plates facing down, in air at atmospheric pressure and at room temperature are:

Turbulent range: $\quad h_c = 0.22 \, \Delta t^{1/3} \qquad (7\text{-}8c) \star$
Laminar range: $\quad h_c = 0.27 (\Delta t / L)^{0.25} \qquad (7\text{-}8d) \star$

For cooled plates facing up, or heated plates facing down:

Laminar range: $\quad h_c = 0.12 (\Delta t / L)^{0.25} \qquad (7\text{-}8e) \star$

The alignment chart[4] for vertical surfaces in the turbulent range (Fig. 7-9) is based on C of 0.13 and n of $\frac{1}{3}$. Since, in the turbulent range C is 0.14 for heated horizontal plates facing upward, the value of h will be

† The constant has been changed from 0.25 to 0.27, to fit the data better.

NATURAL CONVECTION

8 per cent higher than given by Fig. 7-9. In the laminar range the values of h for heated horizontal plates are only 2 per cent above those read from the alignment chart for horizontal cylinders (Fig. 7-11), using plate width instead of cylinder diameter.

IV. ENCLOSED SPACES

Introduction. It is convenient to define a special heat-transfer coefficient h'_c, based on the temperature difference between the two surfaces:

$$h'_c = \frac{q_c}{A(t_{s1} - t_{s2})} \tag{7-9}$$

where q_c is obtained from the total measured q by deducting the q_r transferred by radiation, instead of employing $h_c = q_c/A(t_{s1} - t_G)$ or $h_c = q_c/A(t_G - t_{s2})$. If the clearance x between the plates were so small that natural convection were suppressed, heat would be transferred only by conduction:

$$q_k = \frac{kA(t_{s1} - t_{s2})}{x}$$

The ratio of the actual q_c to q_k is

$$\frac{q_c}{q_k} = \frac{h'_c A(t_{s1} - t_{s2})}{kA(t_{s1} - t_{s2})/x} = \frac{h'_c x}{k} \tag{7-9a}$$

This special Nusselt number, formed with clearance, is used in the correlation of natural convection coefficients in enclosed air spaces and is plotted against the special Grashof number, $x^3 \rho^2 g \beta (t_{s1} - t_{s2})/\mu^2$.

Vertical Enclosed Air Spaces. Based on the extensive data of Mull and Reiher[28] (L/x of 5.2 to 42.2) and including data of Schmidt (L/x of 3.1 to 6.3) and of Nusselt (L/x of 11.5), Jakob[15] plotted $h'_c x/k$ as ordinate Y vs. $x^3 \rho^2 g \beta \, \Delta t/\mu^2$ as abscissa X, with L/x as a parameter.

The data for air could be correlated by equations of the form

$$\frac{h'_c x}{k_f} = \frac{C}{(L/x)^{1/9}} \left[\frac{x^3 \rho_f^2 g \beta_f \, \Delta t}{\mu_f^2} \left(\frac{c_p \mu}{k} \right)_f \right]^n \tag{7-9b} \star$$

For Grashof numbers (based on clearance) below 2×10^3, natural convection is suppressed, and consequently conduction controls ($h'_c = k/x$), and Eq. (7-9b) does not apply. For $N_{Gr,x}$ from 2×10^4 to 2.1×10^5, C is 0.20, and n is $\frac{1}{4}$; for a given L/x, $h_c \propto x^{-1/4}$. For $N_{Gr,x}$ from 2.1×10^5 to 1.1×10^7, C is 0.071, and n is $\frac{1}{3}$; for a given L/x, h'_c is independent of x.

Equation (7-4a) for a single vertical plane involves the temperature difference $t_s - t_G$, both in the Grashof number and in the definition of h_c:

$$\frac{h_c L}{k_f} = 0.13 \left[\frac{L^3 \rho_f^2 g \beta_f \, \Delta t}{\mu_f^2} \left(\frac{c_p \mu}{k} \right)_f \right]^{1/3} \tag{7-9c}$$

If this be applied to an enclosed air space, $q_c/A = h_c(t_{s1} - t_G) = h_c(t_G - t_{s2}) = h'_c(t_{s1} - t_{s2})$; the predicted h'_c is the same as that from Eq. (7-9b) for L/x of 17.5.

Horizontal Enclosed Air Spaces. For air enclosed between horizontal parallel plates, with heat flow *upward*, the extensive data of Mull and Reiher[28] have been correlated by Jakob.[15] For Grashof numbers below 10^3, conduction controls, and $h'_c = k/x$. For limited ranges of Grashof numbers the data are correlated by equations of the form $Y' = CX^n$:

$$\frac{h'_c x}{k_f} = C\left[\frac{x^3 \rho_f^2 g \beta_f \, \Delta t}{\mu_f^2}\left(\frac{c_p \mu}{k}\right)_f\right] \tag{7-9d} \star$$

For Grashof numbers from 10^4 to 3.2×10^5, C is 0.21, and n is $\frac{1}{4}$; for Grashof numbers from 3.2×10^5 to 10^7, C is 0.075, and n is $\frac{1}{3}$. In the latter range, x cancels, and Eq. (7-9d) reduces to the equation for single heated plates facing upward in air. For fluid enclosed between horizontal parallel plates, in the turbulent range the values of h'_c may be taken as 52 per cent of the value of h_c read from Fig. 7-9 for single vertical surfaces; for the laminar range the values of h'_c may be taken as 40 per cent of the value of h_c based on Fig. 7-11 for horizontal cylinders, with clearance replacing tube diameter in both cases.

Horizontal Enclosed Annular Air Spaces. Consider a horizontal annulus formed by concentric tubes having diameters D_1 and D_2, heated at the smaller surface and cooled at the larger. The enclosed fluid will flow upward past the smaller heated cylinder and downward inside the larger cooled cylinder, thus setting up natural convection currents. Let h'_{c1} be based on the area of the smaller tube:

$$h'_{c1} = \frac{q_c}{A_1(t_{s1} - t_{s2})} \tag{7-9e}$$

The heat flow q_k by conduction alone is given by $q_k = 2\pi L k(t_{s1} - t_{s2})/\ln(D_2/D_1)$. The ratio of q_c to q_k is

$$\frac{q_c}{q_k} = \frac{h'_{c1}(\pi D_1 L)(t_{s1} - t_{s2})}{\dfrac{2\pi L k(t_{s1} - t_{s2})}{\ln(D_2/D_1)}} = \frac{h'_{c1} D_1}{2k}\ln\frac{D_2}{D_1}$$

Kraussold[22] used water and two oils (N_{Pr} from 7 to 4000) and D_2/D_1 from 1.2 to 3. Taking x as the clearance $(D_2 - D_1)/2$, it was found that conduction was controlling when the product of the Grashof and Prandtl number was below 10^3. Fishenden and Saunders[10] state that these data of Kraussold for liquids agree with those for enclosed vertical planes, taking x as the clearance in both cases. This correlation should not be extrapolated to large values of D_2/D_1. Beckmann[2a] reports data for three gases in horizontal annuli.

PROBLEMS

1. A bare horizontal steam pipe with an o.d. of 12 inches carries saturated steam at 240°F. The temperature of the ambient air is 70°F. Calculate the rate of heat loss by natural convection, expressed as Btu per hour per foot of pipe length.

2. The parallel wooden outer and inner walls of a building are 15 ft long, 10 ft high, and 4 inches apart. The inner surface of the inner wall is 70°F, and the inner surface of the outer wall is at 0°F.

   a. Calculate the total heat loss, expressed as Btu per hr, neglecting any leakage of air through the walls.

   b. Repeat the calculation for the case in which the air space is divided in half by a sheet of aluminum foil, 0.001 inch thick, stretched parallel to the walls.

CHAPTER 8

INTRODUCTION TO FORCED CONVECTION

Abstract. The over-all and individual coefficients of heat transfer are defined, and the resistance concept is applied to the important problem of transmission of heat through a series of resistances. The effect of deposits of scale on the heating surface is discussed, and a table of coefficients of heat transfer for scale deposits is given. Mean temperature difference is treated in detail, several illustrative problems are solved, and measurement of tube temperature is considered. Correlation and prediction of heat-transfer coefficients and application to design are reserved for subsequent chapters.

Local Over-all Coefficient of Heat Transfer. In the majority of heat-transfer cases met in industrial practice, heat is being transferred from one fluid through a solid wall to another fluid.

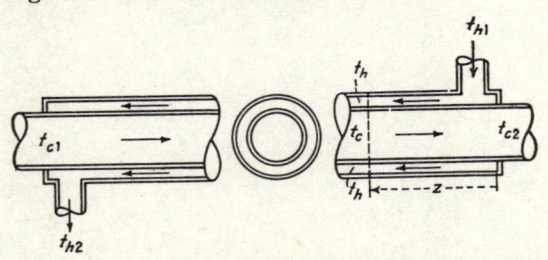

Fig. 8-1. Diagram of counterflow heat exchanger; drawings of typical exchangers are shown in Chap. 15.

In an apparatus such as is shown diagrammatically in Fig. 8-1, hot fluid flows through the jacket, and a cold fluid flows through the tube. All mass flow rates are constant, and the steady state has been attained; hence the temperature at each point in the apparatus is independent of time. Consider any cross section located at a distance z from the point of entry of the hot fluid. If the hot stream were drawn off at this section and mixed, it would have a temperature t_h, called the *bulk temperature*; t_c is the corresponding bulk temperature of the colder fluid at z. For such conditions Newton suggested that the rate dq of heat transfer per unit surface area was directly proportional to the over-all difference between the temperatures of the warmer and colder fluids $t_h - t_c$ or Δt_o and to the heat transfer surface dA:

$$dq = U\, dA\, \Delta t_o \qquad (8\text{-}1) \star$$

and the proportionality factor U is called the *local over-all coefficient of heat transfer* or merely the *local over-all coefficient*. In computing U, one

TABLE 8-1. NOMENCLATURE

A Area of heat transfer surface, square feet; A_h on warmer side, A_c on colder side, A_w for mean area of wall
a' Area of heat-transfer surface per unit length of tube, square feet per foot
b Temperature coefficient in the relation $U = U_o(1 + bt)$, 1/deg F
c_p Specific heat, Btu/(lb fluid)(deg F); c_{ph} for warmer fluid, c_{pc} for colder fluid
d Prefix indicating differential, dimensionless
E Dimensionless ratio in Eq. (8-20), defined in text
F_G Dimensionless ratio, $\Delta t_{om}/\Delta t_{ol}$
F_t Dimensionless ratio in Fig. 8-4. $F_t = (t_{cx} - t_{c1})/\text{rise}$, or $= (t_{hx} - t_{h2})/\text{fall}$
h Individual coefficient between fluid and surface, Btu/(hr)(sq ft)(deg F); h_c is based on A_c and Δt_c; h_h is based on A_h and Δt_h; h_d is for scale deposit, h_{dc} when on A_c, h_{dh} when on A_h
k Thermal conductivity, Btu/(hr)(sq ft)(deg F per ft); k_d for scale deposit; k_w for tube wall
L Tube length, feet
q Rate of heat transfer, Btu per hour; q_t for entire apparatus
R Thermal resistance (hr)(sq ft)(deg F)/Btu; R_c for colder side; R_d for scale deposit; R_h for warmer side; R_t for total, $R_t = \Sigma R$; R_w for tube wall
t Bulk fluid temperature, degrees Fahrenheit; t_c of colder fluid, t_h of warmer fluid; t_{c1} and t_{h1} for entering temperatures, t_{c2} and t_{h2} for exit temperatures; t_{cA} and t_{cB} for colder fluid in first and second passes, respectively; t_x for mean fluid temperature in Fig. 8-4, t_{cx} for colder fluid, t_{hx} for warmer fluid
U Over-all coefficient of heat transfer, Btu/(hr)(sq ft)(deg F); U_c based on A_c, U_h based on A_h, U_w based on A_w; U_d for dirty or fouled apparatus; U_I and U_II at terminals, U_x at average temperature
w Mass rate of flow of fluid, pounds per hour; w_c for colder fluid, w_h for warmer fluid
x Thickness, feet; x_d for dirt or scale deposit, x_w for tube wall
z Distance, feet
Z Dimensionless ratio; $Z = w_c c_{pc}/w_h c_{ph} = (t_{h1} - t_{h2})/(t_{c2} - t_{c1}) = \text{fall/rise}$

Greek

Δt Temperature drop through individual resistances, degrees Fahrenheit; Δt_c for colder side, Δt_d for dirt or scale deposit, Δt_h for warmer side, Δt_w for tube wall
Δt_o Over-all temperature difference between bulk temperatures of warmer and colder fluids, degrees Fahrenheit; $\Delta t_{o\mathrm{I}}$ and $\Delta t_{o\mathrm{II}}$ at terminals; Δt_{oa} is arithmetic-mean value, Δt_{ol} is logarithmic-mean value, Δt_{om} is true-mean value
η Dimensionless ratio; $\eta_C = (t_{h1} - t_{h2})/(t_{h1} - t_{c1}) = \text{fall}/(t_{h1} - t_{c1})$, the "cooling effectiveness"; $\eta_H = (t_{c2} - t_{c1})/(t_{h1} - t_{c1}) = \text{rise}/(t_{h1} - t_{c1})$, the "heating effectiveness"

may use the area of heating surface dA_h, the area of the cooling surface dA_c, or the logarithmic-mean surface dA_w (see Chap. 2). Since in a given case $dq/\Delta t_o$ is fixed, one can obtain three values of local over-all coefficients U_h, U_w, U_c, which are related by means of area ratios: $U_h\, dA_h = U_w\, dA_w = U_c\, dA_c$. It is immaterial which heat-transfer surface is chosen so long as it is specified. The subscripts o for outside, w for wall, and i for inside are often used on U to indicate which area was used in the equation $U = dq/dA\,\Delta t_o$.

As will be shown later, the numerical value of U in a given set of units may vary 10,000-fold, depending on the nature of the fluids, their velocities, and other factors discussed elsewhere. Equation (8-1) is the basic relation for heat transfer between fluids separated by a retaining wall, and the remaining chapters deal with means of predicting the over-all coefficient of heat transfer. This requires consideration of the nature of the thermal resistances met as the heat flows from the warmer fluid through the retaining wall to the colder fluid.

Diagram of Temperature Gradients at a Section of a Heat Exchanger. Continuing the discussion of the conditions at position z of Fig. 8-1, temperatures will be plotted as ordinates against distance from the hottest point in the warmer fluid, as in Fig. 8-2. It is noted that the highest temperature of the hot fluid is slightly above the bulk temperature t_h of the hot fluid and that the outer surface is at a lower temperature.

Local Individual Coefficient of Heat Transfer. As a result of the difference in temperature between the hot fluid and wall, Δt_h, the heat flow rate dq is proportional to Δt_h and to the heat-transfer surface dA_h.

$$dq = h_h\,dA_h\,\Delta t_h \qquad (8\text{-}2)$$

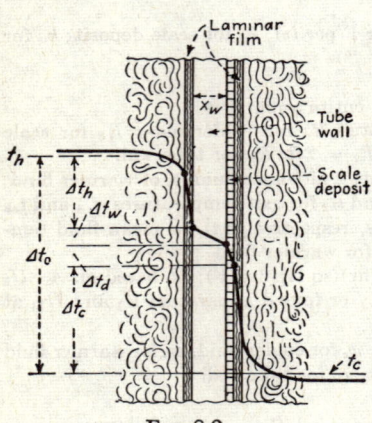

Fig. 8-2.

The proportionality factor h is called the *local individual coefficient of heat transfer* or merely *local individual coefficient* and, depending on the fluid, its velocity, and other factors, could have a numerical value ranging from, say, 0.5 to 50,000 Btu/(hr)(sq ft)(deg F). In flowing from the hot fluid to the surface, heat is transferred both by convection and by conduction in the moving fluid.

Heat Transfer through Solid Conductors. As shown in Chap. 2, the rate of conduction through the retaining wall of thickness x_w having the logarithmic mean area dA_w is given by the equation

$$dq = k_w\,dA_w\,\Delta t_w/x_w \qquad (8\text{-}3)$$

The heat flows by conduction at the same rate dq through the deposit of dirt or scale having thickness x_d on the cooling surface dA_c.

$$dq = k_d\,dA_c\,\Delta t_d/x_d = h_d\,dA_c\,\Delta t_d \qquad (8\text{-}4)$$

INTRODUCTION TO FORCED CONVECTION

The ratio of the thermal conductivity to the thickness of the deposit k_d/x_d is designated as h_d. The thickness of the dirt deposit is usually quite small, and for convenience h_d is based on the area of the heat-transfer surface. In many cases there are deposits of foreign materials on both heat-transfer surfaces.

The heat now flows at the rate dq by conduction through the dirt deposit and thence by convection and also by conduction in the moving fluid to the bulk cold fluid at t_c. As a result of the difference between the temperature of the outer surface of the deposit and the bulk temperature of the cold fluid Δt_c, the local individual coefficient to the cold fluid is defined by the Newton relation

$$dq = h_c \, dA_c \, \Delta t_c \qquad (8\text{-}5)$$

Relation between Local Over-all and Local Individual Coefficients. This relation is readily obtained by solving Eqs. (8-2) to (8-5) for the local individual temperature drops:

$$\Delta t_h = dq/h_h \, dA_h$$
$$\Delta t_w = dq \, x_w/k_w \, dA_w$$
$$\Delta t_d = dq/h_d \, dA_c$$
$$\Delta t_c = dq/h_c \, dA_c$$

Since the rate dq of heat flow is constant and the sum of the individual drops Δt_h, Δt_w, Δt_d, and Δt_c is the over-all temperature difference Δt_o, addition gives

$$\Delta t_o = dq \left(\frac{1}{h_h \, dA_h} + \frac{x_w}{k_w \, dA_w} + \frac{1}{h_d \, dA_c} + \frac{1}{h_c \, dA_c} \right) \qquad (8\text{-}6)$$

Comparison of Eq. (8-6) with Eq. (8-1), $\Delta t_o = dq/U \, dA$, gives

$$\frac{\Delta t_o}{dq} = \frac{1}{U \, dA} = \frac{1}{h_h \, dA_h} + \frac{x_w}{k_w \, dA_w} + \frac{1}{h_d \, dA_c} + \frac{1}{h_c \, dA_c} \qquad (8\text{-}7)$$

where $U \, dA$ may be written as $U_h \, dA_h$, $U_w \, dA_w$, or $U_c \, dA_c$. Equation (8-7) is often solved for a particular U, giving

$$\frac{1}{U} = \frac{dA}{h_h \, dA_h} + \frac{x_w \, dA}{k_w \, dA_w} + \frac{dA}{h_d \, dA_c} + \frac{dA}{h_c \, dA_c} \qquad (8\text{-}7a) \star$$

Where the thickness of the wall is small compared with the diameter of the tube, as an approximation one may use the equation for series flow through a plane wall, with scale deposits on both sides:

$$\frac{1}{U} = \frac{1}{h_h} + \frac{1}{h_{dh}} + \frac{x_w}{k_w} + \frac{1}{h_{dc}} + \frac{1}{h_c} \qquad (8\text{-}7b)$$

Illustration 1. What would be the over-all coefficient for a surface condenser made from 18-gauge Admiralty metal 1-in. tubes, if the steam-side coefficient is 2000, the deposit factors are 2000 on each side, and the water-side coefficient is 1800? What would U_h be if the tubes were cleaned on both sides, and only on the inside?

Solution. As shown in the Appendix, the tube has a wall thickness of 0.049 in. and an i.d. of 0.902 in.; the mean tube diameter is 0.951 in., and k is 63. Noting that the area ratios of Eq. (8-7a) can be replaced by diameter ratios, on a basis of 1 sq ft of *outer* surface, one obtains

$$\frac{1}{U_h} = \frac{1}{2000} + \frac{1}{2000} + \frac{0.049/12}{(63)(0.951/1.00)} + \frac{1}{(2000)(0.902/1.00)} + \frac{1}{(1800)(0.902/1.00)}$$
$$= 0.0005 + 0.0005 + 0.000068 + 0.000554 + 0.000615 = 0.00224$$

Hence $U_h = 1/0.00224 = 446$ Btu/(hr)(sq ft of outside surface)(deg F).

If the scale deposits were removed from both sides, the other coefficients remaining unchanged, U_h would rise to $1/(0.0005 + 0.000068 + 0.000615) = 845$. With tubes clean inside and fouled outside, U_h would be 595.

Because of the fact that the individual coefficients h_h and h_c depend upon fewer variables than the over-all coefficient U, it simplifies correlation of data to study, wherever possible, individual rather than over-all coefficients.† With methods available for predicting values of h, values of U are readily calculated from Eq. (8-7a).

Resistance Concept. By defining a total resistance R_t for unit area dA as $1/U$ and the various individual resistances R_h as $dA/h_h \, dA_h$, R_w as $x_w \, dA/k_w \, dA_w$, R_d as $dA/h_d \, dA_c$, and R_c as $dA/h_c \, dA_c$, Eq. (8-7a) becomes $R_t = R_h + R_w + R_d + R_c$, showing that the individual thermal resistances are additive for steady flow of heat through a series of resistances. Summarizing,

$$dq = \frac{\Delta t_o}{\dfrac{1}{U \, dA}} = \frac{\Delta t_o}{\dfrac{1}{h_h \, dA_h} + \dfrac{x_w}{k_w \, dA_w} + \dfrac{1}{h_d \, dA_c} + \dfrac{1}{h_c \, dA_c}} \qquad (8\text{-}8)$$

It is noted that Eq. (8-8) is analogous to equations for series conduction of heat through several solids.

Coefficients for Scale Deposits. The resistance of a scale deposit is usually obtained from the relation

$$\frac{1}{h_d} = \frac{1}{U_d} - \frac{1}{U} \qquad (8\text{-}9)$$

where U_d is the over-all coefficient for the apparatus with the deposit present and U is the over-all coefficient after cleaning. Apparent thermal conductivities of various boiler scale deposits are given by Partridge[23] and by Biskamp.[4],‡ Table 8-2 shows that values of h_d for water depend on the kind of water, the temperatures of water and heating medium, and the water velocity. Values of h_d are also given in Chap. 13.

† As brought out in Chap. 13, when tests have been run under suitable conditions, it is possible by the use of a graphical method to resolve the over-all thermal resistance for unit area $1/U$ into the component thermal resistances. This method does not involve measurement of surface temperatures in collecting test data and is therefore particularly helpful in the analysis of heat-transfer data from plant equipment.

‡ Biskamp shows the effect of increased porosity in decreasing the apparent thermal conductivity of boiler scale.

TABLE 8-2. HEAT-TRANSFER COEFFICIENTS h_d FOR SCALE DEPOSITS FROM WATER, FOR USE IN EQ. 8-8[a]

Temperature of heating medium	Up to 240°F		240–400°F	
Temperature of water	125°F or less		Above 125°F	
Water velocity, feet per second	3 and less	Over 3	3 and less	Over 3
Distilled	2000	2000	2000	2000
Sea water	2000	2000	1000	1000
Treated boiler feed water	1000	2000	500	1000
Treated make-up for cooling tower	1000	1000	500	500
City, well, Great Lakes	1000	1000	500	500
Brackish, clean river water	500	1000	330	500
River water, muddy, silty[b]	330	500	250	330
Hard (over 15 gm/gal)	330	330	200	200
Chicago Sanitary Canal	130	170	100	130

[a] From "Standards of Tubular Exchanger Manufacturers Association, 3d ed., 1952."
[b] Delaware, East River (New York), Mississippi, Schuylkill, and New York Bay.

TABLE 8-3. h_d FOR MISCELLANEOUS CASES[a]

Organic vapors, liquid gasoline	2000
Refined petroleum fractions (liquid), organic liquids, refrigerating liquids, brine, oil-bearing steam	1000
Distillate bottoms (above 25°API), gas oil or liquid naphtha below 500°F, scrubbing oil, refrigerant vapors, air (dust)	500
Gas oil above 500°F, vegetable oil	330
Liquid naphtha above 500°F, quenching oils	250
Topped crude below 25° API, fuel oil	200
Cracked residuum, coke-oven gas, illuminating gas	100

[a] From "Standards of Tubular Exchanger Manufacturers Association," 3d ed., 1952.

Optimum Operating Conditions. In heating or cooling a liquid or a gas flowing without change in phase, an increase in the mass velocity of the fluid past the surface is accompanied by an increase in the individual coefficient. If corresponding individual resistance $1/h$ is a substantial fraction of the total resistance $1/U$, the over-all coefficient will increase and the total surface required for a given heat-transfer rate q will decrease, thus reducing fixed charges on the investment. On the other hand the use of higher velocity increases the pressure drop and power cost. The optimum velocity, at which total costs are a minimum, is treated in Chap. 15.

Where cooling water is purchased at sufficient pressure to force it through the cooler at any desired velocity, power costs for the water need not be considered and it is possible to calculate the optimum ratio of water to hot fluid or the corresponding over-all temperature difference at the hot end of the cooler (Chap. 15).

In recovering waste heat with an exchanger, as the amount of heat-transfer surface is increased the amount of heat recovered increases, but the fixed charges on the exchanger also increase. Consequently, there is an optimum mean temperature for the particular operation, as shown in detail in Chap. 15.

MEAN TEMPERATURE DIFFERENCE

In a steadily operated heat exchanger the bulk temperatures of both fluids are fixed at a given position, but usually the temperature of one or both of the fluids changes as the fluid flows through the apparatus, and consequently it is necessary to integrate Eq. (8-1), which holds at any point. If the cross section is constant, the velocities are fixed and each individual coefficient and consequently the over-all coefficient U depend on physical properties of the fluids, which, in turn, depend on temperature. Since both temperature and temperature difference are related to q by means of energy and material balances, U and Δt_o depend on q, and hence the variables are separated by writing the equation in the form

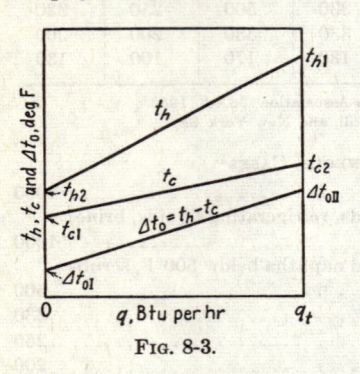

$$\int \frac{dq}{U\,\Delta t_o} = \int dA \tag{8-10}$$

Fig. 8-3.

If necessary, this equation may be integrated graphically, but in many cases the integration may be made algebraically, as illustrated below.

Case I. *Constant U, Parallel or Counterflow Operation.* First consider steady heat exchange between two fluids flowing either in a countercurrent (Fig. 8-1) or in a parallel-flow apparatus. Neglecting kinetic-energy changes relative to changes in enthalpy (Chap. 6) and assuming no heat losses, the energy balance gives†

$$dq = w_h c_{ph}\, dt_h = \pm w_c c_{pc}\, dt_c \tag{8-11}$$

where w is the mass rate of flow, c_p is the specific heat, and t is the bulk temperature. If the specific heats are substantially constant, as is often the case, integration of the heat balance shows that q is linear in each temperature (see Fig. 8-3). Consequently the over-all difference in temperature ($\Delta t_o = t_h - t_c$) is also linear in q. The slope of the plot of Δt_o vs. q is

$$\frac{d(\Delta t_o)}{dq} = \frac{\Delta t_{oII} - \Delta t_{oI}}{q_t} \tag{8-12}$$

† The plus sign applies to counterflow and the minus sign to parallel flow. If one of the temperatures remains constant, because of change in phase, this is equivalent to an infinite value of the specific heat; in such a case the direction of fluid flow is immaterial.

Eliminating dq from Eqs. (8-1) and (8-12), one obtains

$$\frac{d(\Delta t_o)}{U \Delta t_o} = \frac{(\Delta t_{oII} - \Delta t_{oI}) \, dA}{q_t} \qquad (8\text{-}13)$$

If, as when a gas-side resistance is controlling, U is substantially constant, integration, from 0 to A and from Δt_{oI} to Δt_{oII}, gives

$$\frac{1}{U} \ln \frac{\Delta t_{oII}}{\Delta t_{oI}} = \frac{(\Delta t_{oII} - \Delta t_{oI}) A}{q_t} \qquad (8\text{-}14)$$

Upon comparing this with the arbitrarily written equation

$$q_t = UA \, \Delta t_{om} \qquad (8\text{-}15)$$

one finds that in this case†

$$\Delta t_{om} = \frac{\Delta t_{oII} - \Delta t_{oI}}{\ln (\Delta t_{oII}/\Delta t_{oI})} \equiv \Delta t_{ol} \qquad (8\text{-}16) \; \star$$

which is the logarithmic-mean over-all temperature difference. It is strictly correct for constant U, steady operation, constant specific heats, and parallel or counterflow adiabatic operation. If the temperature of one of the fluids is constant, direction of fluid flow is immaterial. Equation (8-16) is often used as an approximation where the percentage variation in U is moderate.

Case II. *Variable U, Parallel or Counterflow Operation.* If U varies substantially with temperature, one can consider the exchanger to consist of a number of exchangers in series, in each of which U is linear in temperature, and hence in Δt_o: $U = a + b \, \Delta t_o$. Eliminating U from Eqs. (8-1) and (8-12), integrating and rearranging, one obtains[8]

$$q_t = A \, \frac{U_I \, \Delta t_{oII} - U_{II} \, \Delta t_{oI}}{\ln (U_I \, \Delta t_{oII}/U_{II} \, \Delta t_{oI})} \qquad (8\text{-}17) \; \star$$

which involves the logarithmic mean of the $U \, \Delta t_o$ products. It is important to note that each product contains Δt_o at one end and the U at the other. If U_I equals U_{II}, this reduces to Eq. (8-14).

Illustration 2. Dry saturated steam, condensing at 227°F, is to be used to heat 13,400 lb/hr of a hydrocarbon oil from 80 to 217°F while flowing inside straight tubes. The specific heat of the oil is substantially constant at 0.47 Btu/(lb)(deg F); the over-all coefficient U_c from steam to oil varies with the bulk temperature of the oil as shown below.

Calculate the square feet of inside heat-transfer surface required.

t, deg F	80	95	110	130	160	190	217
U_c	29.5	38	47	56	70	81.5	91.7

† If $w_h c_{ph}$ equals $w_c c_{pc}$, for counterflow the temperature difference is constant and hence $\Delta t_{om} = \Delta t_{oI} = \Delta t_{oII}$.

Solution. Since heat loss to the surroundings does not affect the temperature of the condensing steam, Eq. (8-17) is not vitiated by heat loss. Upon plotting U vs. t, three straight lines are drawn from 80 to 110°, 110 to 160°, and 160 to 217°F. Equation (8-17) is then applied separately to each of the three zones. For example, for the zone in which t rises from 80 to 110°F, Eq. (8-17) gives

$$A = \frac{(13{,}400)(0.47)(110 - 80)\ln[(29.5)(227 - 110)/(47)(227 - 80)]}{(29.5)(227 - 110) - (47)(227 - 80)} = 37.9 \text{ sq ft}$$

Similar calculations give 59.8 and 143 sq ft for the other two zones, giving a total inside surface of 241 sq ft for all three zones. An alternative method would have been to plot $1/U\,\Delta t$ as ordinates against t and to obtain the area under the curve, which would equal A/wc_p.

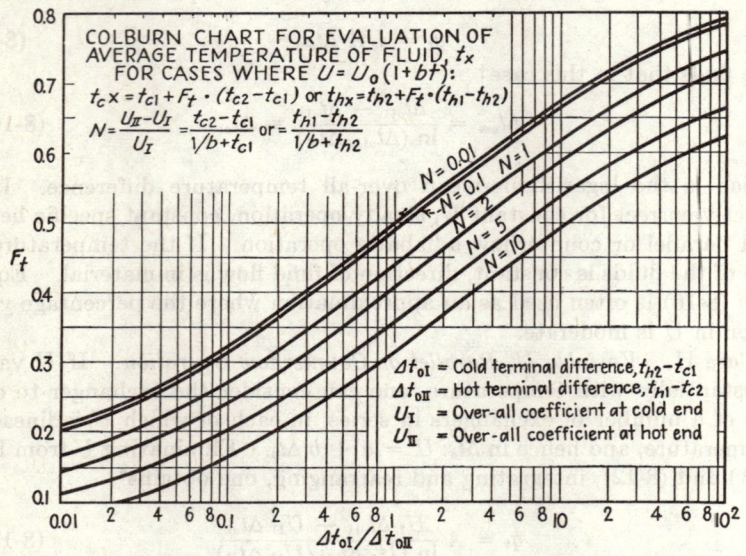

FIG. 8-4. Chart for evaluation of average temperature t_x of fluid in a counterflow heat exchanger.

Even in cases in which the percentage change in U is substantial it is sometimes customary to report U on the basis on the logarithmic-mean over-all temperature difference [Eq. (8-16)], giving

$$U_x = q/A\,\Delta t_{ol} \qquad (8\text{-}17a)$$

If U is linear in t or Δt_o this practice is allowable if U_x is associated with the temperature t_x obtained[8] by eliminating q/A from Eqs. (8-17) and (8-17a), noting that U equals $U_o(1 + bt)$; the value of t_x for *counterflow* is readily obtained from Fig. 8-4.

Illustration 3. Given a counterflow exchanger in which the warmer fluid is cooled from 300 to 105°F and the colder fluid is heated from 100 to 250°F. Estimates of U at the terminals give U_{II} of 120 and U_I of 60, and U of 90 at the section at which the colder stream is at 175°F. Since U is linear in the temperature of the colder stream,

Eq. (8-17) gives the correct flux:

$$\frac{q}{A} = \frac{(60)(50) - (120)(5)}{2.3 \log_{10}(60)(50)/(120)(5)} = \frac{2400}{(2.3)(0.699)} = 1490 \frac{\text{Btu}}{(\text{hr})(\text{sq ft})}$$

Alternatively t_x may be found from Fig. 8-4 as follows: $N = (120 - 60)/60 = 1$, $\Delta t_{oI}/\Delta t_{oII} = 5/50 = 0.10$, $F_t = 0.274$, and $t_{cx} = 100 + (0.274)(250 - 100) = 141°\text{F}$; $U_x = 60 + (141 - 100)(120 - 60)/(250 - 100) = 76.4$. Since the logarithmic-mean difference Δt_{ol} is $(50 - 5)/\ln 10 = 19.5°\text{F}$, the flux is $q/A = (76.4)(19.5) = 1490$ Btu/(hr)(sq ft), checking the value given by Eq. (8-17). Had one arbitrarily evaluated U_x as $(120 + 60)/2 = 90$ and multiplied this by Δt_{ol}, the flux so calculated would have been 18 per cent higher than the correct value. The error would have been further increased by arbitrarily using an arithmetic mean.

Case III. *Constant U, Multipass Flow.* Many exchangers consist of a bundle of tubes inside a suitable shell (Chap. 15). To obtain economic velocities and corresponding heat-transfer coefficients, it is often necessary to arrange the flow paths so that either or both fluids must reverse directions one or more times in passing through the exchanger. In such cases there results a combination of parallel and counterflow, called *reversed flow*. A simple case is shown diagrammatically in Fig. 8-5; this is called a 1-2 *exchanger*, since it contains one well-baffled shell pass and two equal tube passes.

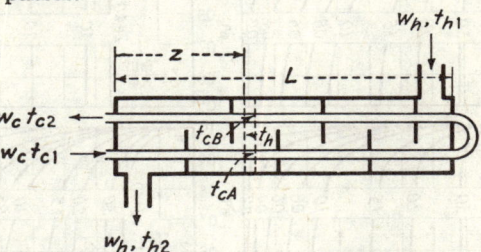

Fig. 8-5. Diagram of 1-2 exchanger, one well-baffled shell pass and two tube passes.

Cold fluid at t_{c1}, having constant specific heat c_{pc}, enters the first pass at the steady mass rate w_c and leaves the second pass at t_{c2}. Warmer fluid at t_{h1} having constant specific heat c_{ph}, enters at a steady mass rate w_h at *either* end of the shell and leaves the other end of the shell at t_{h2}. At any distance z from the inlet to the first pass the tube-side fluid has bulk temperatures t_{cA} in the first pass and t_{cB} in the second pass. Assuming no losses to the surroundings, the heat balance is $\pm w_h c_{ph} \, dt_h = w_c c_{pc}(dt_{cA} - dt_{cB})$, the sign depending on the direction of flow of the shell-side fluid. Designating the heat-transfer area per unit length as a', the rate equations are

$$w_c c_{pc} \, dt_{cA} = U (a' \, dz)(t_h - t_{cA})$$
$$-w_c c_{pc} \, dt_{cB} = U (a' \, dz)(t_h - t_{cB})$$

Upon assuming U constant, integration[35] gives the value of Δt_{om} defined by the equation $q_o = w_h c_h(t_{h1} - t_{h2}) = Ua'2L \, \Delta t_{om} = UA \, \Delta t_{om}$.

$$\Delta t_{om} = \frac{\sqrt{(t_{h1} - t_{h2})^2 + (t_{c2} - t_{c1})^2}}{\ln \dfrac{t_{h1} + t_{h2} - t_{c1} - t_{c2} + \sqrt{(t_{h1} - t_{h2})^2 + (t_{c2} - t_{c1})^2}}{t_{h1} + t_{h2} - t_{c1} - t_{c2} - \sqrt{(t_{h1} - t_{h2})^2 + (t_{c2} - t_{c1})^2}}} \quad (8\text{-}18)$$

Equation (8-18) also applies to cases in which the cold fluid flows through the shell.

Instead of using this cumbersome equation, one defines a dimensionless correction factor F_G in the relation $F_G \equiv \Delta t_{om}/\Delta t_{ol}$, where Δt_{ol} is the logarithmic-mean difference calculated for counterflow:

$$\Delta t_{ol} = \frac{(t_{h1} - t_{c2}) - (t_{h2} - t_{c1})}{\ln\left[(t_{h1} - t_{c2})/(t_{h2} - t_{c1})\right]} \tag{8-19}$$

This purely geometrical factor F_G is often plotted in terms of two dimensionless ratios: the "hourly heat-capacity ratio" $Z = w_c c_{pc}/w_h c_{ph} = (t_{h1} - t_{h2})/(t_{c2} - t_{c1}) = \text{fall/rise}$, and the "heating effectiveness" $\eta_H = \text{rise}/(t_{h1} - t_{c1})$ or the "cooling effectiveness" $\eta_C = \text{fall}/(t_{h1} - t_{c1}) = Z\eta_H$.

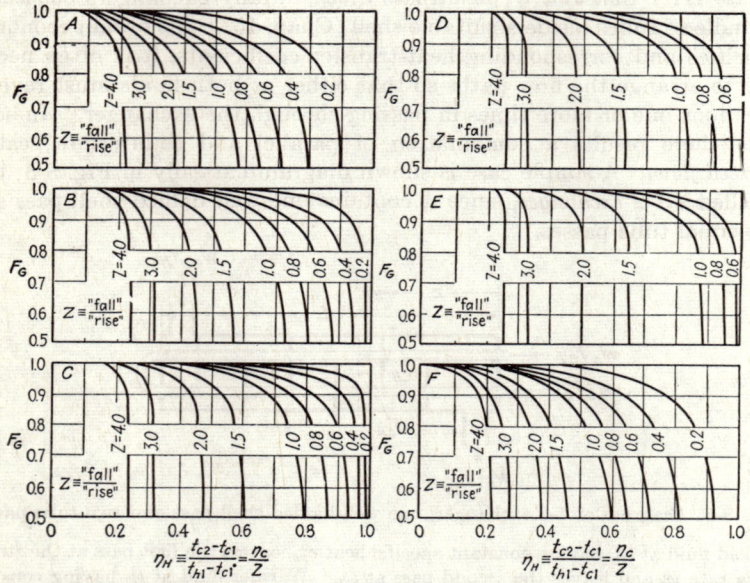

Fig. 8-6. Mean temperature difference in reversed-current exchangers. (Shell-side fluid well mixed at a given cross section.) (A) One shell pass and 2, 4, 6, etc., tube passes. (B) Two shell passes, and 4, 8, 12, etc., tube passes. (C) Three shell passes, and 6, 12, 18, etc., tube passes. (D) Four shell passes, and 8, 16, 24, etc., tube passes. (E) Six shell passes, and 12, 24, 36, etc., tube passes. (F) One shell pass, and 3, 6, 9, etc., tube passes. It is immaterial whether the warmer fluid flows through the shell or the tubes. (*Courtesy of Bowman, Mueller, and Nagle, Trans. ASME.*[6])

With a *counterflow* exchanger containing infinite heat-transfer surface and for Z of 1 or less the colder fluid would be heated from t_{c1} to t_{h1}; since this "maximum rise" in temperature is equal to $t_{h1} - t_{c1}$, η_H can be visualized as the ratio of the actual rise for finite A and Z to that for infinite A and Z of 1 or less. With Z above 1, η_H would be less than 1 even with infinite A in a counterflow exchanger. Figures 8-6 and 8-7 show graphs of F_G vs. η_H for various values of Z for each of a number of types

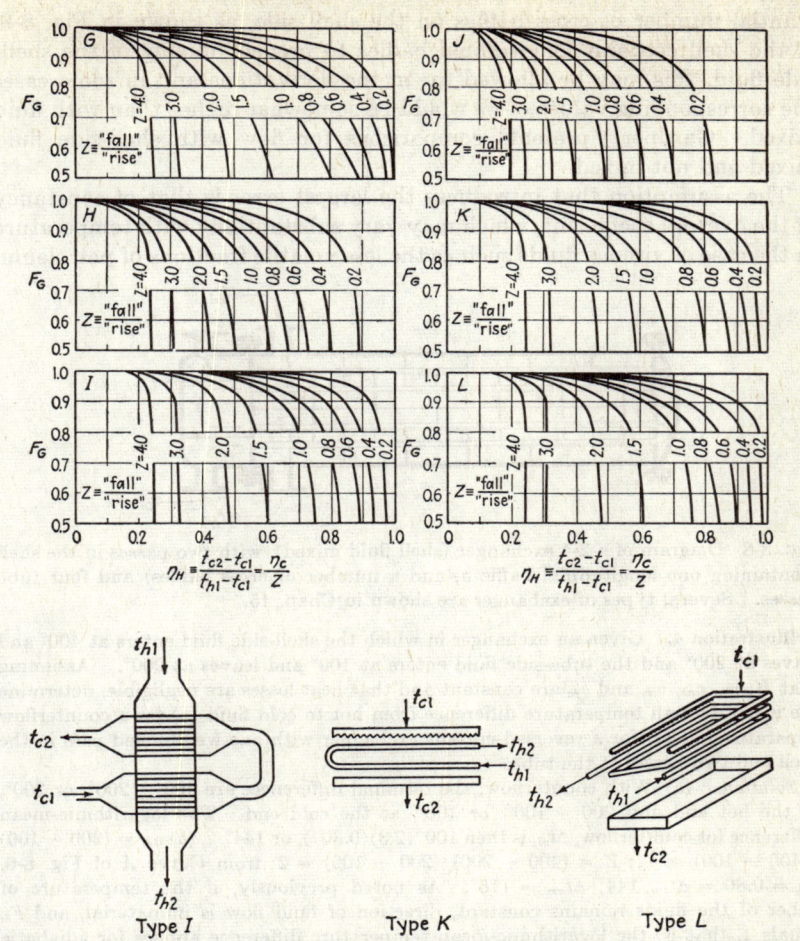

FIG. 8-7. Mean temperature difference in crossflow exchangers. (*G*) Crossflow, both fluids unmixed, one tube pass. (*H*) Crossflow, shell fluid mixed, one tube pass. (*I*) Crossflow, shell fluid mixed, two tube passes, shell fluid flows across second and first passes in series. (*J*) Crossflow, shell fluid mixed, two tube passes, shell fluid flows over first and second passes in series. (*K*) Crossflow (drip type), two horizontal passes with U-bend connections (trombone type). (*L*) Crossflow (drip type), helical coils with two turns. (*Courtesy of Bowman, Mueller, and Nagle, Trans. ASME.*[6])

of exchangers, from the comprehensive paper of Bowman, Mueller, and Nagle.[6,†] Values of F_G are available[12] for exchangers having unequal numbers of tubes per pass.

In the derivations that led to most of the charts the shell-side fluid was assumed to be mixed at a given cross section. This requires a sub-

† Charts *A, B, C*, and *D* are shown to a larger scale in the Appendix.

stantial number of cross baffles on the shell side, as shown in Fig. 8-8. If the shell contains longitudinal baffles to prevent mixing of the shell-side fluid, this may be allowed for in the derivations and in some cases the corresponding values of F_G would be somewhat higher than with fluid mixed. Gardner[13] presents comparisons for flow with shell-side fluid mixed and not mixed.

The assumption that introduces the largest error is that of constancy of the over-all coefficient, which may vary substantially with temperature in the case of viscous fluids such as the less volatile fractions of petroleum.

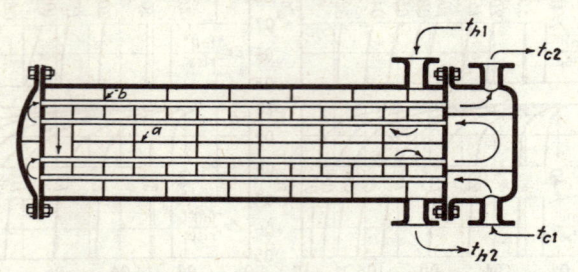

FIG. 8-8. Diagram of a 2-4 exchanger (shell fluid mixed), with two passes in the shell (containing one longitudinal baffle a, and a number of cross baffles) and four tube passes. Several types of exchanger are shown in Chap. 15.

Illustration 4. Given an exchanger in which the shell-side fluid enters at 400° and leaves at 200° and the tube-side fluid enters at 100° and leaves at 200°. Assuming that U, w_h, c_{ph}, w_c, and c_{pc} are constant and that heat losses are negligible, determine the mean over-all temperature difference from hot to cold fluid (a) for a counterflow apparatus and (b) for a reversed-current exchanger with one well-baffled pass in the shell and two passes in the tube.

Solution. (a) With counterflow, the terminal differences are 400 − 200°, or 200°, at the hot end and 200 − 100° or 100°, at the cold end. The logarithmic-mean difference for counterflow, Δt_{ol}, is then $100/(2.3)(0.301)$, or 144°. (b) $\eta_H = (200 - 100)/(400 - 100) = \frac{1}{3}$; $Z = (400 - 200)/(200 - 100) = 2$; from Curve A of Fig. 8-6, $F_G = 0.80 = \Delta t_{om}/144$; $\Delta t_{om} = 115°$. As noted previously, if the temperature of either of the fluids remains constant, direction of fluid flow is immaterial, and F_G equals 1, that is, the logarithmic-mean temperature difference applies for adiabatic operation with constant U, w_h, w_c, c_{ph}, and c_{pc}.

Hurd[17] notes that instead of using F_G charts it is often possible to use the dimensionless equation

$$\frac{w_c c_{pc}}{UA} = \frac{\Delta t_{om}}{t_{c2} - t_{c1}} = \frac{(\Delta t_{oa} + E) - (\Delta t_{oa} - E)}{(t_{c2} - t_{c1})[\ln (\Delta t_{oa} + E)/(\Delta t_{oa} - E)]} \quad (8\text{-}20)$$

where Δt_{om} is the true mean Δt_o, $\Delta t_{oa} = 0.5(t_{h1} + t_{h2}) - 0.5(t_{c1} + t_{c2})$, and E depends only on the geometry of the exchanger. For the 1:2 exchanger $E/(t_{c2} - t_{c1}) = 0.5\sqrt{Z^2 + 1}$. For example, if Z equals 1, $E/(t_{c2} - t_{c1}) = 0.5\sqrt{2} = 0.707$.

In estimating the performance of a given exchanger for fixed values of A, w_h, c_{ph}, t_{h1}, $w_c c_{pc}$, and t_{c1}, U can be estimated; but in solving for t_{h2} or

t_{c2}, use of Fig. 8-6 or 8-7 requires a trial-and-error procedure even with the aid of the over-all heat balance. In such situations the ratios $UA/w_c c_{pc}$ and $UA/w_h c_{ph}$ are known; consequently F_G can be replaced by one of these and the trial-and-error procedure eliminated. The drawback of this modification is that a larger chart is required than with the F_G procedure. Figure 8-9 shows a chart for the crossflow exchanger with

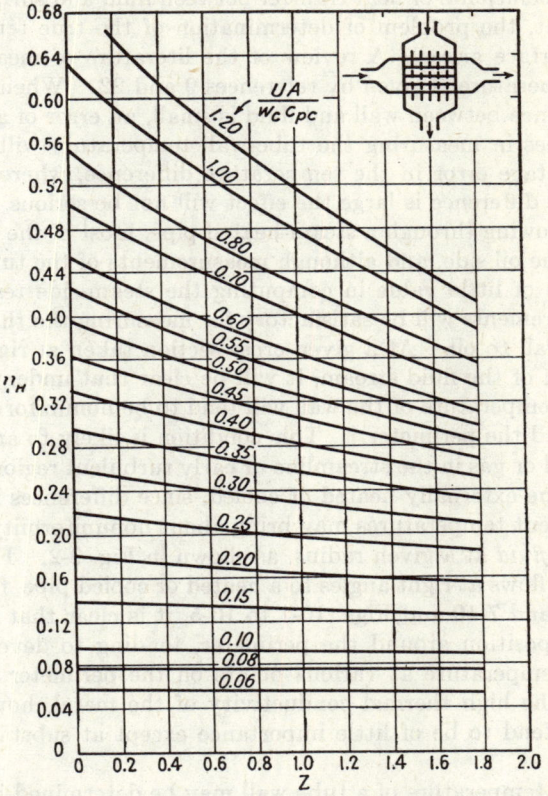

Fig. 8-9. Effectiveness of a crossflow exchanger, both fluids unmixed, one tube pass. (*Courtesy of M. Tribus.*)

both fluids unmixed and one tube pass (case G of Fig. 8-7) in terms of η_H, Z, and $UA/w_c c_{pc}$. Reference 33 presents charts of a similar type and tables for several classes of exchanger.

Kays, London, and Johnson[19] report effectiveness for stepwise countercurrent arrays of two, three, and four crossflow multitubular exchangers with Z of 1; streams are not mixed within each exchanger but are mixed in passing between exchangers. The effectiveness for Z of 1 is plotted against the larger of the ratios $UA/w_c c_{pc}$ and $UA/w_h c_{ph}$, called the "maximum number of transfer units," which is the dimensionless ratio of the larger over-all change in temperature to the true mean Δt_o for the array of exchangers. For various arrays of identical exchangers, Gardner[14] relates an over-all F_G for the array to F_G for an individual component exchanger.

Measurement of Surface Temperature. Where it is desired to obtain individual coefficients of heat transfer between fluid and surface by direct measurement, the problem of determination of the true temperature of the solid surface arises. A review of the literature of measurement of surface temperature is given by references 9 and 22. When the temperature difference between wall and fluid is small, an error of a given number of degrees in measuring the tube-wall temperature will introduce a large percentage error in the temperature difference, whereas when the temperature difference is large the effect will not be serious. For example, for oil flowing through a steam-heated pipe, most of the temperature drop is on the oil side, and although measurements of the tube-wall temperature are of little value in computing the steam-side resistance, the same measurements will be satisfactory for measuring the thermal resistance from wall to oil. At a given cross section taken at right angles to the direction of the fluid stream, it will be clear that under certain conditions the temperature of the wall will tend to be nonuniform at various points around the perimeter.† This condition is likely to arise with the flow of liquid or gas in the streamline or early turbulent regions through a *horizontal* pipe externally heated or cooled, since differences in fluid density at different temperatures may bring about nonuniformity in temperature of the *fluid* at a given radius, as shown in Fig. 9-2. Furthermore, when a fluid flows at right angles to a heated or cooled pipe, from a study of Figs. 7-8 and 7-10 and Figs. 10-1 to 10-5, it is clear that the velocity varies with position around the perimeter, tending to develop nonuniformity in temperature at various points on the perimeter of the wall. Because of the high thermal conductivity of the metal, however, these inequalities tend to be of little importance except at substantial values of q/A.

The mean temperature of a tube wall may be determined by using the tube as a resistance thermometer;[7] the change in tube length with change in mean tube temperature may also be used.[30,31] These methods are suitable only for research work. References 32, 11 present comprehensive reviews of methods of temperature measurement and their application. Thermocouples are widely used in research and industry.‡ Optical methods are also available.

† References 28, 2, 3, and 24.
‡ Morgan and Danforth[21] calibrated thermocouples of tungsten-tantalum up to 5400°F and of tungsten-molybdenum up to 4700°F.

The presence of a thermocouple wire on a surface tends to disturb the flow of the fluid in the zone near the surface, and even if the correct surface temperature were measured, the nature of the fluid flow would be abnormal and the results misleading. For this reason, where the thermocouple leads are brought out through the fluid stream it is desirable, whenever possible, to attach the leads to that side of the surface whose individual coefficient is not under investigation. Where thermocouple leads are brought out through the fluid stream, heat transfer between the fluid and the wire causes a flow of heat along the wire to or from the junction, tending to introduce error (Chap. 10).

Bailey[1] discusses the effects of temperature gradients in thermojunctions on the error introduced in the measurement of surface temperature, the effects of heat capacity upon time lag with varying temperature, and methods for predicting the performance of cylindrical couples. In order to minimize the error due to heat conduction by the leads, it is advisable to submerge the leads from the thermal junction in an isothermal zone[5,26,27] in the tube wall. This matter is discussed by Elias,[10] who employed a number of types of thermocouple. In research work, it is possible to bring the leads out axially through the metal wall itself rather than radially through the fluid stream,[20] although this method has not been used often.[18] Roeser[25] discusses the errors introduced by the diffusion of gases through the walls of pyrometer protection tubes. Spear and Purdy[29] discuss the difficulties involved in making temperature measurements in rubber and insulating materials by means of thermocouples. Various types of thermocouple errors are treated in Chap. 10.

The following methods of installing thermocouples in walls are satisfactory:

1. A groove is cut in that part of the outer surface later to be located in a substantially isothermal zone,[9,24] the bare junction is placed in direct contact with the metal wall of the tube, the electrically insulated leads are installed in the groove so that at least 1 in. of each lead is in the groove,† and the groove is filled with suitable material. If surface conditions are important, as in boiling or condensing, it is advisable to plate the assembly with a suitable coating of metal.

2. The junction is threaded through a chordal hole,[15,16] each lead is submerged in a circumferential groove for at least 1 in., and the groove is filled as described above. This method does not disturb the surface of the metal near the junction. A modification[2] of this method involves placing the junction in a hole drilled at an angle to the axis of the tube.

3. Traveling thermocouples, readily adaptable with short tubes, have been used.

† The lengths of leads submerged in the tube wall should be as large as convenient; this is particularly important if the wire diameter is relatively large.

PROBLEMS

1. Hot oil having a specific heat of 0.5 flows through a reliable meter indicating a steady rate of 50,000 pounds/hr, enters a well-insulated counterflow heat exchanger at a gauge pressure of 80 pounds/sq inch and at a temperature of 380°F, and leaves the exchanger at a gauge pressure of 70 pounds/sq inch and at a temperature of 150°F. Cold oil, having a specific heat of 0.5, flows through a reliable meter indicating a steady rate of 80,000 pounds/hr, enters the exchanger at a gauge pressure of 95 pounds/sq inch and at a temperature of 100°F, and leaves the exchanger at a gauge pressure of 85 pounds/sq inch and a temperature of 300°F. The total cooling surface is 8000 sq ft. It is agreed to overlook the fact that the over-all heat-transfer coefficient may vary with temperature.

Both oils are nonvolatile, even at atmospheric pressure, no chemical changes occur, and the steady state prevails. All thermometers are properly located to measure bulk-stream temperatures, and the thermometers have been recently calibrated.

What sound conclusions should be drawn from these facts? Make conclusions quantitative if possible.

2. An oil is being cooled by water in a double-pipe parallel-flow heat exchanger. The water enters the center pipe at a temperature of 60°F and is heated to 120°F. The oil flows in the annulus and is cooled from 270 to 150°F. It is proposed to cool the oil to a lower final temperature by increasing the length of the exchanger. Neglecting heat loss from the exchanger to the room, and making any reasonable simplifying assumptions which are necessary, but stating all such assumptions clearly, determine:

a. The minimum temperature to which the oil may be cooled.
b. The exit-oil temperature as a function of the fractional increase in the exchanger length.
c. The exit temperature of each stream if the existing exchanger were switched to counterflow operation.
d. The lowest temperature to which the oil could be cooled with counterflow operation.
e. The ratio of the required length for counterflow to that for parallel flow as a function of the exit-oil temperature.

3. Sulfuric acid in the amount of 10,000 lb/hr (specific heat of 0.36) is to be cooled in a two-stage countercurrent cooler of the following type: Hot acid at 174°C is fed to a tank, where it is stirred in contact with cooling coils; the continuous discharge from this tank at 88°C flows to a second stirred tank and leaves the second tank at 45°C. Cooling water at 20°C flows into the cooling coil of the second tank and from there to the cooling coil of the first tank. The water is at 80°C as it leaves the coil in the hot-acid tank.

Calculate the total area of cooling surface necessary, assuming U of 200 and 130 for the hot and cold tanks, respectively, and neglecting heat losses. (Answer is 165 sq ft.)

4. It is desired to heat a cold fluid stream from 20 to 180°F with hot water at a temperature of 200°F. A countercurrent double-pipe heat exchanger is to be used, with the hot water in the inner pipe and the cold fluid in the annulus. The specific heat of the water may be assumed to be constant at 1.0. The true specific heat of the cold fluid is given by the equation

$$c_p = 0.5 + .014(t - 20)$$

a. What is the minimum quantity of water, per pound of cold fluid, that may be used to accomplish this heating? What is the corresponding exit-water temperature?

INTRODUCTION TO FORCED CONVECTION

b. If the water rate is set at 2 pounds of water per pound of cold fluid and the overall coefficient of heat transfer is 200, what is the required area of the exchanger, expressed as square feet of surface area per pound of cold fluid flowing per hour? What is the mean Δt?

5. A vertical shell-and-tube heat exchanger is to be used to condense steam. The exchanger consists of 86 eight-foot lengths of standard 1-inch 18 BWG copper condenser tubing. The steam is to be condensed on the outside of the tubes by water flowing through the inside of the tubes in one pass. The water is available at a rate of 600,000 lb/hr and a temperature of 60°F, and the steam will be condensed at a pressure of 1 atm.

Estimate the capacity of the condenser, expressed as pounds of steam condensed per hour.

Data. The mean individual coefficients of heat transfer may be assumed to be:

$$\begin{aligned}
\text{Water side} &= 1250 \\
\text{Steam side} &= 1750 \\
\text{Scale on water side} &= 1800 \\
\text{Scale on steam side} &= 1800
\end{aligned}$$

6. It is desired to design an adiabatic heat exchanger to cool continuously 110,000 pounds/hr of a solution from 150 to 103°F, using 100,000 pounds/hr of cooling water, available at a temperature of 50°F. The specific heat of the solution is 0.91, and the over-all coefficient of heat transfer may be assumed to be 400. Calculate the heat-transfer area required for each of the following proposals:

a. Parallel flow.
b. Counter flow.
c. Reversed current exchanger with two shell passes and four tube passes, with the hot solution flowing through the shell and the cold water flowing through the tubes.
d. Cross flow, with one tube pass and one shell pass.

7. A steam-heated oil preheater consists of standard 1-in. condenser tubes heated externally by steam condensing at a gauge pressure of 15 lb/sq in. The oil enters the tubes at a gauge pressure of 10 lb/sq in. The volatility of the oil is such that there is no substantial vapor generation in the preheater.

The unit was recently by-passed and shut down for cleaning the inside of the tubes. When the preheater was again returned to operation, the capacity was at first appreciably greater than that which existed before cleaning but within a few hours was far below that which existed just prior to cleaning. The preheater was again shut down, and inspection revealed that the tubes were still clean. What do these facts mean to you?

8. It is proposed to recover waste heat from the stack flue gases in a certain plant by inserting an economizer coil in the stack. The flue gas is available at 300°F, at the rate of 1800 pounds/hr. It is desired to produce saturated steam at 212°F by feeding water at 60°F to the economizer. It is agreed that the economizer coil can be arranged as desired and can be very extensive if necessary. The dew point of the flue gas is 50°F, and the specific heat is 0.26. Calculate the maximum rate of steam generation in the economizer. In which direction should the water flow through the coil for greatest thermal efficiency?

CHAPTER 9

HEATING AND COOLING INSIDE TUBES

Abstract. This chapter is divided into four sections. The *first* deals with turbulent flow in tubes. The mechanisms of transfer of momentum and heat are illustrated by Figs. 9-1 to 9-4. Analogies between the transfer of momentum and heat are shown by Figs. 9-5 and 9-7. Heat-transfer coefficients are correlated in terms of mass velocity, physical properties, length, and diameter. Recommended relations, a design chart, and illustrative problems are included. The *second* section deals with streamline flow, and the *third* with the transition region. The *fourth* treats flow in annular spaces, gravity flow in layer form, flow in rectangular passages, and flow past plane surfaces.

For DG/μ exceeding 10,000 and L/D larger than 60, and $c_p\mu/k$ from 0.7 to 120, heat-transfer coefficients for flow in tubes are correlated within an average deviation of ± 20 per cent by Eqs. (9-10b) and (9-10c). For common gases, these relations reduce to Eq. (9-15), which is the basis of the design chart, Fig. 9-15; the simplified relation for water is Eq. (9-19). For L/D less than 60, the values of h_m are multiplied by the term $1 + (D/L)^{0.7}$. For Prandtl numbers above 120 and below 0.7, Eq. (9-6) and Fig. 9-5 apply; for $c_p\mu/k$ below 0.1, Eq. (9-6) reduces to Eq. (9-7).

For streamline flow ($DG/\mu < 2100$) Eqs. (9-27) and (9-28) are recommended for uniform wall temperatures; for small tubes and moderate Δt the results are approximated by Eq. (9-28a), which is the basis of the simplified design relation, Eq. (9-28b). Theoretical relations are also given for rodlike flow with uniform wall temperature and uniform flux.

For the transition region ($2100 < DG/\mu < 10,000$) Figs. 9-21 and 9-22 are available.

For turbulent flow of liquids in layer form down the walls of vertical tubes, Eq. (9-37a) [which reduces to Eq. (9-37) for water] is recommended; data for laminar flow are given in Fig. 9-26. For laminar flow of water in layer form over horizontal tubes, Eq. (9-39) is recommended. For turbulent flow in rectangular passages, use of Eq. (9-10b), with D_e replacing D, is suggested; data for laminar flow are given in Fig. 9-28. For flow parallel to plates at a uniform temperature, Eq. (9-12a) is used for length Reynolds numbers below 80,000, and Eq. (9-41) is employed for length Reynolds numbers above 500,000; data for the transition region are lacking. Flow normal to plates is treated briefly.

Introduction. Many types of industrial heat-transfer equipment involve heat transfer between a surface, usually metallic or refractory, and a fluid that is heated or cooled without evaporation or condensation. In the power-plant field illustrations include boilers, superheaters, economizers, preheaters, and condensers. In addition to fluids such as air, combustion gases, water, and steam, the petroleum industry involves a variety of products ranging from fixed hydrocarbon gases to the very viscous liquids such as lubricating oils and asphalts. Other industries involve heat transfer to molten metals and slags, acids, and organic liquids.

There is a distinct difference between the mechanism of heat transfer for fluids flowing in turbulent motion on the one hand and streamline motion on the other. Consequently, certain factors, notably average velocity of the fluid past the heat-transfer surface, in general have a more marked effect upon the rate of heat transmission for fluids flowing in turbulent motion than in streamline motion. Other factors, such as tube length, often have greater importance for streamline motion than for turbulent flow. Hence these two cases are treated separately, first consideration being given to the more common turbulent flow.

In many cases, the pipes are smooth, such as drawn tubes of steel, copper, brass, nickel, lead, aluminum, special alloys, and glass, or of only moderate roughness, as cast iron, cast steel, wrought iron, etc., and this section deals largely with heat transfer for turbulent flow of fluids in relatively clean smooth metal tubes. The importance of deposits of slag, scale, and like encrustations on the heat-transfer surface depends upon the thickness and nature of the deposits and also upon the other thermal resistances involved. Such matters are treated in Chap. 8. The effect of roughness is treated on page 223. In the following section, it is assumed that heat transfer by radiation is absent or has been allowed for by the methods of Chap. 4. In other words, this section deals with heat transfer by the combined mechanisms of conduction and convection for fluids flowing inside tubes and miscellaneous cases for flow of fluid parallel to the transfer surface.

Heat-transfer problems ordinarily fall into one of two classes:

1. The use of quantitative relations to design apparatus for proposed installations.

2. The prediction of the effect of changes in operating conditions upon the performance of existing equipment.

As pointed out in Chap. 8, such problems are simplified by resolving the over-all thermal resistance into the various individual resistances: from the warmer fluid to the tube, through the tube wall and scale deposits, and from solid to colder fluid. Consequently the following section deals largely with the quantitative relations between the individual coefficient of heat transfer h between fluid and solid and the various factors influencing the magnitude of the coefficient.

Table 9-1. Nomenclature

- A Area for heat transfer at any radius; $A_w = 2\pi r_w L$ at the wall
- a_1 Constant
- B Thickness of layer of liquid, feet; B_f at film temperature t_f
- b Breadth, wetted perimeter normal to direction of fluid flow, feet
- b_1 Constant
- C Dimensionless factor; see Table 9-4
- c_p Specific heat at constant pressure, Btu/(lb)(deg F)
- D Diameter, feet; D_{Ad} of adiabatic surface; D_e, equivalent diameter equals $4r_h$; D_H of heated wall; D_{He} of helix; D_i, inside diameter of pipe; D_1 and D_2 for sides of rectangular section and for smaller and larger diameters of annulus; D', inches
- d Prefix, indicating differential, dimensionless
- E Eddy diffusivity, square feet per hour; E_H for heat and E_M for momentum
- F_1, F_2 Functions of Z; see Eq. (9-27a) and Table 9-5
- $\mathfrak{F}$ Dimensionless factor in equation for net radiation
- f Friction factor, dimensionless, defined in Chap. 6
- G Mass velocity, lb/(hr)(sq ft of cross section) equals w/S
- g Acceleration due to gravity, usually taken as standard value of 4.17×10^8 ft/(hr)(hr)
- g_c Conversion factor, 4.17×10^8 (lb of fluid)(ft)/(hr)(hr)(pounds force)
- h Coefficient of heat transfer between fluid and surface, Btu/(hr)(sq ft)(deg F); h_a is based on arithmetic mean Δt; h_{ann} for an annulus; h_c for natural convection; h_L is based on logarithmic mean Δt; h_m is based on length mean Δt; h_{m1} is based on initial Δt; h_x is local coefficient based on local Δt, and h_{x1} is h_x based on initial Δt; h_∞ is at large L/D where the local h is constant
- j Product of dimensionless groups; $j = (h/c_p G)(c_p \mu / k) f^{2/3}$; $j' = (h/c_p G)(c_p \mu / k)^{2/3} (\mu_w / \mu)^{0.14}$
- k Thermal conductivity of fluid, Btu/(hr)(sq ft)(deg F per ft)
- L *Heated* length of straight tube, feet; L' is unheated length of calming section; L_c is "critical" or "starting" length; L_{He} is length of helix
- $\ln$ Natural logarithm = $2.303 \log_{10}$
- N Dimensionless group; N_{DR}, ratio of thermal diffusivities (Table 9-2); N_{Gr}, Grashof number = $(D^3 \rho^2 g / \mu^2)(\beta \Delta t)$; N_{Pe}, Peclet number = $DG c_p / k$; N_{Re}, Reynolds number = DG / μ; $N_T = dq_w / dA_w\, c_p \rho T_w u^\star$
- P Absolute pressure, pounds force per square foot
- q Heat-transfer rate, Btu per hour; q_w at the wall
- r Radius, feet; r_w at the wall
- r_h Hydraulic radius, feet, equals cross section divided by total wetted perimeter
- S Cross section of stream in a tube, square feet
- T Absolute temperature, degrees Fahrenheit absolute ($t + 460$); T_b for bulk and T_w for wall
- t Temperature, degrees Fahrenheit; t_c at center of stream; t_s of surface; t_{sv} of saturated vapor; t_w of wall; t_y, local temperature at local distance y from wall; bulk temperatures t_1 and t_2 at stations 1 and 2, respectively; t_0 at base of fin.
- U Over-all coefficient of heat transfer between two streams, Btu/(hr)(sq ft)(deg F); U_i based on inside area and U_o based on outside area
- u Local velocity, feet per hour; u_{max} at axis
- u^+ A dimensionless velocity; $u^+ = u/u^\star$; u_1^+, u_2^+ and u_c^+ at y^+ of 5, 30, and center of stream, respectively

TABLE 9-1. NOMENCLATURE.—(Continued)

- $u^\star$ Friction velocity, feet per hour, $u^\star = V\sqrt{f/2}$
- V Average velocity, feet per hour, $V = G/\rho$; V_1, velocity of deep stream; V' in feet per second
- w Mass rate of flow per tube, pounds fluid per hour
- X Dimensionless abscissa
- x Distance from leading edge, feet
- x_v Fraction dry solids by volume in a slurry
- Y Dimensionless ordinate
- y Distance from wall, feet; y_1, y_2, and y_c at y^+ of 5, y^+ of 30, and at center of pipe
- y^+ Local Reynolds number, dimensionless; $y^+ = y\rho u^\star/\mu = y\rho\, V\sqrt{f/2}/\mu$; $y_1{}^+$ at y^+ of 5, $y_2{}^+$ at y^+ of 30, $y_c{}^+$ at center of pipe
- Z Dimensionless temperature parameter, $Z = (t_2 - t_1)/\Delta t_a$, always positive
- z Dimensionless term, $z = \sqrt{\mathcal{X} + 20\lambda}$

Subscripts

- b For bulk
- f For film; $t_f = (t_b + t_w)/2$
- f', f'' For special film temperatures; $t_{f'} = 0.4t_w + 0.6t_b$; $t_{f''} = 0.6t_w + 0.4t_b$
- w For wall

Greek

- β Coefficient of volumetric expansion, reciprocal degrees Fahrenheit; $\beta = dv/v\,dt$; for perfect gases, $\beta = 1/T$; β_f at T_f
- Γ Mass flow rate per unit breadth, pounds fluid/(hr)(ft); for a vertical tube, Γ equals $w/\pi D$; for a horizontal tube, Γ_H equals $w/2L$
- Δt Temperature difference between bulk temperature of fluid and temperature of surface, degrees Fahrenheit; Δt_a is arithmetic mean, Δt_L is logarithmic mean, Δt_m is length mean, $\Delta t_{\max}$ equals $t_c - t_w$, Δt_y is local value at y
- ϵ Emissivity of a surface, dimensionless; values are tabulated in Appendix
- λ A dimensionless term,
$$\lambda = \frac{\sqrt{2/f}}{DGc_p/k}$$
- μ Viscosity of fluid, lb/(hr)(ft); μ is evaluated at t_b, μ_f at t_f, μ_w at t_w; $\mu = 2.42 \times \mu'$, where μ' is expressed in centipoises
- π A pure number, 3.1416 . . .
- ρ Density, pounds fluid per cubic foot; ρ_f at t_f, ρ_1 at start
- τ Shear stress, force pounds/square foot; τ at any radius and τ_w at wall
- Φ A dimensionless term, equal to $4wc_p/\pi kL$
- ϕ A dimensional term, $\phi = \sqrt[3]{k^3\rho^2 g/\mu^2}$, Btu/(hr)(sq ft)(deg F)
- ψ, ψ_1, ψ_2 Functions defined by context

I. TURBULENT FLOW

A. MECHANISM

Data relative to the isothermal flow of fluids have been presented in Chap. 6, and it was shown that, for isothermal turbulent motion, velocity explorations taken from the wall out into the main body of fluid indicate a thin layer next to the wall where the flow is streamline, a buffer layer beyond the film, and finally a turbulent zone in the main body of the fluid. Explorations of velocity and temperature in a fluid stream are of considerable interest, both in revealing the mechanisms by which the heat is transferred from wall to fluid and in investigating the assumptions made in deriving theoretical relations involving rates of heat transfer from tube to fluid.

Air. In a pioneer investigation in 1916, Pannell[76,†] made velocity and temperature explorations across an *air* stream flowing upward in turbulent motion in an electrically heated *vertical* brass tube having an inside diameter of 1.92 inch. Figure 9-1 shows the results of an experiment made with the wall at 109.4°F and the air at the axis at 75.9°F. The ratio of local to maximum velocity u/u_{max} is plotted vs. the position ratio y/r_w, where u is the local velocity at distance y from the wall and r_w is the radius of the tube. Although heat was being transferred, the velocity-distribution curve resembles those for isothermal runs with air. Figure 9-1 also shows the ratio of local to maximum temperature difference,

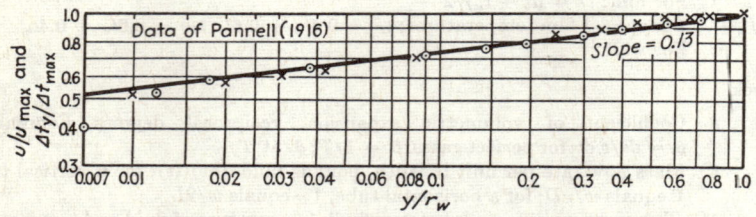

Fig. 9-1. Pannell's explorations[76] of velocity and temperature for air flowing upward in a heated brass pipe, $D' = 1.92$ in., $u_{max} = 87.4$ ft/sec.

$(t_w - t_y)/(t_w - t_c)$, plotted vs. y/r_w, and it is noted that the temperature and velocity fields are substantially identical. The slopes of both curves are roughly 1/7.

Liquids. Woolfenden[99,‡] measured local temperatures t_y in a vertical plane across a stream of water flowing in turbulent motion in a long horizontal copper pipe, 2.06 inches inside diameter, heated by condensing steam. In Fig. 9-2 the ratios of local to maximum temperature difference from the pipe to water are plotted vs. the reduced positions, expressed as fractions of the radius r/r_w. The experiment made at a Reynolds number DG/μ of 11,200, represented by curve AA of Fig. 9-2, brings out the important point that the *temperature distribution* was not symmetrical with respect to the axis. Because of its reduced density, the heated water rose to the upper portion of the horizontal pipe. Certain semi-

---

† Explorations were made across the stream at the middle of the heated section, 46.1 in. long, which was preceded by 197 in. of unheated section.

‡ Explorations were made in a vertical plane at the end of a steam-heated length of 12 ft, preceded by an unheated length of 18 ft. The temperature of the top side of the heated pipe was measured by thermocouples. Immediately upon leaving the heated length, the water was mixed and its bulk temperature determined.

HEATING AND COOLING INSIDE TUBES

theoretical equations ignore this lack of symmetry in the temperature distribution. Referring to the two experiments on water compared in Fig. 9-2, the value of Δt_m, which is the difference between the temperature of the surface and the bulk temperature obtained by mixing all the fluid leaving the section, is 86 to 89 per cent of Δt_{max} from surface to axis. In measuring bulk temperature, it is clear that the omission of a mixing

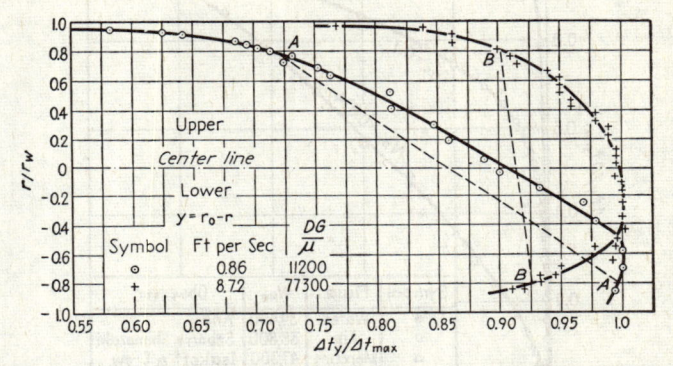

FIG. 9-2. Explorations by Woolfenden,[99] of temperature of water flowing inside a horizontal steam-heated pipe, $D' = 2.06$ in.

device would introduce a substantial error in the calculation of the coefficient of heat transfer.

For Curve BB, where the Reynolds number is 77,300, the dotted tie line is nearly vertical, and the temperature distribution is more nearly symmetrical. For DG/μ of 77,000 to 80,000, Fig. 9-3 shows the temperature and velocity-distribution ratios plotted vs. y/r_w, using logarithmic

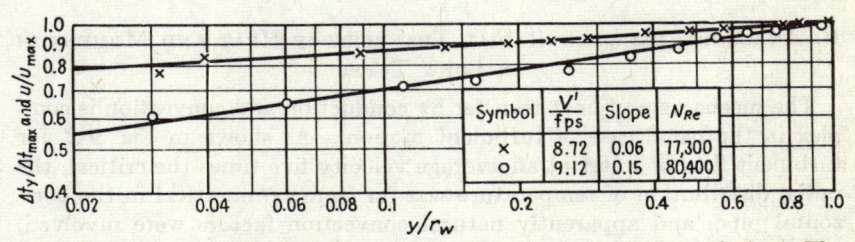

FIG. 9-3. Data of Woolfenden[99] for the upper portion of the horizontal pipe. The upper curve is the temperature field and the lower one is the velocity field.

coordinates. It is noted that the temperature and velocity fields are not the same, since the slopes of the curves are 0.06 and 0.15, respectively. A larger discrepancy was found in slopes for experiments at lower values of DG/μ. The Prandtl number $c_p\mu/k$ ranged from 1.8 to 7.0. For air ($c_p\mu/k$ of 0.7) the temperature and velocity fields are substantially identical (Fig. 9-1).

Effect of Prandtl Number. Figure 9-4 shows the dimensionless temperature fields for water (Koo[54]), for air (Seban and Shimazaki[86]), and for mercury (Isakoff and Drew[43]) flowing upward in externally heated tubes, for Reynolds numbers of 39,400, 38,800 and 47,300. In the turbu-

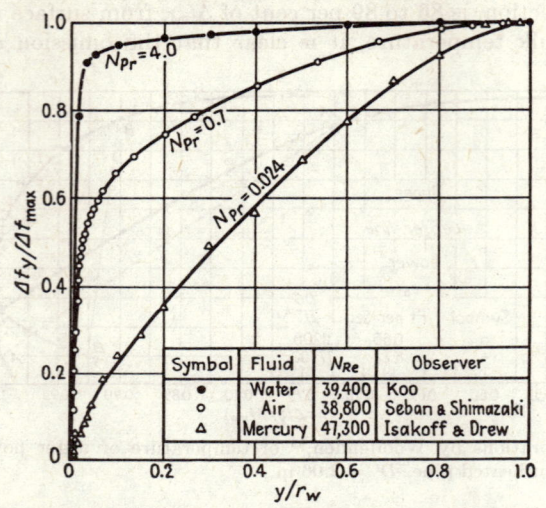

FIG. 9-4. Temperature distribution for upward flow of fluids in vertical heated pipes.[43,54,86]

lent core (y/r_w greater than 0.014, or y^+ greater than 30), as the Prandtl numbers decrease, the slopes of the curves increase. It will be shown later that the slope of the temperature field in the turbulent core is decreased by increasing the Reynolds number.

B. ANALOGIES BETWEEN RADIAL TRANSFER OF HEAT AND MOMENTUM IN LONG TUBES

The mechanism of heat transfer by conduction and convection is complex in the usual case of turbulent motion. As shown in Fig. 9-2, for turbulent flow of water at an average velocity five times the critical, the radial distribution of temperature was far from symmetrical in the horizontal pipe, and apparently natural convection factors were involved, which are ignored in the semi-theoretical treatments.

In the limiting case of the isothermal flow of a liquid or gas in turbulent motion inside a long straight tube with well-developed incompressible flow, radial exploration of local velocity suggests three distinct zones, as shown in Fig. 6-7: (1) a turbulent zone *BC* in the central portion or core, in which eddies are always present; (2) a "buffer layer," or transition zone, *AB*; (3) a laminar sublayer *OA* next to the wall, in which eddies are absent. In the laminar sublayer heat transfer is by molecular

HEATING AND COOLING INSIDE TUBES

transport (conduction), while in the buffer layer and turbulent core the warmer portions of the fluid are mechanically mixed with colder portions by eddy diffusion (convection), thus supplementing the conduction.

The eddy diffusivity of heat E_H is defined by the equation

$$\frac{q}{A} = -(k + \rho c_p E_H)\frac{dt}{dy} \qquad (9\text{-}1)$$

In the laminar sublayer, in which eddies are absent, this reduces to the familiar equation for steady conduction. The eddy diffusivity of momentum E_M is defined by a similar equation

$$\tau g_c = (\mu + \rho E_M)\frac{du}{dy} \qquad (9\text{-}2)$$

In the laminar sublayer ($E_M = 0$) this reduces to the familiar equation for viscous shear.

With well-developed flow the local shear stress is linear in the radius: $\tau/\tau_w = r/r_w$; as an approximation it is assumed that a similar relation applies to the heat flux:

$$\frac{q/A}{(q/A)_w} = \frac{r}{r_w}$$

Equations (9-1) and (9-2) become†

$$\frac{q_w}{A_w \rho c_p}\left(1 - \frac{y}{r_w}\right) = -\left(\frac{k}{\rho c_p} + E_H\right)\frac{dt}{dy} \qquad (9\text{-}1a)$$

$$\frac{g_c \tau_w}{\rho}\left(1 - \frac{y}{r_w}\right) = \left(\frac{\mu}{\rho} + E_M\right)\frac{du}{dy} \qquad (9\text{-}2a)$$

In the right-hand side of Eq. (9-1a), the term in parentheses is the sum of the molecular and eddy diffusivities of heat; the similar term in Eq. (9-2a) is the sum of the molecular and eddy diffusivities of momentum.

1. Reynolds Analogy for $c_p\mu/k$ of 1.0. If the sum of the thermal diffusivities equals the sum of the momentum diffusivities, the ratio of Eqs. (9-2a) and (9-1a) gives

$$\frac{q_w\,du}{A_w c_p g_c \tau_w} = -dt$$

Integration from $u = 0$ to $u = V$ and from $t = t_w$ to $t = t_b$ gives

$$q_w V/A_w c_p g_c \tau_w = t_w - t_b$$

† With uniform heat flux along the tube, if the local velocities were the same at all radii, the radial heat flux would be linear in the radius; allowance for variation of local velocity with radius is introduced later. In reality Eq. (9-1a) is merely the usual arbitrary definition of E_H and is an approximation for uniform heat flux along the tube.

But since by definition $h \equiv q_w/A_w(t_w - t_b)$ and $\tau_w \equiv f\rho V^2/2g_c$, one obtains

$$\frac{h}{c_p V \rho} = \frac{f}{2} \qquad (9\text{-}3) \; \star$$

which is the equation called the Reynolds analogy.[82] The sum of the diffusivities are equal (1) where the eddy diffusivities are controlling and equal, (2) the molecular diffusivities are controlling and equal (Prandtl modulus $c_p\mu/k$ of 1), and (3) the molecular diffusivities of heat and momentum are equal ($c_p\mu/k$ of 1) and the eddy diffusivities are also equal ($E_H = E_M$). Equation (9-3) is a fairly good approximation for common gases (where $c_p\mu/k$ is near 1) and where the temperature potential Δt is moderate.

Since the time of Reynolds, Prandtl[81] (1910), Taylor[89] (1916), von Kármán[48] (1934, 1939), Hoffmann[40b] (1940), and Boelter et al.[6] (1941) have proposed modified analogies; the most recent modifications are discussed below.

2. Martinelli Analogy for Uniform Heat Flux.[66] In order to integrate Eqs. (9-1a) and (9-2a), the following assumptions are made:

1. The temperature potential Δt is negligible; or, alternatively, the physical properties are independent of temperature, and consequently the temperature gradients have no effect on the velocity gradients.

2. The fluid is incompressible and is flowing in well-developed† turbulent motion, and the velocity fields are given by the following equations,‡ based on the data of Nikuradse[72] for water:

In the *laminar sublayer* (y^+ from 0 to 5):

$$u^+ = y^+ \qquad \text{or} \qquad u/u^\star = y\rho u^\star/\mu \qquad (9\text{-}4)$$

where $u^\star$, the "friction velocity," equals $V\sqrt{f/2}$. It is noted that y^+ is also equal to $\frac{1}{2}\frac{DV\rho}{\mu}\sqrt{\frac{f}{2}}\frac{y}{r_w}$; consequently for a fixed Reynolds modulus, y^+ is proportional to y/r_w, the reduced position.

In the *buffer layer* (y^+ from 5 to 30):

$$u^+ = -3.05 + 5 \ln y^+ \qquad (9\text{-}4a)$$

In the *turbulent core* (y^+ above 30):

$$u^+ = 5.5 + 2.5 \ln y^+ \qquad (9\text{-}4b)$$

3. The eddy diffusivities are equal ($E_H = E_M$); other assumptions are cited as they are introduced.

In the laminar sublayer E_H is zero and Eq. (9-1a) is integrated from $y = 0$ and $t = t_w$

---

† The pressure gradient along the tube is due only to shear stress at the wall.

‡ As pointed out by Miller,[68] these equations were fitted to the data by Nikuradse after deducting a correction of 7 from the experimental value of y^+; if this had not been done, a single equation similar to Eq. (9-4b) would have fitted the uncorrected data. Also Miller pointed out that the velocities were measured at a short distance beyond the end of the pipe. Nevertheless, subsequent data of Deissler[19] on air agree closely with the corrected data of Nikuradse.

to $y = y_1$ and $t = t_1$; since this layer is very thin, the term $1 - (y/r_w)$ is taken as 1:

$$q_w = kA_w(t_w - t_1)/y_1 \qquad (9\text{-}5)$$

For the buffer layer the derivative of Eq. (9-4a) is $du/dy = 5u^\star/y$, which is combined with Eq. (9-2a) to give

$$\frac{\mu}{\rho} + E_M = \frac{u^\star}{5}\left(1 - \frac{y}{r_w}\right) y \qquad (9\text{-}5a)$$

This value of E_M is substituted into Eq. (9-1a), again neglecting y/r_w compared with 1; integration from y_1, corresponding to y_1^+ of 5, to y_2, corresponding to y_2^+ of 30, and from t_1 to t_2, gives

$$\frac{5q_w}{A_w \rho c_p u^\star} \ln\left[1 + N_{Pr}\left(\frac{y_2^+}{5} - 1\right)\right] = t_1 - t_2 \qquad (9\text{-}5b)$$

For the turbulent core the derivative of the velocity field [Eq. (9-4b)] is $du/dy = 2.5u^\star/y$, which is combined with Eq. (9-2a) to give

$$\frac{\mu}{\rho} + E_M = \frac{u^\star}{2.5}\left(1 - \frac{y}{r_w}\right) y \qquad (9\text{-}5c)$$

This equation predicts that the sum of momentum diffusivities reaches a maximum at y/r_w of $\tfrac{1}{2}$ and becomes zero at the axis, requiring E_M to be *negative*. This latter unreasonable behavior is due to the fact that the computed du/dy at the axis is finite ($2.5u^\star/r_w$), whereas du/dy is actually zero at the axis, but the error in the final result is small. Since $u^\star$ equals $V\sqrt{f/2}$, Eq. (9-5c) can be rearranged to give

$$1 + \frac{E_M}{\mu/\rho} = \sqrt{\frac{f}{2}}\frac{N_{Re}}{5}\left(1 - \frac{y}{r_w}\right)\frac{y}{r_w} \qquad (9\text{-}5d)$$

Since μ/ρ is small compared with E_M in the turbulent core, it is omitted when combining Eq. (9-5c) with Eq. (9-1a), which combination is integrated from t_2 to t_3 at the axis, and from y_2 corresponding to y_2^+ of 30 to y, giving

$$t_2 - t_3 = 1.25 \frac{q_w}{A_w \rho c_p u^\star} \ln \frac{5\lambda + \dfrac{y}{r_w}\left(1 - \dfrac{y}{r_w}\right)}{5\lambda + \dfrac{y_2}{r_w}\left(1 - \dfrac{y_2}{r_w}\right)}$$

$$+ \frac{1.25 \dfrac{q_w}{A_w \rho c_p u^\star}}{z} \ln \frac{\left[\left(1 - \dfrac{2y}{r_w}\right) - z\right]\left[\left(1 - \dfrac{2y_2}{r_w}\right) + z\right]}{\left[\left(1 - \dfrac{2y}{r_w}\right) + z\right]\left[\left(1 - \dfrac{2y_2}{r_w}\right) - z\right]} \qquad (9\text{-}5e)$$

where

$$\lambda = \frac{\sqrt{2/f}}{DGc_p/k} \qquad \text{and} \qquad z = \sqrt{1 + 20\lambda}$$

By definition of a diffusivity ratio

$$N_{DR} = \frac{E_H}{E_H + (k/\rho c_p)}$$

which is given by the equation†

$$N_{DR}\, 2\ln \frac{N_{Re}}{60}\sqrt{\frac{f}{2}} = \ln \frac{1}{1 + \dfrac{y_2}{5\lambda r_w}\left(1 - \dfrac{y_2}{r_w}\right)} + \frac{1}{z}\ln\left[\left(\frac{z+1}{z-1}\right)\frac{z + \left(1 - \dfrac{2y_2}{r_w}\right)}{z - \left(1 - \dfrac{2y_2}{r_w}\right)}\right] \qquad (9\text{-}5f)$$

where y_2/r_w is based on y_2^+ of 30. Values of N_{DR} are also given in Table 9-2.

† In the published reference 66, typographical errors caused confusion in both Eqs. (9-5e) and (9-5f).

The bulk stream temperature t_b, used in defining $h = dq/dA(t_w - t_b)$, is given by

$$\frac{t_w - t_c}{t_w - t_b} = \frac{\int_0^{r_w} \frac{u}{u_{\max}} \frac{t_w - t}{t_w - t_c} r\, dr}{\int_0^{r_w} (u/u_{\max}) r\, dr}$$

in which $t_w - t_c$ is obtained from Eqs. (9-5), (9-5b), and (9-5e). Computations yield the predictions shown in Table 9-3. The final equation of Martinelli[66] is

$$\frac{h}{c_p G} = \frac{\sqrt{f/2}}{\frac{t_w - t_b}{t_w - t_c}(5)\left[N_{Pr} + \ln(1 + 5N_{Pr}) + 0.5 N_{DR} \ln \frac{N_{Re}}{60}\sqrt{\frac{f}{2}}\right]} \qquad (9\text{-}6) \star$$

In using the Martinelli equation, the friction factor f is graphically related to the Reynolds number (Fig. 6-11), the diffusivity ratio is related to both the Reynolds modulus and the product of the Reynolds modulus and the Prandtl modulus (Table 9-2),† and the temperature-drop ratio is

TABLE 9-2

The values of N_{DR} depend on both N_{Re} and N_{Pe} (= $N_{Re} \times N_{Pr}$), as shown:

N_{Pe}	$N_{Re} \rightarrow$	10^4	10^5	10^6
10^2	....	0.18	0.098	0.052
10^3	....	0.55	0.45	0.29
10^4	....	0.92	0.83	0.65
10^5	....	0.99	0.985	0.980
10^6	....	1.00	1.00	1.00

TABLE 9-3

The ratio $(t_w - t_b)/(t_w - t_c)$ depends upon both N_{Re} and N_{Pr}, as shown:

N_{Pr}	$N_{Re} \rightarrow$	10^4	10^5	10^6	10^7
0	....	0.564	0.558	0.553	0.550
10^{-4}	....	0.568	0.560	0.565	0.617
10^{-3}	....	0.570	0.572	0.627	0.728
10^{-2}	....	0.589	0.639	0.738	0.813
10^{-1}	....	0.692	0.761	0.823	0.864
1.0	....	0.865	0.877	0.897	0.912
10	....	0.958	0.962	0.963	0.966
10^2	....	0.992	0.993	0.993	0.994
10^3	....	1.00	1.00	1.00	1.00

related to the moduli of both Reynolds and Prandtl (Table 9-3).† Thus only three dimensionless ratios (Stanton, Reynolds, and Prandtl or, alternatively, Nusselt, Reynolds, and Prandtl), are controlling.

† The revised values given in these tables and graphs differ slightly from those given in the original reference[66] and were furnished by Martinelli in a personal communication.

HEATING AND COOLING INSIDE TUBES 213

The consequences of the Martinelli equation can be seen from Fig. 9-5. At higher values of the Prandtl modulus the slope of the plot of the Nusselt modulus vs. the Reynolds modulus is substantially 0.8, but as the Prandtl modulus approaches zero, the curve is concave upward, the slope being very small at moderate values of the Reynolds modulus. For

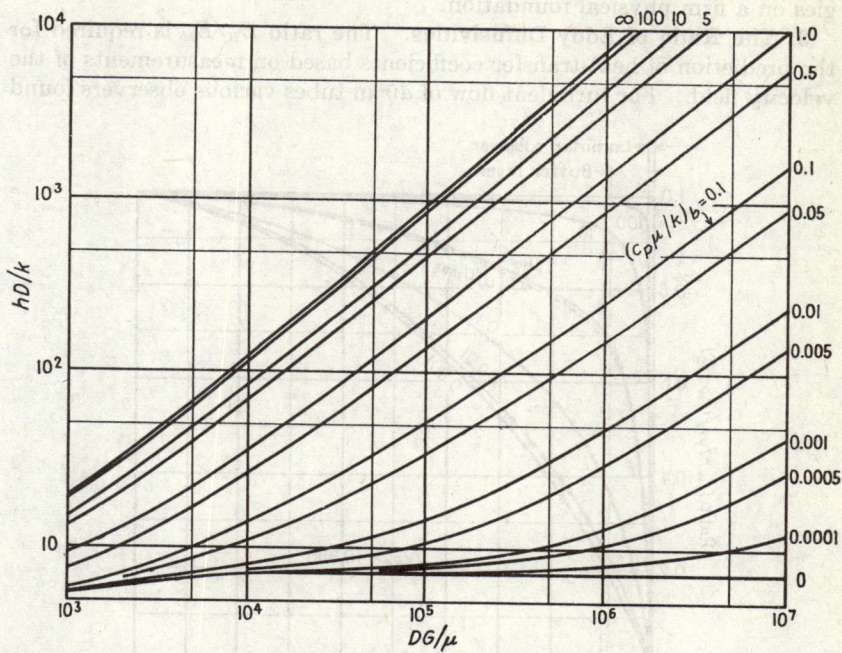

FIG. 9-5. Relations between Nusselt, Reynolds, and Prandtl numbers, predicted by the Martinelli analogy[66] between radial transfer of heat and momentum, for constant heat flux and small Δt.

Prandtl numbers below 0.1, the prediction of Martinelli has been approximated by Lyon[61] within 10 per cent by multiplying the Prandtl and the Reynolds moduli to form the Peclet number:

$$\frac{hD}{k} = 7 + 0.025 \left(\frac{DV\rho c_p}{k}\right)^{0.8} \qquad (9\text{-}7) \star$$

Despite the simplifying assumptions made, the Martinelli equation predicts heat-transfer coefficients of the right order of magnitude, which is encouraging. Such analogies have the advantage over wholly empirical equations in showing how the radial distribution of temperature depends on the Prandtl number. Figure 9-6 shows the dimensionless temperature field for a Reynolds number of 10,000. With a viscous oil (N_{Pr} of 100) 95 per cent of the temperature drop occurs in the laminar sublayer, compared with roughly 3 per cent for liquid mercury; with air (N_{Pr} of 0.7) the temperature and velocity fields are substantially identical.

The theoretical development is handicapped by lack of information on the eddy diffusivities and the effect of the temperature field on the velocity field. It will require an extensive research program to put these analogies on a firm physical foundation.

3. The Ratio of Eddy Diffusivities. The ratio E_H/E_M is required for the prediction of heat-transfer coefficients based on measurements of the velocity field. For turbulent flow of *air* in tubes various observers found

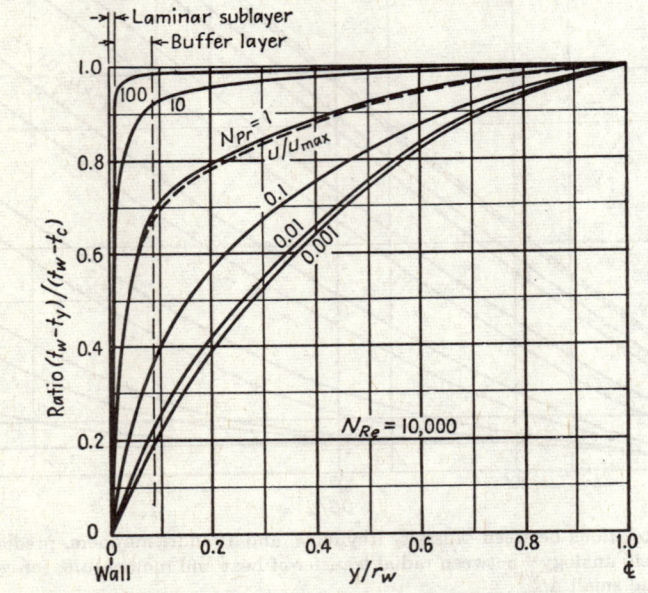

Fig. 9-6. Radial distribution of temperature in a tube, according to the Martinelli analogy[66] between the transport of heat and momentum. Velocity distribution for turbulent flow is shown for comparison. For N_{Pr} of 1 it is seen that the temperature and velocity fields are substantially identical. (*Trans. ASME.*)

E_H/E_M from 0.9 to 1.7. For heating *mercury* at 81°F, Isakoff and Drew[43] find that $E_{M,\max} = 6.3 \times 10^{-5} N_{Re}^{0.83}$, as compared with $3.3 \times 10^{-5} N_{Re}^{0.9}$ from Eq. (9-5d), taking $f = 0.046 N_{Re}^{-0.2}$ and μ/ρ as 0.00434; the Martinelli values are 12 to 25 per cent above those of Isakoff.[43] The local values of both E_H and E_M for mercury reached maxima at y/r_w from 0.35 to 0.4 compared with 0.5 from Eq. (9-5d). The ratio E_H/E_M of the eddy diffusivities, instead of being constant, depended on both N_{Re} and y/r_w as shown in Fig. 9-7, passing through maxima at y/r_w of 0.2. The term $[(E_H/E_M)(N_{Re}^{-0.46})]$ was a unique function of y/r_w.

For cooling air flowing at 30 ft/sec between parallel plates, Schlinger et al.[85] quote data showing that the total diffusivity ($E_H + k/\rho c_p$) reaches a maximum at y/r_w of $\frac{1}{2}$ and at the axis is 20 times $k/\rho c_p$.

Using a modification of the mixing-length theory of Prandtl, Jenkins[45] predicts that the ratio E_H/E_M depends on the Prandtl number and the dimensionless ratio of E_M to μ/ρ. When the latter ratio is very large, E_H/E_M equals 1, but at lower values of $\rho E_M/\mu$, E_H/E_M decreases with decrease in the Prandtl number.

Data for Liquid Metals. Lyon[61] measured over-all coefficients U for sodium-potassium alloy in concentric double-pipe exchangers of nickel.

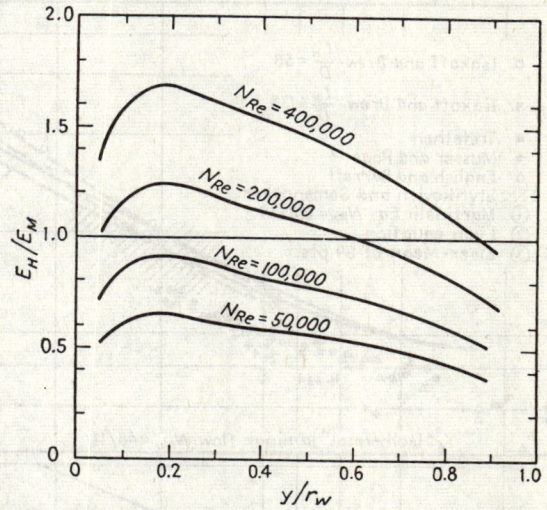

Fig. 9-7. Data of Isakoff and Drew[43] for ratio of eddy diffusivities for mercury flowing vertically upward in an electrically heated tube of stainless steel, 1.5-in. i.d.

Based on the theory of Harrison and Menke[39] for heat transfer between parallel plates with only one plate heated, the following equation was used to predict h in the annulus:

$$\frac{h_{ann}D_e}{k} = 4.9 + 0.0175 \left(\frac{D_e G c_p}{k}\right)^{0.8} \qquad (9\text{-}7a) \star$$

The Martinelli analogy was used to predict h for the tube. The measured values of U fell on both sides of the predicted values, with a maximum deviation of 45 per cent.

Lubarsky[60] measured over-all coefficients U for lead-bismuth eutectic alloy in a vertical double-pipe exchanger. The warmer alloy flowed downward in the annulus (D_2' of 0.625 inch, D_1' of 0.500 inch, and L_H/D_e of 322), and the colder alloy flowed upward inside the central tube (D_1' of 0.402 inch, L_H/D_i of 100). Addition of 0.04 per cent of magnesium as a wetting agent had no effect. It was assumed that the *ratio* of the coefficients (h_{ann} and h_i) could be predicted by using

$$\frac{h_{ann}D_e}{k} = 5.8 + 0.020\left(\frac{D_e G c_p}{k}\right)^{0.8} \tag{9-7b}$$

for the annulus and Eq. (9-7) for the tube. The resulting coefficients, plotted as $h_{ann}D_e/k$ vs. $D_e G c_p/k$ and as hD/k vs. DGc_p/k, fell some 45 per cent below Eqs. (9-7b) and (9-7). These Peclet numbers ranged from 250 to 3800. As the circulation rate increased from 4100 to 17,900

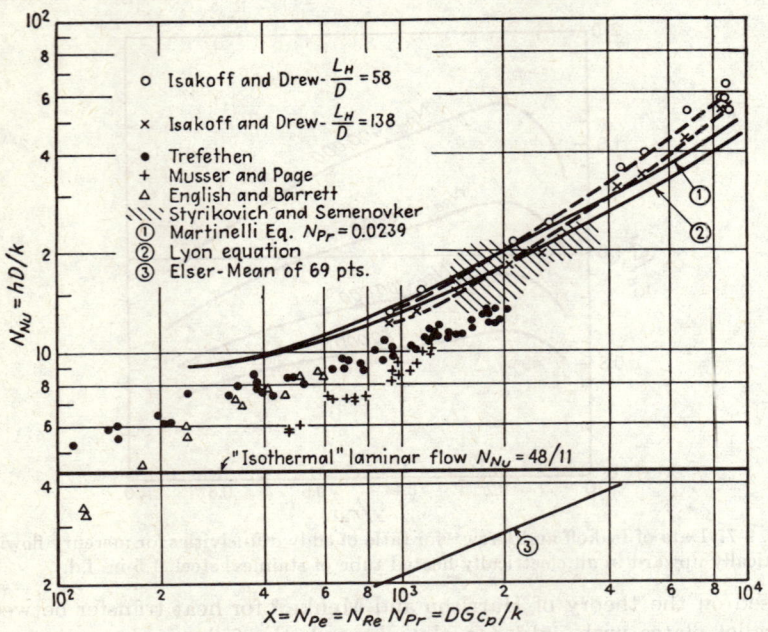

FIG. 9-8. Data of various observers for heating of mercury, taken from Isakoff and Drew.[43]

pounds/hr, h_{ann} increased from 2850 to 5300 and h_i increased from 1330 to 3340.

Data of a number of observers heating mercury in tubes are shown in Fig. 9-8. Some of the observers used uniform flux, and others used constant wall temperature. The data of Isakoff and Drew[43] are in good agreement with the predictions of Martinelli[66] and of Lyon[61] for uniform flux; most of the other data are lower. Trefethen[92] reports that substantially the same results were obtained in heating mercury in tubes whether or not the tube wall was wetted by the mercury.

4. Radial Variation in Physical Properties of Gases, Uniform Heat Flux. Deissler[20,22] treats the cases of $c_p\mu/k$ of 1 and 0.73, with variations in physical properties: $\mu \alpha T^{0.68}$, $k\alpha T^{0.68}$, and $\rho \alpha 1/T$; c_p is assumed constant. The velocity field from y^+ of 0 to 26 is described by a single equation, and the turbulent core is described by the equation $u^+ = 3.8 + 2.78 \ln y^+$, based on data for adiabatic flow of air.[19] Assuming E_H/E_M of 1, the derivation† predicts that hD/k_b depends on DG/μ_b and upon a novel temperature parameter:‡

$$N_T = \frac{(dq/dA)_w}{c_p \rho T_w u^\star} \tag{9-8}$$

For heating (N_T positive) the velocity profile (u^+ vs. y^+) is flattened, and, for a given DG/μ_b, hD/k_b is reduced; for cooling, the reverse effect is predicted. The effect of the parameter N_T was eliminated[20] by evaluating physical properties at special film temperatures:

For heating: $\quad t_{f'} = 0.4 t_w + 0.6 t_b \tag{9-9}$

For cooling: $\quad t_{f''} = 0.6 t_w + 0.4 t_b \tag{9-9a}$

As a close approximation for both heating and cooling, the usual film temperature may be employed:

$$t_f \equiv (t_w + t_b)/2 \tag{9-9b}$$ ★

The predicted heat-transfer coefficients for heating were correlated by plotting

$$Y = \frac{hD}{k_{f'}} \left(\frac{c_p \mu}{k}\right)^{-0.4}_{f'} \text{ vs. } X = DV_b \left(\frac{\rho}{\mu}\right)_{f'}$$

The length-mean coefficients of heat transfer for subsonic heating of air in an Inconel tube were correlated by these coordinates; T_w/T_b ranged from 1.1 to 1.9, and N_T ranged from 0.006 to 0.028. When compared with the theory of Deissler[22] the data fell close to the predicted curve (Fig. 9-9) for $DV_b \rho_{f'}/\mu_{f'}$ above 20,000.

5. Analogy for Uniform Wall Temperature. The derivation of Martinelli for well-developed turbulent flow and uniform flux has been modified by Seban and Shimazaki[86] for uniform wall temperature and small Prandtl numbers. The theory predicts somewhat lower values for uniform wall temperature than for constant flux, particularly at the lower values of the Prandtl modulus, as shown by Fig. 9-10. For Prandtl

---

† The assumptions were made that both the local heat flux and the local shear stress were uniform across the radius, which amounts to omitting the term $[1 - (y/r_w)]$ in Eqs. (9-1a) and (9-2a). It was found that elimination of these assumptions made little difference for adiabatic flow.[19]

‡ This may be rearranged to give

$$N_T = \frac{D}{4} \frac{dt_b}{T_w dL} \sqrt{\frac{2}{f}} \tag{9-8a}$$

in which dt_b/dL is constant since the heat flux is constant.

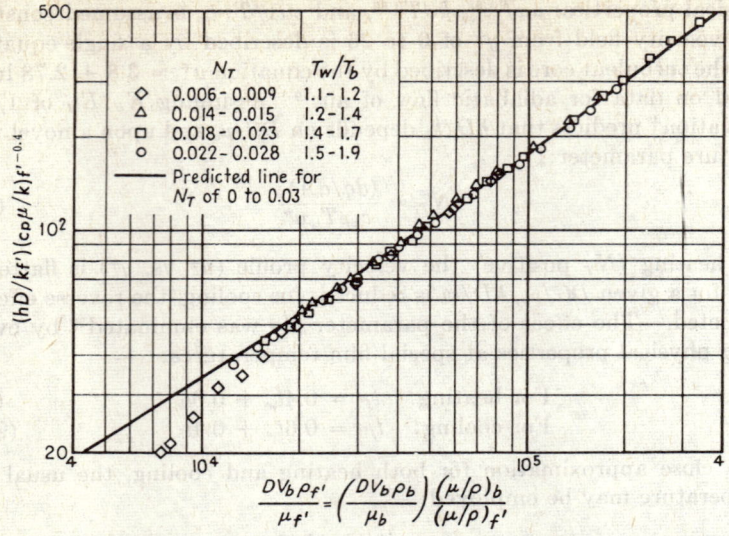

FIG. 9-9. Data of Deissler[22] for subsonic heating of air in an Inconel tube, compared with his theory for $c_p\mu/k$ of 0.73, with variation of k, μ, and ρ with temperature.

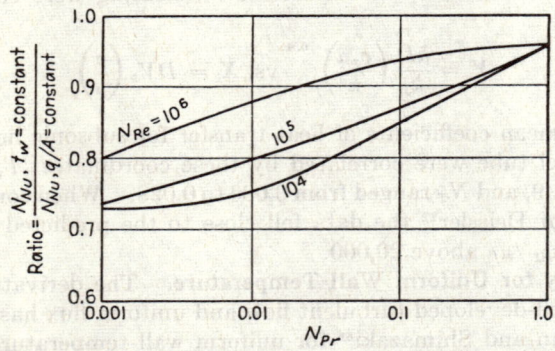

FIG. 9-10. Comparison of theories for uniform flux and uniform wall temperature, Seban and Shimazaki.[86] (*Trans. ASME.*)

moduli of 0.1 or less, the Lyon form of the equation becomes

$$\frac{hD}{k} = 5.0 + 0.025\left(\frac{DGc_p}{k}\right)^{0.8} \quad (9\text{-}9c) \star$$

6. Vorticity Theory. Cope[16] has developed a theoretical relation for heat transfer for turbulent flow in tubes, using the vorticity theory instead of the momentum theory in the turbulent core. For mercury, air, and water, the vorticity theory predicts Stanton numbers which are too low, especially at high Reynolds numbers; at N_{Re} of 10^5 the vorticity theory predicts h/c_pG of 0.00052 for air, compared with the conventional value of 0.0025.

C. Correlation of Data for Turbulent Flow

As shown in Chap. 5, for turbulent flow in tubes, dimensional analysis gives

$$\frac{hD}{k} = \psi\left(\frac{DG}{\mu}, \frac{c_p\mu}{k}\right) \quad \text{or} \quad \frac{h}{c_pG} = \psi_1\left(\frac{DG}{\mu}, \frac{c_p\mu}{k}\right)$$

where the functions are to be determined experimentally. If the tube length is important, as with short tubes, the ratio L/D must be included. With flow of gases at high velocities, the effective Δt involves another dimensionless group (the recovery factor, Chap. 12). For a given fluid whose physical properties are functions of temperature one could include a function of a ratio of two characteristic temperatures, such as wall and bulk stream.

1. Moderate Δt. A number of observers report data for heating or cooling various fluids with Prandtl numbers ranging from 0.7 to 120 for Reynolds numbers from 10,000 to 120,000 and for L/D of 60 or more.†
As shown in reference 62 these data for moderate Δt have been correlated by three *types* of equations. One[23] evaluates all physical properties at the bulk temperature:

$$\frac{h_L D}{k_b} = 0.023\left(\frac{DG}{\mu_b}\right)^{0.8}\left(\frac{c_p\mu}{k}\right)_b^{0.4} \qquad (9\text{-}10a)$$

For viscous liquids where $d\mu/\mu\, dt$ is large there is some difference in opinion as to the temperature at which the viscosity should be evaluated.

Another[14] evaluates all physical properties, except c_p in the Stanton modulus, at the film temperature, $t_f = (t_w + t_b)/2$:

$$\frac{h_L}{c_{pb}G}\left(\frac{c_p\mu}{k}\right)_f^{2/3} = \frac{0.023}{(DG/\mu_f)^{0.2}} \qquad (9\text{-}10b) \star$$

A third[88] evaluates all physical properties at the bulk temperature, except μ_w in a viscosity-ratio term:‡

$$\frac{h_L}{c_{pb}G}\left(\frac{c_p\mu}{k}\right)_b^{2/3}\left(\frac{\mu_w}{\mu_b}\right)^{0.14} = \frac{0.023}{(DG/\mu_b)^{0.2}} \qquad (9\text{-}10c) \star$$

As pointed out by Colburn,[14] the *form* of the last two equations has several advantages over that of Eq. (9-10a):

† As shown on pages 224 to 228, h_L is substantially independent of L/D, for L/D of 60 or more.
‡ In 1937 a correlation[88] of data for petroleum fractions employed a constant of 0.027 in Eq. (9-10c), instead of 0.023, but data for air were correlated with a constant of 0.023. When employing a constant of 0.027 and data for petroleum fractions and for water, the maximum deviations were[62] +43 and −33 per cent. A recent correlation[29a] of new data for air employed a constant of 0.021. It is clear that equations of the forms of the products of the powers of the three dimensionless groups, each with a constant exponent, are inadequate.

220   HEAT TRANSMISSION

1. In evaluating the Stanton number, h_L/c_pG, from experimental data, it is noted that $h_L A \Delta t_L = SGc_p(t_2 - t_1)$, whence

$$\frac{h_L}{c_p G} = \frac{S}{A} \frac{t_2 - t_1}{\Delta t_m} \quad (9\text{-}10d) \star$$

and consequently no physical properties are involved.

2. A plot on logarithmic paper of

$$j \equiv \frac{h_L}{c_p G} \left(\frac{c_p \mu}{k}\right)_f^{2/3}$$

vs. DG/μ_f, yields a line having a slope of only -0.2, whereas a plot of

$$\frac{h_L D}{k_b} \left(\frac{c_p \mu}{k}\right)_b^{-1/3}$$

vs. DG/μ_b yields a line having a slope of 0.8, thus requiring a greater range of ordinates.

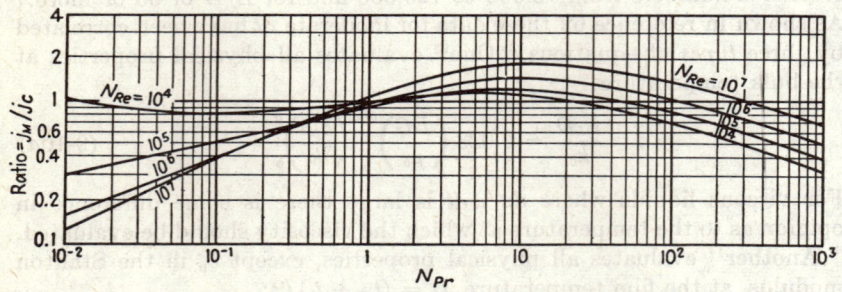

Fig. 9-10a. Ratio of j_M predicted by Martinelli analogy [Eq. (9-6)] to j_C predicted by Colburn relation [Eq. (9-10b)].

3. For Reynolds numbers from 5000 to 200,000, the friction factor for smooth tubes (Chap. 6) is given by the relation $f = 0.046(DG/\mu_f)^{-0.2}$, and consequently one may write[14]

$$j = f/2 \quad (9\text{-}10e) \star$$

The available data are correlated within a *maximum* deviation of ± 40 per cent by the three equations. For extrapolating results for liquids to high Δt, Eq. (9-10c) contains the bulk Reynolds number and the ratio of the viscosities at the wall and bulk temperatures; Eq. (9-10b) involves μ_f and $(c_p\mu/k)_f$ based on the arithmetic average of wall and bulk temperatures. The available data are inadequate[62] to show whether or not one of these two equations is preferable to the other, and use of either is recommended.

Figure 9-10a shows the ratio of the j factors $(h/c_pG)(c_p\mu/k)^{2/3}$ predicted by the Martinelli analogy (j_M) to those predicted by the Colburn relation

[Eq. (9-10b)]:
$$j_C = 0.023/N_{Re}^{0.2} \quad \text{for small } \Delta t$$

While divergence is greatest at small N_{Pr}, significant variation occurs at high N_{Pr}. Values of f for smooth tubes were employed in calculating j_M from Eq. (9-6).

2. Recent Data for Gases: High Δt. Desmon and Sams[24] report data for air flowing through an electrically heated horizontal platinum tube, having an inside diameter of 0.525 inch and a heated length of 24 inches, following a calming length of 24 inches. Figure 9-11 shows the data

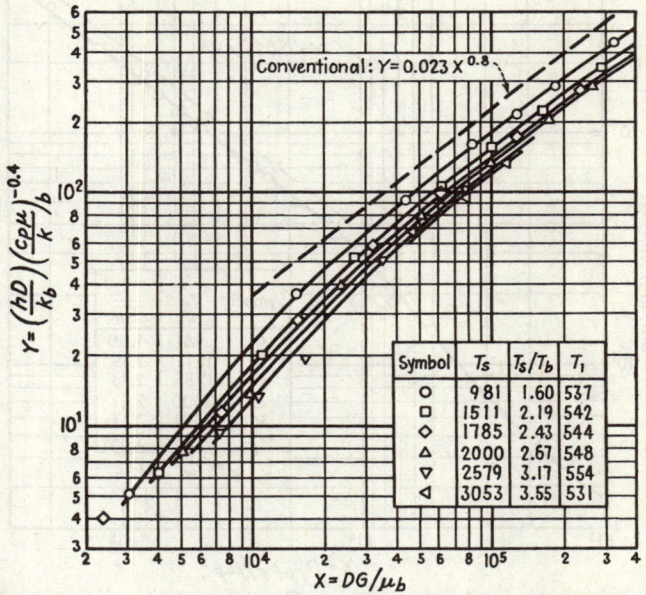

FIG. 9-11. Heat-transfer data[24] at inlet-air temperature of 531 to 554°R. Physical properties of air evaluated at bulk temperature.

plotted with all physical properties evaluated at the average bulk temperature, and the data fall progressively lower than Eq. (9-10a) as the temperature potential increases. The same data are replotted in Fig. 9-12 with all physical properties evaluated at the conventional film temperature, $t_f = (t_w + t_b)/2$, except the velocity which is evaluated at the bulk temperature; it is seen that this method correlates the data better than that used in Fig. 9-11 and the data fall closer to a conventional curve, $Y = 0.023X^{0.8}$. In changing the coordinates from those of Fig. 9-11 to those of Fig. 9-12, the points are shifted both downward and to the left. At the highest value of T_w/T_b, the modified Reynolds number, $DV\rho_f/\mu_f$, is only one-fourth of the conventional Reynolds number,

$DV_b\rho_b/\mu_b$. The data of Lowdermilk and Grele[59] for an Inconel tube agreed with those for the platinum tube. The data of Sams and Desmon[84] for a silicon carbide tube, although well correlated by the coordinates of Fig. 9-12, fell 60 per cent higher than the data for platinum or Inconel. Some of these data are further discussed by Humble *et al.*[42]†

3. Recent Data for Liquids. Bernardo and Eian[5] measured heat-transfer coefficients for water, ethylene glycol, butanol, and aqueous solutions of ethylene glycol, flowing inside an electrically heated tube of 18-8 stainless steel, having an inside diameter of $7/16$ inch. Coefficients for

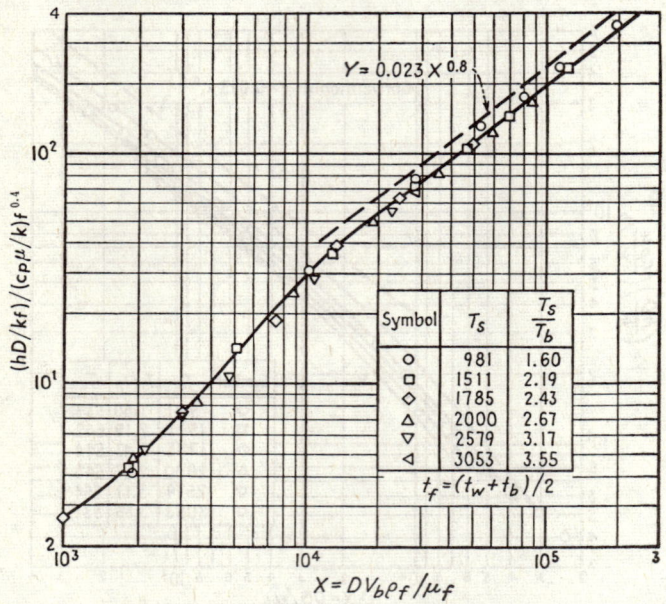

FIG. 9-12. Correlation of heat-transfer data[24] at inlet-air temperature of 531 to 553°R using modified film Reynolds number.

the central 1-ft length were correlated with a maximum deviation of 20 per cent by Eq. (9-10a). The Prandtl numbers ranged from 1.9 to 60, and the Reynolds numbers ranged from 5000 to 300,000; the transition zone extended to DG/μ_b of 10,000. Data for liquid metals are given in Fig. 9-8.

Molten Salts. Where temperatures of 300 to 1000°F are desired, one may use a molten salt mixture (40 per cent $NaNO_2$, 7 per cent $NaNO_3$, and 53 per cent KNO_3 by weight). Physical properties (c_p, μ, and ρ) are available.[53] Values of h_m for flow inside a 6-ft length of standard $3/8$-inch

† In analyzing these data for high Δt, one handicap is the paucity of data for some of the physical properties of air at high temperatures, and the other is lack of correction for axial conduction of heat along the wall of the tube and thence into the long air-cooled bus bars.

iron pipe ranged from 133 to 2660 as V' varied from 0.8 to 6 ft/sec and Reynolds numbers ranged from 2340 to 23,500; bulk temperatures ranged from 580 to 960°F.

Effect of Pulsations. With water flowing in a 2-inch pipe at Reynolds numbers from 35,000 to 80,000, West and Taylor[96] found that the maximum increase (50 to 100 per cent) in heat-transfer coefficient occurred at a pulsation ratio of 1.4, defined as the ratio of maximum to minimum volume of air in the surge tank preceding the test section, over the cycle of one pulsation.

Slurries in Tubes. Bonilla et al.[8] heated aqueous slurries of precipitated chalk flowing in turbulent motion inside a horizontal copper pipe having an inside diameter of 1.61 inches, heated externally by condensing steam. An unheated calming section 46.5 diameters long preceded the test section, which had a length of 58.5 diameters. The viscosity μ of the slurry was calculated from the Hatschek equation:

$$\frac{\mu}{\mu_w} = \frac{1}{1 - x_v^{1/3}}$$

where μ_w is the viscosity of water and x_v is the volume fraction of dry solids, which ranged from 0 to 0.075. The thermal conductivity of the slurry was taken as that of water, and the specific heat c_p and density ρ of the slurry were taken as the mean of the components on a weight-fraction basis; Reynolds numbers ranged from 25,000 to 280,000. For a given Reynolds number, the data for the slurries fell below those for water. The data were roughly correlated by the equation: $Y_w - Y = 555x_w$, where x_w is the weight fraction of solids, which ranged from 0 to 0.18:

$$Y = \frac{hD/k_f}{(c_p\mu/k)_f^{1/3}}$$

and Y_w is the corresponding term for water. At x_w of 0.18, $Y_w - Y = 100$; at $DV\rho/\mu_f$ of 35,000, a conventional equation, $Y_w = 0.023(DV\rho/\mu_f)^{0.8}$, gives Y_w of 100, and consequently the predicted value of Y is zero.

4. Surface Conditions. Cope[17] studied cooling of water in tubes with three degrees of artificial roughness, in which the height of pyramid ranged from one forty-fifth to one-seventh of the radius of the pipe. Although in the turbulent region the friction ran as high as six times that for smooth tubes, the heat transfer ran only 20 to 100 per cent greater than for smooth tubes.† It was concluded that, for the same power loss or pressure drop, greater heat transfer was obtained from a smooth than from a rough tube.

The effect of a given scale deposit is usually far less serious for a gas than for water because of the higher thermal resistance of the gas film

† Because of very small Δt, these values of h are somewhat uncertain.

compared with that of water film. However, layers of dust or of materials that sublime, such as sulfur, may seriously reduce heat transfer between gas and solid. Certain crudes and residual stocks may foul the tubes, thus reducing the over-all coefficient of heat transfer; scale-deposit coefficients are given in Chap. 8. When the primary heating coils in such plants are operated at high rates of heat transfer, a deposit is quite likely to be found in the tubes at the end of a long run and this is usually removed by a mechanical cleaner. With certain hydrocarbon oils, at low temperatures the deposit of a waxlike layer on the cooling surface will give a lower heat-transfer coefficient than predicted for a clean surface.

5. Effect of Length-Diameter Ratio. *Laminar Boundary Layer on a Flat Plate.* Before considering the conditions in a tube near the entrance, it is helpful to consider a thin plate immersed in a deep stream of uniform velocity V_1, density ρ_1, and temperature t_1; the plate is maintained at uniform wall temperature t_w. Owing to the action of the viscous drag, a portion of the stream adjacent to the plate is retarded, thus forming a laminar-flow boundary layer, whose thickness B increases as the distance x from the leading edge is increased (Fig. 9-13). The Pohlhausen theory[78] leads to the dimensionless relation

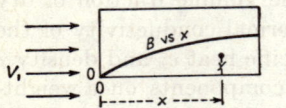

$$\frac{B}{x} = \frac{5.83}{(xV_1\rho_1/\mu_1)^{1/2}} \quad (9\text{-}11) \star$$

Fig. 9-13. Diagram of laminar boundary layer on upper side of a flat plate in a deep stream.

Heat is transferred by conduction through the laminar layer. As x is increased, the length Reynolds number $xV_1\rho_1/\mu_1$ increases until a "critical" value is reached, where transition to a turbulent boundary layer starts, and thereafter the above theory is invalid. The critical value of $xV_1\rho_1/\mu_1$ depends on the disturbances but often ranges from 80,000 to 500,000 or even higher.

Pohlhausen also obtained theoretical equations for both the local coefficient of heat transfer h_{x1} at the length x and the mean coefficient h_{m1} from x of 0 to x of L, both based on $t_w - t_1$:

$$\frac{h_{x1}}{c_p V_1 \rho_1}\left(\frac{c_p \mu}{k}\right)^{2/3} = \frac{1}{3}\left(\frac{xV_1\rho_1}{\mu}\right)^{-1/2} \quad (9\text{-}12) \star$$

$$\frac{h_{m1}}{c_p V_1 \rho_1}\left(\frac{c_p \mu}{k}\right)^{2/3} = \frac{2}{3}\left(\frac{LV_1\rho_1}{\mu_1}\right)^{-1/2} \quad (9\text{-}12a) \star$$

In a recent discussion Tribus[93] used the Chapman-Rubesin analysis[11] to develop equations for a laminar boundary layer of air (N_{Pr} of 0.72) on a flat plate in a deep stream of uniform velocity. For uniform heat flux (q/A) the constants in Eqs. (9-12) and (9-12a) should be multiplied by 1.36. But if the temperature gradient along the plates is linear, the constants in Eq. (9-12) should be multiplied by the dimensionless factor

$$\frac{1 + 1.65\,(x/\Delta t_1)(dt/dx)_w}{1 + (x/\Delta t_1)(dt/dx)_w} \qquad (9\text{-}12b)$$

Where the rise in surface temperature is large compared with the initial temperature difference, the corresponding local coefficient would be 1.65 times that for uniform wall temperature.

Inlet Region of a Tube. As the fluid flows down the tube the thickness of the laminar-flow boundary layer on the inner wall increases and fills the tube (Fig. 9-14); consequently, in this region the local heat-transfer coefficient decreases with increase in x. At the length at which the velocity profile ceases to change, the flow is "fully developed" and the local h is independent of length. The value of L/D at which the flow becomes fully developed depends[7] on the type of entry and the Reynolds number DG/μ.

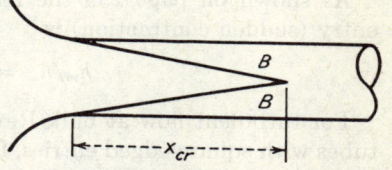

FIG. 9-14. Boundary layer in inlet region of a tube with flared inlet.

According to the theory of Latzko[56,7] for turbulent flow in a tube with a bellmouthed entry, laminar boundary layers build up on the wall in the inlet region of the tube and meet at the "critical," or "starting," length, given by

$$\frac{L_c}{D} = 0.693 \left(\frac{DG}{\mu}\right)^{1/4} \qquad (9\text{-}13)$$

Thus for DG/μ of 40,000, the predicted starting length is 9.8 diameters.

For L/D *less* than given by Eq. (9-13), Latzko[56] predicts that the mean coefficient h_m (from L of 0 to L) is related to the asymptotic value h_∞ by the equation

$$\frac{h_m}{h_\infty} = 1.11 \left[\frac{(DG/\mu)^{1/5}}{(L/D)^{4/5}}\right]^{0.275} \qquad (9\text{-}13a)$$

This prediction is closely substantiated by recent data[7] for a bellmouthed entry.

For L/D *greater* than given by Eq. (9-13), Latzko predicts that the mean h from 0 to L is

$$\frac{h_m}{h_\infty} = 1 + \frac{C}{L/D} \qquad (9\text{-}13b)$$

where the dimensionless factor C is a weak function of DG/μ:

$$C = 0.144(DG/\mu)^{1/4} \qquad (9\text{-}13c)$$

This prediction is confirmed by recent data[7] for a bellmouthed entry, with DG/μ of 26,000 to 56,000, where C was found to be 1.4 compared with 1.83 to 2.22 predicted by Eq. (9-13c). For L/D of 10, a substantial change in C produces only a small change in h_m/h_∞.

Table 9-4 gives experimental values[7] as a function of the type of entry.

TABLE 9-4. VALUES OF C IN EQ. (9-13b) FOR DG/μ FROM 26,000 TO 56,000 FOR $L/D > 5$

Bellmouth with one screen.....................	1.4
Long calming section, $L_c/D = 11.2$.............	1.4
Short calming section, $L_c/D = 2.8$.............	Ca. 3.0
45-deg-angle bend entrance....................	Ca. 5.0
90-deg-angle bend entrance....................	Ca. 7.0

As shown on page 239 the recommended relation for a sharp-edged entry (sudden contraction) is

$$h_m/h_\infty = 1 + (D/L)^{0.7} \qquad (9\text{-}13d) \star$$

For turbulent flow at bulk Reynolds numbers DG/μ_b above 10,000, in tubes with square-edged entries, for Prandtl numbers from 0.7 to 120, the **recommendation** is to employ a combination of Eq. (9-13d) and Eq. (9-10c) or Eq. (9-10b):†

$$\frac{h_L}{c_p G}\left(\frac{c_p \mu}{k}\right)_b^{2/3}\left(\frac{\mu_w}{\mu_b}\right)^{0.14} = \frac{0.023[1 + (D/L)^{0.7}]}{(DG/\mu_b)^{0.2}} \qquad (9\text{-}14) \star$$

or

$$\frac{h_L}{c_p G}\left(\frac{c_p \mu}{k}\right)_f^{2/3} = \frac{0.023[1 + (D/L)^{0.7}]}{(DG/\mu_f)^{0.2}} \qquad (9\text{-}14a) \star$$

6. Simplified Equations, $L/D > 60$. For purposes of rough evaluation, the designer of heat exchangers often desires some approximate equations for moderate Δt.

Common Gases. Since the Prandtl numbers of common gases do not vary widely and since viscosity enters only as $\mu^{0.2}$, a simplified equation is obtained by substituting $(c_p\mu/k)_b$ of 0.78 and μ_b of 0.0455 lb/(hr)(ft) into Eq. (9-10a), giving the dimensional equation‡

$$h_L = 0.0144\, c_p G^{0.8}/D^{0.2} \qquad (9\text{-}15) \star$$

provided the effects of natural convection (page 172) are not important.

Design Chart for Gas Heaters or Coolers. The heat transferred can be equated to that absorbed by the gas:

$$\frac{\pi}{4} D^2 G c_p (t_2 - t_1) = h_L \pi D L\, \Delta t_L \qquad (9\text{-}16)$$

† For turbulent flow of air in steam-heated tubes recent data[29a] indicate that 0.023 should be reduced to 0.021 for air.

‡ Alternatively, this simplified equation may be obtained from Eq. (9-10b) by substituting $(c_p\mu/k)_f$ of 0.78 and μ_f of 0.0426.

or

$$\frac{t_2 - t_1}{\Delta t_L} = \frac{4L}{D}\frac{h_L}{c_p G}$$

For design purposes, h_L may conservatively be predicted from Eq. (9-15). By combining Eqs. (9-15) and (9-16), there results

$$\frac{t_2 - t_1}{\Delta t_L} = \frac{(0.0576)(L/D)}{(DG)^{0.2}} \qquad (9\text{-}17) \star$$

and it is noted that the ratio of rise to mean temperature difference depends mainly upon L/D and slightly upon DG.

The convenient alignment chart[12a] shown in Fig. 9-15 is based on Eq. (9-17).

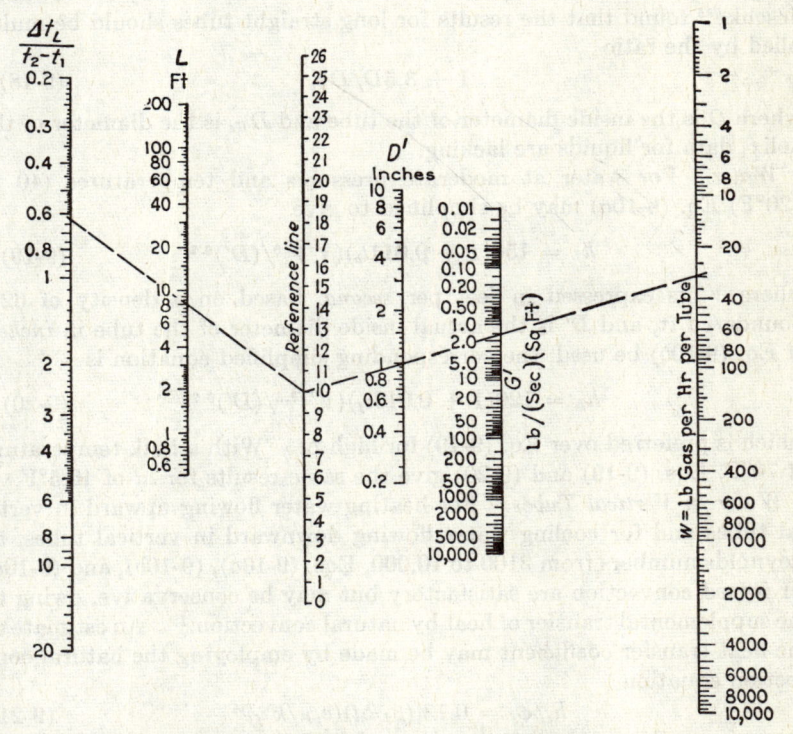

Fig. 9-15. Design chart[12a] for tubular gas heaters or coolers based on Eq. (9-17), which gives conservative results for cases where natural convection or small L/D is important (Chap. 7).

Illustration 1. It is desired to heat 600 lb/hr of a gas from 70 to 190°F, with steam condensing at 220°F outside tubes, maintaining the inner walls at 220°F. It is agreed to use tubes having an i.d. of 0.902 inch and a gas velocity near the optimum value. Calculate the number of tubes in parallel and the length of each tube.

Solution. Employing the methods of Chap. 15 and suitable cost data, the optimum value of G' is 1.88 lb/(sec)(sq ft of cross section). Figure 9-15 shows that the corresponding gas rate per tube is 30 lb/hr; hence 600/30, or 20, tubes in parallel are required.

The ratio of the logarithmic-mean temperature difference to the temperature rise is

$$\frac{(t_w - t_1) - (t_w - t_2)}{(t_2 - t_1)\ln\frac{t_w - t_1}{t_w - t_2}} = \frac{1}{\ln\frac{t_w - t_1}{t_w - t_2}} = \frac{1}{2.3\log_{10}\frac{150}{30}} = 0.621$$

A straight line is now drawn on Fig. 9-15, passing through D' of 0.902 and G' of 1.88 (also through w of 30), and intersects the reference line. Alignment of this intersection with 0.621 on the left-hand line marked $\Delta t_L/(t_2 - t_1)$ gives a reading of 7.3 ft on the L scale. To introduce a factor of safety, each tube should be made 9 ft long. It is seen that the tube length is independent of the specific heat of the gas.

Effect of Curvature. For turbulent flow of air in helically coiled tubes, Jeschke[46] found that the results for long straight tubes should be multiplied by the ratio

$$1 + 3.5 D/D_{He} \qquad (9\text{-}18) \star$$

where D is the inside diameter of the tube and D_{He} is the diameter of the helix; data for liquids are lacking.

Water. For water at moderate pressures and temperatures (40 to 220°F) Eq. (9-10a) may be simplified to give

$$h_L = 150(1 + 0.011 t_b)(V')^{0.8}/(D')^{0.2} \qquad (9\text{-}19) \star$$

where V' is expressed in feet per *second*, based on a density of 62.3 pounds/cu ft, and D' is the actual inside diameter of the tube in *inches*. If Eq. (9-10b) be used, the corresponding simplified equation is

$$h_L = 120(1 + 0.013 t_f)(V')^{0.8}/(D')^{0.2} \qquad (9\text{-}20) \star$$

which is preferred over Eq. (9-19) for high Δt. With a bulk temperature of 70°F, Eqs. (9-19) and (9-20) give the same results for Δt of 46.5°F.

Water in Vertical Tubes. For heating water flowing upward in vertical tubes and for cooling water flowing downward in vertical tubes, at Reynolds numbers from 2100 to 10,000, Eqs. (9-10a), (9-10b), and (9-10c) for forced convection are satisfactory but may be conservative, owing to the supplemental transfer of heat by natural convection.[15] An estimate of the heat-transfer coefficient may be made by employing the natural convection equation

$$h_c/\phi_f = 0.13[(\beta_f \Delta t)(c_p \mu/k)_f]^{1/3} \qquad (9\text{-}21)$$

for the turbulent natural convection range, where the product of the Grashof and Prandtl numbers exceeds 8×10^7. Figure 7-9 is an alignment chart based on Eq. (7-4a) and shows the predicted relation between the fluid, the film temperature, the temperature potential, and h_c; for use with gases, a pressure term appears on the chart.

For products of the Grashof and Prandtl numbers below 10^9, Eq. (9-21)

HEATING AND COOLING INSIDE TUBES

and Fig. 7-9 give conservative results, and Fig. 7-7 should be used. For the combined effects of forced and natural convection with streamline flow, see Eqs. (9-27) and (9-28).

II. STREAMLINE FLOW

As mentioned previously, when a fluid is flowing in fully developed *isothermal* streamline or laminar motion at constant rate through a long tube, there is a parabolic velocity gradient over any cross section, with the maximum velocity at the axis and zero velocity at the wall (Curve *A*, Fig. 9-16). There is no appreciable mixing of the various layers of fluid, and the motion may be visualized as a series of concentric shells slipping past each other.

Consider a long *vertical* tube with a liquid at room temperature flowing at low velocity through it, the first section being bare and the second surrounded by a jacket in which vapor is condensing at constant temperature. In the unheated section the velocity distribution will be parabolic, but, shortly after the liquid enters the heated section, a temperature gradient will be established in the liquid, with a high temperature at the wall and a low temperature at the axis. Since the viscosity of a liquid falls as the temperature rises, there will be a viscosity gradient, with a low viscosity at the wall and a high viscosity at the axis. As a result, the layers of liquid near the wall will flow at a greater velocity than in the unheated section. Since the total flow through both sections is the same, it is clear that some of the liquid from the center of the pipe must flow toward the wall in order to maintain the increased velocity near the wall. Thus, the heating of the liquid develops a *radial* component of the velocity, distorting the parabola (Curve *B*, Fig. 9-16).

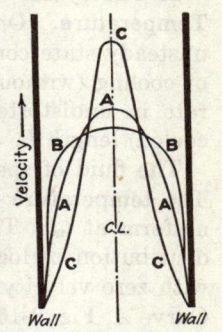

Fig. 9-16. Effect of heat transmission on velocity distribution in streamline motion. Curve *A*, isothermal flow; Curve *B*, heating of liquid or cooling of gas; Curve *C*, cooling of liquid or heating of gas.

If the liquid were cooled, radial flow would again develop, but in this case the flow would be toward the center, changing the velocity distribution as shown diagrammatically by Curve *C*, Fig. 9-16. Where change in density with temperature is appreciable, there may occur additional disturbances as in natural convection (Chap. 7). For liquids the viscosity decreases as the temperature rises, but for gases the viscosity increases as the temperature rises. Therefore, Curve *B* for the heating of a liquid applies also to the cooling of a gas and Curve *C* to the cooling of a liquid and the heating of a gas. Since, in general, the rate of change of viscosity with temperature is much less for a gas than for a liquid, it is expected that the effect of heat transmission on velocity distribution in laminar flow probably would be correspondingly less for a gas than for a liquid.

Since in the presence of heat transmission the velocity distribution of isothermal laminar motion is distorted, the term *modified laminar motion* is sometimes used to describe the nonisothermal type of streamline flow. Data are not available to demonstrate the diagram presented, but the qualitative conclusions seem inescapable. Consequently theoretical equations, which ignore the distortion of the parabola, would be expected to apply only when the temperature differences are small or for fluids whose physical properties vary but little with temperature. The mathematical theory is outlined briefly below.

1. Theory for Parabolic Distribution of Velocity and Uniform Wall Temperature. One of the earliest applications of the equation for unsteady-state conduction was given in 1885 by Graetz[37,38],[†] for heating or cooling (without change in phase) of a fluid flowing at constant mass rate in undistorted laminar motion *inside* a pipe having a heated (or cooled) length L.

The fluid of specific heat c_p and thermal conductivity k enters at t_1. The temperature of the inner surface of the heated length is assumed uniform at t_w. The flow is assumed laminar in character, so that the distribution of local mass velocity over any cross section is parabolic, with zero velocity at the wall and a maximum velocity at the axis (see Curve A, Fig. 9-16). This is the only way in which viscosity is assumed to enter the problem. Heat is assumed to be transferred by radial conduction only; the thermal conductivity of the fluid is assumed uniform. With this assumption, Graetz integrated the Fourier-Poisson equation[79] for radial conduction in a moving fluid and obtained the relation

$$\frac{t_2 - t_1}{t_w - t_1} = 1 - 8\psi(n_1) \qquad (9\text{-}22)$$

In this equation, $\psi(n_1)$ represents the convergent infinite series[‡]

$$\psi(n_1) = 0.10238 e^{-14.627 n_1} + 0.01220 e^{-89.22 n_1} + 0.00237 e^{-212 n_1} + \cdots \qquad (9\text{-}22a)$$

where n_1 represents the dimensionless term $\pi k L / 4 w c_p$. It is noted that a coefficient of heat transfer is not involved in Eqs. (9-22) and (9-22a).

Curve JK of Fig. 9-17 shows this theoretical relation with $(t_2 - t_1)/(t_w - t_1)$ plotted as ordinate against $w c_p / k L$ as abscissa.

From the definition of the individual average coefficient of heat transfer h_m and the usual heat balance, $h_m D / k$ becomes

$$\frac{h_m D}{k} = \frac{1}{\pi} \frac{w c_p}{k L} \frac{t_2 - t_1}{(t_s - t)_m} \qquad (9\text{-}23)$$

[†] An excellent digest of the literature on the mathematics of conduction in moving fluids is given by Drew.[26]

[‡] Drew[26] tabulates n_1 and ψn_1.

HEATING AND COOLING INSIDE TUBES

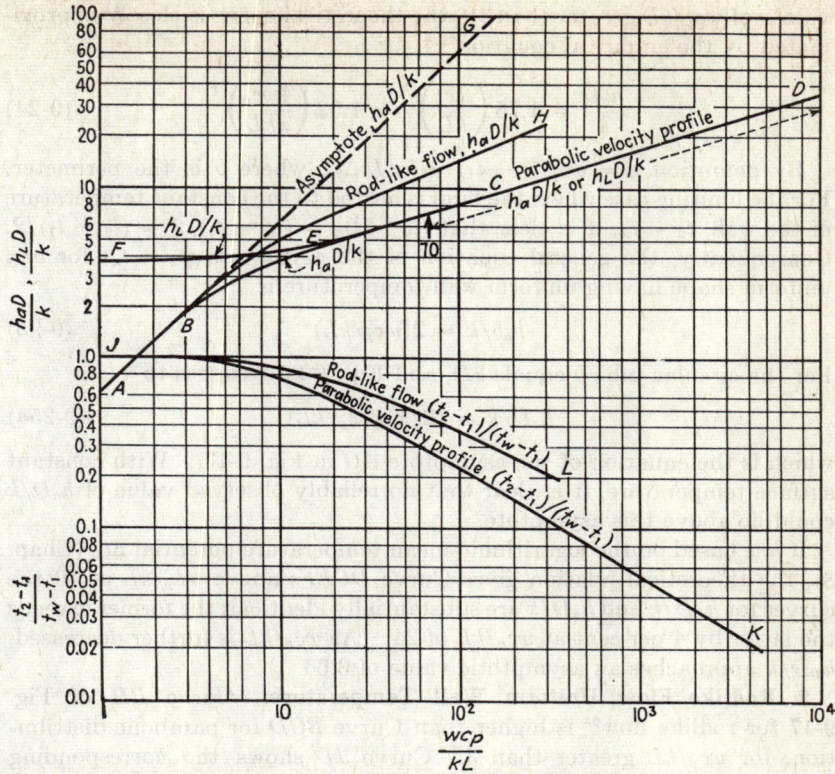

FIG. 9-17. Theoretical relations for laminar and for rodlike flow in circular tubes having uniform wall temperature; invariant physical properties.

Hence equations of the type of Eq. (9-22) can be rearranged to involve h_m, and the average h may be based upon *any* type of mean temperature difference desired.† For design purposes the arithmetic mean of the terminal values

$$(t_w - t)_a \equiv \frac{(t_w - t_1) + (t_w - t_2)}{2} \qquad (9\text{-}23a)$$

is convenient. These four equations may be combined to give

$$\frac{h_a D}{k} = \frac{2}{\pi} \frac{w c_p}{kL} \frac{1 - 8\psi(n_1)}{1 + 8\psi(n_1)} \qquad (9\text{-}23b)$$

Curve ACD of Fig. 9-17 represents the theoretical relation based on parabolic distribution of mass velocity and is plotted from Eq. (9-23b) and extended above wc_p/kL of 400 by an equation of Lévêque.[57] For

† Norris and Streid[73] tabulate the theoretical values of hD/k vs. $4wc/\pi kL$ with h based on initial Δt, arithmetic mean Δt, and logarithmic Δt.

values of wc_p/kL greater than 10, the theoretical curve is closely approximated by the empirical equation[29,†]

$$\frac{h_a D}{k} = 1.75 \left(\frac{wc_p}{kL}\right)^{1/3} = 1.62 \left(\frac{4wc_p}{\pi kL}\right)^{1/3} \qquad (9\text{-}24)$$

By definition $q = wc_p(t_2 - t_1) = h_a bL \Delta t_a$, where b is the perimeter. For the limiting case where the fluid is heated to the constant temperature of the wall, $t_2 = t_w$, it is clear that $(t_w - t)_a = (t_w - t_1)/2 = (t_2 - t_1)/2$. Consequently, the general equation of the asymptote $(t_2 = t_w)$ for *any* uniform shape having uniform wall temperature is

$$h_a b/k = 2(wc_p/kL) \qquad (9\text{-}25)$$

For the circular tube b equals πD, and Eq. (9-25) reduces to

$$h_a D/k = (2/\pi)(wc_p/kL) \qquad (9\text{-}25a)$$

which is the equation of the asymptote AG in Fig. 9-17. With constant surface temperature, it is clear that no reliably observed value of $h_a D/k$ could lie above this asymptote.

If h is based on the logarithmic-mean temperature potential Δt_L (Chap. 8), the theoretical relation gives Curve $DCEF$; above wc_p/kL of 70 the curves for $h_L D/k$ and $h_a D/k$ are substantially identical; the former exceeds the latter by 4 per cent at wc_p/kL of 24. As wc_p/kL is further decreased, $h_L D/k$ approaches an asymptotic value of 3.66.

2. Rodlike Flow, Uniform Wall Temperature. Curve BH of Fig. 9-17 for rodlike flow[26] is higher than Curve BCD for parabolic distribution, for wc_p/kL greater than 3. Curve JL shows the corresponding ratio of temperature change to the initial Δt.

Brinn et al.[9] heated various sands flowing downward in steam-heated vertical tubes, kept full by restricting the outlet. The sand was found to travel in rodlike flow, and the data were well correlated by the theoretical relation[26] for this type of flow:

$$\frac{(t_2 - t_1)}{(t_w - t_1)} = 1 - \psi_1(X) \qquad (9\text{-}26)$$

where

$$\psi_1(X) = 0.692 e^{-5.78X} + 0.131 e^{-30.4X} + 0.0534 e^{-74.8X} + \cdots \qquad (9\text{-}26a)$$

and X equals $\pi kL/wc_p$. For rise/Δt_1 greater than 0.55, only the first term of the infinite series is significant. For a given wc_p/kL, it is noted that tube diameter has no effect on rise/Δt_1. Equation (9-26a) also applies to the transient heating of a long solid cylinder, initially at uniform temperature, by maintaining a uniform surface temperature. It is noted that

† In references 14 and 88 the constants of 1.75 and 1.62 were incorrectly given as 1.65 and 1.5.

$\pi k L/wc_p$ reduces to the Fourier modulus $(k\theta/\rho c_p r_m{}^2)$ of Chap. 3, since w equals $\pi r_m{}^2 V\rho$ and θ equals L/V.

The corresponding local Nusselt numbers $h_x D/k$ are shown in Curve AB of Fig. 9-18, based on the equation given by Drew.[26]

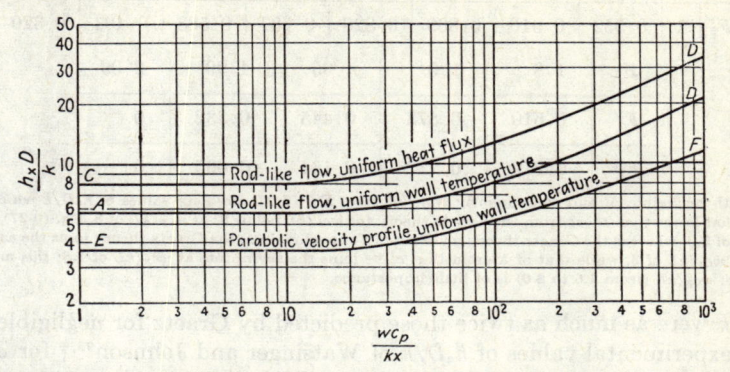

Fig. 9-18. Predicted local Nusselt numbers plotted vs. local Graetz numbers for rodlike flow and invariant physical properties, for two methods of supplying heat.

3. Natural and Forced Convection in Vertical Tubes Having Uniform Wall Temperatures.

Martinelli et al.[66a] analytically treated the cases of heating fluids flowing vertically upward and cooling fluids flowing vertically downward in tubes. In both cases the velocity near the wall is increased by the buoyant force owing to the difference in densities caused by difference in temperatures.† The resultant dimensionless relation is as follows:

$$\frac{h_a D}{k} = 1.75 F_1 \sqrt[3]{\frac{wc_{pb}}{k_b L} + 0.0722 \left[\frac{D}{L} N_{Gr,1} N_{Pr}\right]_w^n} F_2 \qquad (9\text{-}27) \star$$

where $N_{Gr,1}$ is based on tube diameter and *initial* Δt, which is positive for both heating and cooling.

The predicted value of the exponent n was 0.75.

The factor F_2 allows for the decrease in the buoyant force as the fluid approaches the temperature of the tube wall. The dimensionless factors F_1 and F_2 are unique functions of $Z = (t_2 - t_1)/\Delta t_a$ as shown in Table 9-5. The factor F_1 is equal to the ratio $\Delta t_L/\Delta t_a$:

$$F_1 \equiv \frac{Z}{\ln\left[(2+Z)/(2-Z)\right]} \qquad (9\text{-}27a)$$

Martinelli et al.[67] compared this prediction with data for *heating* oils flowing *upward* in vertical steam-heated tubes (L/D from 126 to 602 and D' of 0.422 to 0.494 inch) and for *heating* water flowing *upward* with L/D of 126 and D' of 0.422 inch. With the oils the measured values of

† Density is taken as a linear function of temperature; other pertinent physical properties are considered constant.

TABLE 9-5. FOR F_1 AND F_2 IN EQ. (9-27)[a]

Z	0	0.1	0.2	0.3	0.4	0.5	1.0	1.5	1.7
F_1	1	0.997	0.995	0.990	0.985	0.978	0.912	0.770	0.675
F_2	1	0.952	0.910	0.869	0.828	0.787	0.588	0.403	0.320

Z	1.8	1.9	1.95	1.99	2.00
F_1	0.610	0.573	0.445	0.332	0
F_2	0.272	0.212	0.164	0.095	0

[a] With negligible Δt and consequently $N_{Gr,1}$ of zero, Eq. (9-27) predicts values of $h_a D/k$ which are somewhat lower than those from the Graetz theory for wc_p/kL below 8; at wc_p/kL of 8, Eq. (9-27) gives $h_a D/k$ of 3.0, whereas the Graetz theory gives 3.4. For negligible Δt the Graetz theory joins the asymptote at wc_p/kL of 3, while that of Martinelli et al.[66a] joins the asymptote at wc_p/kL of 1.6; this narrow range of wc_p/kL (from 1.6 to 8.0) is of little importance.

$h_a D/k$ were as much as twice those predicted by Graetz for negligible Δt. The experimental values of $h_a D/k$ of Watsinger and Johnson[95],† for *cooling* water flowing *downward* (L/D of 20 and D' of 1.98 inch) were as much as 4.7 times those predicted for negligible Δt. The data were correlated

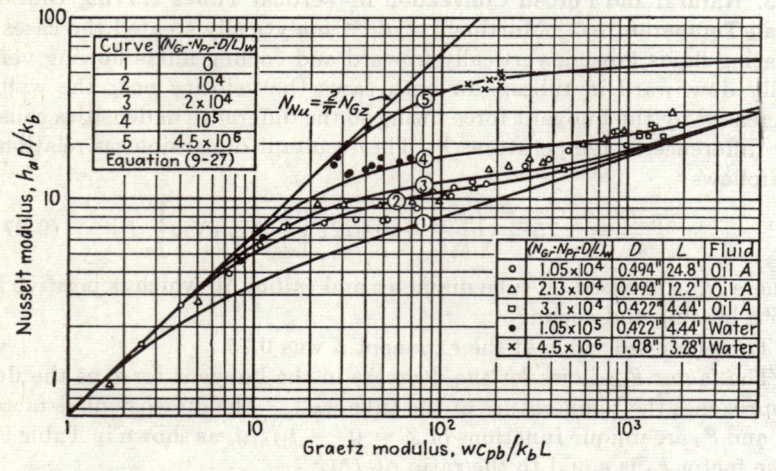

FIG. 9-19. Superimposed effects of forced flow and natural convection in vertical circular tubes. The curves were predicted theoretically by Martinelli and Boelter[66a] for linear variations of density with temperature; based on Eq. (9-27) with n of 0.84.

with a *maximum* deviation of 25 per cent by Eq. (9-27) with n of 0.84, instead of 0.75 (Fig. 9-19). For *heating with downflow* and for *cooling with upflow*, the natural convective forces oppose those due to forced circulation, and consequently $+0.0722$ is changed to -0.0722.

† No comparison was made with the theory for rodlike flow, which might be pertinent with small L_H/D with no calming section.

HEATING AND COOLING INSIDE TUBES

Recently Pigford[77a] analytically treated the combined effects of forced and natural convection for modified streamline flow in vertical tubes, assuming both ρ and $1/\mu$ linear in *local* temperature.† The resultant dimensionless parameters were Nusselt numbers, Graetz numbers, Grashof numbers, and viscosity ratios μ_s/μ_o. For constant viscosity the results were substantially the same as those predicted earlier by Martinelli and Boelter[66a] for linear variation of ρ with temperature, and other physical properties constant. But when heating a very viscous oil with high Δt, the ratio μ_s/μ_o approached 0, and, for given Graetz and Grashof numbers, the Nusselt number increased substantially above that for constant viscosity. The results are shown in Fig. 9-19a for four values of μ_s/μ_o, where μ_s is the viscosity at the constant wall temperature and μ_o is the viscosity at the uniform entrance temperature.

4. Natural and Forced Convection in Horizontal Tubes Having Uniform Wall Temperatures. Eubank and Proctor[31] critically surveyed the available data for streamline flow of petroleum oils in horizontal steam-heated tubes and employed the data shown in Table 9-6. When heating liquids flowing in horizontal tubes, the buoyant force causes a circulation upward at the sides and downward at the center of the tube.

TABLE 9-6. HEATING OR COOLING PETROLEUM OILS FLOWING AT DG/μ_b BELOW 2100, IN HORIZONTAL TUBES[a]

Observers	Diam, inches	$\dfrac{L}{D}$	$\dfrac{L'}{D}$	$\dfrac{wc_{pb}}{k_b L}$	$N_{Gr}N_{Pr}\dfrac{D}{L}$	N_{Pr}	$\dfrac{N_{Pr}N_{Gr}}{10^3}$
Holden[41,28]	0.494	120	50	12–500	2,800–5,300	140–355	330–640
White[97,28]	0.494	112	110	950–4800	900–3,200	345–1,520	100–350
Sherwood et al.[87]	0.593	61–235	None	140–4800	4,000–15,200	150–260	855–1,100
Sieder and Tate[88]	0.620	99	39	280–4900	135–32,000	160–15,200	13–1,670
Kern and Othmer[52]	0.622–2.47	193–49	58–15	115–2800	25,000–17.5 × 10⁶	37–165	4,800–855,000

[a] The heated length L is preceded by an unheated calming length of L'.

This effect and that due to the forced flow sets up forward-moving spirals; the consequent mixing increases the heat-transfer coefficient. The *tentative* empirical equation is similar to that for vertical tubes but does not contain F_1 or F_2:

$$\frac{h_a D}{k}\left(\frac{\mu_w}{\mu_b}\right)^{0.14} = 1.75 \sqrt[3]{\frac{wc_{pb}}{k_b L} + 0.04\left[\frac{D}{L}N_{Gr}N_{Pr}\right]_b^{0.75}} \qquad (9\text{-}28)\;\star$$

where N_{Gr} is based on tube diameter and Δt_a. The data of Table 9-6 are correlated by Eq. (9-28) with a maximum deviation of 30 per cent. Equation (9-28) should not be extrapolated to the left-hand side of the asymptote, Eq. (9-25) on Fig. 9-17. The product $N_{Gr} \cdot N_{Pr}$ ranged from 3.3×10^5 to 8.6×10^8, and Eq. (9-28) was based only on data for petroleum oils with Prandtl numbers ranging from 140 to 15,200. Equation

† See also K. Yamagata, *Kyushu Imp. Univ. Faculty Eng. Mem.*, **8**, 365–449 (1940).

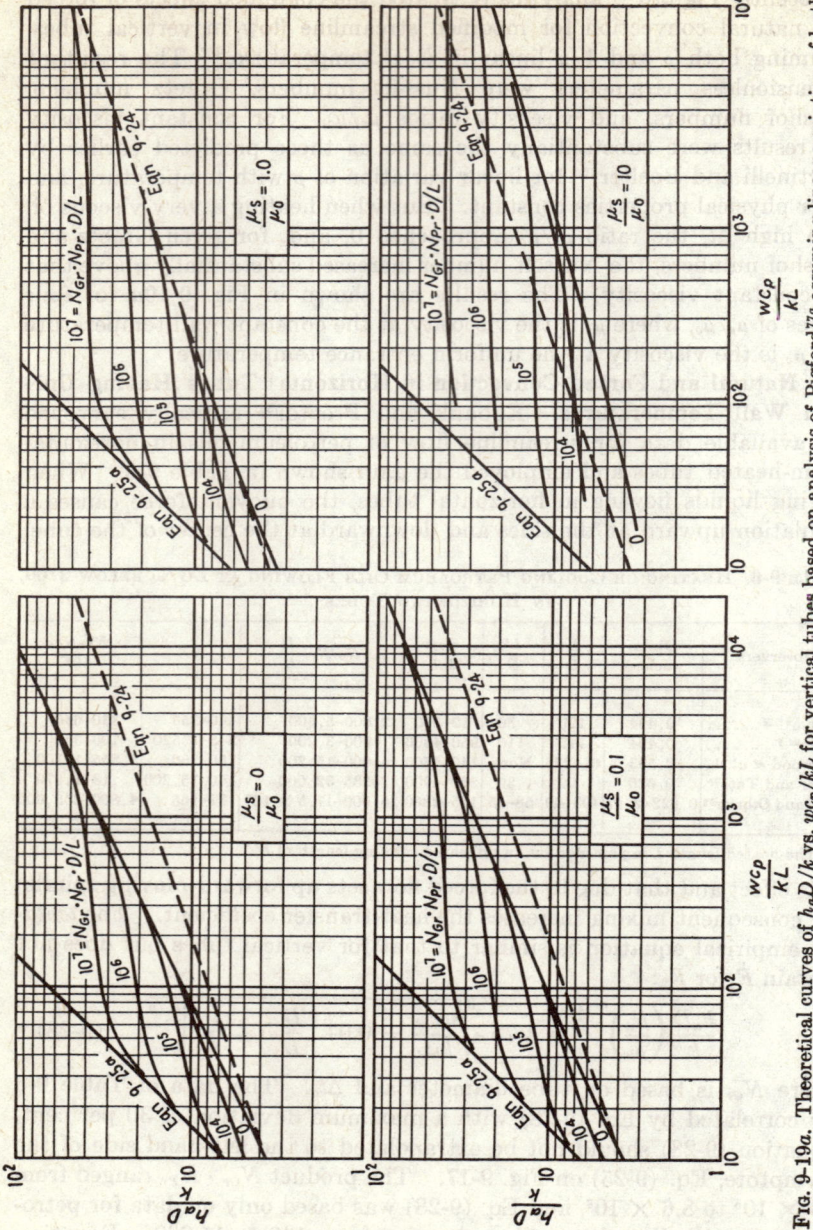

FIG. 9-19a. Theoretical curves of $h_a D/k$ vs. wc_p/kL for vertical tubes based on analysis of Pigford[77a] assuming linear variation of both ρ and $1/\mu$. In viscosity ratio, μ_s is evaluated at constant surface temperature and μ_o at uniform inlet temperature of fluid.

(9-28) is probably conservative for small L/D with no calming section, where the distribution of velocity might be uniform rather than parabolic (see Curve BH, Fig. 9-17).

Small D and Δt. Before the data for large values of D and Δt_a were available,[52] the other data of Table 9-6 were correlated[88] by the relation

$$\frac{h_a D}{k_b}\left(\frac{\mu_w}{\mu_b}\right)^{0.14} = 1.86\left(\frac{4wc_{pb}}{\pi k_b L}\right)^{\frac{1}{3}} = 1.86\left[\frac{DG}{\mu_b}\left(\frac{c_p\mu}{k}\right)_b\frac{D}{L}\right]^{\frac{1}{3}} \qquad (9\text{-}28a) \star$$

with a maximum deviation of 60 per cent, but when the data for large D and Δt are included, the maximum deviation of data *above* Eq. (9-28a) increases to 250 per cent.

For moderate Δt and large μ/ρ or small D, the natural convection effects are small, and Eq. (9-28a) is used as an approximation. By eliminating $h_a D/k$ from Eqs. (9-23), (9-23a), and (9-28a), the design equation is obtained:

$$\left(\frac{wc_p}{kL}\right)^{\frac{2}{3}}\frac{t_2 - t_1}{-\Delta t_a}\left(\frac{\mu_w}{\mu}\right)^{0.14} = 6.33 \qquad (9\text{-}28b) \star$$

For a given value of w/L, the temperature rise is independent of diameter.

Data[71] for heating asphalt at 307 to 395°F, flowing in streamline motion inside a horizontal 1-inch steel tube, give values of h from 30 to 55 at velocities from 2 to 13 ft/sec.

Illustration 2. It is desired to heat oil from 68 to 140°F while it is flowing through horizontal ⅝-in. o.d. steel tubes, No. 16 BWG, each 8 ft long, in a multipass preheater, jacketed by steam condensing at 220°F. It is agreed that each tube shall handle 600 lb/hr of oil. Calculate the necessary number of passes.

Data and Notes. The thermal conductivity of the oil may be assumed constant at 0.080, and the specific heat is constant at 0.50; the viscosity varies with temperature as shown below.

t, deg F	68	80	90	100	110	120	130	140	220
μ', centipoises	23	20	18	16.2	15	13.5	12	11	3.6

Solution. The largest value of DG/μ will correspond to the highest bulk temperature, 140°F, where $\mu = (2.42)(11) = 26.6$ lb/(hr)(ft); since $D = 0.495/12 = 0.0412$ ft, $DG/\mu = 4w/\pi D\mu = (4)(600)/(3.14)(0.0412)(26.6) = 694$; since DG/μ is less than 2100, Eq. (9-28a) applies to all passes. The dimensionless term $4wc/\pi kL$ is $(4)(600)(0.5)/\pi(0.08)(8) = 597$, which is larger than the asymptotic value, below which Eq. (9-28a) does not apply.

Equation (9-28a) must be applied to each pass in turn. If the oil leaves the first pass at 88°F, $t_a = (68 + 88)/2 = 78$, μ' is 20.5 centipoises. If t_s is 215°F, a little colder than the outer surface at 220°F, μ'_s is 4.25 centipoises and $(\mu'/\mu'_s)^{0.14}$ is 1.25. From Eq. (9-28a)

$$h_a = \frac{1.86k}{D}\left(\frac{\mu}{\mu_s}\right)^{0.14}\left(\frac{4wc}{\pi kL}\right)^{\frac{1}{3}} = \frac{(1.86)(0.08)(1.25)(597)^{\frac{1}{3}}}{0.0412} = 38$$

$$\Delta t_a = (t_s - t)_a = 215 - 78 = 137°\text{F}$$

and hence

$$t_2 - t_1 = \frac{h_a A_w (\Delta t)_a}{w c_p}$$

$t_2 - t_1 = (38)(\pi \times 0.0412 \times 8)(137)/(600)(0.5) = 18°F$, which is sufficiently close to the assumed value.

The resistance of the steam side and scale deposit, based on h of 1000, equals $1/(1000)(8 \times \pi \times 0.625/12) = 0.000478$, and that of the tube wall is $(0.065/12)/(26)(8 \times \pi \times 0.56/12) = 0.000176$, giving a total resistance, excluding that on the oil side, of 0.00065; hence $220 - t_s = q\Sigma R = (600 \times 0.5 \times 18)(0.00065) = 3.5°F$ vs. 5° assumed. The oil enters the second pass at 86°F and leaves at a new value of t_2, which is calculated as shown above. The procedure is repeated until the outlet temperature from the last pass equals or exceeds 140°F.

Data for Gases. Figure 9-20 shows recent data† for air flowing in laminar motion ($DG/\mu_b < 2500$) in *steam-heated* tubes where the values of

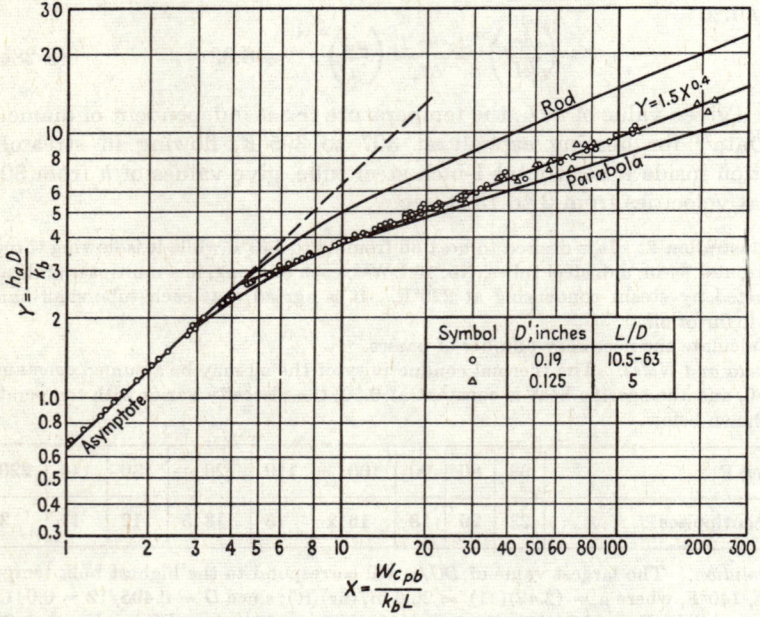

FIG. 9-20. Data of Cholette[13] and of Kroll[55] for air flowing through steam-heated tubes, compared with theory for rodlike and for laminar flow with invariant physical properties.

$N_{Gr} N_{Pr} D/L$ are small. It is seen that these results lie between the curve for parabolic distribution of velocity and that for rodlike flow and are correlated within a maximum deviation of 4 per cent by the equation

$$h_a D/k_b = 1.5(w c_{pb}/k_b L)^{0.4} \qquad (9\text{-}28c) \;\star$$

† Cholette,[13] Margolin,[65] Karim and Travers,[49] and Kroll.[55]

All these data are for multitubular heaters following a well-rounded nozzle, so that all tubes would have their equal shares of air.

5. Rodlike Flow, Uniform Heat Flux. Table 9-7 compares the minimum asymptotic values of $h_x D/k$ for four cases. For example, with rodlike flow, constant flux has a minimum $h_x D/k$ of 8.0, which is 1.38 times the value of 5.8 for constant wall temperature. With uniform wall temperature, the minimum asymptotic value of $h_x D/k$ for rodlike flow is 1.59 times that for parabolic distribution of velocity.

Table 9-7. Minimum Asymptotic Values[73] of $h_x D/k$

	Velocity distribution	
	Parabolic	Uniform
Constant wall temperature, t_w	3.66	5.80
Constant heat flux, q/A	4.36	8.00

6. Radial Variation in Physical Properties, Uniform Heat Flux. For constant flux and well-developed laminar flow of a *gas*, assuming that both c_p and $c_p\mu/k$ are independent of temperature and that both μ and k vary as $T^{0.68}$, as for air and common gases, and that $\rho \propto T^{-1}$, Deissler[21] shows that the asymptotic value of $h_x D/k_b$ varies but little with T_w/T_b from 0.6 to 1.8; in this range his predictions may be represented by

$$h_x D/k_b = 5.14 - 0.78 T_w/T_b \tag{9-29}$$

This agrees with previous theory for isothermal conditions ($h_x D/k_b = 4.36$). The predicted velocity field qualitatively supports the conclusions from Fig. 9-16.

For liquid *metals*, assuming all physical properties uniform except viscosity ($\mu \propto T^{-1.6}$) Deissler predicts that the asymptotic value of $h_x D/k_b$ increases with increase in T_w/T_b. For T_w/T_b from 1 to 1.8, his theoretical curve can be represented by

$$h_x D/k_b = 3.56 + 0.8 T_w/T_b \tag{9-29a}$$

III. TRANSITION REGION

The transition region lies between the end of the region of laminar flow and the start of the region of well-developed turbulent flow. The Reynolds number DG/μ_b corresponding to the lower limit of the transition zone is frequently taken as 2100 to 2500.

1. Gases, Small L/D. Kroll finds that data of a number of observers† for turbulent flow of *air* for L/D from 1 to 68 indicate that

$$h_L/h_\infty = 1 + (D/L)^{0.7} \tag{9-30}$$

† Boelter et al.,[7] Cholette,[13] Karim and Travers,[49] and Margolin.[65]

where the mean value of h from x of 0 to x of L, based on Δt_L, called h_L, is related to h_∞ at large L; when L/D reaches 60, the ratio h_L/h_∞ has fallen to 1.057 and decreases but little as L/D is further increased. These data are plotted in Fig. 9-21 as

$$Y = \frac{h_L/c_p G}{1 + (D/L)^{0.7}}$$

vs. $X = DG/\mu_b$. The lower line† marked $Y = 0.026X^{-0.2}$ is based on revised physical properties[51] of air; the corresponding constant C of Eq. (9-10a) is 0.021; the upper line is based on C of 0.023. For L/D of 63, the value of Y at X of 2500 is only 0.0032, and as DG/μ_b increases, the

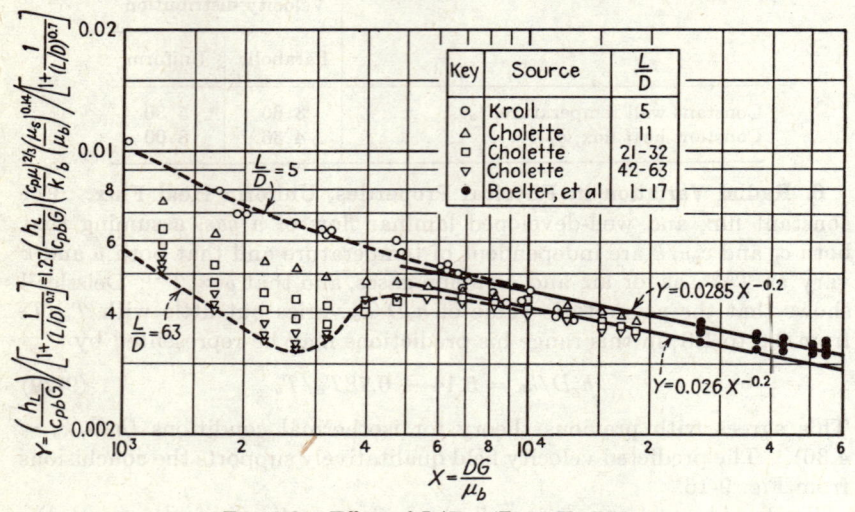

Fig. 9-21. Effect of L/D. (*From Kroll.*[55])

curve loops upward to join the curve for well-developed turbulent flow at DG/μ_b of 10,000. In the laminar range (DG/μ_b below 2500) the ordinate at a given Reynolds number increases as L/D decreases; consequently no dip occurs for L/D of 5. Data for the transition region are inadequate to give a more detailed picture for gases, even for low Δt. At high Δt, the heat transfer is complicated by the superimposed effects of natural convection and by the effect of L/D in the turbulent region.

2. Liquids, $L/D > 60$, Uniform Wall Temperatures. For *streamline* flow of petroleum oils having large kinematic viscosities (μ/ρ) in tubes of small diameters with moderate temperature potentials, the natural convection or buoyant terms on the right-hand sides of Eq. (9-27) for vertical tubes and Eq. (9-28) for horizontal tubes become nearly negligible compared with the Graetz modulus $wc_{pb}/k_b L$. In consequence, as an approxi-

† When the data of Fig. 9-21 were replotted with $1 + (D/L)^{0.7}$ replaced by $1 + 1.1 D/L$, the maximum deviation was increased from 17 to 27 per cent.

mation, Eq. (9-28a) may be used, in which the constant of 2.0 is 14 per cent higher than the theoretical value of 1.75 for negligible Δt. For well-developed turbulent flow, Eq. (9-10c) may be used. The looped curves in Fig. 9-22 for the transition region roughly follow the available data[88] for petroleum oils; more data are needed.

Hausen[40a] gives an interesting form of equation,

$$j'' = (h_m/c_{pb}G)(c_p\mu/k)^{2/3}{}_b(\mu_s/\mu_b)^{0.14} \Big/ \left\{1 + \left(\frac{D}{L}\right)^{2/3}\right\} = 0.116\,\frac{N_{Re}{}^{2/3} - 125}{N_{Re}}$$

which requires a *maximum* value of j'' of 0.0082 at N_{Re} of 7100, *equal* values of j'' of 0.0035 at Reynolds numbers of 3000 and 20,700, and a value of j'' of 0.00234 at N_{Re} of 10^5 where Eq. (9-10c) predicts j of 0.00230.

IV. MISCELLANEOUS SHAPES

1. Concentric Annular Spaces. For turbulent motion the usual procedure involves substituting an equivalent diameter D_e, as is customary in calculation of fluid friction:

$$D_e \equiv 4r_h = \frac{\pi(D_2{}^2 - D_1{}^2)}{\pi(D_2 + D_1)}$$
$$= D_2 - D_1 \quad (9\text{-}31)$$

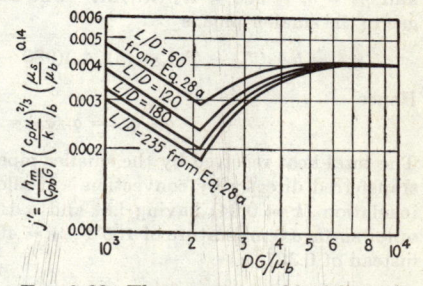

FIG. 9-22. The curves roughly follow the data for heating and cooling petroleum oils flowing through small tubes with moderate Δt. For DG/μ below 2100 and with high Δt use Eq. (9-28) for horizontal tubes and Fig. 9-19 for vertical tubes.

Consider the flow of hot air through an annular space between concentric pipes, with cooling water inside the smaller pipe and insulation outside the larger pipe. Heat will be transferred by convection to both the smaller and larger pipes. The heat received by the larger pipe will in part be lost through the insulation, and the remainder will be radiated to the smaller pipe. The latter may be substantial compared with that received directly by convection, as shown in the following example.

Illustration 3. Assume that the air is at 520°F, the outer surface of the smaller pipe is at 100°F, and the value of h from gas to metal is 5 Btu/(hr)(sq ft)(deg F). For illustration, assume that the i.d. of the larger pipe is 2.07 in. and that the o.d. of the smaller is 1.66 in. Calculate the heat radiated to the smaller pipe expressed as a per cent of that received directly by convection, neglecting heat losses through the insulation. To simplify the problem, it is agreed to ignore the effects of axial gradients in temperature.

Solution. The area ratio A_2/A_1 is $2.07/1.66 = 1.246$. The rate of heat transfer by convection q_{c2} from gas to the larger pipe having absolute temperature T_2 is

$$q_{c2} = hA_2(T_g - T_2) = (5)(1.246A_1)(980 - T_2) = \text{Btu/hr}$$

As shown in Chap. 4, the heat radiated from A_2 to A_1 is given by

$$q_r = 0.173 A_1 \mathfrak{F}_{12} \left[\left(\frac{T_2}{100} \right)^4 - \left(\frac{T_1}{100} \right)^4 \right]$$

where

$$\frac{1}{\mathfrak{F}_{12}} = \frac{1}{\bar{F}_{12}} + \frac{1}{\epsilon_1} - 1 + \frac{A_1}{A_2}\left(\frac{1}{\epsilon_2} - 1\right)$$

Assuming that both emissivities are 0.9,

$$q_r = \frac{0.173 A_1 [(T_2/100)^4 - (560/100)^4]}{\frac{1}{1} + \frac{1}{0.9} - 1 + \left(\frac{1}{1.246}\right)\left(\frac{1}{0.9} - 1\right)} = \text{Btu/hr}$$

By a heat balance on the larger surface, $q_{c2} = q_r$, whence $T_2 = 871°F$ abs, $t_2 = 411°F$, and $q_{c2} = q_r = 686 \times A_1$ Btu/hr. The heat transferred directly by convection from gas to the smaller pipe is

$$q_{c1} = hA_1(T_g - T_1) = (5)(A_1)(980 - 560) = 2100 \times A_1 \quad \text{Btu/hr}$$

Hence

$$q_{c2}/q_{c1} = q_r/q_{c1} = {}^{686}\!/_{2100} = 0.327$$

The total heat received by the smaller pipe by radiation is then 32.7 per cent of that transferred directly by convection. If allowance be made for heat loss through the insulation ($k = 0.04$) having i.d. and o.d. of 6.37 and 2.37 in., respectively, and an outer surface temperature of 105°F, $t_2 = 401°F$, $q_r = 640$, and the ratio q_r/q_{c1} is 0.304 instead of 0.327.

Turbulent Flow in Concentric Annuli. The value of h at the *outer* wall of diameter D_2 is usually correlated in terms of the dimensionless equation

$$\frac{h_L}{c_{pb}G}\left[\frac{c_p\mu}{k}\right]_b^{2/3} = \frac{0.023\psi}{(D_eG/\mu_b)^{0.2}} \tag{9-32}$$

or the equivalent. For D_2/D_1 of 1.85 Monrad and Pelton[70] found

$$\psi = 1.17(\mu_b/\mu_w)^{0.14} \tag{9-32a}$$

Using data for air and water of several observers† Davis[18] found

$$\psi = \left(\frac{D_2}{D_1} - 1\right)^{0.1}\left(\frac{\mu_b}{\mu_w}\right)^{0.14} \tag{9-32b}$$

which gives an unreasonable prediction for D_2 equal to D_1. Based on equations from various sources, for D_2/D_1 from 1 to 10, Wiegand[98] proposed $\psi = 1.0$. For D_2/D_1 from 1 to 10, the *recommended* relation for the *outer* wall is

$$j' = \frac{h_L}{c_{pb}G}\left(\frac{c_p\mu}{k}\right)_b^{2/3}\left(\frac{\mu_w}{\mu_b}\right)^{0.14} = \frac{0.023}{(D_eG/\mu_b)^{0.2}} \tag{9-32c} \star$$

For h at the *inner* wall of diameter D_1, Eq. (9-32) is again employed. For D_2/D_1 of 1.65, 2.45, and 17, Monrad and Pelton[70] found

$$\psi = 0.87(D_2/D_1)^{0.53} \tag{9-32d}$$

† References 34, 70, 71a, and 100.

For various data, Davis[18] found

$$\psi = 1.35 \left(\frac{D_2}{D_1}\right)^{0.15} \left(\frac{D_2}{D_1} - 1\right)^{0.2} \left(\frac{\mu_b}{\mu_w}\right)^{0.14} \quad (9\text{-}32e)$$

At D_2/D_1 of 1.2, this ψ reduces to $(\mu_b/\mu_w)^{0.14}$, but for D_2 equal to D_1, an unreasonable result is predicted. For D_2/D_1 from 1 to 10, Wiegand[98] proposed

$$\psi = (D_2/D_1)^{0.45} \quad (9\text{-}32f)$$

For D_2/D_1 of 1.33, Carpenter et al.[10] found

$$\psi = (\mu_b/\mu_w)^{0.14} \quad (9\text{-}32g)$$

for heating water flowing upward in the vertical annulus.

Recent data[69] for air in an annulus (D_2/D_1 of 2.19) can be well correlated by Eq. (9-32c).

In a recent investigation[64] superheated steam flowed upward through a vertical annulus having an outer diameter of 0.383 inch and an inner diameter of 0.25 inch. Local heat-transfer coefficients h_x were measured for heated lengths from 15 to 80 equivalent diameters. The range of variables was as follows: pressure, 115 to 3500 lb/sq in. abs; wall temperatures, 607 to 1288°F; bulk temperatures, 429 to 1012°F; Δt, 100 to 624°F; $D_e G/\mu_f$, 7000 to 40,000; and h_x, 91 to 354. The local coefficients were correlated by

$$\left(\frac{h_L}{c_{pb}G}\right)\left(\frac{c_p\mu}{k}\right)_f = \frac{0.0214(1 + 2.3 D_e/L)}{(D_e G/\mu_f)^{0.2}} \quad (9\text{-}32h) \ \star$$

with a maximum deviation of 17 per cent, using the values of k of reference 52a.

The **recommendation** is to use Eq. (9-32c) for *both* the outer and inner walls of concentric annuli.†

Laminar Flow in Concentric Annuli. Carpenter et al.[10] heated water, flowing upward in streamline motion in a vertical annulus (D_2' of 0.834 inch and D_1' of 0.625 inch). When compared with the empirical relation for streamline flow in small tubes [Eq. (9-28a), with D replaced by $D_e = D_2 - D_1$], the values of $h_a D/k$ averaged 33 per cent higher than those for tubes.

For *streamline* flow of water in annuli (D_2/D_1 from 1.09 to 2.0) Chen et al.[12] measured the value of h from the larger surface to the water. The results may be roughly correlated[43a] by

$$\frac{h_a D}{k_b} = 1.0 \left(\frac{D_2}{D_1}\right)^{0.8} \left(\frac{w c_p}{kL}\right)_b^{0.45} (N_{Gr})_b^{0.05} \quad (9\text{-}33)$$

† For small L/D_e with gases, use Eq. (9-32c) with 0.023 multiplied by the term $1 + (2.3 D_e/L)$.

where the Grashof number N_{Gr} equals $\beta \Delta t \cdot D_e^3 \rho_b^2 g/\mu_b^2$. Jakob and Rees[44] heated three gases in laminar flow through an annulus and derived a theoretical relation for the difference in temperature between the larger and smaller surfaces.

Rodlike Flow in Long Annuli with Uniform Flux at One Wall. Trefethen[92] gives the asympotic values of $h_x D_e/k$ for rodlike flow in long annuli with uniform flux at one wall, in terms of the ratio D_{Ad}/D_H, where D_{Ad} is the diameter of the adiabatic wall and D_H is the diameter of the heated wall:

$$\frac{h_x(D_{Ad} - D_H)}{k} = \frac{8\left(\dfrac{D_{Ad}}{D_H} - 1\right)\left[\left(\dfrac{D_{Ad}}{D_H}\right)^2 - 1\right]^2}{4\left(\dfrac{D_{Ad}}{D_H}\right)^2 - 3\left(\dfrac{D_{Ad}}{D_H}\right)^4 - 1 + 4\left(\dfrac{D_{Ad}}{D_H}\right)^4 \ln \dfrac{D_{Ad}}{D_H}} \quad (9\text{-}34)$$

2. Gravity Flow of Liquid in Layer Form, Vertical Tubes. In heating or cooling water flowing downward inside vertical tubes, the economical water rate may be such that the tube will not be filled and a layer of water will flow down the vertical wall. By using upward flow, the water will fill the tube. It is interesting to predict which type of flow would give the larger coefficient for a given water rate, tube diameter, and temperature. Equation (9-10a) for turbulent flow will be used for both cases, with D replaced by D_e, equal to four times the hydraulic radius r_h, where r_h is the cross section S filled by the fluid, divided by the perimeter b wetted by the fluid. Since by definition

$$\frac{D_e G}{\mu} = \frac{4 r_h G}{\mu} = \frac{4 S w}{\mu b S} = \frac{4w}{\mu b} = \frac{4\Gamma}{\mu} \quad (9\text{-}35)$$

it is seen that S disappears, and hence $D_e G/\mu$ will be the same whether the fluid partially or completely fills the tube. Since the physical properties are the same, h should vary inversely as D_e for each type of flow. For flow of a layer of thickness B, S equals $\pi(D - B)(B)$, b equals πD, and D_e equals $4S/b$ equals $4B(D - B)/D$; for the full tube, D_e equals D. Hence

$$\frac{h \text{ for layer flow}}{h \text{ for the full tube}} = \frac{D^2}{4B(D - B)} \quad (9\text{-}36)$$

Since layer flow inside a tube can be obtained only under conditions in which B is less than $D/2$, it can be seen from Eq. (9-36) that h for layer flow should exceed that for the full tube.

The extensive data of Bays[3] and coworkers for heating *water* flowing in layer form down the inside walls of vertical tubes† having inside diameters of 1.5 and 2.5 inches and heights ranging from 0.407 to 6.08 feet are plotted in Fig. 9-23 and are correlated within ±18 per cent by the dimen-

† To minimize disturbances, a rounded inlet was used.

sional equation[63]

$$h_m = 120\Gamma^{1/3} = 120\left(\frac{w}{\pi D_i}\right)^{1/3} \qquad (9\text{-}37) \star$$

The Reynolds numbers at the film temperatures ranged from 2100 to 51,000.

Although data for other liquids are not available, the following dimensionless equation[27] may be used for estimating h_m:

$$\frac{h_m}{\phi_f} = \frac{h_m}{\sqrt[3]{k_f^3 \rho_f^2 g / \mu_f^2}} = 0.01 \left(\frac{c_p \mu}{k}\right)^{1/3} \left(\frac{4\Gamma}{\mu_f}\right)^{1/3} \qquad (9\text{-}37a) \star$$

This equation reduces to Eq. (9-37) for water at a film temperature of approximately 190°F; values of ϕ_f are given in the Appendix.

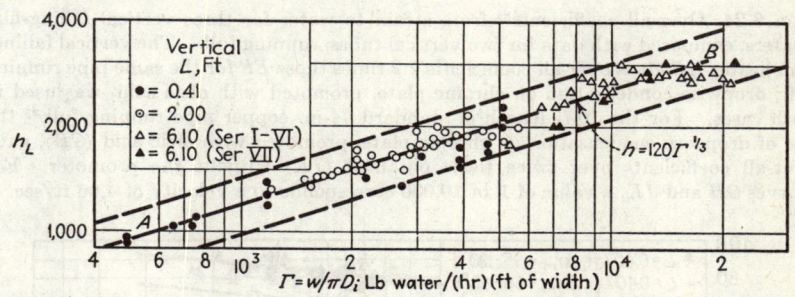

FIG. 9-23. Coefficients for gravity flow of water in layer form inside vertical tubes.[63]

Figure 9-24 shows the over-all coefficients U from steam to water, and the vertical falling-film heater (Curve AB) gives over-all coefficients 2.2 times those for the same pipe (Curve EF) running full with upward flow of water. Curve CD shows data for a falling-film heater 0.41 ft high, having an inside diameter of 0.125 ft.

Using the same apparatus described above, heating data were obtained[4] for two petroleum oils flowing in layer form down the inner walls of vertical tubes. Figure 9-25 shows the data for streamline flow, $4\Gamma/\mu$ less than 2000. It was found that h_a varied directly as $\Gamma^{1/3}$ and inversely as $L^{1/3}$.

Most of these heat-transfer coefficients are higher than predicted by theory[75] for laminar flow and uniform wall temperature, perhaps owing to rippling and consequent mixing in the film. These data were correlated[4] as shown in Fig. 9-26, in which the thickness of the falling film was evaluated from the relation

$$B_f = (3\mu_f \Gamma / \rho_f^2 g)^{1/3} \qquad (9\text{-}38)$$

3. Gravity Flow of Liquid Layers over Horizontal Tubes. In industrial work, various liquids are cooled while they flow through horizontal

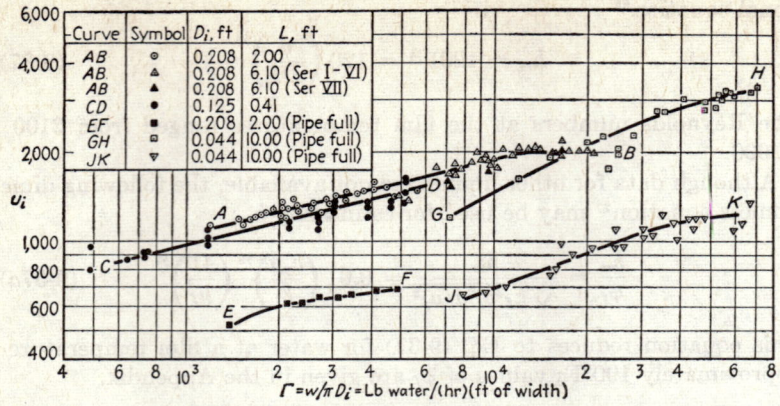

FIG. 9-24. Over-all coefficients[63] from steam to water for three vertical falling-film heaters, compared with data for two vertical tubes, running full. The vertical falling-film heater AB gives over-all coefficients 2.2 times those EF for the same pipe running full; dropwise condensation on chrome plate, promoted with oleic acid, was used in both cases. For the 10-ft length of standard ⅝-in. copper pipe running full,[33] the use of dropwise condensation on chrome plate, promoted with oleic acid (GH), gave over-all coefficients over twice those obtained (JK) without the promoter. For Curves GH and JK, a value of Γ of 10,000 corresponds to a velocity of 4.06 ft/sec.

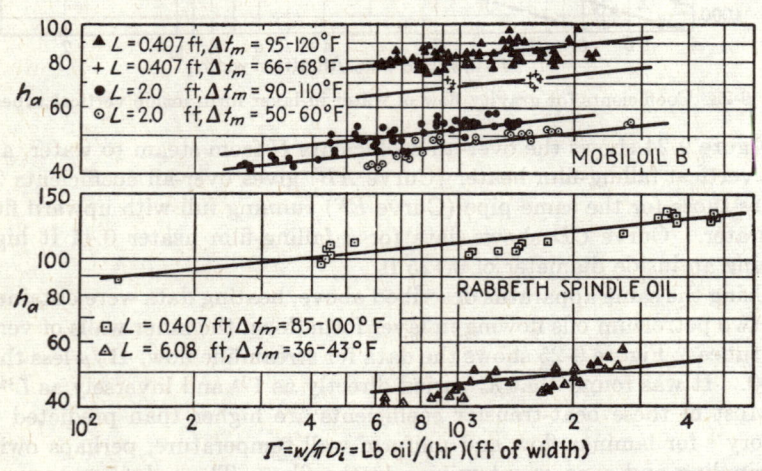

FIG. 9-25. Data of Bays[4] for gravity flow of oils in layer form in streamline motion in vertical tubes.

pipes by allowing a layer of water to trickle over the outer surfaces in the so-called "trickle" or "trombone" coolers. The published data[1,91] were obtained with slightly inclined pipes that caused ripples in the water film, and as a result h_a varied as $(\Gamma_H)^{1/3}$ instead of $(\Gamma_H)^{1/9}$. The symbol Γ_H represents $w/2L$, where w is the water rate over each straight tube of

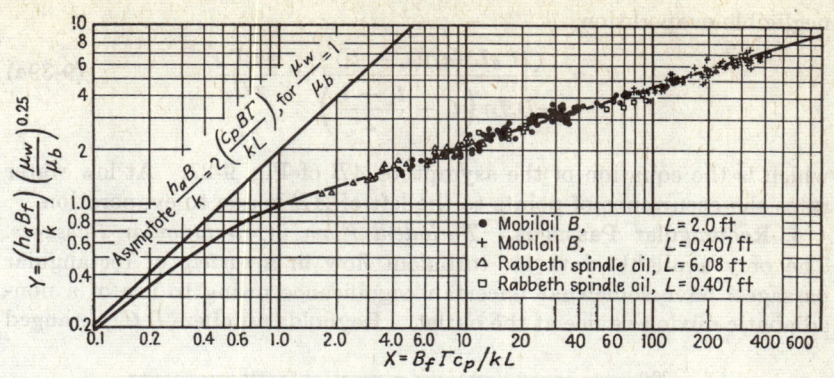

Fig. 9-26. Data for laminar flow in layer form down vertical tubes.[4]

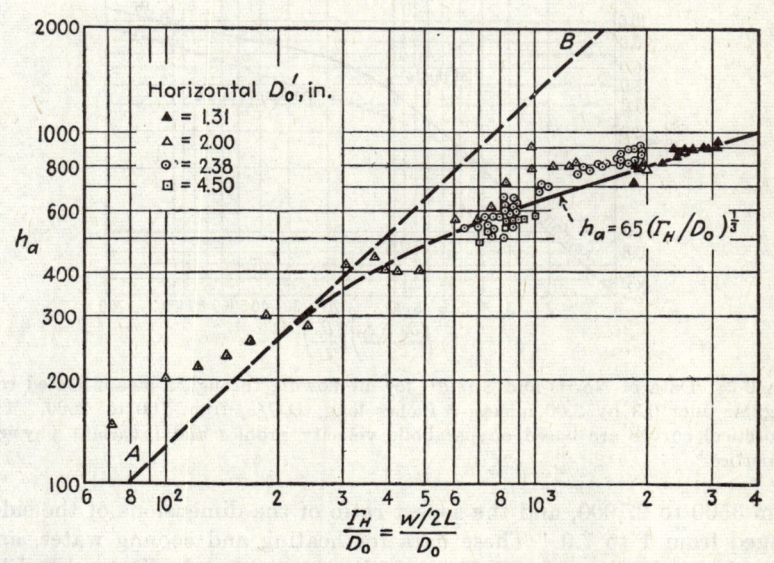

Fig. 9-27. Coefficients for gravity flow of water in layer form over nearly horizontal pipes.[63]

length L. For values of $4\Gamma_H/\mu$ less than 2100, the data are correlated within ± 25 per cent by the dimensional equation[63]

$$h_a = 65(\Gamma_H/D_o)^{1/3} \qquad (9\text{-}39) \star$$

The corresponding data are shown in Fig. 9-27. If the pipe diameter were very large or w were very small, a nonvolatile liquid would leave the bottom of the tube at substantially the temperature of the outer wall of the pipe. For this limiting case, the definition of h_a gives, for negligible evaporation,

$$h_a = \frac{(2\Gamma_H L)(c_p)(t_s - t_1)}{(\pi D_o L)\left(t_s - \dfrac{t_1 + t_s}{2}\right)} = \frac{4c_p \Gamma_H}{\pi D_o} \quad (9\text{-}39a)$$

which is the equation of the asymptote AB of Fig. 9-27. At low water rates the occurrence of points to the left of AB is due to evaporation.

4. Rectangular Passages. *Turbulent Flow in Rectangular Passages.* The only available data for turbulent flow in a *variety* of rectangular passages[2] have somewhat uncertain significance owing to use of a nonadiabatic mixing device at the outlet. Reynolds numbers $D_e G/\mu_f$ ranged

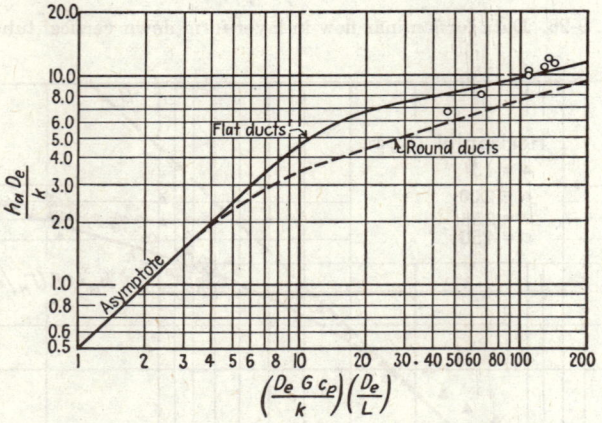

Fig. 9-28. Data of Norris and Streid[73] for air flowing through a steam-heated rectangular duct 0.3 by 2.00 inches, 6 inches long; $D_e G/\mu_f$ from 700 to 2000. The theoretical curves are based on parabolic velocity profiles and invariant physical properties.

from 3500 to 27,000, and the aspect ratio of the dimensions of the sides ranged from 1 to 7.9. These data for heating and cooling water, and those for a circular tube (0.674 inch diameter), were well correlated by plotting $(h_L/c_p G)(c_p \mu/k)_f^{2/3}$ vs. $D_e G/\mu_f$, with a maximum deviation of 24 per cent below Eq. (9-10b).

Kays[50] reports recent data for heating air while flowing in a number of rectangular passages in parallel, each with an aspect ratio of 5.85. The values of $D_e G/\mu_f$ ranged from 3000 to 10,000 and fall 24 per cent below a curve based on Eq. (9-10b).

Laminar Flow in Rectangular Passages. Norris and Streid[73] presented a theoretical relation for laminar flow between broad parallel plates at uniform wall temperature, based on an equation of Nusselt,[75] which is plotted in Fig. 9-28. This equation should not be used on the left-hand side of the asymptote:

$$\frac{h_a D_e}{k} = \frac{1}{2} \frac{D_e G c_p}{k} \frac{D_e}{L} \qquad (9\text{-}40)$$

5. Flow Parallel to a Plane. The theoretical relations of Pohlhausen [Eqs. (9-12) and (9-12a)] for a laminar boundary layer correlate data for air flowing at $LV_1\rho_1/\mu_1$ below 80,000. The transition to a turbulent

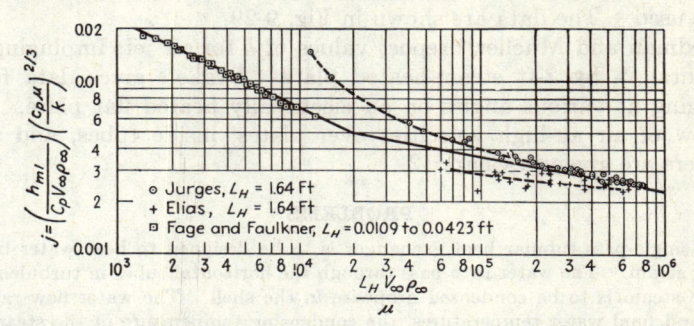

FIG. 9-29. Data for gas flowing parallel to a single plane; solid curves are from Colburn; note that $h_{m1} = q/(A)(t_r - t_\infty)$.

boundary layer may occur at $LV_1\rho_1/\mu_1$ from 80,000 to 500,000, depending on the degree of turbulence in the oncoming air stream and disturbances caused by roughness of the leading edge of the plate. For values of $LV_1\rho_1/\mu_1$ above 500,000 the Colburn relation[14] is suggested:

$$\frac{h_{m1}}{c_p V_1 \rho_1} \left[\frac{c_p \mu}{k}\right]_f^{2/3} = \frac{0.036}{(LV_1\rho_1/\mu_1)^{0.2}} = \frac{f}{2} \qquad (9\text{-}41)$$

The data of Jurges[47] for flow of air at room temperature parallel to a vertical copper plate 1.64 ft square may be represented by Eq. (9-42) and Table 9-8:

$$h_{m1} = a_1 + b_1(V')^n \qquad (9\text{-}42)$$

where h_{m1} is expressed in Btu/(hr)(sq ft)(deg F *initial* Δt) and V' is in feet per second. These tests of Jurges were made with air at room tem-

TABLE 9-8. FACTORS IN EQ. (9-42)

Nature of surface	Velocity less than 16 ft/sec			Velocity between 16 and 100 ft/sec		
	a	b	n	a	b	n
Smooth.........	0.99	0.21	1.0	0	0.50	0.78
Rough..........	1.09	0.23	1.0	0	0.53	0.78

perature and pressure.† Since mass velocity rather than linear velocity is the fundamental variable in forced-convection equations, Schack recommends that V' of Eq. (9-42) be multiplied by $(460 + 70)/(460 + t)$ when the temperature t of the air differs materially from 70°F. For very low velocities the heat transfer is largely by natural convection, in which case the heat-transfer coefficient for natural convection should also be calculated from the procedure given in Chap. 7 and the larger of the two values used. The data are shown in Fig. 9-29.

Friedman and Mueller[36] report values of h for air jets impinging on a horizontal 2- by 2-ft steam-heated plate. Drake[25] gives data for air impinging at various angles on an electrically heated flat plate. Data for flow of air at high velocities over plates, inside tubes, and across cylinders are given in Chap. 12.

PROBLEMS

1. A single-pass tubular heat exchanger is to be designed to heat water by condensing steam. The water is to pass through the horizontal tubes in turbulent flow, and the steam is to be condensed dropwise in the shell. The water flow rate, the initial and final water temperatures, the condensing temperature of the steam, and the available tube-side pressure drop (neglecting entrance and exit losses) are all specified. In order to determine the optimum exchanger design, it is desired to know how the total required area of the exchanger varies with the tube diameter selected. Assuming that the water flow remains turbulent and that the thermal resistance of the tube wall and the steam-condensate film is negligible, determine the effect of tube diameter on the total area required in the exchanger.

2. A tubular air heater has been accurately designed to heat a given stream of air to 170°F. The air passes through 120 tubes, arranged in parallel, with steam condensing on the outside. From Eq. (9-17), it is calculated that the tubes should be 16 ft long.

Because of the deviation of some of the data from Eq. (9-17), it is now decided to build the apparatus with a 20 per cent factor of safety. One engineer proposes to obtain this factor of safety by using 20 per cent more 16-ft tubes of the same diameter. Another argues that the 20 per cent factor of safety should be obtained by making the tube diameters 20 per cent greater, while keeping the same number of tubes and the same tube length.

Discuss the advantages and disadvantages of the foregoing methods. How would you build the heater to have a 20 per cent factor of safety?

3. A liquid metal is flowing in well-developed turbulent motion in a long smooth circular pipe with an i.d. of 1.95 inches. The Prandtl number of the liquid metal is 10^{-4}, the kinematic viscosity μ/ρ is 0.0388 sq ft/hr, and the average velocity of the flow is 3.33 ft/sec.

a. The following equation defines a *special* heat-transfer coefficient h':

$$h' = \frac{q_w}{A_w(t_w - t_{\text{axis}})}$$

where the subscript w refers to the inside surface of the pipe wall. Estimate the numerical magnitude of the special Nusselt number $h'D/k$ by employing the

† Experimental data are also given by Frank,[35] Haucke,[40] Nusselt,[74] Parmelee and Huebscher,[77] Rowley, Algren, and Blackshaw,[83] Taylor and Rehbock,[90] and Wagener.[94]

generalized velocity distribution of Eqs. (9-4), (9-4a), and (9-4b) and by making the assumption that E_H and E_M are equal.

b. What value is predicted by the Martinelli analogy for the Nusselt number based on the standard heat-transfer coefficient h, defined as:

$$h = \frac{q_w}{A_w(t_w - t_m)}$$

where t_m is the "mixing-box" mean temperature?

4. It is planned to construct a horizontal single-pass multitubular preheater to warm 6000 pounds/hr of viscous fuel oil from 80 to 100°F, while flowing in streamline motion inside horizontal tubes jacketed by exhaust steam condensing at 220°F in the shell. For tubes having i.d. of 0.67 inch and lengths of 8, 12, 16, or 20 ft, complete heaters (both shell and tubes) can be obtained at a first cost of $1.80 per foot of each tube. The fixed charges on the heater in dollars per year per foot of length of each tube are $0.36. Mechanical energy, delivered to the fluid, will cost $0.005/million ft-lb. Show which tube length should be selected.

Data and Notes. The oil density is 56 lb/cu ft. The specific heat c_p is constant at 0.48 gm-cal/(gm)(deg C), and k may be assumed constant at 0.08 Btu/(hr)(sq ft)(deg F per ft). Viscosity varies with temperature as shown below:

Deg F.......	80	90	100	110	120	130
Centipoises.....	400	270	185	140	110	82

It may be assumed that the pressure drop through the heater is due only to friction in the tubes. Assume tubes clean on both sides and the $U_i = h_i$.

5. An experimental heat exchanger consists of a vertical copper tube with an i.d. of 1 inch, heated by a steam jacket for a length of 10 ft. The steam condenses at a pressure of 1 atm, and the thermal resistance of the copper tube and steam-condensate film is negligible. A cold oil, entering at 50°F, flows upward through the copper tube.

Plot the exit temperature t of the oil vs. the oil flow rate w for the flow-rate range of 1 to 100,000 pounds/hr. In order to simplify the calculation, the physical properties of the oil will be assumed to be independent of temperature:

$$c_p = 0.5$$
$$k = 0.1$$
$$\mu = 10 \text{ centipoises}$$

6. A single-pass heat exchanger is being used to preheat the oil fed to a distillation column. Heat is supplied by saturated steam condensing on the outside of the tubes while the liquid flows inside the tubes. A "promoter" is added to the steam, which causes dropwise condensation. The tube pass consists of 100 tubes, each 20 ft long with an i.d. of 1 inch.

It is proposed to install an emergency "stand-by" heat exchanger in parallel with the existing exchanger described above. Because of space limitations, the stand-by exchanger will employ tubes only 10 ft in length. The tubes will have an i.d. of ½ inch, and a single tube pass will be used.

How many tubes are required in the stand-by exchanger for it to preheat the feed stock to the same temperature as the original exchanger?

NOTE: It is known that the flow in the original exchanger is in the turbulent range and that the thermal resistance of the tube and the steam-condensate film is negligible.

CHAPTER 10

HEATING AND COOLING OUTSIDE TUBES

Abstract. This chapter is divided into four parts: I, flow normal to single cylinders, spheres, and streamlined shapes; II, flow past extended surfaces (fins); III, flow normal to banks of tubes; and IV, flow through the shell side of cross-baffled exchangers.

The first section treats mechanisms, local and average coefficients of heat transfer, and experimental analogues, such as mass transfer. For flow of gases normal to single cylinders with $D_o G/\mu$ from 0.1 to 250,000, the data are correlated within an average deviation of ± 20 per cent by Fig. 10-7; a simplified relation for common gases, Eq. (10-3c), is included. In measuring the temperature of gases, it is shown that errors owing to both radiation and conduction can be serious. Data for flow of gases normal to noncircular tubes, and past spheres, are correlated in Figs. 10-10 and 10-11, respectively. Data for liquids (Fig. 10-12) are available for a moderate range of Reynolds numbers $D_o G/\mu_f$ but involve a large range of Prandtl numbers, from 6 to 1200.

The second section deals with the "efficiencies" of various types of extended (finned) surfaces, and the corresponding heat-transfer coefficients [Eqs. (10-9) and (10-10)].

The third section treats flow normal to variously arranged banks of tubes. For air, with $D_o G_{max}/\mu_f$ from 2000 to 40,000, Eq. (10-11) and Tables 10-5 and 10-6 apply; simplified dimensional relations for air and water are given by Eqs. (10-11b) and (10-11c), respectively. For both gases and liquids, $D_o G_{max}/\mu_f$ from 1 to 10,000, and for staggered banks 10 rows deep, results can be estimated from Fig. 10-20. With "in-line" arrangements of tubes a "transition region," similar to that for flow inside tubes, is found (Fig. 10-21).

The fourth section considers the complex situation on the shell side of cross-baffled exchangers

I. FORCED FLOW NORMAL TO SINGLE CYLINDERS

Mechanism. Figure 10-1 shows a photograph by Ray[50] taken with air flowing from left to right past a horizontal internally heated hot cylinder. Because of the change of refractive index with change in temperature, the heated air appears brighter than the cold air. Additional data for wires of several diameters are given by Praminik.[48]

Experiments made under isothermal conditions throw light upon the nature of the flow of fluid past cylindrical models. Figure 10-2, by Rubach,[54] shows conditions at an instant when the eddies in the rear of the cylinder were growing in size. Eventually these eddies become too large to be protected by the cylinder and are swept downstream. Similar

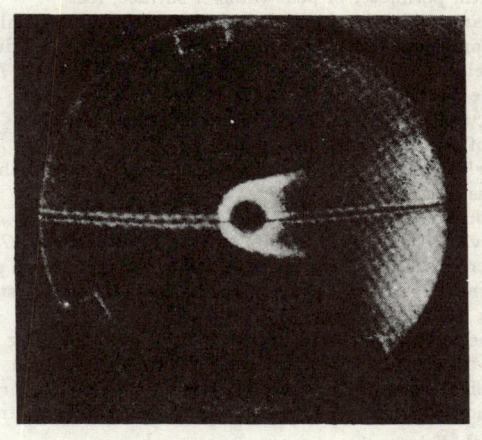

Fig. 10-1. Photograph of air flowing horizontally at a velocity of 0.3 ft/sec at right angles to a heated cylinder. (*Photograph from Ray,*[50] *Proc. Indian Assoc. Cultivation Sci.*)

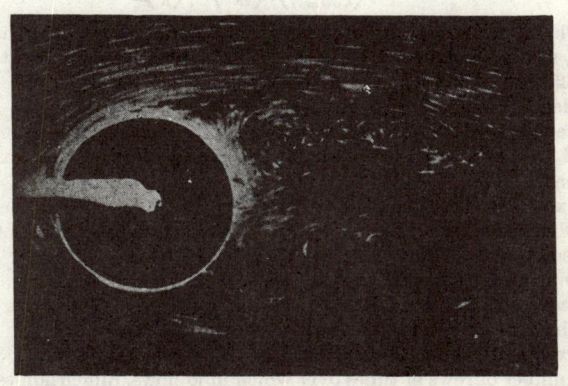

Fig. 10-2. Isothermal flow of liquid normal to the axis of a cylinder. (*Photograph from Rubach,*[54] *Forschungsarb. Metallkunde u. Röntgenmetallog.*)

photographs are shown in papers by Prandtl.[49] Reiher[51] and Lohrisch[37] have published photographs of "smoke" in air currents flowing isothermally past models of various shapes. The models were covered with filter paper wetted with aqueous hydrochloric acid, and the smoke was produced by interaction of ammonia vapor in the air stream with the acid vapor diffusing from the wetted surface. Hence the photographs

Table 10-1. Nomenclature

- A Area of heat-transfer surface, square feet; A_o for outside, A_p for pyrometer, A_s for surface, and A_w for wall
- a Dimensional term, equal to $(hb/kS)^{0.5}$ reciprocal feet; a_p from pyrometer to gas
- b Breadth, perimeter of heat-transfer surface, feet
- b_1, b_2 Constants to be determined experimentally, dimensionless
- C_1, C_2 Constants depending on geometry of fin, C_1 in feet, C_2 dimensionless
- D Diameter, feet; D_e for effective diameter $(D_e = b/\pi)$; D_o for outside; D'_o, outside diameter, inches; D_s for spheres; D_h, hydraulic diameter
- d Prefix, indicating derivative, dimensionless
- e Base of natural logarithms, 2.718
- F_a Factor[19] depending on arrangement of the tubes in the bundle, dimensionless
- f'' Friction factor, dimensionless; defined by $f'' = (2\,\Delta p g_c \rho / 4 G_{\max}^2 N)$
- G Mass velocity of fluid, lb/(hr)(sq ft of cross section); $G_{\max}$ through minimum cross section; G_{app}, G_B, G_{eff}, and G_{ob} are defined on p. 279
- g Acceleration due to gravity, ordinarily taken as standard value of 4.17×10^8 ft/(hr)(hr)
- h Surface coefficient of heat transfer, Btu/(hr)(sq ft)(deg F); h_a for average value to air; h_c for natural convection; h_m for mean value; h_o and h_{ob} for baffled exchangers; h_p from pyrometer to gas; h_r for radiation; h_{rb} for radiation between black bodies; h_x for local value; h_{wa} from thermocouple leads to air
- $j'_{\max}$ A product of dimensionless terms,

$$j'_{\max} = \left(\frac{h}{c_p G_{\max}}\right)\left(\frac{c_p \mu}{k}\right)^{2/3}\left(\frac{\mu_s}{\mu}\right)^{0.14}$$

- k Thermal conductivity of fluids at bulk temperature, Btu/(hr)(ft)(deg F), k_f at film temperature $t_f = (t_s + t)/2$; k_s at surface temperature t_s
- k_e Effective thermal conductivity of thermocouple leads, Btu/(hr)(ft)(deg F); $\sqrt{k_e} = (\sqrt{k_1} + \sqrt{k_2})/2$
- L Geometrical factor, ordinarily height of vertical surface in feet; L', height of horizontal tube, $L' = \pi D'_o/2$, inches; L_c, effective heat exchanger core length
- m Exponent, dimensionless
- m_1, m_2 Dimensionless multipliers, p. 279; m_1 is usually 0.8 to 1.0, and m_2 is usually 0.9 to 1.0
- m' Dimensionless term, $ka/h = (kb/hS)^{0.5}$
- n Exponent, dimensionless
- P Absolute pressure, atmospheres
- p, p_1, p_2 Exponents, dimensionless
- N Number of rows of tubes over which fluid flows, dimensionless
- q Rate of heat transfer, Btu/hr; q_c by convection; q_{gr} by gas radiation; q_k by conduction; q_r by radiation between surfaces
- r_h Hydraulic radius, feet
- S Cross section for flow of fluid, square feet; for S_B and S_{eff}, see p. 279; $S_{\min}$ for minimum value
- s Spacing or pitch between centers of tubes, feet; s_L for longitudinal pitch; s_T for transverse pitch
- T Absolute temperature, degrees Rankine $= t + 460$; T_g for gas; T_p for pyrometer; T_s for surface
- t Thermometric temperature, degrees Fahrenheit; t, bulk temperature of

TABLE 10-1. NOMENCLATURE.—(*Continued*)

fluid; t_a for ambient fluid; t_e for walls of enclosure; t_f for film $= (t_s + t)/2$; t_g for gas; t_p for pyrometer; t_o for outside; t_s for surface; t_x at point x; t_w for wall

U Over-all coefficient of heat transfer, Btu/(hr)(sq ft)(deg F)
V Average velocity, feet per hour; V_{max} through *minimum* free cross section
V_T Volume of heat exchanger, cu ft
w Mass flow rate, pounds fluid per hour; w_A, w_B, w_{B1}, w_C, w_E, w_T, and w_W are defined in Fig. 10-22 and on p. 278
X_{app} A Reynolds number, $D_o G_{app}/\mu$, dimensionless
x Distance along conduction path, feet; x_i, fin immersion; x_f for fin length
x_L, x_T Dimensionless ratio of longitudinal pitch s_L or transverse pitch s_T to outside diameter of tubes D_o; $x_L = s_L/D_o$; $x_T = s_T/D_o$
Y_{app}, Y_{eff} Products of dimensionless ratios, defined on p. 279
y_0 Half thickness of fin at base or root of fin, feet
z Vertical distance, feet
z_p Distance traveled by fluid across fin, feet, Fig. 10-14

Greek

β Coefficient of volumetric expansion, reciprocal degrees Fahrenheit; for perfect gases, $\beta = 1/T$; β_f at t_f
Δt Temperature potential, degrees Fahrenheit; Δt between surface and bulk temperature; Δt_m, length-mean value; Δt_0, at base of fin; Δt_x, local value
ϵ Emissivity, dimensionless (see Appendix); ϵ_p for pyrometer
η Fin efficiency, dimensionless, $\eta = \Delta t_m/\Delta t_0$
μ Absolute viscosity of fluid, lb/(hr)(ft), equal to $2.42 \times$ viscosity in centipoises; μ_f at t_f, μ_s at t_s
π 3.1416, dimensionless
ρ Density of fluid, pounds per cubic foot; ρ_f at t_f; ρ_w at t_w
Σ Prefix, indication of summation, dimensionless
ϕ, ψ Functions to be determined experimentally

show the flow conditions of air that has been in close proximity to the model. Figure 10-3 shows the eddies in the rear of the cylinder. Figure 10-4 shows photographs[22] of isothermal flow of water past a cylindrical model in a closed glass channel. The flow pattern is made visible by the addition of a small concentration of specially prepared colloidal bentonite and is photographed between crossed polarizing plates with transmitted light showing variation in the double refraction of this suspension with velocity gradient.[22]

Local Coefficients. From the foregoing discussion of the mechanism of heat transfer for fluids flowing at right angles to tubes and of the nature of fluid flow over such surfaces, it can be inferred that the coefficient of heat transfer between fluid and pipe will not be uniform around the perimeter. Drew and Ryan[12] have reported measurements of the variation in local rate of heat flow to air at various positions around a vertical steam-heated pipe and have found results similar to those based on the work of Lohrisch[37] † for absorption. The results of an experiment

† See also Thoma[59] and Nusselt.[44]

256 HEAT TRANSMISSION

by Paltz and Starr[45] are shown in Fig. 10-5, with local rates of heat flow plotted radially at various positions around a single steam-heated cylinder. The curve shows the actual distribution of the heat flux, and it is noted that the maximum local rate of heat flow occurred at the front

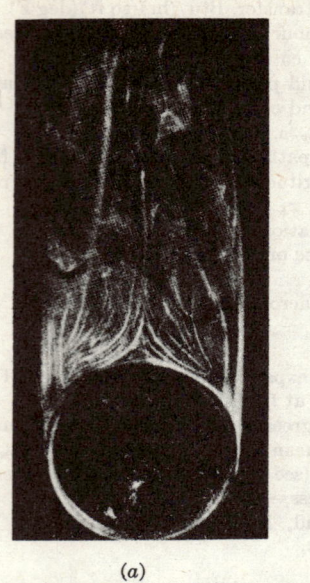

(a) (b)

FIG. 10-3. (a) Photograph of eddies at the rear of a cylinder by air flowing at a Reynolds number of approximately 2700. (b) Idealized drawing from the photograph. (*From Lohrisch,*[37] *Forschungsarb. Metallkunde u. Röntgenmetallog.*)

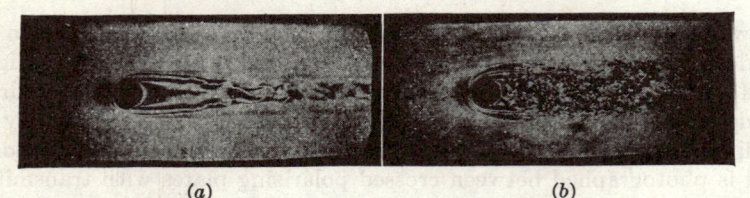

(a) (b)

FIG. 10-4. Shear patterns for isothermal flow of a dilute suspension of bentonite past a cylinder, photographed by Hauser and Dewey.[22] (a) $N_{Re} = 2100$. (b) $N_{Re} = 5300$.

and back, and the minimum rate, approximately 40 per cent of the maximum, was found at the sides.

Experimental Analogues. Since equations for mass transfer are of the same forms as those for heat transfer, it would be expected that mass transfer can be used as an experimental analogue for heat transfer. This proves to be the case, as shown in Fig. 10-6 for flow normal to single tubes.[67] It is seen that the circumferential variation in local Nusselt number is similar for the transfer of both heat and mass. Curves of simi-

HEATING AND COOLING OUTSIDE TUBES

lar shape have been obtained for heat transfer by other workers.[13,14,17,23,55] Analogues between the transfer of mass and of heat have the limitation that a mass-transfer analogue cannot be devised to simulate a heat-transfer case with a large temperature potential.

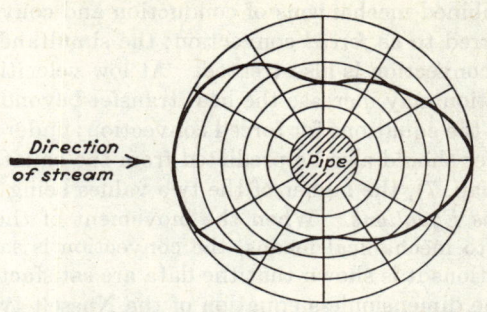

FIG. 10-5. Distribution of heat flow per unit area around a cylinder, the air flowing normal to the axis. (*Drew and Ryan.*[12])

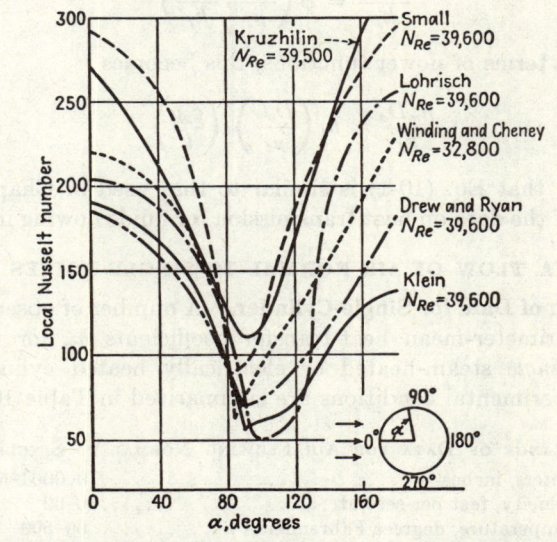

FIG. 10-6. Comparison of local Nusselt numbers for transfer of heat or mass for air flowing normal to a cylinder; data by Winding and Cheney[67] are for evaporation of naphthalene.

Average Coefficients. For practical purposes it is advisable to deal with the average coefficient of heat transfer for a tube, or bank of tubes, rather than a local coefficient at a given point on the perimeter, and the following discussion applies to such average coefficients. The net rate of heat exchange (Σq) between a body and its surroundings is evaluated by adding the rates of heat flow due to convection and radiation, q_c and

258   *HEAT TRANSMISSION*

q_r. Chapter 4 gives details of the evaluation of q_r due to radiation between surfaces and radiation between a surface and certain gases such as carbon dioxide, water vapor, sulfur dioxide, and ammonia. The present chapter deals largely with the evaluation of the rate of heat transfer due to the combined mechanisms of conduction and convection in fluids, ordinarily referred to as *forced* convection; the simultaneous transfer by radiation and convection is also treated. At low velocities, the effect of natural convection may increase the heat transfer beyond what would be predicted from the equations for forced convection; under such conditions the heat transfer should also be predicted from the equations for natural convection (Chap. 7), the higher of the two values being then employed.

Dimensionless Equations. When the movement of the fluid past the surface is due to mechanical means, the convection is said to be *forced*. For such conditions it is shown that the data are satisfactorily correlated by means of the dimensionless equation of the Nusselt type

$$\frac{h_m D_o}{k_f} = \phi\left(\frac{D_o G}{\mu_f}, \frac{c_p \mu_f}{k_f}\right)$$

Expressed in terms of power functions, this becomes

$$\frac{h_m D_o}{k_f} = b_1 \left(\frac{D_o G}{\mu_f}\right)^n \left(\frac{c_p \mu}{k}\right)^m_f \qquad (10\text{-}1)$$

It is recalled that Eq. (10-1) is similar to that used in Chap. 9 for the correlation of the data on heat transmission for fluids flowing inside tubes.

IA. FLOW OF AIR NORMAL TO SINGLE BODIES

Correlation of Data for Single Cylinder. A number of observers[†] have measured perimeter-mean heat-transfer coefficients h_m for flow of air normal to *single* steam-heated or electrically heated cylinders. The ranges of experimental conditions are summarized in Table 10-2. Since

TABLE 10-2. RANGE OF DATA FOR AIR FLOWING NORMAL TO SINGLE CYLINDERS

Diameters, inches	0.0004–5.9
Air velocity, feet per second	0–62
Air temperature, degrees Fahrenheit	60–500
Cylinder temperature, degrees Fahrenheit	70–1,840
Pressure, atmospheres absolute	0.4–4.0
$D_o G/\mu_f$, consistent units	0.2–235,000
hD_o/k_f, consistent units	0.5–500

the Prandtl numbers for air vary but little, only two dimensionless ratios need be considered, the Nusselt and Reynolds numbers, $h_m D_o/k_f$ and $D_o G/\mu_f$. The data of Hilpert[23] cover the largest range of diameters, 0.0079 to 5.9 inches, and the *recommended* curve on Fig. 10-7 of $h_m D_o/k_f$

[†] References 1, 16, 23, 26, 28, 29, 30, 32, and 33.

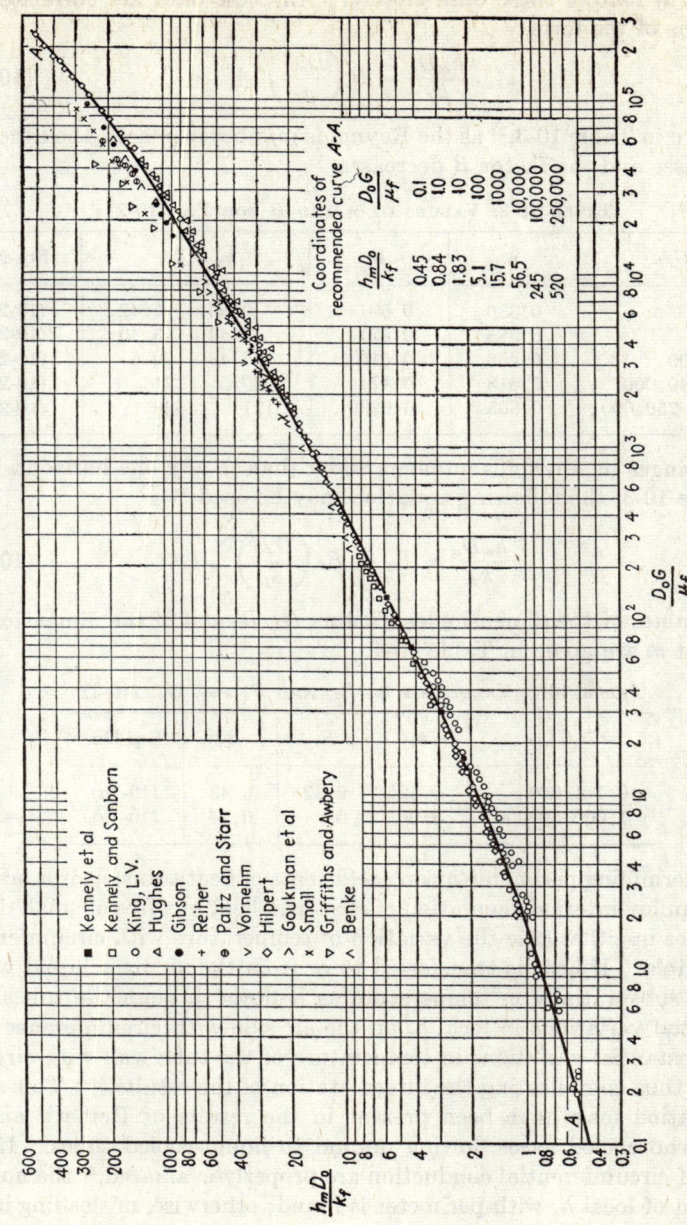

Fig. 10-7. Data for heating and cooling air flowing normal to single cylinders, corrected for radiation to surroundings.

Coordinates of recommended curve A-A	
$\dfrac{D_o G}{\mu_f}$	$\dfrac{h_m D_o}{k_f}$
0.1	0.45
1.0	0.84
10	1.83
100	5.1
1000	15.7
10,000	56.5
100,000	245
250,000	520

Legend:
- ■ Kennely et al
- □ Kennely and Sanborn
- ○ King, L.V.
- △ Hughes
- ● Gibson
- + Reiher
- ⌀ Paltz and Starr
- ▽ Vornehm
- ◇ Hilpert
- > Goukman et al
- × Small
- ▽ Griffiths and Awbery
- ＜ Benke

vs. $D_o G/\mu_f$ follows these data closely. All these data are correlated by equations of the form

$$\frac{h_m D_o}{k_f} = B\left(\frac{D_o G}{\mu_f}\right)^n \qquad (10\text{-}2) \star$$

As shown in Table 10-3,[23] as the Reynolds number increases, the exponent n increases and the factor B decreases.

TABLE 10-3. VALUES OF n AND B FOR EQ. (10-2)

$D_o G/\mu_f$	n	B	$h_m D_o/k_f$	Eq. No.
1–4	0.330	0.891	0.891– 1.42	(10-2a)
4–40	0.385	0.821	1.40 – 3.40	(10-2b)
40–4,000	0.466	0.615	3.43 – 29.6	(10-2c)
4,000–40,000	0.618	0.174	29.5 –121	(10-2d)
40,000–250,000	0.805	0.0239	121 –528	(10-2e)

For ranges in Reynolds numbers wider than in any one horizontal line of Table 10-3, the following equation may be used:[38]

$$\frac{h_m D_o}{k_f} = B_1 + B_2\left(\frac{D_o G}{\mu_f}\right)^m \qquad (10\text{-}3) \star$$

where values of the dimensionless factors B_1, B_2 and of the dimensionless exponent m are given in Table 10-4.

TABLE 10-4. VALUES OF m, B_1, AND B_2 FOR EQ. (10-3)

$D_o G/\mu_f$	m	B_1	B_2	Eq. No.
0.1–1,000	0.52	0.32	0.43	(10-3a)
1,000–50,000	0.60	0	0.24	(10-3b)

In determining perimeter-mean coefficients of heat transfer, it is advisable to employ internal generation of electrical heat in wires of small diameter, since in either case the variation of temperature with circumference is negligible. If heat is transferred to or from the air by a liquid which changes substantially in temperature as it flows through the tubes, the substantial variations in local h_x on the air side with circumference produce substantial variations in temperature of the tube wall with circumference, thus complicating the interpretation of the results.[60] This same complication may have been present in the results of Reiher[51] and of others who cooled gases flowing normal to liquid-cooled tubes. If the effects of circumferential conduction are properly evaluated,[60] the normal variation of local h_x with perimeter is found; otherwise, misleading information is obtained.

Maisel and Sherwood[38b] found that their data for evaporation of several liquids into several gases, from the wetted porous surfaces of single cylinders with transverse flow of gas, were in good agreement with heat-transfer data from dry single cylinders.

Simplified Equation. The simplified equation for air

$$h_m = 0.11 c_p G^{0.6}/D_o^{0.4} \qquad (10\text{-}3c) \star$$

applies for a film temperature of 200°F and $D_o G/\mu_f$ from 1000 to 50,000.

Effect of Turbulence. The effect of increasing the turbulence of the air in increasing the heat-transfer coefficient was studied by Reiher,[51] who obtained as much as 50 per cent increase in h_m by passing the air through a grid to increase turbulence in the air flowing over the single pipe.

For a Reynolds number $D_o G/\mu$ of 5800, Comings *et al.*[9] found that $h_m D_o/k$ increased 31 per cent as the turbulence increased from 2 to 26 per cent. For a Reynolds number of 4900 Maisel and Sherwood[39] found that the mass-transfer equivalent of $h_m D_o/k$ increased by 32 per cent, as the intensity of turbulence increased from 4 to 24 per cent; at $D_o G/\mu$ of 10,000, the increase was 50 per cent; similar effects were found with spheres.

Calculation of True Temperature of a Gas. For computing the true temperature of a gas from the reading of a thermocouple or pyrometer placed in a gas stream and in sight of surrounding walls that may be at various temperatures, a heat balance for the pyrometer is used.

$$q_{gr} + q_c = q_r + q_k$$

where q_{gr} is the rate of heat flow between gas and pyrometer by the mechanism of *gas radiation* (Chap. 4), q_c is the rate of heat flow between gas and pyrometer by *convection* (see Fig. 10-7), q_r is the sum of the various terms representing the radiant heat interchange between the pyrometer and the various surfaces that it "sees," evaluated by the methods of Chap. 4, and q_k is the heat *conducted* from the thermocouple to the walls confining the gas stream. In the simple case of a gas stream having true temperature T_g and flowing through a duct of a diameter large compared with that of the pyrometer at temperature T_p, the inner surfaces of the walls having approximately constant temperature T_s, a heat balance on the pyrometer gives the equation

$$q_{gr} + h_c A_p (T_g - T_p) = (0.171)(\epsilon_p A_p)\left[\left(\frac{T_p}{100}\right)^4 - \left(\frac{T_s}{100}\right)^4\right] + q_k \quad (10\text{-}4)$$

Values of ϵ_p, the emissivity of the surface of the pyrometer, are given in the Appendix; T_p and T_s represent, respectively, the temperatures of the pyrometer and walls, expressed in degrees Fahrenheit absolute. Instead of using Eq. (10-4), which involves the fourth power of the absolute temperature, it is possible to employ temperature in degrees Fahrenheit, as shown in the following alternate procedure. By definition, $q_r = h_r A(t_p - t_s)$, where h_r is the appropriate "radiation coefficient of heat transfer" equal to ϵ times the value h_{rb}, from Eq. (4-44a). Hence Eq. (10-4) may be written as

$$q_{gr} + h_c A_p (t_g - t_p) = h_r A_p (t_p - t_s) + q_k \qquad (10\text{-}4a)$$

Under the conditions in which q_{gr} and q_k are negligible compared with q_c and q_r, Eq. (10-4a) simplifies to

$$t_g - t_p = (t_p - t_s)(h_r/h_c) \qquad (10\text{-}4b)$$

From Eq. (10-4b) it is clear that the difference between the true and apparent temperatures of the gas, $t_g - t_p$, increases with (1) increase in the temperature difference between the pyrometer and the walls, $t_p - t_s$, (2) with increase in the simplified radiation coefficient h_r, and (3) with decrease in the convection coefficient h_c, between gas and pyrometer. Since h_r increases with temperature level far faster than does h_c, the error in measuring gas temperature increases with increase in temperature. In order to reduce the error, several procedures are used.

Illustration 1. Air is flowing steadily through a duct whose inner walls are at 500°F. A thermocouple, housed in a rusted steel well (ϵ of 0.9), inserted at right angles to the air stream, indicates a temperature of 300°F. The mass velocity of the air is 3600 lb/(hr)(sq ft), and the o.d. of the well is 1 in. Estimate the true temperature of the air, neglecting q_k.

Solution. From Eq. (10-3c), $h_c = 9.7$, and, from Chap. 4, $h_{rb} = 4.44$ for a black body; h_r is 0.9×4.44 equals 4.0. From Eq. (10-4b)

$$t_p - t_g = \frac{(t_s - t_p)(h_r)}{h_c} = \frac{(500 - 300)(4.0)}{9.7} = 83°F \qquad t_g = 217°F$$

Reduction of Error. The true gas temperature is sometimes estimated by measuring the apparent gas temperature by several thermocouples of different diameters and extrapolating to zero diameter.[34] Or a sample of the gas stream, flowing at low velocity, is caused to flow at high velocity through an insulated tube past a thermocouple, giving the "high-velocity" thermocouple.[42] These two procedures reduce the error by increasing h_c. The temperature difference between walls and the thermocouple can be reduced by the use of radiation shields, so that the thermocouple "sees" the warmer shields bathed in hot gas rather than the colder walls. The radiation correction also can be reduced by covering the surface of the pyrometer with a tightly fitting surface[65] of polished metal such as aluminum foil. It is interesting to note that the effect of radiation on thermometers was recognized by Wells[65] in 1818. A summary of 26 methods of reducing error in measuring gas temperatures is given by Mullikin.[41]

Moffatt[40] discusses the errors in the calculated gas temperature, owing to uncertainties in predicting heat-transfer coefficients and emissivities, for unshielded probes, shielded probes, aspirated probes, and heated-shield probes. The error due to radiation can be reduced by use of one or more concentric shields. Figure 10-8 shows the error plotted against the true temperature of the gas, for various numbers of shields. With the air stream at 1600°F and the pipe walls at 1100°F, the bare couple is

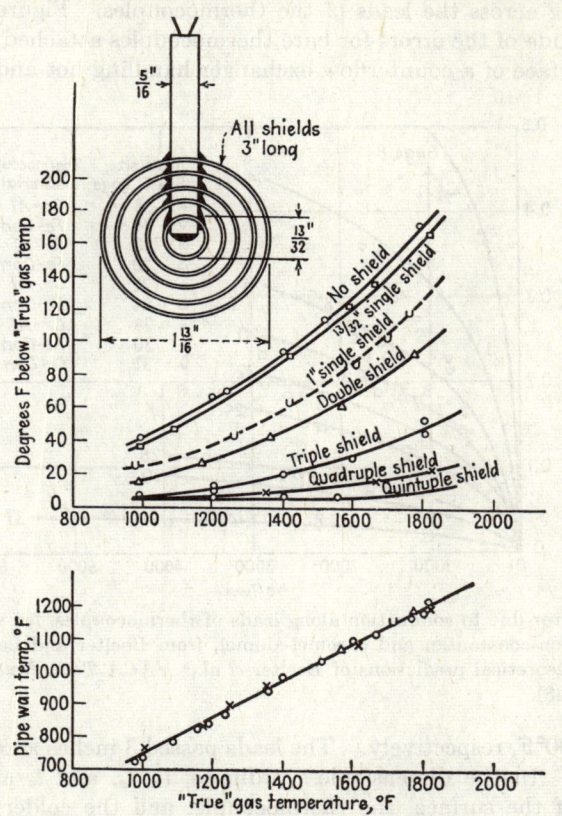

FIG. 10-8. Effect of various numbers of shields upon thermocouple error in hot gas stream, for mass velocity of 16,200 pounds of air/(hr)(sq ft of cross section). (*From W. J. King,*[31] *Trans. ASME.*)

130°F below t_g, and with four shields the error is only 14°F. King[31] gives a detailed description of a recommended quadruple-shielded thermocouple.

With only a single shield in a hot duct transporting colder gas, the temperature of the shield is considerably above that of the gas entering the shield, and consequently the gas is heated before it reaches the shielded thermocouple; it is desirable to measure the temperature of the shield and to correct for the heating of the gas by contact with the shield which precedes the thermocouple. The increase in temperature of a probe, owing to partial conversion of the kinetic energy of the gas to additional enthalpy, is important at high gas velocity (Chap. 12).

Conduction Correction. The temperatures of thermocouples attached to a heat-transfer surface are influenced by the transfer of heat to the fluid flowing across the leads of the thermocouples. Figure 10-9 shows the magnitude of the errors for bare thermocouples attached to the heat-transfer surface of a counterflow exchanger handling hot and cold air, at

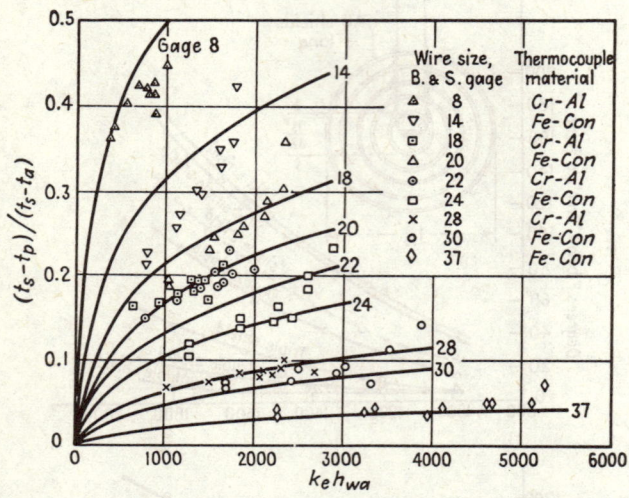

Fig. 10-9. Error due to conduction along leads of thermocouples, for various gauges of wires of iron-constantan and chromel-alumel, from Boelter and Lockhart,[2] compared with theoretical predictions of Boelter *et al.*,[3] *NACA Tech. Notes* 2427 (1951) and 1452 (1948).

1000 and 100°F, respectively. The leads passed 3 inches across the cold-air stream. In the dimensionless ordinate, t_s, t_p, and t_a are the temperatures of the surface, the thermocouple, and the colder air stream, respectively. These data of Boelter and Lockhart[2] are in fairly good agreement with the curves predicted theoretically by Boelter *et al.*[3] The effective thermal conductivity k_e is given by $\sqrt{k_e} = (\sqrt{k_1} + \sqrt{k_2})/2$, and is taken as 16 for chromel-alumel and as 24 for iron-constantan. The heat-transfer coefficient h_{wa} from wire to air increases with increase in mass velocity of the air. For a given thermocouple material, the error increases with increase in wire diameter (decrease in gauge) and with increases in both h_{wa} and k_e. Errors are greatly reduced by electrically insulating the leads and burying them in grooves (4 inches long) in the plate, before bringing the leads out through the colder air stream.

Rizika and Rohsenow[53] derived an equation for calculating the temperature t_g of a fluid from the temperature t_p of a thermocouple in a protecting tube immersed a distance x_i in the fluid:

$$\frac{t_p - t_a}{t_g - t_p} = \cosh a_p x_i - 1 + \sqrt{\frac{h_p}{h_a}} \sinh a_p x_i \qquad (10\text{-}5)$$

In an example a thermocouple junction was housed in the closed end of a metal tube ($\frac{3}{8}$ inch outside diameter and $\frac{1}{4}$ inch inside diameter, k of 15) immersed 2 inches into a gas stream; t_p was 800°F, the room temperature t_a was 65°F, h_p and h_a were 9.4 and 2.4, and t_g was calculated to be 874°F. Charts were included for facilitating the calculation.

Noncircular Tubes. Hilpert[23] and Reiher[51] report data for flow of air normal to noncircular tubes, having oval, square, and hexagonal cross sections;[27] data of the two observers disagree for some of the shapes.

Streamline Shapes. Hughes[26] tested in a wind tunnel a steam-heated copper model having a streamlined contour similar to that of a teardrop. When the air approached the rounded end of the model, higher values of h_m were obtained than when the air approached the pointed end.

For air approaching the rounded end, the data can be brought into close agreement with those for single cylinders (Fig. 10-7) by plotting $h_m D_e/k_f$ vs. $D_e G/\mu_f$, where D_e is the effective diameter, equal to the total perimeter divided by π (Fig. 10-10). When the angle of incidence is changed, the exponent relating h_m and G changes.[4,58] Coefficients for the leading end of a model of an airfoil are available.[20]

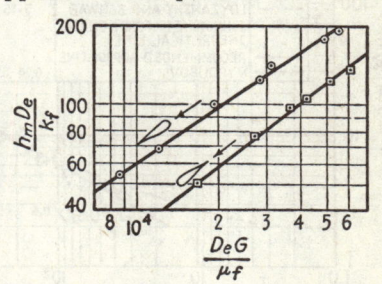

Fig. 10-10. Data of Hughes.[26]

Spheres. Figure 10-11 shows data† for heat transfer between spheres and air, correlated by Williams;[68] subsequent data of Kramers[33] have been added. In the range of $D_s G/\mu_f$ from 17 to 70,000 the **recommended** straight line has the dimensionless equation

$$h_m D_s/k_f = 0.37 (D_s G/\mu_f)^{0.6} \qquad (10\text{-}6) \star$$

In a recent paper Ingebo[26a] reports the effect of air temperature and velocity upon the rates of evaporation of nine organic liquids from the

† In plotting data for irregular particles, where the average surface per particle had been estimated from rates of settling, D_o was taken as the diameter of a spherical particle having the same surface as the irregular particle. The data of Compan[10] and Hartmann[21] would fall roughly 80 to 50 per cent, respectively, below the curve in Fig. 10-11. Part of the data of Vyroubov,[64] plotted in Fig. 10-11, is based on mass-transfer measurements.

surface of a porous sphere having a diameter of 0.27 inch. The heat-transfer coefficients were correlated in terms of Nusselt number, Reynolds number, Schmidt number (Chap. 5), and the ratio of the thermal conductivities of the air and the vapor:

$$\frac{h_m D_o}{k_f} = \left(\frac{k_f}{k_{vf}}\right)^{0.5} \left[2.0 + 0.3 \left(\frac{D_o G}{\mu_f} \cdot N_{Sc}\right)^{0.6}\right] \quad (10\text{-}6a)$$

Local Coefficients. A recent publication[35] reports local coefficients of heat transfer at the equator of a copper sphere to air. Reynolds numbers based on approach velocity ranged from 1.3×10^5 to 10^6, and local Nusselt numbers $h_x D_o/k$ drop slowly as distance from the stagnation

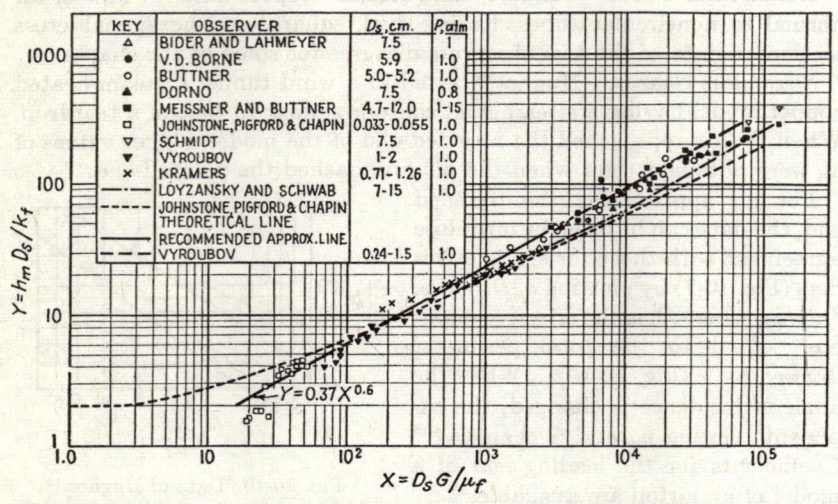

Fig. 10-11. Data for heat transfer between spheres and air, correlated by Williams;[68] data of Vyroubov[64] are based on mass transfer. For conduction to a stagnant medium the theoretically minimum value of $h_m D/k$ is 2 unless the spheres have diameters of the order of the mean free paths of the molecules of the ambient fluid.

point is increased until a minimum is reached at 105°; a maximum occurs at 120°, a second minimum at 150°, and a second maximum at 180° The *average* Nusselt numbers were 40 to 60 per cent above those predicted by $h_m D_o/k_f = 0.33(D_o G/\mu_f)^{0.6}$; the intensity of turbulence in the air stream was not specified.

IB. FLOW OF LIQUIDS NORMAL TO SINGLE BODIES

Single Cylinders. The most comprehensive available data are those of Davis[11a] for several sizes of wires to water and four hydrocarbon oils having a wide range of viscosities; velocities ranged from 0.36 to 2.3 ft sec. Values of c_p, μ, k, and ρ were measured for these oils. Ulsamer[62a] correlated these data for liquids and those of several investigators on air, by Eq. (10-1), with m of 0.31. The values of b_1 and n were 0.91 and 0.385 for $D_o G/\mu_f$ ranging from 0.1 to 50 and 0.6 and 0.5 for $D_o G/\mu_f$

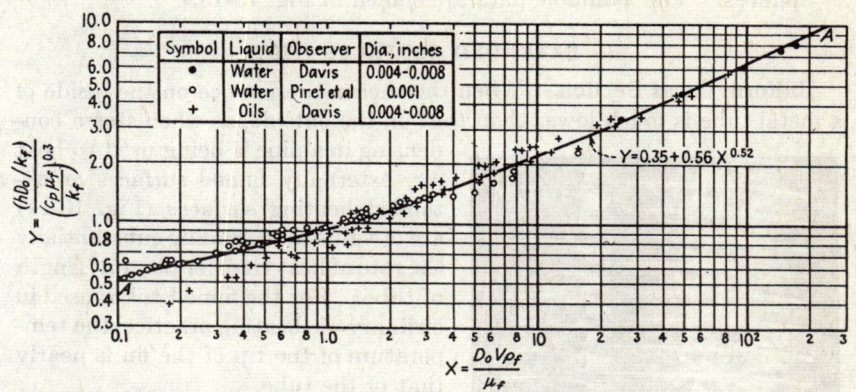

FIG. 10-12. Heating liquids flowing normal to single cylinders. The recommended curve AA is based on Eq. (10-7).

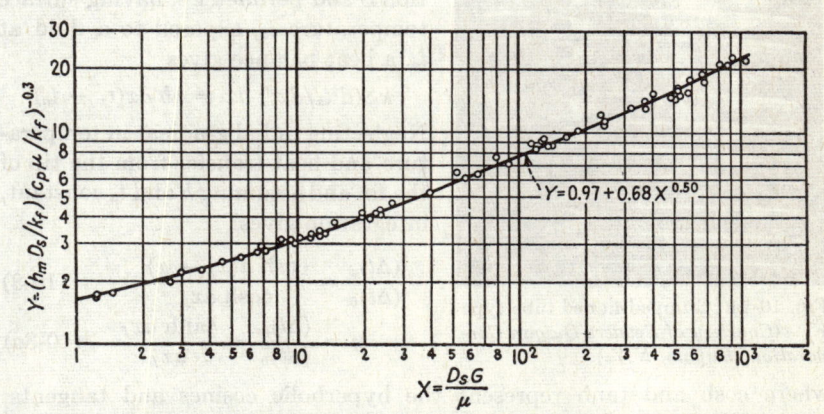

FIG. 10-13. Data of Kramers[33] for flow of water and of spindle oil (Prandtl number from 7.3 to 380 and Δt from 11 to 71°F) past single spheres, having diameters from 0.279 to 0.496 in.

ranging from 50 to 10,000. Later Kramers[33] proposed a more complex relation involving five empirical constants:

$$h_m D_o/k_f = 0.42(c_p\mu/k)_f^{0.2} + 0.57(c_p\mu/k)_f^{0.33}(D_o G/\mu_f)^{0.5} \quad (10\text{-}7)$$

Subsequently Piret[47] reported data for water for Reynolds numbers ranging from 0.8 to 8, correlated by Eq. (10-1), with m of 0.30, b_1 of 0.965, and n of 0.28.

All these data[11a,47] for liquids are plotted in Fig. 10-12. Curve AA is based on the **recommended** relation

$$(h_m D_o/k_f)/(c_p \mu_f/k_f)^{0.3} = 0.35 + 0.56(D_o G/\mu_f)^{0.52} \qquad (10\text{-}7a) \star$$

Spheres. The available data are shown in Fig. 10-13.†

II. EXTENDED SURFACES (FINS)

Uniform Cross Section. When the thermal resistance on the inside of a metal tube is much lower than that on the outside, as when steam condensing in a pipe is being used to heat air, externally finned surfaces or extended heating surfaces (Fig. 10-14) are of value in increasing substantially the rate of heat transfer per unit length of tube. For the finned tubes used in ordinary air-heating practice, the temperature of the tip of the fin is nearly that of the tube.

For finite fins of constant cross section S and perimeter b, having surface temperature t_x, exposed to a fluid at t_a, a heat balance gives

$$kS(d^2 t_x/dx^2)\,dx = hb\,dx(t_x - t_a)$$

Neglecting radial gradient in temperature and heat transfer from the tip of the fin and assuming h and t_a constant, integration gives

$$\frac{(\Delta t)_x}{(\Delta t)_0} = \frac{\cosh a(x_f - x)}{\cosh a x_f} \qquad (10\text{-}8)$$

$$\eta = \frac{(\Delta t)_m}{(\Delta t)_0} = \frac{\tanh a x_f}{a x_f} \qquad (10\text{-}8a)$$

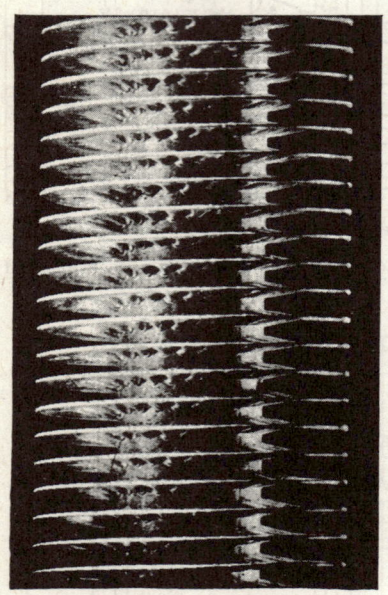

Fig. 10-14. Crimped-finned tube type K. (*Courtesy of Fedders-Quigan Corporation, Buffalo, N.Y.*)

where cosh and tanh represent the hyperbolic cosines and tangents, respectively:

$$\cosh p = \frac{e^p + e^{-p}}{2} \qquad \tanh p = \frac{e^p - e^{-p}}{e^p + e^{-p}}$$

and e is 2.718. In Eqs. (10-8) and (10-8a), a equals $(hb/kS)^{0.5}$, where b is

† The data of Kramers[33] for flow of air, water, and spindle oil past single spheres involved $D_o G/\mu_f$ from 0.4 to 2100 and were correlated by

$$\frac{h_m D_o}{k_f} = 2.0 + 1.3\left[\frac{c_p \mu}{k}\right]_f^{0.15} + 0.66\left[\frac{c_p \mu}{k}\right]_f^{0.31}\left[\frac{D_o G}{\mu_f}\right]^{0.50} \qquad (10\text{-}7b)$$

For small Δt and zero Reynolds number, the value of $h_m D_o/k_f$ should theoretically be 2.0, instead of $2.0 + 1.3(c_p \mu/k)_f^{0.15}$.

the exposed perimeter, x is the length from the base of the fin of length x_f, and Δt is the temperature difference between fin and air, $(\Delta t)_0$ at the base, $(\Delta t)_x$ at distance x, and $(\Delta t)_m$ for the entire fin. The measured[57] temperature gradients along bar fins are well correlated[38a] by Eq. (10-8).

If the tip of the fin is bare, a satisfactory approximation is to increase the actual x_f by the amount S/b. The theoretical equation of Jakob[27] for a fin having a uniform cross section and perimeter, allowing for heat loss from the tip, can be rearranged to give

$$q = \frac{h_m b (\Delta t)_0}{a}\left(\frac{2}{1 + \frac{m'-1}{m'+1}\frac{1}{e^{2ax_f}}} - 1\right) \qquad (10\text{-}8b)$$

where $m' = ka/h = \sqrt{kb/hS}$. When one calculates the ratio of the heat loss q_f from a pin fin to the heat loss q_o for a bare wall having the same surface area as the cross section of the fin, granting that h and t_o are the same in both cases, the q ratio is given by

$$\frac{q_f}{q_o} = m'\left(\frac{2}{1 + \frac{m'-1}{m'+1}\frac{1}{e^{2ax_f}}} - 1\right) \qquad (10\text{-}8c)$$

For m' of 1, this q ratio equals 1, regardless of the value of ax_f; consequently the fin should be omitted. For m' greater than 1, the q ratio exceeds 1 and increases with increase of ax_f. For m' less than 1, the q ratio is less than 1.

For a given steam temperature the outer tube wall will be somewhat colder with a fin attached to the outside than without a fin. The q ratio is given by Gardner:[15]

$$\frac{q_f}{q_o} = \frac{A_f \eta}{A_o} \frac{1 + (h_f/h_i)}{1 + \eta(h_f/h_i)} \qquad (10\text{-}8d)$$

For a given mass velocity of the air, Norris and Spofford[43] find the highest coefficients for flow normal to cylindrical fins, intermediate values for flow parallel to discontinuous strip fins, and the lowest coefficients for flow between parallel continuous fins. The data are correlated on Fig. 10-15 by the dimensionless equation[43]

$$\frac{h_m}{c_{pb}G_{\max}}\left[\frac{c_p\mu}{k}\right]_f^{2/3} = 1.0 \left[\frac{z_p G_{\max}}{\mu_f}\right]^{-1/2} \qquad (10\text{-}9) \star$$

The values of $z_p G_{\max}/\mu_f$ ranged from 260 to 12,000, and for air, the left-hand side of the equation equals $3.3 h_m/G_{\max}$. The gas-film coefficients h_m were computed from the measured over-all coefficients U by the resistance equation

$$\frac{1}{h_m(\Delta t)_m/(\Delta t)_0} = \frac{1}{U} - \frac{A}{h_s A_s} - \frac{x_w A}{k_w A_w} \qquad (10\text{-}10)$$

The efficiency $\eta\ (= \Delta t_m/\Delta t_0)$ was calculated from Eq. (10-8a), and the vapor-side coefficient was evaluated by assuming film-type condensation (Chap. 13). Friction factors were also reported.

Variable Cross Section. If the cross section of the fin is not constant, the heat balance becomes

$$kS \frac{d^2t}{dx^2} dx + k\, dS \frac{dt}{dx} = h\, dA\, (t - t_a)$$

Gardner[15] notes that the heat balance could be rearranged (by multiplying by $x^2/kS\, dx$) to give the same form as the general Bessel function

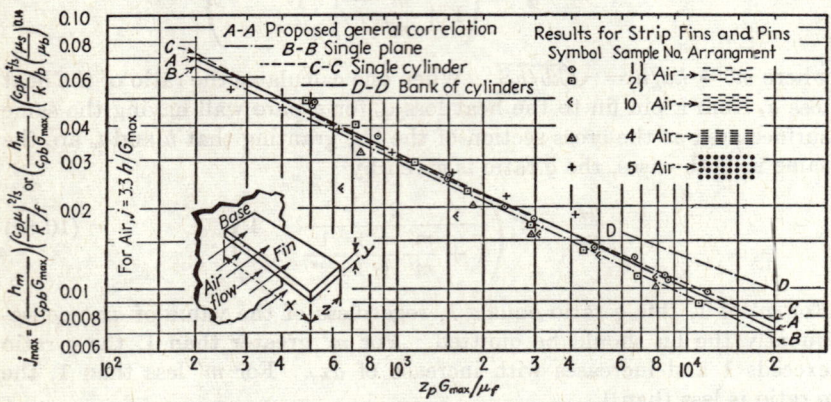

Fig. 10-15. Correlation by Norris and Spofford[43] of their data for heating air by short fins consisting of strips or pins; $z_p = 2(z + y)$; hence z_p is twice the heated length L_H over which the air travels. In the range of $z_p G_{max}/\mu$ involved, this method of plotting also correlates data for air flow normal to smooth single cylinders (Fig. 10-7) and parallel to short planes (Fig. 9-29). The pin fins had diameters ranging from 0.02 to 0.125 in., and the strip fins were 0.125 in. long. The viscosity is evaluated at t_f. See also Chap. 11.

given by Douglass in Sherwood and Reed,[56] provided both cross section and perimeter vary in a specified manner with distance along the x axis:

$$S = C_1 x^{1-2p_1 p_2}$$
$$dA/dx = C_2 x^{2p_1(1-p_2)-1}$$

Various geometrical configurations correspond to definite numerical values assigned to the exponents p_1 and p_2. After making the usual assumptions† and integrating the differential equations, the resulting dimensionless values of fin efficiency, $\eta = (t - t_a)_m/(t_0 - t_a)$, are plotted as ordinates vs. a dimensionless term equal to fin length x_f multiplied by the square root of h/ky_0, where y_0 is the half thickness of the fin at the base or root of the fin; t_0 denotes the temperature at the base of the fin.

† Assuming h, k, t_o, and t_a constant and neglecting both radial temperature gradients and heat transfer from the tip.

Figure 10-16 shows fin efficiencies for various fins, plotted vs. $x_f \sqrt{h_m/ky_0}$. It is interesting to note that the term $x_f \sqrt{h_m/ky_0}$ is in reality the square root of the product of a Nusselt modulus (hx_f/k) and a dimension ratio (x_f/y_0). The original paper[15] gives efficiencies for five

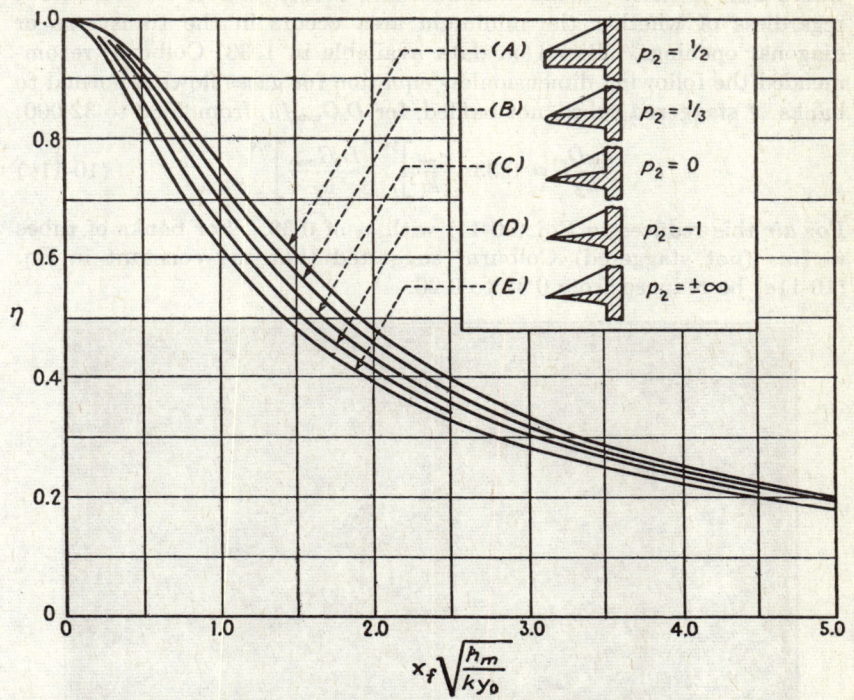

FIG. 10-16. Fin efficiency of several types of straight fins (K.A. Gardner,[15] *Trans. ASME*). This reference also gives curves for annular fins and spine fins.

types of straight fins, four types of spines, six annular fins of constant thickness, and four tapered annular fins having constant cross section of metal.

III. FLOW ACROSS BANKS OF TUBES

The case of flow normal to banks of nonbaffled tubes is of considerable importance, especially in air heaters and heat exchangers. Figures 10-17 and 10-18 show the flow of air at 5 ft/sec through a bank of tubes 0.67 inch in diameter, "staggered" in Fig. 10-17 and "in line" in Fig. 10-18. For the same air velocity, Reiher[51] found that the staggered arrangement, similar to Fig. 10-17, gave substantially higher coefficients of heat transfer than an in-line arrangement similar to Fig. 10-18.

Gases. Many data were available in 1942 for the flow of *air* at right angles to banks of tubes.[38] Since the term $c_p\mu_f/k_f$ was constant, the data were usually correlated by an equation of the type used for air flowing across single tubes:

$$h_m D_o/k_f = b_2(D_o G_{max}/\mu_f)^n \qquad (10\text{-}11) \star$$

where G_{max} is based on the *minimum* free area available for fluid flow, regardless of whether the minimum area occurs in the transverse or diagonal openings. Based on data available in 1933, Colburn[8] recommended the following dimensionless equation for gases flowing normal to banks of **staggered** tubes, not baffled, for $D_o G_{max}/\mu_f$ from 2000 to 32,000:

$$\frac{h_m D_o}{k_f} = 0.33 \left[\frac{c_p \mu}{k}\right]_f^{1/3} \left[\frac{D_o G_{max}}{\mu_f}\right]^{0.6} \qquad (10\text{-}11a)$$

For *air* this reduces to Eq. (10-11) with b_2 of 0.30. For banks of tubes *in line* (not staggered) Colburn[8] suggested that the constant in Eq. (10-11a) be reduced from 0.33 to 0.26.

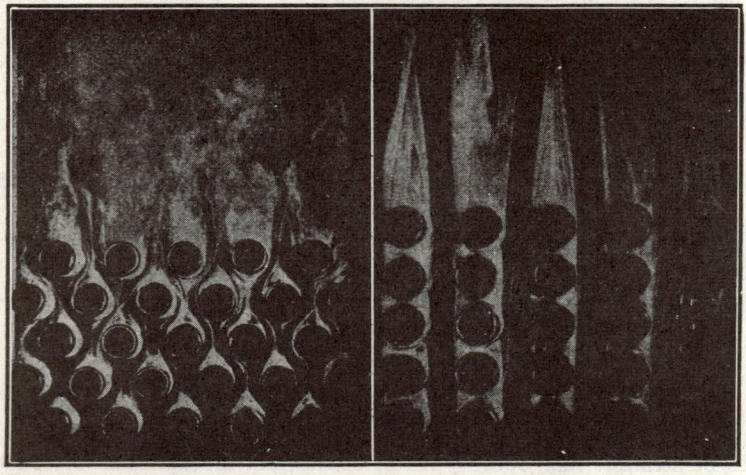

Fig. 10-17. Air flowing normal to a bank of staggered pipes. (*Photograph from Lohrisch,*[37] *Handbuch der Experimental-Physik.*)

Fig. 10-18. Air flowing normal to a bank of pipes in line. (*Photograph from Lohrisch,*[37] *Handbuch der Experimental-Physik.*)

Tucker[62] determined the heat-transfer coefficients to air flowing normal to a single vertical 1-in. tube placed at the center of the third row of a bank of *staggered* unheated tubes 5 rows deep. In each of the 15 bundles used, the tubes were arranged at the apexes of isosceles triangles. Three ratios of longitudinal to transverse spacing were used: s_L/s_T of $0.5\sqrt{3}$, 1.0, and 1.5; with each of these, transverse pitches of 1.25, 1.5, 2, 3, and 4 tube diameters were employed. It was found that the large changes in the transverse and longitudinal clearances had a negligible effect on the value of h_3 at a given mass velocity through the minimum free area,

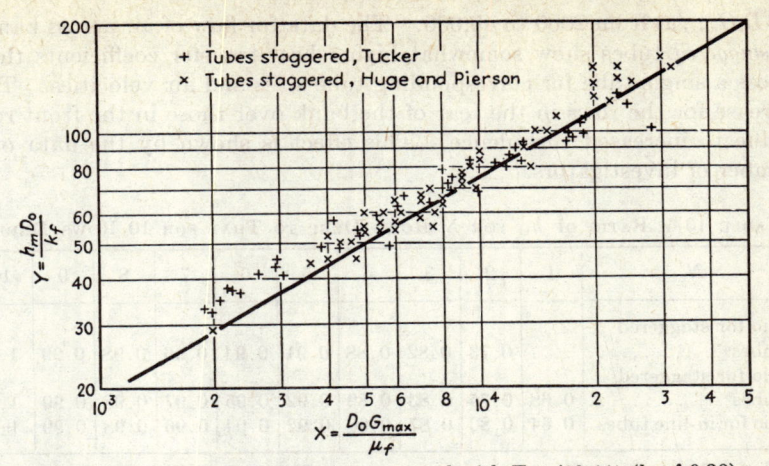

FIG. 10-19. Data for heating air, compared with Eq. (10-11) (b_2 of 0.30).

TABLE 10-5. GRIMISON'S VALUES OF b_2 AND n IN EQ. (10-11)[19]

$$h_m D_o/k_f = b_2(D_o G_{max}/\mu_f)^n; \quad t_f = t_s - (t_s - t)_m/2$$

$x_L = \dfrac{s_L}{D_o}$	$x_T = 1.25$		$x_T = 1.5$		$x_T = 2$		$x_T = 3$	
	b_2	n	b_2	n	b_2	n	b_2	n
Staggered:								
0.600	.....	.....	.....	.....	.....	.....	0.213	0.636
0.900	.....	.....	.....	.....	0.446	0.571	0.401	0.581
1.000	.....	.....	0.497	0.558				
1.125	.....	.....			0.478	0.565	0.518	0.560
1.250	0.518	0.556	0.505	0.554	0.519	0.556	0.522	0.562
1.500	0.451	0.568	0.460	0.562	0.452	0.568	0.488	0.568
2.000	0.404	0.572	0.416	0.568	0.482	0.556	0.449	0.570
3.000	0.310	0.592	0.356	0.580	0.440	0.562	0.421	0.574
In line:								
1.250	0.348	0.592	0.275	0.608	0.100	0.704	0.0633	0.752
1.500	0.367	0.586	0.250	0.620	0.101	0.702	0.0678	0.744
2.000	0.418	0.570	0.299	0.602	0.229	0.632	0.198	0.648
3.000	0.290	0.601	0.357	0.584	0.374	0.581	0.286	0.608

$x_L = s_L/D_o$; $x_T = s_T/D_o$.

which in all cases involved the transverse clearance, $s_T - D_o$; the data are plotted in Fig. 10-19 and are correlated by Eq. (10-11), with b_2 of 0.30 and n of 0.6, with an average deviation of ±5 per cent.

Table 10-5 shows the values of b_2 and n obtained by Grimison[19] from the extensive data of Huge[25] and of Pierson[46] for banks 10 rows deep,†

† The values of b_2 and n of Table 10-5 are in fair agreement[19] with the data of references 5, 6, 36, 51, 52, and 18.

for $D_o G_{max}/\mu_f$ from 2000 to 40,000. The data for flow of air across banks of *staggered* tubes show somewhat higher heat-transfer coefficients than across a single tube for corresponding tube sizes and air velocities. The increase for the rows in the rear of the bank over those in the front row is due to increased turbulence. This effect is shown by the data of a number of investigators.†

TABLE 10-6. RATIO OF h_m FOR N ROWS DEEP TO THAT FOR 10 ROWS DEEP

N	1	2	3	4	5	6	7	8	9	10
Ratio for staggered tubes[a]	....	0.73	0.82	0.88	0.91	0.94	0.96	0.98	0.99	1.0
Ratio for staggered tubes[b]	0.68	0.75	0.83	0.89	0.92	0.95	0.97	0.98	0.99	1.0
Ratio for in-line tubes	0.64	0.80	0.87	0.90	0.92	0.94	0.96	0.98	0.99	1.0

[a] Pierson.[46]
[b] Kays and Lo.[27a]

Kays and Lo[27a] recently constructed banks of *staggered* unheated vertical aluminum tubes, D_o of 0.0313 ft and 0.693 ft high; other details are given in the following table. Air at room temperature and 1 atm flowed

x_T and x_L	1.50, 1.25	1.25, 1.25	1.50, 1.50	2.00, 1.00	2.50, 0.75
N rows deep	15	13	15	19	26
L_o, feet	0.586	0.509	0.702	0.595	0.608
$D_h = L_c S_{min}/A$, feet	0.0256[a]	0.0125[a]	0.0298[a]	0.0327[b]	0.0271[b]
A/V_T, reciprocal feet	53.6	64.4	44.8	50.3	53.6

[a] S_{min} is based on s_T.
[b] S_{min} is based on diagonal clearances.

normal to the tubes at a steady mass rate. By replacing one of the unheated tubes by a transiently heated tube, "a thermal capacitor," heat-transfer coefficients were obtained for small Δt. It was found that the ratio of heat-transfer coefficients h_N/h_∞ (for any row N to that for a row where h no longer changed) was a function of N, as shown in Table 10-7, which also gives the row-mean coefficient h_m for rows 1 through N divided by h_∞. For a staggered bank at least eight rows deep, h_6 and h_7 were equal to h_∞ although h_m was less. The values of DG_{max}/μ_f ranged from 900 to 20,000 and when the values of j_{max} were plotted vs. $D_o G_{max}/\mu_f$, the results were within 10 per cent of Eq. (10-11a) but were some 15 per cent below the correlations of Grimison[19] for both j_{max} and friction factors. For each exchanger the preferred method[27a] was to plot curves of both j_{max} and f (rather than f'') vs. a Reynolds number $D_{v,min} G_{max}/\mu_f$, where $D_{v,min}$ represents the ratio of the *minimum* free volume ($L_c S_{min}$) to the

† References 51, 52, 46, and 66.

total heat-transfer surface:

$$D_{v,\min} = L_c S_{\min}/A$$

When plotted on logarithmic paper with $D_{v,\min}G_{\max}/\mu_f$ as abscissa and with both $j_{\max}$ and f as ordinates, linear relations were always obtained even with the lowest value of $D_{v,\min}G_{\max}/\mu_f$ of 500; consequently the air flow was described as "turbulent." Comparison of the curves of heat transfer and friction power, both per unit surface and per unit gross volume, showed that performance is increased by decreasing x_T, x_L, or both. The results of a steady-state test on a single steam-heated exchanger (D_o of 0.0208 ft, N of 15, x_T of 1.50 and x_L of 1.25) agreed with those of the transient tests within 10 per cent.

TABLE 10-7. ROW-TO-ROW VARIATION IN h/h_∞ AND h_m/h_∞, FROM KAYS AND LO[27a]

N	1	2	3	4	5	6	7	8
Staggered h/h_∞	0.63	0.76	0.93	0.98	0.99	1.0	1.0	1.0
Staggered h_m/h_∞	0.63	0.70	0.77	0.83	0.86	0.88	0.90	0.91

N	9	10	12	15	18	25	35	72
Staggered h/h_∞	1.0	1.0	1.0	1.0	1.0	1.0	1.0	1.0
Staggered h_m/h_∞	0.92	0.93	0.94	0.95	0.96	0.97	0.98	0.99

At a given mass velocity of air, Winding[66] finds that a bank of staggered steel tubes eight rows deep gives substantially the same results whether the tubes are circular, oval or streamline in shape; all three tubes had a perimeter of 4.7 in. Hence these data can be correlated in terms of z_p.

A simplified *dimensional* equation for gases flowing normal to a 10-row bank of staggered tubes is

$$h_m = 0.133 c_p G_{\max}^{0.6}/D_o^{0.4} \qquad (10\text{-}11b) \star$$

Recommended Procedure for Air. For $D_o G_{\max}/\mu$ from 2000 to 40,000, Eq. (10-11) is recommended for staggered banks 10 rows deep; for N less than 10, multiply h_m from Eq. (10-11) by the values of h_m/h_{1-10} from Table 10-6; for N greater than 10, multiply h_m from Eq. (10-11) by the value of h_m/h_∞ from Table 10-7, and divide by 0.93. For in-line banks, use Eq. (10-11), with Tables 10-5 and 10-6. These equations should not be extrapolated to smaller values of $D_o G_{\max}/\mu_f$.

Flow of Liquids Normal to Staggered Banks of Tubes. The conventional relation is given as Fig. 115 of reference 38 (and as Fig. T-8 of reference 56a) and is shown as Curve *ABD* of Fig. 10-20; Curve *BC* is based on data for oil coolers from reference 61; Curve *EF* is for heating liquids flowing normal to single cylinders (Fig. 10-12). Further data are needed.

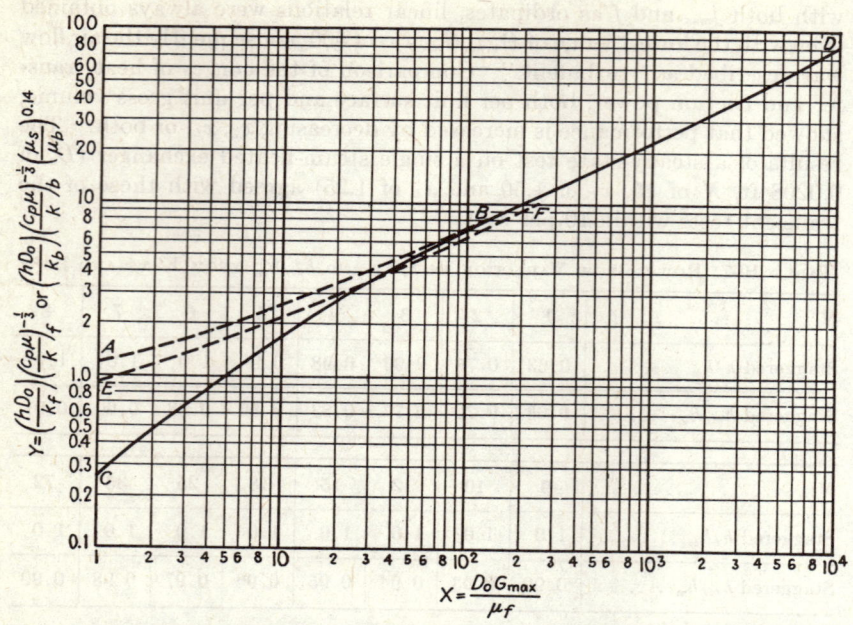

FIG. 10-20. Recommended curve *ABD* for *estimating* h_m for heating or cooling fluids flowing normal to *staggered* banks of tubes 10 rows deep; for $D_o G_{max}/\mu_f$ above 2000 the curve is based on Eq. (10-11a), and for X from 2000 to 150 is drawn parallel to the curve for single tubes (Fig. 10-7) and is supported by unpublished data for liquids. For values of N less than 10, multiply h_m by the factors given in Table 10-6.

A simplified dimensional relation, for $D_o G_{max}/\mu$ above 2000, for water is

$$h_m = (370)(1 + 0.0067 t_f)(V'_{max})^{0.6}/(D'_o)^{0.4} \qquad (10\text{-}11c) \star$$

where t_f is in degrees Fahrenheit, V_{max} is in feet per second, and D'_o is in inches.

IV. CROSS-BAFFLED EXCHANGERS

In various industries, shell-and-tube heat exchangers are widely used. The usual type consists of a bundle of long tubes with one fluid flowing inside the tubes and the other outside the tubes. Cross baffles of various types are inserted at right angles to the tubes to increase velocity and give more crossflow in long heat exchangers. In this design, the flow of the fluid outside the tubes is partly parallel to and partly across the tubes. In short-tube exchangers the fluid in the shell flows across the tubes; baffles are not used; this design is similar to that for air flowing at right angles to banks of tubes, except that the cross section varies where a cylindrical shell is used.

Bergelin, Colburn, and Hull[1b] measured heat-transfer coefficients for streamline flow of petroleum oil normal to 10 to 14 rows of tubes in seven different models of exchangers. For a given $D_o G_{max}/\mu$, the j'_{max} values were found to vary inversely as $N^{0.18}$; if Eq. (9-28a) for streamline flow

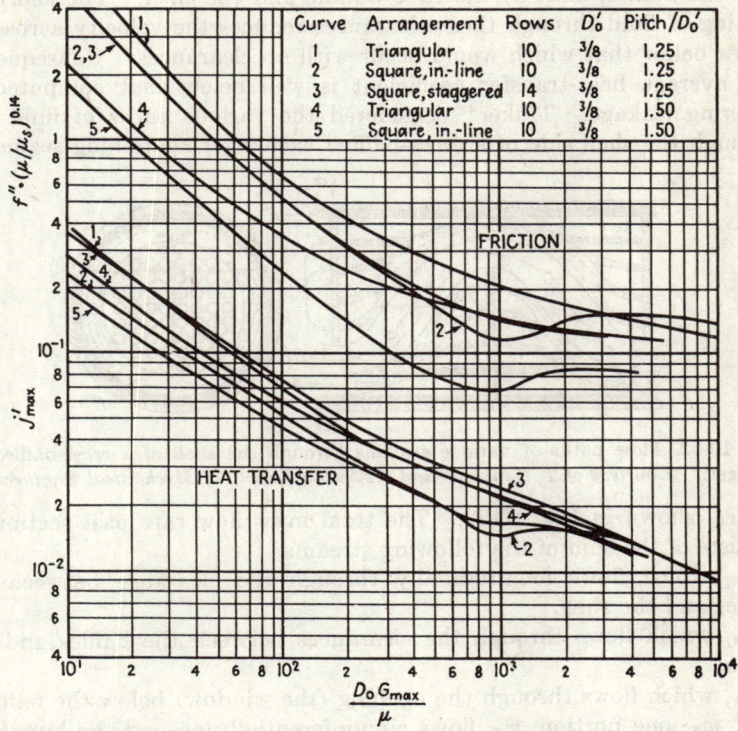

FIG. 10-21. Values of f'' and j_{max} plotted vs. $D_o G_{max}/\mu$. The curves for in-line arrangements show dips in the transition region, while those for the staggered arrangements do not. (*Courtesy of Bergelin, Brown, and Doberstein, and American Society of Mechanical Engineers.*[1a])

inside tubes were applicable, with L replaced by $N\pi D_o/2$, j should vary inversely as $N^{1/3}$. For 10 rows deep and $D_v G_{max}/\mu$ from 1.5 to 70 the data may be approximated by

$$j_{max} = b_3 (D_v G_{max}/\mu)^n$$

where b_3 is 1.75 for staggered banks and 1.20 for in-line banks, and n is -0.71 and -0.66, respectively.

Additional data, by Bergelin, Brown, and Doberstein,[1a] are shown in Fig. 10-21: further experiments are in progress.

278   HEAT TRANSMISSION

Flow through the Shell of a Cross-baffled Exchanger. Figure 10-22 shows diagrammatically a conventional transversely baffled heat exchanger. The baffles are used to increase the mass velocities and consequently the heat-transfer coefficient on the shell side. As shown in Chap. 15, in the construction of commercial equipment it is necessary to employ clearances between (1) the baffles and the shell, (2) the tubes and the baffles, and (3) the tube bundle and the shell. The short circuiting of fluid through these clearances reduces the velocity across the tubes below that which would occur with no clearances. Consequently the average heat-transfer coefficient is well below that computed by ignoring leakage. Tinker[61] considered the various paths of fluid flow through the shell side of a cross-baffled exchanger containing segmental

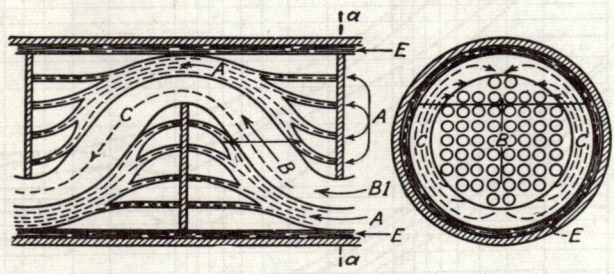

Fig. 10-22. Flow paths of various streams through the shell of a cross-baffled exchanger. (*Courtesy of T. Tinker,*[61] *and American Society of Mechanical Engineers.*)

baffles, shown in Fig. 10-22. The total mass flow rate past section aa consists of the sum of the following streams:

w_E, which flows longitudinally through the clearance between the baffles and the shell.

w_A, which flows through the clearances between the baffles and the tubes.

w_W, which flows through the opening (the window) below the baffle.

Of w_W, one portion, w_C, flows circumferentially around the bundle to the next window, a second portion equivalent to w_A leaks through the next baffle, and the remainder, w_{B1}, flows transversely across the tubes, and out through the next window, along with w_A. For a given mass velocity and tube diameter, flow across tubes gives a substantially higher heat-transfer coefficient than flow parallel to tubes.† Consequently it is desirable to evaluate the average mass rate of flow w_B across the tubes and the corresponding average mass velocity G_{max} and heat-transfer coefficient h_{ob}. Tinker[61] takes w_B as $w_{B1} + 0.5w_A$. Since the pressure drops

† The ratio of Eq. (10-11a) for turbulent flow normal to banks of tubes to Eq. (9-10a) for turbulent flow inside tubes is

$$\frac{h_o}{h_i} = 14.3\mu^{0.2}\frac{D_i^{0.2}}{D_o^{0.4}}\frac{G_{max}^{0.6}}{G_i^{0.8}}$$

between given cross sections must be equal for the various streams, it is possible to evaluate the various streams and h_{ob}.

In various zones lower values of h occur owing to reduced velocities or flow parallel to the tubes. Measured values of h_o, based on the entire outer tube surface A_o, are found to be less than h_{ob}. By definition, $h_o = m_1 m_2 h_{ob}$, where the multiplier m_1 ranged from 1.0 to 0.8, depending upon baffle design. The multiplier m_2 allows for any differences in the velocities in the inlet and outlet compartments and that between baffles. Unless the velocities in the end compartments are unduly low, m_2 is 0.9 or more.

Alternatively, instead of employing $G_B = w_B/S_B$, an effective mass velocity G_{eff} is based on the *total* flow rate w_T and an "effective" cross section, S_{eff}:

$$G_{eff} = \frac{w_B}{S_B} = \frac{w_T}{S_{eff}}$$

Tinker[61] correlated the shell-side coefficients for 11 different baffled oil coolers. When the dimensionless term Y_{app} was taken as

$$Y_{app} = \frac{h_o D_o}{k_b} \left(\frac{c_p \mu}{k_b}\right)^{-\frac{1}{3}} \left(\frac{\mu_b}{\mu_w}\right)^{-0.14}$$

and the Reynolds number was taken as $X_{app} = D_o G_{app}/\mu$, where $G_{app} = w_T/S_B$, the results fell well below a conventional relation for straight crossflow without baffles or leakage. Upon basing Y_{eff} on h_{ob}, and X_{eff} on G_{eff}, the scatter of data was substantially reduced and the points fell on both sides of the curve for crossflow, for X_{eff} from 20 to 7000. Research along these lines is in progress.[63]

For a certain conventional unit of the type shown in Fig. 10-22, Tinker[61] predicted the effect of changing the spacing of the baffles in a shell of fixed diameter and length. As the baffle spacing was reduced from one shell diameter to one-tenth of the shell diameter, the fraction of the total oil rate w_T which flowed across the tubes decreased from 39 to 12 per cent; the values of G_{ob} and h_{ob} increased; the power loss per unit heat-transfer surface increased.

In commercial exchangers, Tinker[61] estimates that the crossflow stream through the bundle is usually only 12 to 60 per cent of the total stream. The resistance encountered by the C stream is very low compared with that of the B stream, and hence its velocity is far higher. As the total rate of flow through the shell is reduced, streamline flow develops in the crossflow stream, while the C stream remains turbulent; consequently, the fraction of the total stream which by-passes the tubes increases at the expense of the B stream. In tests[61] with a viscous oil entering the shell at 180°F, a reduction of the temperature of the inlet water from 120 to 70°F caused a *rise* of more than 20°F in the temperature of the outlet oil. The fluid in the tube nest apparently was chilled to such a high viscosity that the unchilled leakage and by-pass routes transported the bulk of the oil around the tube bundle. In other tests[61] with ⅝-inch tubes on a ¾-inch triangular pitch the water entered at 70°F, and the oil was cooled from 108 to 100°F. When the pitch was increased to ⅞ inch (26 per cent fewer tubes) and the water entered at 70°F, the same oil was cooled from 117 to 100°F; water and oil rates were the same in both cases. The wider pitch reduced the crossflow resistance relative to that of the by-pass routes, thus increasing the crossflow current sufficiently to double the heat-transfer rate.

Reduction in the by-pass C stream can be accomplished by use of side strips.

PROBLEMS

1. An air heater consists of a bank of vertical 2-inch standard steel pipes, each 6 ft long. Each row, normal to the air flow, consists of 30 tubes with center-to-center spacings of 4 inches. The bank is 40 rows deep, and the row centers are 3 inches apart. Successive rows are staggered, every other row being displaced 1½ inches to the side. Exhaust steam, available at 11 lb/sq in. gauge, will be condensed inside the tubes, and air will be heated from 65 to 220°F as it flows past the tube bank.

Estimate the capacity of this heater, expressed as pounds of air per hour.

2. Hot air is flowing inside a steel pipe with an i.d. of 18 inches. Two thermocouples are used to measure the air temperature, one at the center of the pipe and one embedded in the pipe wall. The thermocouple in the center of the pipe is shielded by a cylindrical metal tube, concentric with the large pipe, which is centered around the junction of the thermocouple. The shield is ½ inch in diameter and 12 inches long, and its emissivity is 0.3. The emissivity of the thermocouple junction may be assumed to be 0.7.

The center-line thermocouple indicates a temperature of 917°F, and the temperature of the pipe wall is 631°F. Estimate the true temperature of the gas. What temperature would the center-line thermocouple indicate if the shield were removed? It is agreed to neglect the open ends of the shield and to neglect conduction in the thermocouple leads. Assume $h_c = 8$ for convection to the shield and to the thermocouple.

3. An air heater involves condensation of steam inside horizontal tubes and flow of air at right angles to the tubes in a shell containing a number of vertical segmental baffles. When using exhaust steam condensing at a gauge pressure of 5 lb/sq in. and an air rate of 30,000 lb/hr, the apparatus will heat the air from 80 to 180°F.

If this exchanger were used to heat hydrogen from 80 to 180°F using the same steam pressure, what would be the capacity of the exchanger, expressed as pounds of hydrogen per hour?

4. Air is heated in the annulus of a concentric-tube heat exchanger by steam condensing in the center tube. The o.d. of the inner tube is 1 inch, and its outer surface is finned with 32 longitudinal fins, each of which extends the length of the tube. The fins extend out radially 1 inch from the tube, and each fin is 0.06 inch thick. The fins are made of an alloy steel ($k = 16$). The saturation temperature of the condensing steam is 240°F, the air-side heat-transfer coefficient is estimated to be 4, and the thermal resistance of the pipe wall and steam-condensate film are considered negligible.

Calculate the heat transfer rate, expressed as Btu per hour per foot of pipe length at a point in the exchanger at which the air temperature is 140°F, by the following procedures:

HEATING AND COOLING OUTSIDE TUBES

   a. By using Eq. (10-8*b*).
   b. By using Eq. (10-8*a*) and adding S/b to x_f to allow for the bare tips.
   c. By using Eq. (10-8*a*) for the sides of the fins and evaluating the heat transfer from the fin tips by using Eq. (10-8).

5. An experimental apparatus consists of a solid horizontal polished copper cylinder spanning a duct. The ends of the cylinder, which are embedded in the walls of the duct, are maintained at a temperature of 150°F by electrical heaters, and heat is transferred from the copper cylinder to air flowing in the duct. The width of the duct is 4 ft, and the diameter of the cylinder is ½ inch. The approach velocity of the air is 25.4 ft/sec, and the air is at a temperature of 130°F and a pressure of 1 atm.

   a. Calculate the rate of heat transfer from the cylinder to the air.
   b. Plot the cylinder temperature vs. position along the length of the cylinder.

Note: It is to be assumed that radial temperature gradients in the cylinder are negligible and that the effect of the walls of the duct on the heat-transfer coefficient from the air to the cylinder near the walls may be neglected.

6. A multipass heat exchanger contains 1200 tubes, each 16 ft long, having an o.d. of 1.25 in. and an i.d. of 1.12 in. These tubes are arranged as a 12-6 exchanger in which the fluid in the tubes makes 12 equal single passes and that in the shell makes 6 corresponding passes. Each tube pass contains 100 tubes, and each shell pass is well baffled. It is desired to predict the performance of this apparatus if 693,000 lb/hr of a hot oil, $c = 0.540$ gm-cal/(gm)(deg C), were introduced continuously into the shell at a temperature of 360°F and cold oil at 60°F, $c = 0.470$ gm-cal/(gm)(deg C), $\rho = 56$ lb/cu ft, were fed to the first pass of the tubes at a rate of 796,000 lb/hr. From the data and notes given below, calculate the temperature to which the hot oil will be cooled.

Data and Notes. It is estimated that the average coefficient from hot oil to the outer surface of the tubes, including suitable allowance for scale deposit on both sides, will be 100 Btu/(hr)(sq ft)(deg F), based on the outside surface. The thermal conductivity of the tubes will be taken as 26 Btu/(hr)(sq ft)(deg F per ft). The coefficient h on the inside of the tube is given by Eq. (9-10*b*). The average absolute viscosity of the oil in the tubes may be taken as 0.0416 gm/(sec)(cm); and the thermal conductivity of the oil in the tubes is 0.080 Btu/(hr)(sq ft)(deg F per ft).

CHAPTER 11

COMPACT EXCHANGERS, PACKED AND FLUIDIZED SYSTEMS

Abstract. This chapter covers three different topics. The first section treats heat transfer between fluids in compact exchangers employing extended surface. Heat-transfer coefficients are correlated for surfaces of various designs, and the heat-transfer performance of several typical compact exchangers is compared on the basis of pumping power per unit surface area and pumping power per unit volume of exchanger.

The second section deals with fixed and moving beds of packed solids. Heat exchange is considered between the fluid and the packing material and between the fluid and the wall of a packed bed. A simplified analysis of heat regenerators is presented.

The final part of this chapter treats beds of fluidized solids. The data of numerous observers are presented for dense- and lean-phase beds with internal and external heat-transfer surfaces. Terms dealing with fluidization are defined. Data on the flow characteristics of fluidized beds are given in order to facilitate evaluation of the variables employed in the heat-transfer correlations.

I. COMPACT EXCHANGERS

Introduction. In conventional shell-and-tube heat exchangers one fluid flows inside the tubes, and the other flows across and along the tubes to an extent depending on the arrangement of any baffles in the shell. The ratio of outer surface of the tubes to the over-all volume of the tube bundle depends on the diameter, spacing, and arrangement of the tubes. Consider tubes of outside diameter D_o and length L arranged at the corners of a rectangle as shown in Fig. 11-1. For the quadrants of the four tubes associated with the rectangle having volume $z_1 z_2 L$, the surface area is $4\pi D_o L/4$, and consequently the ratio of heat-transfer surface to total volume is given by

$$\zeta = \frac{\pi D_o L}{z_1 z_2 L} = \frac{\pi D_o}{z_1 z_2} \qquad (11\text{-}1)$$

For a square arrangement ζ equals $\pi/s^2 D$, where s is the pitch ratio z/D_o. With a small pitch ratio of 1.25, ζ equals $2.01/D_o$; thus for tubes of ordinary sizes (0.5 to 2.0 inches) the value of ζ would range from 48 to 12 sq ft/cu ft. With tubes at the apexes of equilateral triangles (Fig.

11-1b), the value of ζ is given by

$$\zeta = \frac{3(\pi D_o/6)L}{(\sqrt{3}/4)z^2 L} = \frac{3.63}{s^2 D_o} \tag{11-2}$$

For the same tube diameter and center-to-center distance, this value of ζ is 15 per cent larger than for the square arrangement, but the latter is easier to clean mechanically. With a circular shell there is some waste space near the wall and the actual values of ζ are consequently less than these maximum values, particularly when the number of tubes is small.

Where greater compactness is desired, as in aircraft and certain other applications, additional heat-transfer surface per unit volume is obtained by the use of fins of very small cross section brazed or otherwise attached in good thermal contact with the primary heat-transfer surface.[43] The two principal arrangements employing extended surfaces are the "plate-fin" exchanger and the "finned-tube" exchanger.

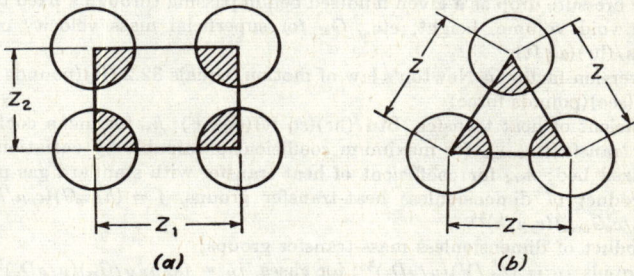

FIG. 11-1. Tube arrangements. (a) Rectangular. (b) Equilateral triangular.

The primary heat-transfer surface of the plate-fin design consists of multiple parallel plates structurally connected by fins; the space between each pair of plates comprises a fluid passage. Alternate fluid passages are connected in parallel by suitable headers to form the two "sides" of the exchanger. Figure 11-2 shows one type of plate-fin exchanger employing plain draped fins brazed at the contacts with the plates, giving a rigid structure of light weight. In the "pin-fin" type, the fins are solid metal rods (pins) of small diameter, brazed to the plates, so that the fluid flows across a bank of small cylinders, either staggered or in-line. In the "strip-fin" type, the round pins are replaced by metal strips of rectangular cross section, arranged either staggered or in-line. The plate spacing may be optimized independently for each side of the plate-fin exchanger, resulting in great flexibility of design.

Where fins are desired on only one side of the primary surface, tubes (round or flattened) are placed through holes m thin metal plates and are brazed, giving one form of finned-tube design (Fig. 11-3). Finned tubes are frequently advantageous when the tube fluid is either a liquid or a high-pressure gas and the se ond fluid is a gas at ordinary pressure.

Table 11-1. Nomenclature

A Area of heat transfer surface, square feet; A_f for area of extended surface (fins); A_p for surface area of particle

a Surface area of packing per unit volume of packed bed, square feet per cubic foot

b_{avg} Average perimeter of flow passage, feet; $b_{\text{avg}} = A/L$

c_p Specific heat of fluid at constant pressure, Btu/(lb)(deg F)

c_s Specific heat of solid, Btu/(lb)(deg F)

D Diameter, feet; D_e for equivalent diameter; D_o for outside diameter of tube; D_p for particle diameter; D_i for inside diameter of tube

D_v Molecular diffusivity of mass, square feet per hour

E Eddy diffusivity of heat, square feet per hour

e Base of Napierian or natural logarithms

f Friction factor, dimensionless, defined by Eq. (11-3)

G Mass velocity of fluid, lb/(hr)(sq ft); $G_{\max}$ for maximum mass velocity at point of minimum free cross section; G_o for superficial mass velocity based on total cross section without particles; G_{oe} for the mass velocity which would cause the same pressure drop as a given fluidized bed in passing through a fixed bed of the same void volume, height, etc.; G_M for superficial mass velocity of gas, lb moles/(hr)(sq ft)

g_c Conversion factor in Newton's law of motion, equals 32.2 (ft)(pounds matter)/(sec)(sec)(pounds force)

h Coefficient of heat transfer, Btu/(hr)(sq ft)(deg F); h_m for mean coefficient of heat transfer; $h_{m,\max}$ for maximum coefficient obtainable by variation of G_o in fluidized bed; h_{std} for coefficient of heat transfer with standard gas properties

j A product of dimensionless heat-transfer groups, $j = (h/c_p G)(c_p\mu/k)^{2/3}$; $j_{\max} = (h/c_p G_{\max})(c_p\mu/k)^{2/3}$

j_D A product of dimensionless mass-transfer groups: for liquids $j_D = (k_L'/V)(\mu/\rho D_v)^{2/3}$; for gases, $j_D = (k_G' p_{BM}/G_m)(\mu/\rho D_v)^{2/3}$

K Thermal conductivity of a fluid-solid system, Btu/(hr)(sq ft)(deg F per ft); K_a for apparent thermal conductivity defined by Eq. (11-11); K_B for thermal conductivity of packed bed without fluid flow; K_F for apparent thermal conductivity of a fluidized bed; K_r for thermal conductivity of packed bed due to radiation; K_s for thermal conductivity of packed bed due to conduction through solids and fluid fillets at points of contact

k True thermal conductivity, Btu/(hr)(sq ft)(deg F per ft); k_g for the gas, k_s for the solid particles. See below for k'

k_G' Mass transfer coefficient in gas system, lb moles/(hr)(sq ft)(atm)

k_L' Mass transfer coefficient in liquid system, feet per hour

L Length of exchanger or packed bed, feet; L_f for height of fluidized bed; L_{mf} for height of bed at condition of minimum fluidization

M Dimensionless group, $2k_s/hD_p$

m Exponent defined by Eq. (11-38)

N_{Pe}' Modified Peclet number, dimensionless; $N_{Pe}' = DV/E$

N_v Dimensionless bed expansion ratio, equals L_f/L_{mf}

P Friction power, foot-lb (force) per second; or horsepower

p_{BM} Mean pressure of nondiffusing fluid, atmospheres

Q Volumetric flow rate, cubic feet per second

q Heat-transfer rate, Btu per hour

r Radius, feet; r_B for volume of packing per unit surface of packing; r_h for hydraulic radius, $S_{\min}/b_{\text{avg}}$

TABLE 11-1. NOMENCLATURE.—(*Continued*)

S Cross section for flow, superficial, square feet; S_{min} for minimum cross section for flow

s Pitch ratio, dimensionless; $s = z/D_o$

T Temperature, degrees Fahrenheit absolute; $T = t + 460$

t Temperature, degrees Fahrenheit; t_1 at inlet; t_2 at outlet; t_b for fluidized mass; t_g for temperature of a gas at a point in a fluidized bed; t_s of solids; t_w at wall; t'' for the hot fluid in a regenerator; t' for the cold fluid in a regenerator

V Velocity, feet per second; V_o for superficial velocity; V_{mf} for superficial velocity through bed at point of minimum fluidization

W Weight of solid and fluid in a fluidized bed, pounds

w Mass rate of flow of fluid, lb per hour; w_s for solids

X Dimensionless group, $4\alpha_s x/VD_p^2$

x Axial position from inlet, feet

Y Dimensionless group, $(t - t_s)/(t_1 - t_s)$

Z Dimensionless group, $(w_s c_s)/(w c_p)$

z Tube spacing, center-to-center, feet

Greek

α Molecular thermal diffusivity of fluid, square feet per hour; α_s for a solid particle

β Heat-transfer surface per unit of volume associated with that side of the exchanger, square feet per cubic foot

Δp Pressure drop, pounds force per square foot; Δp_f for pressure drop due to friction

Δt Temperature difference, degrees Fahrenheit; Δt_{lm} for log-mean temperature difference

Δx Distance in direction of heat flow, feet

∂ Prefix indicating partial derivative

δt Maximum fluctuation in regenerator outlet temperature, $t_{2,max} - t_{2,min}$; $\delta t''$ for hot stream, $\delta t'$ for cold stream

ϵ Average fraction void in bed, dimensionless; ϵ_{mf} for a bed at point of incipient fluidization

ζ Heat transfer area per unit total volume, square feet per cubic foot

η Fractional effectiveness of surface, dimensionless (from 0 to 1); η_0 for average efficiency of entire surface; η_f for average efficiency of extended surface (fins)

θ Time period, hours; θ'' for passage of hot fluid through regenerator; θ' for passage of cold fluid through regenerator

λ Generalized regenerator size, dimensionless; λ'' for the hot fluid, λ' for the cold fluid, λ_s for a symmetrical cycle

μ Viscosity of fluid, lb/(hr)(ft); μ_f at the film temperature; μ_{std} with the standard gas properties

ξ Dimensionless fluidization efficiency

π 3.1416

ρ Density of fluid, lb per cubic foot; ρ_b for the average bulk density of dumped bed; ρ_b for the average bulk density of the settled bed; ρ_{bf} for the average bulk density of the fluidized bed; ρ_c for the average solids concentration in a fluidized bed; ρ_g for the density of the gas; ρ_m for the average density of the fluid; ρ_s for the true density of the solid particles

Σ Prefix indicating summation

τ Generalized time for heating or cooling period of a regenerator, dimensionless; τ'' for hot fluid, τ' for cold fluid; τ_s for a symmetrical cycle

ϕ Temperature effectiveness of a regenerator; ϕ'' for hot fluid, ϕ' for cold fluid; ϕ_s for a symmetrical cycle

Pressure-drop Analysis. The pressure drop due to friction Δp_f is calculated by deducting the usual end losses (Chap. 6) from the over-all pressure drop measured with flow of air through the core without heat addition. The friction factor f is then computed from the Fanning

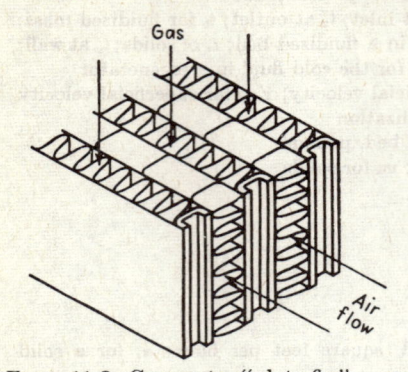

FIG. 11-2. Compact "plate-fin" exchanger with plain continuous fins.

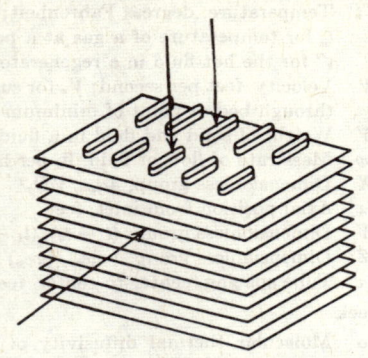

FIG. 11-3. Compact "finned-tube" exchanger with flattened tubes and continuous fins.

equation, based on the maximum velocity $G_{\max}$ occurring at the minimum cross section of the flow passage:

$$\Delta p_f = \frac{fLG_{\max}^2}{2g_c r_h \rho_m} \tag{11-3}$$

where r_h is the *minimum* hydraulic radius based on minimum cross section and average perimeter, defined as total heat-transfer surface divided by the length of the gas passage

$$r_h = \frac{S_{\min}}{b_{\text{avg}}} = \frac{S_{\min} L}{A} \tag{11-4}$$

The resulting values of f for isothermal conditions are plotted against a Reynolds number $D_e G_{\max}/\mu_f = 4 r_h G_{\max}/\mu_f$, where $G_{\max}$ is based on the minimum cross section and viscosity μ_f is evaluated at the film temperature $t_f = 0.5(t_w + t)$. The available data cover the latter part of the streamline region, the transition region, and in some cases the early part of the turbulent region. Separate curves of f vs. $D_e G_{\max}/\mu_f$ are required for each exchanger geometry, *i.e.*, each side of each exchanger. In using these isothermal friction factors to predict pressure drop while heating or cooling a gas, it is necessary to allow for the change in kinetic energy.

Heat-transfer Analysis. The heat-transfer performance is expressed as a mean heat-transfer coefficient h_m, based on the average temperature efficiency η_0 for the particular side of the exchanger under consideration.

The average efficiency is computed from the efficiency η_f of the extended surface (Chap. 10) by the relation

$$\eta_0 = 1 - \frac{A_f}{A}(1 - \eta_f) \qquad (11\text{-}5)\star$$

which assumes 100 per cent efficiency for the primary surface. In many designs A_f/A is 0.8 or higher; hence η_f should be near 1.0 in order to secure acceptable performance. Values of η_f depend strongly on the thermal conductivity of the fin, as discussed in Chap. 10; copper and aluminum are most frequently employed in practice.

The coefficients are frequently plotted vs. $D_e G_{max}/\mu_f$ as a Stanton modulus $h_m/c_p G_{max}$ multiplied by the two-thirds power of the Prandtl modulus:

$$j_{max} \equiv \frac{h_m}{c_p G_{max}}\left[\frac{c_p \mu}{k}\right]_f^{2/3} \qquad (11\text{-}6)$$

Since the values of j_{max} for compact exchangers usually fall well below the values of f, both curves are usually shown on the same logarithmic graph (Fig. 11-4). A separate curve of j_{max} vs. $D_e G_{max}/\mu_f$ is required for each side of each compact exchanger; for a given Reynolds number the j_{max} values vary substantially, depending on the geometry of the exchanger.

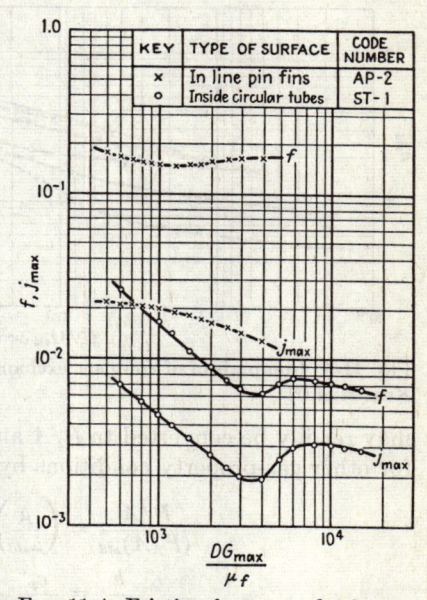

FIG. 11-4. Friction factors and j factors for several compact exchanger surfaces. (*Data of Kays et al.*[30])

Comparison of Compact Exchangers. The performance of various compact exchangers may be compared in several ways, depending on the objective. If only investment charges and power costs need be considered, the comparison is made per unit area of heat-transfer surface. Consider a particular exchanger operating with a given fluid and fixed terminal temperatures; D_e and μ_f are thereby fixed, and hence $D_e G_{max}/\mu_f$ is a unique function of the mass flow rate, G_{max}. Since for this case c_p and k_f are also fixed, the heat-transfer rate per unit area is proportional to $(h/c_p)(c_p\mu/k)_f^{2/3}$, which is equal to $j_{max} G_{max}$ and hence is a unique function of G_{max}. The friction power per unit surface is

$$\frac{P}{A} = \frac{\Delta p_f Q}{A} = \frac{1}{A}\frac{fLG_{max}^2}{2g_c(S_{min}/A)L\rho_m}\frac{G_{max}S_{min}}{\rho_m} = \frac{fG_{max}^3}{2g_c\rho_m^2} \qquad (11\text{-}7)$$

which indicates that $\rho^2 P/A$ is also a unique function of G_{max}. Curves of $(h/c_p)(c_p\mu/k)_f{}^{2/3}$ may therefore be plotted vs. $\rho^2 P/A$, as suggested by Colburn.[8]

An alternate method[42] is to plot the heat-transfer coefficient vs. the friction power per unit area for specified fluid properties, h_{std} vs. $(P/A)_{std}$, as in Fig. 11-5.[30] These values of h_{std} and $(P/A)_{std}$ are based on the following properties for air at 1 atm abs having a film temperature of 500°F: $c_p = 0.248$, $\mu = 0.0678$, $(c_p\mu/k) = 0.671$, and $\rho = 0.0413$. They

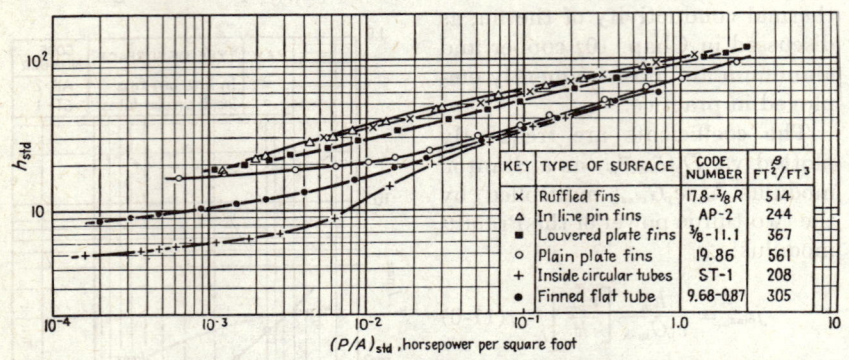

KEY	TYPE OF SURFACE	CODE NUMBER	β FT²/FT³
×	Ruffled fins	17.8-3/8 R	514
△	In line pin fins	AP-2	244
■	Louvered plate fins	3/8-11.1	367
○	Plain plate fins	19.86	561
+	Inside circular tubes	ST-1	208
●	Finned flat tube	9.68-0.87	305

FIG. 11-5. Comparison of compact exchanger surfaces on equal area basis. (*Data of Kays et al.*[30])

may readily be converted to P/A and h at the same Reynolds number but for other gas-property conditions by the following relations:

$$\frac{P/A}{(P/A)_{std}} = \left(\frac{\mu}{\mu_{std}}\right)^3 \left(\frac{\rho_{std}}{\rho}\right)^2 \tag{11-8}$$

$$\frac{h}{h_{std}} = \frac{c_p}{c_{p,std}} \frac{\mu}{\mu_{std}} \left[\frac{(c_p\mu/k)_{f,std}}{(c_p\mu/k)_f}\right]^{2/3} \tag{11-9}$$

If compactness is most important, one compares the heat transfer and friction power per unit volume. For this purpose it is convenient with plate-fin exchangers to employ the volume of the one side of the exchanger under consideration; hence one plots[28] $(\beta)(h_{std})$ vs. $(\beta)(P/A)_{std}$, as in Fig. 11-6.[30] For finned-tube exchangers one plots $(\zeta)(h_{std})$ vs. $(\zeta)(P/A)_{std}$ or $(\beta)(h_{std})$ vs. $(\beta)(P/A)_{std}$.

Design Data. Kays et al.[30,29] present data for 34 cores representing plate-fin surfaces and finned-flat-tube surfaces. Data of Kays[27] are also available for other surfaces, including circular tubes. Data of Norris and Spofford[49] for several sizes of pin and strip fins are given in Chap. 10. Typical results for the six cores described in Table 11-2 are shown in Figs. 11-5 and 11-6.

Study of Fig. 11-4 reveals that the ratio of f/j_{max}, at a given Reynolds number, varies with the Reynolds number and the geometry of the

COMPACT EXCHANGERS

exchanger. Thus f/j_{max} ranged from 7.5 to 13.0 for in-line pin fins and from only 2.5 to 3.8 for inside circular tubes. Inspection of the curves in Fig. 11-5 shows that, for a given friction power per unit surface, the in-line pin fins give heat-transfer coefficients which are 1.7 to 5.5 times

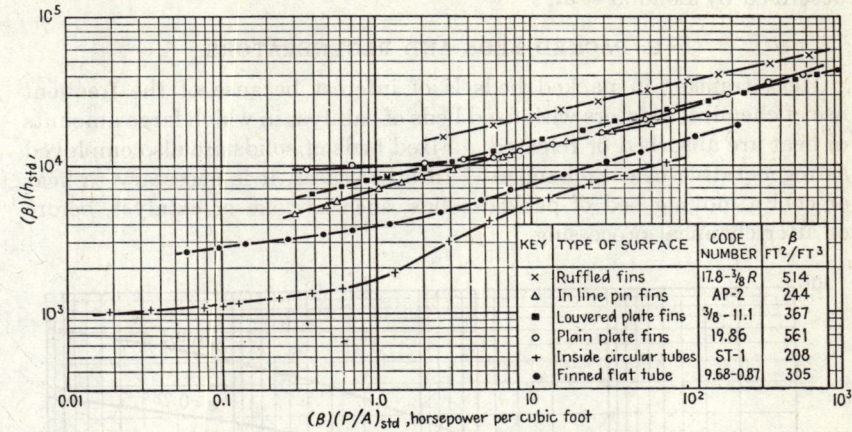

FIG. 11-6. Comparison of compact exchanger surfaces on equal volume basis. (*Data of Kays et al.*[30])

those for inside circular tubes. For a given heat-transfer coefficient the values of P/A for inside circular tubes are 10 to 20 times those for the in-line pin fins. An economic comparison may be made by comparing heat-transfer coefficients for the several types of exchanger, each

TABLE 11-2. DATA FOR COMPACT EXCHANGERS

Type of surface	Code number	Fins per inch	Plate spacing, inches	Fin thickness, inches	Minimum r_h, ft	β, sq ft/cu ft	A_f/A
Plain plate fins[30]	19.86	19.86	0.25	0.006	0.00150	561	0.849
Ruffled fins[30]	17.8-⅜ R	17.8	0.413	0.006	0.00174	514	0.892
Louvered plate fins[30]	⅜-11.1	11.1	0.25	0.006	0.00253	367	0.756
In-line pin fins[30]	AP-2	8.33	0.358	0.040	0.00350	244	0.686
Finned flat tubes[30]	9.68-.87	9.68	....	0.004	0.00295	a	0.795
Inside circular tubes[27]	ST-1	None	None	None	0.00481	208	None

a ζ = 229 sq ft/cu ft.

at the optimum friction power per unit surface, where the sum of fixed charges and pumping costs is a minimum (Chap. 15). Taking compactness into account, the curves of Fig. 11-6 show that the heat-transfer coefficient per unit volume $(h\beta)$, for a given friction power per unit volume $(\beta)(P/A)$, for the in-line pin fins is 1.8 to 4.0 times those for inside circular tubes. Of the data shown in Fig. 11-6, the highest curve is for ruffled fins.

The optimization of compact exchangers and radiators is discussed by Aronson[3,4] and Fong.[15] A compact liquid-coupled exchanger is described by London et al.[41]

II. PACKED BEDS AND REGENERATORS

Heat transfer in packed beds is of interest because of the frequent use of chemical reactors with fixed beds of catalyst in which large amounts of heat are absorbed or released. Fixed beds of solids are also employed as regenerative heat exchangers. In other cases, it is necessary to heat or cool a moving bed of packed solids, such as coke or catalyst, before or after chemical processing.

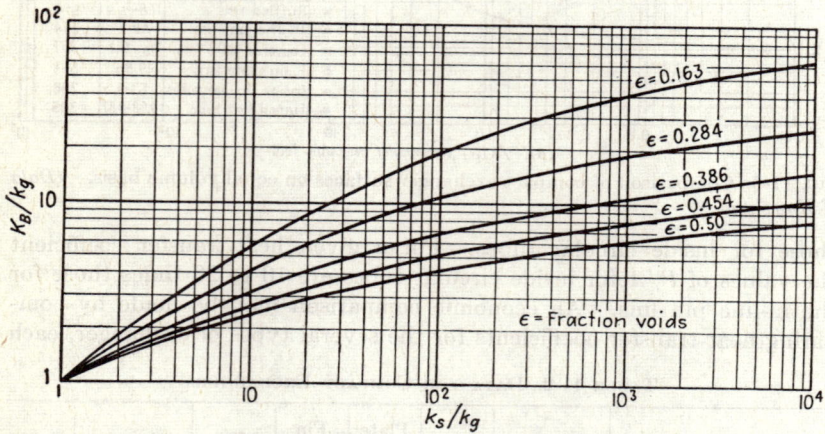

Fig. 11-7. Thermal conductivity of a packed bed of spherical particles with stagnant interstitial fluid. (*Correlation of Schumann and Voss.*[60])

Conductivity of Stagnant Packed Beds. The thermal conductivity of packed beds without fluid flow was investigated experimentally by Polack[51] and by Schumann and Voss.[60] Polack varied the pressure and temperature level in order to discover the relative importance of conduction, natural convection, and radiation. The contribution of radiation was found to be in fair agreement with the prediction of Damköhler,[10]

$$K_r = 0.00692 \epsilon\epsilon H D_p \left(\frac{T}{100}\right)^3 \qquad (11\text{-}10)$$

In the absence of radiation and natural convection, the residual bed conductivity K_B is correlated by plotting K_B/k_g against k_s/k_g for various void fractions. Data of reference 60 are shown in Fig. 11-7.

The conductivity of fine beds of magnesium oxide was measured and treated theoretically by Deissler and Eian.[11] An electric analogue gave theoretical results for pure conduction in packed spheres.[67]

Packed Cylinders. In some cases a packed cylinder is used as a heater or cooler. A fluid flows at steady mass rate through the voids in the column, and heat is supplied or removed at the wall. An apparent thermal conductivity K_a is defined by the conduction equation

$$\frac{\delta t}{\delta x} = \frac{K_a}{G_o c_p}\left(\frac{\delta^2 t}{\delta r^2} + \frac{1}{r}\frac{\delta t}{\delta r} + \frac{\delta^2 t}{\delta x^2}\right) \quad (11\text{-}11)\star$$

where G_o is the superficial mass velocity of the fluid, *i.e.*, the mass rate of flow divided by the cross section of the empty tower.

If axial conduction is neglected, $\delta^2 t/\delta x^2$ disappears from Eq. (11-11), which then becomes integrable for constant wall temperature and uniform mass velocity. The integrated expression is identical with that for rodlike flow [Eq. (9-26a)]; an appropriate change of variable gives

$$\frac{t_w - t_2}{t_w - t_1} = 0.692 e^{-23.14X} + 0.1312 e^{-121.9X} + 0.0535 e^{-299.6X} + \cdots \quad (11\text{-}12)$$

in which

$$X = \frac{K_a L}{c_p G_o D_t^2} \quad (11\text{-}13)$$

Figure 11-8 shows Eq. (11-12) for X up to 0.1; for values of X greater than 0.04, or for $(t_w - t_2)/(t_w - t_1)$ less than 0.28, only the first term of Eq. (11-12) need be employed for accuracy within 1 per cent.

Alternatively, the results for packed cylinders may be expressed as a mean heat-transfer coefficient h_m based on the area of the wall of the tube and the logarithmic mean of the terminal temperature differences,

$$h_m = \frac{q}{A(t_w - t)_m} \quad (11\text{-}14)$$

where t is the bulk stream temperature. Thus for $(t_w - t_2)/(t_w - t_1)$ less than 0.28, Eq. (11-12) may be expressed in terms of h_m as follows:

$$h_m = 5.79\frac{K_a}{D_t} + 0.0912\frac{c_p G_o D_t}{L} \quad (11\text{-}15)$$

Apparent Conductivity of Bed. Bunnell *et al.*[6] measured radial temperature profiles at several depths in a 2-inch-diameter vertical tower packed with ⅛-inch cylinders of alumina. Hot air flowed upward, and heat was removed at the wall by water boiling in an external jacket. The values of K_a computed with Eq. (11-11) from the temperature profiles in the bed were correlated by the dimensionless relation

$$\frac{K_a}{k_g} = 5 + 0.061 \frac{D_p G_o}{\mu} \qquad (11\text{-}16)$$

The Reynolds number ranged from 30 to 110.

Felix and Neill[14] employed a variety of packings having thermal conductivities from 0.1 to 100 in 3- and 5-inch tubes and flowed heated

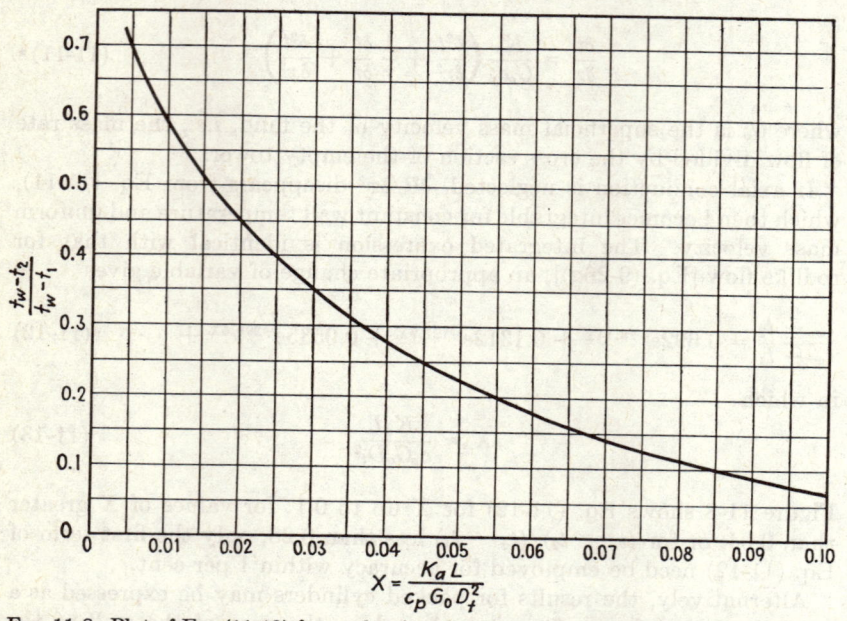

FIG. 11-8. Plot of Eq. (11-12) for evaluating the temperature change of a fluid flowing through a packed bed with constant wall temperature; unaccomplished temperature ratio vs. generalized bed size.

(y-axis: $\frac{t_w - t_2}{t_w - t_1}$; x-axis: $X = \frac{K_a L}{c_p G_o D_t^2}$)

and cooled air upward through the voids. They neglected $\delta^2 t/\delta x^2$ in Eq. (11-11) and obtained a dimensional correlation:

$$\frac{K_a}{k_g} = \frac{1}{D}\left(\frac{k_s}{k_g}\right)^{0.12}\left(C_1 + C_2 \frac{D_p G_o}{\epsilon \mu}\right) \qquad (11\text{-}17)$$

The values of C_1 and C_2 were 3.65 and 0.0106 for cylindrical packing and 3.4 and 0.00584 for spherical packing.

Hougen and Piret[24] cooled air flowing downward in tubes containing Celite packing. The values of K_a computed by the integrated relation Eq. (11-12) were correlated by the dimensionless equation

$$\frac{K_a}{k_g} = \frac{2.74}{\epsilon}\left(\frac{G_o \sqrt{A_p}}{\mu}\right)^{1/3} \qquad (11\text{-}18)$$

PACKED BEDS

where A_p is the surface area of one piece of packing; $G_o \sqrt{A_p}/\mu$ ranged from 130 to 2800.

Vershoor and Schuit[64] obtained data for heating of air in tubes containing glass beads, lead and steel balls, crushed pumice, and terrana tablets. The values of K_a computed from the data were correlated by

$$\frac{K_a}{k_g} = 1.72 \left(\frac{k_s}{k_g}\right)^{0.26} + 0.1(aD_t)^{0.5}\left(\frac{G_o}{\mu a}\right)^{0.69} \quad (11\text{-}19)$$

within 16 per cent. For spherical particles, this gives

$$\frac{K_a}{k_g} = 1.72 \left(\frac{k_s}{k_g}\right)^{0.26} + \frac{0.071(D_p G_o/\mu)^{0.69}}{(D_p/D_t)^{0.5}(1-\epsilon)^{0.19}} \quad (11\text{-}20)$$

Leva[32,33,35] presents results for a large number of heat-transfer experiments in terms of the average heat-transfer coefficient defined by Eq. (11-14). Air was heated while flowing downward through a steam-jacketed tube (½-, ¾-, and 2-inch standard pipe) filled with a wide variety of packing. The results were correlated by

$$\frac{h_m D_p}{k_g} = 0.125 \left(\frac{D_p G_o}{\mu}\right)^{0.75} \quad \text{for } \frac{D_p}{D_t} \text{ from 0.35 to 0.60} \quad (11\text{-}21)$$

$$\frac{h_m D_t}{k_g} = \frac{0.813}{e^{6D_p/D_t}} \left(\frac{D_p G_o}{\mu}\right)^{0.9} \quad \text{for } \frac{D_p}{D_t} \text{ less than 0.35} \quad (11\text{-}21a)$$

Singer and Wilhelm[61] postulated the various heat-flow mechanisms which contribute to the value of K_a:

1. Heat flow through the solid phase.
    a. With transfer of heat between particles through the flowing fluid, with allowance for film resistance at the particle surface.
    b. With transfer of heat between particles through the points of contact and the adjoining fillets of stagnant fluid.
2. Heat flow through the continuous fluid phase.
    a. By molecular conduction.
    b. By turbulent eddy conduction.

When mechanism 1a is negligible, as often occurs, the derivation gives

$$\frac{K_a}{k_g} = \frac{K_s}{k_g} + \epsilon + \frac{E}{\alpha} \quad (11\text{-}22)$$

in which the terms represent mechanisms 1b, 2a, and 2b, respectively. It is noted that K_s, the effective thermal conductivity of the *solid*, is a function of Reynolds number because of the variation in the gas film at the fillets between particles. Molecular conduction is represented by ϵ, the fraction voids in the bed. The term E/α is the ratio of eddy diffusivity of heat to the molecular diffusivity of heat, $\alpha \equiv (k/\rho c)_g$.

With particles of low conductivity, K_s/k_g in Eq. (11-22) is unimportant, thus permitting determination of the term E/α. Singer and Wilhelm[61] plotted a modified Peclet number

$$N'_{Pe} \equiv \frac{D_p G_o}{\mu} \frac{c_p \mu}{k} \frac{\alpha}{E} \qquad (11\text{-}23)$$

against the particle Reynolds number $(D_p G_o/\mu)$ for heating of gas, as shown in Fig. 11-9. For D_p/D_t less than 0.1 the wall effect becomes small. The subsequent theoretical treatment and measurements of radial eddy diffusion by Latinen and Wilhelm[31] show that N'_{Pe} approaches

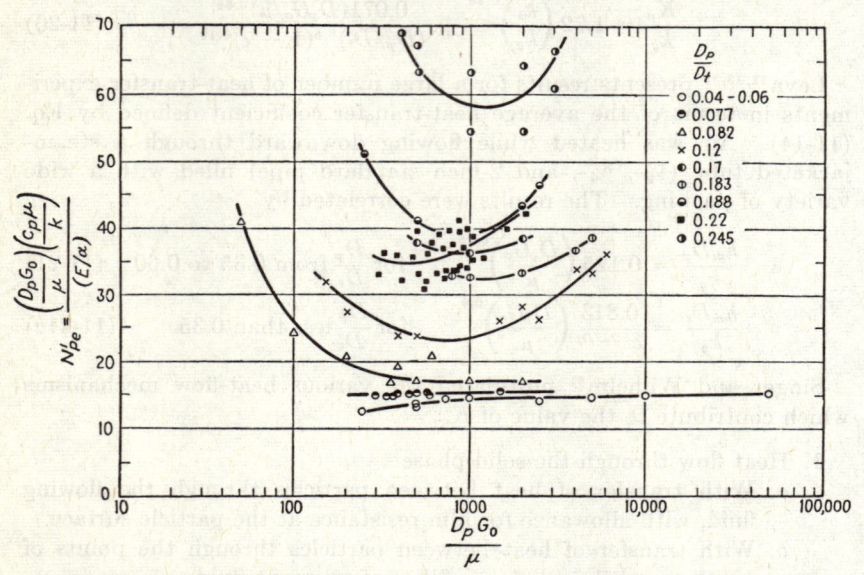

Fig. 11-9. Modified Peclet number vs. particle Reynolds number for a packed bed. (*Correlation of Singer and Wilhelm.*[61])

a constant value of 11 at high Reynolds numbers, as indicated by the trends of Fig. 11-9. Below an abscissa of 200 there is evidence of a change in mechanism, and the data are less consistent.

With particles of high conductivity K_s/k_g is not always negligible. Singer and Wilhelm[61] present values of K_s/k_g as a function of the Reynolds number for lead and steel balls, but a general correlation was not obtained.

Film Coefficient between Fluid and Packed Solids. The heat-transfer coefficient between a flowing fluid and the particles of a packed bed cannot usually be measured by direct steady-state techniques because of difficulty in continuously heating or cooling the particles themselves. Heat-transfer coefficients are obtained by a transient heat-transfer technique as in regenerators, or, alternatively, mass-transfer experiments are performed from which heat-transfer coefficients may be predicted by employing the analogy between the transfer of heat and mass.

Gamson et al.[16] report transient mass-transfer experiments in an adiabatic tower filled with wet porous particles; Reynolds numbers range from 60 to 4000. The results are reported in terms of j and are correlated by

$$j \equiv \frac{h}{c_p G_o}\left[\frac{c_p \mu}{k}\right]_f^{2/3} = 1.06 \left[\frac{D_p G_o}{\mu}\right]^{-0.41} \qquad (11\text{-}24)\star$$

for turbulent flow, with Reynolds numbers from 350 to 4000.

Wilke and Hougen[68] report values of the mass transfer group j_D for flow of air over wet porous pellets. For Reynolds numbers from 40 to 350 the data yield

$$j_D \equiv \frac{k'_G p_{BM}}{G_M}\left[\frac{\mu}{\rho D_v}\right]_f^{2/3} = 1.82 \left[\frac{D_p G_o}{\mu}\right]^{-0.51} \qquad (11\text{-}25)\star$$

For turbulent flow with Reynolds numbers from 350 to 4000, j_D averages 6 per cent below the values of j given by Eq. (11-24).

Mass-transfer experiments with liquids are reported by McCune and Wilhelm,[45] who give

$$j_D = 1.625 \left(\frac{D_p G_o}{\mu}\right)^{-0.507} \qquad (11\text{-}26)\star$$

for Reynolds numbers from 15 to 120 and

$$j_D = 0.687 \left(\frac{D_p G_o}{\mu}\right)^{-0.327} \qquad (11\text{-}26a)\star$$

for Reynolds numbers from 120 to 2000.

Equations (11-24) to (11-26) are compared in Fig. 11-10. In the turbulent range the Colburn analogy $j = j_D$ is found to apply within 25 per cent (Chap. 9).

Heat-transfer coefficients for stacked wire screens, dimpled plates, brick checkers, and other solid shapes are given in the following section on heat regenerators.

Blast-furnace Stoves and Heat Regenerators. With extremely high temperatures or with gases carrying suspended ash and solids, the use of the conventional heat exchanger becomes impractical. Under these conditions the transfer of heat from one gas to another is accomplished more economically by the alternate heating and cooling of a refractory solid, as in the blast-furnace stove or heat regenerator. At lower temperatures, metal packing is frequently employed, as in rotary preheaters or in regenerators for very low temperature service.

When hot and cold gases flow alternately through a chamber partly filled with solids the mathematical relations become very complex for the general case.† Several methods of approximate calculations are available.‡ Axial conduction within the matrix has usually been neglected; Schultz,[59] however, considers the effect of axial conduction by subdividing the matrix into thermally separated parts.

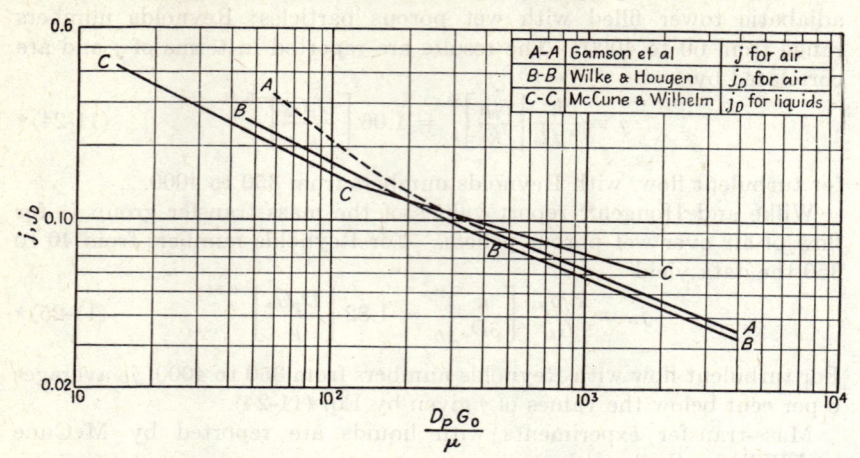

Fig. 11-10. Comparison of data for mass transfer and heat transfer between packed solids and flowing fluids.

For a counterflow regenerator in which the thermal resistance of the packing is negligible, a simpler treatment developed by Hausen[20] and Hottel[23] is applicable. The method is summarized below.

Employing ('') to denote the hot gas stream or gas blow and (') to denote the cold air stream or air blow, the dimensionless regenerator size is defined by

$$\lambda'' \equiv \frac{h''}{c_p'' G_o''} \frac{(1-\epsilon)L}{r_B} \qquad (11\text{-}27)$$

$$\lambda' \equiv \frac{h'}{c_p' G_o'} \frac{(1-\epsilon)L}{r_B} \qquad (11\text{-}27a)$$

and the dimensionless regenerator period by

$$\tau'' \equiv \frac{h'' \theta''}{c_s \rho_s r_B} \qquad (11\text{-}28)$$

$$\tau' \equiv \frac{h' \theta'}{c_s \rho_s r_B} \qquad (11\text{-}28a)$$

For a symmetrical cycle ($\tau'' = \tau' = \tau_s$ and $\lambda'' = \lambda' = \lambda_s$), the temperature efficiency of both streams is the same and is given by

$$\phi_s = \frac{t_2' - t_1'}{t_1'' - t_1'} \equiv \frac{t_1'' - t_2''}{t_1'' - t_1'} = \frac{\lambda_s}{2 + \lambda_s + (\tau_s/\Gamma)} \qquad (11\text{-}29)\ast$$

† Heiligenstadt,[22] Nusselt,[50] Hausen,[21] Lubojatsky,[44] Ackermann,[2] and others.
‡ For example, see Schack,[57,58] Rummel and Schack,[55] Rummel,[53] Hausen[20] and Trinks.[62]

PACKED BEDS

in which t_2'' and t_2' are the average outlet temperatures of the hot and cold streams and t_1'' and t_1' are the uniform inlet temperatures of the hot and cold streams. It is apparent that ϕ_s is the ratio of the actual temperature change of the stream divided by the temperature change which would occur in a regenerator of infinite size. For the case of negligible thermal resistance within the solids, the term Γ is a function only of τ_s and λ_s; hence ϕ_s may be plotted as shown in Fig. 11-11. The selection of τ_s and λ_s as the coordinates is advantageous since the design limitations for a particular regenerator may often be superimposed directly on this generalized plot; in addition, cycle times and regenerator lengths may be shown for different packings and mass velocities by the use of auxiliary scales along the coordinates.[23] In Fig. 11-11, the intercepts

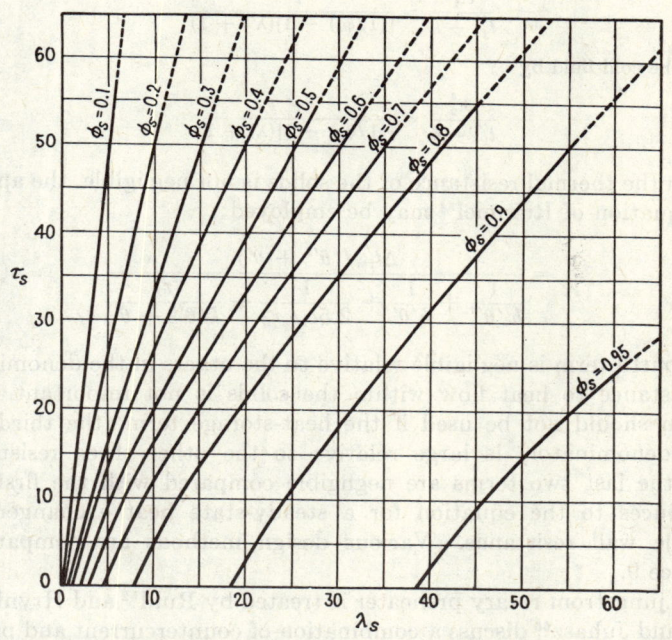

FIG. 11-11. Generalized regenerator plot for the case of symmetrical cycles and negligible thermal resistance within the solid.[20,23] (*Courtesy of Hottel.*[23])

for an infinitely short period correspond to the temperature efficiency which would be obtained in a conventional heat exchanger involving heat transfer through a wall of negligible thermal resistance, $\phi_s = \lambda_s/(2 + \lambda_s)$.

For a nonsymmetrical cycle, the rise or fall in temperature of each stream is given approximately by the relations

$$\frac{t_1'' - t_2''}{\Delta t_{lm}} = \frac{\lambda''/\tau''}{\dfrac{1}{\tau''} + \dfrac{1}{\tau'} + \dfrac{2}{\tau'' + \tau'}\left[\lambda_s\left(\dfrac{1}{\phi_s} - 1\right) - 2\right]} \quad (11\text{-}30a)$$

and

$$\frac{t_2' - t_1'}{\Delta t_{lm}} = \frac{\lambda'/\tau'}{\dfrac{1}{\tau''} + \dfrac{1}{\tau'} + \dfrac{2}{\tau'' + \tau'}\left[\lambda_s\left(\dfrac{1}{\phi_s} - 1\right) - 2\right]} \quad (11\text{-}30b)$$

For this purpose ϕ_s is evaluated from Fig. 11-11 using

$$\tau_s = \frac{\tau'' + \tau'}{2} \tag{11-31}$$

and

$$\lambda_s = \frac{2}{(1/\lambda'') + (1/\lambda')} \tag{11-32}$$

The maximum fluctuation in the outlet temperature of each stream during the cycle is of interest and is given for the hot gas by

$$\frac{\delta t_2''}{t_2'' - t_1'} = \frac{\tau''}{[(1/\phi_s) - 1](\lambda'' + 1)} \tag{11-33}$$

and for the cold fluid by

$$\frac{\delta t_2'}{t_1'' - t_2'} = \frac{\tau'}{[(1/\phi_s) - 1](\lambda' + 1)} \tag{11-34}$$

When the thermal resistance of the solids is not negligible, the approximate equation of Rummel[54] may be employed:

$$\frac{q}{A} = \frac{\Delta t_{lm}/(\theta'' + \theta')}{\dfrac{1}{h''\theta''} + \dfrac{1}{h'\theta'} + \dfrac{1}{2.5 c_s \rho_s r_B} + \dfrac{r_B}{k(\theta'' + \theta')}} \tag{11-35}$$

If the fourth term is negligible relative to the others in the denominator, the resistance to heat flow within the solids is not important. This equation should not be used if the heat-storage term (the third term in the denominator) is large relative to the other three resistances. Where the last two terms are negligible compared with the first two, this reduces to the equation for a steady-state heat exchanger with negligible wall resistance. Various design methods are compared in Reference 9.

The Ljungstrom rotary preheater is treated by Ruhl[52] and Hrynizak.[25] Widell and Juhasz[66] discuss a combination of countercurrent and parallel flow to increase the temperature of the matrix at the cold end of the rotary regenerator and thus to increase its life by preventing condensation.

Heat-transfer coefficients for matrices of brick checkerwork are given by Rummel.[53] Saunders and Smoleniec[56] present experimental results for matrices of wire gauze having 320 to 1530 sq ft of surface/cu ft of gross volume; Nusselt numbers ranged from 1 to 10, as Reynolds numbers increased from 4 to 300, and were of the same order as for flow normal to single cylinders. Ambrosio *et al.*[1] report heat-transfer coefficients and friction factors for laminar flow of air through compact regenerators packed with stacked wire screens, dimpled flat plates, or small round tubes.

Moving Bed. In some cases a counterflow heat exchanger consists of an insulated vertical container of uniform cross section in which solid particles are introduced at the top and move downward as a bed while gas flows upward through the voids. The case of uniform velocity is treated analytically by Munro and Amundson[48] for spherical particles. Allowance is made for thermal gradients in the spheres; the surface coefficient of heat transfer between the spheres and the gas is assumed constant. The results involve four dimensionless ratios.

$$Y \equiv \frac{t - t_{s1}}{t_1 - t_{s1}}$$

$$X \equiv \frac{4\alpha_s x}{V D_p{}^2} = \frac{4\alpha\theta}{D_p{}^2}$$

$$Z \equiv \frac{w_s c_s}{w c_p}$$

$$M \equiv \frac{2k_s}{h D_p}$$

Figure 11-12 shows a graph of Y vs. X, for three combinations of Z and M: 2.5 and 0.5, 1 and 0.2, and 0.5 and 0.1. This paper also considers various cases of internal heat generation.

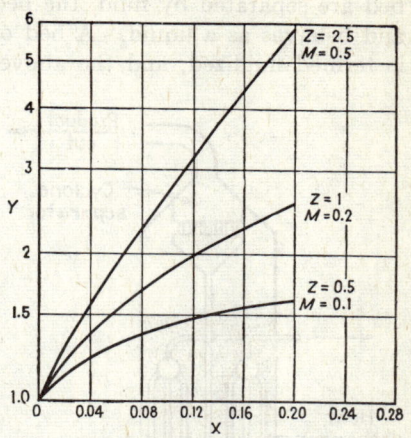

Fig. 11-12. Generalized plot for obtaining the fluid outlet temperature for a counterflow moving bed. (*Courtesy of Munro and Amundson.*[48])

Downflow of packed solids without gas flow is treated by Brinn et al.[5a] (Chap. 9).

III. FLUIDIZED SYSTEMS

Introduction. Following the construction during World War II of giant fluidized reactors for the catalytic cracking of petroleum, wide interest in the fluidization phenomenon has arisen. Numerous experimental studies of gas[18] and solids[17] flow and of heat transfer in fluidized beds have been reported; however, the field is still in the stage of development, and the various effects which have been observed are incompletely understood. In the following sections the heat-transfer characteristics of fluidized beds are discussed, and the results of investigations which have been reported to date are summarized. A fluidized bed with internal heat-transfer surface is shown in Fig. 11-13.

Reference 26 presents nomenclature and symbols suggested for fluidization. The recommended terminology and in most cases the symbols are used throughout this section.

Types of Fluidization. Consider upward flow of a gas through a bed of finely divided solid material. With low superficial velocity the gas flows through the interstices of the bed, the solid remaining fixed in position. At some higher velocity there is a shifting of solid with a slight increase in bed volume, and the frictional drag on each solid particle becomes high enough to support it without contact with other particles. The pressure drop at this point is essentially equal to the weight of the bed per unit cross section. Since the particles forming the bed are separated by fluid, the bed is unable to resist a shearing force and behaves as a liquid. A bed of solids exhibiting this characteristic is termed fluidized, and the above condition is the point of *minimum fluidization*.

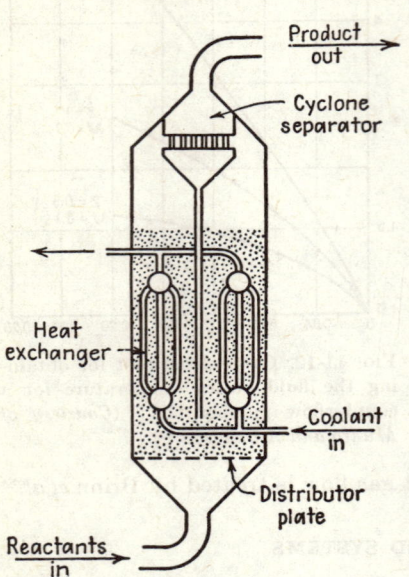

At the lowest fluidizing velocities the *quiescent* bed has a well-defined upper surface, and there is little mixing of solids. Over a range of gas velocities higher than those for quiescence, solid mixing increases, and there is pronounced circulation of the fluidized mass. The bed at this point is very similar in appearance to a liquid: objects less dense than the bed will float on the surface, and disturbance of the surface causes ripples; the fluidized mass possesses hydrostatic head and will flow between containers. The condition in which there is a well-defined surface and smooth flow of gas through the bed is described as *dense-phase* fluidization.

BATCH FLUIDIZED POWDER SYSTEM

FIG. 11-13. Batch fluidized bed with internal heat-transfer surface. (*Courtesy of E. R. Gilliland.*[17])

Higher gas velocities result in an important change in the character of fluidization. The bed becomes highly turbulent, and rapid mixing of the solids occurs. The surface is no longer well defined but is diffuse and boils violently, and bubbles of gas similar in appearance to those in boiling liquids rise through the bed.

At still higher gas velocities the bubbles become large enough to force up masses of particles above them; in small beds these bubbles may extend from wall to wall. This condition of fluidization is called *slugging*. A fluidized system in which the solids concentration is low and in which there is no boundary defining the upper surface of the bed (a condition more analogous to a gas than a liquid) is termed *disperse*.

In a bed fluidized with a gas, the solid is thought to move in clusters of particles, or aggregates, which are continuously broken up and re-formed; hence fluidization of this nature is designated *aggregative*. When solids are fluidized with liquid, however, the individual solid particles are thought normally to be separated by fluid, rather than to form clusters. Fluidization in which the solid remains separated is termed *particulate*.

The boundaries between the conditions of fluidization described above are not sharp; hence the various regimes are difficult to distinguish.

Incipient Fluidization. The "critical" velocity V_{mf}, corresponding to the transition from a fixed bed to a quiescent fluidized bed, has been studied by several observers.[5,38] Miller and Logwinuk[47] found that the following dimensionless equation correlated their data for small particles, D_p from 0.0035 to 0.0098 in., having true densities ρ_s from 70 to 243 lb/cu ft, fluidized by gases having densities ρ_g from 0.01 to 0.112 lb/cu ft, independent of weight or depth of bed.

$$V_{mf} = \frac{D_p{}^2(\rho_s - \rho_g)^{0.9}\rho_g{}^{0.1}g}{800\mu} \quad (11\text{-}36)$$

The mean particle diameter D_p is the weighted average of the effective diameters of individual sieve fractions

$$D_p = \sum_1^n \chi d_p \quad (11\text{-}37)$$

in which $d_p = \sqrt{d_1 d_2}$ and d_1 and d_2 are openings of adjacent standard sieve sizes used in separating a fraction, χ is the weight fraction of material of nominal diameter d_p, and n is the number of fractions into which the material is separated.

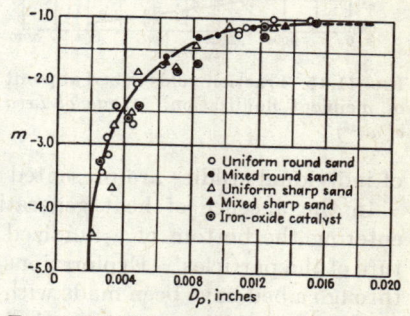

FIG. 11-14. Values of exponent m in Eq. (11-38) determined in 2.5- and 4-in.-diameter beds; note that m is always negative. (*Data of Leva et al.*[38])

Correlation of Fraction Voids. Leva *et al.*[38] relate the fraction voids ϵ in a fluidized bed to the superficial velocity V_o by the following equation:

$$\frac{V_o}{V_{mf}} = \left(\frac{1 - \epsilon}{1 - \epsilon_{mf}}\right)^{2m+1} \left(\frac{\epsilon_{mf}}{\epsilon}\right)^{3m} \quad (11\text{-}38)$$

The exponent m in Eq. (11-38) is a function of the particle size; values for several sharp and round sands and an iron oxide catalyst are plotted in Fig. 11-14 vs. D_p as obtained from Eq. (11-37). The bed condition in which the void fraction is the same as that at incipient fluidization can be obtained by gradually reducing to zero the mass velocity through a fluidized bed. Leva *et al.*[38] found that for sands this condition is approximated by slowly pouring the material into an open vessel. Values of ϵ_{mf} for several materials are given in Fig. 11-15.

Pressure Drop. Since the frictional drag at the wall of a fluidized bed is small compared with the weight of the bed per unit cross section, the latter is normally taken as the pressure drop between bottom and top,

$g_c \Delta p/g = W/S = L_f \rho_{bf}$. For gases at ordinary pressures the weight of the bed is essentially equal to the weight of the solids alone,

$$g_c \Delta p/g = L_f(1 - \epsilon)\rho_s \qquad (11\text{-}39)$$

With smooth fluidization Eq. (11-39) very closely predicts the pressure drop, independent of G_o; however, in slugging beds or with electrostatic effects, values of Δp as high as 20 per cent greater than predicted by Eq. (11-39) have been measured.[40]

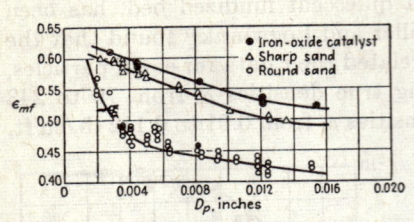

FIG. 11-15. Fraction voids in bed at point of incipient fluidization. (*Data of Leva et al.*[38])

General Heat-transfer Characteristics. The over-all temperature pattern and heat-transfer characteristics of a fluidized bed are briefly outlined below; the detailed results of individual studies are presented later under the appropriate headings.

In the absence of heat generation in or on the solid particles, gas entering the bottom of a fluidized system rapidly attains the temperature of the particles. Explorations through a bed have been made with a probe formed from two identical thermocouples.[37] One couple was bare and contacted the solid particles and fluid in the bed, whereas the other couple was shielded from the solid by a bag of fine wire mesh and thus contacted only the fluid. The temperature difference between the thermocouples is plotted as a function of bed height in Fig. 11-16. At heights greater than 1 inch from the bottom of the bed the temperature difference between the couples was always less than 1°F, indicating that solid and gas were at nearly the same temperature. A similar effect was observed by Levenspiel and Walton.[39] The high rate of heat exchange between the gas and the particles is caused by the large surface area of solid per unit volume of bed and by the high gas-to-particle heat-transfer coefficients resulting from the small particle sizes (see Fig. 10-11).

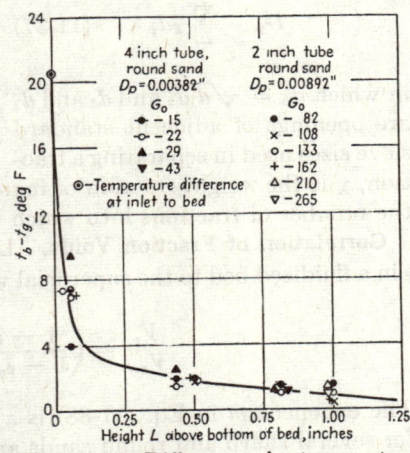

FIG. 11-16. Differences in temperature between bare thermocouple and thermocouple shielded from solid particles by screen, at various levels above fluid inlet. (*Data of Leva, Weintraub, and Grummer*[37] *for gas entering an externally heated fluidized bed.*)

With moderate rates of heat addition to a turbulent dense-phase bed, the heat transfer within the bed is so rapid that the fluidized mass is essentially at uniform temperature; the temperature drop between the heat-transfer surface and fluidized mass takes place across a small distance near the surface[5,46] (Fig. 11-17). Even with a high rate of heat transfer to the bed the temperature gradients within the fluidized material are very small relative to the temperature drop at the heat-transfer surface.[19] Data on heat transfer *within* fluidized beds are presented in the discussion of apparent thermal conductivity.

Heat-transfer coefficients, $h = q/(t_w - t_b)A$, between fluidized materials and surfaces in fluidized beds have been reported by a number of investigators, but the effects of different variables are not yet conclusively determined. Increased gas flow rate in a *dense-phase* bed increases the heat-transfer coefficient. However, higher flow rates appreciably reduce the solids concentration, and *lean-phase* fluidization occurs: in this region the effect of reduced concentration may cause the coefficient to fall off. Hence a maximum value of h occurs as flow rate is varied,[5] and different effects are observed in dense and lean beds. The thermal conductivity of the gas has an important influence in determining h, whereas manyfold variation in the conductivity of the solid does not appear to have effect; increase in particle size is found to cause a reduction in h. The manner in which other variables affect heat transfer in fluidized beds is uncertain.

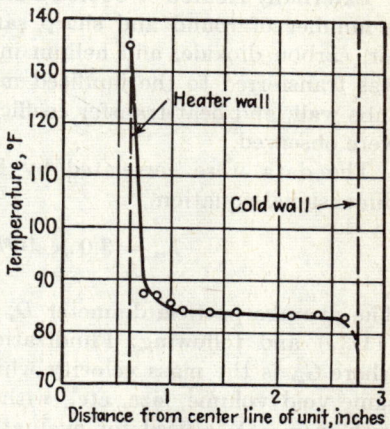

Fig. 11-17. Typical horizontal temperature profile inside a fluidized bed with heat at axis and water-cooled wall. (*Data of Baerg, Klassen, and Gishler.*[5])

Apparent Thermal Conductivity. The rate at which heat is transferred between points within a fluidized bed has been studied by Girouard[19] with 2- and 3-in.-diameter beds. Heat was supplied through the tube wall to the bottom of the bed and removed near the top at rates high enough to cause appreciable temperature gradients in the fluidized mass. For a given radial position the axial temperature gradient was found to be essentially the same at any elevation between the heating and cooling sections and was directly proportional to the heat flux. In addition the rate at which heat was transferred vertically was proportional to the cross-sectional area of the column.

The above relations are identical with those for conduction in a solid, and an *apparent thermal conductivity* K_F may be defined such that

$$q/S = K_F \Delta t/\Delta x \qquad (11\text{-}40)$$

Equation (11-40) was found to apply with the same value of K_F even when the positions of heater and cooler were reversed and heat was transferred *downward* through the bed. The apparent thermal conductivity was also found to be independent of height and diameter for the columns tested. The values of K_F in some cases were thousands of times greater than the apparent conductivity of the static bulk material; conductivities as high as 22,000 Btu/(hr)(sq ft)(deg F per ft) were obtained with glass beads at a superficial air velocity of 1.7 ft/sec.

Externally Heated or Cooled Dense-phase Beds. Leva *et al.*[37] fluidized a number of round and sharp sands and an iron oxide catalyst with air, carbon dioxide, and helium in 2- and 4-inch-diameter beds. Heat was transferred to the fluidized mass by steam condensing outside the tube wall, and heat-transfer coefficients up to 80 Btu/(hr)(sq ft)(deg F) were observed.

The data were correlated by Leva[34] within ±14 per cent by the dimensional equation

$$h_m = 3.0 \times 10^6 k_g D_p \left(\frac{D_p G_o \xi}{N_v \mu} \right)^{0.6} \qquad (11\text{-}41)$$

The average particle diameter D_p is obtained using the method of Eq. (11-37) and following: Fluidization efficiency ξ equals $(G_o - G_{oe})/G_o$, where G_{oe} is the mass velocity which would occur in a fixed bed of the same void volume, size, etc., with the same pressure drop; Eq. (11-38) with $m = -1$ is used for evaluation of G_{oe}. The bed expansion ratio N_v equals $L_f/L_{mf} = (1 - \epsilon_{mf})/(1 - \epsilon)$. The following range of variables was covered by the data: G_o from 10 to 1200, V_o from 0.05 to 2.50, pressure from atmospheric to 80 lb/sq in. abs, fluid thermal conductivity from 0.016 to 0.103, and particle size from 0.006 to 0.018 in.

Leva and Grummer[36] point out that although h is proportional to $D_p^{1.6}$ in Eq. (11-41), the fluidization efficiency ξ is inversely proportional to D_p to the 2.5 to 5.2 power; thus the net effect of increase in D_p is a reduction in h, as observed by other investigators.

Dow and Jakob[12] fluidized Aerocat catalyst particles, finely divided coke, and powdered iron with air in 2- and 3-inch-diameter beds; heat-transfer coefficients from 20 to 174 Btu/(hr)(sq ft)(deg F) were obtained. Heat was supplied to the bed by steam condensing on the outer surface of the wall. Operation was entirely in the range of smooth dense-phase fluidization, with mass velocity of air varied from 60 to 305 lb/(hr)(sq ft) and fraction of solids, $1 - \epsilon$, from 0.47 to 0.31. Average particle sizes ranged from 0.0027 to 0.0067 inches as determined using sieves; fluidized bed height L_f ranged from 1.7 to 14.1 inches. The data are correlated by the following dimensionless equation to within ±20 per cent:

$$\frac{h_m D_t}{k_g} = 0.50 \left(\frac{D_t}{L_f}\right)^{0.65} \left(\frac{D_t}{D_p}\right)^{0.17} \left(\frac{\rho_s c_s}{\rho_g c_p}\right)^{0.25} \left(\frac{D_t G_o}{\mu}\right)^{0.80} \quad (11\text{-}42)$$

Van Heerden et al.[63] cooled beds of powders of coke, carborundum, Devarda's alloy, iron oxide, and lead, fluidized with various gases in a 3.3-inch-diameter bed; particle diameters ranged from 0.002 to 0.03 in. A 3.9-in.-high section of the wall was cooled with water and coefficients between the wall and the fluidized mass determined; the entering air was heated by an electrical coil located about 4 inches below the cooler. The mean heat-transfer coefficient at the wall was correlated by the following dimensionless relation,

$$\frac{h_m D_p}{k_g} = 0.58 \left(\frac{\rho_b'}{\rho_g}\right)^{0.18} \left(\frac{c_s}{c_p}\right)^{0.36} \left(\frac{c_p \mu}{k_g}\right)^{0.5} \left(\frac{B D_p G_o}{\mu}\right)^{0.45} \quad (11\text{-}43)$$

for $BD_p G_o/\mu$ from 0.1 to 5.0, where ρ_b' is the apparent density of the bed as the air rate is slowly reduced to zero and B is a generalized shape factor which deviates very little from 1.0. For given solid and gas properties h_m is a unique function of G_o ($h_m \sim G_o^{0.45}$). As the Reynolds number is increased beyond 5, the data fall below the curve for the above correlation and h_m probably reaches a maximum value.

Internally Heated or Cooled Dense-phase Beds. Baerg et al.[5] measured heat-transfer coefficients for a 1.25-in.-diameter electric heater centrally located in a 5.5-in.-diameter fluidized bed from which heat was removed at the wall by a cooler; 10 per cent of the heat passed out in the air stream. Seven types of materials (iron powder, round sand, foundry sand, jagged silica, glass beads, cracking catalyst, and alumina) with diameters ranging from 0.0024 to 0.035 in. and bulk densities from 65 to 148 lb/cu ft were fluidized with air; mass velocities up to 600 lb/(hr)(sq ft) were employed.

The heat-transfer coefficient for the heater was found to pass through a maximum as the mass velocity was increased; for all materials except glass beads the maximum coefficient was correlated by

$$h_{m,\text{max}} = 49 \log (0.00037 \rho_b/D_p) \quad (11\text{-}44)$$

The values of $h_{m,\text{max}}$ for glass beads were about 18 per cent higher than predicted by Eq. (11-44); this difference was attributed to freer movement of the beads resulting from their smooth surfaces. Lower values of $h_{m,\text{max}}$ were found for glass beads with etched surfaces, but no effect on the minimum velocity required for fluidization was observed.

For fluidizing velocities below those yielding $h_{m,\text{max}}$ the heat-transfer coefficient was roughly correlated by Eq. (11-45) with deviations increasing with larger particle sizes,

$$\log (h_{m,\text{max}} - h_m) = 1.74 - 0.0052(G_o - 0.71 \rho_b) \quad (11\text{-}45)$$

No effect of specific heat or thermal conductivity of the solid was detected.

In a large bed of 22 inches diameter, using air, Vreedenberg[65] fluidized various types of sand with static bed heights of 47 and 67 inches. The fluidized mass was cooled by water flowing through a single tube, 1.3 inches outside diameter, positioned horizontally and vertically at various locations in the bed. Heat-transfer coefficients ranged from 20 to 70 Btu/(hr)(sq ft)(deg F) with mass velocities from 65 to 180 lb/(hr)(sq ft); for the same gas flow, rate values of h_m measured with various tube positions differed by as much as 65 per cent.

Campbell and Rumford[7] investigated the effect of solid properties on heat transfer to a cooling coil immersed in a 2.3-in.-diameter fluidized bed. No effect of thermal conductivity of the solid material was observed with solids ranging in conductivity from 0.8 to 220.

Miller and Logwinuk[47] fluidized silicon carbide and aluminum oxide powders with air, carbon dioxide, and helium in a 2-inch steel pipe with low gas flow rates. The bed was heated electrically through the column wall and simultaneously cooled by water flowing through a vertical $3/8$-inch copper tube at the axis. Coefficients determined for the copper tube surface ranged from 40 to 200. The development of electrostatic charges in fluidized beds and their effects were qualitatively investigated.

Fairbanks[13] investigated the effects of gas and solid properties on heat transfer in a turbulent dense-phase bed. Glass beads and Aerocat catalyst particles of 0.003 in. diameter were fluidized using six different gases in a 4-in. bed with a $1/4$-in.-diameter axial heater. For gas velocities up to 1 ft/sec, coefficients ranged from 50 to 180. For a given gas, between 0.1 and 1.0 ft/sec, there was only slight increase in h_m with increasing velocity. Erratic behavior found with ammonia was attributed to condensation on the particles.

Fairbanks proposed that clusters of particles having the density of a quiescent bed contact the heater for short periods. Heat flows into each cluster with a limited depth of penetration, and the mathematical treatment of a semi-infinite solid (Chap. 3) is applicable, yielding for fixed dynamical conditions

$$h \sim \sqrt{k_q \rho_q c_s}$$

where k_q and ρ_q are the thermal conductivity and density which the bed would have in the quiescent state. The values of k_q were obtained from the measured conductivity of the packed bed by correcting for the change in ϵ according to the correlation of Fig. 11-7. The experimental values of h_m were found to be proportional to the 0.5 power of k_q, in agreement with the postulated mechanism.

Lean-phase Fluidization. Mickley and Trilling[46] determined local heat-transfer coefficients in beds of spherical glass beads fluidized with air. Operation was largely in a range of air flow rates yielding low bed densities; the fraction of solids, $1 - \epsilon$, ranged from 0.53 to 0.016. Average particle sizes ranged from 0.002 to 0.02 in. and superficial air velocities from 0.8 to 15 ft/sec. Experiments were conducted with internally and externally heated beds with electrical generation of heat; batch operation was employed. The beds were found to be at nearly uniform temperature throughout, with all temperature gradients occurring at the heater and wall surfaces; no temperature measurements were made at the bottom of the bed, where initial contact between air and solid takes place. Heat-transfer coefficients ranged from 10 to 120 Btu/(hr)(sq ft)(deg F); coefficients predicted for open tubes for the same flow rates would range between 1 and 4.

The internally heated bed was formed from a steel tube 2.88 in. inside diameter by 50 in. long and contained an axially located heater 0.49 in. in diameter and 34.5 in. long. The heat-transfer coefficient was found to be a function of the bead size and solids concentration only; for a given bead size and solids concentration there was no effect of gas flow rate. Since the properties of the gas and solid were not varied, they were omitted from the correlation. The data for the four larger bead sizes are correlated by Eq. (11-46) within ±25 per cent:

$$h = 0.0433 \left(\frac{\rho_c^2}{D_p^3}\right)^{0.238} \quad (11\text{-}46)$$

Values for the smallest beads were considerably less than predicted by the correlation. The solids concentration, $\rho_c = (1 - \epsilon)\rho_s$, ranged from 4 to 80 lb/cu ft for the data correlated by Eq. (11-46).

Mickley and Trilling also operated 1- and 4-in.-diameter fluidized beds with heat supplied through the wall of the column. In contrast to the results for the internally heated bed there was an appreciable effect of air flow rate. The data were tentatively c rrelated by Eq. (11-47):

$$\lambda = 0.0118 \left(\frac{\rho_c G_o}{D_p^3}\right)^{0.263} \quad (11\text{-}47)$$

No effect of bed diameter was detected. Solids concentrations varied from 3 to 70 lb/cu ft.

PROBLEMS

1. A fluidized bed, 3 in. in diameter, is being heated through the external walls. At a given time during the heating operation it is found that the wall temperature is 266°F and the temperature of the fluidized bed is 150°F. The air flow rate was 914 lb/(hr)(sq ft), and the heat flux was 3970 Btu/(hr)(sq ft). The solid being fluidized was glass beads having an average diameter of 14.8×10^{-4} ft. The absolute density of the glass beads was found to be 159 lb/cu ft, and the dense-phase density of the fluidized bed was 51.7 lb/cu ft. Using these data, calculate the heat-transfer coefficient from the wall to the fluidized bed.

2. A carbon-steel ball, 0.975 in. in diameter, is heated and then plunged into a fluidized bed. The temperature at the center of the steel ball is measured with a thermocouple. In one test after the ball had been in the fluidized bed for 4 min., the center temperature was found to be 96°F. The original temperature of the steel ball was 210°F, and the temperature of the fluidized bed remained constant at 78°F. The fluidized bed was 3 in. in diameter and the solid employed was finely divided activated alumina. In this particular test the air flow rate was 4.0 cu ft/min, measured at standard temperature and pressure. Using these data, calculate the average heat-transfer coefficient between the steel ball and the fluidized bed.

3. In order to measure the ability of a fluidized bed to transmit heat from one region to another, measurements were made in which heat was added at one level and removed at some higher level. Temperature gradients were measured in the section between the heater and the cooler. They were found to be essentially linear with distance, and the results for the intermediate section could be expressed as an apparent thermal conductivity by using the usual Fourier conduction equation. In one such investigation, a 3-in. column was employed fluidizing microspherical cracking catalyst with particle sizes from 74 to 160 microns. The superficial air velocity at the average fluidized-bed conditions was 0.230 ft/sec. The average bed pressure was 30.8 in. Hg abs. The air entered the fluidized bed at 75°F and left the top of the unit at 144°F. The water rate to the cooler was 186 lb/hr, and it entered at a temperature of 14.4°C and left at a temperature of 24.0°C. The temperature in the bed between the heater and cooler was measured at two levels 29 in. apart and found to be 209 and 167°F. The pressure drop per foot through the dense phase was 4.25 in. of water. Neglecting the heat losses from the column,

  a. Calculate the apparent thermal conductivity.
  b. Assuming that the heat is carried by solid at 209°F flowing up and being cooled to 167°F, estimate the pounds of solid that must flow per second.
  c. For the density prevailing, estimate the solid velocity corresponding to the answer for (*b*).

4. The thermal conductivity of a packed bed was measured in the annular space between a center heater and the outer cooling surface. The internal core, which was 0.84 in. in diameter, was heated with steam at atmospheric pressure. The outer tube, which served as a cooling system, had an internal diameter of 3.06 in. The length of the test section was 24.1 in. In one experiment the annular space between the heater and the cooler was packed with glass beads 0.158 in. in diameter. During a 10-min. run, 14.4 cm^3 of condensate was obtained. The temperature of the cooling surface was 69.8°F.

  a. Neglecting any temperature drop through the metal wall of the heater, calculate the thermal conductivity of the packed section.
  b. During the experiment given above, the voids around the glass beads were filled with air at atmospheric pressure. The same bed was tested when the absolute pressure was 0.4 mm Hg and the cubic centimeters of condensate obtained in 10 min. decreased to 5.4. Estimate the thermal conductivity in this case.

CHAPTER 12

HIGH-VELOCITY FLOW; RAREFIED GASES

Abstract. This chapter is divided into two sections. The first deals with flow of gases at supersonic and high subsonic speeds. The effect of frictional heating of the wall is considered, and recovery factors and heat-transfer coefficients are given for various shapes.

The second section treats heat transfer and friction with gases at pressures so low that the fluid no longer behaves as a continuous medium. Accommodation coefficients are discussed; correlations of heat-transfer coefficients are presented for both forced and natural convection.

Introduction. Because of the recent advent of supersonic airplanes, rockets, and missiles which can operate at high altitudes, considerable research has been undertaken on fluid flow and heat transfer at high Mach numbers and, in some cases, with rarefied gases. Some interesting findings have been made, but these fields are still in the process of exploration and clarification.

HIGH-VELOCITY FLOW

Mechanism. With flow of gases at high velocities the frictional heating of the wall of the duct becomes appreciable and necessitates a new definition of the heat-transfer coefficient. Also, with flow velocities higher than approximately one-third the velocity of sound, the changes in pressure and density along the flow path are often considerable, requiring integration along the length of the duct to compute the exit conditions. When the density changes significantly owing to high velocities, the flow is termed *compressible*.

RECOVERY FACTORS

If a gas stream of uniform temperature is brought to rest adiabatically, as at the true stagnation point of a blunt body, the temperature rise for an ideal gas will be

$$t_s - t_m = V^2/2g_cJc_p \tag{12-1}$$

where t_s is the "stagnation" or "total" temperature and t_m is the local bulk temperature t of the moving stream having velocity V. At every other point on the body, the gas is brought to rest by viscous effects in the boundary layer. The temperature of the gas near the wall is increased

Table 12-1. Nomenclature

- A Heat transfer area, square feet
- c_p Specific heat at constant pressure, Btu/(lb of fluid)(deg F)
- c_v Specific heat at constant volume, Btu/(lb of fluid)(deg F)
- D Diameter, feet
- D_o Diameter (outside) of tube or wire, feet
- d Prefix indicating differential, dimensionless
- E Internal energy, Btu per pound of fluid
- f Fanning friction factor, dimensionless
- G Mass velocity, equals w/S, lb of fluid/(sec)(sq ft of cross section); G equals $V\rho = V/v$
- g_c Conversion factor in Newton's law of motion, equals 32.2 (ft)(pounds mass)/(sec)(sec)(pounds force)
- h Coefficient of heat transfer between fluid and surface, Btu/(hr)(sq ft)(deg F); h_c, based on $t_w - t_{aw}$; h_m, based on $t_w - t_m$; h_s, based on $t_w - t_s$
- i Enthalpy, $E + (pv/J)$, Btu per pound of fluid
- J Mechanical equivalent of heat, 778 ft-pounds force per Btu
- j Dimensionless factor $= (N_{St})(N_{Pr})^{2/3}$
- k Thermal conductivity, Btu/(hr)(sq ft)(deg F per ft); k_0 evaluated at the stagnation temperature
- L Length of duct, feet
- M Average molecular weight
- N_{Kn} Knudsen number, dimensionless; $N_{Kn} = \lambda_m/D$ for a circular duct
- N_{Ma} Mach number, dimensionless; $N_{Ma} = V/V_a$
- N_{Nu} Nusselt number, dimensionless; $N_{Nu} = hD/k$
- N_{Pr} Prandtl number, dimensionless; $N_{Pr} = c_p\mu/k$; $N_{Pr,0}$ evaluated at the stagnation temperature, $N_{Pr,m}$ evaluated at the moving-stream temperature
- N_{Re} Reynolds number, dimensionless; $N_{Re} = DV\rho/\mu$; $N_{Re,x}$ based on the characteristic length x; $N_{Re,0}$ evaluated at the stagnation temperature
- N_{rf} Recovery factor, dimensionless
- N_{St} Stanton number, dimensionless; $N_{St} = h/c_p G$
- N_v Molecular speed ratio, dimensionless; $N_v = V/\bar{u}$
- p Absolute pressure, pounds force per square foot
- q Heat transfer rate, Btu per hour
- R_G Gas constant in mechanical units equals pv/T, equals $1544/M$ (ft)(pounds force)/(pounds mass)(deg F abs)
- R'_G Gas constant in heat units equals $1.987/M$ (Btu)(pounds mass)/(deg F abs)
- T Absolute temperature, degrees Fahrenheit absolute; T_{aw} for adiabatic wall, T_m for moving stream, T_s for stagnation, T_w for wall temperature
- t Temperature, degrees Fahrenheit; t_{aw} for adiabatic wall, t_m for moving stream, t_s for stagnation, t_w for wall temperature
- u_m Mean molecular speed, feet per second
- $\bar{u}$ Most probable molecular speed, feet per second
- v Specific volume, cubic feet per pound of fluid, equals $1/\rho$
- w Mass rate of flow, pounds of fluid per second
- V Velocity, feet per second; V_a for acoustic velocity, equals $\sqrt{\gamma g_c R_G T}$ for an ideal gas; V_∞ for velocity of free stream
- x Distance from leading edge or entrance, feet
- z Ratio of absolute stagnation and wall temperatures, T_s/T_w

Greek

- α Accommodation coefficient defined by Eq. (12-18)
- β Fraction of molecules striking surface which are diffusely reflected, dimensionless

HIGH-VELOCITY FLOW

TABLE 12-1. NOMENCLATURE.—(*Continued*)

γ Ratio of specific heats, $\gamma = c_p/c_v$
Δt Temperature difference, degrees Fahrenheit; Δt_e, effective temperature difference
λ_m Mean free path of the molecules, feet
μ Viscosity, lb mass/(ft)(sec); μ_0 evaluated at the stagnation temperature
ρ Density, pounds fluid per cubic foot; ρ_0 evaluated at the stagnation temperature
Φ Correction factor to Poiseuille law defined by Eq. (12-26)
ψ Relative pressure, dimensionless, defined by Eq. (12-11b)

by stagnation and in addition by transfer of momentum toward the wall resulting from the velocity gradient. In an adiabatic system no heat is transferred through the body itself; however, the rise in temperature of the wall above the moving-stream temperature causes conduction of heat back through the gas layers near the wall into the bulk stream. Consequently the adiabatic wall assumes a temperature t_{aw}, higher than the moving stream by an amount depending primarily on the ratio of the molecular diffusivity of momentum to the molecular diffusivity of heat:

$$\frac{\mu/\rho}{k/\rho c_p} = \frac{c_p \mu}{k} \equiv N_{Pr}$$

For convenience, a temperature recovery factor N_{rf} is defined:

$$N_{rf} \equiv \frac{t_{aw} - t_m}{t_s - t_m} = \frac{t_{aw} - t_m}{V^2/2g_c J c_p} \qquad (12\text{-}2)$$

The ratio V/V_a is termed the *Mach number* N_{Ma}, and for an ideal gas the acoustic velocity V_a equals $\sqrt{g_c \gamma R_G T_m}$. Equation (12-1) can thus be rewritten for an ideal gas:

$$T_s = \left(1 + \frac{\gamma - 1}{2} N_{Ma}^2\right) T_m \qquad (12\text{-}1a)$$

Substitution for $t_s - t_m$ in Eq. (12-2) yields

$$N_{rf} = \frac{T_{aw} - T_m}{N_{Ma}^2 [(\gamma - 1)/2] T_m} \qquad (12\text{-}2a)$$

Flat Plate. Johnson and Rubesin[13,27] review both theory and data for recovery factors. For the flat plate with a *laminar* boundary layer (see Chap. 9), the recovery factor is independent of Mach number from 0 to 10 and of the Reynolds number; for Prandtl number from 0.7 to 1.2:

$$N_{rf} = \sqrt{N_{Pr}} \qquad (12\text{-}3) \; \star$$

For a turbulent boundary layer, reference 13 predicts

$$N_{rf} = N_{Pr}^{1/3} \qquad (12\text{-}4) \; \star$$

As shown in Fig. 12-1, the data of Eckert and Weise[7] for flow *parallel* to a cylindrical probe confirm Eq. (12-3) for a laminar boundary layer; at $xV_\infty\rho_\infty/\mu$ of 5×10^5, the transition to a turbulent boundary layer causes a rise in the recovery factor, as predicted by Eq. (12-4).

For air (N_{Pr} of 0.72) Seban[30] predicted N_{rf} for a turbulent boundary layer:

$$N_{rf} = 1 - 0.869(N_{Re,x})^{-0.2} \qquad (12\text{-}5)$$

Slack[33] reports local recovery factors for air at a Mach number of 2.4 flowing along a flat plate. Below $xV_\infty\rho_0/\mu_0$ of 80,000, the recovery factors agreed with those predicted for a laminar boundary layer ($N_{rf} = \sqrt{N_{Pr}}$); at $xV_\infty\rho_0/\mu_0$ of 300,000, the recovery factor agreed with that predicted for a turbulent boundary layer ($N_{rf} = N_{Pr}^{1/3}$); the recovery factor went through a maximum of 0.975 in the intermediate (transition) range.

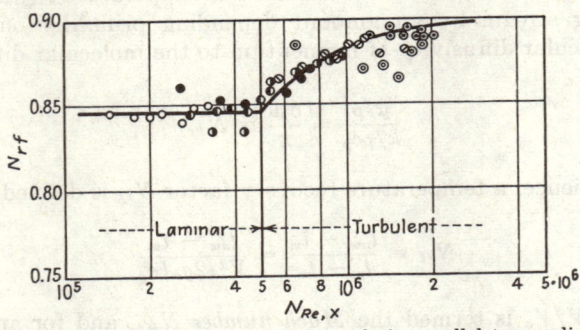

FIG. 12-1. Data of Eckert and Weise[7] for flow of air parallel to a cylindrical probe. (*Courtesy of E. R. G. Eckert.*)

Higgins and Pappas[10] studied the transition from a laminar to a turbulent boundary layer on a flat plate in an air stream with a Mach number of 2.4. The length Reynolds numbers corresponding to the start and end of the transition were found to decrease with increase in the ratio T_w/T_m of the absolute temperatures of the wall and the free stream.

Inside Tubes. Reference 22 reports recovery factors for subsonic flow of air near the outlet end of a long smooth tube; the results were in good agreement with those predicted from Eq. (12-4). Kaye *et al.*[14,15,16] report local recovery factors for supersonic flow of air in a short smooth tube.

Other Shapes. Hottel and Kalitinsky[11] give recovery factors for subsonic flow of air at 140 to 315 ft/sec past various probes as follows: axial flow past butt-welded junction (0.01 inch diameter), N_{rf} of 0.86; axial flow past ball junction (0.07 inch diameter), N_{rf} of 0.79; flow normal to butt-welded junction, N_{rf} of 0.66; flow normal to ball junction,

N_{rf} of 0.72. With a special stagnation probe (Fig. 12-2) N_{rf} was 0.93 at 250 ft/sec and 0.98 at 1000 ft/sec.

Stine and Scherrer[37] report local recovery factors for air in a turbulent boundary layer on a cone having a total angle of 10 deg, pointed upstream; the values of N_{rf} fell between those predicted for a laminar boundary layer ($N_{rf} = N_{Pr}^{1/2}$) and a turbulent boundary layer ($N_{rf} = N_{Pr}^{1/3}$). The transition from laminar to turbulent boundary layer occurred at a lower value of the local length Reynolds number as the Mach number

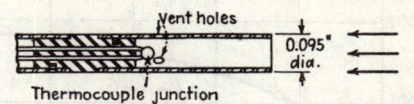

FIG. 12-2. Probe for measurement of stagnation temperature of high-velocity gas stream. (*Hottel and Kalitinsky.*[11])

increased from 1.97 to 3.77. Scherrer[29] reports similar findings for a 20-deg cone; the effects of heating and cooling on the transition from a laminar to a turbulent boundary layer are also reported.

HEAT-TRANSFER COEFFICIENTS

Effective Δt. In an adiabatic tube the effect of the stagnation phenomenon is to make the conventional Δt (wall temperature t_w less bulk temperature t_m of the moving stream) finite, although the heat-transfer rate q through the wall is zero; under these conditions, the conventional equation ($q = h_m A \, \Delta t$) has no utility. This dilemma is solved by defining a new potential, the *effective* temperature difference, $\Delta t_e = t_w - t_{aw}$, which is zero when the heat transfer through the wall is zero. If the wall temperature t_w is raised above t_{aw}, heat is transferred according to the equation

$$dq/dA = h_e(t_w - t_{aw}) \qquad (12\text{-}6)\;\star$$

Figure 12-3 summarizes results of a number of subsonic heat-transfer experiments.[22] The effective heat-transfer coefficient h_e (expressed in terms of the Stanton number, $h_e/c_p G$) is seen to be independent of temperature potential, while that based on the conventional Δt is not. When a coefficient is based on the excess of the wall temperature over the stagnation temperature,

$$dq = h_s \, dA(t_w - t_s) \qquad (12\text{-}7)$$

the corresponding heat-transfer coefficient h_s is more nearly independent of temperature potential than the coefficient h_m. At high temperature potentials h_s/h_e approaches unity, and consequently Eq. (12-7) may be used.

Figure 12-4 shows results[22] of a heat-transfer experiment with a mass velocity of 43,500 lb of air/(hr)(sq ft) in a smooth tube. Air entered

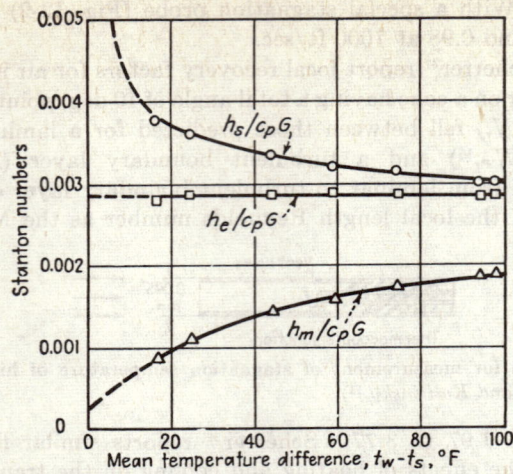

FIG. 12-3. Comparison of Stanton moduli defined in terms of different temperature potentials:[22] $h_s = dq/(t_w - t_s) \, dA$; $h_e = dq/(t_w - t_{aw}) \, dA$; $h_m = dq/(t_w - t_m) \, aA$.

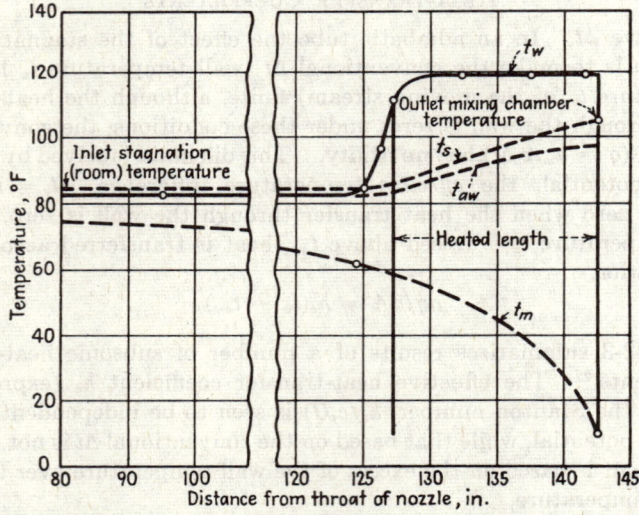

FIG. 12-4. Results of heating air at high velocity in a tube, with sonic velocity at the outlet.[22] Despite heat addition sufficient to increase the stagnation temperature 20°F, the temperature of the moving stream fell some 50°F.

at room temperature, and the wall temperature in the test section was maintained at 120°F by means of condensing steam. Although enough heat was transferred through the wall to increase the stagnation temperature by some 20°F, the temperature of the moving stream of air fell roughly 50°F (from 60°F to 10°F), in accordance with the total energy balance [Eq. (6-4)]. The Mach number increased from 0.46 at the beginning of the heated section to 1.0 at the outlet.

Flat Plates. Brown and Donoughe[4] mathematically treated laminar compressible flow of air along a flat plate, allowing for variation in physical properties with temperature. For ratios of the absolute temperatures of the stream to those of the plate ranging from $\frac{1}{4}$ to 4, the predicted local coefficients of heat transfer were in close agreement with the theoretical equation of Pohlhausen (based on initial conditions, Chap. 9) for invariant properties and incompressible flow. The effect of passing air inward or outward through a porous wall was also treated.

Wedges. Garbett[9] found that the measured average heat-transfer coefficients for 20-deg wedges to air agreed closely with the Pohlhausen equation (Chap. 9) up to length Reynolds numbers of 2×10^5; Mach numbers ranged from 0.01 to 1.6.

Cones. Fischer and Norris[8] report local coefficients of heat transfer at various points on the conical nose of a V-2 rocket in supersonic flight. The local values of h were plotted as ordinates in terms of $j = N_{St}N_{Pr}^{2/3}$ vs. the length Reynolds number, $xV_\infty\rho_\infty/\mu$, which ranged from 5×10^4 to 8×10^6. The data were bracketed by the equation for a laminar boundary layer on a cone, $j = 0.58/N_{Re,x}^{0.5}$, and that for a turbulent boundary layer, $j = 0.028/N_{Re,x}^{0.2}$.

Scherrer[29] finds that the heat-transfer relation of Johnson and Rubesin[13] for a laminar boundary layer and uniform surface temperature is supported by data for 20-deg cones in a supersonic stream of air (N_{Ma} of 1.5).

Flow Normal to Cylinders. Kovasznay[20] determined heat transfer from small wires (0.00015 and 0.00030 inch) to air flowing at Mach numbers from 1.15 to 2.03. For small Δt the results were correlated by

$$\frac{h_e D_o}{k_0} = -0.795 + 0.58 \sqrt{\frac{D_o V_\infty \rho_\infty}{\mu_0}} \qquad (12\text{-}8)$$

At high Δt, the results fell below this equation.

For flow of air at Mach numbers from 0.12 to 0.94 normal to a cylinder having a diameter of $\frac{3}{8}$ inch, Garbett[9] finds

$$\frac{h_e D_o}{k} = 0.664 \left(\frac{D_o G}{\mu}\right)^{1/2} \left(\frac{c_p \mu}{k}\right)^{1/3} \qquad (12\text{-}9)$$

which is lower than the conventional curve for incompressible flow (Fig. 10-7); the stream was only one-third wider than the model; $D_o G/\mu$ ranged from 20,000 to 165,000.

Scadron and Warshawsky[28] measured the time constants for four pairs of common thermocouple materials mounted normal to a stream of air flowing at high velocity. Errors due to radiation, and conduction in the leads, were considered, and calculations were facilitated by graphs and nomograms. The Mach number ranged from 0.1 to 0.9 and the Reynolds number $D_0\rho_0 V/\mu_0$ from 250 to 30,000. The Nusselt number $h_e D_0/k_0$ was correlated in terms of $N_{Re,0}$ and $N_{Pr,m}$ (based on the temperature of the moving stream) within ± 7.4 per cent by

$$\frac{h_e D_o}{k_0} = 0.478 \left(\frac{D_o V \rho_0}{\mu_0}\right)^{0.5} \left(\frac{c_p \mu}{k}\right)_m^{0.3} \quad (12\text{-}10)$$

This correlation falls within the spread of data for incompressible flow (Fig. 10-7).

COMPRESSIBLE-FLOW CALCULATIONS

Heat-transfer and pressure-drop calculations in the range of compressible flow are distinguished by two factors which are unimportant at lower velocities. First, the use of average values of the flow variables may no longer be permissible when the density changes significantly; hence integration is often required in the direction of flow. Second, in accordance with the entropy considerations of the second law of thermodynamics, the maximum mass flow rate may be limited or "choked" under certain conditions when the linear velocity of the fluid approaches the acoustic velocity. For example, with a given inlet pressure, the discharge rate of a duct of uniform cross section reaches a maximum when the exit velocity is equal to the sonic velocity; further reduction in downstream pressure does not result in increased flow. A generalized review of the thermodynamics of one-dimensional compressible flow is given by Shapiro and Hawthorne,[31] who develop simplified equations for several important cases. The thermodynamics of flow in ducts, with attention to the maximum velocity at the outlet, is discussed also by Keenan[17] and by Hunsaker and Rightmire.[12]

Under certain restrictions and with the aid of simplifying assumptions, tedious stepwise numerical calculations for the temperature and pressure of a gas undergoing compressible flow can be eliminated by use of special integrated equations or generalized design charts. As an example, consider the steady flow of a perfect gas through a horizontal tube of uniform diameter and wall temperature. By employing the Reynolds analogy (Chap. 9), Nielsen[25] related the ratio of stagnation and wall temperatures, $z = T_s/T_w$, to the Mach number for heated and cooled tubes with uniform wall temperatures. The basic relations are:

1. The relations for perfect gases:
$$pv = R_G T$$
$$c_p - c_v = R_G/J$$
$$V_a = \sqrt{g_c \gamma R_G T_m}$$

2. The definitions of the Mach number and the stagnation temperature:
$$N_{Ma} = V/V_a$$
$$T_s - T_m = V^2/2g_c J c_p$$

HIGH-VELOCITY FLOW

3. The equation of continuity for constant cross section:

$$V = Gv$$
$$dV = G\,dv$$

4. The total energy balance:

$$dq = wc_p\,dt_s$$

5. The rate equation:

$$dq = h_s\,dA(t_w - t_s)$$

6. The Reynolds analogy based on h_s:

$$h_s/c_p G = f/2$$

7. The force balance for compressible flow:

$$-v\,dp = \frac{V\,dV}{g_c} + \frac{4fV^2\,dL}{2g_c D}$$

Relations 1, 2, and 3 give

$$\frac{G^2 R_G T_s}{g_c p^2} = \gamma N_{Ma}^2\left(1 + \frac{\gamma-1}{2}N_{Ma}^2\right) \tag{12-11}$$

Since G is constant, this gives

$$\frac{T_{s1}}{T_{s2}}\left(\frac{p_2}{p_1}\right)^2 = \left(\frac{N_{Ma,1}}{N_{Ma,2}}\right)^2 \frac{1 + \dfrac{\gamma-1}{2}N_{Ma,1}^2}{1 + \dfrac{\gamma-1}{2}N_{Ma,2}^2} \tag{12-11a}$$

which may be rearranged as follows:

$$\frac{p_2}{p_1} = \frac{\left[\dfrac{z_2}{N_{Ma,2}^2\left(1 + \dfrac{\gamma-1}{2}N_{Ma,2}^2\right)}\right]^{1/2}}{\left[\dfrac{z_1}{N_{Ma,1}^2\left(1 + \dfrac{\gamma-1}{2}N_{Ma,1}^2\right)}\right]^{1/2}} \equiv \frac{\Psi_2}{\Psi_1} \tag{12-11b}$$

Upon assuming f and T_w constant, relations 4, 5, and 6 give

$$\ln\frac{1-z_1}{1-z_2} = \frac{2f_m L}{D} \tag{12-12}$$

Introducing relation 7, the various relations combine to give

$$\frac{dz}{z(z-1)} = \frac{(N_{Ma}^2 - 1)\,dN_{Ma}^2}{N_{Ma}\left(1 + \dfrac{\gamma-1}{2}N_{Ma}^2\right)[\gamma N_{Ma}^2 + 1 + z(\gamma N_{Ma}^2 - 1)]} \tag{12-13}$$

318   HEAT TRANSMISSION

Since the variables are not separated, Nielsen[25],† integrated Eq. (12-13) by numerical methods, obtaining the path curves of Fig. 12-5 for subsonic heating of air ($\gamma = 1.4$). As shown by Eq. (12-11b), Ψ is a function only of z and N_{Ma}; hence lines of constant Ψ are plotted in Fig. 12-5.

For a run with acoustic velocity at the outlet, Fig. 12-5, shows that dz/dN_{Ma} is zero at N_{Ma} of 1. For example, consider the path curve starting at any of the following points: $N_{Ma,1}$ of 0.25, z_1 of 0.15; $N_{Ma,1}$ of 0.35, z_1 of 0.26; $N_{Ma,1}$ of 0.45, z_1 of 0.35.

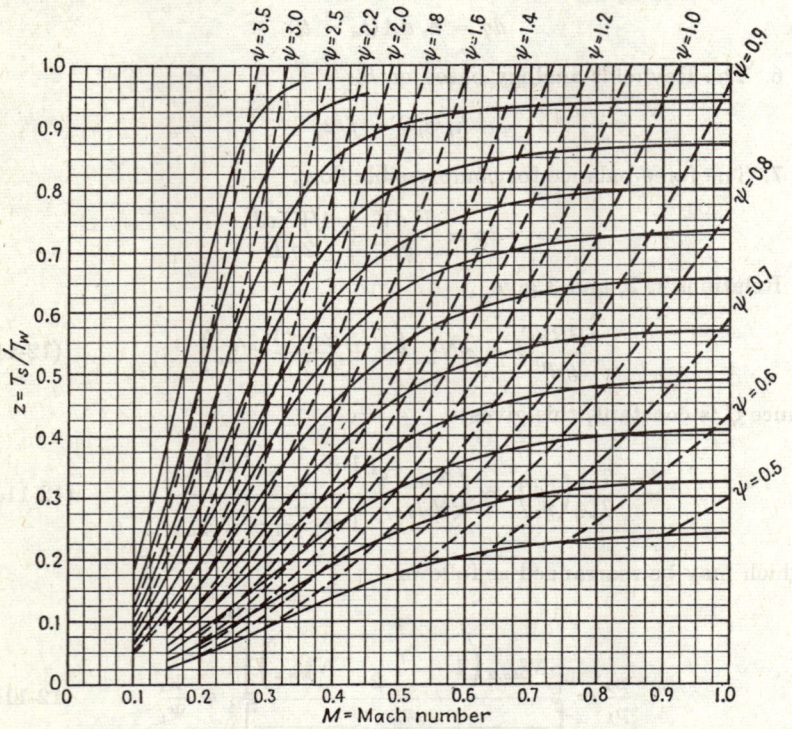

FIG. 12-5. Nielsen plot for paths of subsonic heating of air in a tube with uniform wall temperature.[25]

In all cases, the maximum z_2 is 0.485 at $N_{Ma,2}$ of 1.0. Equation (12-12), based only on heat-transfer considerations, can predict values of z greater than are attainable, since other relations would thus be ignored. From $N_{Ma,1}$ of 0.45 and z_1 of 0.35 to $N_{Ma,2}$ of 1 and z_2 of 0.485 the lines of constant Ψ give $\Psi_2/\Psi_1 = 0.64/1.30 = 0.492 = p_2/p_1$, permitting evaluation of p_2 for any value of p_1. Stepwise calculations can be made for nonuniform wall temperature using Fig. 12-5 if the tube temperature pattern is approximated by a series of uniform temperature sections.

Valerino and Doyle[40] present charts for estimating the pressure drop for subsonic flow of *monatomic* gases through heated smooth tubes

† Nielsen also gives curves for subsonic cooling.

having uniform wall temperatures; other modes of heat addition are considered. Sibulkin and Koffel[32] prepared a chart for subsonic flow of a compressible fluid through a *cooled* tube having an exponential longitudinal temperature distribution.

Valerino[39] prepared generalized charts for determination of pressure drop and heat transfer of air ($\gamma = 1.40$) flowing in tubes of constant wall temperature, for various values of L/D_e. The utility of the Mach number in developing generalized differential equations for subsonic and supersonic gas dynamics is demonstrated by Shapiro and Hawthorne.[31]

RAREFIED GASES

In the conventional treatment of the high-speed flow of gases at ordinary pressures, the flow phenomena are satisfactorily described in terms of Reynolds numbers and Mach numbers. However at very small absolute pressures the mean free path λ_m of the gas molecules may become large compared with a characteristic dimension D of the body; in this case an additional dimensionless ratio, the Knudsen number ($N_{Kn} = \lambda_m/D$), or its equivalent in terms of other ratios, must be included in the analysis.

According to the kinetic theory of gases,[18] the viscosity μ is given very closely in terms of the mean molecular velocity u_m and the mean free path λ_m by the relation

$$\mu = 0.5 u_m \lambda_m \rho \qquad (12\text{-}14)$$

where

$$u_m = \sqrt{\frac{8}{\pi} g_c R_G T} \qquad (12\text{-}15)$$

Since the acoustic velocity equals $\sqrt{\gamma g_c R_G T}$, the Knudsen number may be expressed in terms of the Reynolds and Mach moduli:

$$N_{Kn} \equiv \frac{\lambda_m}{D} = \sqrt{\frac{\pi \gamma}{2}} \cdot \frac{N_{Ma}}{N_{Re}} \qquad (12\text{-}16)$$

Consequently the ratio N_{Ma}/N_{Re} may be used instead of the Knudsen number. Another useful dimensionless group is the molecular speed ratio N_v, defined as the ratio of the stream velocity to the most probable molecular speed $\bar{u}$. Since $\bar{u} = \sqrt{2 g_c R_G T}$,

$$N_v = \frac{V}{\bar{u}} = \sqrt{\frac{\gamma}{2}} N_{Ma} \qquad (12\text{-}17)$$

Three regimes of flow of gases may be distinguished: (1) continuum flow, in which the gas behaves as a continuous medium, (2) slip flow, in which the mean free path becomes significant and the gas velocity at the wall is no longer zero, and (3) free-molecule flow, in which the mean free path is so great that intermolecular collisions are comparatively rare and are therefore less important than collisions with the wall. The criteria bounding these regimes are not yet clearly defined but are indicated in Table 12-2.

TABLE 12-2. FLOW REGIMES FOR GASES

	Reference a	Reference b
Continuum flow (Chap. 6)	$N_{Kn} < 0.001$	$\dfrac{N_{Ma}}{N_{Re}} < 0.01$
Slip flow	$0.001 < N_{Kn} < 2$	$0.01 < \dfrac{N_{Ma}}{N_{Re}} < 10$
Free-molecule flow	$N_{Kn} > 2$	$N_{Kn} > 10$

Reference a: Stalder et al.[36]
Reference b: Tsien.[38]

Accommodation Coefficients. Consider the transfer of heat from a solid surface to a rarefied gas in which the mean free path is large. Molecules at temperature t_i striking the wall at temperature t_w are found to leave the surface at some intermediate temperature t_r. This effect was recognized quantitatively by Knudsen,[19] who defined the accommodation coefficient α as the ratio of the actual energy interchange to the maximum possible energy interchange. Theoretically for nonatomic perfect gases, but with good approximation in most practical cases,

$$\alpha = \frac{t_r - t_i}{t_w - t_i} \qquad (12\text{-}18)$$

With diffuse surfaces on which multiple collisions occur at each incidence, α approaches unity. For any surface, the effects of the accomnodation coefficient and the resulting temperature discontinuity are negligible in the continuum regime because the mean free path is small.

Dushman[6] summarizes accommodation coefficients from a number of observers; values for several gases are presented in Table 12-3. Wied-

TABLE 12-3. VALUES OF ACCOMMODATION COEFFICIENTS FOR SEVERAL GASES[6]

Surface	H_2	O_2	CO_2	N_2
Tungsten	0.20	....	....	0.57
Ordinary Pt	0.36	....	....	0.89
Polished Pt	0.36	0.84	0.87	
Pt slightly coated with black	0.56	0.93	0.94	
Pt heavily coated with black	0.71	0.96	0.97	

mann and Trumpler[41] determined values of α for air on flat black lacquer and on bronze, cast iron, and aluminum surfaces with various finishes. All measured values were between 0.87 and 0.97. At extremely low pressures, Amdur[2] finds the accommodation coefficient to be a strong function of the amount of adsorbed gas and proposes a correlation in terms of the Langmuir isotherm

$$\alpha = \alpha_0 + \frac{b(p)}{1 + b(p)} (\alpha_\infty - \alpha_0) \qquad (12\text{-}19)$$

where p is the gas pressure in millimeters of mercury, α_0 the accommodation coefficient at zero pressure, α_∞ the accommodation coefficient at infinite pressure, and b a constant having units of reciprocal millimeters. For various gases and metal surfaces values of α_0 averaged about 0.05 and α_∞ ranged from 0.4 to 0.88 and b from 126 to 702.

Heat Transfer to Cylinders. Stalder et al.[34] determined the equilibrium temperature T_{aw} for a small cylinder (0.0031 inch diameter) placed transversely in a high-velocity stream of nitrogen or helium. The molecular speed ratio N_v ranged from 0.5 to 2.3, Knudsen numbers from 2 to 92, Mach numbers from 0.55 to 2.75, and Reynolds numbers from 0.005 to 0.9. The ratios of T_{aw}/T_m ranged from 1.2 to 3.6 and agreed fairly well with values predicted theoretically for free-molecule flow, in which the recovery factor may exceed unity even when N_{Pr} is less than 1. Drag coefficients were measured and found to be a function only of the molecular speed ratio N_v, as predicted theoretically.

In the range of slip flow, Stalder et al.[35] determined heat-transfer coefficients for rarefied air flowing normal to single cylinders having diameters from 0.001 to 0.126 inch. Knudsen numbers ranged from 0.025 to 11.8, Mach numbers from 2.0 to 3.3, and Reynolds numbers from 0.15 to 70. The data were correlated by the equation

$$\frac{h_c D}{k_0} = 0.132 \left(\frac{D_0 V_\infty \rho_\infty}{\mu_0} \right)^{0.73} \qquad (12\text{-}20) \star$$

where subscript 0 refers to stagnation temperature and subscript ∞ refers to the free stream. At the lowest Reynolds number the Nusselt number was only 0.06 of that for continuum flow at the same value of N_{Re}; at the highest Reynolds number this ratio was 0.64. The recovery factor exceeded 1.0 when the Knudsen number exceeded 0.2.

Madden and Piret[23] studied natural convection from heated wires of 0.00276 and 0.00988 inch diameter in air and helium at pressures from 760 to 0.05 mm Hg. The temperature discontinuity which occurs at the gas-solid interface in the noncontinuum regime is treated theoretically by defining a fictitious Nusselt number $N_{Nu}{}^\star$

$$\frac{2}{N_{Nu}} = \frac{2}{N_{Nu}{}^\star} + \frac{8\gamma \lambda_m}{\alpha N_{Pr}(\gamma + 1)D} - \ln\left(1 + 2\frac{\lambda_m}{D}\right) \qquad (12\text{-}21) \star$$

The observed Nusselt number N_{Nu} becomes equal to $N_{Nu}{}^\star$ when free-molecule effects are absent. As expected, the minimum asymptotic values of $N_{Nu}{}^\star$ were correlated by the conduction equation for hollow cylinders,

$$\frac{2}{N_{Nu}{}^\star} = \ln \frac{2r}{D} \qquad (12\text{-}22)$$

in which r is the distance from the center of the wire to the point at which the gas temperature is measured. In the presence of natural convection, values of $N_{Nu}{}^\star$ computed from Eq. (12-21) were correlated by

$$\frac{2}{Nu^\star} = \ln\left[1 + \frac{6.82}{(N_{Gr}N_{Pr})^{1/3}}\right] \qquad (12\text{-}23)$$

for $N_{Gr}N_{Pr}$ from 10^{-8} to 10^{-1}.

Parallel Flat Plates. At low pressures, when convection is absent, the rate of heat transfer by conduction between two broad, parallel flat plates is given theoretically by

$$q = \frac{kA\,\Delta t}{x\left[1 + \frac{4}{N_{Pr}}\frac{\lambda_m}{x}\frac{2-\alpha}{\alpha}\frac{\gamma}{\gamma+1}\right]} \qquad (12\text{-}24)$$

in which x is the distance between the plates.[6,18]

Heat Transfer to Other Shapes. Analytical approximations for the Nusselt number with slip flow over flat plates are given by Drake and Kane[5] and by Stalder et al.[36] Slip flow over spheres is treated in reference 5, which correlates Nusselt numbers and recovery factors for data on air in terms of $\sqrt{N_{Re}}/N_{Ma}$.

In the region of free-molecule flow, Oppenheim[26] presented a generalized technique for computing the heat-transfer characteristics of flat plates, cylinders, spheres, and cones; complex shapes were treated by subdivision into fundamental shapes. At low Mach numbers the limiting value of the Stanton number is given by

$$N_{St} = \left[\frac{\gamma-1}{\gamma}\right]\frac{\alpha}{4\sqrt{\pi}\,N_v} \qquad (12\text{-}24a)$$

At higher Mach numbers, Oppenheim finds a greater effect of shape but, for $N_{Ma} > 2$, predicts a minimum Stanton number of $\alpha(\gamma+1)/2\gamma$. Recovery factors are discussed.

Pressure Drop inside Tubes. The flow of gases at low pressures is normally in the range of Reynolds numbers which characterize the streamline-flow regime. However, with Knudsen numbers of approximately 0.001 or higher, pressure drop in tubes is less than that predicted by the Hagen-Poiseuille law,[4]

$$-\frac{dp}{dL} = \frac{32\mu V}{g_c D^2} \qquad (12\text{-}25)$$

which assumes the velocity of the fluid to be zero at the tube wall (Chap. 6). Since at low pressures this assumption is not valid, Kundt and

RAREFIED GASES

Warburg[21] applied a correction factor Φ to Eq. (12-25) to allow for slip flow and free-molecule flow:

$$-\frac{dp}{dL} = \frac{32\mu V}{g_c D^2}\left[\frac{1}{\Phi}\right] \qquad (12\text{-}26)\,\star$$

For a given pressure drop, the resulting flow rate is Φ times the value predicted by Eq. (12-25); flow rates up to 40,000 times greater than predicted by the Hagen-Poiseuille law have been reported for free-molecule flow.

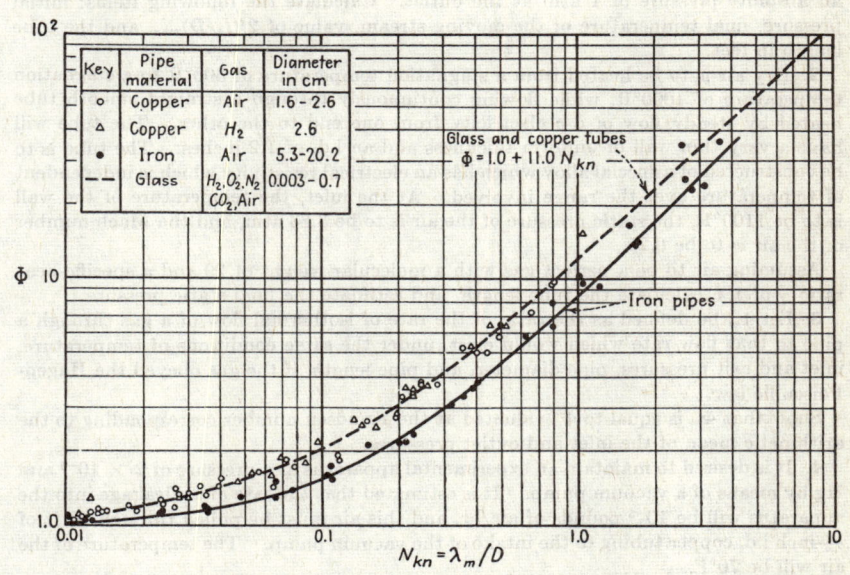

Fig. 12-6. Correction factor to the Hagen-Poiseuille law for slip flow and free-molecule flow; data of Brown et al.[3]

Maxwell[24] derived a theoretical relationship for Φ in terms of the Knudsen number,

$$\Phi = 1 + 8\left(\frac{2}{\beta} - 1\right)N_{Kn} \qquad (12\text{-}27)$$

in which β is the fraction of molecules striking the wall which are diffusely reflected (and $1 - \beta$ the fraction specularly reflected).

Data of five observers for flow of various gases and air in glass capillaries ranging from 0.003 to 0.7 cm in diameter as reported by reference 3 are correlated by Eq. (12-27) with β of 0.84 for N_{Kn} from 0.01 to 10, yielding

$$\Phi = 1.0 + 11 N_{Kn} \qquad (12\text{-}28)$$

For N_{Kn} from 10 to 3400 better correlation was obtained with $\beta = 0.77$.

Data of Brown et al.[3] for air and hydrogen in copper tubes are also correlated by Eq. (12-28) as shown in Fig. 12-6. With air in iron pipes, lower values of Φ were obtained. Slip flow in annuli is discussed by Alancraig and Bromley.[1]

PROBLEMS

1. It is desired to heat air from a stagnation temperature of 400°R to a stagnation temperature of 2600°R while flowing at the maximum possible mass velocity through a single smooth horizontal tube having an actual i.d. of 0.10 ft, with the temperature of the inner wall maintained at a uniform temperature of 4000°R. The air is to have an absolute pressure of 1 atm at the outlet. Calculate the following items: initial pressure, final temperature of the moving stream, value of $2f(L/D)_{max}$, and the tube length in feet.

2. Dry air is to be heated from a stagnation temperature of 500°R to a stagnation temperature of 1000°R, while flowing continuously through a straight smooth tube heated by steady flow of d-c electricity from one end to the other. The tube will have a very thin wall of uniform thickness and an i.d. of 1.2 inches. The tube is to be constructed of a special alloy which has an electrical resistivity which is independent of temperature over the range involved. At the inlet, the temperature of the wall is to be 1100°R, the static pressure of the air is to be 3.83 atm, and the Mach number of the air is to be 0.25.

Assuming air to be a perfect gas with a molecular weight of 29 and a specific-heat ratio γ of 1.4, calculate the tube length, and estimate the final static pressure.

3. Let Φ_m be defined as the ratio of the rate of isothermal flow of a gas through a pipe to that flow rate which would exist, under the same conditions of temperature, inlet and exit pressures, pipe diameter, and pipe length, if the gas obeyed the Hagen-Poiseuille law.

Show that Φ_m is equal to Φ evaluated at the Knudsen number corresponding to the arithmetic mean of the inlet and outlet pressures.

4. It is desired to maintain an experimental apparatus at a pressure of 2×10^{-2} mm Hg by means of a vacuum pump. It is estimated that the rate of air leakage into the apparatus will be 10^{-5} pounds of air/hr, and this air must be pulled through 10 ft of ½-inch i.d. copper tubing to the intake of the vacuum pump. The temperature of the air will be 70°F.

Calculate the required intake pressure of the vacuum pump.

CHAPTER 13

CONDENSING VAPORS

Abstract. For film-type condensation of pure saturated vapor on *vertical* tubes the data for $4\Gamma/\mu_f$ below 1800 average 20 per cent above the Nusselt theory for negligible vapor velocity (Fig. 13-7); for $4\Gamma/\mu_f$ above 1800 the data are correlated by an empirical equation; the recommended curves for both cases are shown in Fig. 13-6. The effect of noncondensable gas is shown by Eqs. (13-6) and (13-6a). The effect of vapor velocity is given by Eq. (13-8). Subcooling of condensate is treated on page 338.

With film-type condensation on *horizontal* tubes the Nusselt theoretical equation for negligible vapor velocity is recommended [Eqs. (13-12) and (13-13)]. Data are summarized in Table 13-4. The effect of air is shown in Fig. 13-9.

Over-all coefficients for condensers are given on page 343. The graphical method of Wilson is presented on page 343.

Values of h_m for dropwise condensation of steam are much higher than for film-type condensation (pages 347 to 351).

With pure superheated vapor, if the temperature of the tube wall is below the saturation temperature, condensation occurs and the flux is controlled by the potential $t_{sv} - t_s$, as with saturated vapor. If t_s exceeds t_{sv}, the gas is merely desuperheated. Condensation of mixtures of vapors is treated on pages 351 to 354.

Dehumidification of vapor-gas mixtures by cooled tubes is outlined on page 355. Direct contact of water with moist air, as in humidifiers, dehumidifiers, and water coolers, is treated on pages 356 to 365. Calculations are facilitated by combining the humidity and temperature potentials in an enthalpy potential which controls the rate of enthalpy transfer.

I. CONDENSATION OF PURE SATURATED VAPOR

Mechanism: Dropwise vs. Film-type Condensation. In a study[93] of the condensation of steam on a *vertical* water-cooled wall, photographs were taken of the condensate side of the plate. On a rusty or etched plate, the steam condensed in a continuous *film* over the entire wall. With a polished surface of chromium-plated copper, the condensate was formed in *drops*, which rapidly grew in size (up to 3 mm in diameter)

TABLE 13-1. NOMENCLATURE

A Area of heat-transfer surface, square feet; A_i for inside, A_L for coolant side, A_o for outside, A_v for vapor side, A_w for tube wall (mean of A_L and A_v)

A_p Surface area of a particle, square feet

a Total transfer surface divided by total volume of tower, $a = A/S_o z$, square feet per cubic foot; a_H for heat transfer and a_M for mass transfer

b Breadth of condensing surface, feet; for a vertical tube, b equals πD; for a horizontal tube, b equals L_H

C_0 Dimensional constant in Eq. (13-17), equal to h_L for a water velocity of 1 ft/sec but based on outside surface

C_1 Dimensionless constant in Eq. (13-15)

C_2 $\lambda_0 - 0.45 t_0$; with t_0 of 32°F, $C_2 = 1061.4$ Btu/lb of steam; with t_0 of 0°F, $C_2 = 1094$ Btu/(lb of steam)

c Specific heat, Btu/(lb fluid)(deg F); c_L for liquid coolant, c_p for constant pressure, c_{pv} for vapor

c_s Humid heat, Btu/(lb dry gas)(deg F); for air at ordinary temperatures, c_s equals $0.24 + 0.45H$

D Diameter of tube, feet; D_o for o.d., D_i for i.d., D_i' for inches

D_v Diffusivity of vapor (volumetric units), square feet per hour (see $\mu_v \rho_v / D_v$)

d Prefix, indicating derivative, dimensionless

e Base of Napierian or natural logarithms

F_c Dimensionless factor to allow for variation in tube temperature; see Table 13-2

F_V Dimensionless factor to allow for finite velocity of vapor; see Table 13-4

G_f Mass velocity of film of condensate, lb/(hr)(sq ft of cross section of film)

G_G Superficial mass velocity of dry gas; equals w_G/S_o, lb/(hr)(sq ft of total cross section)

G_L Superficial mass velocity of coolant, equals W/S_o, lb/(hr)(sq ft of gross cross section)

g Acceleration due to gravity, ordinarily taken as 4.17×10^8 ft/(hr)(hr), or 32.2 ft/(sec)(sec)

g_c Conversion factor, 4.17×10^8 (lb of fluid)(ft)/(hr)(hr)(pounds force)

H "Absolute" humidity, pounds of water vapor per pound of dry gas; H_i is saturation value at t_i; H_G is H of bulk of vapor-gas stream, H_{G1} at gas inlet, and H_{G2} at gas outlet; H_L is saturation value of H at t_L

h Local coefficient of heat transfer, Btu/(hr)(sq ft)(deg F)

h_d Mean coefficient for dirt deposit, Btu/(hr)(sq ft)(deg F); see Chap. 8

h_G Mean coefficient for transfer of sensible heat through gas film, Btu/(hr)(sq ft)(deg F); see Chaps. 9 and 10

h_L Mean coefficient of heat transfer to coolant, Btu/(hr)(sq ft)(deg F); see Chaps. 9 and 10

h_m Mean value of h with respect to height of condensing surface, Btu/(hr)(sq ft)(deg F)

h_N Coefficient for any horizontal tube in a vertical tier of N tubes, h_1 for top tube, h_2 for second, etc.

h_{Nu} Coefficient predicted by Nusselt equation for zero vapor velocity, based on Δt_m employed

h_r Coefficient of heat transfer due to radiation, Btu/(hr)(sq ft)(deg F); see Chap. 4

i Enthalpy, Btu per pound; for gas-vapor mixture, i_G equals $c_s(t - t_0) + \lambda_0 H_G$; i_i is value of i_G at t_i, i_L equals i_G based on t_L

TABLE 13-1. NOMENCLATURE.—(*Continued*)

K_G Mass transfer coefficient through gas film, lb of vapor condensed/(hr)(sq ft) (unit potential difference expressed in atmospheres)

K' Mass transfer coefficient, expressed in lb of vapor condensed/(hr)(sq ft) (unit humidity potential difference); K'_G for **gas film**, based on $H_i - H_G$; K'_o for **over-all** value based on $H_L - H_G$.

k_f Thermal conductivity of condensate at t_f, Btu/(hr)(sq ft)(deg F per ft)

k_G Thermal conductivity of gas-vapor mixture

k_w Thermal conductivity of tube wall, Btu/(hr)(sq ft)(deg F per ft)

L Length of a straight tube, feet; L_H for horizontal and L_v for vertical tube

M Molecular weight; M_n for noncondensable gas (28.97 for air), M_v for vapor (18.02 for steam)

N Number of rows of tubes in a vertical tier; N_H is a horizontal plane

$N_{Re,f}$ Reynolds number, $4r_h G/\mu_f$, dimensionless; for a horizontal tube $N_{Re,f} = 2\Gamma'/\mu_f$; for a vertical tube $N_{Re,f} = 4\Gamma/\mu_f$; $r_h = S/b$

P Total absolute pressure, *atmospheres*

p Partial pressure, *atmospheres* absolute, p_n for noncondensable gas, p_{nm} for logarithmic mean of p_{ni} at gas-liquid interface, and p_{nv} in main body of gas stream; p_v for condensable vapor in main body of gas stream; p_{vi} is vapor pressure of liquid at t_i, p_L is vapor pressure of condensate at temperature t_L of coolant

q Rate of heat transfer, Btu per hour; in general, q or Σq is total by all mechanisms; q_G is based on transfer of sensible heat by means of h_G, q_λ of latent heat by means of conduction through film of condensate, q_L total transfer to coolant

R Thermal resistance, degrees Fahrenheit times hours times square feet divided by Btu; R_d for dirt deposit, R_L for coolant, R_v for vapor side, R_w for tube wall, ΣR for total resistance $(= 1/U)$

S_o Total cross section (ground area) of vertical tower, square feet

S Cross section of film of condensate, square feet

t Temperature, degrees Fahrenheit; $t_f = t_{sv} - 3\,\Delta t/4$ for condensate film; t_G for gas stream (bulk temperature); t_{G1} and t_{G2} for inlet and outlet gas, respectively; t_i for liquid-gas interface; t_L for coolant (bulk temperature); t_{L1} and t_{L2} for inlet and outlet coolant; t_s for condensing surface; t_{sv} for saturation temperature or dew-point temperature of vapor; t_{wb} for wet-bulb temperature; t_v for actual temperature of vapor

U Over-all coefficient of heat transfer, Btu/(hr)(sq ft)(deg F); U_i is based on A_i and U_o and A_o

U' Defined in Eq. (13-20)

V Average velocity, feet per hour

V' Average velocity of coolant, feet per second

W Coolant rate, pounds per hour; W_1 at inlet and W_2 at outlet

w_G Dry-gas rate, pounds per hour

w Mass rate of flow of condensate from *lowest* point on condensing surface, pounds per hour; w_z at distance z from top of condensing surface

x Length of conduction path, feet

x_i Mole fraction of more volatile component in condensate at t_i

x_w Thickness of tube wall, feet

X Abscissas of graphs

Y Ordinates of graphs

y Thickness of condensate film, feet

TABLE 13-1. NOMENCLATURE.—(*Continued*)

- z Distance of a point from top of a vertical tube; height of a vertical tower, feet
- z_G Height or length of a "mass-transfer unit," based on gas-film resistance; $z_G = G_G/K_G'a$
- z_{OG} Height of an over-all "mass-transfer unit," feet, based on over-all humidity difference; $z_{OG} = G_G/K_O'a$
- z_{ot} Height of an over-all "heat-transfer unit," feet; $z_{ot} = G_G c_s/U_G a$
- z_t Height of a "heat-transfer unit," feet; $z_t = G_G c_s/h_G a$

Greek

- α Angle between any radius of a horizontal tube and a vertical plane at the axis
- Γ Mass rate of flow of condensate from lowest point on condensing surface, divided by the breadth, lb/(hr)(ft); for a vertical tube, Γ equals $w/\pi D$; for a horizontal tube, Γ' equals w/L; in general, Γ is used to designate both Γ and Γ'; Γ_z is the local value of Γ at position z
- ΔP Pressure drop, force pounds per square foot of cross section
- Δt Temperature difference, degrees Fahrenheit; Δt_l for logarithmic mean. For condensing pure vapors, saturated or superheated, Δt equals $t_{sv} - t_s$
- Δt_o Over-all temperature difference, $t_{sv} - t_L$, degrees Fahrenheit; Δt_{ob} and Δt_{ot} refer to bottom and top, respectively; Δt_{om} refers to mean
- λ Enthalpy change, latent heat of condensation, Btu per pound; λ at saturation temperature, λ_{32} at 32°F, and λ_0 at base temperature
- μ_f Absolute viscosity of condensate film at t_f, lb/(hr)(ft); μ equals $2.42 \times \mu_f'$ in centipoises. If μ_F is viscosity in force pounds × hours per square foot, $\mu_f = \mu_F g_c$
- μ_v Absolute viscosity μ of gas-vapor stream, lb/(hr)(ft)
- $\mu_v/\rho_v D_v$ Schmidt number, dimensionless; for air-steam mixtures at ordinary pressures, $\mu_v/\rho_v D_v$ equals 0.71
- π 3.1416 . . .
- ρ_f Density of condensate film at t_f, pounds per cubic foot
- ρ_v Density of gas-vapor mixture, pounds per cubic foot
- τ_w Shear stress at the wall, evaluated for a dry tube, pounds force per square foot of wall
- Φ A dimensionless group $(k^3 \rho^2 g/\mu^2)^{1/3}$; Φ for liquid water is a function of temperature (see Appendix); $\Phi_f = (k_f^3 \rho_f^2 g/\mu_f^2)^{1/3}$
- ϕ Prefix, indicating function; $\phi(V') = C_1(V')^{0.8}$
- ψ Angle of inclination from horizontal

and ran down the plate. Dropwise condensation could also be brought about, for a while, by coating the surface with a thin petroleum oil that later washed away. Figure 13-1 shows dropwise condensation on the polished chromium-plated surface of copper; Fig. 13-2 shows dropwise condensation on clean areas and film condensation on areas coated with rust brought in by the steam. Dropwise condensation gave coefficients of heat transfer four to eight times as high as film condensation; results for the latter type agreed with the theoretical equation of Nusselt given later. A more complete discussion of dropwise condensation is also given later.

1. Film-type Condensation on Vertical Surfaces

Theoretical Relations for Laminar Flow of Condensate. In 1916, Nusselt[85]† derived theoretical relations for predicting the coefficient of heat transfer between a pure saturated vapor and a colder surface. It was assumed that streamline motion exists throughout the thickness of the continuous film of condensate on the cooling surface. In deriving the simple relations given below, it was further assumed that the force of gravity alone causes the flow of condensate over the surface, thus neglecting the possible effect of vapor velocity upon the thickness of the condensate film. By employing the definition of viscosity and assuming zero velocity of the condensate at the wall and a maximum velocity at

Fig. 13-1.   Fig. 13-2.

Fig. 13-1. *Dropwise* condensation of steam on plate.
Fig. 13-2. Film and dropwise condensation of steam on vertical plate. (*Photographs from Schmidt, Schurig, and Sellschopp,*[93] *Tech. Mech. u. Thermodynam.*)

the liquid-vapor interface, theoretical equations were obtained for the thickness of the film of condensate at a given point on the surface. The local coefficient of heat transfer may then be calculated upon assuming that the total thermal resistance lies in the film of condensate, through which the latent heat of condensation is conducted, neglecting the cooling of the condensate. Assuming that the temperature difference between vapor and wall is constant at all points, the resulting theoretical equations for the mean coefficients of heat transfer involve the thermal conductivity, viscosity, and density of the condensate, the temperature difference between vapor and solid, and certain dimensions of the apparatus.

Upon increase of the difference in temperature between saturated

† The derivations are given in English in references 28, 80, and 55 and in German in references 103, 101, 44, and 77. In reference 67, it is shown that the accelerating force in the film is negligible compared with the viscous drag.

vapor and surface, it is clear that the rate of heat transmission q would be increased and the average thickness of the condensate layer would increase. Since heat is assumed to flow through the condensate film only by conduction, it follows that the coefficient of heat transfer should decrease with increase in the temperature difference. Since the physical properties involved, especially viscosity, depend upon temperature, changes in film temperatures should theoretically affect the coefficient of heat transfer.

At any distance z below the top of the vertical tube where condensation starts, heat is transferred solely by conduction through the condensate film having local thickness y:

$$h = \frac{k_f}{y} = \frac{dq}{dA\,\Delta t} = \frac{\lambda\,dw_z}{b\,dz\,\Delta t} = \frac{\lambda\,d\Gamma_z}{dz\,\Delta t} \qquad (13\text{-}1)$$

Let $d\Gamma_z$ equal dw/b, where b is the perimeter πD. For the entire tube, of length L by definition,

$$h_m = \frac{q}{A\,\Delta t} = \frac{\lambda w}{bL\,\Delta t} = \frac{\lambda \Gamma}{L\,\Delta t} \qquad (13\text{-}1a)$$

Combining Eqs. (13-1) and (13-1a) to eliminate Δt, which is assumed constant,

$$\frac{k_f}{y} = \frac{h_m L}{\Gamma}\frac{d\Gamma_z}{dz} \qquad (13\text{-}1b)$$

From the theory of isothermal streamline flow down vertical wetted walls,†

$$\Gamma_z = \rho_f^2 g y^3 / 3\mu_f \qquad (13\text{-}1c)$$

Elimination of y from Eqs. (13-1b) and (13-1c), neglecting ρ_v relative to ρ_f, gives‡

$$k_f \left(\frac{\rho_f^2 g}{3\mu_f}\right)^{1/3} dz = \frac{h_m L}{\Gamma} \Gamma_z^{1/3} d\Gamma_z \qquad (13\text{-}1d)$$

Integration from 0 to L and from 0 to Γ gives

$$h_m = \frac{4}{3} k_f \left(\frac{\rho_f^2 g}{3\mu_f \Gamma}\right)^{1/3} = 0.925 k_f \left(\frac{\rho_f^2 g}{\mu_f \Gamma}\right)^{1/3} \qquad (13\text{-}1e)$$

Equation (13-1e) may be rearranged to involve the Reynolds number based on μ_f:

$$N_{Re,f} = \frac{4 r_h G}{\mu_f} = \frac{4}{\mu_f} \frac{S}{b} \frac{w}{S} = \frac{4w}{\mu_f b} = \frac{4\Gamma}{\mu_f}$$

giving

$$h_m \left(\frac{\mu_f^2}{k_f^3 \rho_f^2 g}\right)^{1/3} = 1.47 \left(\frac{4\Gamma}{\mu_f}\right)^{-1/3} = 1.47 (N_{Re,f})^{-1/3} \qquad (13\text{-}1f)$$

† Assuming a linear gradient in temperature through the film, and that $1/\mu$ is linear in t, Drew[32] shows that μ_f should be evaluated at $t_f = t_{sv} - 3\,\Delta t/4$.

‡ In the literature the density ρ_v of the vapor has been neglected relative to that of the liquid; strictly, the term ρ_f^2 should be replaced by $\rho_f(\rho_f - \rho_v)$. Nevertheless, even with steam at 1133 lb/sq in. and a saturation temperature of 560°F, omission of ρ_v causes the predicted value of h_m, for a given Γ, to be only 2.5 per cent higher than were ρ_v included; as the critical pressure is approached, the predicted value of h_m approaches zero.

Elimination of Γ from Eqs. (13-1a) and (13-1e) gives

$$h_m = 0.943 \left(\frac{k_f{}^3 \rho_f{}^2 g \lambda}{L \mu_f \Delta t}\right)^{1/4} \quad (13\text{-}1g)$$

The local coefficient h at any z is obtained by eliminating y from Eqs. (13-1) and (13-1c).

As shown above, the theoretical dimensionless equations of Nusselt for film-type condensation on vertical surfaces are

$$h = (k_f{}^3 \rho_f{}^2 g / 3 \mu_f \Gamma_z)^{1/3} \quad (13\text{-}2) \star$$

$$h_m = 0.943 (k_f{}^3 \rho_f{}^2 g \lambda / \mu_f L \Delta t)^{1/4} \quad (13\text{-}3) \star$$

$$h_m \left(\frac{\mu_f{}^2}{k_f{}^3 \rho_f{}^2 g}\right)^{1/3} = 1.47 \left(\frac{4\Gamma}{\mu_f}\right)^{-1/3} \quad (13\text{-}4) \star$$

To allow for the fact that the condensate is cooled from the saturation temperature to the film temperature, for both vertical and horizontal tubes Bromley[15] showed that the latent heat of condensation λ should be multiplied by the term

$$\left(1 + 0.4 \frac{c_p \Delta t}{\lambda}\right)^2$$

This correction becomes important at high pressure, where c_p increases and λ decreases.

Simplified Relation for Steam at 1 Atm. For steam condensing at atmospheric pressure with Δt ranging from 10 to 150°F, Eq. (13-3) reduces to the dimensional equation

$$h_m = \frac{4000}{L^{1/4} \Delta t^{1/3}} \quad (13\text{-}5) \star$$

where L is the height in feet. Similar relations for other vapors can readily be obtained by plotting readings from the alignment chart,[18] Fig. 13-7. These equations for streamline flow should not be used for $4\Gamma/\mu_f$ exceeding 2100 or for dropwise or mixed condensation.

Comparison of Eqs. (13-2) and (13-4) shows that the local coefficient at the end of the condensing section is three-fourths of the mean value for the condensing section. For *planes*† inclined at an angle ψ with the horizontal, the values of h and h_m are to be multiplied by $(\sin \psi)^{1/4}$.

When a liquid flows isothermally down vertical tubes,[30] measurements of the average thickness of the liquid film substantiate Eq. (13-1c) for values of the Reynolds number $4w/b\mu$ ranging from 1 to 2000. Wave motion in the film may cause the *maximum* thickness at the crest of a ripple to be as much as 2.5 times that predicted by Eq. (13-1c), although the *mean* thickness of the film is found to check the theoretical value.[38] For a given amount of liquid on the wall, the occurrence of ripples causes

† Not for inclined tubes.

a variation in y and gives a larger *average* value[41] of k_f/y than when no ripples occur and y is constant. Consequently, ripples should cause the coefficient of heat transfer to exceed that predicted from theory.

Variation in Tube-wall Temperature. In the Nusselt derivations, it was assumed that the temperature difference was constant. In the usual condenser, cooled by a liquid that rises considerably in temperature, the temperature difference varies substantially. The heat balance shows that the over-all temperature difference, Δt_o, from condensing vapor to coolant, is linear in q_z or Γ_z. Modification[32,52] of the derivation[107] for condensation of pure saturated vapors shows that the value of h_m should be taken as F_c times those obtained from Eq. (13-4), where the dimensionless factor F_c varies slightly with the ratio of over-all temperature differences at bottom and top of the vertical condenser, $\Delta t_{ob}/\Delta t_{ot}$, as shown in Table 13-2; $q = UA\ \Delta t_{om}$, where U is obtained from Eq. (8-7a), p. 187, with $F_c h_m$ replacing h_k.

TABLE 13-2

$\Delta t_{ob}/\Delta t_{ot}$	0.5	1.0	2.0	5.0
F_c	0.96	1.0	1.06	1.15

Data. A number of experiments have been published on heat transfer between pure saturated vapor and single horizontal or vertical tubes.

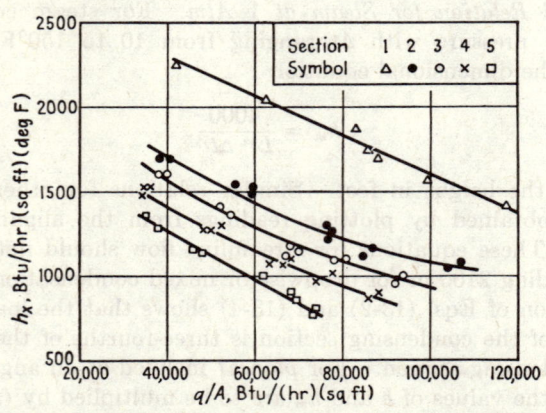

FIG. 13-3. Local coefficients for film-type condensation on five sections of a vertical condenser with downflow of steam.[95]

Because the coefficients of heat transfer are so high for vapors such as saturated steam, temperature differences between vapor and solid are relatively small. The measurement of small differences in temperature offers considerable difficulty, and, in general, the available coefficients of heat transfer between pure saturated vapors and single horizontal tubes are not so precise as might be desired.

CONDENSING VAPORS 333

Local Coefficients. Shea and Krase[95] measured local coefficients of heat transfer for film-type condensation of air-free steam at substantially atmospheric pressure flowing downward past a *vertical* flat water-cooled copper plate 4 inches wide and 23 inches high. Local coefficients were measured for each of five sections, each 4.6 inches (0.393 ft) high. In Fig. 13-3 the upper curve is for the top section, the lowest curve is for the bottom section, and the intermediate curve is the average for the entire plate, 1.92 ft high. The results for the top section were 10 per cent above the Nusselt theoretical equation for zero vapor velocity, while those for the bottom section were 25 per cent higher. When vapor velocity for a given heat flux was increased fourfold by reducing the width of the jacket from 4 inches to 1 inch, the coefficients increased by 5 to 30 per cent.

Average Coefficients. Table 13-3 describes the apparatus and operating conditions employed by a number of different investigators and gives deviations from Eq. (13-3) for vertical tubes. Average coefficients for

TABLE 13-3. CONDENSING PURE SATURATED VAPORS ON VERTICAL TUBES

Vapor	Range of h_m	Range of Δt, deg F	Range Obs. h_m / h_{Nu} [a]	Diameter, in.	Length, ft	Observers
Steam.........	1900–680	5–39	Avg 1.5	1.0	12.0	Hebbard and Badger[50]
Steam.........	1800–840	5–48	Avg 1.5	1.0	12.0	Meisenburg et al.[76]
Steam.........	2500–1300	2–23	*Ca.* 1.7	0.875	8.0	Fragen[5]
Steam.........	1500–700	3–34	Avg 1.2	2.0	20.0	Baker et al.[5]
Steam.........	2280–1420	17–83	Avg 1.1	Plate	0.39	Shea and Krase[95]
Diphenyl......	430–120	23–130	Turbulent	0.75	12.0	Badger et al.[3]
Dowtherm A...	545–118	35–72	Turbulent	0.875	11.7	Badger[2]
Ethanol.......	340–200	20–98	0.78–1.03	0.405	0.47	Hagenbuch[46]

[a] Based on observed Δt_m.

film-type condensation of steam on vertical tubes are shown in Fig. 13-4 and, for a given $4\Gamma/\mu_f$ at the bottom, average roughly 75 per cent above those predicted by Eq. (13-4). If compared on the basis of the Δt_m employed, these values of h_m average roughly 50 per cent above Eq. (13-3).† The increase above theory is probably due to the effect of the downward-flowing vapor in reducing the thickness of the condensate film and to the mixing action of ripples in this film. The values of h_m for a vertical copper plate[47] 5 inches high were 20 per cent above theoretical

† The data were taken from the paper by Baker, Kazmark, and Stroebe,[5] which gave data of Stroebe, Hebbard, and Fragen for tube lengths of 20, 12, and 8 ft, respectively.

[Eq. (13-3)]; values for a vertical copper plate 4.6 inches high[95] were 10 per cent above theoretical.

Effect of Noncondensable Gas. Using a vertical copper plate 5 inches high to condense steam at atmospheric pressure, Hampson[47] found that h_m decreased linearly with increase in the weight ratio X of nitrogen to steam from 0 to 0.02:

$$h_m/h_{Nu} = 1.2 - 20X \qquad (13\text{-}6)$$

The heat flux was 100,000 Btu/(hr)(sq ft). With steam condensing outside a vertical oxidized copper tube 12 ft long, Meisenburg et al.[76] found higher results for a given concentration of air,

$$h_m/h_{Nu} = 1.17/C^{0.11} \qquad (13\text{-}6a)$$

for air concentration C ranging from 0.2 to 4.0 per cent air by weight. For higher concentrations of air, see page 351.

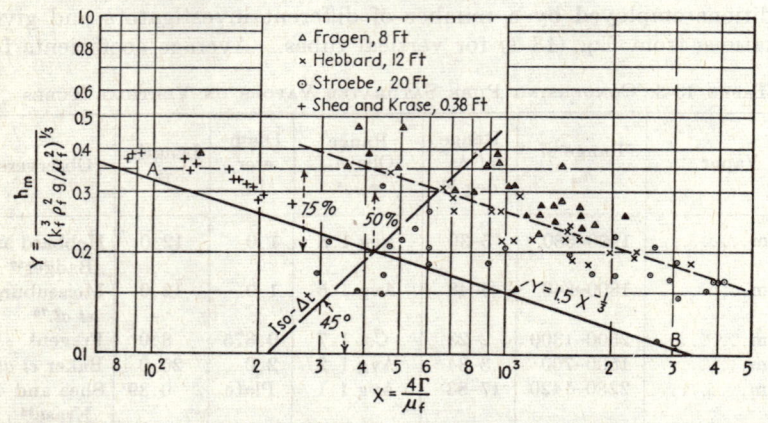

Fig. 13-4. Data for steam condensing on vertical tubes or plates, compared with the theoretical relation for film-type condensation, Eq. (13-15).

Effect of Turbulence. With film condensation of vapor on a tall vertical tube, one can easily obtain condensation rates such that the Reynolds number $4\Gamma/\mu_f$ exceeds the critical value (ca. 1800) at which turbulence begins;[80] this is not the case with horizontal tubes where the condensing height ($\pi D_o/2$) is inherently small. Because of turbulence in the layer of condensate on the lower part of the tube, the mean coefficient for the entire tube should lie above the line predicted from Eq. (13-4). Figure 13-5 shows data[3,105] for the condensation of diphenyl oxide and Dowtherm A on vertical tubes, correlated by Kirkbride[64] and Badger,[2] giving the recommended dimensionless equation for $4\Gamma/\mu_f$ exceeding 1800,

$$h_m\left(\frac{\mu_f^2}{k_f^3 \rho_f^2 g}\right)^{1/3} = 0.0077\left(\frac{4\Gamma}{\mu_f}\right)^{0.4} \qquad (13\text{-}7)\ \star$$

which is plotted as line CE in Fig. 13-5.

CONDENSING VAPORS 335

A semitheoretical relation of Colburn[24] is available for the condensate film for the case where $4\Gamma/\mu_f$ exceeds 2100 before reaching the bottom of a vertical tube. At the top of the tube the local coefficient for the streamline region is given by Eq. (13-2), and that for the turbulent

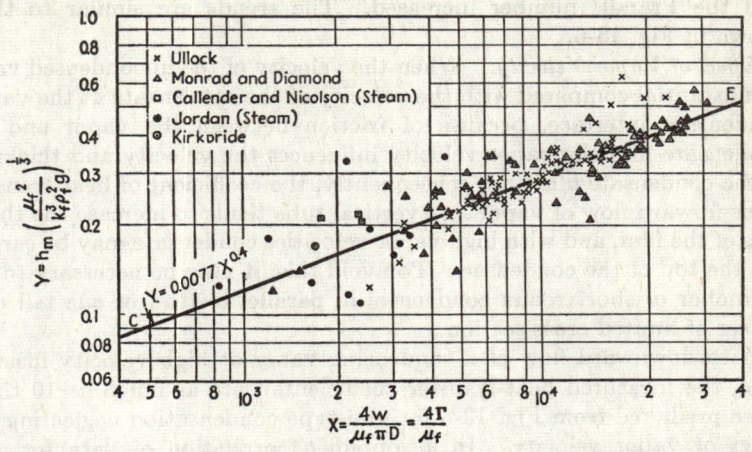

FIG. 13-5. Kirkbride-type plot[64] of data for single vapors condensing on vertical tubes; the increase in Y with increase in X is attributed to turbulence in the condensate layer on the lower portion of the tube.

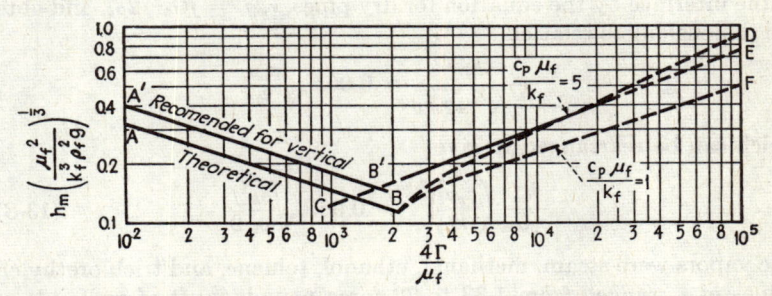

FIG. 13-6. Recommended curves $A'B'$ and CE for film-type condensation of single vapors on vertical tubes or plates.

region is based on an equation similar to Eq. (9-10c) with G_f replaced by Γ_z/y, DG/μ_f replaced by $4\Gamma/\mu_f$, and omitting the term $(\mu/\mu_w)^{0.14}$, giving

$$\frac{hy}{c_p\Gamma_z}\left(\frac{c_p\mu_f}{k_f}\right)^{2/3} = 0.027\left(\frac{4\Gamma_z}{\mu_f}\right)^{-0.2} \qquad (13\text{-}7a)$$

The resulting integrated relation, involving the mean coefficient for the entire tube and Γ, is shown in Fig. 13-6 as curves BD and BF for two values of $c_p\mu_f/k_f$.

Grigull[43] condensed several different vapors on the outside of vertical tubes having heights from 3.2 to 22 ft. The Reynolds numbers $4\Gamma_b/\mu$ ranged from 800 to 38,000, and the Prandtl numbers of the condensate film ranged from 1.5 to 5.0. The values of h_m ran progressively above the Nusselt equation for laminar flow as the condensation rates increased and the Prandtl number increased. The trends are similar to those shown in Fig. 13-6.

Effect of Vapor Velocity. When the velocity of the uncondensed vapor is substantial compared with the velocity of the condensate at the vapor-condensate interface, because of friction between the vapor and the condensate film the vapor velocity influences the velocity and thickness of the condensate film and, consequently, the coefficient of heat transfer. Thus upward flow of vapor in a vertical tube tends to increase the thickness of the film, and with high vapor velocities condensate may be carried out the top of the condenser. To avoid this, it may be necessary to use a number of short reflux condensers in parallel instead of one tall condenser of limited cross section.†

With downward flow of a condensing vapor at high velocity inside a tube, the measured heat-transfer coefficients[16] are as much as 10 times those predicted from Fig. 13-6 for film-type condensation neglecting the effect of vapor velocity. In a simplified correlation of data for condensation of a saturated vapor flowing downward in a water-cooled tube at high velocity, Carpenter and Colburn[16] compute the shear stress τ_w at the interface by the equation for dry pipes, $\tau_w g_c = fG_a^2/2\rho_v$, and obtain the dimensionless relation

$$\frac{h\mu}{k\sqrt{\tau_w g_c \rho_L}} = 0.065 \sqrt{\frac{c_p \mu}{k}}$$

which can be rearranged to give

$$\frac{h_m}{c_p G_m} \left(\frac{c_p \mu}{k}\right)^{0.5} = 0.065 \sqrt{\frac{\rho_L}{\rho_v} \frac{f}{2}} \qquad (13\text{-}8) \star$$

The vapors were steam, methanol, ethanol, toluene, and trichlorethylene; values of τ_w ranged from 1.33 to 72 force pounds/sq ft of wall. It was assumed that q/A was linear in L. For incomplete condensation G_m is taken as $\sqrt{(G_1^2 + G_1 G_2 + G_2^2)/3}$; for complete condensation this gives $G_m = 0.58 G_1$. All physical properties, except ρ_v, are taken for the condensate film, $t_f = t_v - 0.75 \, \Delta t$.

The data of Jakob *et al.*[57] for complete condensation of saturated steam flowing downward in a tube having an inside diameter of 1.57 inches and a cooled length of 4 ft are represented by the dimensional equation

$$h_m = \frac{1100 + 9.9 V_1'}{L^{1/4}} \qquad (13\text{-}9)$$

† Data on the loading velocities are given in references 106 and 14.

CONDENSING VAPORS 337

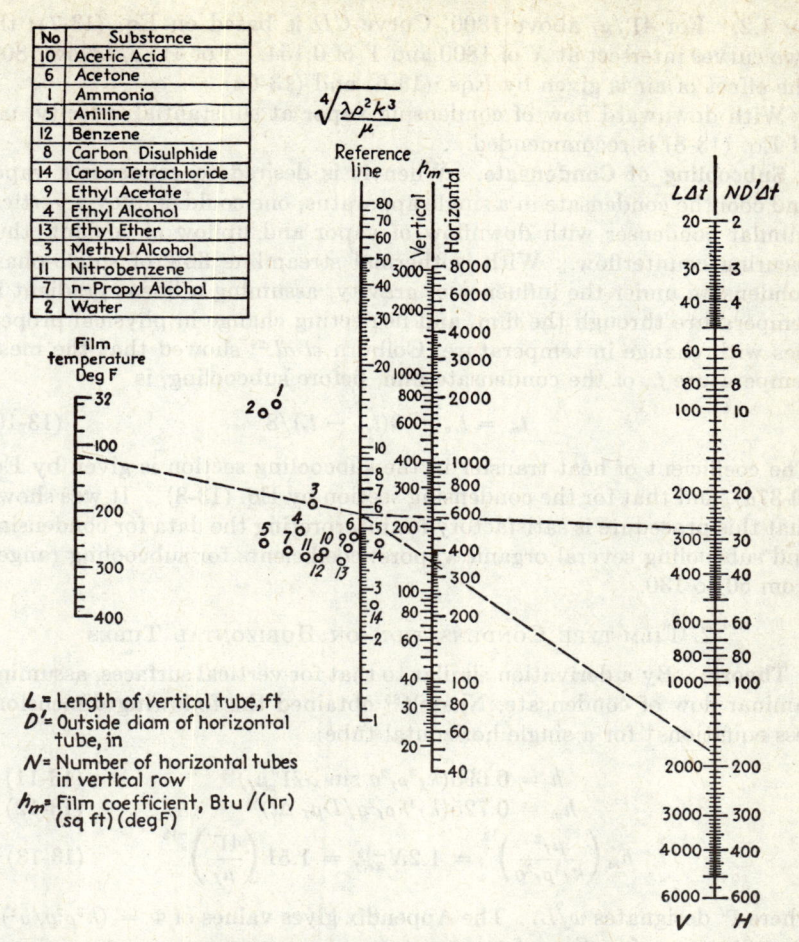

No	Substance
10	Acetic Acid
6	Acetone
1	Ammonia
5	Aniline
12	Benzene
8	Carbon Disulphide
14	Carbon Tetrachloride
9	Ethyl Acetate
4	Ethyl Alcohol
13	Ethyl Ether
3	Methyl Alcohol
11	Nitrobenzene
7	n-Propyl Alcohol
2	Water

L = Length of vertical tube, ft
D' = Outside diam of horizontal tube, in
N = Number of horizontal tubes in vertical row
h_m = Film coefficient, Btu/(hr)(sq ft)(deg F)

FIG. 13-7. Alignment chart for film-type condensation of single vapors, with streamline flow of condensate, based on Nusselt's theoretical relations [Eqs. (13-3) and (13-12)]. For vertical tubes, multiply h_m by 1.2. For $4\Gamma/\mu_f$ exceeding 1800, use Fig. 13-6 or Eq. (13-7). (*Chilton, Colburn, Genereaux, and Vernon.*[18])

The entering velocity V'_1 of the steam ranged from 33 to 260 ft/sec. The data were in fair agreement with an extended theory for laminar flow of condensate, allowing for the effect of vapor velocity.

Recommended Relations for Film-type Condensation of Saturated Vapors, Vertical Tubes. At low vapor velocities use of Fig. 13-6 is recommended. For $4\Gamma/\mu_f$ below 1800 the line $A'B'$ is based on Eq. (13-4) multiplied by 1.28, which is equivalent to Eq. (13-3) multiplied by 1.2; alternatively h_m from the alignment chart (Fig. 13-7) is multiplied by 1.2. For $4\Gamma/\mu_f$ above 1800, Curve CD is based on Eq. (13-7); the two curves intersect at X of 1800 and Y of 0.154. For $4\Gamma/\mu_f$ below 1800, the effect of air is given by Eqs. (13-6) and (13-6a).

With downward flow of condensing vapor at substantial velocity, use of Eq. (13-8) is recommended.

Subcooling of Condensate. When it is desired to condense a vapor and cool the condensate in a single apparatus, one could employ a vertical tubular condenser with downflow of vapor and upflow of coolant, thus securing counterflow. With isothermal streamline flow of single-phase condensate under the influence of gravity, assuming a linear gradient in temperature through the film, and neglecting change in physical properties with change in temperature, Colburn et al.[29] showed that the mean temperature t_m of the condensate film, before subcooling, is

$$t_m = t_{sv} - 3(t_{sv} - t_s)/8 \qquad (13\text{-}10)$$

The coefficient of heat transfer in the subcooling section is given by Eq. (9-37a), and that for the condensing section by Eq. (13-3) It was shown that this procedure is satisfactory for interpreting the data for condensing and subcooling several organic vapors; coefficients for subcooling ranged from 50 to 130.

2. Film-type Condensation on Horizontal Tubes

Theory. By a derivation similar to that for vertical surfaces, assuming laminar flow of condensate, Nusselt[85] obtained the following dimensionless equations† for a single horizontal tube:

$$h = 0.693(k_f{}^3\rho_f{}^2 g \sin \alpha / \Gamma'_s \mu_f)^{1/3} \qquad (13\text{-}11) \star$$

$$h_m = 0.725(k_f{}^3 \lambda \rho_f{}^2 g / D \mu_f \Delta t)^{1/4} \qquad (13\text{-}12) \star$$

$$h_m \left(\frac{\mu_f{}^2}{k_f{}^3 \rho_f{}^2 g}\right)^{1/3} = 1.2 N_{Re,f}^{-1/3} = 1.51 \left(\frac{4\Gamma'}{\mu_f}\right)^{-1/3} \qquad (13\text{-}13) \star$$

where Γ' designates w/L. The Appendix gives values of $\Phi = (k^3 \rho^2 g / \mu^2)^{1/3}$ for water, as a function of temperature.

For a horizontal tube,

$$N_{Re,f} = \frac{4 r_h G}{\mu_f} = \frac{4}{\mu_f} \frac{S}{L} \frac{w/2}{S} = \frac{2\Gamma'}{\mu_f} \qquad (13\text{-}13a)$$

Hence the critical Reynolds number of 2100 corresponds to $4\Gamma'/\mu_f$ of 4200 for a horizontal tube.

In the derivation, it was shown that 59.4 per cent of the condensate is formed on the upper half of the tube. Since the local coefficient h decreases as the thickness of the condensate layer increases in flowing

---

† In Eq. (13-12) by assuming that the local rate of condensation was uniform with perimeter, instead of assuming constant Δt, Parr[89] obtained a constant of 0.75 instead of the 0.725 obtained by Nusselt.

downward around the perimeter of the tube, it would be expected that the temperature of the outer wall of the tube would be highest at the top and lowest at the bottom. This was confirmed by Baker and Mueller[6] in an experimental study in which wall temperatures were measured at 30° intervals around the entire circumference by rotating the tube. No one location was satisfactory in giving the perimeter mean temperature; consequently, with a horizontal tube in a fixed position, temperatures should be measured at a number of points around the perimeter of the tube.

For N horizontal tubes arranged in a vertical plane so that the condensate from one tube flows directly onto the top of the tube directly below without splashing, h_m in Eq. (13-12) theoretically depends on the inverse fourth root of N. This effect of N is automatically allowed for in Eq. (13-13), which contains Γ', based on the total condensate from the lowest tube in the tier. Consider measured coefficients $h_1, h_2, \ldots, h_N$ for each of N rows of horizontal pipes[110] in a vertical tier or for successive sections of a vertical surface.[95] The coefficients for the top tube or section may properly be compared with Eqs. (13-12) or (13-3) for the coefficients on a surface having no condensate fed to the highest point. Since the second row or section receives condensate from the first, the coefficients for these sections should not be compared with Eqs. (13-12) or (13-3). However, the mean coefficient for the two highest sections may correctly be compared with Eq. (13-15), involving Γ leaving the second section. If the two sections are equal in size,

$$h_m = \frac{(q_1 + q_2)/2A_1}{(\Delta t_1 + \Delta t_2)/2}$$

Hence $h_m \sqrt[3]{\mu_f^2/k_f^3 \rho_f^2 g}$ should be plotted vs. $4\Gamma/\mu_f$ and compared with Eq. (13-15), where $\Gamma'_2 = (q_1 + q_2)/\lambda L$ for two horizontal tubes in a vertical tier and $\Gamma_2 = (q_1 + q_2)/\lambda \pi D$ for the two highest sections of a vertical tube. A similar procedure applies for any number of sections, so long as the Reynolds number is less than 2100 for the condensate leaving the bottom section.

The "local" coefficient h_N for any row N is theoretically related to that for the top row by the equation

$$h_N/h_1 = N^{0.75} - (N - 1)^{0.75}$$

Thus h_2/h_1 is 0.683, h_3/h_1 is 0.60, and $(h_1 + h_2 + h_3)/3h_1$ is 0.762, as is also predicted by $h_m/h_1 = 3^{-0.25} = 0.762$. As shown later, the coefficients for the lower tubes run higher than predicted, owing to the effect of turbulence caused by dripping and splashing as the condensate falls intermittently from one tube to the next.

For film-type condensation of steam at *atmospheric* pressure on N horizontal tubes in a vertical tier, the Nusselt relation reduces to the

TABLE 13-4. DATA OF VARIOUS OBSERVERS CONDENSING PURE SATURATED VAPORS OUTSIDE SINGLE HORIZONTAL TUBES

Vapor	Range of h_m	Range of Δt, deg F	Range[a] Obs. h_m / h_{Nu}	Diameter, in.	Length, ft	Observers
Steam........	3400–2100	22–43	1.0–1.3	0.675	4.0	McAdams and Frost[73]
Steam........	5000–1700	2–19	0.7–1.2	3.0	3.9	Othmer[87]
Steam........	4100–2800	3–13	....	1.31	3.67	Baker and Mueller[6]
Steam........	4300–3800	9–11	1.5–1.7	0.84	2.5	Wallace and Davison[108]
Benzene......	370–310	23–28	0.8–0.9	0.675	4.0	McAdams and Frost[73]
Benzene......	381–242	42–67	0.84–1.22	1.31	8.2	Kirkbride[63]
Benzene......	262–235	31–62	....	1.31	3.67	Baker and Mueller[6]
Benzene......	289–264	56–79	0.98–1.07	0.84		Wallace and Davison[108]
Benzene......	315–307	114	1.07–1.10	0.625	3.67	Baker and Tsao[7]
Diphenyl.....	400–225	8–27	....	1.7	9.4	Montillon et al.[81]
Naphtha......	361–174	22–85	0.84–1.16	1.31	8.2	Kirkbride[63]
Oleum spirits.	362–212	17–38	0.81–1.05	1.31	8.2	Kirkbride[63]
Oil vapor.....	260–190	55–60	0.80–1.27	1.31	8.2	Kirkbride[63]
Tetrachloroethylene[b]...	190–170	58–135	0.64–0.90	1.00	3.67	Tsao[104]
Tetrachloromethane (CCl₄).....	280	27	0.9	0.675	4.0	McAdams and Frost[73]
Trichloroethylene[b]...	262–244	54–85	0.84–0.99	1.00	3.67	Tsao[104]
Toluene[b].....	300–203	21–115	0.76–1.04	1.00	3.67	Tsao[104]
Toluene......	241–193	55–72	0.77–0.96	0.84	2.5	Wallace and Davison[108]
Turpentine[c]..	514–326	24–84	....	1.31	6.0	Patton and Feagen[90]
Methanol[d]...	600–500	14–28	1.06	2.0	1.63	Othmer and White[88]
Ethanol[d].....	450–320	11–39	1.02	2.0	1.63	Othmer and White[88]
Propanol[d]....	300–250	23–47	0.87	2.0	1.63	Othmer and White[88]
Butanol[d].....	300–250	22–50	0.85	2.0	1.63	Othmer and White[88]
i-Propanol[d]...	263–235	17–45	0.73–0.94			Othmer and White[88]
i-Butanol[d]....	210–194	26–56	....	2.0	1.63	Othmer and White[88]
Sec-Butanol[d].	213–191	21–58	....	2.0	1.63	Othmer and White[88]
Tert-Butanol[d]	192–154	14–46	....	2.0	1.63	Othmer and White[88]

[a] Based on observed Δt_m.
[b] Data made available through the courtesy of Prof. E. M. Baker.
[c] Condensate (turpentine and water) boils at 204°F.
[d] Figure 6 of reference 88 shows an alignment chart based on data of reference 88.

following dimensional relation:

$$h_m = \frac{5800}{(ND_o')^{1/4}(\Delta t_m)^{1/3}} \tag{13-14} \star$$

where D_o' is the outside diameter in inches. Thus with 16 horizontal 1-inch tubes and Δt_m of 27°F, h_m would be $(5800)/(2)(3) = 967$, and $q_m/A = h_m \Delta t_m = (967)(27) = 26{,}200$ for the average tube, while the top tube (N of 1) would give twice this flux.

These equations should not be employed for $2\Gamma'/\mu_f$ above 1800.

Vertical vs. Horizontal Tubes. Comparison of Eqs. (13-3) and (13-12) for a given Δt shows that vertical and horizontal tubes should give the same h_m if the length of the vertical tube is 2.87 times the outside diameter of the horizontal tube. If a tube had a ratio of length to diameter of 100:1, theoretically, h_m for the horizontal position would be 2.44 times that for the vertical position.

The term Γ of the equation for vertical surfaces equals $w/\pi D$, and Γ' of the equation for horizontal tubes equals w/L_H, where w is the total condensation rate per tube, D is the diameter of the vertical tube, and L_H is the length of the straight horizontal tube. If Γ is used to designate both Γ for vertical tubes and Γ' for horizontal tubes, Eqs. (13-4) and (13-13) become

$$h_m/\Phi_f = a_1 \left(\frac{4\Gamma}{\mu_f}\right)^{-1/3} \tag{13-15}$$

where the purely numerical constant a_1 is 1.47 for a vertical and 1.51 for a horizontal tube. Hence little error is introduced by using Eq. (13-15) for both vertical and horizontal tubes, with a_1 taken as 1.5. To facilitate calculations, the term Φ for water is shown as a function of temperature in the Appendix.

Data. Table 13-4 shows data for film-type condensation of steam and also for 18 organic vapors, outside single horizontal tubes. The data have been compared with the theoretical equation of Nusselt for laminar flow of condensate, neglecting any effect of vapor velocity. The measured values of h_m ran from 36 per cent below to 70 per cent above those predicted from the measured values of Δt_m. The average of the ratio of measured to predicted coefficients was 1.23 for steam and 0.94 for the organic vapors.

Figure 13-8 shows data of several observers for condensation of single pure saturated vapors outside single horizontal tubes and on a short vertical plate; it is seen that the deviations from Eq. (13-15) range from -50 to $+25$ per cent. Short and Brown[98] determined local coefficients for filmwise condensation of Freon-11 vapor on the outer surfaces of a single bank of twenty ⅝-inch horizontal brass tubes, arranged in a vertical tier. Values of h for each row were obtained from Wilson plots (page 343). Data for the top tube gave values of h increasing from 500 to 940 as the values of $10^{-7}k^3\rho^2 g/\mu\Gamma'$ increased from 12 to 80, somewhat above the values predicted by the Nusselt equation for the top tube. Coefficients for the lower rows were higher than predicted by theory, probably owing to turbulence induced by the intermittent dripping of condensate. In consequence, the average h for the tier was substantially equal to the value predicted for the top tube.

Effect of Air. Othmer[87] determined values of h_m for film-type condensation of steam on a 3-inch horizontal tube, as a function of the volume per cent of air in the ambient vapor and the value of Δt_m based on the temperature of the vapor (not the dew-point temperature). Figure 13-9 shows the measured values of h_m plotted vs. Δt_m, for two vapor temperatures, for various concentrations of air in the steam. An alternative method is given on page 355.

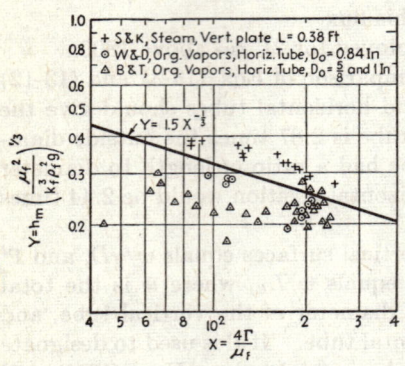

Fig. 13-8. Film-type condensation of single vapors on horizontal tubes or short vertical plates, compared with Eq. (13-15).

Recommended Relations. For film-type condensation of a pure saturated vapor outside of a vertical tier of N horizontal tubes, the use of Eq. (13-16) is recommended, depending on whether it is more convenient to estimate Δt or w, giving

$$h_m = 0.725\left(\frac{k_f^3 \rho_f^2 g \lambda}{N D_o \mu_f \Delta t}\right)^{1/4} = 0.95\left(\frac{k_f^3 \rho_f^2 g L}{\mu_f w}\right)^{1/3} \quad (13\text{-}16) \star$$

This recommendation is probably conservative, since it contains no special allowance for turbulence due to high vapor velocity or splashing of condensate. This equation is solved graphically on the alignment chart, Fig. 13-7. In the unlikely event that $2\Gamma'/\mu_f$ exceeds 2100, use of Eq. (13-7) (equivalent to Curve CD of Fig. 13-6) is suggested. The effect of noncondensable gas may be estimated from Fig. 13-9.

Illustration 1. Dry saturated steam at 212°F is condensing outside a bank of horizontal tubes, 16 tubes high, and the average temperature of the outer surface of the tubes is 200°F. The o.d. of the tubes is 1 in. Estimate the value of h_m from steam to metal.

Solution. The value of t_f is $(212 + 200)/2 = 206$°F. In Fig. 13-7, align 206°F with point 2, thus obtaining an intersection with the reference line. The product $D'\Delta t$ is $1 \times 12 = 12$. Aligning 12 on the right-hand scale and the point already found on the reference line, read h_m of 2500 Btu/(hr)(sq ft)(deg F) for a single horizontal pipe. For 16 pipes high, $h_m = 2500/(16)^{0.25} = 1250$. Approximate equation (13-14) gives h_m of 1270.

CONDENSING VAPORS

In case $2\Gamma'/\mu_f$ exceeds, say, 2100 because of turbulence, Fig. 13-7 will be too conservative, and one should use Eq. (13-7) or the right-hand part of Fig. 13-6. For a single tier of 16 tubes each L ft long, the corresponding A equals $(16)(\pi)(D_o)(L)$ = $(16)(\pi)(\frac{1}{12})(L)$ = $4.18L$ sq ft. The corresponding value of w is equal to $(h_m)(A_m)(t_{sv} - t_s)/\lambda = (1250)(4.18L)(212 - 200)/(970 + 6) = 64.3L$ lb/hr, and Γ' equals $w/L = 64.3$ lb/(hr)(ft), from the lowest tube. Since, from the Appendix, μ_f' equals 0.294 centipoise, μ_f equals $2.42\mu_f' = (2.42)(0.294) = 0.711$ lb/(hr)(ft), and $2\Gamma'/\mu_f = (2)(64.3)/0.711 = 181$; since this is less than the critical value of $ca.$ 2100, the value of h_m of 1250 from Fig. 13-7 is satisfactory unless a promoter be used to secure drop-wise condensation.

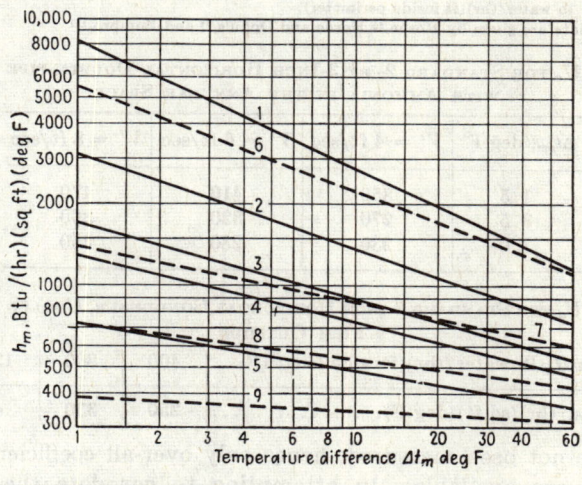

FIG. 13-9. Coefficients for steam condensing on a horizontal 3-in. tube. (Othmer.[87]) (*Courtesy W. J. King.*)

Side Lines ($t_v = 230°F$)
1 = 0 per cent air by vol.
2 = 1.07 per cent air
3 = 1.96 per cent air
4 = 2.89 per cent air
5 = 4.53 per cent air

Broken Lines ($t_v = 212°F$)
6 = 0 per cent air
7 = 1.42 per cent air
8 = 3.47 per cent air
9 = 6.21 per cent air

3. OVER-ALL COEFFICIENTS IN CONDENSERS

Condensing Ammonia. Kratz et al.[66] condensed ammonia vapor at a gauge pressure of 145 lb/sq in. in several types of condensers, using cooling water at 60 to 75°F, and obtained the over-all coefficients given in Tables 13-5 to 13-7.

Graphical Method of Wilson.[109] In surface condensers, a vapor is being condensed on the outer surface of a tube, while a liquid, usually cooling water, flows in turbulent motion inside the tube. Although in some cases tube temperatures have been measured, especially in testing single-tube experimental apparatus, in the majority of cases tube temper-

TABLE 13-5. U_o FOR VERTICAL SHELL-AND-TUBE CONDENSER, WATER FILM INSIDE TUBES[a]

Δt_{om} deg F	$\Gamma = 400$	$\Gamma = 800$	$\Gamma = 1200$	$\Gamma = 1600$	$\Gamma = 2000$	$\Gamma = 2400$
1.5	220	275	310	350		
3.5	170	225	270	315	390	430
7.0	150	215	260	300	340	370

$\Gamma = w/\pi D_i$ = lb water/(hr)(ft inside perimeter).
[a] Additional data are given by Horne,[53] Horne and Ophuls,[54] and Zumbro.[111]

TABLE 13-6. U_o FOR STANDARD 2- BY 3-INCH HORIZONTAL DOUBLE-PIPE CONDENSER WITH AMMONIA IN THE ANNULAR SPACE

Δt_{om}, deg F	$V' = 4$ ft/sec	$V' = 6$ ft/sec	$V' = 8$ ft/sec
1.5	350	410	470
3.5	270	320	390
7.0	230	280	350

TABLE 13-7. U_i FOR STANDARD 2-INCH HORIZONTAL CONDENSER, COOLED BY A WATER FILM OUTSIDE

$\Gamma' = w/L$, lb water/(hr)(ft tube length)....	400	800	1200
U_i, Btu/(hr)(sq ft)(deg F)................	250	330	400

atures have not been measured; hence only over-all coefficients of heat transfer U are available. In attempting to correlate these over-all coefficients, many engineers have failed to take advantage of the fact that the over-all resistance to heat flow ($\Sigma R = 1/U$) is numerically equal to the sum of the individual series resistances, namely, the resistance on the vapor side R_v, that of the wall itself R_w, that of the dirt deposit R_d, and that on the water side R_L. In 1915, Wilson[109] employed a valuable graphical analysis of the over-all coefficient of heat transfer. The following treatment presents the method and illustrates the application to several sets of data.

Consider a series of runs made in condensing substantially air-free vapor at a given temperature, employing different water velocities. From the concept of resistances in series (page 188) it is clear that the total resistance is equal to the sum of the individual resistances $1/U = R_v + R_w + R_d + R_L$. According to the theoretical equation of Nusselt, the resistance on the vapor side depends upon the temperature difference and the temperature of the condensate film; hence R_v should vary somewhat as water velocity is changed. Also, the thermal resistance of any scale on the heating surface might differ in the various runs; and changes in tube-wall temperature, brought about by changes in water velocity, would cause minor variations in the thermal conductivity of the tube wall. However, except where very high water velocities are used, the water-side resistance is usually the major resistance from condensing steam to water, and, under these conditions, serious error would not be introduced by assuming that the sum of the first three individual resistances $R_v + R_w + R_d$ is approximately constant. As is well known, the water-side resistance is an inverse function of the water velocity V' through the tubes, and, neglecting the effects of changes in water temperature, due to changes in water velocity, the water-side resistance R_L could be taken as a function of the water velocity alone. As Wilson points out, a plot of $1/U$ vs. $1/\phi(V')$ should give a straight line when plotted to ordinary rectangular coordinates. For turbulent flow of the water in a given apparatus, $\phi(V')$ may be taken as $C_1(V')^{0.8}$, and hence

$$\frac{1}{U_o} = R_v + R_w + R_d + \frac{1}{C_1(V')^{0.8}} \qquad (13\text{-}17)$$

where C_1 is an empirical constant and may be considered as the apparent individual coefficient of heat transfer from tube to water, based on the *outside* surface for water velocity of 1 ft/sec. Figure 13-10 shows experimental data of Orrok,[36] plotted as suggested by Wilson. The crosses represent data for a new, clean tube; and, for a given abscissa, the total thermal resistances are lower for the new tube than for the old tube. Empirical equations for the two sets of data are as follows:

For the old tube:

$$\frac{1}{U_o} = 0.00092 + \frac{1}{268(V')^{0.8}}$$

For the clean tube:

$$\frac{1}{U_o} = 0.00040 + \frac{1}{268(V')^{0.8}}$$

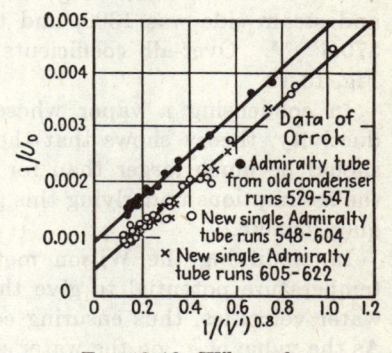

FIG. 13-10. Wilson plot.

where U_o is expressed as Btu/(hr)(sq ft on the steam side)(deg F over-all temperature difference) and $V' =$ water velocity in the tubes in feet per second.

The reciprocal slope C_1 for each set of tests was 268, indicating that the water-side coefficient, at a water velocity of 1 ft/sec, was 268 Btu/(hr) (sq ft)(deg F, difference from tube to water). Since the tube had outside and inside diameters of 1.00 and 0.902 in., respectively, the water-side coefficient per square foot of inside surface was $h_i = 268/0.902 = 297$. The tube-wall thickness was 0.00408 ft, and taking $k_w = 63$ Btu/(hr)(sq ft)(deg F per ft) for the Admiralty-metal tube, $R_w =$

0.000068. Referring to the curve for the clean tube, R_w may be deducted from the intercept of 0.00040, giving by difference

$$R_v = 0.00040 - 0.000068 = 0.000332 = \frac{1}{3010}$$

i.e., the vapor-side coefficient of heat transfer was 3010.

Thus, by the aid of this graphical method, the observed over-all resistance has been subdivided into the three component resistances $R_v = 0.000332$, $R_w = 0.000068$, and $R_L = 1/268(V')^{0.8}$. By comparing the empirical equations for the old and new tubes, the resistance of the scale is found by difference to be $0.00092 - 0.00040 = 0.00052 = 1/1920$, that is, the apparent coefficient of heat transfer through the scale is 1920 based on the steam side, or $1920/0.902 = 2130$/sq ft of water side.

Even if one does not wish to evaluate the individual thermal resistances as outlined above, the graphical method gives a straight line, thereby facilitating interpolation and extrapolation. When the water velocity becomes too low, the assumption of an 0.8 exponent will be incorrect; hence this relation will break down at low water velocities. The graphical method is applied to a number of sets of data in reference 75. Graphical analysis of over-all coefficients[92] for surface condensers gives[23] scale-deposit coefficients h_d of 760 before cleaning the inside surface, 1200 after cleaning with rubber plugs, 2000 after using a mechanical tube cleaner, and 4200 after sandblasting; the combined coefficient for the tube wall and steam side was 1300, and the water-side coefficient was equal to $370(V')^{0.8}$. Over-all coefficients for condensing steam are given in Fig. 13-12.

In condensing a vapor whose condensate has a low thermal conductivity, theory shows that the thermal resistance on the vapor side should be much larger than for condensing steam, and hence some of the assumptions underlying this graphical method will appear somewhat questionable.

In modifying the Wilson method, Chu *et al.*[19] adjust the over-all temperature potential to give the same average heat flux $U_o \Delta t_o$ at all water velocities, thus ensuring constant value of h_m on the vapor side. As the value of h_m on the water side is proportional to $(1 + 0.011t)(V')^{0.8}$ [see Eq. (9-19)], $1/U_o$ is plotted vs. $1/(1 + 0.011t)(V')^{0.8}$.

If refinements in the method be desired, it is possible to introduce factors for water viscosity and tube diameter into the abscissa. The graphical method may also be applied to cases in which neither fluid changes phase, as in liquid-to-liquid or gas-to-gas heat exchangers, provided that the velocity of one of the fluids is held constant. If the liquid being heated or cooled flows in streamline motion, one should plot $1/U$ vs. $1/(V')^{1/3}$ [see Eq. (9-23a)].

Finned Tubes. Katz and Geist[62] measured over-all coefficients U_o for a bank of finned copper tubes containing six horizontal tubes in a vertical tier. The tubes had outside diameters of 0.75 inch and 15 integral copper fins per inch, 0.0625 inch high. Mean condensing coefficients, based on the total outer surface (0.5 sq ft per ft) obtained from a Wilson plot were 309 for Freon-12, 395 for butane, and 610 for acetone and exceeded the Nusselt theory by 25, 46, and 53 per cent, owing to the effect of splashing, so that part of the condensate did not fall on the tubes below; the over-all values of Δt_{om} were 40, 80, and 80°F, respectively. With mixed condensation of water, and Δt_{om} of 77°F, h_m was 1500, 20 per cent higher than predicted for film-type condensation.

Cleaning Condensers. The tubes of surface condensers may be cleaned by turbine-driven cleaners, rubber plungers, sandblast, or in some cases chemical agents.[39] Cooling water is sometimes treated with 4 parts of chlorine per million to prevent algae growth and consequent reduction in U. A number of articles† deal with the economic intervals between cleaning of condenser tubes.

4. Dropwise Condensation of Pure Vapor

Mechanism. Dropwise condensation has been obtained with mixtures of steam and other vapors, but steam is the only pure vapor for which conclusive evidence of dropwise condensation is available. The subject has been studied by numerous investigators,‡ and a number of erroneous conclusions were drawn by the early workers. The situation was investigated and clarified in papers by Nagle[82] and coworkers, and the following summary is based on the paper by Drew, Nagle, and Smith.[33]

1. Film-type condensation is always obtained with clean steam condensing on clean surfaces, whether rough or smooth, regardless of the presence of simple noncondensable gases.

2. Dropwise condensation of steam is obtained only when the condensing surface is contaminated with a suitable promoter that prevents the condensate from wetting the surface.

3. Although many substances (including hydrocarbon oils) will make the surface nonwettable temporarily, only those which are adsorbed or otherwise firmly held are significant as promoters. Some of the important promoters are specific for certain surfaces (*e.g.*, mercaptans on copper and its alloys); others are effective with a number of surfaces (*e.g.*, oleic acid on copper, brass, nickel, and chromium).§ Octyl thiocyanate is effective on copper and is less noxious than benzyl mercaptan.

If the surface contaminant reduces the interfacial tension sufficiently to render the surface nonwettable, the condensate will collect in drops that grow in size until downward forces cause them to roll down the surface.

† References 96, 25, 92, and 1.
‡ References 93, 100, 11, 33, 58, 34, and 35.
§ The use of promoters in inducing dropwise condensation is disclosed in U.S. Patent 1,995,361 granted to W. M. Nagle, Mar. 26, 1935.

Since at any moment a substantial fraction of the condensing surface is free of condensate, much higher rates of condensation are obtained with a given temperature difference than with a nonpromoted or wettable surface that is insulated with a continuous film of condensate.† Emmons[34] used the Blodgett technique to form a monomolecular layer of molecules and found that two layers of suitable promoter on the condensing surface were adequate to produce dropwise condensation of steam.

Data. Table 13-8 summarizes the results of several investigators who report film coefficients for the dropwise condensation of steam. With filtered steam condensing on well-promoted surfaces from 2 to 6 ft high, the data of references 83 and 37 show that h_m averages 13,000 and is independent of heat flux in the range from 70,000 to 250,000 Btu/(hr)(sq ft); the lower values were obtained with insufficient promoter.[37]

TABLE 13-8. DROPWISE CONDENSATION OF STEAM

Reference number	Height, ft	Steam filter used	Range h_m	Range Δt, deg F	Range q/A, Btu/(hr)(sq ft)
95[a]	1.92	Yes	7,000–75,000	1.36–28.8	74,000–276,000
83[b]	2.0	Yes	11,000–17,000	5.2 –13.9	77,000–170,000
37[c]	6.08	Yes	7,000–16,000	....	68,000–252,000
37[d]	10.0	No	4,200		

[a] The steam was passed through an entrainment separator, and the condensing surface was not fouled.
[b] Based on thesis of Bays and Blenderman, reference 9, using oleic acid on chrome plate.
[c] Based on thesis of Baum, reference 8, using benzyl mercaptan on copper.
[d] Based on thesis of Fitzpatrick, reference 36; benzyl mercaptan was used on a fouled surface of copper.

Shea and Krase[95] measured "local" coefficients for dropwise condensation of steam at atmospheric pressure flowing downward past five vertical water-cooled sections each 0.384 ft high. The vertical surface was of copper, well promoted with benzyl mercaptan to induce dropwise condensation. Heat flux could be increased by increasing the velocity of the coolant in all water jackets. The heat-transfer coefficient was highest for the top section, where the steam velocity was greatest, and the rate of flow of condensate was least. As the heat flux was increased, the value of h_m for any section increased owing to the increase in steam velocity, but with further increase in heat flux and steam velocity h_m passed through a maximum and then decreased, since the detrimental effect of increased condensate rate was more important than the beneficial effect of steam velocity. The results are shown on Fig. 13-11; the numbers on

† Unless the layer of contaminant is so thick as to introduce a substantial resistance to conduction.

the curves are the average downward steam velocities in feet per second at the center of each section. The steam had passed through an entrainment separator and must have been free of rust, as otherwise such high coefficients would not have been obtained. In contrast, data of reference 37 show much lower coefficients for a 10-ft vertical tube where no steam filter was used, but even so very substantial coefficients (4200) were obtained.

Hayes and Bartol[49] heated water flowing at 32 to 58 ft/sec inside a horizontal copper tube having inner and outer diameters of 0.54 and 0.67 inch, and 5 ft long. The outer surface of the tube was chrome-plated and heated by condensing steam, promoted with oleic acid to give dropwise condensation. Over-all coefficients ranged from 5000 to 6700,

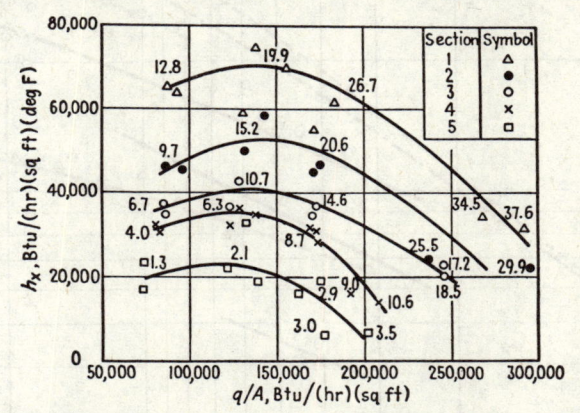

FIG. 13-11. Local coefficients for dropwise condensation of steam flowing downward past a vertical copper plate promoted with benzyl mercaptan. Numbers on curves give average steam velocities in feet per second. (*Shea and Krase.*[95])

and flux density ranged from 700,000 to 2,000,000 Btu/(hr)(sq ft). A conventional equation gave a satisfactory correlation of the data on the water side, where the Reynolds moduli ranged from 200,000 to 500,000. Steam-side coefficients were of the order of 20,000 to 40,000 Btu/(hr)(sq ft)(deg F).

The following examples[37] are based on the results with commercial steam (not filtered) condensing on a 10-ft vertical No. 18 BWG tube having an outside diameter of ⅝ in., comparing chrome-plated copper promoted with benzyl mercaptan vs. copper mildly promoted with oleic acid. In order to transfer 200,000 Btu/hr with each surface, the former would require 2000 lb of cooling water/hr as compared with 4200 for the latter. When using 9400 lb/hr of water in each case, the former transferred 580,000 Btu/hr and the latter, 270,000. In order to preheat water from 65 to 150°F, with steam condensing at 216°F, the former would heat 5100 lb/hr of water and the latter, 1090.

Figure 13-12 shows over-all coefficients from steam condensing at atmospheric pressure outside a vertical 10-ft length of ⅝-in. condenser tube having a wall thickness of 0.049 in., internally cooled by water flowing upward. Since the steam used contained a small concentration of oleic acid introduced in the lubricant for the boiler feed-water pump, the reference curve D-I for film condensation is based on work by Drew[31]

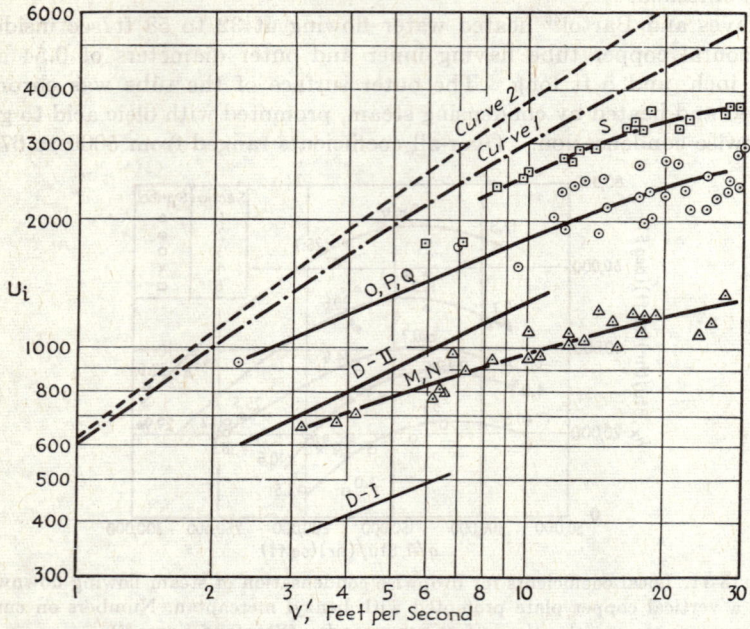

Fig. 13-12. Condensation of steam at 1 atm on a vertical 10-ft length of ⅝-in. copper tube.[37]

Curve 2, h_m of ∞ on vapor side.
Curve 1, h_m of 14,000 on vapor side.
Curve S, chrome-plated on steam side, and promoted with oleic acid or benzyl mercaptan.
Curve OPQ, copper tube promoted with benzyl mercaptan.
Curve MN, copper tube, trace of oleic acid present.
Curve D-I, film-type condensation on vertical Admiralty tube (6 ft by ⅝ in.) not promoted; Curve D-II, same tube, with benzyl mercaptan.

using uncontaminated steam. For a water velocity of 7 ft/sec, the over-all coefficient is 2100 with promoted chrome plate as compared with 520 for film condensation. The water-side coefficient was $630(V')^{0.8}$ and k/x_w for the tube wall 54,000; Curve 1 is based on h_m of 14,000 (as obtained in reference 78), and Curve 2 is based on h_m of infinity. In the range of economical water velocity, 6 to 10 ft/sec, it is seen that U is increased but little as h_m on the vapor side increases from 14,000 to infinity.

With a single horizontal-tube condenser, over-all coefficients of reference 58 analyzed by plotting $1/U$ vs. $1/(V')^{0.8}$ indicate an approximate value of h_m of 10,000 with a dropwise condensation of steam on a tube of Muntz metal.

Because of the very high coefficients obtained with fine-grained dropwise condensation of steam, in many cases the resistance on the steam side will be negligible compared with other resistances involved, thus eliminating the necessity for using thermocouples in the tube wall; if the latter are used, because of the very small temperature drop from steam to thermocouple junction, the error due to thermal conduction through the leads is materially reduced.

II. CONDENSING OR COOLING OF PURE SUPERHEATED VAPOR

Case 1. The mechanism of film-type condensation of superheated vapor differs from that of saturated vapor. Nevertheless, with corresponding surface temperatures and steam pressures, superheated steam has been found to transfer heat at a rate only slightly higher than saturated steam. For example, Merkel[77] showed that, in condensing steam with 180°F superheat, the rate of heat transfer q/A was only 3 per cent more than for saturated steam at the same pressure and with the same wall temperature. Similar findings have been reported by others.† Hence, for film-type condensation of superheated steam, little error is made in computing the rate of heat flow per unit area q/A by multiplying the value of h_m for a saturated vapor (Fig. 13-6) by the difference between the saturation temperature of the steam t_{sv} and the surface temperature t_s:

$$q = hA(t_{sv} - t_s)$$

Case 2. If the pipe wall is above the saturation temperature of the vapor, no condensation will take place; the vapor will merely lose some of its superheat. The heat-transfer mechanism is the same as that for cooling a noncondensable gas; the recommended equation is given in Chap. 9.

Case 3. It may happen that the temperature of the tube wall is above the saturation temperature of the pure vapor near the vapor inlet and below the saturation temperature at the outlet. Under these conditions, the vapor will be cooled without condensation in the first part of the apparatus and will condense in the latter part. It should be remembered that the coefficient of heat transfer in the desuperheating section is very small compared with that in the condensing section, and such problems should be divided into two stages: desuperheating and condensing.

III. CONDENSATION OF MIXTURES OF VAPORS

Mechanism. If several vapors are condensing simultaneously in the absence of noncondensable gas, the results differ, depending on the nature

† References 101, 56, 60, 65, and 13.

of the condensates. Thus, if the various components, when condensed, are not miscible in all proportions, such as condensed hydrocarbons and water, dropwise condensation occurs.

For a condensate forming a true solution, h_m should be predictable from the physical properties of the solution and Fig. 13-6.†

When a mixture of condensable vapors at temperature t_v is exposed to a surface having a temperature t_s below the dew point of the vapor mixture, condensation occurs. The temperature t_i of the interface between the condensate film and the vapor film will be intermediate between t_v and t_s. Under the influence of the temperature difference $t_v - t_i$, sensible heat dq_G flows through the gas film to the interface. Simultaneously, vapor condenses on the element dA of condensing surface, and the corresponding enthalpy change dq_λ (largely latent heat) is liberated at the interface. The sum of $dq_G + dq_\lambda$ is transferred through the film of condensate, flows through the tube wall and dirt deposit, and is absorbed by the cooling medium.

With film-type condensation, as is usually obtained with a *single-phase* condensate, the coefficient h for the condensate film is predicted by the methods given above for pure vapors, and the corresponding rate dq of heat transfer is calculated from the equation

$$dq = h \, dA (t_i - t_s) \qquad (13\text{-}18)$$

The interfacial temperature is evaluated by assuming equilibrium at the interface, *i.e.*, that t_i is the boiling temperature corresponding to the total pressure P and the composition of the liquid at the interface.‡

Colburn and Drew[27] give the complex equations from which the mole fraction in the condensate of the more volatile component x_i may be computed for a binary mixture. In an example of the differential condensation of a saturated vapor containing 70 mole per cent methanol and 30 per cent steam at 1 atm, they show that the local condensate varies in composition from 50 mole per cent methanol when the cooling surface is at 140°F, t_i is 165.2°F, and $t_i - t_s$ is 25.2°F to 67 mole per cent methanol when the cooling surface is at 32°F, t_i is 158°F, and $t_i - t_s$ is 126°F.§

To avoid the laborious computations necessary to calculate t_i, one assumes that the condensate has the same composition as the original vapor and hence computes too small a temperature drop through the condensate layer. This conservative approximation involves values of

† If special heat effects, such as heat of solution, are important compared with latent heat of condensation, due allowance should be made for such effects.

‡ In the analogous fields of absorption and distillation, equilibrium at the interface is a satisfactory assumption; in heat transfer, this assumption is employed for condensation of mixtures in references 63, 6, 27, and 7.

§ A coefficient of 500 was taken for the condensate film, independent of condensate composition.

$t_i - t_s$ of $158 - 140$, or $18\,°\text{F}$, and $158 - 32$, or $126\,°\text{F}$, respectively, instead of the correct values of 25.2 and $126°\text{F}$. These approximate values are used in Eq. (13-18), to obtain h from Eq. (13-2) or (13-11), evaluating the physical properties at the film temperature $t_f = t_{sv} - 3\,\Delta t/4$.

Single-phase Condensate. Two sets of data[108,46] are available for the condensation of binary mixtures; in one case,[108] nearly all the vapor was condensed; in the other,[46] only a small fraction was condensed in the test condenser. In both cases, the condensate from the test condenser was

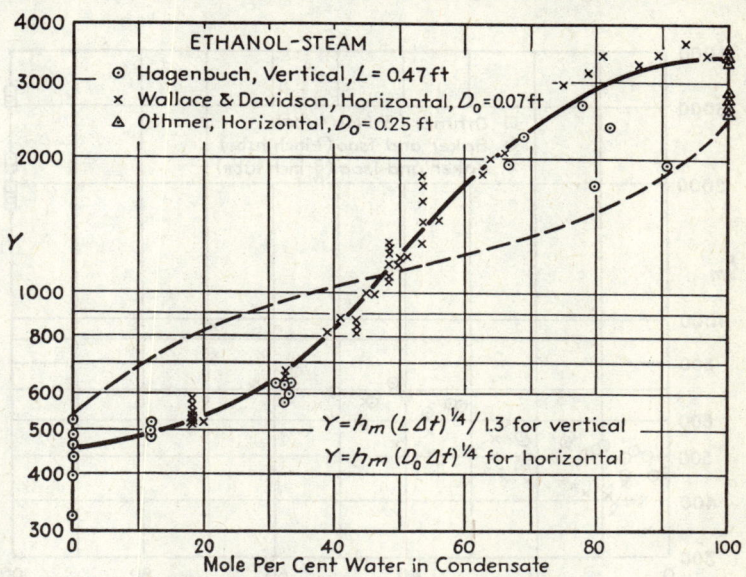

Fig. 13-13. Condensation of mixtures of ethanol and steam at 1 atm; the dashed curve shows the approximate location of the theoretical curve for film-type condensation; $h_m = q/(A)(t_i - t_s)_m$, in which t_i is the boiling point of the condensate.

collected and analyzed. Taking t_i as the boiling point of the condensate, the values of h_m were computed from the equation

$$h_m = q/A(t_i - t_s)_m \tag{13-19}$$

Figure 13-13 shows the data of both observers, plotted vs. the mole per cent water in the condensate. To facilitate comparison of the results with theory, the ordinates are $h_m(D_o\,\Delta t)^{1/4}$ for the horizontal tubes and $h_m(L\,\Delta t)^{1/4}/1.3$ for the vertical tube. The data of these observers for steam was omitted, since dropwise or mixed condensation was obtained, and those of Othmer[87] for film-type condensation of steam were included instead. The maximum deviations of h_m, from the theoretical values, are $+80$ and -25 per cent.†

† Haselden and Prosad[48] condensed oxygen and nitrogen on the outer surfaces of small vertical tubes; the data supported the Nusselt equation.

Immiscible Condensates. Cogan[22] noticed that mixtures of steam and benzene, when condensed, gave partial dropwise condensation. A film of benzene preferentially wet the tube, and drops of water were present in the film. This observation was confirmed by Baker and coworkers,[6,7] who determined values of h_m for condensation of binary mixtures of steam and benzene, toluene, chlorobenzene, heptane, and trichloroethylene outside horizontal tubes having diameters of 0.625, 1.00, and 1.31 in. Although the vapor was superheated ($t_v > t_i$) in a number of

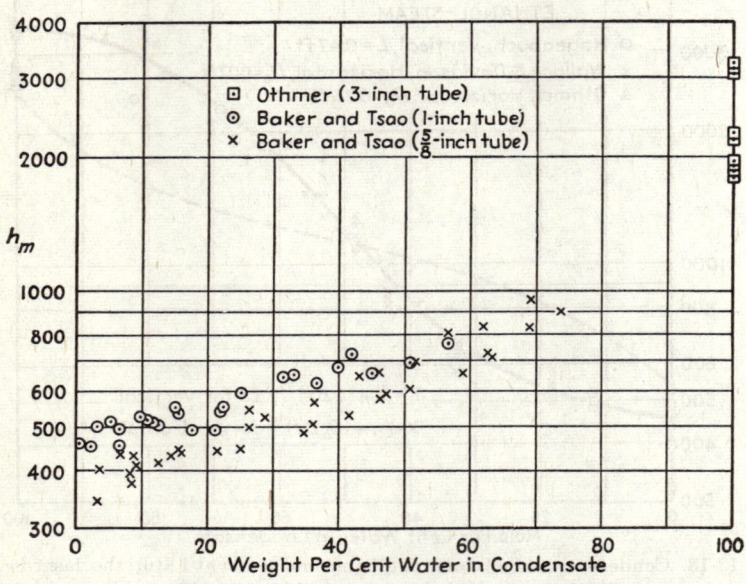

Fig. 13-14. Condensation of mixtures of steam and organic vapors (benzene, toluene, chlorobenzene, or trichloroethylene) giving two-phase condensate; $h_m = q/(A)(t_i - t_s)_m$; t_i is the temperature at which sum of the vapor pressures of the water and the substantially immiscible organic liquid equals the total pressure.

cases, as suggested by Kirkbride[63] h was based on $t_i - t_s$, where t_i is the equilibrium temperature at which the sum of the vapor pressures of water and organic substance equals the total pressure (1 atm) and t_s is the mean temperature of the outer surface of the tube. It was found that h_m for the different mixtures and a given diameter could be correlated in terms of the weight per cent water in the condensate. Since, in most runs, nearly all the entering vapor was condensed, the composition of the withdrawn condensate differed little from that of the entering vapor and equally good correlation could be obtained in terms of composition of entering vapor. The effect of diameter, if any, is difficult to determine; the data are plotted in Fig. 13-14.

IV. EFFECT OF NONCONDENSABLE GAS IN CONDENSATION OF VAPORS ON TUBES

Mechanism. When a mixture of a condensable vapor and a noncondensable gas is exposed to a surface colder than the dew point of the mixture some condensation occurs. In the absence of dropwise condensation, a layer of condensate collects on the cooling surfaces, and a film of a mixture of noncondensable gas and vapor forms next to the condensate layer, the concentration of vapor in the gas film being lower than in the main body of the mixture. As pointed out by Lewis,[68] because of the difference in partial pressure of the vapor between the main body of the mixture and that at the interface between gas and liquid films, the vapor diffuses from the main body through the gas film to liquefy at the interface. Thus both the latent heat of condensation and the sensible heat lost by the vapor are transferred through the condensate layer. Nevertheless, the latent heat is not transferred through the gas film unless, under special conditions, with a very cold surface, the dew point might be reached in the gas film, causing "fogging" in the gas film. As the main body of the mixture flows past the cooling surface, it is cooled and the sensible heat so removed is transferred through the gas film by conduction and convection, later to pass through the condensate layer and metal wall to the cooling medium on the other wide. The rate of condensation is thus governed by the laws of diffusion of vapor through a film of noncondensable gas, whereas sensible heat transmission is governed by the usual laws of heat transfer by conduction and convection.

Data[69,28,87] are available for dehumidifying air-steam mixtures, but these are strictly applicable only for the conditions of the experiments.

Method of Calculation. The following design procedure is that recommended by Colburn and Hougen.[28] A partial resistance $1/U'$ is computed that includes only the resistances of the water side, the dirt deposit, the tube wall, and the condensate film:

$$\frac{1}{U'} = \frac{1}{h_L A_L/A_v} + \frac{1}{h_d A_w/A_v} + \frac{L}{k_w A_w/A_v} + \frac{1}{h_m} \quad (13\text{-}20)$$

Since the heat transferred to the cooling water, neglecting subcooling of the condensate, equals that delivered to the gas-liquid interface, for an element of surface dA_v the following heat balance is employed:

$$W c_L \, dt_L/dA = U'(t_i - t_L) = h_G(t_v - t_i) + \lambda[K_G(p_v - p_i)] \quad (13\text{-}21)$$

One assumes a value of t_i and substitutes t_i, together with the corresponding vapor pressure p_i, into Eq. (13-21) and repeats the procedure until the equation balances, thus giving the correct value of t_i. The total area required is then determined by graphically or otherwise integrating the following equation:

$$A = \int_0^q \frac{dq}{U'(t_i - t_L)} \tag{13-22}$$

A detailed example of the use of these equations, in designing a tubular dehumidifier, is given elsewhere.[28] The mass-transfer coefficient K_G is obtained by substituting the dimensionless terms

$$\frac{K_G p_{nm}}{G}\left(\frac{\mu_v}{\rho_v D_v}\right)^{2/3} \quad \text{or} \quad \frac{K_G p_{nm}}{G_{\max}}\left(\frac{\mu_v}{\rho_v D_v}\right)^{2/3}$$

for the corresponding dimensionless terms

$$\frac{h_G}{c_{pv} G}\left(\frac{c_p \mu}{k_v}\right)^{2/3} \quad \text{or} \quad \left(\frac{h_G}{c_{pv} G_{\max}}\right)\left(\frac{c_p \mu}{k_v}\right)^{2/3}$$

appearing in Eq. (9-10b).

V. DIRECT CONTACT OF LIQUID AND GAS

Introduction. In a previous section, methods of calculation were given for an indirect type of dehumidifier, where the gas and cooling medium were separated by a metal wall through which the heat was conducted. In some instances, the cooling water is brought into direct contact with the gas, as in a spray chamber or in a packed tower. Nevertheless, the same general method of calculation applies, except that it may be difficult to evaluate the square feet of cooling surface in which case performance is reported on a basis of unit volume of the apparatus. Thus, let a represent the square feet of cooling surface per cubic foot of chamber, having height z, gross cross section S_o, and consequently gross volume $S_o z$; the total cooling surface A is equal to the product $a S_o z$.

In apparatus such as humidifiers, water coolers, and dehumidifiers, the partial pressures of the water vapor are low relative to those of the air, and it is convenient to employ absolute humidity (pounds of water per pound of dry air), rather than partial pressure of water vapor, in the potential for mass transfer.

Basic Equations. Consider steady operation of a vertical tower in which air and water are brought into direct countercurrent contact. Air of dry-bulb temperature t_{G1} enters at constant mass rate w_G at the bottom and leaves at t_{G2} at the top, and water enters the top at mass rate W_1 and temperature t_{L1} and leaves the bottom at W_2 and t_{L2}. Let H_G represent the absolute humidity of the gas stream, expressed as pounds of water vapor per pound of dry air. The water balances are

$$W - W_2 = w_G(H - H_{G1}) \tag{13-23}$$
$$dW = w_G \, dH \tag{13-23a}$$

CONDENSING VAPORS

The enthalpy i_v of 1 lb of low-pressure steam at any temperature t equals the enthalpy change λ_0 due to vaporization at the base temperature t_0 plus the enthalpy corresponding to the superheat: $i_v = \lambda_0 + 0.45(t - t_0)$. The enthalpy i_a of 1 lb of dry air having a specific heat of 0.24 is $i_a = 0.24(t - t_0)$. The enthalpy i_G of 1 lb of dry air plus H_G lb of water vapor is consequently

$$i_G = i_a + i_v H_G = 0.24(t - t_0) + H_G[\lambda_0 + 0.45(t - t_0)] \quad (13\text{-}24)$$

As suggested in 1908 by Grosvenor,[45] the term $0.24 + 0.45 H_G$ is designated as the humid heat c_s, and hence

$$i_G = c_s(t - t_0) + \lambda_0 H_G \quad (13\text{-}24a)$$
$$di_G = c_s\, dt + \lambda_0\, dH_G \quad (13\text{-}24b)$$

The enthalpy of 1 lb of liquid of specific heat c_L is $c_L(t_L - t_0)$. For adiabatic operation, an enthalpy balance (above a base temperature t_0 for both gas and liquid) for the upper part of the tower gives

$$W_1 c_L(t_{L1} - t_0) + w_G[\lambda_0 H_G + c_s(t_G - t_0)] = W c_L(t_L - t_0) + w_G[\lambda_0 H_{G2} + c_{s2}(t_{G2} - t_0)] \quad (13\text{-}25)$$

Differentiating Eq. (13-25) and neglecting unimportant terms, one obtains

$$w_G\, di_G = W c_L\, dt_L \quad (13\text{-}25a) \star$$

The rate of transfer of heat from the stream of liquid at t_L through the liquid film to the liquid-gas interface at t_i is

$$W c_L\, dt_L = h_L a_H S_o\, dz(t_L - t_i) \quad (13\text{-}26)$$

and the rate of transfer of sensible heat from the interface through the gas film to the gas stream at t_G is

$$w_G c_s\, dt_G = h_G a_H S_o\, dz(t_i - t_G) \quad (13\text{-}27)$$

The rate of diffusion of water vapor from the interface through the gas film to the gas stream is proportional to the humidity H_i at the interface (based on the saturation humidity at t_i) less the humidity H_G of the gas stream:†

$$w_G\, dH_G = K'_G a_M S_o\, dz(H_i - H_G) \quad (13\text{-}28)$$

Since, fortuitously, for air and water vapor at ordinary temperatures, the ratio of thermal and mass-transfer diffusivities is such that h_G is

† The mass transfer coefficients K'_G of Eq. (13-28) and K_G of Eq. (13-21) are related by the relation $K'_G(H_i - H_G) = K_G(p_i - p_G)$, and H_G and p_v are related by the definition (based on the gas laws)

$$H_G = \frac{p_v}{P - p_v} \frac{M_v}{M_a}$$

As an approximation, $K'_G = P K_G M_v / M_a$.

substantially equal† to $K_G'c_s$, neglecting any difference between the surface areas a_M and a_H for the transfer of mass and of heat, Eq. (13-27) becomes

$$w_G c_s \, dt_G = K_G' a S_o \, dz(c_s t_i - c_s t_G) \qquad (13\text{-}29)$$

Upon multiplying both sides of Eq. (13-28) by λ_0 and adding to Eq. (13-29), defining i_i as $\lambda_0 H_i + c_{si} t_i$ and neglecting the differences between c_s and c_{si}, the following relation is obtained:‡

$$w_G \, di_G = K_G' a S_o \, dz(i_i - i_G) \qquad (13\text{-}30) \star$$

Figure 13-15 shows values of i_i vs. t_i for air saturated with water vapor at standard barometric pressure,§ based on the relation $i_i = 1061.4 H_i + c_s t_i$.

Let the wet-bulb temperature of the entering air be represented by t_{wb}. Since for air and water the usual equation, relating the wet- and dry-bulb temperatures and the corresponding humidities H_{wb} and H_G, takes the form $i_{wb} = i_G$, the enthalpy i_G of the vapor stream is fixed by t_{wb}, and hence the wet-bulb lines start on the saturation curve and run horizontally to the right on Fig. 13-15.

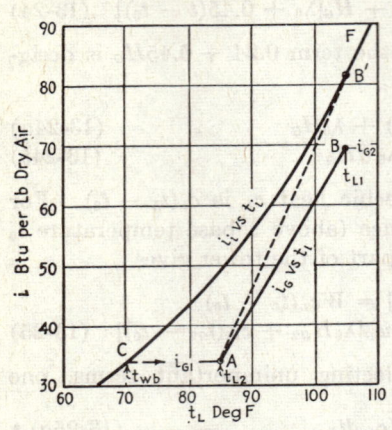

Fig. 13-15. Graphical method[97] for design of cooling towers.

The relation between i_G and t_L is given by Eq. (13-25a), but ordinarily the fractional change in W is so small that little error is made by assuming W constant, and for these conditions Eq. (13-25a) shows that t_L is linear in i_G, the slope di_G/dt_L being equal to Wc_L/w_G.

Since the enthalpy i_{G1} and water temperature t_{L2} are fixed at the bottom of the tower, point A of Fig. 13-15, the enthalpy balance line is a straight line[97] passing through point A. In order to reach the abscissa t_{L1}, corresponding to the temperature of the water entering the top of the tower,

† Based on pp. 587 to 592 of reference 107.
‡ Merkel,[78] Goodman,[42] Boelter,[12] and Sherwood.[97]
§ In such calculations, one usually employs base temperatures of 0°F for the dry air and 32°F for the steam. Since the latent heat of vaporization at 32°F is 1075.8 Btu/lb of steam and the specific heat of low-pressure steam is 0.45, $i_v = 1075.8 + 0.45(t - 32) = 1061.4 + 0.45t$. Per pound of dry air, containing H lb of water vapor, the enthalpy is $i_G = 0.24t + i_v H = c_s t + 1061.4 H$.

If t_0 is taken as 0°F for *both* air and steam, and since the extrapolated value of λ_0 is 1094, $i = 0.24t + (1094 + 0.45t)H = c_s t + 1094 H$. Each of the two procedures gives $i = C_2 H + c_s t$; C_2 is 1061.4 when 0 and 32°F are used as base temperatures for air and steam, respectively, and C_2 is 1094 when 0°F is used for both air and steam. Since both the equilibrium curve and the operating line will be based on the same value of C_2, it makes little difference which value of C_2 is used.

the steepest possible line is AB', which contacts the saturation curve CF at point B'. The corresponding maximum slope is $(i_{G2} - i_{G1})/(t_{L2} - t_{L1})$ $= (Wc_L/w_G)_{\max}$, but since the potential difference $i_{G2} - i_{i2}$ is zero, Eq. (13-30) shows that the corresponding height of tower is infinite. Consequently, it will be necessary to decrease the ratio of water to dry air, using a flatter operating line such as AB.

Approximate Method. Since it is difficult to measure the interfacial temperature, it usually is assumed equal to the bulk temperature of the liquid, and Eq. (13-30) is written to involve an over-all coefficient K'_{OG}:

$$w_G \, di_G = K'_{OG} a S_1 \, dz (i_L - i_G) \tag{13-30a}$$

If $h_L a$ were infinite, i_i would equal i_L and consequently K'_G would equal K'_{OG}. Actually $h_L a$ is finite; consequently, t_i is lower than t_L, $i_i - i_G$ is less than $i_L - i_G$, and K'_G is larger than K'_{OG}. If in the range involved the equilibrium curve is linear, comparison[99] of Eqs. (13-30) and (13-30a) shows that

$$\frac{i_L - i_G}{i_i - i_G} = \frac{K'_G}{K'_{OG}}$$

Because of the constant ratio of the apparent to the true enthalpy potential, the ratio of the true and apparent enthalpy-transfer coefficients is consequently constant. Apparent coefficients should not be extrapolated from one range of temperatures to another, owing to the change in the slope of the equilibrium curve. With this limitation, use of apparent coefficients is helpful in the design of cooling towers.

At any t_L, the potential difference $i_L - i_G$ corresponds to the vertical distance[97] between the saturation curve (i_L vs. t_L) and the operating line (i_G vs. t_L), and the height z of the tower is found by integration:

$$z = \frac{G_G}{K'_{OG} a} \int \frac{di_G}{i_L - i_G} = z_{OG} \int \frac{di_G}{i_L - i_G} \tag{13-30b} \star$$

It is noted that the ratio w_G/S_o, or G_G, is the mass velocity based on the dry-gas rate and the total cross section. Since for a given G_L of the liquid ($G_L = W/S_o$) in a given type of contacting device the over-all mass-transfer coefficient $K'_{OG} a$ is a function of G_G, the ratio $G_G/K'_{OG} a = z_{OG}$ is sometimes called[17] the *height of an over-all mass-transfer unit*. For a given type of contacting device and a fixed value of G_L, one may plot either $K'_{OG} a$ or z_{OG} vs. G_G. Values of z_{OG} are given in Table 13-9.

Illustration 2. It is desired to design a forced-draft tower to cool 7450 lb/hr of water from 105° to 85°F, using air entering with a wet-bulb temperature of 70°F at the bottom. In order to avoid entrainment of liquid by the gas stream, it is agreed to use G_G of 600 lb/(hr)(sq ft of total cross section). In order that the height of the tower be not excessive, it is planned to use a ratio of water to air equal to three-fourths of the maximum value.

Calculate the total cross section and number of over-all mass-transfer units required.

Solution. The enthalpy i_{G1} of the entering air is obtained from Fig. 13-15 by reading 34 Btu/lb from the saturation curve at $t_{wb} = 70°F$. This is plotted in Fig. 13-15 as the ordinate of point A, corresponding to the bottom of the tower, the abscissa being $t_{L2} = 85°F$. The maximum value of W/w_G corresponds to equilibrium at the top of the tower, reference to Fig. 13-15 shows that at t_{L1} of 105°F, i_{G2} is 81 Btu/lb of dry air. From the approximate enthalpy balance

$$\frac{W}{w_G} = \frac{i_{G2} - i_{G1}}{c_L(t_{L1} - t_{L2})} = \frac{81 - 34}{(1)(105 - 85)} = 2.35 \frac{\text{lb water}}{\text{lb dry air}}$$

This is the slope of the dotted line AB'. Then the actual ratio of water to air will be $(2.35)(\frac{3}{4}) = 1.76$, and the air rate will be $(7450)/(1.76) = 4230$ lb/hr of dry air, which corresponds to a tower having gross cross section S_o of $4230/600 = 7.05$ sq ft.

From point A in Fig. 13-15, the actual operating line AB is drawn with a slope of 1.76, and the point B is located at the intersection of this line with t_L of 105°F at an ordinate of i_{G2} of 69.3. From Eq. (13-30b),

$$z = z_{OG} \int_{34}^{69.3} \frac{di_G}{i_L - i_G}$$

where i_L corresponds to saturation at t_L. At any point the over-all driving force $i_L - i_G$ is the difference between the ordinates of the saturation line and the operating line, and, in general, the preceding equation for z must be solved by a graphical integration.† Since, over the range of i_L from 34 to 81, the curvature of the saturation line is not great, a sufficiently accurate approximation may be made by using the arithmetic average of the driving forces at five points:

$$(i_L - i_G)_a = \frac{15 + 13 + 11 + 11 + 11.7}{5} = 12.3$$

$$\int \frac{di_G}{i_L - i_G} = \frac{i_{G2} - i_{G1}}{(i_L - i_G)_a} = \frac{69.3 - 34.5}{12.3} = 2.83 \text{ over-all mass-transfer units}$$

Incidentally, the wet-bulb temperature of the exit air is found to be 99°F.

The value of $G_L = 7450/7.05 = 1060$ lb of water/(hr)(sq ft of total cross section). The value of z_{OG} may depend upon both G_G and G_L, as well as the type of contacting device used; values of z_{OG} are given in Table 13-9. The height z of the tower is $2.83 z_{OG}$.

Performance data are available for several types of cooling towers: atmospheric type with cross flow of air,‡ forced draft with counterflow of air,§ and spray ponds.|| Table 13-9 summarizes data for cooling water by air in several types of contacting devices. Cost data are available.[91,99]

Allowance for Liquid-film Resistance. Comparison[72] of Eqs. (13-26) and (13-30) shows that

$$\frac{i_i - i_G}{t_i - t_L} = -\frac{h_L}{K'_G} \qquad (13\text{-}31) \star$$

† A simple geometrical method for determining the integral has been devised by Baker.[4]

‡ Coffey and Horne,[21] Geibel,[40] and Perry.[91]

§ Geibel,[40] Johnstone and Singh,[59] London et al.,[70] Niederman et al.,[84] Perry,[91] and Simpson and Sherwood.[99]

|| Coffey and Dauphinee[20] and Perry.[91]

CONDENSING VAPORS

TABLE 13-9. DATA FOR FORCED-DRAFT COOLING TOWERS

Observer	Type of contacting device	z_{OG}, ft	G_G	G_L
Niederman et al.[84]	Vertical spray tower	10–2.5	420	200–600
		4.7	300–700	420
Geibel[40]	Grid packing	14.4	550	720
Robinson[92a]	Grid packing	19	1020–2280	3180
Johnstone and Singh[59]	Crossed grids[a]			
	Spacing 0.625 in.	1.1	2460	1100
	Spacing 1.25 in.	2.5	2460	1100
	Spacing 1.75 in.	3.7	2460	1100
	Spacing 2.25 in.	3.2–5.2	2460	1100
London et al.[70]	Ovate slats			
Parekh[72]	1-in. Raschig rings	0.4–2.0	180–700	500–3000
Hooker and Sackheim[72]	Spined tubing, horizontal, staggered	3.5–3.9	430–760	700

[a] Air was cooled by feeding water at wet-bulb temperature, and hence the values are for z_G, not z_{OG}. Wooden grids were 0.25 in. thick, 4.0 in. high and were crossed to form an "egg-crate" pattern.

and hence a "tie line" having a negative slope equal to h_L/K'_G may be used to relate i_i and t_i and t_L, as shown by the dotted lines in Fig. 13-16. The potential differences $i_i - i_G$ may be read from Fig. 13-16 and the tower height calculated from Eq. (13-30):

$$z = \frac{G_G}{K'_G a} \int \frac{di_G}{i_i - i_G} = z_G \int \frac{di_G}{i_i - i_G} \qquad (13\text{-}32) \star$$

where z_G is the height of a mass-transfer unit, based on the resistance of the gas phase itself. The application of the tie line is illustrated in Fig. 13-16.

Since it is difficult to measure t_i, the film coefficients h_G, h_L, and K'_G are best obtained by measuring K'_{OG} and U_G for cases in which the liquid-film resistance is inoperative, i.e., in an adiabatic humidifier in which water is fed at the wet-bulb temperature and consequently the change in water temperature is negligible, and substantially all the latent heat for evaporation is furnished by the cooling of the air. Since the heat lost or gained by the water is negligible, t_i is substantially equal to t_L; hence K'_{OG} equals K'_G and U_G equals h_G. Then, with the same values of G_G and G_L, the same apparatus is operated adiabatically as a water cooler or water heater (dehumidifier), and the new over-all values of K'_{OG} and U_G, particularly the former, should be smaller than before.

Reference 74 reports data for cooling air by water at constant temperature in an adiabatic tower packed with 1-inch carbon Raschig rings. The sump at the bottom was found equivalent to 0.6 ft of packing. The heat-transfer coefficients were correlated by

$$h_G a_H = 2.1 G_G^{0.7} G_L^{0.07}$$

for G_G ranging from 350 to 1020 and G_L ranging from 540 to 2620. The corresponding mass-transfer coefficients were somewhat smaller than expected from the relation $h_G a_H / K'_G a_M = c_s$, since the packing was not fully wetted except at the higher values of G_G, where the values of K'_G

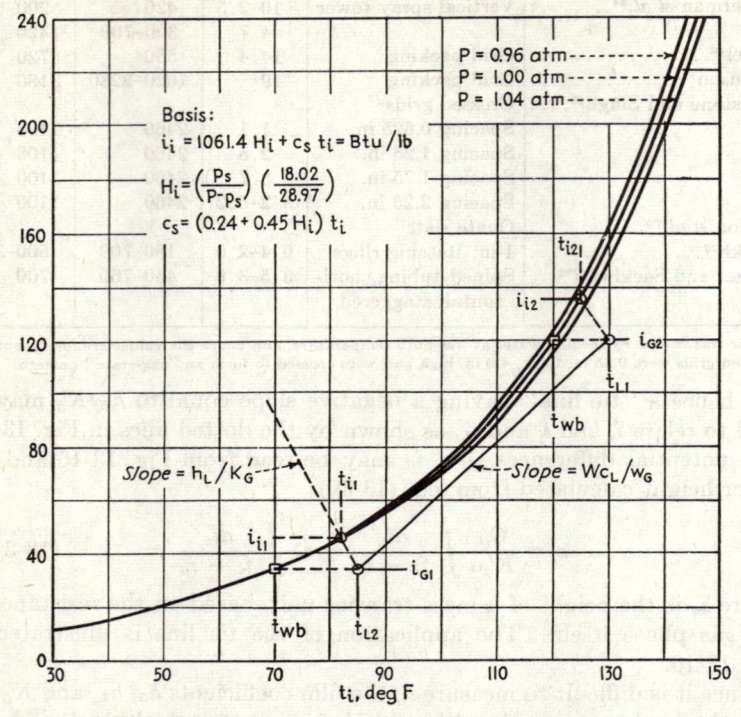

FIG. 13-16. Method of allowing for thermal resistance of the liquid film. In the example shown, water is to be cooled from 130 to 85°F by air initially having a wet-bulb temperature of 70°F, using 1.91 lb water per lb of dry air. Assuming $h_L a/K'_G$ of 3.4, the terminal values of t_i are found to be 125 and 82°F, and the gas-film driving forces are $136 - 120 = 16$ and $46 - 34 = 12$ Btu/lb dry air; if the liquid-film resistance is neglected, the corresponding values are $156 - 120 = 36$ and $49 - 34 = 15$.

were apparently equal to h_G/c_s. Cooling-tower runs were than made and gave

$$h_L a / K'_G a = 0.092 G_L^{0.43}$$

from G_L from 500 to 2000 and G_G from 350 to 1000. Inlet water temperatures ranged from 128 to 138°F, and outlet water temperatures ranged from 97 to 104°F; the ratio $h_L a/K'_G a$ ranged from 1.8 to 2.7. Considering a case with di_i/dt_i of 3 and $h_L a/K'_G a$ of 2.25, the ratio of the resistance of the gas phase to that of the liquid phase, both in the same units, is $3/2.25 = 1.33$.

In cooling air flowing up a vertical tower cooled by water at uniform temperature flowing downward over Berl saddles, at the lower values of G_G and G_L, Hensel and Treybal[51] found that values of $h_G a_H / K'_G a_M$ were as high as twice c_s, since a_H was apparently twice a_M; at high values of G_G and G_L, $h_G a_H / K'_G a_M$ was substantially equal to c_s.

By measuring the constant rate of evaporation of water into warmer air from porous 1- and 2-inch Raschig rings, Taecker and Hougen[102] correlated the heat-transfer coefficients from air to wetted rings by the dimensionless equation

$$\frac{h_G}{c_p G_0} \left[\frac{c_p \mu}{k} \right]_f^{2/3} = 1.15 \left[\frac{G_0 A_p}{\mu} \right]^{-0.41}$$

where A_p is the total surface area of one particle and G_0 is the superficial mass velocity.

Schulman and DeGouff[94] estimate that the wetted area for 1-inch ceramic Raschig rings increases with increase in G_L and was rather insensitive to variations in G_G.

The method based on the use of the tie line and Eq. (13-32) gives the final enthalpy of the gas and consequently its wet-bulb temperature, but not the dry-bulb temperature. A method of finding the latter by a graphical procedure was recently developed by Mickley.[79]

The key to the method is the division of Eq. (13-30) by Eq. (13-29), giving a relation between the changes in gas enthalpy and gas temperature and the corresponding potentials for the transfer of enthalpy and sensible heat:

$$\frac{di_G}{dt} = \frac{i_i - i_G}{t_i - t_G} \qquad (13\text{-}33) \ \star$$

In using Eq. (13-33) graphically the bulk temperature t_{G1} of the entering air is plotted as abscissa vs. the corresponding value i_G as ordinate, and an incremental "dog-chase-rabbit" procedure is used, as illustrated in Fig. 13-17. Upon assuming that i_i and t_i remain constant for a small incremental change in gas enthalpy, a straight-path line is drawn from the point t_G, i_G (E) toward the point t_i, i_i (D) ending after a small increment in i_G, thus giving the new values of t_G and i_G (F). A horizontal line is then drawn from the new i_G to the operating line (i_G vs. t_L), giving the new water temperature t_L. The new tie line [based on Eq. (13-31)] is then drawn from the point i_G and t_L (J) to the equilibrium curve, giving the new values of t_i and i_i (K). This procedure is repeated until the final value of i_G is reached, thus giving the final gas temperature (M).

If, in a test on a cooling tower, data are available on S_o, z, W, w_G, t_{G1}, t_{wb1}, t_{G2}, and t_{wb2}, the Mickley method can be used to solve for the slope of the tie line ($h_L a / K'_G a$) necessary to predict the measured value of t_{G2}, although a trial-and-error procedure is involved. Measurement of the final t_G is difficult if the outlet gas contains liquid spray.

If fog forms in the bulk gas stream, the transfer equations given above are not applicable. Consequently, designs wherein the gas-temperature curve crosses the equilibrium curve should be avoided; such cases are readily detected by use of the Mickley graphical method.

The enthalpy potential method can also be adapted to cases in which the cooling medium is inside tubes, instead of being in direct contact with

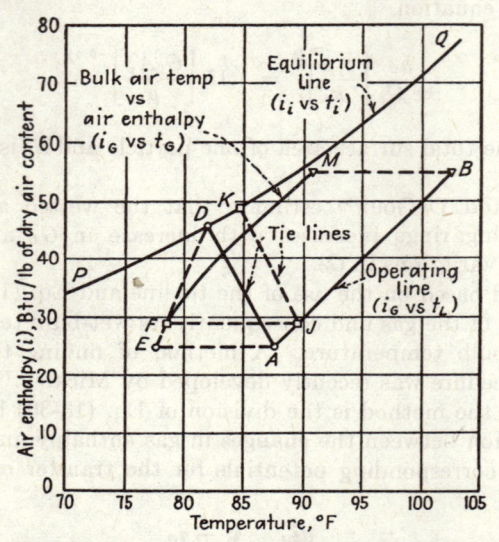

Fig. 13-17. Mickley[79] graphical method of predicting temperature of air leaving a forced-draft counterflow vertical cooling tower.

the gas. In such a case, $aS\,dz = dA_i$, and an over-all heat-transfer coefficient U'' in Eq. (13-26) is used instead of h_L:

$$\frac{1}{U''} = \frac{A''}{h'A'} + \frac{x_w A''}{k_w A_w} + \frac{1}{h''}$$

where h' and h'' refer to the heat-transfer coefficients on the inside and outside surfaces of the tube, respectively, and A' and A'' are the corresponding surface areas.

Frost Formation. When a mixture of air and water vapor is in contact with a cold surface having a temperature below 32°F, frost may form if the dew point of the air exceeds 32°F. References 61, 10, and 47a deal with frost formation.

Constant Water Temperature. In an adiabatically operated humidifier wherein the unevaporated water is recirculated, the water reaches an equilibrium temperature t_e, substantially equal to t_{wb}. The heat given up by the cooling of the air is transferred to the water and is largely consumed in evaporating water that humidifies the air. In this case, where the rates of heat transfer and vapor diffusion are substantially equal, the performance of the apparatus may be expressed in terms of either $h_G a$ or $K'_G a$, obtained by integrating Eqs. (13-27) and (13-28):

$$\ln \frac{t_{G1} - t_i}{t_{G2} - t_i} = \frac{h_G a_H S_o z}{w_G c_s} = \frac{h_G a z}{G_G c_s} = \frac{z}{z_t} \qquad (13\text{-}34) \star$$

$$\ln \frac{H_i - H_{G1}}{H_i - H_{G2}} = \frac{K'_G a_M S_o z}{w_G} = \frac{K'_G a z}{G_G} = \frac{z}{z_G} \qquad (13\text{-}35)$$

Since for air and water h_G equals $K'_G c_s$, Eqs. (13-34) and (13-35) show that H_G is linear in t_G, as is also found by integrating the heat balance, calling c_s constant.

PROBLEMS

1. It is proposed to heat 10,000 pounds/hr of process water from 60 to 80°F by direct countercurrent contact with air discharged from a drier. The hot air is available at the rate of 10,000 pounds/hr of bone-dry air with a dry-bulb temperature of 120°F and a wet-bulb temperature of 99°F.

For a tower with a cross-sectional area of 10 sq ft, the superficial mass velocities are 1000 lb bone-dry air/(hr)(sq ft) and 1000 lb water/(hr)(sq ft). For the proposed packing and the above superficial mass velocities, available data indicate that the following coefficients should apply: $h_G a = 120$ Btu/(hr)(cu ft)(deg F); $h_L a = 460$ Btu/(hr)(cu ft)(deg F).

a. Calculate the local rate of evaporation of water at the bottom of the tower in pounds of water per hour per cubic foot of tower volume.
b. What are the dry- and wet-bulb temperatures of the air leaving the top of the tower?
c. *Estimate* the depth of the packing required.
d. Calculate the maximum amount of process water which can be heated from 60 to 80°F by countercurrent contact with the air from the drier if the tower height were infinite.

2. The following data were obtained in tests on a packed column being used to cool water by countercurrent contact with atmospheric air. Water was fed to the top of the tower at a rate of 10,000 lb/hr and a temperature of 103°F and left the tower at a temperature of 82°F. Air was fed to the tower at a rate of 21,300 lb of wet air/hr, with an entering temperature of 105°F and a dew point of 65°F. The temperature of the exit air was 90.5°F.

a. Assuming that the flow rates of *dry* air and water remain the same and that the water still enters at 103°F, calculate the temperature to which the water will be cooled in this tower on a day when the atmospheric air is at a temperature of 100°F and has a dew point of 81.5°F. What would be the temperature of the exit air?
b. Estimate the quantity of water evaporated per hour under the test conditions and under the new conditions.

3. It is proposed to use an existing tubular heat exchanger to condense exhaust steam. The steam, saturated at a gauge pressure of 5 lb/sq in., is to condense on the outside of the ¾-inch 18 BWG copper condenser tubes. The water, available at a temperature of 75°F, is to flow through the tubes in a single pass. The tubes in the exchanger are 12 ft long, and the exchanger is presently installed in a horizontal position. The exchanger could be installed in a vertical position if extra piping were provided. Assuming the tubes to be clean and neglecting any effect of vapor velocity, calculate the *maximum* capacity (at infinite water flow rate) of this exchanger, expressed as pounds of steam condensed per hour, for:

a. The horizontal position.
b. The vertical position.

Note. The tube layout for the horizontal position is given in the table below, where n is the number of vertical tiers having N tubes per tier (thus the total number of tubes is 385):

N	3	9	13	15	17	19	21	23
n	2	2	2	2	2	4	6	3

4. Tests have been made to determine whether or not heat-transfer coefficients for organic liquids with high surface tension would follow the prediction of Nusselt when condensing on horizontal finned tubes, described below. The over-all heat-transfer coefficients were measured at a series of water velocities and were plotted by the graphical method of Wilson to obtain the condensing-film coefficients.

Calculate the experimentally determined heat-transfer coefficient for acetone condensing at 154°F. Compare the calculated coefficient with the coefficient predicted by the Nusselt equations.

Physical dimensions of copper condenser tube. Fifteen fins per in.
Root diameter, 0.621 inch. Length of finned section, 34⅜ inches.
Diameter over fins, 0.746 inch. Length of cooled section, 35⅞ inches.
Diameter of plain ends, 0.750 inch. Actual outside area, 1.452 sq ft.

Physical properties of acetone:

   Latent heat of vaporization at 154°F = 121.4 cal/gm
                  Viscosity at 94°F = 0.30 centipoise
              Thermal conductivity = 0.1 Btu/(hr)(sq ft)(deg F per ft)

Test Data.

           Avg bulk-water temperature = 74.0°F
       Avg condensing-vapor temperature = 154.0°F
        Avg mean temperature difference = 80°F

Run number	Avg water velocity, ft/sec	U, Btu/(hr)(sq ft)(deg F)(based on outside area)	Water rate, lb/hr	Temperature rise of water, deg F
48	4.60	254.5	1709	17.28
51	6.43	289.6	2394	14.08
53	8.60	324.3	3208	11.75
55	10.68	349.1	3982	10.17
57	12.40	367.5	4620	9.25

5. Ultimately it will be necessary to design a multitubular *total* condenser for an ammonia-refrigeration plant to condense 8000 pounds of ammonia/hr, but because of uncertainties in some of the design data, it is planned to construct a pilot unit, con-

CONDENSING VAPORS

sisting of a single tube of steel, having an o.d. of 1.000 inch and a wall thickness of 0.049 inch, provided with a suitable external water jacket. In the pilot unit, ammonia vapor at an absolute pressure of 300 pounds/sq inch and a temperature of 300°F is to be introduced continuously at the top of the vertical tube at a rate of 20 pounds/hr. The condensate will leave at the bottom, and brackish clean river water at 80°F will flow vertically upward through the jacket, leaving at 105°F. Because of limitations of headroom, it is desired to attain complete condensation at the bottom, thus using the *shortest possible* length of condenser tube. Other data are given below.

a. Estimate the tube length.
b. Estimate the temperature of the condensate leaving the bottom of the tube.
c. Should the pilot plant differ from the one proposed?

Enthalpy of liquid ammonia:

t, deg F	80	90	100	110	123
i, Btu/lb	131.5	143	154.5	166	181

Properties of NH_3 vapor at 300 psia;

t, deg F	Saturated vapor at 123°	Superheated vapor at 300°
i, Btu/lb	634	758
v, cu ft/lb	1.003	1.496

The critical pressure is 1638 lb/sq in. abs, and critical temperature is 270°F.

It will be assumed that the ammonia side will remain clean, but the water side will become fouled. The coefficient on the water side when clean will be 2000 Btu/(hr)(sq ft)(deg F). Film-type condensation will be obtained. The saturation temperature is 123°F; $\rho_L = 35.2$ lb per cu. ft.

CHAPTER 14

BOILING LIQUIDS

Abstract. The first section of this chapter discusses boiling of saturated liquids on submerged heat-transfer surfaces. The factors which influence the heat-transfer film coefficient in nucleate boiling are considered, and data for a number of liquids are presented in graphical form. The peak heat flux $(q/A)_p$ for nucleate boiling is then treated in detail, and several correlations are shown. A theoretical treatment of film boiling is shown to agree with data.

The second part of the chapter is concerned with surface boiling of subcooled liquids. In this phenomenon, often called local boiling, a liquid below its saturation temperature is in contact with a surface hotter than the saturation temperature of the liquid. Extremely high heat-transfer rates are obtained as a result of the local generation and collapse of bubbles on the surface.

The third section starts with a discussion of horizontal two-phase flow of air-water mixtures, giving data for the pressure drop, liquid holdup, and heat-transfer coefficient. The more complex case of vaporization with forced circulation inside horizontal tubes is then considered, bringing in the additional factors of bubble formation inside the tube and acceleration of the stream as the quantity of vapor is increased. Vaporization with forced circulation inside vertical tubes is considered, introducing the variation in hydrostatic head as an important factor. Finally the performance of natural-circulation evaporators is discussed.

I. BOILING OF SATURATED LIQUIDS ON SUBMERGED SURFACES

Heat transfer to liquids boiling on submerged surfaces in pools is treated first because the factors involved are less complex than those encountered when boiling occurs inside tubes with natural or forced circulation. Horizontal tubes internally heated by condensing vapor are employed in one form of commercial evaporator to boil liquids surrounding the tubes. Liquids are also boiled on the submerged wall surface of heated kettles of various types. Submerged tubes of small dimensions have been used with condensing vapor in a number of experimental studies of boiling. Recently the use of electric-resistance heating in submerged tubes, wires, and strips has permitted the range of data to

Table 14-1. Nomenclature

- A Area of heat-transfer surface, square feet; unless otherwise specified, A refers to the boiling side
- a Empirical constants, a_1, a_2, etc.
- c_p Specific heat of liquid, Btu/(lb)(deg F)
- D Diameter of tube, feet
- e Base of Napierian or natural logarithms
- f Prefix denoting functional relationship
- G Mass velocity, lb/(hr)(sq ft of cross section), $G = w/S$
- g Acceleration due to gravity, ordinarily taken as 4.17×10^8 ft/(hr)(hr)
- h Coefficient of heat transfer, Btu/(hr)(sq ft)(deg F); h_B for mean value in the boiling section; h_c for conduction and convection through the vapor film in presence of radiation; h_{co} for conduction and convection through the vapor film in absence of radiation; h_L for heating liquid without phase change; h_m for mean value in entire apparatus
- k Thermal conductivity of liquid, Btu/(hr)(sq ft)(deg F per ft); k_v for vapor
- L Tube length, feet
- m An exponent in Eq. (14-8)
- P Pressure, lb per square inch absolute; P_c for critical pressure at which liquid and vapor have same properties
- q Rate of heat transfer, Btu per hour
- q/A Heat flux, Btu/(hr)(sq ft on boiling side); $(q/A)_p$ for peak heat flux in range of nucleate boiling
- R_L Volume fraction of the tube filled with liquid, dimensionless
- S Cross section for flow, square feet; for round pipe $S = \pi D^2/4$
- t Temperature of liquid, degrees Fahrenheit; t_0 for temperature in vapor-liquid separator; t_{sat} for saturation temperature of liquid; t_{sv} for saturation temperature of vapor; t_w for surface temperature
- U Over-all coefficient of heat transfer, Btu/(hr)(sq ft)(deg F); U_B for boiling section; U_o based on $t_{sv} - t_0$; U_m for mean value
- V Average velocity, feet per second; V_G for the gas in two-phase flow; V_L for the liquid in two-phase flow; V' for velocity entering tube
- w Mass rate of flow, lb per second; w_G for gas; w_L for liquid
- X Two-phase parameter defined by Eq. (14-11), the square root of the ratio of the pressure drop for the liquid alone to the pressure drop for the gas phase alone
- y Weight fraction of vapor in stream, dimensionless

Greek

- Δt Temperature difference between surface and fluid, degrees Fahrenheit; Δt_c at peak heat flux; Δt_i for initiation of nucleate boiling in tubes; Δt_m for mean value along tube, Δt_o for $t_w - t_0$
- $\Delta p/\Delta L$ Pressure gradient, lb/(sq in. abs)(ft); $(\Delta p/\Delta L)_G$ for gas alone; $(\Delta p/\Delta L)_L$ for liquid alone; $(\Delta p/\Delta L)_{TP}$ for co-current flow of two-phase mixture
- λ Latent heat of vaporization, Btu per pound
- μ Viscosity, lb/(hr)(ft); μ_G of the gas; μ_L of the liquid, μ_v of the vapor
- π 3.1416
- ρ Density, lb per cubic foot; ρ_G of gas; ρ_L of liquid; ρ_v of vapor
- Φ_L Two-phase parameter defined by Eq. (14-10); the square root of the ratio of the pressure drop for two-phase flow to the pressure drop for the liquid alone

be extended to extremely high heat-transfer rates and temperature differences.

Regimes of Boiling. The existence of several regimes of boiling was first clearly discussed by Nukiyama[73] in 1934, although Leidenfrost[59] in 1756 and Lang[56] in 1888 had been aware of the existence of minimum and maximum rates of evaporation. Drew and Mueller[35] presented further data and reviewed the theory in 1937.

Consider, for example, the boiling of water on an electrically heated horizontal platinum wire. The heat flux q/A may be computed from the measured input of electrical energy and the measured surface area of the wire. The temperature difference Δt between the surface temperature t_w of the wire and the saturation temperature t_{sat} of the liquid may be

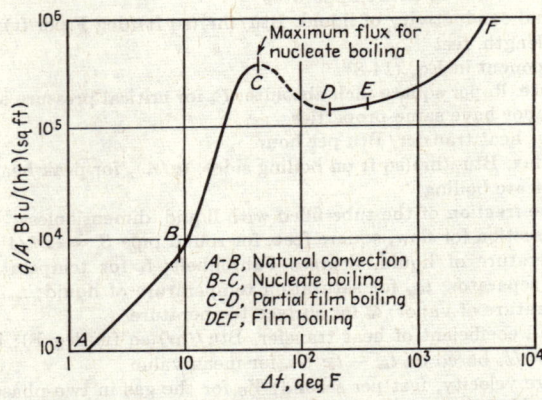

FIG. 14-1. Boiling of water at 212°F on an electrically heated platinum wire, showing different regimes.[73]

obtained by the techniques of resistance thermometry with suitable correction for radial gradients in temperature within the wire. The resulting graph of q/A vs. Δt is shown in Fig. 14-1 for water boiling at 1 atm. It is similar *in shape* to the boiling curves which are available for other fluids and other pressures.

In the range AB, the liquid water is being superheated by natural convection, and evaporation occurs only at the surface of the pool. As might be expected with this mechanism, q/A is proportional to $\Delta t^{5/4}$. In the range BC, bubbles form at active nuclei on the heat-transfer surface and rise through the pool, setting up natural circulation currents. In this range of *nucleate boiling*, q/A varies as Δt^n, where n ranges from 3 to 4. At point C the heat flux goes through a maximum, or *peak heat flux*, at a temperature difference called the "critical Δt." In the range CD, where part of the surface is insulated with a vapor film, q/A decreases as Δt increases. At point D, the heat flux goes through a minimum at

BOILING LIQUIDS 371

the Leidenfrost point.[59] In the *film-boiling* region EF, heat is transferred through the vapor film by conduction and radiation. At point F, the Δt corresponds to the melting point of the wire; a further increase in heat generation would cause the wire to burn out. If the ordinate at point F were lower than that of C, any increase in heat supply above the peak flux would cause the electric heater to melt unless the heat input were

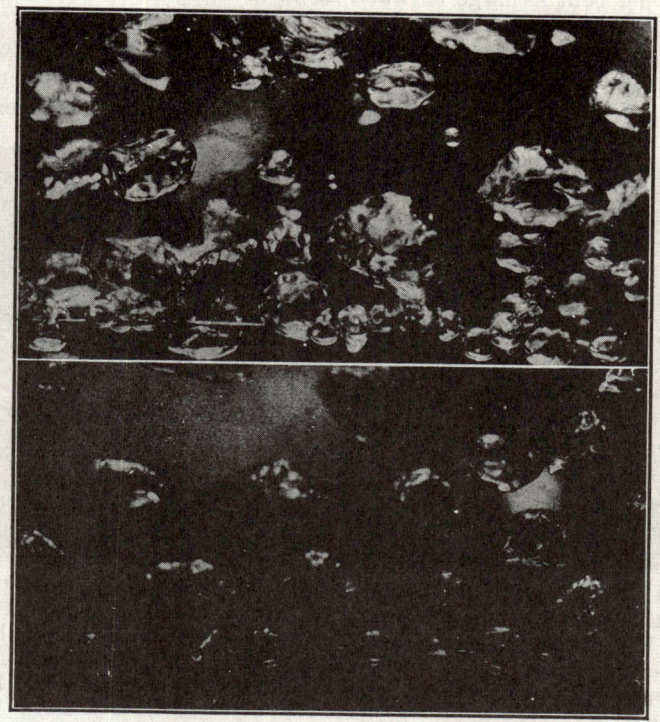

Fig. 14-2 (*top*). Nucleate boiling of a pool of water at 212°F by a horizontal wire of Chromel A, at a density of heat flux of 80 per cent of the peak value; Δt is small. (*Edgerton photograph by Castles.*[22])

Fig. 14-3 (*bottom*). Film boiling of a pool of water at 212°F by a horizontal wire of Chromel A, at a density of heat flux of 72 per cent of peak value; wire is red hot and Δt is large. (*Edgerton photograph by Castles.*[22])

immediately reduced as soon as the critical Δt was exceeded. The curve of Fig. 14-1 is reversible under steady conditions, unless, while operating in the range CF, irreversible changes occur in the heating surface.

Figures 14-2 and 14-3 show Edgerton high-speed photographs by Castles[22] of water boiling at 212°F on a horizontal wire of Chromel C, at a heat flux of 72 to 80 per cent of the peak value. In the region of nucleate boiling (Fig. 14-2) Δt is small, while in the region of film-boiling

(Fig. 14-3) the wire is red hot and Δt is large. The continuous film of vapor which virtually insulates the wire is visible in Fig. 14-3.

Figure 14-4 shows two high-speed photographs of ethyl acetate boiling at atmospheric pressure in a pool heated by steam condensing inside a horizontal tube of aluminum having an outside diameter of 0.5 in. In using an over-all Δt (from steam to boiling liquid) of 104°F, the flux is only 14 per cent as large as when Δt_o is 73°F. The critical Δt will be discussed in detail later.

Mechanism of Nucleate Boiling. At a given temperature, the vapor pressure over a very small concave liquid surface, such as that generated by a steam bubble in water, is less than the vapor pressure over a flat

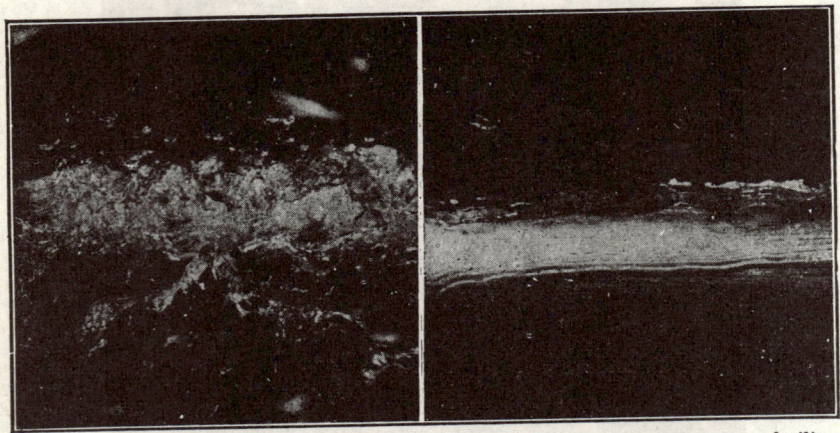

FIG. 14-4. Photographs by E. T. Sauer[81] and W. B. Tucker of ethyl acetate boiling at 1 atm. Nucleate boiling is shown in the left-hand picture; $\Delta t_o = 73°F$, $q/A = 41,000$. Film boiling, typical of a vapor-bound surface, is shown in the right-hand picture, where an excessive Δt_o of 104°F was used, giving q/A of 5800 and U_o of 53. Photographs were taken with an exposure of 1/100,000 sec, using the Edgerton technique.

liquid surface. Hence, for a given pressure, a liquid must be hotter to evaporate into a small bubble of vapor than into the vapor space above the liquid. The difficulty in forming vapor bubbles then becomes apparent, since the curvature of the liquid surface of a very small newly formed bubble is very great and the vapor pressure is reduced substantially, requiring a higher temperature than the saturation temperature corresponding to the pressure. Therefore it is not surprising that bubbles form preferentially at those active nuclei on the heating surface where the temperature and the nature of the surface are favorable. On some surfaces the vapor bubbles form readily and quickly detach themselves, whereas on others they form only when the liquid has a higher degree of superheat.[46]

Jakob and Fritz[46] studied the effect of roughness of the heating surface on heat-transfer rates and made temperature explorations in the boiling liquid and in the vapor space above it. In Fig. 14-5, some of these temperature explorations are plotted as degrees of superheat of liquid (temperature of liquid minus saturation temperature) vs. the vertical distance from the heated plate. For the rough heating surface, Curve C, the Δt required to transfer 14,500 Btu/(hr)(sq ft) was 10.8°F, whereas for the smooth surface, Curve A, the corresponding temperature drop

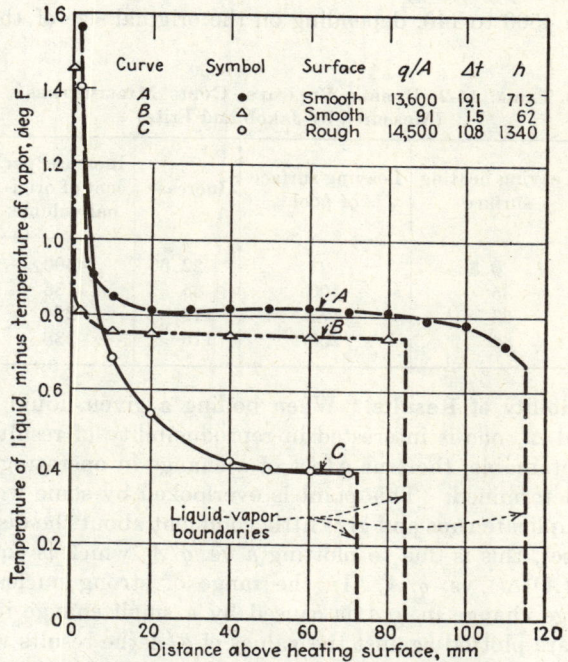

Curve	Symbol	Surface	q/A	Δt	h
A	●	Smooth	13,600	19.1	713
B	△	Smooth	91	1.5	62
C	○	Rough	14,500	10.8	1340

FIG. 14-5. Temperature explorations in water boiling at 1 atm on a horizontal plate. (*Data of Jakob and Fritz.*[46])

from plate to vapor required to transfer 13,600 Btu/(hr)(sq ft) was 19.1°F. In these two runs, the rough plate gave a coefficient of 1340, and the smooth plate gave 713; this was attributed to the larger number of vaporization nuclei on the rough plate. The liquid boiling on the rough plate had about half as much superheat as the liquid boiling on the smooth plate. Curve B, with a very low rate of heat transfer, 91 Btu/(hr)(sq ft), and Δt of 1.46°F, shows, nevertheless, approximately the same liquid superheat as Curve A. In the region near the heating surface, these curves show a very sharp increase in the temperature of the liquid, indicating substantial superheat in the liquid film.

At low flux, since a substantial portion of the heating surface is covered with liquid, a considerable fraction of the heat transferred from the surface is first used to superheat liquid. As a bubble leaves the heating surface and rises through the pool, the superheat is consumed in vaporizing additional liquid, thus increasing the volume of the bubble. By means of stroboscopic photography, Jakob and Fritz[46] measured the increase in volume of typical bubbles. As shown in Table 14-2, the increase in volume of the bubble after leaving the heating surface, expressed as a percentage of its volume leaving the heating surface, ranged from 4500 to 140, depending on the original size of the liberated bubble.

TABLE 14-2. BUBBLE VOLUMES, CUBIC MILLIMETERS
(Measured by Jakob and Fritz)

Leaving heating surface	Leaving surface of pool	Increase	Increase, per cent of original volume
0.5	23	22.5	4500
35	100	65	185
53	172	119	224
84	200	116	138

Reproducibility of Results. When boiling a given liquid at a given pressure and Δt, one is interested in reproducibility of results, for until these are obtainable, the true effect of a change in operating conditions cannot be determined. This point is overlooked by some workers, who make few duplicate runs and give little comment about the discrepancies. In some cases, this is due to plotting h vs. q/A, which is equivalent to plotting $q/(A)(\Delta t)$ vs. q/A. In the range of strong nucleate boiling, where a large change in q/A is caused by a small change in Δt, if the values of h are plotted against the values of q/A the results will seem to be quite reproducible, despite the fouling that may have occurred. The true situation would be disclosed by plotting q/A vs. Δt. Both methods of plotting a given set of data are shown in Fig. 14-6. Since one is often interested in the value of h rather than that of q/A, the same data are also plotted as h vs. Δt in Fig. 14-6. If, owing to fouling, Δt must be larger than before to deliver the same q/A, the point will be moved to the right and downward, thus bringing it vertically far below the curve for the clean surface; consequently this is the most revealing method of plotting the data.

Nature of Surface. Jakob and Fritz[46] found that a grooved copper surface adsorbed air and gave initially much higher coefficients in the range of moderate flux, although with continued boiling the coefficients decreased and approached more closely those for the smooth chromium-

BOILING LIQUIDS

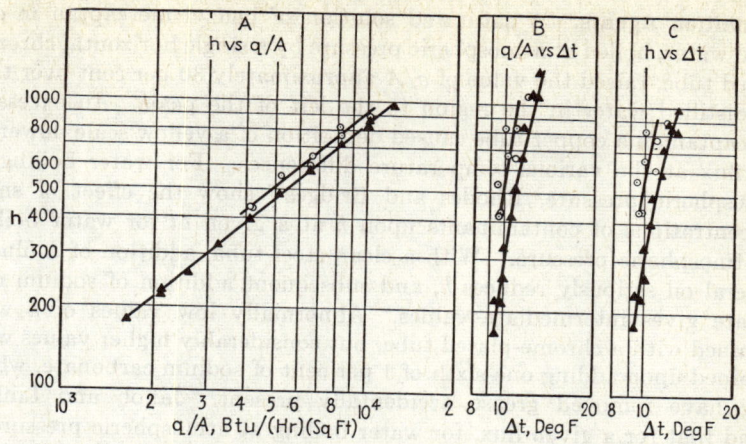

FIG. 14-6. Three methods of plotting. (*Data of Insinger and Bliss.*[44])

plated surface. In both cases, A was taken as the projected area of the plate, regardless of the fact that the grooved surface had an actual area 1.8 times that of the smooth plate. At Δt of 10°F the coefficient based on the *actual* area was less for the grooved than for the smooth plate. In boiling ethanol at atmospheric pressure with high flux, Sauer[81] found that the over-all coefficient U, based on the projected area, was increased by grooving the copper, but the percentage of increase varied with the Δt_o employed. Deutsch and Rhode[32] boiled distilled water at atmospheric pressure with high flux and for a given Δt_o found that U based on the projected area was not increased by roughening the surface and that U based on the total surface was less than for the smooth tube.

Figure 14-7 shows data[15] for boiling ethanol at atmospheric pressure on a horizontal flat plate provided with several different surfaces. For a Δt of 40°F, h is 4800 for freshly polished copper, 2200 for fresh gold plate, 1300 for fresh chrome plate, and 580 for aged chrome plate. With water or methanol boiling at atmospheric pressure in a small submerged tube evaporator, Cooper[29,82] found that U was larger with iron than with copper tubes, indicating that the increase in the number of vaporization nuclei more than compensated for the decrease in thermal conductivity.

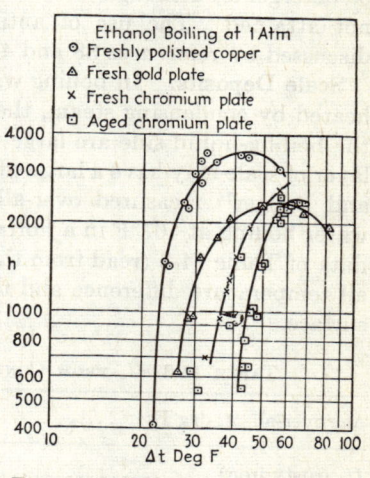

FIG. 14-7. Effect of nature of surface. (*Data of Bonilla and Perry.*[15])

Addition Agents. A saturated solution of benzyl mercaptan in distilled water, boiled at atmospheric pressure by a single horizontal chrome-plated tube, raised the value of q/A approximately 30 per cent over that for distilled water in the region to the left of the peak. An excess of mercaptan on a copper tube caused deposition of a yellow scale, lowering the flux at the various temperature differences. For water boiling at atmospheric pressure, Rhodes and Bridges[77] show the effect of small concentrations of contaminants upon h at a given Δt for water boiling at atmospheric pressure. With a clean steel tube, addition of a film of mineral oil seriously reduces h, and subsequent addition of sodium carbonate gives intermediate values. Abnormally low values of h were obtained with a chrome-plated tube, but considerably higher values were obtained upon adding one-sixth of 1 per cent of sodium carbonate, which may have removed grease accidentally present. Jakob and Linke[47] found that for a given flux, for water boiling at atmospheric pressure, h was increased 23 per cent by adding a wetting agent (Nekal BX), which reduced the surface tension 45 per cent. At a constant flux of 12,800, Insinger and Bliss[44] found that a wetting agent (Triton W-30) increased h 20 per cent while reducing surface tension 27 per cent.

Morgan et al.[70] report increased boiling coefficients in the presence of detergents, although the equilibrium surface tensions were probably not attained. The use of antifoaming agents in industrial boilers is discussed by references 38 and 45.

Scale Deposits. In boiling water or aqueous solutions in evaporators heated by condensing steam, the coefficients on both the steam side and the boiling-liquid side are large with clean surfaces. Hence even a thin layer of scale may have a large effect on the over-all coefficient. Pridgeon and Badger[76] measured over-all coefficients from condensing steam to water boiling at 167°F in a horizontal submerged-tube evaporator. The data of Table 14-3 (read from their curves) show the importance of over-all temperature difference and the cleanliness and nature of the heating surface.

TABLE 14-3. U_o FROM CONDENSING STEAM TO BOILING WATER[76]

Δt_o, over-all Δt, deg F	18	27	36	45	54
U_o (rusty iron)	280	300	325	350	375
U_o (clean iron)	300	385	460	535	610
U_o (slightly dirty copper)	580	780	950	1120	
U_o (polished copper)	820	1110	1470	1810	2120

Later work on the copper tubes, after acid cleaning, gave U_o as high as 3070 with $\Delta t_o = 38°F$, but it was difficult to reproduce these very high values of U_o, owing to the collection of a thin deposit on the tubes. The small amount of scale necessary to reduce a high coefficient by a substantial amount is not generally realized. By taking the thermal conductivity k of scale as 1.3 Btu/(hr)(sq ft)(deg F per ft), the calculated thickness of scale required to reduce the coefficients in Table 14-3 for polished copper to those for slightly dirty copper are shown in Table 14-4.

Webre and Robinson[91] show the effect of cleanliness in tests by Kerr on a quadruple-effect evaporator in a sugar mill. Tests were made $3\frac{1}{2}$ days and 6 days after the evaporators were cleaned. The effect of

TABLE 14-4

Δt_o, deg F	18	27	36	45
U (polished copper)	820	1110	1470	1810
U (slightly dirty copper)	580	780	950	1120
Necessary scale thickness, inches	0.0079	0.0059	0.0058	0.0053

increasing scale formation is illustrated by Table 14-5. Webre and Robinson also give data on an experimental apparatus in which the over-all coefficient U dropped from 1550 to 900 in less than 2 hr, then less rapidly to a value of 700.

In boiling saturated solutions of sodium sulfate or sodium chloride at atmospheric pressure Coffey[27] found that scale was rapidly deposited on the heating surface. Starting with clean surfaces, a number of runs were made, each with a constant over-all temperature difference, in the

TABLE 14-5. U, BTU/(HR)(SQ FT)(DEG F)

Evaporator	1	2	3	4
After $3\frac{1}{2}$ days	321	508	203	90
After 6 days	239	270	97	79

range 50 to 150°F. Comparison of these results, 1 hr after boiling started, showed that the largest flux of 75,000 was obtained with Δt_o of 75°F, whereas with water a maximum flux of 330,000 was found with Δt_o of 95°F. With a solute of low solubility, calcium sulfate, Bulkley[21] found that the fractional increase in U with increase in Δt_o is less than for water boiling in the same apparatus; at Δt_o of 25°F, the solution gave the same U as water, but with Δt_o of 120°F, U was only 64 per cent of that for water.

The use of an organic agent for prevention of scale deposition on the heating surface of crystallizers for sodium sulfate is discussed by Vener and Thompson.[88]

Effect of Δt. It is evident from a study of the various regimes of boiling that Δt is an important variable controlling the nature and rate of the heat-transfer process. While the effect of Δt is significant in all regimes of boiling, it is most important in the range of strong nucleate boiling, for which the data of many observers may be expressed by

$$q/A = a_1 \Delta t^n \qquad (14\text{-}1)$$

where n is a constant ranging from 3 to 4 and a_1 is a particular constant for each liquid, surface, and pressure.

The influence of Δt in the nucleation process has been treated in several papers which discuss the thermodynamics of metastable states[57,88] and review the nucleation theory.[12] A general correlation of boiling data in terms of theory is not yet available because of the strong influence of minor experimental variables which are not readily controlled, such as the amount of dissolved and adsorbed gases, impurities and contamination on the surface, and differences in experimental technique. There is considerable evidence of hysteresis in obtaining boiling curves, indicating that more nuclei are active at a given Δt if the Δt has first been increased to a greater value for an initial period.

Data on the effect of Δt in nucleate boiling are given in the graphs for the various liquids (Fig. 14-7 to 14-14). In normal applications the heat-transfer surface should clearly be operated at Δt less than the critical value, beyond which q/A decreases.

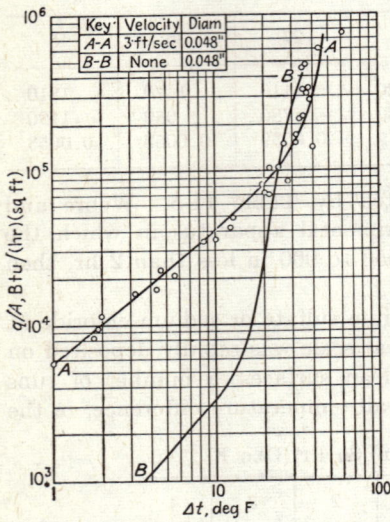

Fig. 14-8. Effect of increased velocity (data of Beecher[9] for water boiling at 212°F with 3 ft/sec velocity, compared with data[2] for water boiling without mechanical agitation).

Effect of Increased Velocity. In the range of low Δt, the use of forced convection in a boiling system results in an increase in the heat flux for a given Δt, as shown in Fig. 14-8. Line B represents data for water boiling on a 0.048-in. platinum wire in an unstirred pool at 1 atm. Line A shows data of Beecher[9] for water at 212°F and 1 atm flowing at 3 ft/sec normal to an electrically heated 0.050-inch-diameter stainless-steel tube. In the region of strong nucleate boiling the curves differ very little, Curve A falling to the right perhaps because of a slight scale deposit. For Δt less than 10°F, Curves A and B are in agreement with the respective nonboiling correlations for forced convection and natural convection. The use of velocities higher than 3 ft/sec resulted in cavitation.

Mechanical agitation has been used in commercial equipment.[91]

BOILING LIQUIDS

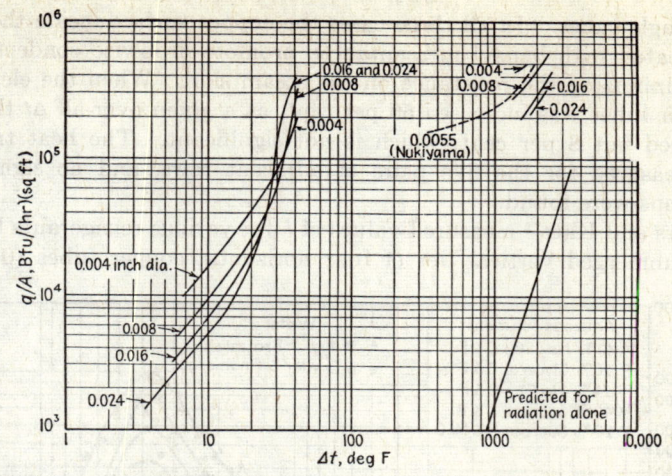

FIG. 14-9. Summary of data of Rinaldo[63] for four wire diameters, compared with data of Nukiyama[73] for 0.0055 in. Water boiling at 1 atm.

Drew and Mueller[35] find that stirring did not prevent vapor binding and that the critical over-all Δt was not affected substantially.

Tube Size and Arrangement. Figure 14-9 shows the data of Rinaldo[63] for four horizontal platinum wires with diameter ranging from 0.004 to 0.024 in., submerged in a pool of distilled water at 212°F. At a given Δt, wire diameter has a substantial effect on the flux in the region of natural-convective superheating (Δt from 4 to 15°F) and in the film-boiling region (Δt from 1350 to 2350°F). In the region of nucleate boiling (Δt from 20 to 40°F) wire diameter has little effect.

Most of the data in the literature for laboratory apparatus are for a single tube or a horizontal plate. Abbott and Comley[1] tested a model evaporator containing 60 nearly horizontal chrome-plated copper tubes, each 10 in. long, having inside and outside diameters of 0.438 and 0.500 in., respectively. When the tubes were placed on 1-in. centers, giving horizontal and vertical clearances of 0.50 in., with 10 rows deep, the results shown in Fig. 14-10 were obtained, which are approximately the same as those obtained by Kaulakis and Sherman[51] and by Dresely[34]

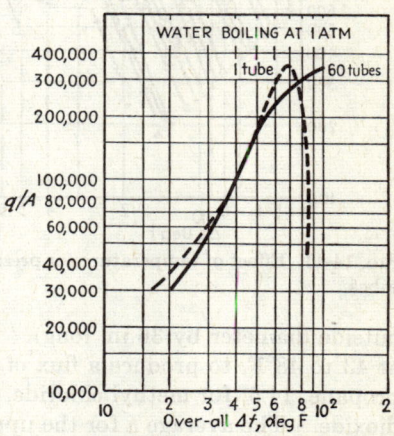

FIG. 14-10. Good agreement is shown between data[1] for small evaporators containing 60 tubes and 1 tube.

with single tubes. In all three cases the copper surface inside the tube was treated with benzyl mercaptan to promote dropwise condensation, thus minimizing the resistance on the steam side. When the clearance between tubes was increased 50 per cent, at a given over-all Δt the flux decreased but 8 per cent, which is not significant. The heat transfer was measured for the five pairs of adjacent rows, and no significant variations were found.

Myers and Katz[72] measured values of h for various refrigerants boiling on a submerged vertical tier of four horizontal copper tubes (0.75 in.

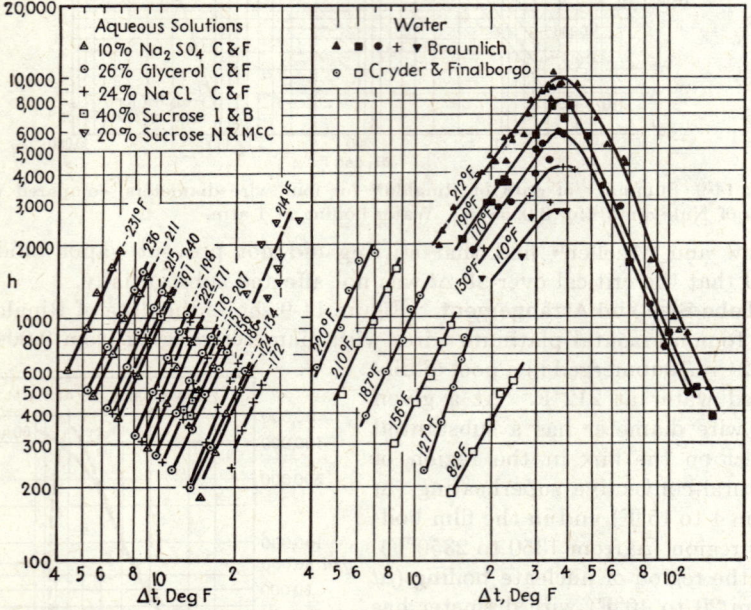

Fig. 14-11. Effect of temperature (or pressure) upon h, for a given Δt; single horizontal tubes.

outside diameter by 36 in. long). For the bottom tube, and refrigerant at 43 to 48°F, to produce a flux of 16,000 Btu/(hr)(sq ft) Δt was 10° for propane, 14.5° for methyl chloride, 15.5° with F-12, and 20°F with sulfur dioxide. The average h for the upper three rows was given by the equation $h_{2-4}/h_1 = 14.3 h_1^{0.66}$; h_1 ranged from 350 to 2000.

Jakob and Linke[47] obtained data for a given fluid boiled with horizontal and vertical surfaces and found that the horizontal plate gave nearly as good results as the vertical surface for both water and carbon tetrachloride boiling at atmospheric pressure.

Effect of Pressure. For a large number of liquids boiling in pools in the nucleate region at pressures of 1 atm and less, the data show that a decrease in saturation pressure gave a lower heat flux for a given Δt. Figure 14-11 shows the data of Cryder and Finalborgo[30] and Braunlich[17] for water boiling at various temperatures. The peak heat flux and the boiling coefficient are less for the lower pressures. The heat-transfer

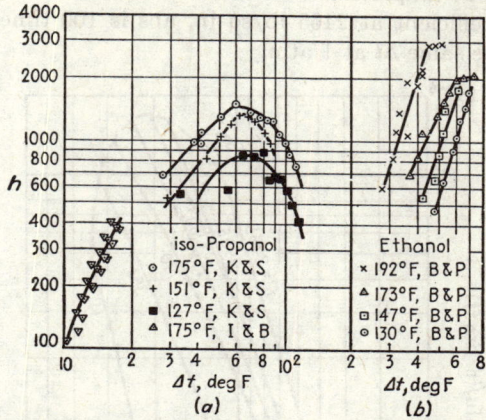

FIG. 14-12. (a) Data of Kaulakis and Sherman[51] and of Insinger and Bliss[44] for i-propanol; (b) data of Bonilla and Perry[15] for ethanol at several pressures.

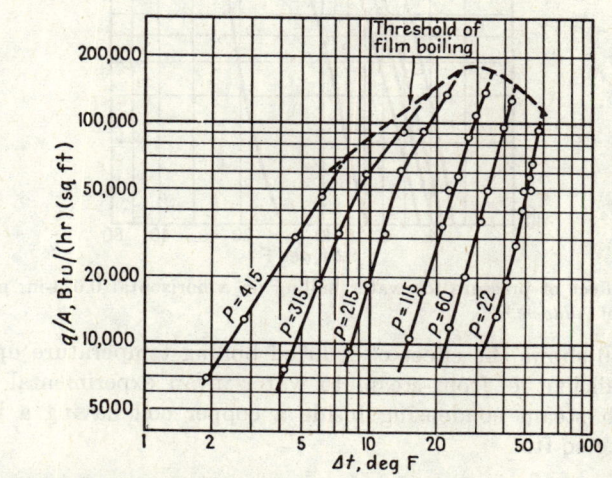

FIG. 14-13. Nucleate boiling of n-pentane (90 per cent pure) on a clean horizontal chrome-plated disk. (*Data of Cichelli and Bonilla.*[23])

coefficient shown for five solutions boiling at various pressures follow the same trend. Figure 14-12 shows a similar effect of pressure in the data of Bonilla and Perry[15] for ethanol and the data of Kaulakis and Sherman[51] and Insinger and Bliss[44] for isopropanol.

Cichelli and Bonilla[23] boiled a number of organic liquids on a horizontal chrome-plated copper disk at pressures ranging from 1 atm to as high as 95 per cent of the critical pressure. Typical results are shown in Fig. 14-13, for pentane. The data of Addoms[2] for water boiling at elevated pressure on a 0.024-in. platinum wire are summarized in Fig. 14-14. The heat-transfer coefficient at 2465 lb/sq in. abs is 100 times greater than the value for the same Δt at 1 atm.

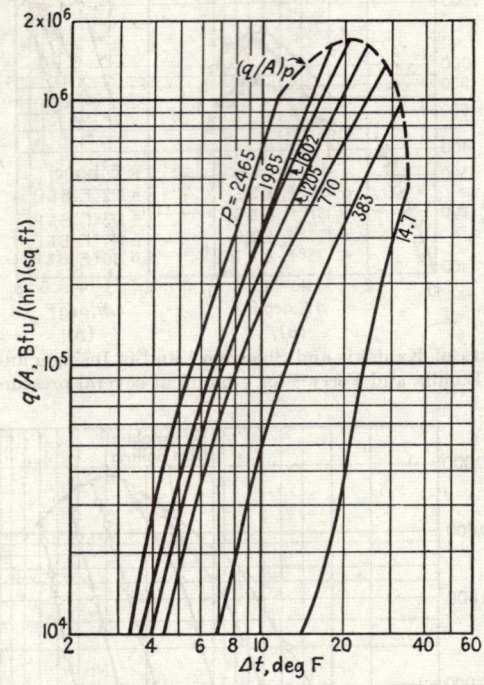

Fig. 14-14. Effect of pressure for water boiling on a horizontal 0.024-in. platinum wire. (*Data of Addoms.*[2])

Table 14-6 shows the effect of reduced boiling temperature upon the over-all coefficient U from steam to water in an experimental evaporator,[24] with steam condensing inside a copper coil having a heating surface of 5.4 sq ft.

TABLE 14-6. EVAPORATION OF WATER BY SUBMERGED COPPER COIL (Claassen)

Δt_o, °F	20	30	40	50	60	70	80	90
U for 212°F	390	490	560					
U for 187°F	...	360	440	520	600			
U for 158°F	...	...	...	...	510	600	660	720

BOILING LIQUIDS 383

Nature of Liquid. Data of different observers, for the boiling of a number of liquids, have been shown in Figs. 14-7 to 14-14. Figure 14-15 shows the data of Cryder and Finalborgo,[30] who boiled a number of liquids in the range of moderate flux. For a given Δt, the value of h for methanol is only 12 per cent of that for water, but for a given heat flux, shown in the right-hand part of Fig. 14-15, the value of h is 55 per cent of that for water.

As shown in Fig. 14-11, aqueous solutions of salts or organic substances usually give lower coefficients than water at the same pressure for a given Δt or flux. A few data for miscible binary mixtures are available.[15] Data for the immiscible mixtures styrene-water and butadiene-water are reported by Bonilla and Eisenberg.[14]

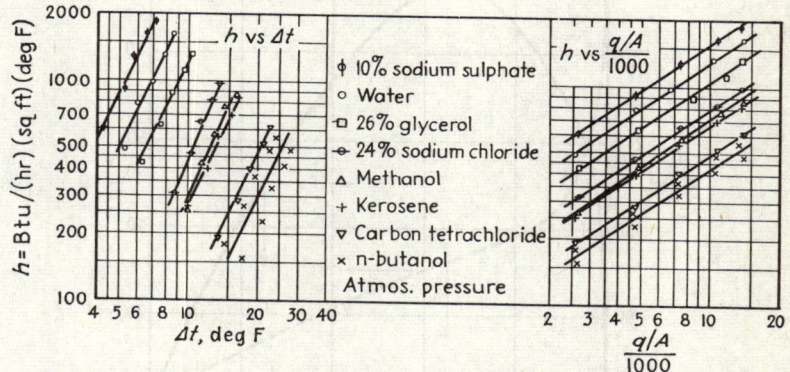

FIG. 14-15. Data of Cryder and Finalborgo[30] for a single horizontal tube.

In the range of nucleate boiling, a general correlation of the data for various liquids remains to be discovered. Since the slopes of the curves of heat flux vs. Δt are roughly parallel for all liquids, Lukomskii[62] proposed that the ratio of the heat flux to the peak heat flux might be a unique function of the ratio of the Δt to the critical Δt,

$$\frac{q/A}{(q/A)_p} = f\left(\frac{\Delta t}{\Delta t_c}\right) \qquad (14\text{-}2)$$

While this relationship is useful for estimating when only the peak heat flux and corresponding Δt_c are known, it is oversimplified and gives poor correlation of the available data.

Peak Heat Flux and Critical Δt**.** The values of the peak flux for a number of organic liquids boiling at pressures up to 95 per cent of the critical pressure were correlated by Cichelli and Bonilla,[23] as shown in Fig. 14-16, by plotting the peak flux $(q/A)_p$ divided by the critical pressure in pounds per square inch absolute as ordinate vs. reduced pressure as abscissa. When the heat-transfer surface was slightly fouled, the peak heat flux was found to increase somewhat, in which case the ordinates were divided by a factor of 1.15. With organic liquids the highest peak density was 350,000 for ethanol on a clean surface. From Fig. 14-16 it is noted that the maximum in the peak flux occurred at an ordinate of 380, at an abscissa of one-third of the critical pressure. It is surprising that such a simple dimensional relation correlates the data for a number of organic liquids, for which P_c ranged from 474 to 928 lb/sq in. abs; perhaps a number of the more important physical factors are functions of reduced pressure. Since the buoyancy term $\rho_L - \rho_v$ becomes zero at the critical pressure, it is not surprising to note that the curve of

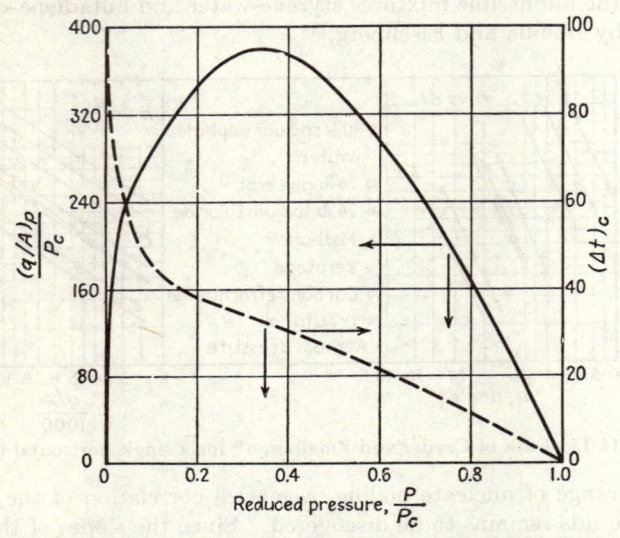

FIG. 14-16. Correlation of Cichelli and Bonilla[23] for peak flux and critical Δt, based on data for organic liquids.

peak flux is approaching zero at the critical pressure. The critical Δt was roughly correlated by the dotted curve in Fig. 14-16.

The peak flux for water boiling at elevated pressures on a 0.024-in. platinum wire was measured by Addoms[2] as shown by the broken line in Fig. 14-14. These values of peak flux to water under pressure ran somewhat above those predicted by the curve of Cichelli and Bonilla based on organic liquids; the same is true of the data of Kazakova[52] for water boiling on a platinum wire (0.059 in.) at pressures from 14.7 to 2880 lb/sq in. abs shown in Fig. 14-17. Data of Braunlich[17] for water boiling under reduced pressures were shown in Fig. 14-11. The peak-flux values for water[2,17] and organic liquids[23] were correlated by Addoms[2] as shown in Fig. 14-18. The dimensionless abscissa contains a buoyancy term, $(\rho_L - \rho_v)/\rho_v$, and the dimensionless ordinate contains the average volu-

BOILING LIQUIDS 385

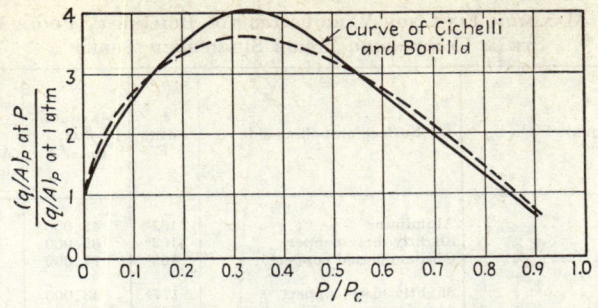

FIG. 14-17. Data of Kazakova[52] for water boiling on a horizontal platinum wire (0.059 in.) compared with the prediction of Cichelli and Bonilla for organic liquids.

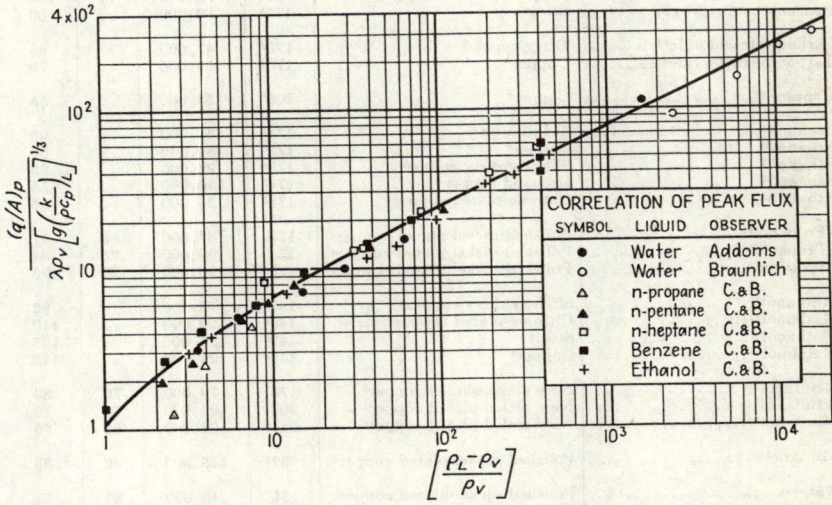

FIG. 14-18. Dimensionless correlation[2] of peak flux for water boiling on a single horizontal platinum wire and a chrome-plated tube, and for organic liquids boiling on a horizontal chrome-plated copper disk.

metric vapor disengaging rate per unit surface,

$$\frac{(q/A)_p}{\lambda \rho_v}$$

divided by the cube root of the product of the thermal diffusivity in the liquid $(k/\rho c_p)_L$ and the acceleration due to gravity,

$$\sqrt[3]{g \left(\frac{k}{\rho c_p} \right)_L}$$

Table 14-7 summarizes values of the peak flux for a number of liquids

TABLE 14-7. MAXIMUM FLUX FOR VARIOUS LIQUIDS BOILING IN POOLS HEATED BY STEAM CONDENSING INSIDE SUBMERGED TUBES[a]

Liquid	Surface on boiling side	t, deg F	Maximum flux, q_{max}/A	Critical Δt Δt_L,[†] deg F	Critical Δt Δt_o, deg F
Ethyl acetate[82]	Aluminum[e]	162[b]	42,000	..	80
Ethyl acetate[82]	Slightly dirty copper[e]	162[b]	62,000	..	57
Ethyl acetate[82]	Chrome-plated copper[e]	162[b]	77,000	..	70
Benzene[82]	Slightly dirty copper[e]	177[b]	43,000	..	100
Benzene[82]	Aluminum[e]	177[b]	50,000	..	80
Benzene[82]	Copper[e]	177[b]	55,000	..	80
Benzene[82]	Chrome-plated copper[e]	177[b]	69,000	..	100
Benzene[82]	Copper[e]	177[b]	72,000	..	60
Benzene[82]	Chrome-plated copper[e]	177[b]	70,000	..	100
Benzene[82]	Steel[e]	177[b]	82,000	..	100
Carbon tetrachloride[82]	Dirty copper[e]	170[b]	47,000	..	83
Carbon tetrachloride[82]	Copper[e]	170[b]	58,000	..	79
Heptane[82]	Copper[e]	209[b]	53,000	..	55
Ethanol[82]	Aluminum[e]	173[b]	54,000	..	90
Ethanol[82]	Copper[e]	173[b]	80,000	..	66
Ethanol[82]	Slightly dirty copper[e]	173[b]	93,000	..	65
Ethanol[82]	Grooved copper[e]	173[b]	120,000	..	55
Ethanol[82]	Chrome-plated copper[e]	173[b]	126,000	..	65
i-Propanol[51]	Polished nickel-plated copper[e]	127	67,000	85	91
i-Propanol[51]	Polished nickel-plated copper[e]	151	90,000	75	84
i-Propanol[51]	Polished nickel-plated copper[e]	175[b]	110,000	90	96
Methanol[82]	Slightly dirty copper[e]	149[b]	78,000	..	92
Methanol[82]	Chrome-plated copper[e]	149[b]	120,000	..	110
Methanol[82]	Steel[e]	149[b]	123,000	..	105
Methanol[82]	Copper[e]	149[b]	124,000	..	115
n-Butanol[51]	New nickel-plated copper[e]	173	79,000	70	83
n-Butanol[51]	New nickel-plated copper[e]	207	92,000	70	79
n-Butanol[51]	New nickel-plated copper[e]	241[b]	105,000	60	70
i-Butanol[51]	Polished nickel-plated copper[e]	222[b]	115,000	80	85
Water[51]	Polished nickel-plated copper[c]	131	115,000	42	53
Water[17]	Chrome-plated copper[d]	110	150,000	45	
Water[17]	Chrome-plated copper[d]	130	175,000	45	65
Water[51]	New nickel-plated copper[c]	155	190,000	50	
Water[17]	Chrome-plated copper[d]	150	220,000	45	64
Water[17]	Chrome-plated copper[d]	170	243,000	45	64
Water[51]	Polished nickel-plated copper[c]	171	250,000	50	72
Water[51]	New nickel-plated copper[c]	191	260,000	50	
Water[17]	Chrome plated copper[d]	190	300,000	45	70
Water[e]	Chrome-plated copper[c]	212[b]	330,000	45	80
Water[51]	New nickel-plated copper[c]	212[b]	360,000	52	68
Water[51]	Polished nickel-plated copper[c]	212[b]	370,000	45	72
Water[17]	Chrome-plated copper[d]	212[b]	390,000	45	72
Water[82]	Steel[e]	212[b]	410,000	..	150

[a] Additional data are given in reference 35, obtained with a very small apparatus.
[b] Boiling at atmospheric pressure.
[c] Steam side was promoted with benzyl mercaptan.
[d] Steam side was promoted with octyl thiocyanate.
[e] Steam probably contained a trace of oleic acid.
[†] ΔT_L is the temperature drop from the surface of the tube to the liquid. ΔT_o is the over-all temperature drop from the steam to the liquid.

boiled at atmospheric or reduced pressure and includes the corresponding values of the temperature potential.

Table 14-8 shows the peak flux for horizontal wires submerged in pools of water boiling at 1 atm. The data of the two observers do not agree for a given metal, but the ratio of the values for Chromel C and nickel are 2.6 and 2.2.

TABLE 14-8. PEAK FLUX FROM HORIZONTAL WIRES TO WATER BOILING AT 1 ATM, $q/A \times 10^{-6}$

Metal	Ref. a	Ref. b
Nickel	0.380	0.294
Platinum	.....	0.295
Aluminum	.....	0.418
Chromel A	0.216	0.453
Tungsten	0.618	
Chromel C	0.993	0.651

Reference a, Farber and Scorah,[36] used wires 0.040 inch in diameter.
Reference b, Castles,[22] used wires 0.010 to 0.016 inch in diameter.

Film Boiling. Bromley[18] derived a theoretical equation for film boiling assuming upward laminar flow of vapor and heat transfer by conduction through the vapor film. The resulting equation is of the same form as that of Nusselt for film-type condensation of the pure vapor with downward laminar flow of condensate, except that $\rho_v(\rho_L - \rho_v)$ replaces ρ_L^2,

$$h_{co} = a_2 \left[\frac{g \lambda k_v^3 \rho_v (\rho_L - \rho_v)}{D \mu_v \Delta t} \right]^{1/4} \qquad (14\text{-}3)\star\dagger$$

Upon assuming that the liquid rises freely with the vapor, the constant a_2 is 0.724 as in the Nusselt equation; if the liquid is assumed stagnant, the constant is reduced to 0.512. Since any heat transferred by radiation results in an increase in the thickness of the vapor film, the coefficient h_c for conduction and convection in the presence of radiation is less than the coefficient h_{co} in the absence of radiation; the resulting film coefficient for both mechanisms is given by

$$h = h_c + h_r = h_{co} \left(\frac{h_{co}}{h} \right)^{1/3} + h_r \qquad (14\text{-}4)\star$$

Bromley obtained data on film boiling of benzene, carbon tetrachloride, ethanol, nitrogen, and water, boiling on a single horizontal submerged carbon tube having a diameter of 0.35 in. With pentane, data were obtained with carbon tubes having diameters of 0.238, 0.352, and 0.469 in. and with a stainless-steel tube having a diameter of 0.188 in. Values of Δt ranged from 150 to 2250°F, and values of h_{co} ranged from 20 to 52.

† The properties k_v, ρ_v, and μ_v were evaluated at $t_v = (t_\omega + t_{sv})/2$; λ represents the latent heat of the saturated vapor plus $c_p(t_v - t_{sv})$.

Figure 14-19 shows the values of h_r, h_{co}, and h for nitrogen on a 0.35-in. carbon tube, and it is seen that the contribution of radiation (based on emissivity of 0.8) was substantial at high Δt. Upon solving for h_{co} by use of Eq. (14-4), it was found that h_{co} for all the data could be predicted by using a_2 of 0.62, ± 0.04:

$$h_{co} = 0.62 \left[\frac{g \lambda k_v^3 \rho_v (\rho_L - \rho_v)}{D \mu_v \Delta t} \right]^{1/4}$$
(14-5)★

Equation (14-5) is also supported by data of Kane[50] for methanol, ethanol, toluene, and benzene boiling on horizontal stainless-steel tubes of 0.25 in. diameter. Data of reference 63 for film boiling of water at 1 atm on platinum wires (0.004 to 0.024 in.) indicated that the theory is not applicable to such small wires, where the coefficients are 30 to 100 per cent higher than predicted (Fig. 14-9).

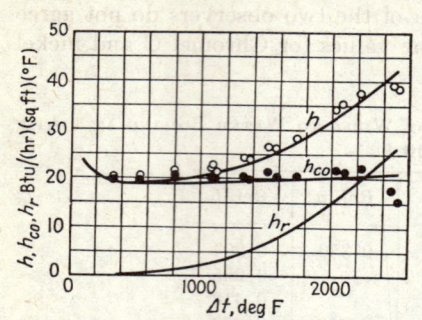

FIG. 14-19. Film boiling of nitrogen on 0.35-in. carbon tube. (*Data of Bromley.*[18])

SUMMARY, SATURATED LIQUIDS BOILING ON SUBMERGED SURFACES

In boiling a liquid at a given pressure, as the Δt is increased the heat flux at first increases slowly in the region of natural convective superheating, then increases rapidly in the region of nucleate boiling, passes through a maximum at the critical Δt, reaches a minimum as the heat-transfer surface becomes fully insulated with a vapor film at the Leidenfrost point, and subsequently increases again in the region of film boiling at high Δt (Fig 14-1).

The maximum value of the peak flux generally occurs with a reduced pressure of substantially one-third, at a moderate value of Δt (Fig. 14-16); the peak flux at a given pressure is best estimated from Fig. 14-18. Table 14-7 gives the peak flux and corresponding critical Δt at substantially atmospheric pressure for steam-heated submerged horizontal tubes. In using exhaust steam to supply the heat to vaporize organic liquids of moderately low boiling points, to avoid exceeding the critical Δt it may be necessary to boil under elevated pressure or to condense the steam under vacuum. Water or aqueous solutions are often boiled in multiple-effect evaporators, wherein a total temperature drop of 100 to 150°F is shared among the several evaporators connected in series with respect to vapor, and under such conditions there is little or no chance of exceeding the critical temperature difference.

The effects of addition agents, scale deposits, and artificial roughening of the surface may be significant in nucleate boiling, but the results have not yielded to a general correlation. The effect of freshly adsorbed air is important but decreases with time of boiling. Whereas a very thin layer of scale on the heat-transfer surface may increase h, a thicker deposit seriously decreases h (Tables 14-3 and 14-5).

In nucleate boiling of various liquids at atmospheric pressure h may vary widely from one liquid and surface to another when comparisons are made at a given Δt, but the variation is considerably less at a given flux; data for pure liquids and solutions are given in Fig. 14-9 to 14-12. The effect of tube diameter is unimportant. Substantially the same coefficients have been obtained with horizontal and vertical surfaces, and meager data indicate little effect of substantial variation in the spacing of horizontal tubes in a bundle. Agitation with stirrers is unimportant, except at low flux (Fig. 14-8).

In the region of film boiling the heat-transfer coefficients for several liquids range from 20 to 52 as Δt ranges from 150 to 2250°F; h is predicted from Eqs. (14-4) and (14-5).

II. SURFACE BOILING OF SUBCOOLED LIQUIDS

Introduction. Surface, or local, boiling is a form of nucleate boiling which occurs when a liquid at a temperature below saturation is brought into contact with a metal surface hot enough to cause boiling at the surface of the heater. The vapor bubbles condense in the cold liquid, and no net generation of vapor is realized with degassed liquid. Extremely high film coefficients and peak fluxes may be obtained; hence local boiling is advantageous in high-duty applications.

Mosciki and Broder[71] were probably the first investigators to study local boiling. They used an electrically heated vertical platinum wire submerged in water at atmospheric pressure. The water temperature was varied from 68 to 212°F, and the critical wire temperature (266°F) at the peak heat flux was essentially independent of the temperature of the water. With subcooled water at 68°F the estimated peak heat flux was 1,200,000 Btu/(hr)(sq ft), which was 6.7 times that with water at the boiling point. The same general effect of subcooling was noted by Nukiyama,[73] who immersed an electrically heated horizontal platinum wire in water at 1 atm.

In the preheating section of an experimental horizontal-tube evaporator heated externally by dropwise condensation of steam, Woods[65] found that the local over-all coefficients of heat transfer were four times those predicted for nonboiling conditions. This was explained as possibly being due to local boiling at the surface of the tube with subsequent condensation of the vapor in the cold liquid.

Tibbetts and Cohen[86] connected a high-amperage d-c generator to a german silver tube (0.08 in. inside diameter) through which tap water flowed at a velocity of 53.5 ft/sec. They attained a peak heat flux of 4,700,000 Btu/(hr)(sq ft) before the tube melted. Taylor[85] used an electrically heated stainless-steel tube, 4.75 inches long with an inside diameter of 0.305 in., with tap water entering at a velocity of 37 ft/sec; audible noise occurred at a heat flux of 2,000,000, and burnout occurred at 3,400,000 Btu/(hr)(sq ft). In a pioneer paper, Knowles[54] studied surface boiling of river water in a centrally heated vertical glass annulus, obtaining peak flux as high as 2.4×10^6 Btu/(hr)(sq ft) but had difficulty with fouling of the heater.

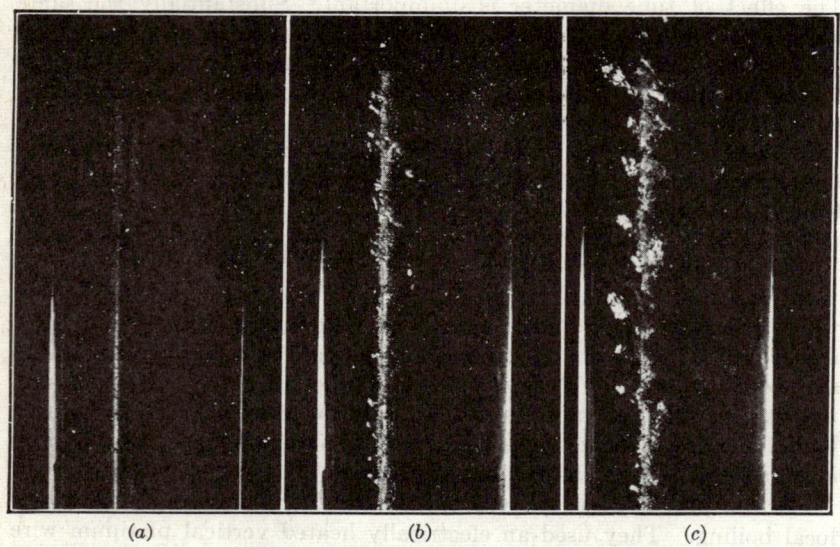

FIG. 14-20. High-speed photographs[64] showing surface boiling of degassed distilled water flowing upward at 4 ft/sec, 30 lb/sq in. abs, 50°F below saturation: (a) $q/A = 295{,}000$, $\Delta t = 83.6$; (b) $q/A = 510{,}000$, $\Delta t = 89.7$; (c) $q/A = 830{,}000$, $\Delta t = 96.7$.

Surface Boiling in Annuli. Colburn, Schoenborn, and Sutton[28] obtained data for local boiling of tap water flowing upward at velocities from 0.3 to 6 ft/sec through an annulus having a central heated tube 1.66 inches in outside diameter and an unheated outer jacket 2.00 inches inside diameter. The heat-transfer coefficients for local boiling were higher than predicted for nonboiling conditions, and the increase was more pronounced with low velocity and high temperature difference. The highest heat flux obtained was 69,000 Btu/(hr)(sq ft).

Reference 64 reports data for heat transfer for degassed distilled water flowing upward in a vertical annulus containing a central electrically heated stainless-steel tube in glass jackets having diameters of 0.77, 0.73, and 0.43 in. Pressures ranged from 30 to 90 lb/sq in. abs, and water velocities ranged from 1 to 12 ft/sec.

BOILING LIQUIDS 391

As shown by Fig. 14-20, at a given instant bubbles are present on certain parts of the surface but later condense in the nearby cold liquid. Heat is transferred from other parts of the tube directly to the cold liquid by the mechanism of forced convection without phase change. Owing to the disturbances set up in the boundary layer by the growth and collapse of the bubbles, the heat-transfer rates are higher than without phase change. High-speed motion pictures of the surface were taken

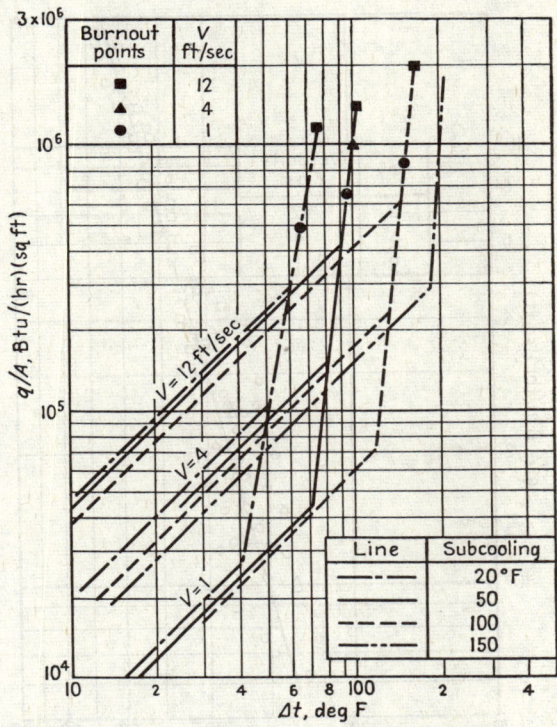

FIG. 14-21. Effect of velocity and degree of subcooling on surface boiling of water.[64]

by Dew;[33] from a study of two of the reels Rohsenow and Clark[79] estimated that the transfer of latent heat was only a small fraction of the total heat transfer and attributed the high heat-transfer coefficients to the disturbances in the boundary layer, confirming earlier conclusions of Gunther and Kreith[40] in heating a pool of subcooled water with an electrically heated horizontal strip.

Figure 14-21 shows a logarithmic graph of the heat flux q/A plotted as ordinate vs. the total Δt from the heater to the degassed water. In the nonboiling region, the results for each velocity agree with those expected from conventional equations for forced convection without change in phase. In the local-boiling region, the curves are steep and are displaced horizontally for each subcooling by values of Δt corresponding closely to the differences in subcooling. It is seen that the steep portions of the curves start at various values of Δt, depending on the temperature of the water.

The data for surface boiling are plotted in Fig. 14-22 vs. the surface boiling potential, $t_w - t_{sat}$, where t_{sat} is the saturation temperature of the

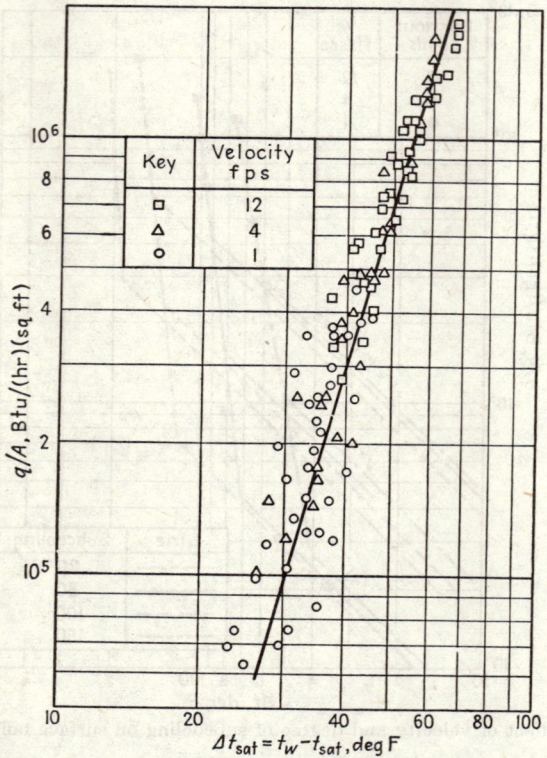

Fig. 14-22. Approximate correlation of surface boiling of degassed distilled water in terms of $t_w - t_{sat}$.

liquid. When plotted in this fashion, results are insensitive to water temperature and velocity. The slope of the curve in Fig. 14-22 is similar to that for the boiling of a pool of saturated liquid. With surface boiling the temperature of the heater was independent of length.

The various values of the peak heat flux for the 0.77-in. jacket flux are correlated by the dimensional equation:

$$(q/A)_p = 400{,}000 V^{1/3} + 4800(t_{sat} - t)V^{1/3} \qquad (14\text{-}6)$$

Surface Boiling in Tubes. Kreith and Summerfield[55] made experiments on surface boiling of water at moderate pressure in both a horizontal and a vertical tube of stainless steel. In plotting peak flux for upflow in the vertical tube vs. the product of water velocity and the degree of subcooling, it was noted that a higher peak flux was obtained by decreasing the subcooling and/or the water velocity than by increasing the flux, since the latter method gave a premature burnout, owing to lack of temperature equilibrium. These workers also report data for *n*-butyl alcohol and impure aniline.

Jens and Lottes[48] summarized experiments[20,80] on surface boiling of water flowing upward in vertical electrically heated tubes of stainless steel or nickel, having inside diameters ranging from 0.143 to 0.226 in. and L/D from 21 to 168. Pressures ranged from 100 to 2500 lb/sq in. abs, water temperatures from 229 to 636°F, mass velocities from 0.008×10^6 to 7.7×10^6 pounds/(hr)(sq ft of cross section), and q/A was as high as 4×10^6 Btu/(hr)(sq ft). The data were correlated by the dimensional equation:

$$\Delta t_{sat} = t_w - t_{sat} = \frac{1.9(q/A)^{1/4}}{e^{P/900}} \quad (14\text{-}7) \star$$

where P is the absolute pressure in pounds per square inch absolute. The peak flux was correlated by the dimensional equation:

$$\frac{(q/A)_p}{10^6} = a_3 \left(\frac{G}{10^6}\right)^m (t_{sat} - t)^{0.22} \quad (14\text{-}8)$$

As pressure increased from 1000 to 2000 lb/sq in. abs, the average value of a_3 decreased from 0.76 to 0.50 and m increased linearly from 0.27 to 0.50.

Horizontal Rectangular Duct. Gunther[39] made studies of surface boiling of degassed subcooled water flowing through a horizontal glass passage ½ in. high, ³⁄₁₆ in. wide, and 6 in. long, containing a horizontal electrically heated metal strip at the axis. By high-speed photography, the size of a vapor bubble could be determined as it formed. The bubble then grew while sliding downstream in contact with the heater, and finally disappeared owing to condensation in the subcooled liquid. In an experiment at 1.6 atm, with water subcooled 155°F flowing at 10 ft/sec and a flux of 2.33×10^6 Btu/(hr)(sq ft), there were 1.4×10^6 bubbles per second per square inch, the bubble lifetime was 220 microseconds, the maximum bubble radius was 0.01 in., and 4 per cent of the surface was covered by bubbles; the peak flux for these conditions was 3.86×10^6. Burnout data were taken in 38 experiments with water velocities from 5 to 40 ft/sec, pressures from 1 to 11.1 atm, and subcooling from 22 to 282°F and were correlated by the dimensional equation:

$$(q/A)_p = 7000(t_{sat} - t)(V)^{0.5} \quad (14\text{-}9)$$

The peak flux ranged from 420,000 to 11,400,000 Btu/(hr)(sq ft).

III. BOILING INSIDE TUBES

Progressive vaporization of a fluid as it flows inside a heated tube occurs in both forced- and natural-circulation evaporators, in tubular water boilers, and in pipe stills of various types. Despite the wide industrial application of this operation over a period of many years, the basic heat-transfer and fluid-flow phenomena are not yet fully understood, owing to the large number of experimental variables which must be incorporated in the analysis. The mechanical design and operation of evaporators and boilers has been extensively studied,[5,6,75,90] largely with the goal of minimizing scale deposits or facilitating their removal.[78,43]

Mechanism. Consider a single long vertical tube in which a pure liquid is being vaporized as it flows upward at a steady total mass rate of flow. The liquid feed to the tube is obtained from a separator at the top of the tube; hence the entering feed is below the saturation temperature which prevails at the bottom of the tube because of the increased hydrostatic pressure. The liquid flowing upward in the heated tube is brought to the saturation condition by two mechanisms: (1) the temperature of the liquid increases because of heat transfer, and (2) the saturation temperature decreases because the pressure falls owing to wall friction and reduction in hydrostatic head. In this lower part of the tube where the liquid is below saturation, the heat-transfer rate is governed by the relationships for flow inside tubes (Chap. 9), unless the wall temperature is sufficient to cause local boiling, in which case the relations of the previous section of this chapter would be applicable.

In the upper portion of the tube the fluid is saturated, vaporization occurs progressively, and a two-phase mixture emerges from the top. The rate of heat transfer at any point in this section of the tube depends upon the local temperature difference and upon the relative flow rates of liquid and vapor. As the weight fraction of vapor becomes larger, the velocity of the mixture increases because of the greater volume occupied by the vapor. The pressure gradient is increased because of greater wall drag and because of the force necessary to accelerate the stream to the higher velocities; hence the saturation temperature usually falls more rapidly near the outlet of the tube.

From the foregoing it is evident that a sound analysis of boiling inside tubes involves consideration of the variation in local conditions along the tube, including the tube wall temperature, the fluid temperature, the pressure, and the relative fractions of vapor and liquid. However, in many studies, only the inlet and outlet conditions and mean temperature difference have been measured. The early workers reported only the over-all effects on performance resulting from change in the independent variables controlled by the operator. However, several studies of local conditions in evaporator tubes have become available,[31,92] and local pressure-drop and heat-transfer measurements have been reported for two-phase mixtures of gas and liquid.[61,49]

Two-phase Flow without Vaporization

In order to provide background for the subsequent discussion of forced-circulation and natural-circulation evaporators, it is advantageous at this point to examine the phenomena which occur in two-phase flow without vaporization. Boiling inside tubes can then be discussed in the light of these factors which affect two-phase flow, as well as the factors which were previously shown to affect boiling on submerged surfaces.

Pressure Drop. Lockhart and Martinelli[61] gave an empirical dimensionless correlation for the frictional pressure drop in isothermal two-phase flow of air and several liquids in horizontal tubes. The flow was substantially incompressible; hence the pressure drop was that due to friction. The basic assumptions were that the static pressure drop for the liquid phase equals that for the gas phase, regardless of the flow pattern for the two-phase flow, and that the volume occupied by the liquid plus that occupied by the gas at any instant equals the volume of the pipe. These assumptions eliminate consideration of stratified flow and slug flow. The parameters used in correlating the data were defined as follows:

$$\Phi_L = \sqrt{\frac{(\Delta p/\Delta L)_{TP}}{(\Delta p/\Delta L)_L}} \qquad (14\text{-}10)\ \star$$

$$X = \sqrt{\frac{(\Delta p/\Delta L)_L}{(\Delta p/\Delta L)_G}} \qquad (14\text{-}11)\ \star$$

where $(\Delta p/\Delta L)_{TP}$ is the pressure gradient for two-phase flow, $(\Delta p/\Delta L)_L$ is that which would exist were the liquid flowing alone, and $(\Delta p/\Delta L)_G$ is that if the gas were flowing alone. Four types of flow were arbitrarily defined, as shown by the Reynolds numbers for single-phase flow in Table 14-9, where the first subscript refers to the flow of liquid alone

Table 14-9

Flow type	$t-t$	$v-t$	$t-v$	$v-v$
$4w_L/\pi D\mu_L$	> 2000	< 1000	> 2000	< 1000
$4w_G/\pi D\mu_G$	> 2000	> 2000	< 1000	< 1000

and the second to the gas alone. Thus $v-t$ designates viscous flow for the liquid alone and turbulent flow for the gas alone. From experimental data the functional relations between Φ_L and X were determined for substantially incompressible isothermal flow of air with water and air with various organic liquids in horizontal pipes having diameters ranging from 0.0586 to 1.017 in. The curves of Φ_L vs. X are shown in Fig. 14-23. Vertical downflow has been investigated by Bergelin.[11]

Liquid Holdup. In two-phase flow the liquid and gas ordinarily possess different average velocities, owing to the slip of the liquid relative to the gas. Hence the average density of the mixture existing at a given section of the tube is not identical with that computed from the respective mass flow rates and densities of the two phases. The equation of continuity gives

$$\frac{w_L}{V_L \rho_L R_L} = \frac{w_G}{V_G \rho_G (1 - R_L)} = S \tag{14-12}$$

in which R_L is the average volumetric fraction of liquid in the tube at a given point and $1 - R_L$ is the corresponding fraction of gas or vapor.

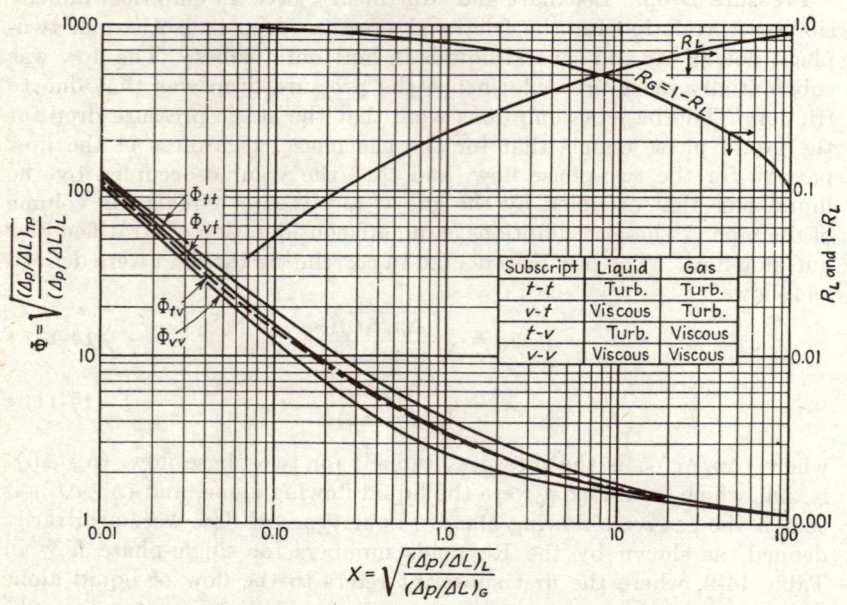

Fig. 14-23. Curves of Lockhart and Martinelli[61] for horizontal two-phase flow without vaporization.

Lockhart and Martinelli[61] measured the volume fraction R_L of liquid for horizontal flow by means of quick-acting valves and obtained a single curve of R_L vs. X as shown in Fig. 14-23. Bergelin and Gazley[10] and Johnson and Abou-Sabe[49] also measured R_L for horizontal flow of air-water mixtures and obtained results within 50 per cent of those predicted by Fig. 14-23, for X from about 0.2 to 20.

For upflow of water-air and oil-air mixtures, R_L was reported by Moore and Wilde;[69] temperatures and viscosities were not measured, and hence analysis of the data is incomplete.

Heat Transfer. The data of Verschoor and Stemmerding[89] and Johnson and Abou-Sabe[49] indicate that the heat-transfer coefficient inside a tube for a given mass flow rate of liquid first increases as the co-current gas flow rate is made larger but then in some cases reaches a maximum and declines. Data of the former workers for upward flow of air and water in a 14-mm tube with a 400-mm test section are shown in Fig. 14-24. The results of the latter investigation employing a horizontal

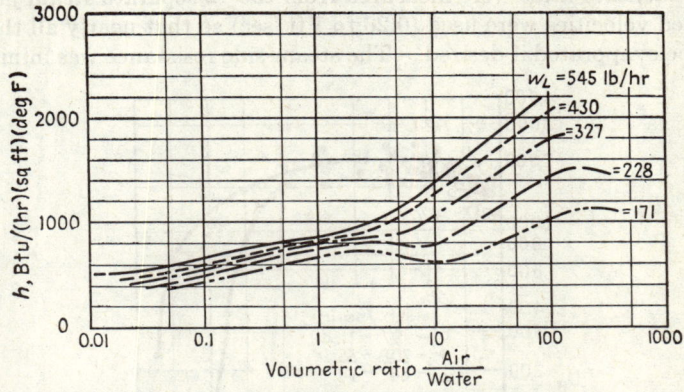

FIG. 14-24. Data of Vershoor and Stemerding[89] for heat transfer to upward flowing co-current air-water mixtures.

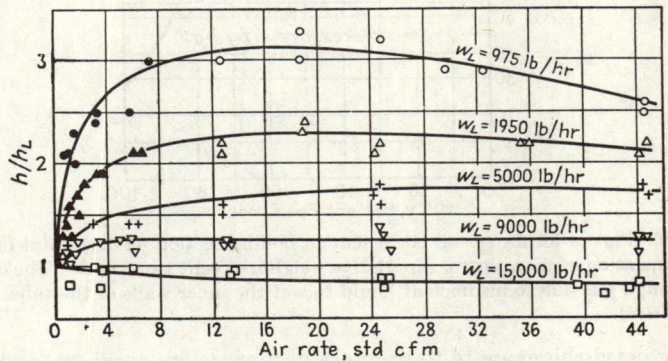

FIG. 14-25. Data of Johnson and Abou-Sabe[49] for horizontal flow of air-water mixtures.

1-in. 16-gauge tube are shown in Fig. 14-25, which exhibits a maximum for the lowest water rate.

Vaporization with Forced Circulation

In this application the total mass flow rate through the tubes is usually high and is set by the capacity of the feed pump. The flow rate is usually known and reported since it is under the control of the experimenter or designer. The analysis of a forced-circulation system is therefore simpler than that of a natural-circulation evaporator, in which

398   *HEAT TRANSMISSION*

the interaction of heat transfer and pressure drop may become extremely complex.

Horizontal Tubes. Woods et al.[65] vaporized water or benzene flowing through a four-pass horizontal steam-jacketed standard 1-in. copper pipe. Temperatures of the fluid were determined at the end of each 12-ft pass, and condensate rates were measured from the 12 separate steam jackets. Low feed velocities were used (0.26 to 1 ft/sec) so that nearly all the feed could be evaporated if desired. The steam-side resistance was minimized

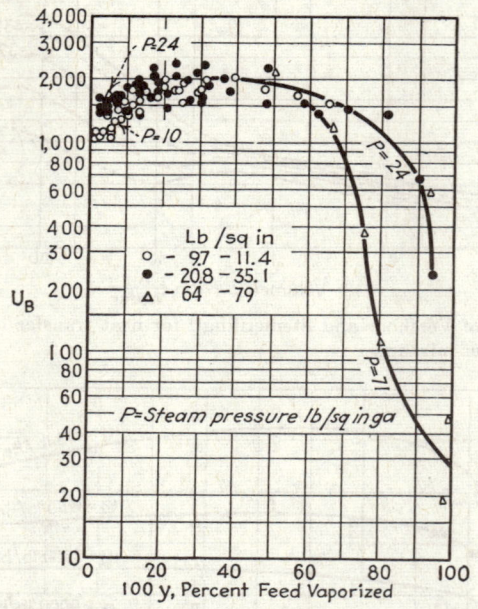

Fig. 14-26. Plot of local over-all coefficient in boiling section of horizontal four-pass forced-circulation evaporator vs. cumulative weight per cent vaporized. The decrease in U_B at high y is due to insufficient liquid to wet the inner walls of the tube. (*Data of Woods.*[92,65])

by using octyl thiocyanate to promote dropwise condensation, and hence over-all coefficients should be nearly as large as the inside coefficients. In the benzene runs the local over-all coefficients in the preheating section ran from two to four times those predicted for forced convection without phase change, which may be due to surface boiling in the film near the wall. As shown in Fig. 14-26, with moderate temperature differences, as the fluid is progressively vaporized the local over-all coefficient at first increases, goes through a maximum, and then decreases sharply toward values typical of superheating dry vapor. This dry-wall vapor binding is attributed to insufficient liquid to wet the walls. With high temperature differences, there was encountered the type of hot-wall vapor binding typical of boiling pools of liquid heated with submerged tubes. Figure 14-27 shows the average flux in the boiling section plotted vs. the length-mean over-all temperature difference for various per cent vaporizations of the feed.

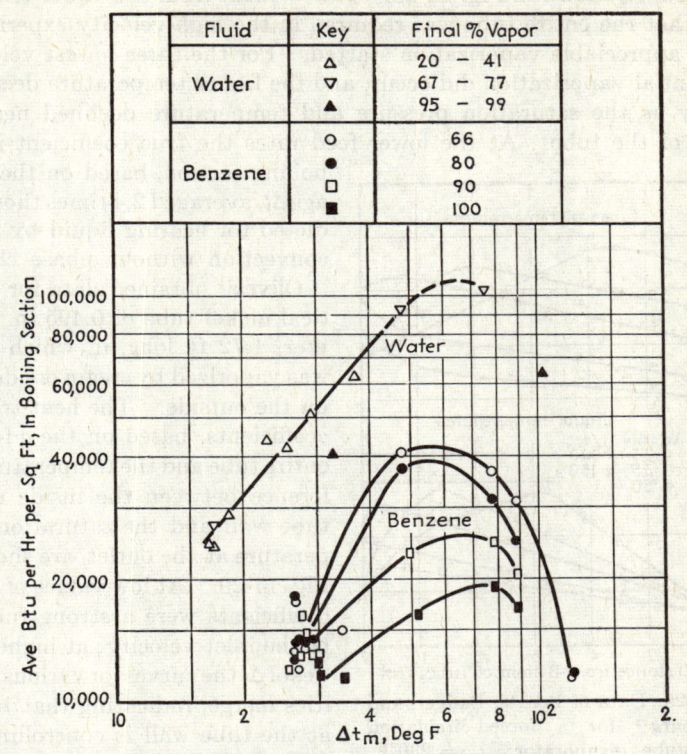

FIG. 14-27. Average flux in boiling section of a horizontal-tube evaporator vs. length-mean temperature difference, for various values of y. In the benzene runs, hot-wall vapor binding (due to excessive Δt) is clearly shown for several values of y; at a given Δt, dry-wall vapor binding is indicated by the decrease in flux with increase in y. Owing to use of a promoter of dropwise condensation on the steam side, the Δt on the boiling side should be substantially the same as the measured over-all Δt. (*Data of Woods*[92,66] *and Bryan*.)

The pressure-drop data of Woods[66] were analyzed by Martinelli and Nelson,[67] who modified the technique of Lockhart and Martinelli[61] by including an approximate allowance for the pressure drop resulting from acceleration of the fluid. The data fell within 30 per cent of the mean value predicted by reference 67, indicating the general validity of the theoretical treatment; pressures ranged from 18 to 25 lb/sq in. abs with outlet qualities from 4 to 95 per cent vapor by weight.

Vertical Tubes. Boarts et al.[13] evaporated water at boiling temperatures of 212 to 139°F inside a vertical copper tube (0.76 in. by 12 ft) jacketed with condensing steam. By use of a traveling thermocouple and tube-wall thermocouples, temperature-distribution curves were obtained, as shown in Fig. 14-28. It is evident from the liquor temperatures that the entire tube was required in the high-velocity experiments before appreciable vaporization started. For the three lowest velocities substantial vaporization did occur, and the liquor temperature decreased rapidly as the saturation pressure and temperature declined near the outlet of the tube. At the lower feed rates the true coefficient in the boiling section, based on the average Δt, averaged 2.4 times those predicted for heating liquid by forced convection without phase change.

Oliver[74] obtained data for a vertical nickel tube of 0.495 in. diameter, 1.72 ft long, in which water was vaporized by steam condensing on the outside. The heat-transfer coefficients, based on the q for the entire tube and the temperature difference between the inside of the tube wall and the saturation temperature at the outlet, are shown in Fig. 14-29. At low values of Δt the coefficients were a strong function of the inlet velocity; at higher values of Δt the curves for various velocities merge, indicating that boiling at the tube wall is controlling the film coefficient. A similar indication of the combined effect of velocity and Δt was shown for submerged cylinders in Fig. 14-8. The corresponding over-all coefficients for runs with the short tube at entering velocities of 3.6 ft/sec are plotted vs. the steam pressure in Fig. 14-30, and it is seen that the oleic acid was a better promoter of dropwise condensation on nickel than was octyl thiocyanate.

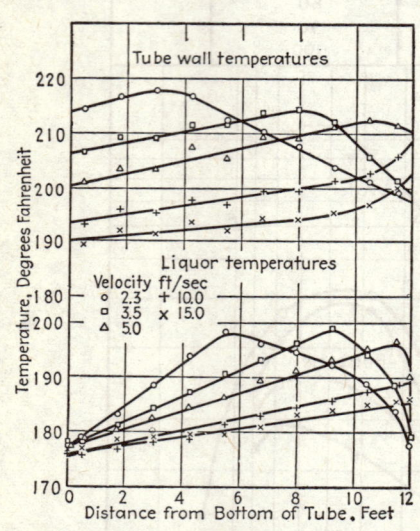

Fig. 14-28. Data of Boarts, Badger, and Meisenburg[13] for a forced-circulation vertical-tube evaporator; $t_{sv} = 230°F$, $t_0 = 176°F$, overall $\Delta t = 54°F$.

Local film heat-transfer coefficients and local pressure gradients were measured by Dengler[31] for vaporization of water flowing upward in a 1-in. 20-ft vertical copper tube. Five steam jackets were closely spaced along the tube, and 21 thermocouples were embedded in the tube wall. Local pressure drops and saturation temperatures and pressures were determined by means of a manometer system connected to pressure taps located between the jackets. The feed was preheated and introduced

BOILING LIQUIDS 401

into the test section at the boiling point. Total mass flow rate was varied from 240 to 5500 lb/hr, corresponding to entrance velocities of 0.2 to 4.8 ft/sec and entering Reynolds numbers of 5200 to 120,000. The weight fraction of vapor varied from 0 to 100 per cent depending upon flow rate. The separator pressure was varied from 7.2 to 29 lb/sq in. abs and the film temperature difference from 0 to 40°F.

The local heat-transfer coefficient was controlled by the combined influence of boiling and forced convection. At low flow rates and low weight fractions of vapor the coefficient increased with increase in Δt and increase in pressure, as in nucleate boiling on submerged surfaces. At

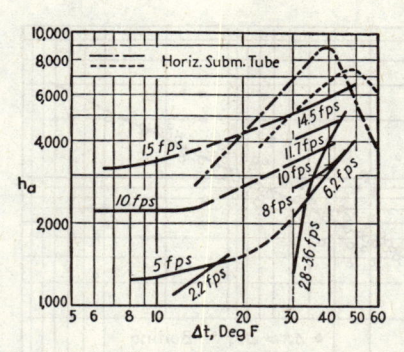

FIG. 14-29. Coefficients of Oliver[74] for water in a vertical forced-circulation evaporator, based on $t_m - t_0$, compared with coefficients for boiling in pools heated by a submerged surface.

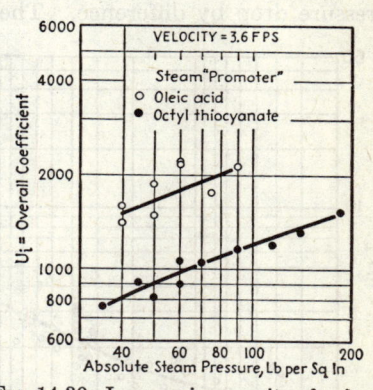

FIG. 14-30. Increase in capacity of a short vertical nickel-tube forced-circulation evaporator (0.495 in. by 1.72 ft) by promoting dropwise condensation by use of oleic acid; film coefficients are given in Fig. 14-29. (*Data of Oliver.*[74])

high mass flow rates and large weight fractions of vapor the coefficient became independent of Δt and was reduced by increase in pressure. Under these conditions the effect of nucleate boiling was apparently suppressed by the increased heat-transfer coefficient obtained with strong two-phase circulation, and the results were correlated by

$$\frac{h}{h_L} = 3.5 \left(\frac{1}{X}\right)^{0.5} \tag{14-13}$$

from $1/X$ from 0.25 to 70, as shown in Fig. 14-31. When Δt exceeded the value Δt_i required for initiation of boiling,

$$\Delta t_i = 11(V_L)^{0.3} \tag{14-14}$$

the coefficients ranged up to 2.5-fold greater than those given by Eq. (14-13).

The observed pressure drops were analyzed by Dengler[31] in terms of the components due to acceleration, wall friction, and change in elevation. For this purpose, the local volumetric fraction of liquid R_L was determined by adding a radioactive tracer to the liquid. A Geiger-Mueller counter was moved up and down outside the test section, and the resultant counting-rate measurement with suitable corrections yielded local values of R_L, given in Fig. 14-32. The gravitational head was computed from the average density based on R_L, and the acceleration pressure drop was computed from the change in momentum between pressure taps, assuming a uniform, but different, velocity for each phase. The frictional pressure drop was then obtained from the observed total pressure drop by difference. The values of Φ_L so obtained were about

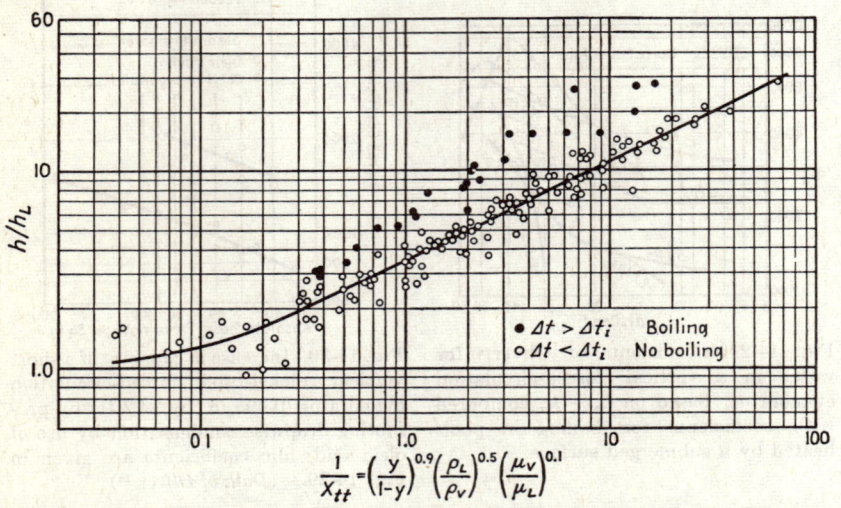

Fig. 14-31. Correlation of Dengler[31] for upflow of steam-water mixture; when Δt exceeded Δt_i, the ratio of h_B/h_L was up to 2.5 times the values shown.

20 per cent above the curve of Lockhart and Martinelli given in Fig. 14-23; when Δt exceeded Δt_i, the observed values of Φ_L were up to three times the predicted values. A pressure profile is presented in Fig. 14-33.

Critical Velocity. For flow of a vapor-liquid mixture in a duct of uniform cross section there exists a maximum or critical mass flow rate, analogous to the limitation of sonic velocity in compressible flow of gases. The critical vapor-liquid velocity is reached when further increase in upstream pressure produces no further increase in mass flow rate. While not yet fully explored, the conditions for critical velocity are discussed by Schweppe and Foust[83] and Harvey and Foust.[41] Data were taken for upflow of water, but the thermodynamic treatment should be applicable for horizontal flow. The theory was compared with data for water in a vertical stainless-steel evaporator (1.12 in. inside diameter

BOILING LIQUIDS 403

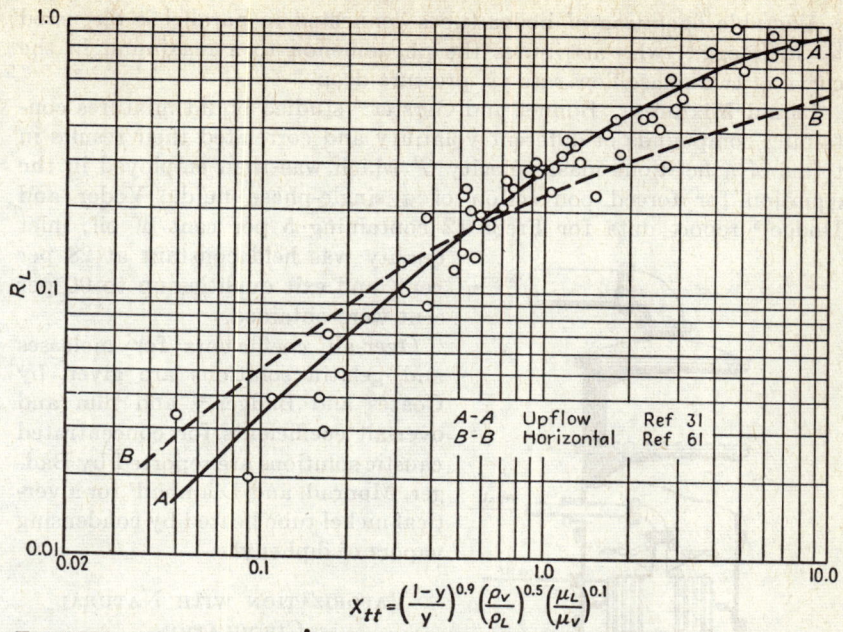

FIG. 14-32. Data of Dengler[31] for the volumetric fraction of liquid, R_L, with upflow of steam and water, as determined by a radioactive tracer.

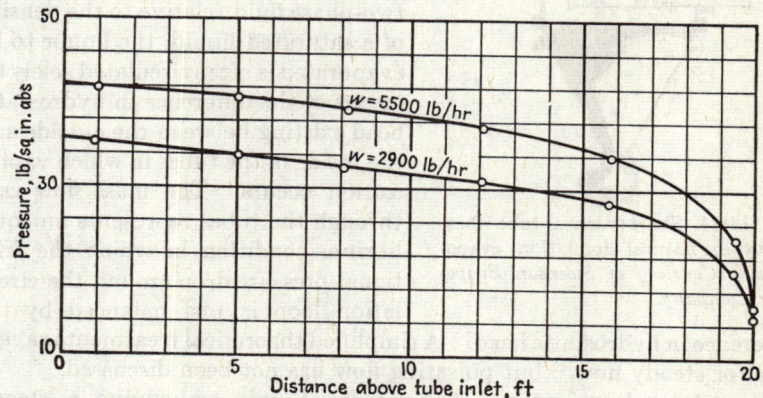

$$X_{tt} = \left(\frac{1-y}{y}\right)^{0.9}\left(\frac{\rho_v}{\rho_L}\right)^{0.5}\left(\frac{\mu_L}{\mu_v}\right)^{0.1}$$

FIG. 14-33. Pressure profiles in 20-ft 1-in. vertical evaporator, showing steep gradient at discharge end.[31]

and 20 ft) with vapor-head temperatures of 125 to 200°F, a variety of wall temperatures, and entering velocities ranging from 0.36 to 1.3 ft/sec. It was concluded that critical flow was frequently reached at the end of the tube and that the stream then expanded adiabatically to the pressure and temperature in the vapor head.

Unstable operation of boiler tubes connected in parallel is discussed by Ledinegg,[58] who attributes the phenomenon to a maximum in the curve of two-phase flow rate vs. pressure drop.

Liquid Mixtures. Bonnet and Gerster[16] studied liquid mixtures containing compounds of different volatility and correlated their results in terms of a fictitious mass velocity G' which was then employed in the equation for forced convection of a single-phase fluid. Yoder and Dodge[93] report data for Freon-12 containing 5 per cent of oil; inlet quality was held constant at 28 per cent, and exit qualities up to 90 per cent were obtained.

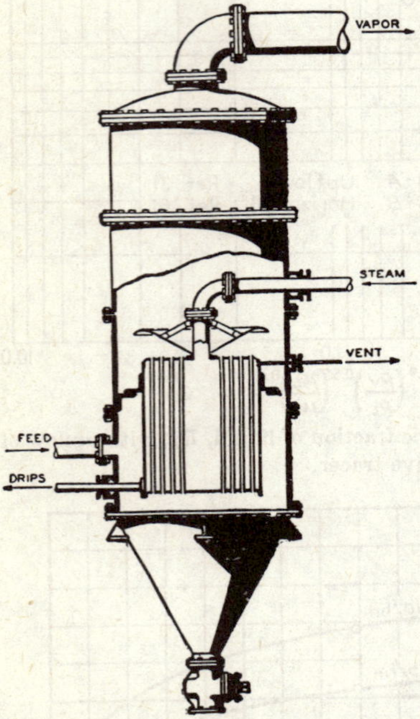

Over-all coefficients for molasses and gelatin solutions are given by Coates and Badger,[26] and film and over-all coefficients for concentrated caustic solutions are reported by Badger, Monrad, and Diamond[7] for a vertical nickel tube heated by condensing vapors of diphenyl.

Vaporization with Natural Circulation

Owing to the reduced density of a two-phase fluid relative to the density of a saturated liquid, the liquor to be evaporated is often circulated solely by means of the difference in hydrostatic head existing between the outside and the inside of the tubes in which vaporization occurs. The mass flow rate through the tubes represents an equilibrium condition in which the frictional pressure drop around the circulation loop is just balanced by the

Fig. 14-34. Short vertical-tube (basket-type) natural-circulation evaporator. (*Courtesy of Swenson Evaporator Company.*)

difference in hydrostatic head. A simplified theoretical treatment is available for steady flow,[42] but pulsating flow has not been discussed.

Short-tube Evaporators. A commercial unit embodying a steam-heated bundle of short vertical tubes is shown in Fig. 14-34. Data of Cleve[25] for evaporation of water in an evaporator of this type are shown in Fig. 14-35. The mass flow rate of liquid reached a maximum at a Δt of 9°F; the subsequent decline in velocity indicates that for greater vaporization the wall friction increased more rapidly than the difference in hydrostatic head, which serves as a driving force for circulation. The heat-transfer film coefficient continued to increase, but at a lower rate.

Foust, Baker, and Badger[37] studied the evaporation of water under reduced pressure in a basket-type evaporator containing 31 ten-gauge steel tubes (2.5 in. outside diameter and 4 ft long), heated by condensing steam. The liquor velocity entering the tubes, measured by a pitot tube, ranged from 1 to 4 ft/sec and for each value of the over-all temperature drop went through a maximum with respect to liquor level. The over-all coefficients increased with lowered level of the liquor, with increased Δt_o, and with increased boiling point. Values of U ranged from 78 to 340.

Long-tube Evaporators. In a pioneer investigation Kirschbaum et al.[53] studied the performance of a natural-circulation evaporator containing a vertical copper tube having an inside diameter of 1.58 in., heated externally for a length of 77.5 in. by steam condensing in a concentric jacket.

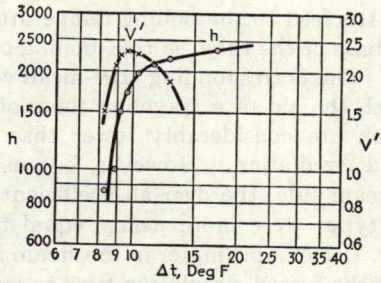

Fig. 14-35. Data of Cleve[25] for heating and boiling water in a short vertical-tube natural-circulation evaporator; V' is liquor velocity entering tube, expressed in feet per second.

Water entered the bottom of the tube at the saturation temperature corresponding to the pressure in the overhead vapor-liquid separator. The mass rate of liquid discharged to the separator, when plotted against the mass rate of vapor generation, always passed through a maximum, which increased both with increase in saturation temperature and with increase in the submergence. Radial and axial traverses of temperature inside the test section gave a curve of mean fluid temperature vs. heated length, which passed through a maximum, indicating a nonboiling region in the lower portion of the tube and a boiling region in the upper portion. The average heat flux increased roughly in proportion to the square of the apparent Δt_a, based on the mean temperature of the inner wall of the tube less the saturation temperature in the separator, and increased with increase in saturation temperature. For a given Δt_a and saturation temperature the apparent heat-transfer coefficient passed through a maximum at a submergence of 25 per cent. Other things being equal, tubes having inside diameters of 1.18 in. gave higher apparent coefficients than tubes having inside diameters of 0.59 in.

In some experiments with forced circulation, Kirschbaum et al.[53] found that the apparent coefficient of heat transfer was less sensitive to Δt_a than with natural circulation and increased with increase in entering velocity and saturation temperature. A much smaller fraction of the tube was used for boiling than with natural circulation. Brooks and Badger[19] measured the distribution of temperature in a single vertical tube of copper (1.76 in. by 20 ft) jacketed by condensing steam. It was found that a substantial portion of the length was used for preheating, and true over-all coefficients U_B for the boiling section were determined for distilled water boiling at temperatures ranging from 150 to 200°F. The data show that U_B increased with increase in over-all temperature difference.

Stroebe et al.[84] used the same apparatus to boil water at 150 to 200°F, with feed entering at velocities of 0.065 to 0.58 ft/sec and injected enough live steam to bring the feed to the boiling temperature corresponding to the pressure at the base of the tube, so that boiling occurred through the length of the tube. The corresponding true-mean coefficients h_m (based on Δt_m obtained with the aid of a traveling thermocouple) ranged from 1160 to 2640, which are considerably lower than those obtained by Boarts[13] with forced circulation. However, because of the substantial resistance on the steam side, the over-all coefficients in the forced- and natural-circulation types were more nearly equal than the coefficients on the boiling side. Use of a promoter in the steam for both evaporators would have allowed the forced-circulation type to take advantage of the high velocity inside the tubes.

Inclined-tube Evaporators. Table 14-10 shows the dimensions of two evaporators having tubes inclined 45 deg, in which the feed enters at the bottom and the unevaporated liquid returns by gravity to recirculate through the tubes. Table 14-11 gives smoothed values of the over-all

TABLE 14-10

Observer	Number of tubes	O.D., inches	I.D., inches	Heated length, ft	A_i, sq ft
Linden and Montillon[60]	1	1.32	1.06	4.08	1.14
Van Marle[87]	7	3.00	2.83	4.88	25.3

TABLE 14-11

Boiling temperature, deg F	120	130	140	180	195	210
U, $\Delta t_o = 10°F$, reference 60	...	...	...	...	...	300
U, $\Delta t_o = 20°F$, reference 60	...	...	...	550	700	900
U, $\Delta t_o = 60–79°F$, reference 87	650	850	1050			

BOILING LIQUIDS 407

coefficients of heat transfer for the evaporation of water at several temperatures. The inside coefficients, determined by Linden and Montillon,[60] were correlated by the equation $h = a_4 \Delta t^{2.5}$, where a_4 is 0.63 at 180°, 1.0 at 195°, and 1.56 at 210°F.

Effect of Liquid Level. As expected from the dependence of h upon mass velocity and weight fraction of vapor, the depth of liquid over the heating surface of a natural-circulation evaporator strongly affects the heat-transfer rate. As shown by Webre and Robinson in Curves 2, 3, and 4 of Fig. 14-36, short vertical-tube evaporators show an increase in the apparent over-all coefficient U (based on the difference between the saturation temperature on the steam side and the saturation temperature in the separator) with increase in depth of liquid up to a certain point

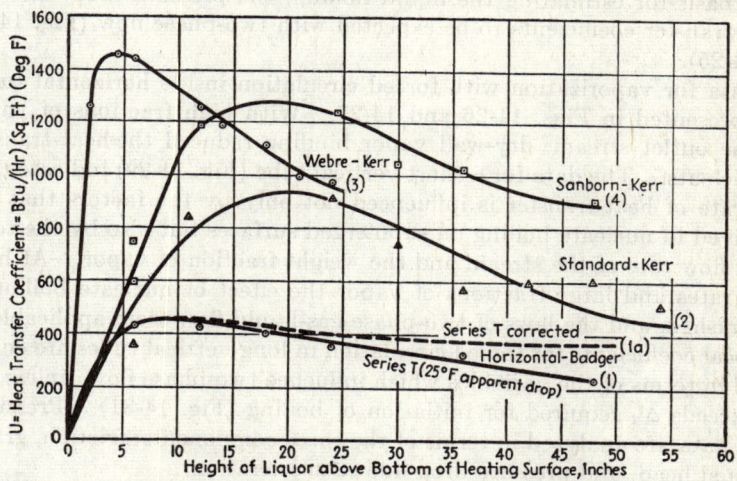

FIG. 14-36. Effect of liquor level on U in natural-circulation evaporators.[91]

and then a decrease with increase in depth. The low coefficients at very low liquid levels may be due to insufficient liquid in contact with the heating surface. At high levels the decrease with increase in liquid level is attributed both to the effect of hydrostatic head on the temperature difference and to a change in circulation. Badger and Shepard[8] had previously studied the variation with liquid level of the over-all coefficient from steam to boiling water in a vertical-tube evaporator and observed a behavior analogous to that illustrated in Fig. 14-36.

In 1920, Badger[4] published tests on a 138-sq-ft horizontal-tube evaporator operated with various liquid levels. Water was evaporated by condensing steam, and the effect of the water level on the over-all coefficient is shown as Curve 1 in Fig. 14-36. The solid line is based on the difference between the saturation temperature corresponding to the pressure in the steam chest and the temperature corresponding to the pressure in the vapor space. The dotted line, Curve 1a, is based on the temperature difference, corrected for hydrostatic head, taken as the difference between the saturation temperature in the steam chest and the arithmetic mean of the saturation temperatures of the water corresponding to the pressures at the top and at the bottom of the heating surface. It is evident that the effect of liquor level is not critical for the horizontal evaporator. Nevertheless, there is an optimum liquid level at which the capacity of either type of evaporator is at a maximum.

Summary, Boiling Inside Tubes

Data for the co-current flow of air-water mixtures are first presented as a basis for estimating the liquid holdup, the pressure drop, and the heat-transfer coefficients to be expected with two-phase flow (Fig. 14-23 to 14-25).

Data for vaporization with forced circulation inside horizontal tubes are presented in Figs. 14-26 and 14-27. With high fractions of vapor in the outlet stream, dry-wall vapor binding reduced the heat-transfer coefficients. The data for a short vertical tube (Fig. 14-29) indicate that the rate of heat-transfer is influenced not only by the factors that are involved in nucleate boiling on submerged surfaces but also by the total mass flow rate of the stream and the weight fraction of vapor. At high flow rates and large fractions of vapor the effect of nucleate boiling is diminished, and the laws of two-phase gas-liquid flow seem applicable.

Local coefficients for forced circulation in long vertical tubes are correlated in terms of the variables which influence two-phase flow, unless the Δt exceeds Δt_i required for initiation of boiling (Fig. 14-31). Pressure-drop data are analyzed in terms of the three components: friction, gravitational head, and pressure drop due to acceleration.

With natural circulation there is interaction between pressure drop and heat transfer; hence the results are sensitive to change in liquid level. The boiling-side and over-all coefficients increase with increase in temperature difference and with increase in temperature of boiling, as is also true for submerged-tube evaporators.

PROBLEMS

1. An experimental evaporator shell consists of a horizontal cylinder approximately 1 ft in length by 8 in. in diameter, provided with an overhead water-cooled total condenser and a submerged horizontal chromium-plated copper tube that is heated by steam condensing on the inside surface. The submerged tube is 12.1 in. long and has an o.d. of 0.840 in. and a wall thickness of 0.109 in. Thermocouples are installed at points 4 in. from each end of the tube and 0.025 in. below the outer surface.

In a typical run, the heating surface is covered with water to a depth of 1.25 in.; steam is passed through a vapor-liquid separator, thence through a glass-wool filter, and is throttled to the desired condensing pressure before it enters the heating tube. Cooling water is supplied to the condenser.

BOILING LIQUIDS

When steady-state conditions have been obtained, the following readings are taken with the water boiling at 212°F:

Run number	1	2	3	4
Steam gauge pressure, lb/sq in	99	13	43	78
Average temperature of thermocouples, deg F	329	234	258.5	317
Temperature of inlet condenser water, deg C	16.2	17.2	12.8	18
Temperature of outlet condenser water, deg C	54.5	58.5	57	58.5
Condenser-water rate, lb/hr	322	305	1132	317

From these data, compute the over-all coefficient and the two surface coefficients.

2. It is desired to boil methanol at 151°F in a natural-convection evaporator heated by steam condensing inside horizontal submerged tubes. In view of the data below, what should be the condensing temperature of the steam?

t_{av}, deg F	221	226	236	246	256	266	272	286	296
U	1450	1510	1540	1500	1420	1300	1200	1040	950

3. Estimate the pressure drop for isothermal co-current flow of water and air in a horizontal standard 1-inch iron pipe 10 ft long. The water rate will be 1200 lb/hr and the air rate 30 lb/hr. The temperature is 70°F, and the discharge pressure is 1 atm.

Repeat for co-current upflow of the same mixture.

4. The outer surface of a long horizontal tube with an o.d. of 0.5 inch is maintained at a constant temperature of 1000°F by electrical heat input. The tube is immersed in boiling water at 212°F. What is the average heat flux from the tube to the water? If the tube is 16 BWG and has a thermal conductivity of 15 Btu/(hr)(sq ft)(deg F), what is the temperature inside the tube? Use $\epsilon = 0.7$.

CHAPTER 15

APPLICATIONS TO DESIGN

Abstract. The first section of the chapter deals with the general case of heat loss through insulated surfaces, the critical radius of insulation, and the optimum thickness of insulation in reducing heat loss.

The second section treats the design of heat-transfer equipment. Various types of tubular exchangers are described, alternate methods of routing the streams are discussed, and unit-cost data are included. The basic quantitative relations among the various design factors are reviewed and summarized. In designing to obtain predetermined terminal temperatures with tubes of fixed diameters, the design procedure is discussed for cases in which the design is fixed by virtue of preselected values of velocities, heated length, or pressure drops. A procedure is given for calculating the outlet temperatures which a given heat exchanger should produce for fixed inlet temperatures and mass flow rates. The section dealing with optimum operating conditions treats the complex general case of optimum velocities in exchangers with fixed terminal temperatures and mass flow rates. Equations and procedures are developed for cases where only one power cost need be considered, where additional design factors are fixed, or where flow of a given fluid either inside or outside the tubes is optional. A quantitative method for comparing fluids as heat-transfer media is presented. Equations and graphs are included for determining the optimum amount of cooling water for condensers and coolers, and equations are developed for the optimum temperature difference to be used in recovering waste heat.

I. HEAT TRANSMISSION THROUGH INSULATION

Consider the general case of a hot fluid at t' inside a pipe having length L, inside and outside diameters D_i and D_o, and mean thermal conductivity k_w, insulated with any thickness x of insulant having mean thermal conductivity k_m, exposed to colder surroundings at t_a. Under steady conditions the same heat flows in turn through a number of resistances: by conduction and convection from the warm fluid to the inside surface, by conduction through the tube wall and the insulation to the outer surface, and thence by conduction, convection, and radiation to the surroundings. In view of the relations developed in preceding chapters, the following general relation may be written,

$$\frac{dq}{dL} = \frac{t' - t_a}{\dfrac{1}{h_i(\pi D_i)} + \dfrac{x_w}{k_w(\pi D_w)} + \dfrac{x}{k_m(\pi D_m)} + \dfrac{1}{(h_c + h_r)(\pi D_s)}} = \frac{t' - t_a}{\Sigma R} \quad (15\text{-}1)$$

which applies to the usual case in which both the air and the surfaces of the surroundings have the same temperature. Values of the thermal conductivities are given in the Appendix, values of h_i are given in Chap. 9, and values of $h_c + h_r$ are given in Table 7-2.

Illustration 1. Calculate the heat loss per 100 ft of standard 1-in. steel pipe carrying saturated steam at a gauge pressure of 150 lb per sq in., insulated with a 2-in. layer of magnesia pipe covering, exposed to a room at 80°F.

Solution. From the steam tables (Appendix), it is found that the saturation temperature corresponding to an absolute pressure of 165 lb per sq in. is 366°F, and consequently the over-all Δt is $366 - 80 = 286$°F. From the Appendix $D_i' = 1.049$ in., $D_o' = 1.315$ in., and $x_w' = 0.133$ in.; since $x' = 2$ in., $D_s' = 5.315$ in. Since the resistance on the steam side will be small relative to that of the other resistance, h need not be predicted accurately; consequently the combined film coefficient for the condensate film and dirt deposit will be taken as 1000. Then

$$R_1 = \frac{1}{h_i(\pi D_i)} = \frac{1}{(1000)(3.14 \times 1.049/12)}$$
$$= 0.0036$$

From the Appendix, k for steel is 26.

$$R_w = \frac{x_w}{k_w \pi D_w} = \frac{0.133/12}{(26)(3.14 \times 1.182/12)} = 0.0014$$

If the temperature drop in the insulation is 90 per cent of the over-all drop, the mean temperature of the insulation will be 238°F, and, from the Appendix, k_m is interpolated as 0.0418. The logarithmic mean of 1.32 and 5.32 is 2.86 in.

$$R_x = \frac{x}{k_m A_m} = \frac{2/12}{(0.0418)(3.14 \times 2.86/12)} = 5.33$$

With 10 per cent (29°) of the total drop through the surface resistance, reference to Table 7-2, at Δt_s of 29°F and D_s' of 5.32 in., gives $h_c + h_r$ of 1.9.

$$R_s = \frac{1}{(h_c + h_r)(\pi D_s)} = \frac{1}{(1.9)(3.14 \times 5.32/12)} = 0.38$$

The total resistance ΣR is then $0.0036 + 0.0014 + 5.33 + 0.38 = 5.72$, and hence the resistances R_1 and R_w were, as usual, negligible compared with the total resistance. The estimated value of R_s is $(0.38/5.72)(100) = 6.6$ per cent of the total resistance instead of the 10 per cent assumed above, but the discrepancy is too small to warrant a second calculation based on Δt_s of $(0.066)(286) = 19$°F. The predicted heat flow per foot is then $(t - t_a)/\Sigma R = 286/5.72 = 50$ Btu/hr, or 5000 Btu/hr from 100 ft.

Optimum Thickness of Insulation. The optimum thickness of insulation to be used in a particular case can be determined by straightforward calculation for a number of standard thicknesses and suitable cost data. The procedure is to calculate the heat loss for various thicknesses of

TABLE 15-1. NOMENCLATURE

A Area of heat-transfer surface, square feet; A_i for inside, A_o for outside, A_s for outer surface of insulation, A_w for tube wall, A_1 and A_2 for surfaces 1 and 2

a Constant, dimensionless; $f_i = a_i(\mu/DG)_i{}^m$; $f_o = a_o(\mu/DG)_o{}^{m_o}$; $j_i = a_3/F_s(DG/\mu)_i{}^{1-n}$; $j_o = a_4/F_s(DG/\mu)_o{}^{1-n_o}$.

B_i, B_o Correction factors, dimensionless, in friction relations [Eqs. (15-8) and (15-11)]

C Unit costs: C_A is first cost of apparatus, dollars per square foot of heat-transfer surface (C_{Ai} is based on A_i, C_{Ao} on A_o); C_e is cost of mechanical energy supplied to the fluid, dollars per foot-pound (C_{ei} for fluid inside tubes and C_{eo} for fluid outside tubes); C_H is cost of heat in dollars per Btu used ($C_{H'}$ available at low temperature and $C_{H''}$ at high); C_w is cost of cooling water in dollars per pound

C_a Hourly fixed charges on apparatus, dollars/(hr)(sq ft); C_{ai} is based on A_i and C_{ao} on A_o; $C_a = C_A F_A/\theta$

C_p Hourly power cost, dollars/(hr)(sq ft of heat-transfer surface); C_{pi} for fluid in tubes and C_{po} for fluid in shell

c Mean specific heat of fluid, Btu/(lb)(deg F); c'' is value for colder fluid, c' for warmer, c_i for fluid inside tubes, c_o for fluid outside tubes

D Diameter of tube in feet; D_i for inside, D_m for mean diameter of insulant, D_o for outside, D_s for surface of insulation, D_w for tube wall

d Prefix, indicating derivative, dimensionless

E Power loss per unit of heat-transfer surface, ft-lb/(hr)(sq ft)

F_c, F_e, F_i, F_r Friction, foot-pounds per pound of fluid, due to sudden contraction, sudden enlargement, friction in tubes, and reversal in direction of flow, respectively

F_s Factor of safety, dimensionless; value by which h in equations recommended in Chap. 9 to 14 should be divided to allow for deviation of data from recommended equation

F_A Fraction of first cost of apparatus charged off annually

F_G Correction factor, dimensionless; $F_G = \Delta t_m/\Delta t_l$ (see Chap. 8)

f_i, f_o Friction factor, dimensionless, f_i for flow inside tubes; f_o for flow normal to a bank of tubes

G Mass velocity, lb/(hr)(sq ft of cross section); for flow inside tubes, $G_i = w/S = 4w/\pi D_i^2$; for flow across tubes, $G_o = w/S_o$, where S_o is based on the minimum free area (previously G_o was designated as $G_{\max}$); G_{io}, G_{oo} for optimum velocity inside tubes and outside tubes, respectively

g_c Conversion factor in Newton's law; $g_c = 4.17 \times 10^8$ (lb fluid)(ft)/(hr)(hr)(pounds force)

h Surface, film, or individual coefficient of heat transfer, Btu/(hr)(sq ft)(deg F); h_c by natural convection, h_d for dirt film, h_i for inside tubes, h_o for outside tubes, h_r by radiation

j Dimensionless factor, $(h/cG)(c\mu/k)^{2/3}$; j_i for inside tubes, j_o for outside tubes (flow normal to axis)

K Dimensional term, K_i defined in Eq. (15-25), K_o defined in Eq. (15-26)

K_1 Dimensionless factor in definition of B_i; see Eq. (15-9)

K_2 Sum of power costs and fixed charges, dollars/(hr)(sq ft); $K_2 = C_a(X_a + X_p)/X_a$

k Thermal conductivity, Btu/(hr)(sq ft)(deg F per ft); k_i for fluid in tubes, k_m for insulant, k_o for fluid outside tubes, k_w for tube wall

Table 15-1. Nomenclature.—(Continued)

L Length of each tube, feet; L_H is total heated length, $L_H = N_{TP}L$, where N_{TP} is number of tube passes

ln   Operator, dimensionless, denoting natural logarithm to base e; $\ln = 2.303 \times \log_{10}$

m_i, m_o Exponents, dimensionless, in dimensionless equations: $f_i = a_i(\mu/DG)_i{}^{m_i}$, and $f_o = a_o(\mu/DG)_o{}^{m_o}$

N Number of rows of tubes across which shell fluid flows, dimensionless

N_{TP} Number of tube passes, dimensionless

N_w Number of tubes in each tube pass

n_i, n_o Exponents, dimensionless, in dimensional equations: $h_i = \alpha_i G_i{}^{n_i}$, $h_o = \alpha_o G_o{}^{n_o}$

P Power delivered to the fluid allowing for over-all efficiencies of motors, pumps, etc., foot-lb per hour

Q Quantity of heat, Btu, transferred through a tube wall

q Rate of heat transfer, Btu per hour

R Individual thermal resistance, feet $\times$ degrees Fahrenheit $\times$ hours per Btu; R_i on inside of pipe, R_w of pipe wall, R_{wd} of pipe wall and dirt film, R_x of insulation, R_s on outer surface of insulant; ΣR is total resistance

r Radius, feet; r_1 is inside radius of insulant; r_c is critical outer radius at which dq/dL reaches maximum value

S Cross section normal to flow of fluid, square feet; S_i for flow inside tubes, S_o for flow outside tubes, S_H for cross section of header

t Bulk temperature of fluid, degrees Fahrenheit; t_a for air, t' for warmer fluid, t'' for colder fluid

U Over-all coefficient of heat transfer, Btu/(hr)(sq ft)(deg F); U_i is based on A_i and U_o on A_o; in a given case $U_i A_i = U_o A_o$

V Average velocity, feet per hour; $V = G/\rho$; V_i for inside tubes, V_H in headers

w Mass rate of flow, pounds per hour: w' for warmer fluid and w'' for colder fluid

X Cost, dollars per Btu transferred; X_a represents fixed charges on apparatus and equals $C_a A/q$ (X_{ai} is for a given fluid inside the tubes and X_{ao} for a given fluid outside the tubes); $(X_{ai})_o$ represents optimum value of X_{ai}. X_p represents cost of mechanical power delivered to the fluid, allowing for over-all efficiencies of motors, pumps, etc., and equals $C_p A/q$ (X_{pi} is for a given fluid inside the tubes, and X_{po} is for a given fluid outside the tubes)

X_o Dimensionless ratio defined in Eq. (15-39)

x Thickness of conductor, feet, measured in direction of heat flow; x_w for tube wall, x for insulant covering tube

Y_o Dimensionless ratio defined in Eq. (15-39)

y_o Clearance, feet, between outer surfaces of tubes in a bundle, taken to correspond to the minimum free area

Z Dimensionless ratio defined in Eq. (15-39)

z Dimensionless ratio, $w'c'/w''c''$

z' Dimensionless ratio, $w''c''/w'c'$

Greek

α Dimensional factors in the equations $h_i = \alpha_i G_i{}^n$, $h_o = \alpha_o G_o{}^n$

Δp Pressure drop, pounds per square foot; Δp_i for inside tubes and Δp_o for outside tubes

TABLE 15-1. NOMENCLATURE.—(*Continued*)

Δt Mean value of temperature drop across an individual thermal resistance, degrees Fahrenheit; Δt_i for inside fluid film, Δt_o for outside fluid film, Δt_{wd} for tube wall and dirt deposits

Δt_m Mean value of over-all temperature drop $(t' - t'')$ from warmer to colder fluid, degrees Fahrenheit; $\Delta t_m = F_G \Delta t_l$, where Δt_l is the logarithmic mean of terminal values for counterflow

θ Hours of operation per year

μ Viscosity of fluid, lb/(hr)(ft); μ_i for fluid inside tubes, μ_o for fluid outside tubes

π 3.1416 . . .

ρ Density of fluid, pounds per cubic foot; ρ_i for fluid inside tubes, ρ_o for fluid outside tubes, ρ_m for mean density, ρ_p for fluid in pump

ϕ Dimensionless term, defined by Eq. (15-14)

insulation from Eq. (15-1). The annual cost of the heat lost can be determined for each thickness from the dollar per Btu value of the heat. The annual cost of the installed insulation can be determined from the initial cost and the annual depreciation rate. Typical results of such a calculation are shown in Fig. 15-1. As the thickness of the insulation is increased, the cost of the heat lost decreases but the fixed charges on the insulation increase. The optimum thickness is determined by the minimum of the resulting curve of total variable costs (annual fuel value of heat lost plus fixed charges on insulation), corresponding to approximately a 2-in. layer in this case.

Many articles† have appeared giving charts or equations to facilitate the determination of the optimum thickness from the various pertinent data.

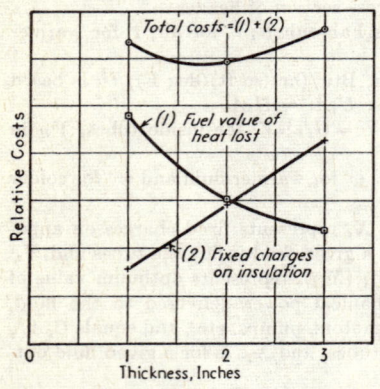

FIG. 15-1. Determination of optimum thickness of insulation on a steam main.

Critical Radius. Under special conditions an increase in heat loss is caused by insulating, since the effect of the added surface more than offsets the reduced temperature of the surface. Thus, in 1910, Porter and Martin[20] showed that as the thickness x of the insulation is increased the heat loss may reach a maximum value and then decrease with further increase in x. The maximum loss occurs when the outside radius r_c of the insulation is equal to the ratio of the thermal conductivity to the surface coefficient:‡

---

† See references 1, 9, 10, 14 to 17, 19, 21, 23, and 25 to 28.

‡ Equation (15-2) may be derived by setting the derivative of the sum of R_x and R_s with respect to the outside radius of insulation equal to zero. Let r_1 be the inside

$$r_c = \frac{k_m}{(h_c + h_r)_c} \quad (15\text{-}2) \ \star$$

This principle is used by the electrical engineer in insulating wires to secure a combination of electrical insulation with an increased cooling effect.

Thus a tube having an outside diameter of ½ in. and a temperature of 248°F, exposed to a room at 68°F, would lose approximately 60 Btu/hr, if bare. If insulated with dense asbestos (k of 0.15), the heat loss would be 76 Btu/hr with a covering ½-in. thick and 60 Btu/hr with a 2.8-in. layer. Had a really good insulation been used, the heat loss would not have been increased by adding the covering.

Heat loss from a bare surface depends on h_c (Chap. 7) and upon h_r, which is directly proportional to the emissivity (Appendix) as shown in Chap. 4. Hence for a surface having a low emissivity, such as galvanized iron, the heat loss may be increased by adding a thin layer of asbestos paper having a high emissivity, as shown by Day;[7] eight layers of paper were necessary to make the heat loss as low as that from the bare pipe.

II. DESIGN OF HEAT-TRANSFER EQUIPMENT

In this section the design of heat-transfer equipment is discussed both qualitatively and quantitatively. Selection of the best type of heat exchanger for a given case is considered, including the factors that influence construction and operation of the unit. The quantitative relations developed in previous chapters are reviewed and summarized, and their application to design problems is demonstrated. Equations are developed for predicting the optimum operating conditions for various applications of heat-transfer equipment.

Structural and Cost Factors

The designer of exchangers has to specify a number of factors. The following section describes various types of exchangers† and discusses temperature strains, thickness of shells and tubes, types of baffles, and unit costs of surface as affected by length, diameter, and number of tubes and material of construction. Usually the apparatus is designed to meet the requirements of the safety codes of the ASME or the API-ASME. Specifications governing tubular exchangers and the materials to be

---

radius and r the outside radius of the insulation, having mean thermal conductivity k_m. Now $R_x = x/k_m 2\pi r_m = \ln (r/r_1)/2\pi k_m$, and $R_s = 1/(h_c + h_r)(2\pi r)$. Setting $d(R_x + R_s)/dr = 0$ gives $r_c = k_m/(h_c + h_r)$, neglecting change in k_m and $h_c + h_r$ with change in t, and r.

† For the sake of brevity in the descriptive section the term exchanger is used in the broad sense, regardless of whether the object is to cool one stream, to heat the other, or to do both. Several types of evaporators are described in Chap. 14.

used in fabrication are given in reference 24. The names of typical heat-exchanger parts are given in Figure 15-2.

Basic Types. Three basic types of heat exchanger are illustrated in Fig. 15-3 to 15-5. Figure 15-3 shows a typical exchanger with fixed tube sheets. Since the tube bundle is not removable, this construction is suitable only for those cases in which the fluid in the shell does not foul

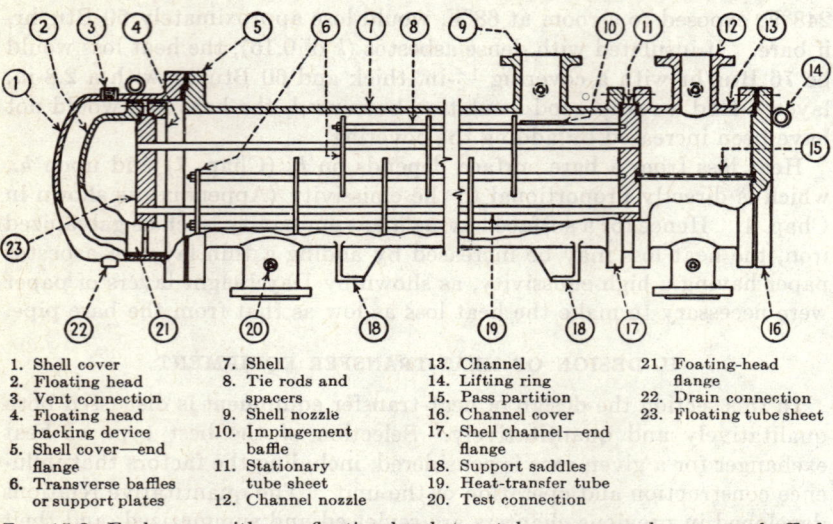

1. Shell cover
2. Floating head
3. Vent connection
4. Floating head backing device
5. Shell cover—end flange
6. Transverse baffles or support plates
7. Shell
8. Tie rods and spacers
9. Shell nozzle
10. Impingement baffle
11. Stationary tube sheet
12. Channel nozzle
13. Channel
14. Lifting ring
15. Pass partition
16. Channel cover
17. Shell channel—end flange
18. Support saddles
19. Heat-transfer tube
20. Test connection
21. Floating-head flange
22. Drain connection
23. Floating tube sheet

FIG. 15-2. Exchanger with one floating head, two tube passes, and one cross-baffled shell pass. (*Adapted from Tubular Exchanger Manufacturers Association.*[24])

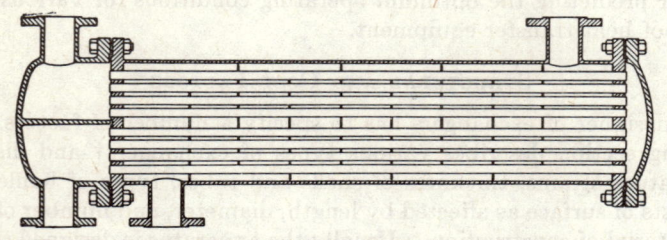

FIG. 15-3. Heat exchanger with fixed tube sheets, two tube passes, and one cross-baffled shell pass. (*Courtesy of Struthers-Wells Company.*)

the tubes. Large differences in temperature between the shell and tubes may produce severe temperature stresses, and when this is encountered, a type of construction that will allow independent expansion of either the shell or the tubes should be provided. Figure 15-4 shows an exchanger with the expansion joint in the shell. Figures 15-5 and 15-6 show the more common type of arrangement in which the tubes are free to expand by means of a floating head. Sieder[22] gives an empirical formula for

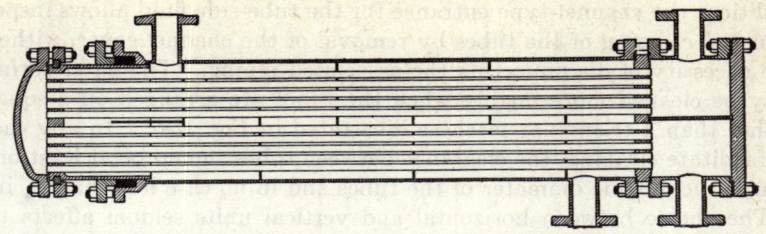

FIG. 15-4. Heat exchanger with an expansion joint in the shell, two tube passes, and one cross-baffled shell pass. (*Courtesy of Struthers-Wells Company.*)

FIG. 15-5. Heat exchanger with one floating head, four tube passes, and one cross-baffled shell pass. (*Courtesy of M. W. Kellogg Company.*)

(a)   (b)

FIG. 15-6. Photographs from Sieder[22] of portions of exchangers; (*a*) shows split floating heads, with shell cover removed, and (*b*) shows the method of assembling tubes and baffles in an unfinished exchanger.

calculating the number of floating heads that should be used to avoid failure due to severe temperature strains. Alternatively, temperature stresses may be avoided by the use of U-shaped tubes, which, however, are more difficult to clean inside by mechanical means.

If frequent cleaning is required, the construction employed should facilitate this operation. For this reason many exchangers are provided with removable tube bundles as shown in Figs. 15-2 and 15-5. In addition, the channel-type entrance for the tube-side fluid allows inspection and cleaning of the tubes by removal of the channel cover, without the necessity of disconnecting the associated piping. The outer surface may be cleaned more readily when the tubes are arranged on a square rather than a triangular pitch as illustrated in Fig. 15-7. In any case, to facilitate cleaning, the clearance between tubes should be at least one-fourth the outside diameter of the tubes and in no case less than ¼ in.

The choice between horizontal and vertical units seldom affects the first cost of an exchanger of a given size, and the decision will depend on plant layout or process requirements and in some cases on installation

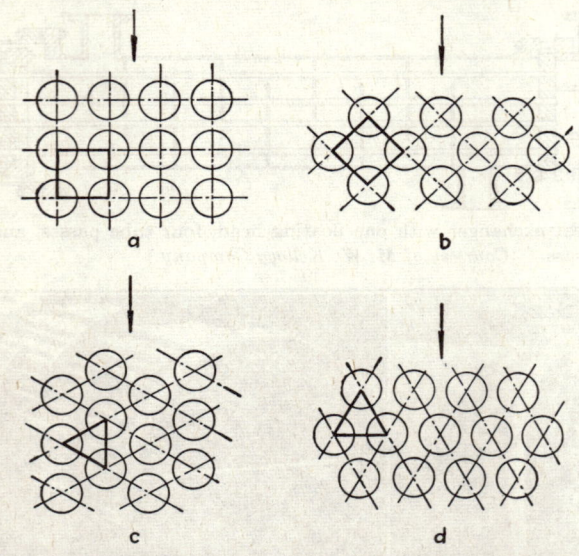

Fig. 15-7. Several arrangements of tubes in bundles: (a) An in-line arrangement with a square pitch; (b) a staggered arrangement with a square pitch; (c) and (d) staggered arrangements with triangular pitches.

costs. As shown in Chap. 13, with film condensation, higher coefficients are obtained with horizontal than with vertical tubes. However, if it is desired to condense the vapor and substantially subcool the condensate in a single apparatus, vertical tubes are used.

Routing of Fluids. If one of the fluids fouls the surface much more rapidly than the other, it should be routed through the tubes, since the inside surface may be cleaned without removing the tube bundle from the shell. If both fluids are equally nonfouling and only one is under high pressure, it should flow inside the tubes to avoid the expense of a high-pressure shell. Where only one of the fluids is corrosive, it should flow inside the tubes to avoid the expense of special metal for both shell and tubes. If one of the fluids is much more viscous than the other, it may be routed through the shell to increase the over-all coefficient. At times, because of a limitation in available pressure drop, it is necessary to route a given stream through the shell.

Tube Size. Although tube lengths of 4 to 22 ft are readily obtainable and longer tubes are available, one often selects[24] a standard length of 8, 12, or 16 ft, using more than one pass where necessary. The shorter lengths are used when the exchanger is located well above grade in order to minimize the cost of platforms and equipment for removing the bundles for cleaning, where length is limited by available space, or where very large shell diameters are required; the longer lengths are used when the equipment is near grade and consequently platforms are not needed. Where maintenance work must be done frequently, location at or near grade is preferred.

In exchangers the tubes ordinarily used have outside diameters of $5/8$, $3/4$, 1, $1\tfrac{1}{4}$, or $1\tfrac{1}{2}$ in.; the larger diameters tend to be used for fluids that foul the tubes rapidly. In some cases fouling may be reduced by use of high velocity, thus permitting the use of tubes of moderate diameter. Where a number of exchangers for different duty are to be installed in a given plant, service and replacement costs are reduced by standardizing on a minimum number of tube diameters and lengths. The thickness of tube walls should be selected not only to withstand working pressures and extreme temperatures and to provide allowance for corrosion but also to facilitate expanding the tubes into the tube sheets. Manufacturers of tubes specify the outside diameter and nominal thickness of the wall. The variation in tube-wall thickness from the nominal value may be ± 10 per cent for "average-wall" tubes and $+22$ per cent for "minimum-wall" tubes. The effects of these variations should not be overlooked, especially in tubes of small diameter, since pressure drop varies inversely as the diameter raised to an exponent ranging from 4 to 4.8.

Baffle Arrangements. Figure 15-8 shows three types of transverse baffles used to increase velocity on the shell side. With a bored shell the clearance between baffles and shell is often $1/32$ to $3/64$ in., but the clearance and consequently the leakage will increase because of corrosion. With unbored shells the clearance may be considerably larger because of greater tolerances.[24] The orifice type of baffle should fit the shell closely to prevent leakage, and the baffles are spaced fairly closely to give frequent changes in the velocity; this type should not be used for fluids that rapidly foul the outer surface or where corrosion or erosion is likely to cut the tubes. In a variation of the orifice type, alternate baffles have orifices for one-half the number of tubes and support the other half. With disk-and-doughnut baffles, the disks offer no support for the central tubes unless braced to the shell. Sometimes the minimum clearance between the tubes and the edges of the holes in the segmental or disk-and-doughnut baffles is 1 per cent of the tube diameter, but since the tubes may have a diameter tolerance of 2 per cent, the maximum clearance may be 3 per cent; consequently, some fluid will flow through these clearances. In reference 24 it is recommended that the tube holes in the baffles and those in the support plates be drilled $\frac{1}{32}$ and $\frac{1}{64}$ in. larger than the outside diameter of the tubes. To avoid vibration of the baffles

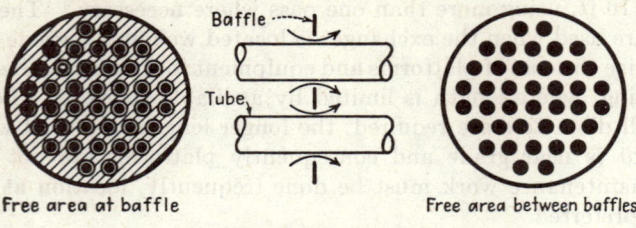

ORIFICE BAFFLE

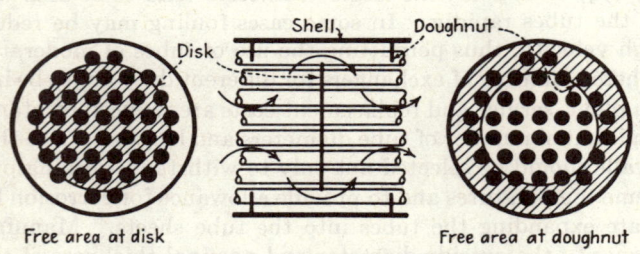

Free area at baffle — Free area between baffles

DISK-AND-DOUGHNUT BAFFLE

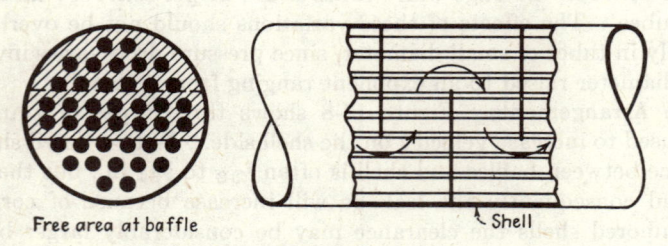

Free area at disk — Free area at doughnut

SEGMENTAL BAFFLE

Free area at baffle — Shell

Fig. 15-8. Three types of transverse baffles. (*Sieder*.[22])

and scoring of the tubes, the baffles should have a thickness of $\frac{1}{8}$ in. and preferably $\frac{3}{16}$ or $\frac{1}{4}$ in., and the edges of the tube holes should be chamfered. The baffle is a fixture, whereas the tubes can be replaced individually, and the baffles may be subjected to considerable abuse when the tube bundle is being withdrawn; consequently the baffle should be at least twice as thick as the wall of the tube. Sufficient space between baffles should be provided to facilitate cleaning. The thickness of tube sheets should be at least $\frac{7}{8}$ in. and should be not less than the outside diameter of the tubes.[24] Longitudinal baffles, if used, may be welded to the shell or may be of the removable type, with arrangement to prevent leakage. Some users of exchangers prefer two single-pass shells in series rather than one shell with a longitudinal baffle and a split floating head. Where a spare tube bundle is to be provided, it may be more economical to employ several smaller units in multiple instead of a single large unit. To avoid erosion of the tubes near the point of entry of the shell fluid, impingement baffles and flaring nozzles are sometimes installed.

Venting. Adequate vents and drains should always be provided, and relief valves or rupture disks are required for high pressure. Since the removal of condensate becomes difficult when the gauge pressure is as

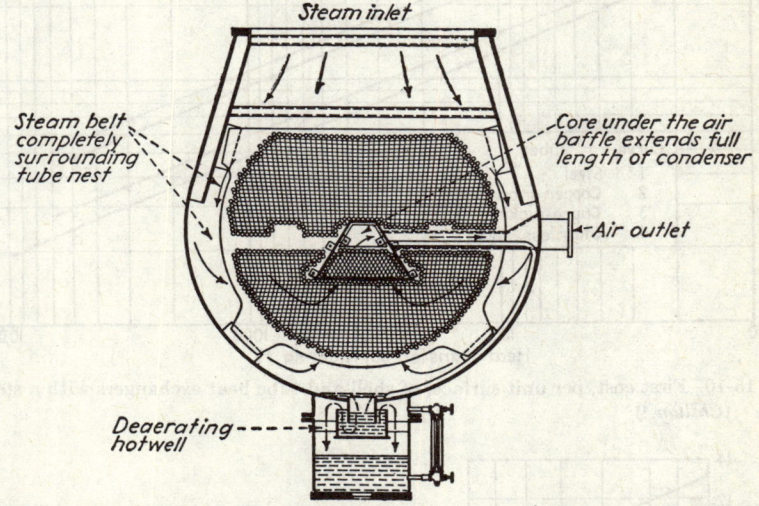

Fig. 15-9. Cross section of a small radial steam-flow type of surface condenser, with two tube passes. (*Courtesy of Westinghouse Electric Corporation.*)

low as 10 lb/sq in., gauge glasses should be provided to show the condensate level so that the operator may avoid flooding the tubes, with consequent reduction in capacity. Steam traps should be provided with by-passes so that the equipment may be operated with hand control in case of failure of the traps. With vapor-heated equipment, it is advisable to provide suitably located vents for the removal of noncondensable gas rather than to rely wholly on devices to remove both condensate and permanent gas.

Figure 15-9 shows a two-pass surface condenser in which noncondensable gases are collected along the axis and pass over cold tubes before entering the air-outlet line. Single-pass steam condensers have been built that contain more than 100,000 sq ft of condensing surface in a single shell.

Costs. The initial cost of a tubular heat exchanger, expressed as dollars per square foot of heat-transfer area, depends upon a large number of factors: the total heat-transfer surface, the tube size and length, the material of construction of the tubes and the shell, the working pressure, the degree of baffling, and the cost of special features, if any. As expected, costs for standard units are considerably below those for the

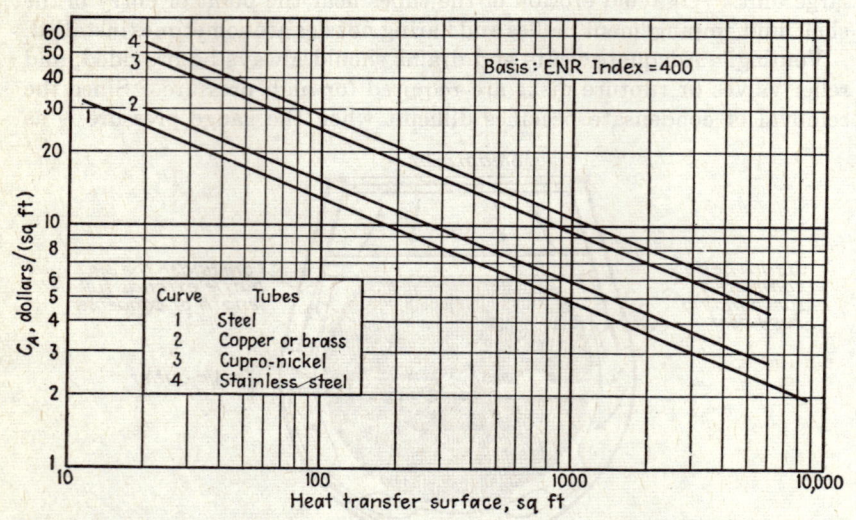

FIG. 15-10. First cost, per unit surface, of shell-and-tube heat exchangers with a steel shell. (*Chilton*.[4])

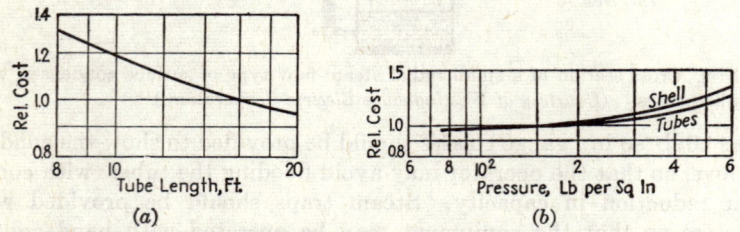

Curve	Tubes
1	Steel
2	Copper or brass
3	Cupro-nickel
4	Stainless steel

Basis: ENR Index = 400

FIG. 15-11. Relative costs of shell-and-tube heat exchangers as affected by (a) tube length and (b) working pressure. (*Sieder*.[22])

less common types. For industrial exchangers of normal size the cost per square foot decreases as the total surface increases, as shown in Fig. 15-10, in which the slope of the curve is -0.4. The curve is based on the cost index of the *Engineering News-Record* relative to a value of 100 for the year 1913.

The use of two floating heads in a single shell instead of one may increase the cost 30 per cent, and four floating heads instead of one may increase the cost 50 per cent.[22] For the same area, a typical exchanger would be reduced 24 per cent in cost by using tube lengths of 16 instead of 8 ft as shown in Fig. 15-11a. The effect of working pressure upon the relative cost of a typical exchanger of fixed area is shown in Fig. 15-11b.[22]

As the tube diameter is decreased, the coefficient of heat transfer increases somewhat in many cases; hence the required area decreases. However, the pressure drop increases, and consequently there is an optimum diameter corresponding to the minimum sum of fixed charges and power cost. The difficulty of cleaning the tubes increases with decrease in diameter, which may result in higher maintenance costs.

Summary of Quantitative Relations

The following section presents quantitative relations used in the design of tubular heat-transfer apparatus such as heaters, coolers, exchangers, and condensers. In these cases the heat is transferred from fluid to surface by the combined mechanisms of conduction and forced convection. In warming or cooling fluids flowing inside or outside tubes without change in phase, both the coefficient of heat transfer and the pressure drop increase with increase in velocity, and in many cases the designer is free to employ the optimum velocity at which the total costs are at a minimum; however, this may lead to impractical proportions of apparatus. At times, because of process requirements or for other reasons a fixed pressure drop is available, and the exchanger is designed to meet this situation. Alternatively, one may fix the tube length and diameter and use whatever velocity and pressure drop are necessary. Regardless of which factors are fixed in advance and which are computed, the same basic relations are always involved; although these have been derived and discussed in earlier chapters, they are summarized below for convenience.

Heat Balance. As shown in Chap. 6, the heat Q transferred through the wall of a tube may be expended in increasing the enthalpy of the stream, in changing the kinetic energy of the stream, and in doing work against gravity; some of the heat may be lost to the surroundings if the shell fluid receives the heat. Thus if the stream is a gas flowing at high velocity with large percentage change in pressure and small change in temperature, the change in kinetic energy is substantial compared with Q. With a small-scale apparatus involving a small heat flux, a substantial fraction of Q may be lost to the surroundings. Except in such special cases, substantially all the heat transferred through the wall of the tube is utilized to increase the enthalpy of the stream of fluid flowing past the heat-transfer surface, and heat loss to the surroundings is negligible. In the cases discussed below, heat is transferred from a warmer stream at t' through a heat-transfer surface to a colder stream at t'', and at least one of the streams is heated or cooled without change in phase; consequently the heat balance becomes

$$q = wc(t_2 - t_1) = SGc(t_2 - t_1) \qquad (15\text{-}3)$$

In case neither fluid changes in phase, the heat balance is

$$q = w'c'(t_1' - t_2') = w''c''(t_2'' - t_1'') \qquad (15\text{-}3a)$$

In these cases the heat balances have direct utility, since they fix the relation between the terminal temperatures and mass flow rates. As pointed out in Chap. 8, the ratio z' $[= w''c''/w'c' = (t_1' - t_2')/(t_2'' - t_1'')]$ has an important bearing on the mean over-all temperature difference $(t' - t'')_m$ in multipass or cross-flow exchange of sensible heat. If this limitation is overlooked, one might select four terminal temperatures that satisfy the heat balance and that seem plausible, since the lowest temperature of the warmer stream exceeds the highest temperature of the colder stream, yet the value of z' is such that the proposed heat exchange cannot occur in the particular type of apparatus under consideration, although it could readily be obtained with counterflow.

Rate Equation. This takes the familiar form

$$q = UA\,\Delta t_m = U_i(\pi D_i L_H)F_G\,\Delta t_l = h_i(\pi D_i L_H)\,\Delta t_i = h_o(\pi D_o L_H)\,\Delta t_o \qquad (15\text{-}4)$$

in which L_H is the heated length and F_G is the ratio of the true mean-temperature difference for the apparatus in question to the logarithmic mean value Δt_l for counterflow:

$$\Delta t_l = \frac{(t_1' - t_2'') - (t_2' - t_1'')}{\ln\,[(t_1' - t_2'')/(t_2' - t_1'')]} \qquad (15\text{-}4a)$$

Values of F_G, given in Figs. 8-6 and 8-7 depend on two dimensionless parameters as well as on the type of exchanger. If one of the temperatures remains constant, F_G is always unity for any type of exchanger. It will be recalled that the values of F_G were derived on the assumption that U was independent of temperature. If the ratio of the terminal values of U differs substantially from unity, U is evaluated at a special temperature t_x given by Fig. 8-4.

As shown in Chap. 8, the over-all coefficients are related to the individual coefficients by the resistance equation

$$\frac{1}{U_i} = \frac{1}{h_i} + \frac{1}{h_{di}} + \frac{x_w}{k_w D_w/D_i} + \frac{1}{h_{do}D_o/D_i} + \frac{1}{h_o D_o/D_i} \qquad (15\text{-}5)$$

This is sometimes written as

$$\frac{1}{U_i} = D_i \Sigma R + \frac{1}{h_i} \qquad (15\text{-}5a)$$

in which ΣR represents the sum of all resistances except the resistance

$1/h_i$ of the film under consideration; all resistances are based on the inside surface.

Coefficients. Values of the surface coefficients h_i and h_o are given in Chap. 9 to 14 in terms of the pertinent physical properties, dimensions of apparatus, and operating variables; dirt-deposit factors are given in Tables 8-2 and 8-3, and various physical properties and standard diameters are given in the Appendix. For turbulent flow without change in phase, as shown in Chaps. 9 and 10, the values of h_i and h_o are given by the following dimensionless equations:

$$\left(\frac{h}{cG}\right)_i \left(\frac{c\mu}{k}\right)_i^{2/3} = \frac{a_3/F_s}{(DG/\mu)^{1-n}} \tag{15-6}$$

$$\left(\frac{h}{cG}\right)_o \left(\frac{c\mu}{k}\right)_o^{2/3} = \frac{a_4/F_s}{(DG/\mu)_o^{1-n_o}} \tag{15-7}$$

Various refinements and special cases are discussed in Chaps. 9 and 10. The proper value of the factor of safety F_s depends on the case but, in general, may be taken as 1.25 for gases or water and 1.5 for viscous liquids.† In general, Eq. (15-6) applies for Prandtl numbers from 0.7 to 120, Reynolds numbers from 10,000 to 120,000, and L/D ratio greater than 60. In these ranges a_3 is 0.023, and n_i is 0.8. Equation (15-7) is based on flow normal to tubes that are not baffled, at Reynolds numbers between 2000 and 40,000. The recommended value of a_4 is 0.33 for staggered tubes and 0.26 for in-line tubes, and n_o is 0.6. If baffles are used, because of leakage of fluid from one compartment to the next, the velocity and consequently h_o will be less than computed on the basis of no leakage. Furthermore, in baffled exchangers, the flow is not strictly normal to the tubes, tending to reduce the coefficients; also certain portions of the tubes are not swept by the stream, thus reducing the effectiveness of these areas. Data are scarce, but some work has been done for quantitatively predicting the effect of these factors on h_o (see page 278). Alternatively, one may space the baffles more closely so that the desired velocity is obtained despite leakage.

For a given case, Eqs. (15-6) and (15-7) reduce to

$$h_i = \alpha_i G_i^{n_i} \tag{15-6a}$$
$$h_o = \alpha_o G_o^{n_o} \tag{15-7a}$$

Pressure Drop. The pressure drop Δp_i through the tube passes is conveniently expressed as B_i times that due to friction in the straight tubes:

$$\Delta p_i = \frac{B_i 4 f_i L_H G_i^2}{2 g_c \rho_i D_i} \tag{15-8}$$

---

† If a given equation predicts h within ± 20 per cent, to play safe, h is taken as 0.8 times the value from the equation and the corresponding F_s is $1/0.8 = 1.25$.

wherein†

$$B_i = 1 + \frac{K_1}{4f_i L/D_i} \tag{15-9}$$

As shown in Fig. 6-11, f_i depends on the Reynolds number $(DG/\mu)_i$ and somewhat on the roughness of the surface of the wall; f_i may be approximated by the relation

$$f_i = a_i(\mu/DG)_i^{m_i} \tag{15-10}$$

where the exponent m_i is 0.2 for turbulent flow and 1.0 for streamline flow. For $D_i G_i/\mu_i$ in the range of 5000 to 200,000 for isothermal flow, a_i is 0.046 for smooth tubes and 0.055 for steel tubes. Corrections for nonisothermal flow are given on page 149.

For flow across nonbaffled tubes one employs

$$\Delta p_o = \frac{4f_o B_o N G_o^2}{2g_c \rho_o} \tag{15-11}$$

in which f_o (previously designated as f''') depends on a Reynolds number $D_o G_o/\mu_o$ and the arrangement and spacing of tubes and G_o (previously called G_{max}) is the mass velocity of flow through the minimum free area; it will be recalled that f_o is much larger than f_i. For a given arrangement and spacing

$$f_o = a_o(\mu/DG)_o^{m_o} \tag{15-12}$$

where m_o is 0.15; values of a_o depend on the spacing and arrangement of tubes and are given on page 163. If baffles are used, an empirically determined value of B_o is introduced to allow for the effect on ΔP_o of reversal in direction of flow and nonuniformity in cross section. For flow of fluid across banks of tubes that are not baffled, B_o is 1.0.

Application of Quantitative Relations

In the application of the quantitative relations outlined above, two general types of problem are frequently encountered. In the first type,

---

† $B_i = (F_c + F_i + F_e + F_r)/F_i$, in which the friction F_c due to each sudden contraction is $K_c V_i^2/2g_c$, the friction F_i in each tube pass is $4f_i L V_i^2/2g_c D_i$, the friction F_e due to each sudden enlargement is $(V_i - V_H)^2/2g_c$, where V_H is the velocity in the header compartments and F_r is the friction in each header due to reversal in direction of flow. As an approximation the average value of F_r is taken as $0.45 V_i^2/2g_c$, although this depends on the details of construction. Hence

$$B_i = 1 + \frac{K_c + [1 - (S_i/S_H)]^2 + 0.45}{4f_i L/D_i} = 1 + \frac{K_1}{4f_i L/D_i}$$

The over-all change in kinetic energy $(V_{i2}^2 - V_{i1}^2)$ can be included in K_1, so that the Δp_i will represent the over-all pressure drop; otherwise Δp_i represents the over-all pressure drop due to friction, not corrected for any net change in kinetic energy.

For S_i/S_H of 0.5, the term $[1 - (S_i/S_H)]^2$ is 0.25, Fig. 6-13 shows that K_c is 0.3, and K_1 equals $0.3 + 0.25 + 0.45 = 1.0$.

the number of variables that have been fixed is sufficient to cause the design to be inflexible, and it is desired to calculate the resulting operating characteristics and the complete heat-exchanger design from these predetermined values. In the second type of problem, which is discussed in the section on Optimum Operating Conditions, there is at least one independent variable that has not been fixed, and it is desired to determine the most economic design for the given case.

The independent variables that are most frequently used by the designer or fixed by plant requirements are selected from the operating and process variables of terminal temperatures, mass flow rates, and pressure drop inside and outside the tubes and the design variables of tube length, tube diameters, and tube layout.

In many cases the terminal temperatures and mass flow rates are determined by process requirements and the heat balance, and consequently q and Δt_l are known. Values of D_i and D_o are often selected for the reasons given on page 419. Since the physical properties of the fluids are fixed, the designer may specify one additional factor when a single resistance controls or two additional independent factors when both resistances are substantial. The others are then computed from the proper combination of basic relations as shown below for several cases.

Fixed Velocities. In some instances the fluid velocities for a tube of given D_i and D_o are fixed for purposes of minimizing fouling and corrosion. In this case, the corresponding values of q, h_i, h_o, and U are computed, and the product $F_G L_H$ is found from Eq. (15-4). If L_H is not excessive, single pass is used and F_G is 1.0; otherwise F_G is evaluated from Figs. 8-6 and 8-7. The corresponding pressure drops are found from Eqs. (15-8), (15-10), (15-11), and (15-12).

Fixed Heated Length. One may have available or under consideration a certain exchanger which might be used for at least a portion of the total proposed load. If, as before, the desired terminal temperatures are specified, the required value of G_i may be obtained by combining Eqs. (15-3), (15-4), and (15-16), if it is noted that $S = (\pi/4)(D_i{}^2)$:

$$\left(\frac{DG}{\mu}\right)_i = \left(\frac{4a_3 L_H}{D_i \phi_i F_s}\right)^5 \tag{15-13}$$

$$\phi_i = \frac{(t_2 - t_1)_i}{\Delta t_i}\left[\frac{c\mu}{k}\right]_i^{2/3} \tag{15-14}$$

If the inside resistance is controlling, ϕ_i is known and G_i can be obtained directly, although the term involving the exponent of 5 should be known with precision. Thus if the factor of safety F_s is to be 1.25, G_i will be 33 per cent of the velocity based on F_s of 1.0. For common gases, for which Eq. (9-15) is applicable, this case may be solved by the use of Fig. 9-15.

If the inside resistance is substantial but not controlling, Eq. (15-13) may be used by a trial-and-error procedure but in this case ϕ_i must be closely approximated. If the outside thermal resistance controls, Eqs. (15-3), (15-4), and (15-7) may be combined to give

$$\left(\frac{DG}{\mu}\right)_o = \left(\frac{a_4 \pi D_o N}{y_o \phi_o F_s}\right)^{2.5} \tag{15-15}$$

$$\phi_o = \frac{(t_1 - t_2)_o}{\Delta t_o} \left[\frac{c\mu}{k}\right]_o^{2/3} \tag{15-15a}$$

Fixed Pressure Drop inside Tubes. The permissible pressure drop across the heater may be fixed by the process-flow requirements. The corresponding G_i for flow inside tubes can be obtained by eliminating L_H from Eqs. (15-3), (15-4), (15-6), (15-8), and (15-10):

$$G_i = \left(\frac{2a_3 g_c \rho_i \, \Delta p_i}{a_i F_s \phi_i B_i}\right)^{0.5} \tag{15-16}$$

B_i is similarly obtained from Eqs. (15-3), (15-6), (15-8), (15-9), and (15-10):

$$B_i = 1 + \frac{N_{TP} K_1 a_3/a_i}{\phi_i F_s} \tag{15-17}$$

For turbulent flow in tubes in which the average velocity is twice that in the header compartments, K_1 is 0.55 for single pass and approximately 1.0 for multipass; for a given N_{TP}, B_i can be obtained from Eq. (15-17) without computing L_H.

If the thermal resistance on the inside is controlling, as in the case of a vapor-heated gas heater, the procedure is simple. First the term ϕ is computed from Eq. (15-14), and B_i is obtained from Eq. (15-17). Second, G_i is calculated from Eq. (15-16), with turbulent flow assumed. Third, D_i is selected, and the Reynolds number $D_i G_i/\mu_i$ is computed; if, as will usually be the case with gases, $D_i G_i/\mu_i$ exceeds 2100, the value of G_i obtained from Eq. (15-16) applies. Fourth, one computes h_i/cG_i from Eq. (15-6) and obtains† L_H from the following relation, which is based on Eqs. (15-3) and (15-4) (noting that $S_i = \pi D_i^2/4$):

$$\frac{4L_H}{D_i} = \frac{(t_2 - t_1)/\Delta t_i}{(h/cG)_i} \tag{15-18}$$

If this value of L_H is inconveniently long for a single pass, multipass will be used.

Illustration 2. Assume that it is necessary to heat 100,000 lb/hr of air from 70 to 200°F by means of steam condensing at 220°F outside steel tubes, that the air enters

† Alternatively, f_i may be computed from Eq. (15-10) and L_H found from Eq. (15-8).

at an absolute pressure of 10 atm, and that the permissible pressure drop across the heater (exclusive of that required for change in kinetic energy) is ⅓ atm. If the tubes have an i.d. of 0.834 in. and F_s is taken as 1.25, what must be the number of tubes in parallel and the length of each?

Solution. Since the thermal resistance inside the tube is controlling, U_i equals h_i. From Eq. (15-14)

$$\phi_i = \frac{t_2 - t_1}{\Delta t_i} \left(\frac{c\mu}{k}\right)^{2/3} = \left(\ln \frac{220 - 70}{220 - 200}\right)(0.74)^{2/3} = 1.65$$

Assume that the headers have twice the cross sections of the tubes; hence K_1 is 0.55 for single pass, and from Eq. (15-17)

$$B_i = 1 + (1)(0.55)(0.023)/(0.055)(1.65)1.25 = 1.11$$

From Eq. (15-4a), the mean temperature difference is 64.4°F, and ρ_i is evaluated at $t_m = t' - \Delta t_i = 220 - 64 = 156$°F. Since 29 lb of air occupies 359 cu ft at normal pressure and 492°F abs:

$$\rho_i = \frac{(29)(492)(9.83)}{(359)(460 + 156)} = 0.633 \frac{\text{lb}}{\text{cu ft}}$$

If the flow is turbulent, Eq. (15-16) applies:

$$G_i = \sqrt{\frac{2a_3 g_c \rho_i \, \Delta p_i}{a_i F_c \phi_i B_i}} = \sqrt{\frac{(2)(0.023)(32.2)(0.633)(705)}{(0.055)(1.25)(1.65)(1.11)}} = 72.4 \frac{\text{lb}}{(\text{sec})(\text{sq ft})}$$

$$= 260,000 \frac{\text{lb}}{(\text{hr})(\text{sq ft})}$$

In order to handle 100,000 lb/hr of air, one needs

$$N_w = \frac{100,000}{(260,000)(0.785)(0.834/12)^2} = 102 \text{ tubes in parallel}$$

As shown in the Appendix, μ' is 0.020 centipoise, and μ_i is $(2.42)(0.020) = 0.0484$ lb/(hr)(ft).

$$\left(\frac{DG}{\mu}\right)_i = \frac{(0.834/12)(260,000)}{0.0484} = 373,000$$

and since this exceeds 2100 the flow is turbulent, as assumed. One could now save time by using Eq. (9-17) or the corresponding chart (9-15); however, the example is continued to illustrate the general procedure. From Eq. (15-6), with a_3 of 0.023, n of 0.2, and F_s of 1.25,

$$\left(\frac{h}{c_p G}\right)_i = \frac{a_3/F_s}{(DG/\mu)_i^{0.2}(c_p\mu/k)^{2/3}} = \frac{0.023/1.25}{(373,000)^{0.2}(0.74)^{2/3}} = 0.00172$$

$$L_H = \frac{D_i}{4} \frac{(t_2 - t_1)_i/\Delta t_i}{(h/c_p G)_i} = \frac{0.834}{48} \frac{2.02}{0.00172} = 20.4 \text{ ft}$$

If F_s had been taken as 1, one would have obtained N_w of 92, L_H of 16.6, and $N_w L_H$ of 1530 ft, instead of 102, 20.4, and 2080, respectively.

If the thermal resistance inside is substantial but not controlling, one assumes a preliminary value of Δt_i based on an estimate of the resistance ratio U_i/h_i, that is, $\Delta t_i = U_i \, \Delta t_m/h_i$. The value of G_i is then found by the procedure outlined above, and the corresponding ratio U_i/h_i is computed. If this value differs seriously from the assumed value, the procedure is repeated until compatible results are obtained.

Fixed Pressure Drop across Tubes. The relations are similar to those for flow inside tubes; elimination of the number of rows deep from Eqs.

(15-3), (15-4), (15-7), (15-11), and (15-12) gives

$$G_o = \left[\frac{\pi a_4 g_c \rho_o \, \Delta p_o (D_o/y_o)}{2 a_o F_s \phi_o B_o (D_o/\mu_o)^{0.25}}\right]^{0.444} \quad (15\text{-}19)$$

If the thermal resistance on the shell side is controlling ($h_o = U_o$, $\Delta t_o = \Delta t_m$), ϕ_o is found from Eq. (15-15a), G_o is obtained from Eq. (15-19), h_o from Eq. (15-7), A_o from Eq. (15-4), and S_o from Eq. (15-3). Obviously it is necessary that N be made a whole number. If the thermal resistance on the shell side is substantial but not controlling, G_o could be found for a fixed ΔP from Eq. (15-19) by a trial-and-error procedure similar to that outlined on page 429.

Calculation of Outlet Temperatures. Both manufacturers and users of heat exchangers may face the problem of predicting the outlet temperatures t_2' and t_2'' which a given heat exchanger should produce for fixed inlet temperatures, mass flow rates, and physical properties. The procedure is as follows: The over-all coefficient is predicted in the usual manner, and the outlet temperature t_2' is computed from the following equation, which was obtained by eliminating t_2'' from Eqs. (15-3a), (15-4), and (15-4a):

$$t_2' = \frac{t_1'(1-z) + (a_5 - 1)t_1''}{a_5 - z} \quad (15\text{-}20)$$
$$a_5 = e^{UAF_G(1-z)/w'c'}$$

If the ratio $z = w'c'/w''c''$ is unity, the terminal over-all temperature differences are equal and one obtains

$$t_2' = (t_1' + a_6 t_1'')/(1 + a_6) \quad (15\text{-}20a)$$
$$a_6 = UAF_G/w'c'$$

The solutions to the above equations involve trial and error to find F_G except for counterflow exchangers where $F_G = 1$.

Optimum Operating Conditions

As pointed out in the previous section, in case one or more independent variables have *not* been fixed prior to design calculations, there exists a particular design which will yield the lowest total cost per year. This section develops procedures for determining the *optimum* design for a given case from the basic quantitative relations and suitable cost data.

Optimum Velocities. A problem often encountered is the determination of the optimum heat exchanger for a given case with fixed terminal temperatures and mass flow rates. The general case, in which the surface coefficient on each side of the tube varies substantially with velocity, is complex, since the velocities and corresponding power costs can be independently varied within limits. Let C_{af} represent the fixed charges on the exchanger, expressed in dollars/(hr)(sq ft of inside surface); the corresponding fixed charge, expressed in dollars per Btu, is

$$X_a = \frac{C_{ai}A_i}{q} = \frac{C_{ai}}{U_i \Delta t_m} \qquad (15\text{-}21)$$

The power theoretically required to force the fluid through the tubes is $\Delta p_i w_i/\rho_i$, and Δp_i is obtained from Eqs. (15-8) and (15-10). Let C_{ei} represent the cost of supplying mechanical energy to the fluid, expressed in dollars per foot-pound, allowing for the over-all efficiency of the pump. By use of Eqs. (15-3) and (15-4) and the relations given in the preceding section, the corresponding cost of power, expressed in dollars per Btu transferred, is found to be [6,8,13]

$$X_{pi} = \frac{C_{pi}A_i}{q} = C_{ei} \frac{B_i a_i \mu_i^{m_i}}{2 g_c \rho_i^2 D_i^m} \frac{G_i^{3-m_i}}{U_i \Delta t_m} = \frac{C_{ei} K_i G_i^{3-m_i}}{U_i \Delta t_m} \qquad (15\text{-}22)$$

Similarly, using the equations for pressure drop across tubes, the power cost for flow through the shell, expressed in dollars per Btu, is[8]

$$X_{po} = \frac{C_{po}A_i}{q} = C_{eo} \frac{2}{\pi} \frac{B_o a_o \mu_o^{m_o} y_o}{g_c \rho_o^2 D_o^{m_o} D_i} \frac{G_o^{3-m_o}}{U_i \Delta t_m} = \frac{C_{eo} K_o G_o^{3-m_o}}{U_i \Delta t_m} \qquad (15\text{-}23)$$

Since the inside and outside individual coefficients of heat transfer depend on the velocities $h_i = \alpha_i G_i^{n_i}$ and $h_o = \alpha_o G_o^{n_o}$, Eq. (15-5) may be written

$$\frac{1}{U_i} = D_i R_{wd} + \frac{D_i/D_o}{h_o} + \frac{1}{h_i} = D_i R_{wd} + \frac{D_i/D_o}{\alpha_o G_o^{n_o}} + \frac{1}{\alpha_i G_i^{n_i}} \qquad (15\text{-}24)$$

The total cost $\Sigma X = X_a + X_{pi} + X_{po}$ may be made a minimum by using the proper combination of the two velocities. The corresponding optimum relation between the velocities and the unit costs is found by partially differentiating ΣX with respect to both G_i and G_o at constant D_i and D_o and setting the resulting partial derivatives equal to zero. This procedure results in the two following symmetrical equations for optimum velocities:[6]

$$G_{io}^{3-m_i} = \frac{\dfrac{\Delta t_i}{\Delta t_m} \dfrac{n_i}{3-m_i} \dfrac{C_{ai}}{C_{ei}K_i}}{1 - \dfrac{\Delta t_i}{\Delta t_m} \dfrac{n_i}{3-m_i} - \dfrac{\Delta t_o}{\Delta t_m} \dfrac{n_o}{3-m_o}} \qquad (15\text{-}25)\ \star$$

$$G_{oo}^{3-m_o} = \frac{\dfrac{\Delta t_o}{\Delta t_m} \dfrac{n_o}{3-m_o} \dfrac{C_{ai}}{C_{eo}K_o}}{1 - \dfrac{\Delta t_i}{\Delta t_m} \dfrac{n_i}{3-m_i} - \dfrac{\Delta t_o}{\Delta t_m} \dfrac{n_o}{3-m_o}} \qquad (15\text{-}26)\ \star$$

in which†

$$K_i = \frac{B_i a_i \mu_i^{m_i}}{2 g_c \rho_i^2 D_i^{m_i}} \qquad K_o = \frac{2 B_o a_o y_o \mu_o^{m_o}}{\pi g_c \rho_o^2 D_i D_o^{m_o}}$$

The ratio of the power costs is obtained by dividing Eq. (15-25) by Eq. (15-26):

$$\frac{C_{pi}}{C_{po}} = \frac{C_{ei} K_i G_{io}^{3-m_i}}{C_{eo} K_o G_{oo}^{3-m_o}} = \frac{\Delta t_i}{\Delta t_o} \frac{n_i}{3-m_i} \frac{3-m_o}{n_o} \qquad (15\text{-}27)$$

In many cases the resistance of the tube wall is a negligible fraction of the total resistance and $\Delta t_m = \Delta t_o + \Delta t_i$. For values of $D_i G_i/\mu_i$ and $D_o G_o/\mu_o$ exceeding 10,000 and 1000 respectively, the exponents have the following values: $n_i = 0.8$, $m_i = 0.2$, $n_o = 0.6$, and $m_o = 0.15$.‡

Using these values and neglecting the wall resistance, Eqs. (15-25) and (15-26) become

$$G_{io} = \left[\frac{C_{ai}/C_{ei} K_i}{2.5 + 2.76 \dfrac{\Delta t_o}{\Delta t_i}} \right]^{0.357} \qquad (15\text{-}25a) \star$$

$$G_{oo} = \left[\frac{C_{ai}/C_{eo} K_o}{3.75 + 3.39 \dfrac{\Delta t_i}{\Delta t_o}} \right]^{0.351} \qquad (15\text{-}26a) \star$$

A procedure for solving a specific problem is as follows: From the physical properties of the two streams and the proposed diameters§ and spacing of tubes, K_i and K_o are evaluated, and the unit costs C_{ai}, C_{ei}, and C_{eo} are known. From a preliminary estimate of the temperature-drop ratios, trial values of G_{io} and G_{oo} are obtained from Eqs. (15-25) and (15-26) or (15-25a) and (15-26a). Using Eqs. (15-4), (15-6), and (15-7), revised values of the temperature-drop ratios are calculated and used to find new values of G_{io} and G_{oo} from Eqs. (15-25) and (15-26) or (15-25a) and (15-26a). Since substantial variations in the numerical values of the temperature-drop ratios have but little effect on the values of G_{io} and G_{oo}, the values converge rapidly.

Equations (15-25) and (15-26) may be combined to give a relation

---

† It should be noted that the term ρ^2, which appears in the definitions of K_i and K_o, represents the product $\rho_m \rho_p$, wherein ρ_m is the mean density and ρ_p that in the pump.

‡ For flow across tubes there is some doubt as to the exact value of m_o, but apparently it lies between 0 and 0.2.

§ The effect of tube diameter on the total cost per year at the optimum operating conditions is relatively slight[3] as can be seen by substituting Eqs. (15-25a) and (15-26a) in the expression for X_T:

$$X_T = \frac{C_{ai}}{U_i \Delta t_m} \left[1 + \frac{1}{2.5 + 2.76(\Delta t_o/\Delta t_i)} + \frac{1}{3.75 + 3.39(\Delta t_i/\Delta t_o)} \right]$$

The term in brackets containing the temperature drop ratios is only slightly affected by variations in D_i and D_o. As the diameter is decreased, U increases somewhat but this is counteracted by an increase in C_{ai}. [As U increases, A decreases and C_{ai} increases (see Fig. 15-10).]

between the optimum values of the two velocities:

$$G_{oo}^{3-m_o+n_o} = \frac{\alpha_i D_i (3-m_i) n_o C_{ei} K_i}{\alpha_o D_o (3-m_o) n_i C_{eo} K_o} G_{io}^{3-m_i+n_i} \quad (15\text{-}28)$$

Using the values of m_i, m_o, n_i, and n_o given on page 432, this reduces to

$$G_{oo} = 0.915 \left(\frac{\alpha_i D_i C_{ei} K_i}{\alpha_o D_o C_{eo} K_o} \right)^{0.29} G_{io}^{1.043} \quad (15\text{-}28a)$$

Since the exponent is close to 1, the ratio of optimum velocities is practically independent of velocity. With fluids of similar physical properties on both sides of the tubes, G_{oo} tends to be only a fraction of G_{io}.

Illustration 3. It is desired to design a gas-to-gas exchanger for use in a continuous process. One gas stream is to be heated from 180 to 540°F, and the other is to be cooled from 750 to 390°F. Each stream will flow at the rate of 20,000 lb/hr and has the same physical properties as air. The average absolute pressures are to be 10 atm in the tubes and 9.65 atm in the shell. Annual fixed charges on a suitable exchanger will be 20 per cent of the installed cost. The apparatus is to operate 8640 hr/year, and power delivered to the fluid costs $0.02/kw-hr. It is proposed to employ steel tubes arranged in line with a square pitch (center-to-center distance) of $1.50 D_o$, but not less than ¼ in. It is agreed to use Eq. (6-13a), which gives a_o of 0.31 and m_o of 0.15, to employ F_s of 1.25 in Eqs. (15-6) and (15-7), and to neglect all thermal resistances except those of the two gas films.

Solution. The problem will be solved by use of Eqs. (15-25a) and (15-26a). If counterflow is used, F_G is 1 and $\Delta t_m = 210°F$. The densities ρ_o and ρ_i will be evaluated at 360 and 570°F; hence

$$\rho_o = \frac{29}{359} \frac{492}{460 + 360} \frac{9.65}{1} = 0.468 \text{ lb/cu ft}$$

$$\rho_i = 0.468 \frac{460 + 360}{460 + 570} \frac{10.0}{9.65} = 0.385 \text{ lb/cu ft}$$

From the alignment chart in the Appendix, the corresponding viscosities are 0.0247 and 0.0283 centipoises, or $\mu_o = 0.0598$ and $\mu_i = 0.0685$ lb/(hr)(ft). From Eq. (15-14), with $\Delta t_i = 105°F$, $\phi_i = (360/105)(0.74)^{2/3} = 2.8$; from Eq. (15-17), if N_{TP} is 1, $B_i = 1 + (1)(0.55)(0.023)/(0.055)(2.8)1.25 = 1.07$.

For 12-BWG tubes having an o.d. of 1.00 in. and an i.d. of 0.782 in.:

$$K_i = \frac{B_i a_i}{2 g_s \rho_i^2 (D/\mu)_i{}^m} = \frac{(1.07)(0.055)}{(2)(4.17 \times 10^8)(0.385)^2 (0.0652/0.0685)^{0.2}} = 4.78 \times 10^{-10}$$

$$K_o = \frac{2 B_o a_o y_o / D_i}{\pi g_c \rho_o^2 (D/\mu)_o{}^{m_o}} = \frac{(2)(1)(0.31)(0.5/0.782)}{(3.14)(4.17 \times 10^8)(0.468)^2 (0.0833/0.0598)^{0.15}} = 13.1 \times 10^{-10}$$

$$C_{ei} = C_{eo} = (0.02)(0.746)/(33,000)(60) = 7.55 \times 10^{-9}$$

For the evaluation of C_{ai} from the cost data given on Fig. 15-10, assume that $A_i = 675$ sq ft.

$$C_{ai} = 5.45 \frac{0.2}{8640} = 1.27 \times 10^{-4}$$

Assume $\Delta t_i/\Delta t_o = 1.02$; $\Delta t_o/\Delta t_i = 0.98$. From Eqs. (15-25a) and (15-26a),

$$G_{io} = \left[\frac{(1.27 \times 10^{-4})/(7.55 \times 10^{-9})(4.78 \times 10^{-10})}{2.5 + (0.98)(2.76)} \right]^{0.357} = 38,400 \text{ lb/(hr)(sq ft)}$$

$$G_{oo} = \left[\frac{(1.27 \times 10^{-4})/(7.55 \times 10^{-9})(13.1 \times 10^{-10})}{3.75 + (1.02)(3.39)} \right]^{0.351} = 19,800 \text{ lb/(hr)(sq ft)}$$

From Eqs. (15-6) and (15-7),

$$h_i = \frac{(0.25)(38,400)}{(0.74)^{2/3}} \frac{(0.023)/(1.25)}{[(0.0652)(38,400)/(0.0685)]^{0.2}} = 26.2$$

$$h_o = \frac{(0.25)(19,800)}{(0.74)^{2/3}} \frac{(0.26)/(1.25)}{[(0.0833)(19,800)/(0.0598)]^{0.4}} = 21.2$$

From Eq. (15-4),

$$\frac{\Delta t_i}{\Delta t_o} = \frac{h_o D_o}{h_i D_i} = \frac{(21.2)(0.0833)}{(26.2)(0.0652)} = 1.03$$

Since this value differs so little from that assumed, it is unnecessary to make a second trial.

$$\frac{1}{U_i} = \frac{1}{h_i} + \frac{D_i/D_o}{h_o} = \frac{1}{26.2} + \frac{0.782}{21.2} = \frac{1}{13.4}$$

The heated length L_H is found from Eqs. (15-3) and (15-4):

$$F_G L_H = \frac{D_i}{4} \frac{G_i c_p}{U_i} \frac{(t_i - t_2)_i}{\Delta l_m} = \frac{(0.0652)(38,400)(0.25)(360)}{(4)(13.4)(210)} = 20.2 \text{ ft}$$

Since $F_G L_H$ is nearly 20 ft, single pass may be used with counterflow; consequently F_G is 1, and 20-ft lengths would be used.

Since the cross section of one tube is $(\pi/4)(0.9652)^2 = 0.00334$ sq ft, the number of tubes in parallel should be $20,000/(38,400)(0.00334) = 156$. $A_i = N_w(\pi D_i L_H) = (156)(3.14)(0.0652)(20) = 640$ sq ft.

$$C_{ai} = 5.50 \frac{0.2}{8640} = 1.28 \times 10^{-4}$$

The value of C_{ai} used in the original calculation was adequate, and a new calculation is not necessary.

The diameter D_S of a circular shell, containing N_c inscribed circles of diameter D_c, depends on the pitch, as shown by the following equations:†

With square pitch and $N > 25$:

$$D_s/D_c = 1.37 N_c^{0.475}$$

With equilateral triangular pitch and $N > 20$:

$$D_s/D_c = 0.94 + \sqrt{(N_c - 3.7)/(0.907)}$$

In the present case, D_c is 1.5 in., and for 156 tubes, $D_s = (1.37)(1.5)(156)^{0.475}/12 = 1.88$ ft. The necessary free area S_o is $20,000/19,800 = 1.01$ sq ft. Since the maximum clearance is $(1.88)(0.5/1.5) = 0.63$ ft, if the baffles are spaced $1.01/0.63 = 1.6$ ft apart, the average velocity should be approximately as large as the desired value even with substantial leakage. Neglecting the thickness of the baffles, with tubes 20 ft long, the number of baffled compartments would be $20/1.6 = 12.5$; hence 13 compartments would be used.

The total cost per year will be

$$X_T = \frac{C_{ai}}{U_i \Delta t_m}\left[1 + \frac{1}{2.5 + 2.76(\Delta t_o/\Delta t_i)} + \frac{1}{3.75 + 3.39(\Delta t_i/\Delta t_o)}\right]$$

$$= \frac{1.28 \times 10^{-4}}{(13.4)(210)}(1 + 0.192 + 0.139) = 6.05 \times 10^{-8} \text{ dollar/Btu}$$

i.e., \$0.0605/million Btu transferred.

$$q = wc_p(t_i - t_2) = (20,000)(8640)(0.25)(360) = 1.55 \times 10^{10} \text{ Btu/year}$$
$$\text{Cost/year} = (6.05 \times 10^{-8})(1.55 \times 10^{10}) = \$940$$

† "Machinery's Handbook," p. 74, Industrial Press, New York, 1927.

APPLICATIONS TO DESIGN

The above procedure was repeated for proposed tube diameters of 1.25 in. o.d., 1.032 in. i.d.; and 0.75 in. o.d., 0.532 in. i.d. Tabulated results are shown below for the operating conditions and for the costs.

D_o	G_{io}	G_{oo}	U_i	$F_G L_H$	N_w	A_i
0.75	39,100	18,900	15.2	12	328	559
1.00	38,400	19,800	13.4	20	156	640
1.25	35,200	21,500	12.0	28	98	730

D_o	X_a	X_{pi}	X_{po}	Total annual cost
0.75	4.28×10^{-8}	0.86×10^{-8}	0.56×10^{-8}	$885
1.00	4.55	0.87	0.63	940
1.25	4.80	0.93	0.66	990

For this particular problem wherein it was assumed that $C_{ai} \approx A^{-0.4}$ (page 422), the above results indicate that, the smaller the tube, the more economical will be the design. However, this may not be true in actual practice. The decrease in the total annual cost with decrease in diameter is due almost entirely to a decrease in the annual fixed charges, $C_{ai} A_i$. However, two factors that may affect the value of C_{ai} were neglected in this problem. First, the cleaning and maintenance costs for small-tube exchangers would be somewhat higher than those for large-tube exchangers. More frequent cleaning may be required to maintain the higher U_i, and the cost per cleaning may also be higher owing to the greater number of tube feet to be cleaned. This factor may become particularly important in cases where severe fouling is expected. Second, the shell costs for exchangers of equal area were assumed to be equal. However, in many cases, for the same heat-transfer area, a shell having a moderate L/D ratio would be cheaper than one having an extremely low L/D ratio. This factor is particularly important where moderate and high pressures are to be used. Both of these factors would tend to increase C_{ai} for the small-tube exchangers and make the resulting total costs per year more nearly equal.

Optimum Velocity in Tubes: Shell Power Immaterial. This case has been frequently discussed in the literature,[5,8,13] and the corresponding equation can be derived by setting the derivative of $X_{ai} + X_{pi}$ equal to zero, giving

$$G_{io}^{3-m_i} = \frac{C_{ai}/C_{ei} K_i}{\left(\dfrac{3-m_i}{n_i}\right) \dfrac{\Delta t_m}{\Delta t_i} - 1} \qquad (15\text{-}29) \star$$

Thus for $D_i G_i / \mu_i$ exceeding 10,000, n_i is 0.8, m_i is 0.2, and Eq. (15-29) becomes

$$G_{io} = \left[\frac{C_{ai}/C_{ei} K_i}{3.5 \dfrac{\Delta t_m}{\Delta t_i} - 1} \right]^{0.357} \qquad (15\text{-}29a) \star$$

This is readily solved by assuming a value of the temperature-drop ratio, computing G_{io}, and repeating the procedure until compatible results are obtained. Since the exponent 0.357 is small, values of G_{io} converge rapidly. It is interesting to note that at the optimum velocity the ratio of costs of power and fixed charges is

$$\frac{X_{pi}}{X_{ai}} = \frac{C_{pi}}{C_{ai}} = \frac{1}{\frac{\Delta t_m}{\Delta t_i}\left(\frac{3-m_i}{n_i}\right) - 1} \tag{15-30}$$

Thus, for turbulent flow, where n_i is 0.8 and m_i is 0.2, the optimum ratio X_{pi}/X_{ai} is 0.4 if the inside resistance controls and otherwise is less. The minimum total cost in dollars per Btu, except for that of the heat itself, is

$$(X_{pi} + X_{ai})_o = \frac{C_{ai}}{U_{io}\,\Delta t_m}\left(\frac{1}{1 - \frac{\Delta t_i}{\Delta t_m}\left(\frac{n_i}{3-m_i}\right)}\right) \tag{15-31}$$

Optimum Velocity across Tubes: Inside Power Immaterial. The derivation for this case† is similar to that for the previous case, and the corresponding equations are

$$G_{oo}{}^{3-m_o} = \frac{C_{ao}/C_{eo}K_o'}{\left(\frac{3-m_o}{n_o}\right)\frac{\Delta t_m}{\Delta t_o} - 1} \tag{15-32} \star$$

wherein $K_o' = K_o D_i/D_o$.

For $D_o G_o/\mu_o$ between 1000 and 40,000, if n_o is 0.6 and m_o is 0.15, Eq. (15-32) becomes

$$G_{oo} = \left[\frac{C_{ao}/C_{eo}K_o'}{4.75\,\frac{\Delta t_m}{\Delta t_o} - 1}\right]^{0.351} \tag{15-32a} \star$$

The cost expressions are

$$\frac{X_{po}}{X_{ao}} = \frac{C_{po}}{C_{ao}} = \frac{1}{\frac{\Delta t_m}{\Delta t_o}\left(\frac{3-m_o}{n_o}\right) - 1} \tag{15-33}$$

and

$$(X_{po} + X_{ao})_o = \frac{C_{ao}}{U_{oo}\,\Delta t_m}\left(\frac{1}{1 - \frac{\Delta t_o}{\Delta t_m}\left(\frac{n_o}{3-m_o}\right)}\right) \tag{15-34}$$

If the outside resistance controls, $\Delta t_o = \Delta t_m$ and $X_{po}/X_{ao} = 0.267$.

Illustration 4. It is desired to compute the optimum velocity for a gas having ρ of 0.069 lb/cu ft, ρ_p of 0.075, $c_p\mu/k$ of 0.74, and μ of 0.045 lb/(hr)(ft), flowing normal to a bank of in-line tubes laid out with a square pitch, so that the transverse clearance is one-fourth the o.d. of the 1-in. tubes. Assume that power delivered to the fluid costs

† Reference 11 solves this case in terms of optimum pressure drop.

0.02/kw-hr and that the hourly fixed charges are $0.66/7200 = \$9.13 \times 10^{-5}/(\text{hr})(\text{sq ft of outside surface})$. Since baffles are not involved, B_o will be taken as 1.0. What would be the optimum U_o and the optimum total costs $(X_{po} + X_{ao})$ for Δt_m of 105°F?

Solution. From Eq. (6-13a), $a_o = 0.68$. With outside resistance controlling, Eq. (15-32a) will be used.

$$C_{eo} = (0.02)(0.746)/(33{,}000)(60) = 7.55 \times 10^{-9}$$

$$K_o' = \frac{2B_o a_o \mu_o{}^{m_o} y_o}{\pi g_c \rho_o{}^2 D_o D_o{}^{m_o}}$$

$$= \frac{(2)(1.0)(0.68)(0.045)^{0.15}(0.25)}{(3.14)(4.17 \times 10^8)(0.075)(0.069)(\frac{1}{12})^{0.15}} = 4.59 \times 10^{-8}$$

Substitution in Eq. (15-32a), with $\Delta t_o = \Delta t_m$, gives

$$G_{oo} = \left[\frac{(9.13 \times 10^{-5})/(7.55 \times 10^{-9})(4.59 \times 10^{-8})}{4.75 - 1} \right]^{0.351} = 6410 \text{ lb}/(\text{hr})(\text{sq ft})$$

The Reynolds number $(DG/\mu)_o = (0.0833)(6410)/(0.045) = 11{,}860$. Taking F_s as 1.25 in Eq. (15-7) gives

$$h_o = \frac{0.33/1.25}{(11{,}860)^{0.4}} \frac{1}{(0.74)^{\frac{2}{3}}} (0.24)(6410) = 11.6$$

From Eq. (15-34), since $U_o = h_o$,

$$(X_{po} + X_{ao})_o = \frac{9.13 \times 10^{-5}}{(11.6)(105)} \frac{1}{0.79} = 9.49 \times 10^{-8}$$

i.e., the sum of the cost of power and fixed charges is 9.49 cents/million Btu transferred.

Summary of Optimization Procedure. In the general case of heat-exchanger design both resistances are substantial, and the terminal temperatures and mass flow rates are fixed. Four independent variables then remain to be determined.

If none of these variables are specified prior to design calculations, the four variables are subject to optimization but the analytical equations are unwieldy. A simple technique, therefore, is to employ Eqs. (15-25) and (15-26) to find optimum values of G_i and G_o for a series of selected values of D_i and D_o. The combination giving the lowest total annual cost is selected as shown in Illustration 3. Sometimes this procedure is not necessary as the total annual cost at the optimum velocities changes very little with tube diameter.[†,3]

In some instances there are other fixed design requirements in addition to fixed terminal temperatures and mass-flow rates. For example, if the heated length is specified in advance, only three variables remain to be determined. Usually D_i and D_o are preselected (for reasons given on page 419), leaving only one quantity subject to optimization.

Selection of Flow Path. In heating a gas by a condensing vapor the thermal resistance on the gas side controls, and the question arises whether the gas should flow inside or across the tubes. The exponents usually have the following values, $n_i = 0.8$, $m_i = 0.20$, $n_o = 0.6$, and $m_o = 0.15$, and the ratio of the optimum sums of power and fixed charges

† See Illustration 3 and the footnote on page 432.

is readily found to be

$$\frac{(X_{po}+X_{ao})_o}{(X_{pi}+X_{ai})_o} = 0.905 \frac{C_{ao}U_{io}(\Delta t_m)_i}{C_{ai}U_{oo}(\Delta t_m)_o} \tag{15-35}$$

Consequently the choice will be determined by the product of several ratios. If a gas were to flow either outside or inside the tubes of a given type of apparatus, where the hourly fixed charge per foot of tube was the same, C_{ao}/C_{ai} would equal D_i/D_o. If the gas were to be cooled under pressure with water, a considerably less expensive type of construction (trickle cooler) could be used with gas inside the tubes than with gas outside the tubes in a pressure shell. Conversely, if the gas were to be heated at atmospheric pressure with steam condensing under pressure, a sheet-metal tunnel could be used for gas flow across tubes and C_{ao}/C_{ai} would now be a fractional value. The ratio of the optimum values of U_i and U_o may be obtained from equations previously given.

If a fixed pressure drop may be used either inside or across tubes, ΔP_i equals ΔP_o, and if the over-all thermal resistance is controlled by the fluid for which the pressure drop is fixed, Eqs. (15-4), (15-6), (15-7), (15-16), and (15-19) may be combined to give

$$\frac{X_{ao}}{X_{ai}} = \frac{C_{ao}A_o}{C_{ai}A_i} = \frac{C_{ao}a_3}{C_{ai}a_4}\left(\frac{2a^3}{a_iB_i}\right)^{0.4}\left(\frac{D_o}{D_i}\right)^{0.2}\left[\left(\frac{2a_oB_oy_o}{\pi a_4\mu}\right)^2\frac{g_c\rho\,\Delta p}{F_s\phi}\right]^{1/7.5} \tag{15-36}$$

If Δp is small, X_{ao}/X_{ai} will be substantially less than 1 and flow across tubes will be preferred; if Δp is quite large, X_{ao}/X_{ai} will exceed 1 and flow inside tubes is preferable. Although the equation is formidable in appearance it is not very difficult to use in a specific case.

Illustration 5. Consider the case of 1-in. tubes having a wall thickness of 0.109 in., with a transverse clearance y_o of 0.025 in.; assume that the gas is diatomic ($c\mu/k = 0.74$), μ is 0.05 lb/(hr)(ft), ρ is 0.06, and a_o is 0.69. Assume $(t_2 - t_1)/\Delta t_m$ is 2.3, whence ϕ equals $(2.3)(0.740)^{2/3} = 1.88$. From Eq. 15-17, assuming one pass, $B_i = 1 + N_{TP}K_1a_3/a_i\phi F_s = 1 + (1)(0.55)(0.023)/(0.055)(1.88)(1.25) = 1.1$. Substituting values in Eq. (15-36), assuming that the cost per foot of length is fixed (and consequently $C_{ao}/C_{ai} = D_i/D_o$), one obtains

$$X_{ao}/X_{ai} = 0.381\,\Delta p^{0.133}$$

and flow across tubes would be preferred to flow inside tubes until Δp exceeded a value of 1400 lb/sq ft or 9.8 lb/sq in., at which point X_{ao} equals X_{ai}; the corresponding mass velocity G_i would be 96,000 lb/(hr)(sq ft of cross section).

Selection of Heat-transfer Fluid. In some instances only one fluid in a heat exchanger is part of a process stream, and the designer is at liberty to select the remaining fluid to be used as a heat-transfer agent. General considerations in the choice of a fluid include factors such as corrosion and fouling tendencies, vapor pressure, weight, cost, and availability. Assuming these factors to be equal in a given case, comparison of fluids

APPLICATIONS TO DESIGN

as heat-transfer agents is based on the relationship between the heat-transfer coefficient and the energy consumed in circulating the fluid.[5,18] As shown on page 287, the power loss per unit of surface is

$$E = P/A = KG^{3-m} \tag{15-37}$$

and as shown by Eqs. (15-6a) and (15-7a), the heat-transfer coefficient is

$$h = \alpha G^n \tag{15-37a}$$

Eliminating G gives the following relation:

$$h = \alpha K^{n/(m-3)} E^{n/(3-m)} \tag{15-38}$$

For the same power loss per unit surface, fluids of different physical properties will yield different heat-transfer coefficients, the most desirable having a high value of the term $\alpha K^{n/(m-3)}$. This type of analysis has been applied to many fluids of interest by references 2 and 18. Selection of a fluid for heat removal from nuclear reactors is discussed in reference 12.

Selection of Surface Geometry. A method of comparing the performance of heat-transfer surfaces is to plot[5] jG vs. $\rho^2 E$ as shown in Chap. 11, as applied to compact exchangers.

Optimum Water Flow for Condensers or Coolers. In some cases the cooling water in the plant mains is under sufficient pressure to give the desired rate of flow through the heat-transfer apparatus, and the cost of cooling water is directly proportional to the amount used. The optimum water rate corresponds to the minimum annual sum of the cooling-water costs and the fixed charges on the condenser or cooler.

Consider the general case in which w' lb/hr of warmer fluid enters at t_1' and leaves at t_2', and the heat removed $w'c'(t_1' - t_2')$ is absorbed by w'' lb/hr of cooling water entering at t_1'' and having specific heat c''. The cooling water costs C_w dollars/lb and is available in adequate amount. Let U represent the over-all coefficient from warmer fluid to water; preferably this should be evaluated at the optimum velocity for the warmer fluid, to maintain proper balance between fixed charges on the apparatus and pumping costs for the warmer fluid (page 430). The highest useful water velocity should be used, since the water is under adequate pressure. The first cost per square foot of heat-transfer surface is C_A, and the fraction F_A is charged off annually as fixed charges; the apparatus is to be operated θ hr/year. Let t_2'' represent the temperature of the cooling water leaving the apparatus, calculated from the heat balance

$$q = w'c'(t_1' - t_2') = w''c''(t_2'' - t_1'')$$

The annual water cost equals $w''\theta C_w = q\theta C_w/c''(t_2'' - t_1'')$, and the annual fixed charges are $AC_A F_A$. But q equals $UAF_G \Delta t_l$, where F_G

depends on the geometrical arrangement of the shell and tube passes in the exchanger (Chap. 8) and Δt_l is evaluated for counterflow by Eq. (15-4a). By combining the foregoing relations, the annual sum of water costs and fixed charges is differentiated with respect to the temperature difference $t_1' - t_2''$ at the hot end and set equal to zero to find the minimum total costs, giving

$$\left(\frac{Z + 1 - Y_o}{Y_o - 1}\right)^2 \left(\frac{1 - Y_o}{Y_o} + \ln Y_o\right) = X_o \qquad (15\text{-}39) \star$$

wherein $Y_o = \Delta t_H / \Delta t_C = (t_1' - t_2'')_o / (t_2' - t_1'')$, $X_o = UFG_oC_w\theta / C_AF_Ac''$, and $Z = (t_1' - t_2') / (t_2' - t_1'')$.† Trial-and-error calculation of the desired Δt_H is avoided by use of Fig. 15-12.

Illustration 6. It is desired to design a counterflow apparatus to cool 10,000 lb/hr of gas, having c' of 0.24, from 200 to 90°F by use of water entering at 85°F, costing $0.20/1000 cu ft. Annual fixed charges are $0.50/sq ft, θ is 8400, and the optimum value of U is 7.8, after including F_s of 1.25. Calculate the optimum outlet temperature of the water, the corresponding ratio of water to gas, and the square feet required.

Solution.

$$X_o = \frac{UFC_w\theta}{C_AF_Ac''}$$

$$= \frac{(7.8)(1.0)(0.2/62,300)(8400)}{(0.5)(1.0)} = 0.422$$

$$Z = (t_1' - t_2')/(t_2' - t_1'')$$
$$= (200 - 90)/(90 - 85) = 22$$

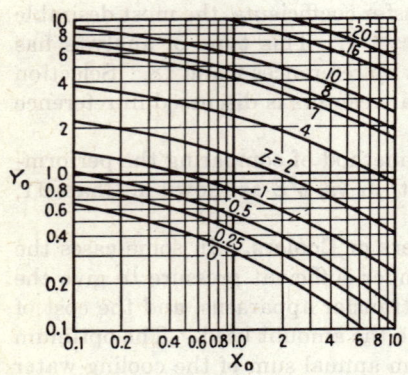

Fig. 15-12. Solution of Eq. (15-39) for ordinary range of Z.

and from Fig. 15-12, $Y_o = 15.9 = (200 - t_2'')/5$, whence $200 - t_2'' = 79.5$ and $t_2'' = 120.5°F$. Per pound of gas, $(200 - 95)(0.24)/(120.5 - 85.0) = 0.71$ lb of water is required. The mean over-all temperature difference is $(79.5 - 5)/\ln 15.9 = 26.9°F$, and $A = q/U \Delta t_m = (10,000)(0.24)(200 - 95)/(7.8)(26.9) = 1200$ sq ft.

Optimum Final Δt for Recovering Waste Heat. Exhaust steam having saturation temperature t' may be used to furnish part of the heat required to warm w lb/hr of a fluid without phase change from t_1 to t, the remainder of the heat being furnished by more expensive high-pressure steam condensing at t''. The problem is to determine the optimum over-all Δt at the outlet of the first heater, in which the low-pressure steam is used. Let the value of exhaust steam be $C_{H'}$ dollars/Btu of latent heat; $C_{H''}$ is the corresponding value for the high-pressure steam; $(C_a + C_p)$ or K_2

† Where the temperature of the warmer fluid is constant, as in a condenser, the equation reduces to

$$\frac{1 - Y_o}{Y_o} + \ln Y_o = X_o \qquad (15\text{-}39a)$$

represents the optimum sum of fixed charges and power required to force the gas or liquid through the heaters, expressed as dollars/(hr)(sq ft) of heat-transfer surface. The total cost y, which depends on the intermediate temperature t, is then

$$\Sigma y = wc(t - t_1)C_{H'} + A_1K_2 + wc(t_2 - t)C_{H''} + A_2K_2 \quad (15\text{-}40)$$

and A equals $q/U \, \Delta t_m$ for each of the two heaters; Δt_m is the logarithmic-mean over-all temperature difference, which applies if U is constant (Chap. 8); F is unity for any arrangement of surface, since the vapors condense at constant temperature. Combining these relations, $d\Sigma y/dt$ is set equal to zero to find the following relation for minimum total cost:

$$(t' - t)(t'' - t) = \frac{(C_a + C_p)(t'' - t')}{U(C_{H''} - C_{H'})} \quad (15\text{-}41)$$

which can be used directly or solved for t by the quadratic rule.

Illustration 7. It is planned to heat air at atmospheric pressure from 70 to 300°F. Steam condensing at 220°F costs $0.05/10⁶ Btu, and steam condensing at 370°F costs $0.20/10⁶ Btu. At the optimum velocity U_o is 8, fixed charges are estimated at $0.45/(year)(sq ft), and the equipment is to be operated 8400 hr/year. What should be the temperature of the air leaving the first heater?

Solution. For the reasons given on page 438 it would be more economical to employ flow across tubes instead of flow inside tubes. Assume that the cost data and other factors are such that Eq. (15-32) gives an optimum velocity that corresponds to U of 8, after including F_s of 1.25. Since the Reynolds number would fall in the region where the exponents n_o and m_o are 0.6 and 0.15, respectively, the ratio $(X_p + X_a)/X_a = (3 - m_o)/(3 - n_o - m_o) = 2.85/2.25 = 1.27$.

$$K_2 = \frac{X_p + X_a}{X_a} \frac{C_A F_A}{\theta} = 1.27 \frac{0.45}{8400} = \frac{\$6.8 \times 10^{-5}}{(\text{hours})(\text{square feet})}$$

Substituting of values in Eq. (15-41) gives

$$(220 - t)(370 - t) = \frac{(6.8 \times 10^{-5})(370 - 220)}{(8)(0.15 \times 10^{-6})}$$

whence t is 176°F and the optimum Δt at the outlet of the first heater is 44°F.

PROBLEMS

1. For process use, 50,000 lb/hr of air is to be heated from 70 to 300°F. In order to utilize exhaust steam, two heaters in series will be used. The air will flow through 1-in. o.d. steel tubes (18 BWG, thickness = 0.049 in.) at a mass velocity of 10,000 lb/(hr)(sq ft). The over-all coefficient of heat transfer in both heaters may be assumed constant at 8 Btu/(hr)(sq ft)(deg F) based on the outside surface area of the tubes.

The exhaust steam condenses at 212°F and is valued at 20 cents/10⁶ Btu of latent heat. The high-pressure steam has a condensing temperature of 350°F and costs 40 cents/10⁶ Btu of latent heat.

The energy cost for pumping the air through the heaters at the given mass velocity is $0.40/(sq ft of heating surface)/year, and the fixed charges on both heaters are $1/(year)(sq ft of heating surface). (By heating surface is meant the outside surface area of the tubes.)

If the heaters operate 8000 hr/year, to what temperature should the air be heated with the exhaust steam in order to give the lowest over-all cost of heating for air from 70 to 300°F?

The specific heat of air is 0.24 Btu/(lb)(deg F).

2. It is necessary to heat 24,000 lb/hr of air from 70 to 350°F while it is flowing under pressure at the optimum mass velocity of 7200 lb/(hr)(sq ft of cross section), inside tubes having an actual i.d. of 0.870 in. Low-pressure exhaust steam (220°F saturation temperature) now being discarded would be available at a cost of $0.05/million Btu of latent heat, and high-pressure steam (370°F saturation temperature) is available at a cost of $0.20/million Btu of latent heat. The heaters must run 8400 hr/year. The annual fixed charges, expressed in dollars/(year)(ft of each tube), will be assumed as 0.15 for the low-pressure heater and 0.25 for the high-pressure heater, independent of the length of tube or the number of tubes. The air is to leave the high-pressure heater at an absolute pressure of 10 atm and a temperature of 350°F. It is agreed to use Eq. (9-15). Calculate the minimum yearly costs and the tube length required for each heater.

3. An air cooler consisting of a bundle of 1-in. 18 BWG copper tubes, enclosed in a well-baffled shell, is being built to cool 45,000 lb of air/hr from 200 to 90°F. The air flows under pressure in a single pass through the tubes, and cooling water at 80°F, under sufficient pressure to force it through at any desired rate, flows countercurrently through the shell. The air flows inside the pipes at a mass velocity of 8600 lb/(hr)(sq ft), which is the optimum velocity; the corresponding U_i will be 7.8. Cooling water costs 20 cents/1000 cu ft, and the annual fixed charges on the cooler are 50 cents/sq ft of heating surface. It is proposed to operate 8400 hr/year. From the standpoint of the lowest total yearly cost, calculate:

a. The optimum pounds of cooling water per pound of air.
b. The over-all temperature difference at the hot end.
c. The length of each tube.
d. The number of tubes in parallel.
e. The total annual cost in dollars.

APPENDIX

		Page
Conversion Factors		
Table A-1a	Thermal Conductivity	444
Table A-1b	Coefficient of Heat Transfer	444
Table A-1c	Heat Flux	444
Table A-1d	Viscosity	444
Thermal Conductivity		
Tables A-2 to A-4	Metals	445–447
Tables A-5 to A-9	Materials and Insulants	448–454
Tables A-10 and A-11	Liquids	455–456
Table A-12	Gases and Vapors	457
Fig. A-1	Gases at High Pressures	459
Table A-13	Soils	460
Specific Heat		
Table A-14	Solids	461
Fig. A-2	Liquids	462
Table A-15	Organic Liquids	463
Fig. A-3	Gases	464
Table A-16	Liquefied Gases	465
Table A-17	Air at High Pressures	465
Viscosity		
Fig. A-4	Liquids	466
Table A-18	Liquids	467
Table A-19	Gases	468
Fig. A-5	Gases	468
Fig. A-6	Gases	469
Table A-20	Steam	471
Table A-21	Efflux Viscosimeters	471
Prandtl Number		
Fig. A-7	Liquids	470
Table A-22	Gases	471
Other Properties		
Table A-23	Emissivity	472
Fig. A-8	Latent Heat	480
Fig. A-9	Surface Tension	481
Special Substances		
Table A-24	Steam Table	482
Table A-25	Air Properties	483
Table A-26	Enthalpy of Saturated Air	483
Table A-27	Liquid Water	484
Table A-28	Molten Metals	485
Table A-29	Standard Atmosphere	486
Standard Dimensions		
Table A-30	Condenser Tubes	487
Table A-31	Iron Pipe	488
Mean Temperature Difference		
Fig. A-10	One Shell Pass, Two or More Tube Passes	489
Fig. A-11	Two Shell Passes, Four or More Tube Passes	489
Fig. A-12	Three Shell Passes, Six or More Tube Passes	490
Fig. A-13	Four Shell Passes, Eight or More Tube Passes	490

Table A-1. Conversion Factors
a. Thermal Conductivity

$\dfrac{\text{Btu}}{\text{hr-ft}^2\text{-}°\text{F/ft}}$	$\dfrac{\text{gm-cal}}{\text{sec-cm}^2\text{-}°\text{C/cm}}$	$\dfrac{\text{watts}}{\text{cm}^2\text{-}°\text{C/cm}}$	$\dfrac{\text{kg-cal}}{\text{hr-m}^2\text{-}°\text{C/m}}$
1	0.004134	0.01731	1.488
241.9	1	4.187	360
57.79	0.2388	1	86
0.672	0.002778	0.01163	1

b. Coefficient of Heat Transfer

$\dfrac{\text{Btu}}{\text{hr-ft}^2\text{-}°\text{F}}$	$\dfrac{\text{gm-cal}}{\text{sec-cm}^2\text{-}°\text{C}}$	$\dfrac{\text{watts}}{\text{cm}^2\text{-}°\text{C}}$	$\dfrac{\text{kg-cal}}{\text{hr-m}^2\text{-}°\text{C}}$
1	0.0001355	0.0005678	4.882
7,373	1	4.187	36,000
1,761	0.2388	1	8,600
0.2048	0.00002778	0.0001163	1

c. Heat Flux

$\dfrac{\text{Btu}}{\text{hr-ft}^2}$	$\dfrac{\text{gm-cal}}{\text{sec-cm}^2}$	$\dfrac{\text{watts}}{\text{cm}^2}$	$\dfrac{\text{kg-cal}}{\text{hr-m}^2}$
1	0.00007535	0.0003154	2.712
13,272	1	4.187	36,000
3,170	0.2388	1	8,600
0.3687	0.00002778	0.0001163	1

d. Viscosity

Centipoises[a]	$\dfrac{\text{lb}}{\text{sec-ft}}$	$\dfrac{\text{lb force-sec}}{\text{ft}^2}$	$\dfrac{\text{lb}}{\text{hr-ft}}$	$\dfrac{\text{kg}}{\text{hr-m}}$
1	0.000672	0.0000209	2.42	3.60
1,490	1	0.0311	3600	5350
47,800	32.2	1	116,000	172,000
0.413	0.000278	0.00000864	1	1.49
0.278	0.000187	0.00000581	0.672	1

[a] 100 centipoises = 1 poise = 1 gm/(sec)(cm).

TABLE A-2. EFFECT OF TEMPERATURE UPON THERMAL CONDUCTIVITY OF METALS AND ALLOYS[a]

[Main body of table is k in Btu/(hr)(sq ft)(deg F per ft)]

t, deg F	32	212	392	572	752	932	1112	Melting point, deg F
t, deg C	0	100	200	300	400	500	600	
Aluminum	117	119	124	133	144	155	...	1220
Brass (70 copper, 30 zinc)	56	60	63	66	67	...	...	1724
Cast iron	32	30	28	26	25	...	...	2192
Copper, pure	224	218	215	212	210	207	204	1976
Graphite (longitudinal)	97	87	76	66	58	53	48	None
Lead	20	19	18	18	...	...	...	621
Nickel	36	34	33	32	...	...	...	2642
Silver	242	238	...	...	...	...	...	1760
Steel, mild	...	26	26	25	23	22	21	2507
Tin	36	34	33	...	...	...	...	450
Wrought iron, Swedish	...	32	30	28	26	23	...	2741
Zinc	65	64	62	59	54	...	...	786

[a] From "International Critical Tables," McGraw-Hill, New York, 1929, and other sources.

TABLE A-3. THERMAL CONDUCTIVITIES OF METALS
[k = Btu/(hr)(sq ft)(deg F per ft)][a]

Substance	Deg F	k	Substance	Deg F	k
Metals:			Metals:		
Antimony	32	10.6	Mercury	32	4.8
Antimony	212	9.7	Nickel alloy (62Ni, 12Cr, 26Fe)	68	7.8
Bismuth	64	4.7			
Bismuth	212	3.9	Platinum	64	40.2
Cadmium	64	53.7	Platinum	212	41.9
Cadmium	212	52.2	Tantalum	64	32
Gold	64	169.0	Alloys:		
Gold	212	170.0	Admiralty metal	86	65
Iron, pure	64	39.0	Bronze, commercial	...	109
Iron, pure	212	36.6	Constantan (60Cu, 40Ni)	64	13.1
Iron, wrought	64	34.9	Constantan (60Cu, 40Ni)	212	15.5
Iron, wrought	212	34.6	Nickel silver	32	16.9
Iron, cast	129	27.6	Nickel silver	212	21.5
Iron, cast	216	26.8	Manganin {84Cu, 4Ni, 12Mn}	64	12.8
Steel (1 per cent C)	64	26.2			
Steel (1 per cent C)	212	25.9		212	15.2
Magnesium	32–212	92.0	Platinoid	64	14.5

[a] From L. S. Marks, "Mechanical Engineers' Handbook," 5th ed., McGraw-Hill, New York, 1951.

TABLE A-4. THERMAL CONDUCTIVITY OF NICKEL-CHROMIUM ALLOYS WITH IRON[a]
[k = Btu/(hr)(sq ft)(deg F per ft)]

American Iron and Steel Institute type number	k at 212°F	k at 932°F
301, 302, 302B, 303, 304, 316[b]	9.4	12.4
308	8.8	12.5
309, 310	8.0	10.8
321, 347	9.3	12.8
403, 406, 410, 414, 416[b]	14.4	16.6
430, 430f	12.5	14.2
501, 502[b]	21.2	19.5

[a] Based on recent information from manufacturers. The variation of k is substantially linear in the range 212 to 932°F.
[b] S. M. Shelton and W. N. Swanger, *Trans. Am. Soc. Steel Treating*, **21**, 1061 (1933).

TABLE A-5. THERMAL CONDUCTIVITIES OF SOME BUILDING AND INSULATING MATERIALS[a]

[k = Btu/(hr)(sq ft)(deg F per ft)]

Material	Apparent density ρ, lb/cu ft at room temperature	Deg F	k
Aerogel, silica, opacified................	8.5	248	0.013
		554	0.026
Asbestos-cement boards...............	120	68	0.43
Asbestos sheets.......................	55.5	124	0.096
Asbestos slate........................	112	32	0.087
	112	140	0.114
Asbestos (see Tables A-7–A-9)			
Aluminum foil, 7 air spaces per 2.5 in.....	0.2	100	0.025
		351	0.038
Ashes, wood..........................		32–212	0.041
Asphalt..............................	132	68	0.43
Bricks:			
Alumina (92–99% Al_2O_3 by weight) fused....		801	1.8
Alumina (64–65% Al_2O_3 by weight)..........		2399	2.7
(See also Bricks, fire clay)..................	115	1472	0.62
	115	2012	0.63
Building brickwork.......................		68	0.4
Carbon.................................	96.7	....	3.0
Chrome brick (32% Cr_2O_3 by weight)........	200	392	0.67
	200	1202	0.85
	200	2399	1.0
Diatomaceous earth, natural, across strata[b]....	27.7	399	0.051
	27.7	1600	0.077
Diatomaceous, natural, parallel to strata[b]....	27.7	399	0.081
	27.7	1600	0.106
Diatomaceous earth, molded and fired[b]........	38	399	0.14
	38	1600	0.18
Diatomaceous earth and clay, molded and fired[b]	42.3	399	0.14
	42.3	1600	0.19
Diatomaceous earth, high burn, large pores[c]	37	392	0.13
	37	1832	0.34
Fire clay, Missouri.....................		392	0.58
		1112	0.85
		1832	0.95
		2552	1.02
Kaolin insulating brick[c]..................	27	932	0.15
	27	2102	0.26

TABLE A-5. THERMAL CONDUCTIVITIES OF SOME BUILDING AND INSULATING MATERIALS.[a]—(*Continued*)

Material	Apparent density ρ, lb/cu ft at room temperature	Deg F	k
Kaolin insulating firebrick[d]...............	19	392	0.050
	19	1400	0.113
Magnesite (86.8% MgO, 6.3% Fe$_2$O$_3$, 3% CaO, 2.6% SiO$_2$ by weight)..................	158	399	2.2
	158	1202	1.6
	158	2192	1.1
Silicon carbide brick, recrystallized[c]..........	129	1112	10.7
	129	1472	9.2
	129	1832	8.0
	129	2192	7.0
	129	2552	6.3
Calcium carbonate, natural..................	162	86	1.3
White marble.............................	......	......	1.7
Chalk....................................	96	......	0.4
Calcium sulfate (4H$_2$O), artificial..............	84.6	104	0.22
Plaster, artificial..........................	132	167	0.43
Building.................................	77.9	77	0.25
Cambric, varnished.........................	......	100	0.091
Cardboard, corrugated......................	......	......	0.037
Celluloid.................................	87.3	86	0.12
Charcoal flakes............................	11.9	176	0.043
	15	176	0.051
Clinker, granular..........................	......	32–1292	0.27
Coke, petroleum...........................	......	212	3.4
	......	932	2.9
Coke, powdered...........................	......	32–212	0.11
Concrete, cinder...........................	......	......	0.20
1:4 dry..................................	......	......	0.44
Stone....................................	......	......	0.54
Cotton wool...............................	5	86	0.024
Cork board................................	10	86	0.025
Cork, ground..............................	9.4	86	0.025
Regranulated.............................	8.1	86	0.026
Diatomaceous earth powder, coarse[b]...........	20.0	100	0.036
	20.0	1600	0.082
Fine[b]...................................	17.2	399	0.040
	17.2	1600	0.074
Molded pipe covering[b].....................	26.0	399	0.051
	26.0	1600	0.088
4 vol. calcined earth and 1 vol. cement, poured and fired[b]...............................	61.8	399	0.16
	61.8	1600	0.23

TABLE A-5. THERMAL CONDUCTIVITIES OF SOME BUILDING AND INSULATING MATERIALS.[a]—(*Continued*)

Material	Apparent density ρ, lb/cu ft at room temperature	Deg F	k
Dolomite	167	122	1.0
Ebonite			0.10
Enamel, silicate	38		0.5–0.75
Felt, wool	20.6	86	0.03
Fiber insulating board	14.8	70	0.028
Fiber, red	80.5	68	0.27
With binder, baked		68–207	0.097
Glass			0.2–0.73
Borosilicate type	139	86–167	0.63
Soda glass			0.3–0.44
Window glass			0.3–0.61
Granite			1.0–2.3
Graphite, longitudinal		68	95.
Powdered, through 100 mesh	30	104	0.104
Gypsum, molded and dry	78	68	0.25
Hair felt, perpendicular to fibers	17	86	0.021
Ice	57.5	32	1.3
Kapok	0.88	68	0.020
Lampblack	10	104	0.038
Lava			0.49
Leather, sole	62.4		0.092
Limestone (15.3 vol. % H_2O)	103	75	0.54
Linen		86	0.05
Magnesia, powdered	49.7	117	0.35
Magnesia, light carbonate	19	70	0.034
Magnesium oxide, compressed	49.9	68	0.32
Marble			1.2–1.7
Mica, perpendicular to planes		122	0.25
Mill shavings			0.033–0.05
Mineral wool	9.4	86	0.0225
	19.7	86	0.024
Paper			0.075
Paraffin wax		32	0.14
Petroleum coke		212	3.4
		932	2.9
Porcelain		392	0.88
Portland cement (see Concrete)		194	0.17
Pumice stone		70–151	0.14
Pyroxylin plastics			0.075
Rubber, hard	74.8	32	0.087
Para		70	0.109
Soft		70	0.075–0.092

TABLE A-5. THERMAL CONDUCTIVITIES OF SOME BUILDING AND INSULATING MATERIALS.[a]—(*Continued*)

Material	Apparent density ρ, lb/cu ft at room temperature	Deg F	k
Sand, dry	94.6	68	0.19
Sandstone	140	104	1.06
Sawdust	12	70	0.03
Silk	6.3	......	0.026
Varnished	......	100	0.096
Slag, blast furnace	......	75–261	0.064
Slag wool	12	86	0.022
Slate	......	201	0.86
Snow	34.7	32	0.27
Sulphur, monoclinic	......	212	0.09–0.097
Rhombic	......	70	0.16
Wallboard, insulating type	14.8	70	0.028
Wallboard, stiff pasteboard	43	86	0.04
Wood shavings	8.8	86	0.034
Wood, across grain			
Balsa	7–8	86	0.025–0.03
Oak	51.5	59	0.12
Maple	44.7	122	0.11
Pine, white	34.0	59	0.087
Teak	40.0	59	0.10
White fir	28.1	140	0.062
Wood, parallel to grain			
Pine	34.4	70	0.20
Wool, animal	6.9	86	0.021

[a] L. S. Marks, "Mechanical Engineers' Handbook," 5th ed., McGraw-Hill, New York, 1951, and "International Critical Tables," McGraw-Hill, New York, 1929, and other sources. For additional data, see pp. 452–454.

[b] B. Townshend and E. R. Williams, *Chem. Met. Eng.*, **39**, 219–222 (1932).

[c] F. H. Norton, "Refractories," McGraw-Hill, New York, 1949.

[d] F. H. Norton, personal communication, 1939.

TABLE A-6. THERMAL CONDUCTIVITIES OF SOME MATERIALS FOR REFRIGERATION AND BUILDING INSULATION[a]

[k = Btu/(hr)(sq ft)(deg F per ft) at approximately room temperature]

Material	Apparent density, lb/cu ft, room temperature	k
Soft, flexible materials in sheet form		
Chemically treated wood fiber	2.2	0.023
Eel grass between paper	3.4–4.6	0.021–0.022
Felted cattle hair	11–13	0.022
Flax fibers between paper	4.9	0.023
Hair and asbestos fibers, felted	7.8	0.023
Insulating hair and jute	6.1–6.3	0.022–0.023
Jute and asbestos fibers, felted	10.0	0.031
Loose materials		
Charcoal, 6-mesh	15.2	0.031
Cork, regranulated, fine particles	8–9	0.025
Diatomaceous earth, powdered	10.6	0.026
Glass wool, curled	4–10	0.024
Gypsum in powdered form	26–34	0.043–0.05
Mineral wool, fibrous	6	0.0217
	10	0.0225
	14	0.0233
	18	0.0242
Sawdust	12	0.034
Wood shavings, from planer	8.8	0.034
Semiflexible materials in sheet form		
Flax fiber	13.0	0.026
Semirigid materials in board form		
Corkboard	7.0	0.0225
Corkboard	10.6	0.025
Mineral wool, block, with binder	16.7	0.031
Stiff fibrous materials in sheet form		
Wood pulp	16.2–16.9	0.028
Sugar-cane fiber	13.2–14.8	0.028
Cellular gypsum	8	0.029
	12	0.037
	18	0.049
	24	0.064
	30	0.083

[a] Abstracted from *U.S. Bur. Standards, Letter Circ.* 227 (Apr. 19, 1927). For additional data, see pp. 448–451, 453–454.

TABLE A-7. THERMAL CONDUCTIVITIES OF INSULATING MATERIALS AT HIGH TEMPERATURES[a]

$[k = \text{Btu}/(\text{hr})(\text{sq ft})(\text{deg F per ft})]$

Material	For temperatures, deg F up to	Mean temperature, deg F								
		100	200	300	400	500	600	800	1000	1500	2000
Laminated asbestos felt (approx. 40 laminations/in.)	700	0.033	0.037	0.040	0.044	0.048					
Laminated asbestos felt (approx. 20 laminations/in.)	500	0.045	0.050	0.055	0.060	0.065					
Corrugated asbestos (4 plies/in.)	300	0.050	0.058	0.069							
85% magnesia (13 lb/cu ft)	600	0.034	0.036	0.038	0.040						
Diatomaceous earth, asbestos and bonding material	1600	0.045	0.047	0.049	0.050	0.053	0.055	0.060	0.065		
Diatomaceous earth brick	1600	0.054	0.056	0.058	0.060	0.063	0.065	0.069	0.073		
Diatomaceous earth brick	2000	0.127	0.130	0.133	0.137	0.140	0.143	0.150	0.158	0.176	
Diatomaceous earth brick	2500	0.128	0.131	0.135	0.139	0.143	0.148	0.155	0.163	0.183	0.203
Diatomaceous earth powder (density, 18 lb/cu ft)		0.039	0.042	0.044	0.048	0.051	0.054	0.061	0.068		
Rock wool		0.030	0.034	0.039	0.044	0.050	0.057				

Asbestos cement, 0.1; 85% magnesia cement, 0.05; asbestos and rock wool cement, 0.075 approx.

[a] L. S. Marks, "Mechanical Engineers' Handbook," McGraw-Hill, New York, 1951.

TABLE A-8. THERMAL CONDUCTIVITIES OF INSULATING MATERIALS AT MODERATE TEMPERATURES (Nusselt)[a]

$[k = \text{Btu}/(\text{hr})(\text{sq ft})(\text{deg F per ft})]$

Material	Density lb/cu ft	Temperatures, deg F						
		32	100	200	300	400	600	800
Asbestos	36.0	0.087	0.097	0.110	0.117	0.121	0.125	0.130
Burned infusorial earth for pipe coverings	12.5	0.043	0.046	0.052	0.057	0.062	0.073	0.085
Insulating composition, loose	25.0	0.040	0.046	0.050	0.053	0.055		
Cotton	5.0	0.032	0.035	0.039				
Silk hair	9.1	0.026	0.030	0.034				
Silk	6.3	0.025	0.028	0.034				
Wool	8.5	0.022	0.027	0.033				
Pulverized cork	10.0	0.021	0.026	0.032				
Infusorial earth, loose	22.0	0.035	0.039	0.045	0.047	0.050	0.053	

[a] L. S. Marks, "Mechanical Engineers' Handbook," McGraw-Hill, New York, 1951.

TABLE A-9. THERMAL CONDUCTIVITIES OF INSULATING MATERIALS AT LOW TEMPERATURES[a]

$[k = \text{Btu}/(\text{hr})(\text{sq ft})(\text{deg F per ft})]$

Material	Density, lb/cu ft	Temperature, deg F				
		32	0	−100	−200	−300
Asbestos................	44.0	0.135	....	0.130	0.125	0.100
Asbestos................	29.0	0.0894	....	0.0820	0.072	0.0545
*Carbon stock............	94	3.59	....	2.00	0.55	
*Corkboard..............	6.9	......	0.24	0.22	0.17	0.12
Cotton.................	5.0	0.0325	....	0.0276	0.0235	0.0198
*Glass blocks, expanded......	10.6	......	0.42	0.39	0.35	0.30
*Glass, fiberboard...........	11	......	0.22	0.17	0.13	0.08
*Mineral wool fiberboard.....	14.3	......	0.25	0.19	0.15	0.09
*Rubber board, expanded....	4.9	......	0.21	0.18	0.13	0.05
*Shredded redwood bark.....	4	......	0.22	0.17	0.14	0.14
Silk....................	6.3	0.0290	....	0.0235	0.0196	0.0155
*Sugar cane fiberboard.......	14.4	......	0.29	0.25	0.21	0.16

[a] Largely from L. S. Marks, "Mechanical Engineers' Handbook," McGraw-Hill, New York, 1951. and G. B. Wilkes, "Heat Insulation," Wiley, New York, 1950.

* $[k = \text{Btu}/(\text{hr})(\text{sq ft})(\text{deg F per inch})]$

TABLE A-10. THERMAL CONDUCTIVITY OF LIQUIDS

[k = Btu/(hr)(sq ft)(deg F per ft)]

A linear variation with temperature may be assumed. The extreme values given constitute also the temperature limits over which the data are recommended.

Liquid	t, deg F	k	Liquid	t, deg F	k
Acetic acid 100%[f]	68	0.099	Benzene[m]	86	0.086
50%[f]	68	0.20		140	0.082
Acetone[e]	86	0.102	Bromide[e]	68	0.070
	167	0.095	Ether[e]	86	0.080
Allyl alcohol[f]	77–86	0.104		167	0.078
Ammonia[i]	5–86	0.29	Iodide[e,f]	104	0.064
Ammonia, aqueous 26%[f]	68	0.261		167	0.063
	140	0.29	Ethylene glycol[h]	32	0.153
Amyl acetate[h]	50	0.083			
Alcohol (n-)[e]	86	0.094	Gasoline[g,m]	86	0.078
	212	0.089	Glycerol 100%[b]	68	0.164
(iso-)[m]	86	0.088	80%	68	0.189
	167	0.087	60%	68	0.220
Aniline[f]	32–68	0.100	40%	68	0.259
			20%	68	0.278
Benzene[m]	86	0.092	100%[b]	212	0.164
	140	0.087			
Bromobenzene[m]	86	0.074	Heptane (n-)[m]	86	0.081
	212	0.070		140	0.079
Butyl acetate (n-)[f]	77–86	0.085	Hexane (n-)[m]	86	0.080
Alcohol (n-)[e]	86	0.097		140	0.078
	167	0.095	Heptyl alcohol (n-)[g]	86	0.094
(iso-)[e]	50	0.091		167	0.091
			Hexyl alcohol (n-)[g]	86	0.093
Calcium chloride brine 30%[f]	86	0.32		167	0.090
15%[f]	86	0.34			
Carbon disulfide[e]	86	0.093	Kerosene[e]	68	0.086
	167	0.088		167	0.081
Tetrachloride[k]	32	0.107			
	154	0.094	Mercury[h]	82	4.83
Chlorobenzene[m]	50	0.083	Methyl alcohol 100%[e]	68	0.124
Chloroform[k]	86	0.080	80%	68	0.154
Cymene (para-)[m]	86	0.078	60%	68	0.190
	140	0.079	40%	68	0.234
Decane (n-)[m]	86	0.085	20%	68	0.284
	140	0.083	100%	122	0.114
Dichlorodifluoromethane[m]	20	0.057	Chloride[i,k]	5	0.111
	60	0.053		86	0.089
	100	0.048			
	140	0.043	Nitrobenzene[m]	86	0.095
	180	0.038		212	0.088
Dichloroethane[k]	122	0.082	Nitromethane[m]	86	0.125
Dichloromethane[k]	5	0.111		140	0.120
	86	0.096	Nonane (n-)[m]	86	0.084
				140	0.082
Ethyl acetate[h]	68	0.101	Octane (n-)[m]	86	0.083
Alcohol 100%[c]	68	0.105		140	0.081
80%	68	0.137	Oils[f,m,a]	86	0.079
60%	68	0.176	Oils, castor[j]	68	0.104
40%	68	0.224		212	0.100
20%	68	0.281	Oils, olive[j]	68	0.097
100%[c]	122	0.087		212	0.095

TABLE A-10. THERMAL CONDUCTIVITY OF LIQUIDS.—(Continued)

Liquid	t, deg F	k	Liquid	t, deg F	k
Paraldehyde[m]	86	0.084	Sulfur dioxide[c]	5	0.128
	212	0.078		86	0.111
Pentane (n-)[m]	86	0.078			
	167	0.074	Toluene[c,m]	86	0.086
Perchloroethylene[k]	122	0.092		167	0.084
Petroleum ether[c]	86	0.075	β-Trichloroethane[k]	122	0.077
	167	0.073	Trichloroethylene[k]	122	0.080
Propyl alcohol (n-)[g]	86	0.099	Turpentine[h]	59	0.074
	167	0.095	Vaseline[h]	59	0.106
Alcohol (iso-)[m]	86	0.091			
	140	0.090	Water[n]	32	0.343
				100	0.363
Sodium	212	49		200	0.393
	410	46		300	0.395
Sodium chloride brine 25.0%[f]	86	0.33		420	0.376
12.5%[f]	86	0.34		620	0.275
Sulfuric acid 90%[j]	86	0.21			
60%	86	0.25	Xylene (ortho-)[h]	68	0.090
30%	86	0.30	(meta-)[h]	68	0.090

[a] For many oils an average value of 0.079 may be used.
[b] O. K. Bates, *Ind. Eng. Chem.*, **28,** 494 (1936).
[c] O. K. Bates, G. Hazzard, and G. Palmer, *Ind. Eng. Chem.*, **10,** 314 (1938).
[d] A. F. Benning, private communication, 1940.
[e] P. W. Bridgman, *Proc. Am. Acad. Arts Sci.*, **59,** 141 (1923).
[f] T. H. Chilton and R. P. Genereaux, personal communication, 1939, based on data selected from the literature.
[g] M. Daniloff, *J. Am. Chem. Soc.*, **54,** 1328 (1932).
[h] "International Critical Tables," McGraw-Hill, New York, 1929.
[i] A. Kardos, *Z. Ver. deut. Ing.*, **77,** 1158 (1933); *Z. ges. Kälte-Ind.*, **41,** 29 (1934).
[j] G. W. C. Kaye and W. F. Higgins, *Proc. Roy. Soc. (London)*, **A117,** 459 (1928).
[k] DuPont Chlorinated Hydrocarbons, *Tech. Bull.*, Electrochemicals Department, du Pont, Buffalo, N.Y., 1938.
[l] S. Shiba, *Sci. Papers Inst. Phys. Chem. Research (Tokyo)*, **16,** 205 (1931).
[m] J. F. D. Smith, *Trans. ASME*, **58,** 719 (1936).
[n] D. L. Timrot and N. Vargaftik, *J. Tech. Phys. (U.S.S.R.)*, **10,** 1063 (1940).

TABLE A-11. THERMAL CONDUCTIVITIES OF PETROLEUM OILS
(J. F. D. Smith)
[k = Btu/(hr)(sq ft)(deg F per ft)]

Designation of hydro-carbon oil	Avg mol wt.[a]	Viscosity, centipoise			k at 86°F	k at 212°F	Sp. gr. at 60°F
		68°F	140°F	212°F			
Light heat-transfer oil	284	62.0	9.5	3.2	0.0765	0.0748	0.925
Rabbeth spindle oil	303	24.5	5.7	2.37	0.0825	0.0805	0.870
Velocite B oil	333	73.0	11.0	4.20	0.0825	0.0800	0.897
Red oil	418	....	44.0	9.90	0.0815	0.0796	0.928

[a] H. O. Forrest and L. W. Cummings.

TABLE A-12. THERMAL CONDUCTIVITIES OF GASES AND VAPORS
$[k = \text{Btu}/(\text{hr})(\text{sq ft})(\text{deg F per ft})]$

The extreme temperature values given constitute the experimental range. For extrapolation to other temperatures, it is suggested that the data given be plotted as log k vs. log T or that use be made of the assumption that the ratio $c_p\mu/k$ is practically independent of temperature (or of pressure, within moderate limits).

Substance	Deg F	k	Substance	Deg F	k
Acetone (ref. 10)[a]	32	0.0057	Chloroform (ref. 10)[a]	32	0.0038
	115	0.0074		115	0.0046
	212	0.0099		212	0.0058
	363	0.0147		363	0.0077
Acetylene (ref. 3)[a]	−103	0.0068	Cyclohexane	216	0.0095
	32	0.0108			
	122	0.0140	Dichlorodifluoromethane	32	0.0048
	212	0.0172		122	0.0064
Air (ref. 7)	−328	0.0040		212	0.0080
	−148	0.0091		302	0.0097
	32	0.0140			
	212	0.0184	Ethane (ref. 1, 3)	−94	0.0066
	392	0.0224		−29	0.0086
	572	0.0260		32	0.0106
Ammonia (ref. 7)	−58	0.0097		212	0.0175
	32	0.0126	Ethyl acetate (ref. 10)[a]	115	0.0072
	212	0.0192		212	0.0096
	392	0.0280		363	0.0141
	572	0.0385	Ethyl alcohol (ref. 10)[a]	68	0.0089
	752	0.0509		212	0.0124
Argon (ref. 7)	−148	0.0063	Ethyl chloride (ref. 10)[a]	32	0.0055
	32	0.0095		212	0.0095
	212	0.0123		363	0.0135
	392	0.0148		413	0.0152
	572	0.0171	Ether (ref. 10)[a]	32	0.0077
				115	0.0099
Benzene (ref. 10)[a]	32	0.0052		212	0.0131
	115	0.0073		363	0.0189
	212	0.0103		413	0.0209
	363	0.0152	Ethylene (ref. 3)[a]	−96	0.0064
	413	0.0176		32	0.0101
Butane (n-) (ref. 9)	32	0.0078		122	0.0131
	212	0.0135		212	0.0161
(iso-) (ref. 9)	32	0.0080			
	212	0.0139	Helium (ref. 7)	−328	0.0338
				−148	0.0612
Carbon dioxide (ref. 7)	−58	0.0064		32	0.0818
	32	0.0084		212	0.0988
	212	0.0128	Heptane (n-) (ref. 10)[a]	212	0.0103
	392	0.0177		392	0.0112
	572	0.0229	Hexane (n-) (ref. 9)	32	0.0072
Carbon disulfide (ref. 3)[a]	32	0.0040		68	0.0080
	45	0.0042	Hexene (ref. 10)[a]	32	0.0061
Carbon monoxide (ref. 7)	−328	0.0037		212	0.0109
	−148	0.0088	Hydrogen (ref. 7)	−328	0.0293
	32	0.0134		−148	0.0652
	212	0.0176		32	0.0966
Carbon tetrachloride (ref. 10)[a]	115	0.0041		212	0.1240
	212	0.0052		392	0.1484
	363	0.0065		572	0.1705
Chlorine (ref. 5)	32	0.0043			

TABLE A-12. THERMAL CONDUCTIVITIES OF GASES AND VAPORS.—(*Continued*)

Substance	Deg F	k	Substance	Deg F	k
Hydrogen and carbon dioxide (ref. 4)[a]	32		Neon	32	0.0256
			Nitric oxide (ref. 7)	−148	0.0089
0% H₂	....	0.0083		32	0.0138
20%	....	0.0165		122	0.0161
40%	....	0.0270	Nitrogen (ref. 7)	−328	0.0040
60%	....	0.0410		−148	0.0091
80%	....	0.0620		32	0.0139
100%	....	0.10		212	0.0181
Hydrogen and nitrogen (ref. 4)	32			392	0.0220
0% H₂	....	0.0133		572	0.0255
20%	....	0.0212		752	0.0287
40%	....	0.0313			
60%	....	0.0438	Nitrogen and carbon dioxide (ref. 8)	122	
80%	....	0.0635			
Hydrogen and nitrous oxide (ref. 4)	32		0% N₂	....	0.0105
			34.06%	....	0.0121
0% H₂	....	0.0092	52.88%	....	0.0130
20%	....	0.0170	66.50%	....	0.0137
40%	....	0.0270	100%	....	0.0161
60%	....	0.0410	Nitrous oxide (ref. 7)	−148	0.0047
80%	....	0.0650		32	0.0088
Hydrogen sulfide (ref. 3)[a]	32	0.0076		212	0.0138
			Oxygen (ref. 7)	−328	0.0038
Mercury (ref. 5)	392	0.0197		−148	0.0091
Methane (ref. 7)	−328	0.0045		32	0.0142
	−148	0.0109		122	0.0166
	32	0.0176		212	0.0188
	212	0.0255			
	392	0.0358	Pentane (n-) (ref. 10)[a]	32	0.0074
	572	0.0490		68	0.0083
Methyl acetate (ref. 10)[a]	32	0.0059	(iso-) (ref. 10)[a]	32	0.0072
	68	0.0068		212	0.0127
Alcohol (ref. 10)[a]	32	0.0083	Propane (ref. 9)	32	0.0087
	212	0.0128		212	0.0151
Chloride (ref. 10)[a]	32	0.0053			
	115	0.0072	Sulfur dioxide (ref. 2)	32	0.0050
	212	0.0094		212	0.0069
	363	0.0130			
	413	0.0148	Water vapor, zero pressure (ref. 7)[b]	212	0.0136
Methylene chloride (ref. 10)[a]	32	0.0039		392	0.0182
	115	0.0049		572	0.0230
	212	0.0063		752	0.0279
	413	0.0095		932	0.0328

[a] Data from Eucken[3] and Moser[10] are measurements relative to air. Data in this table from these sources are based on the thermal conductivity of air at 32°F of 0.0140 Btu/(hr)(sq ft)(deg F per ft).

[b] For saturated water vapor (ref. 6):

lb/sq in. abs	250	500	1000	1500	1750	2000
Deg F	401	467	545	596	617	636
k	0.0212	0.0250	0.0316	0.0380	0.0412	0.0445

Table A-12. Thermal Conductivities of Gases and Vapors.—(*Continued*)

References:
1. T. H. Chilton and R. P. Genereaux, personal communication, 1946.
2. B. G. Dickens, *Proc. Roy. Soc. (London)*, **A143**, 517 (1934).
3. A. Eucken, *Physik. Z.*, **12**, 1101 (1911), **14**, 324 (1913).
4. T. L. Ibbs and A. A. Hirst, *Proc. Roy. Soc. (London)*, **A123**, 134 (1929).
5. "International Critical Tables," McGraw-Hill, New York, 1929.
6. J. H. Keenan and F. G. Keyes, "Thermodynamic Properties of Steam," Wiley, New York, 1950 (22d impression).
7. F. G. Keyes, *Tech. Rept.* 37, Project Squid (Apr. 1, 1952).
8. F. G. Keyes, *Trans. ASME*, **74**, 1303 (1952).
9. W. B. Mann and B. G. Dickens, *Proc. Roy. Soc. (London)*, **A134**, 77 (1931).
10. Moser, Dissertation, Berlin, 1913.
11. L. S. Marks, "Mechanical Engineers' Handbook," 5th ed., McGraw-Hill, New York, 1951.

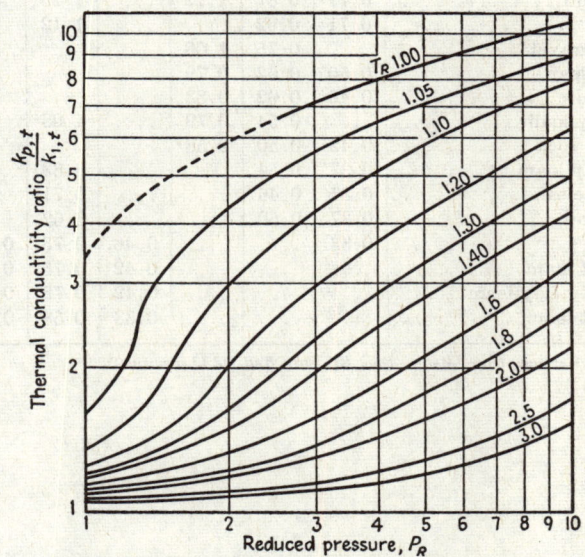

Fig. A-1. Generalized correlation of thermal conductivity of gases at high pressures. [Comings and Nathan, *Ind. Eng. Chem.*, **39**, 964–970 (1947).] For data on a number of gases see Lenoir and Comings, *Chem. Eng. Progr.*, **47**, 223–231 (1951), and F. G. Keyes, *Trans. ASME*, **73**, 589–606 (1951).

TABLE A-13. THERMAL CONDUCTIVITY OF SOME SOILS AT 40°F[a]
[k = Btu/(hr)(sq ft)(deg F per ft)]

Type of soil	Moisture content, per cent						
	4	4	4	10	10	20	20
	Dry density, lb/cu ft						
	100	110	120	90	110	90	100
Fine crushed quartz	1.0	1.33					
Crushed quartz	0.96	1.33	1.83				
Graded Ottawa sand	0.83	1.17					
Fairbanks sand	0.71	0.87	1.12	....	1.25		
Lowell sand	0.71	0.92	....	....	1.12		
Chena river gravel	....	0.75	1.08				
Crushed feldspar	0.50	0.62	0.79				
Crushed granite	0.46	0.62	0.83				
Dakota sandy loam	....	0.54	0.79	....	1.08		
Crushed trap rock	0.42	0.50	0.58				
Ramsey sandy loam	0.37	0.54	....	....	0.83		
Northway fine sand	0.37	0.46	....	....	0.71		
Northway sand	0.37	0.50	....	....	0.62		
Healy clay	0.33	....	....	0.46	0.75	0.67	0.83
Fairbanks silt loam	....	....	....	0.42	0.75	0.62	0.83
Fairbanks silty clay loam	....	....	....	0.42	0.75	0.62	0.79
Northway silt loam	....	....	....	0.33	0.58	0.50	0.58

From M. S. Kersten, *Univ. Minn., Eng. Ex. Sta., Bull.* 28 (June, 1949).

APPENDIX

TABLE A-14. SPECIFIC HEATS OF SOLIDS
[Expressed in Btu/(lb)(deg F) = gm-cal/(gm)(deg C)]

Deg F	Deg C	Pb	Zn	Al	Ag	Au	Cu	Ni	Fe	Co	Quartz
32	0	0.0306	0.0917	0.2106	0.0557	0.0305	0.0919	0.1025	0.1051	0.1023	0.1667
212	100	0.0315	0.0958	0.2225	0.0571	0.0312	0.0942	0.1132	0.1166	0.1079	0.2061
392	200	0.0325	0.0999	0.2344	0.0585	0.0320	0.0965	0.1241	0.1280	0.1138	0.2315
572	300	0.0335	0.1041	0.2463	0.0599	0.0327	0.0988	0.1352	0.1395	0.1192	0.2518
752	400	0.0328	0.1082	0.2582	0.0612	0.0334	0.1011	0.1295	0.1508	0.1249	0.2696
932	500	0.0328	0.1225	0.2702	0.0626	0.0341	0.1034	0.1310	0.1622	0.1305	0.2865
1112	600	0.0328	0.1233	0.2821	0.0640	0.0349	0.1057	0.1326	0.1737	0.1362	0.2624
1292	700	0.0328	0.1242	0.259	0.0654	0.0356	0.1080	0.1341	0.1853	0.1418	0.2714
1472	800	0.0328	0.1250	0.259	0.0668	0.0363	0.1103	0.1356	0.1741	0.1475	0.2806
1652	900	0.0328	0.1259	0.259	0.0682	0.0371	0.1126	0.1372	0.1805	0.1531	0.2897
1832	1000	0.0328	0.1267	0.259	0.076	0.0378	0.1149	0.1387	0.1505	0.1588	0.2989
2012	1100	.....	.....	.....	0.076	0.0355	0.118	0.1403	0.1505	0.1644	0.3080
2192	1200	.....	.....	.....	0.076	0.0355	0.118	0.1418	0.1505	0.1701	0.3172
2372	1300	.....	.....	.....	0.076	0.0355	0.118	0.1434	0.1505	0.1757	0.3264
2552	1400	.....	.....	.....	.....	.....	.....	0.1449	0.1505	0.1814	0.3355
2732	1500	.....	.....	.....	.....	.....	.....	0.1455	0.1790	0.1425	0.3447
2912	1600	.....	.....	.....	.....	.....	.....	0.1455	0.1460	0.1425	0.3538
Melting points, deg F......		621	786	1220	1760	1945	1981	2646	2795	2696	

[a] Calculated values from equations by K. K. Kelley, *U.S Bur. Mines Bull.* 371 (1934). These values are the true specific heats for the particular physical state or allotropic modification existing at the indicated temperatures.

Specific heat = Btu/(Lb)(Deg F) = Pcu/(Lb)(Deg C)

NO.	LIQUID	RANGE DEG C
29	ACETIC ACID 100%	0- 80
32	ACETONE	20- 50
52	AMMONIA	-70- 50
37	AMYL ALCOHOL	-50- 25
26	AMYL ACETATE	0-100
30	ANILINE	0-130
23	BENZENE	10- 80
27	BENZYL ALCOHOL	-20- 30
10	BENZYL CHLORIDE	-30- 30
49	BRINE, 25% CaCl$_2$	-40- 20
51	BRINE, 25% NaCl	-40- 20
44	BUTYL ALCOHOL	0-100
2	CARBON DISULPHIDE	-100- 25
3	CARBON TETRACHLORIDE	10- 60
8	CHLOROBENZENE	0-100
4	CHLOROFORM	0- 50
21	DECANE	-80- 25
6A	DICHLOROETHANE	-30- 60
5	DICHLOROMETHANE	-40- 50
15	DIPHENYL	80-120
22	DIPHENYLMETHANE	30-100
16	DIPHENYL OXIDE	0-200
16	DOWTHERM A	0-200
24	ETHYL ACETATE	-50- 25
42	ETHYL ALCOHOL 100%	30- 80
46	ETHYL ALCOHOL 95%	20- 80
50	ETHYL ALCOHOL 50%	20- 80
25	ETHYL BENZENE	0-100
1	ETHYL BROMIDE	5- 25
13	ETHYL CHLORIDE	-30- 40
36	ETHYL ETHER	-100- 25
7	ETHYL IODIDE	0-100
39	ETHYLENE GLYCOL	-40-200

NO.	LIQUID	RANGE DEG C
2A	FREON-11 (CCl$_3$F)	-20- 70
6	FREON-12 (CCl$_2$F$_2$)	-40- 15
4A	FREON-21 (CHCl$_2$F)	-20- 70
7A	FREON-22 (CHClF$_2$)	-20- 60
3A	FREON-113 (CCl$_2$F-CClF$_2$)	-20- 70
38	GLYCEROL	-40- 20
28	HEPTANE	0- 60
35	HEXANE	-80- 20
48	HYDROCHLORIC ACID, 30%	20-100
41	ISOAMYL ALCOHOL	10-100
43	ISOBUTYL ALCOHOL	0-100
47	ISOPROPYL ALCOHOL	-20- 50
31	ISOPROPYL ETHER	-80- 20
40	METHYL ALCOHOL	-40- 20
13A	METHYL CHLORIDE	-80- 20
14	NAPHTHALENE	90-200
12	NITROBENZENE	0-100
34	NONANE	-50- 25
33	OCTANE	-50- 25
3	PERCHLORETHYLENE	-30-140
45	PROPYL ALCOHOL	-20-100
20	PYRIDINE	-50- 25
9	SULPHURIC ACID 98%	10- 45
11	SULPHUR DIOXIDE	-20-100
23	TOLUENE	0- 60
53	WATER	10-200
19	XYLENE ORTHO	0-100
18	XYLENE META	0-100
17	XYLENE PARA	0-100

FIG. A-2. True specific heats of liquids. (*Chilton, Colburn, and Vernon, personal communication, based mainly on data from "International Critical Tables."*)

TABLE A-15. AVERAGE SPECIFIC HEATS OF ORGANIC LIQUIDS[a]
[Expressed in Btu/(lb)(deg F) = gm-cal/(gm)(deg C)]

Compound	Deg F	c_p	Compound	Deg F	c_p
Acetal	32	0.467	Isobutane	32	0.549
	66–210	0.520	Isoheptane	32–122	0.501
Allyl alcohol	70–205	0.665	Isopentane	32	0.512
Amylene	32	0.282	Isopentane	46.4	0.527
Benzaldehyde	72–342	0.428	Kerosene	32–212	0.50
Bromobenzene	68–212	0.231	Lauric acid	104–212	0.572
Butane (n-, iso-)	32	0.549	Machine oil	32–212	0.40
Butyl chloride (n-)	68	0.451	Methane (saturated liquid)	−280	0.811
Caproic acid	84–221	0.531	Methane (saturated liquid)	−226	0.861
Castor oil	59	0.51	Methane (saturated liquid)	−172	0.992
Chlorophenol	32–68	0.399	Mesityl oxide	70–250	0.521
Coal tar oils	59–194	0.34	Mesitylene	32	0.393
Cresol (o-)	32–68	0.497	Myristic acid	133–212	0.539
Cresol (m-)	70–387	0.551	Naphthol (α)	32	0.388
Cyclohexane	50–64	0.431	Naphthol (β)	32	0.403
Cyclohexanol	59–64	0.416	Nonylene	32–122	0.485
Cyclohexanone	59–64	0.431	Octylene	32–122	0.486
Cymene (o-)	32	0.398	Olive oil	32–212	0.40
Decahydronaphthalene (cis-)	59–64	0.393	Palmitic acid	150–220	0.653
Decylene	32–122	0.467	Paraffin oil	32–212	0.52
Diamylene	78–266	0.543	Pentadecane	32–122	0.497
Dichloroacetic acid	70–385	0.348	Pentadecylene	32–122	0.471
Dichlorobenzene (o-)	32	0.269	Petroleum	32–212	0.50
Dichlorobenzene (m-)	32	0.269	Phenol	57–79	0.561
Dichlorobenzene (p-)	127–210	0.297	Propane	32	0.576
Dichlorodifluormethane	−45	0.21	Pseudocumene	68	0.414
Dihydronaphthalene	64–82	0.345	Stearic acid	167–279	0.550
Di-isoamyl	71–311	0.588	Tetrachlorethane	68	0.268
Dodecane	57–68	0.505	Tetrachlorethylene	68	0.216
Dodecylene	32–122	0.455	Tetradecane	32–122	0.497
Ethylene chloride	−22	0.278	Toluidine (o-)	32	0.454
	+68	0.299		72–383	0.598
	140	0.318	Toluidine (p-)	109	0.524
Formic acid	68–212	0.524		136	0.634
Furfural	68–212	0.416		201	0.533
Fusel oil	32–212	0.56	Trichlorethane	68	0.266
Gasoline	32–212	0.50	Trichlorethylene	68	0.233
Heptylene	32–122	0.486	Tridecane	32–122	0.499
Hexadiene (1, 5-)	32	0.405	Tridecylene	32–122	0.457
Hexahydrocresol (o-)	59–64	0.416	Trinitrotoluene (2, 4, 6-)		0.335
Hexahydrocresol (m-)	59–64	0.420	Turpentine	32–212	0.42
Hexahydrocresol (p-)	59–64	0.421	Undecane	32–122	0.501
Hexylene	32–122	0.504	Undecylene	32–122	0.482

[a] "International Critical Tables," Vol. 5, pp. 107–113, McGraw-Hill, New York, 1929. See also Fig. A-2.

464   HEAT TRANSMISSION

C = Specific heat = Btu/(Lb)(Deg F) = Pcu/(Lb)(Deg C)

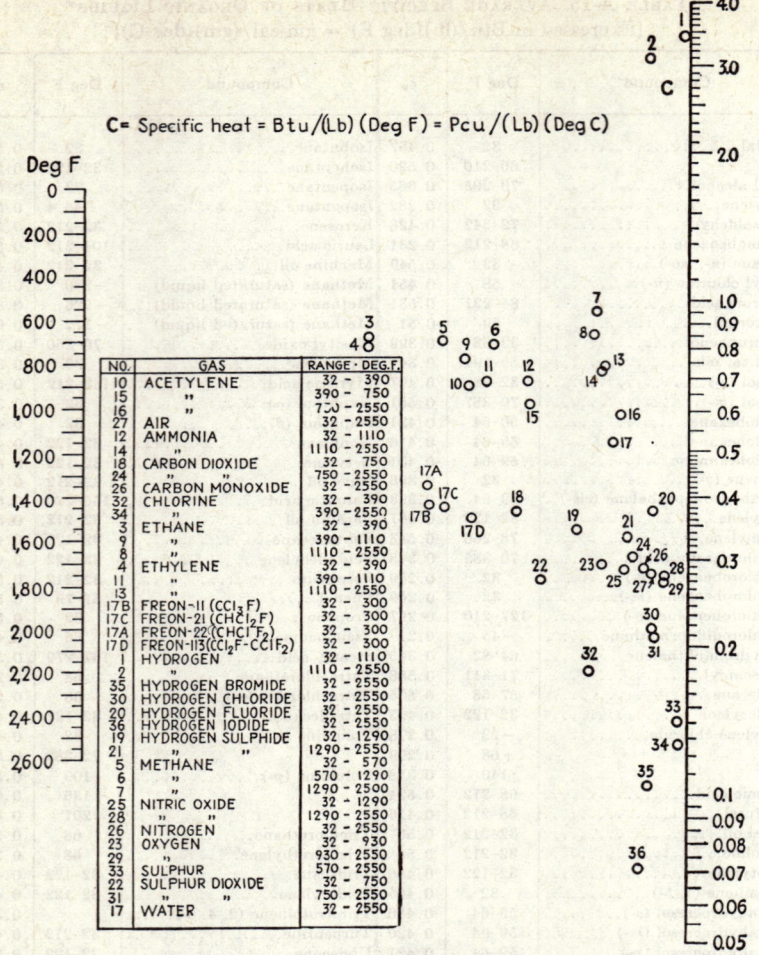

NO.	GAS	RANGE · DEG.F.
10	ACETYLENE	32 – 390
15	"	390 – 750
16	"	750 – 2550
27	AIR	32 – 2550
12	AMMONIA	32 – 1110
14	"	1110 – 2550
18	CARBON DIOXIDE	32 – 750
24	"	750 – 2550
26	CARBON MONOXIDE	32 – 2550
32	CHLORINE	32 – 390
34	"	390 – 2550
3	ETHANE	32 – 390
9	"	390 – 1110
8	"	1110 – 2550
4	ETHYLENE	32 – 390
11	"	390 – 1110
13	"	1110 – 2550
17B	FREON-11 (CCl$_3$F)	32 – 300
17C	FREON-21 (CHCl$_2$F)	32 – 300
17A	FREON-22 (CHClF$_2$)	32 – 300
17D	FREON-113 (CCl$_2$F–CClF$_2$)	32 – 1110
1	HYDROGEN	32 – 1110
2	"	1110 – 2550
35	HYDROGEN BROMIDE	32 – 2550
30	HYDROGEN CHLORIDE	32 – 2550
20	HYDROGEN FLUORIDE	32 – 2550
36	HYDROGEN IODIDE	32 – 2550
19	HYDROGEN SULPHIDE	32 – 1290
21	" "	1290 – 2550
5	METHANE	32 – 570
6	"	570 – 1290
7	"	1290 – 2500
25	NITRIC OXIDE	32 – 1290
28	" "	1290 – 2550
26	NITROGEN	32 – 2550
23	OXYGEN	32 – 930
29	"	930 – 2550
33	SULPHUR	570 – 2550
22	SULPHUR DIOXIDE	32 – 750
31	" "	750 – 2550
17	WATER	32 – 2550

Fig. A-3. True specific heats c_p of gases and vapors at 1 atm pressure. (*Chilton, Colburn, and Vernon, personal communication, based mainly on data from "International Critical Tables."*)

Table A-16. Specific Heats of Liquefied Gases[a]

Liquid	Deg C	Specific heat, Btu/(lb)(deg F)
Ammonia	−60	1.05
	0	1.10
	40	1.16
	80	1.29
	100	1.48
	110	1.61
Carbon dioxide (63 atm)	−50 to −10	0.465 to 0.539
Disulfide	−100 to 150	$0.235 + 0.00046t$
Monoxide	−206 to −190	0.0615
Chlorine	−205	0.229
Hydrogen	−258 to −252	1.75 to 2.33
Nitric oxide	−158 to −156	0.580
Nitrogen	−209 to −197	0.475
Oxygen	−216 to −200	0.398
Sulfur dioxide	−20	0.313
	0	0.318
	20	0.328
	60	0.361
	100	0.419
	150	0.846

[a] From J. H. Perry, "Chemical Engineers' Handbook," 3d ed., McGraw-Hill, New York, 1950.

Table A-17. Specific Heat of Air at High Pressures[a]

Temperature, deg C	c_p, specific heat of air					
	1 atm	10 atm	20 atm	40 atm	70 atm	100 atm
100	0.237	0.239	0.240	0.245	0.250	0.258
0	0.238	0.242	0.247	0.251	0.277	0.298
−50	0.238	0.246	0.257	0.279	0.332	0.412
−100	0.239	0.259	0.285	0.370	0.846	
−150	0.240	0.311	0.505			

[a] For the specific heats of other gases as a function of the pressure, see "International Critical Tables," Vol. 5, pp. 82–83, McGraw-Hill, New York, 1929.

466   HEAT TRANSMISSION

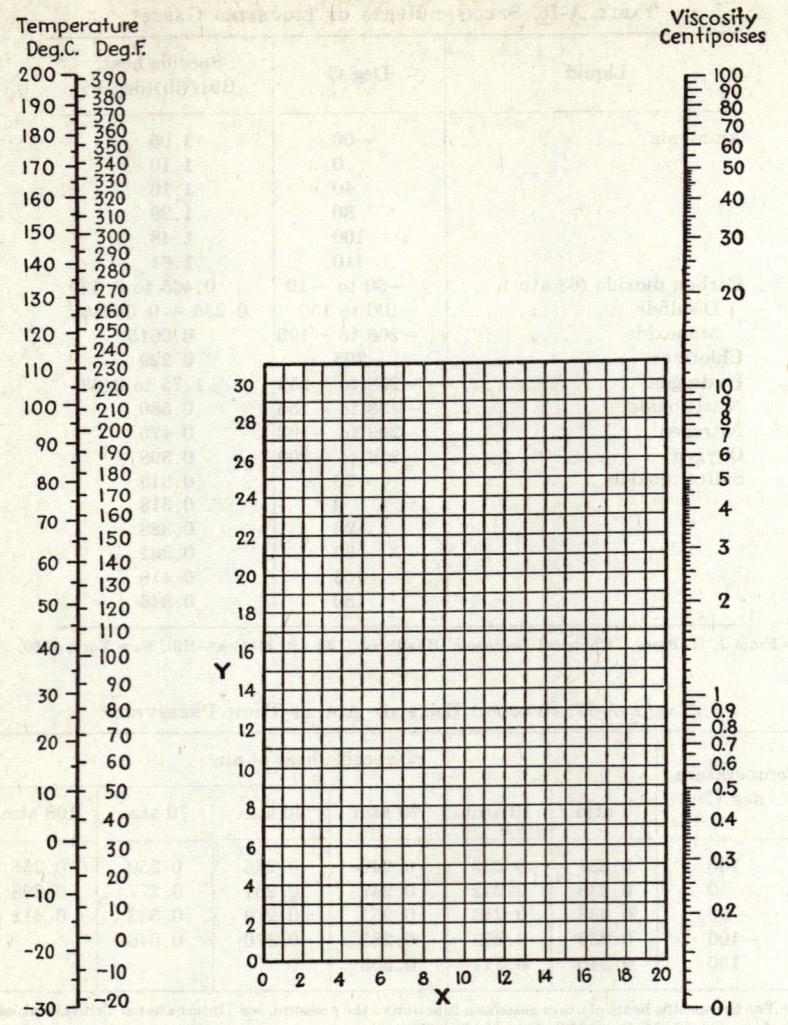

FIG. A-4. Viscosities of liquids at 1 atm. For coordinates, see Table A-18. (*Genereaux, personal communication.*)

TABLE A-18. VISCOSITIES OF LIQUIDS[a]
(Coordinates for Fig. A-4)

No.	Liquid	X	Y	No.	Liquid	X	Y
1	Acetaldehyde	15.2	4.8	56	Freon-22	17.2	4.7
2	Acetic acid, 100%	12.1	14.2	57	Freon-113	12.5	11.4
3	Acetic acid, 70%	9.5	17.0	58	Glycerol, 100%	2.0	30.0
4	Acetic anhydride	12.7	12.8	59	Glycerol, 50%	6.9	19.6
5	Acetone, 100%	14.5	7.2	60	Heptane	14.1	8.4
6	Acetone, 35%	7.9	15.0	61	Hexane	14.7	7.0
7	Allyl alcohol	10.2	14.3	62	Hydrochloric acid, 31.5%	13.0	16.6
8	Ammonia, 100%	12.6	2.0	63	Isobutyl alcohol	7.1	18.0
9	Ammonia, 26%	10.1	13.9	64	Isobutyric acid	12.2	14.4
10	Amyl acetate	11.8	12.5	65	Isopropyl alcohol	8.2	16.0
11	Amyl alcohol	7.5	18.4	66	Kerosene	10.2	16.9
12	Aniline	8.1	18.7	67	Linseed oil, raw	7.5	27.2
13	Anisole	12.3	13.5	68	Mercury	18.4	16.4
14	Arsenic trichloride	13.9	14.5	69	Methanol, 100%	12.4	10.5
15	Benzene	12.5	10.9	70	Methanol, 90%	12.3	11.8
16	Brine, CaCl$_2$, 25%	6.6	15.9	71	Methanol, 40%	7.8	15.5
17	Brine, NaCl, 25%	10.2	16.6	72	Methyl acetate	14.2	8.2
18	Bromine	14.2	13.2	73	Methyl chloride	15.0	3.8
19	Bromotoluene	20.0	15.9	74	Methyl ethyl ketone	13.9	8.6
20	Butyl acetate	12.3	11.0	75	Naphthalene	7.9	18.1
21	Butyl alcohol	8.6	17.2	76	Nitric acid, 95%	12.8	13.8
22	Butyric acid	12.1	15.3	77	Nitric acid, 60%	10.8	17.0
23	Carbon dioxide	11.6	0.3	78	Nitrobenzene	10.6	16.2
24	Carbon disulphide	16.1	7.5	79	Nitrotoluene	11.0	17.0
25	Carbon tetrachloride	12.7	13.1	80	Octane	13.7	10.0
26	Chlorobenzene	12.3	12.4	81	Octyl alcohol	6.6	21.1
27	Chloroform	14.4	10.2	82	Pentachloroethane	10.9	17.3
28	Chlorosulfonic acid	11.2	18.1	83	Pentane	14.9	5.2
29	Chlorotoluene, ortho	13.0	13.3	84	Phenol	6.9	20.8
30	Chlorotoluene, meta	13.3	12.5	85	Phosphorus tribromide	13.8	16.7
31	Chlorotoluene, para	13.3	12.5	86	Phosphorus trichloride	16.2	10.9
32	Cresol, meta	2.5	20.8	87	Propionic acid	12.8	13.8
33	Cyclohexanol	2.9	24.3	88	Propyl alcohol	9.1	16.5
34	Dibromoethane	12.7	15.8	89	Propyl bromide	14.5	9.6
35	Dichloroethane	13.2	12.2	90	Propyl chloride	14.4	7.5
36	Dichloromethane	14.6	8.9	91	Propyl iodide	14.1	11.6
37	Diethyl oxalate	11.0	16.4	92	Sodium	16.4	13.9
38	Dimethyl oxalate	12.3	15.8	93	Sodium hydroxide, 50%	3.2	25.8
39	Diphenyl	12.0	18.3	94	Stannic chloride	13.5	12.8
40	Dipropyl oxalate	10.3	17.7	95	Sulphur dioxide	15.2	7.1
41	Ethyl acetate	13.7	9.1	96	Sulphuric acid, 110%	7.2	27.4
42	Ethyl alcohol, 100%	10.5	13.8	97	Sulphuric acid, 98%	7.0	24.8
43	Ethyl alcohol, 95%	9.8	14.3	98	Sulphuric acid, 60%	10.2	21.3
44	Ethyl alcohol, 40%	6.5	16.6	99	Sulphuryl chloride	15.2	12.4
45	Ethyl benzene	13.2	11.5	100	Tetrachloroethane	11.9	15.7
46	Ethyl bromide	14.5	8.1	101	Tetrachloroethylene	14.2	12.7
47	Ethyl chloride	14.8	6.0	102	Titanium tetrachloride	14.4	12.3
48	Ethyl ether	14.5	5.3	103	Toluene	13.7	10.4
49	Ethyl formate	14.2	8.4	104	Trichloroethylene	14.8	10.5
50	Ethyl iodide	14.7	10.3	105	Turpentine	11.5	14.9
51	Ethylene glycol	6.0	23.6	106	Vinyl acetate	14.0	8.8
52	Formic acid	10.7	15.8	107	Water	10.2	13.0
53	Freon-11	14.4	9.0	108	Xylene, ortho	13.5	12.1
54	Freon-12	16.8	5.6	109	Xylene, meta	13.9	10.6
55	Freon-21	15.7	7.5	110	Xylene, para	13.9	10.9

[a] From J. H. Perry, "Chemical Engineers' Handbook," 3d ed., McGraw-Hill, New York, 1950.

Table A-19. Viscosities of Gases[a]
(Coordinates for use with Fig. A-6)

No.	Gas	X	Y	No.	Gas	X	Y
1	Acetic acid	7.7	14.3	29	Freon-113	11.3	14.0
2	Acetone	8.9	13.0	30	Helium	10.9	20.5
3	Acetylene	9.8	14.9	31	Hexane	8.6	11.8
4	Air	11.0	20.0	32	Hydrogen	11.2	12.4
5	Ammonia	8.4	16.0	33	$3H_2 + 1N_2$	11.2	17.2
6	Argon	10.5	22.4	34	Hydrogen bromide	8.8	20.9
7	Benzene	8.5	13.2	35	Hydrogen chloride	8.8	18.7
8	Bromine	8.9	19.2	36	Hydrogen cyanide	9.8	14.9
9	Butene	9.2	13.7	37	Hydrogen iodide	9.0	21.3
10	Butylene	8.9	13.0	38	Hydrogen sulfide	8.6	18.0
11	Carbon dioxide	9.5	18.7	39	Iodine	9.0	18.4
12	Carbon disulfide	8.0	16.0	40	Mercury	5.3	22.9
13	Carbon monoxide	11.0	20.0	41	Methane	9.9	15.5
14	Chlorine	9.0	18.4	42	Methyl alcohol	8.5	15.6
15	Chloroform	8.9	15.7	43	Nitric oxide	10.9	20.5
16	Cyanogen	9.2	15.2	44	Nitrogen	10.6	20.0
17	Cyclohexane	9.2	12.0	45	Nitrosyl chloride	8.0	17.6
18	Ethane	9.1	14.5	46	Nitrous oxide	8.8	19.0
19	Ethyl acetate	8.5	13.2	47	Oxygen	11.0	21.3
20	Ethyl alcohol	9.2	14.2	48	Pentane	7.0	12.8
21	Ethyl chloride	8.5	15.6	49	Propane	9.7	12.9
22	Ethyl ether	8.9	13.0	50	Propyl alcohol	8.4	13.4
23	Ethylene	9.5	15.1	51	Propylene	9.0	13.8
24	Fluorine	7.3	23.8	52	Sulfur dioxide	9.6	17.0
25	Freon-11	10.6	15.1	53	Toluene	8.6	12.4
26	Freon-12	11.1	16.0	54	2,3,3-trimethylbutane	9.5	10.5
27	Freon-21	10.8	15.3	55	Water	8.0	16.0
28	Freon-22	10.1	17.0	56	Xenon	9.3	23.0

[a] From J. H. Perry, "Chemical Engineers' Handbook," 3d ed., McGraw-Hill, New York, 1950.

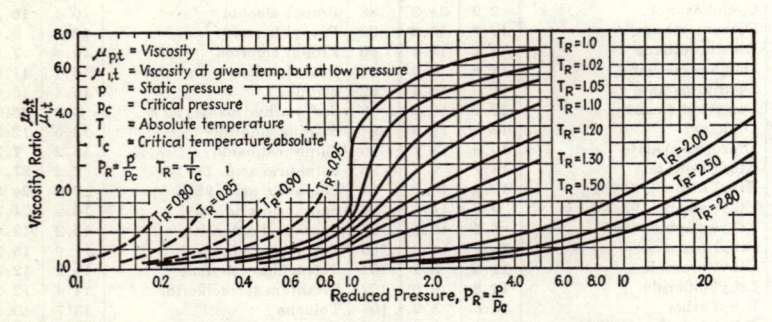

Fig. A-5. Effect of pressure on viscosities of gases at several temperatures. For steam, see Table A-20. [*Comings and Egly, Ind. Eng. Chem.*, **32**, 714–718 (1940).]

APPENDIX

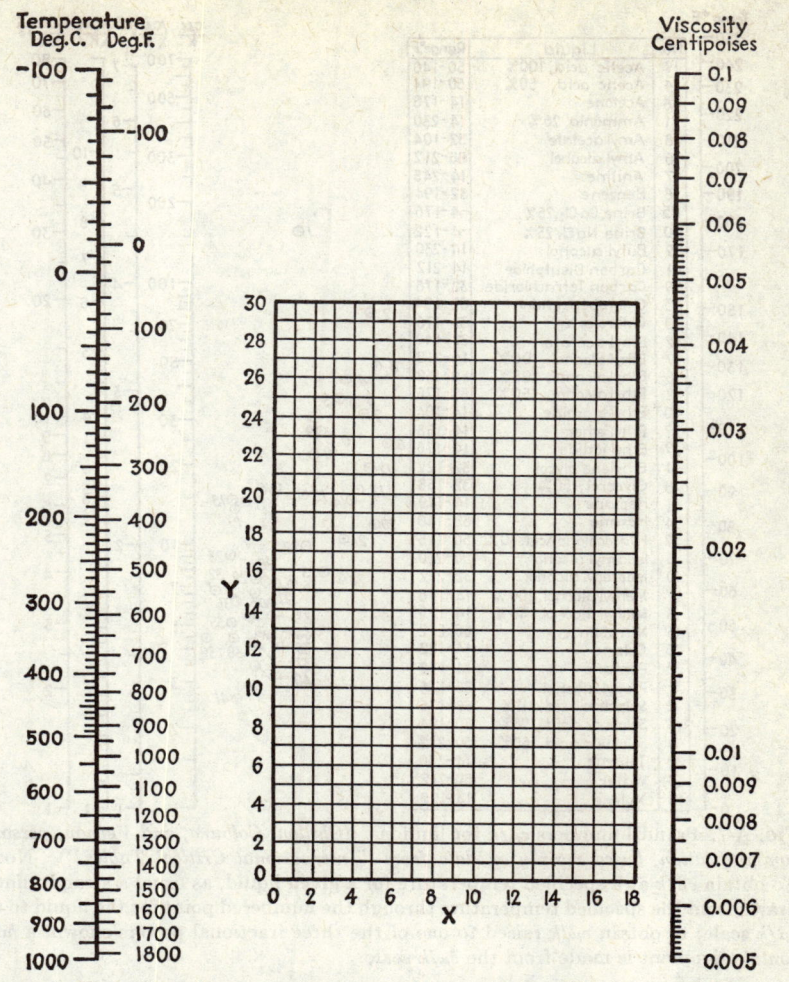

Fig. A-6. Viscosities of gases and vapors at 1 atm; for coordinates, see Table A-19. (*Genereaux, personal communication.*)

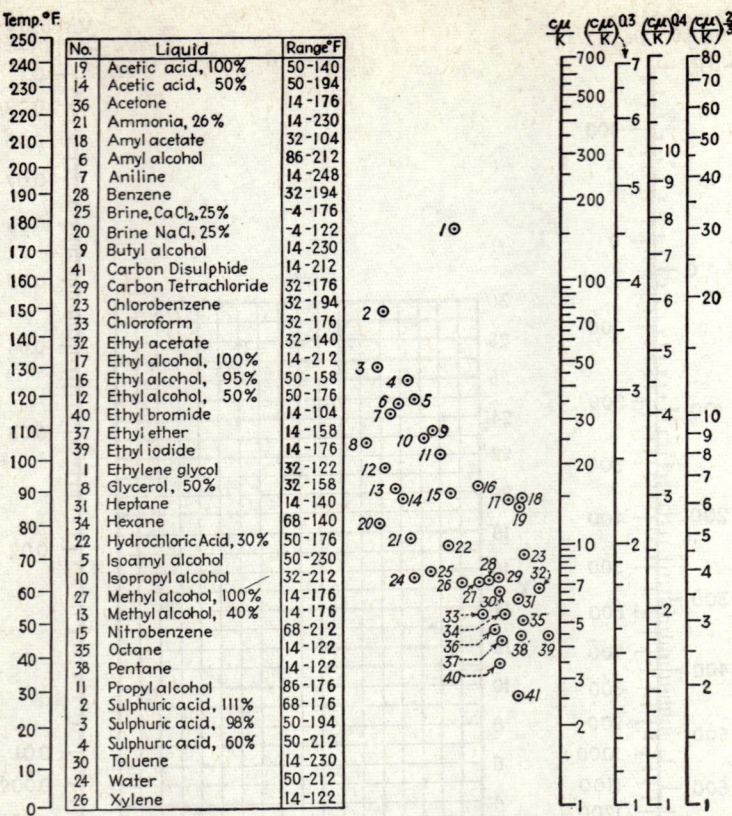

No.	Liquid	Range °F
19	Acetic acid, 100%	50-140
14	Acetic acid, 50%	50-194
36	Acetone	14-176
21	Ammonia, 26%	14-230
18	Amyl acetate	32-104
6	Amyl alcohol	86-212
7	Aniline	14-248
28	Benzene	32-194
25	Brine, CaCl$_2$, 25%	-4-176
20	Brine NaCl, 25%	-4-122
9	Butyl alcohol	14-230
41	Carbon Disulphide	14-212
29	Carbon Tetrachloride	32-176
23	Chlorobenzene	32-194
33	Chloroform	32-176
32	Ethyl acetate	32-140
17	Ethyl alcohol, 100%	14-212
16	Ethyl alcohol, 95%	50-158
12	Ethyl alcohol, 50%	50-176
40	Ethyl bromide	14-104
37	Ethyl ether	14-158
39	Ethyl iodide	14-176
1	Ethylene glycol	32-122
8	Glycerol, 50%	32-158
31	Heptane	14-140
34	Hexane	68-140
22	Hydrochloric Acid, 30%	50-176
5	Isoamyl alcohol	50-230
10	Isopropyl alcohol	32-212
27	Methyl alcohol, 100%	14-176
13	Methyl alcohol, 40%	14-176
15	Nitrobenzene	68-212
35	Octane	14-122
38	Pentane	14-122
11	Propyl alcohol	86-176
2	Sulphuric acid, 111%	68-176
3	Sulphuric acid, 98%	50-194
4	Sulphuric acid, 60%	50-212
30	Toluene	14-230
24	Water	50-212
26	Xylene	14-122

FIG. A-7. Prandtl numbers $c\mu/k$ for liquids. (*Chilton, Colburn, and Vernon, personal communication, based mainly on data from "International Critical Tables."*) NOTE: To obtain $c\mu/k$ at a specified temperature for a given liquid, as usual a straight line is drawn from the specified temperature through the numbered point for the liquid to the $c\mu/k$ scale; to obtain $c\mu/k$ raised to one of the three fractional powers shown, a *horizontal* alignment is made from the $c\mu/k$ scale.

Table A-20. Viscosity of Steam[a]
[Lb/(hr)(ft)]

Absolute pressure, lb/sq in.	Saturated vapor	\multicolumn{7}{c}{Temperature, deg F}
		32	200	400	600	800	1000	1200
0	.....	0.023	0.031	0.041	0.050	0.059	0.067	0.074
500	0.055	.....	.....	.....	0.059	0.066	0.073	0.080
1000	0.070	.....	.....	.....	0.070	0.074	0.080	0.086
1500	0.081	.....	.....	.....	0.081	0.081	0.087	0.092
2000	0.094	.....	.....	.....	.....	0.092	0.094	0.097
2500	0.108	.....	.....	.....	.....	0.101	0.101	0.104
3000	0.116	.....	.....	.....	.....	0.110	0.108	0.110
3500	.....	.....	.....	.....	.....	0.119	0.114	0.116

[a] From J. H. Keenan and F. G. Keyes, "Thermodynamic Properties of Steam," 22d impression, Wiley, New York, 1950.

Table A-21. Efflux Viscosimeters

These instruments are designed to conform to the equation $\mu'/\rho' = A'\theta - B'/\theta$, where μ' is expressed in poises and ρ' in grams per cubic centimeter; θ is the time of efflux expressed in seconds. The ratio μ'/ρ' is called the **kinematic viscosity** and in the preceding equation is expressed in stokes. Such instruments should not be used outside the range of calibration.

	A'	B'
Saybolt Universal viscosimeter	0.0022	1.8
Redwood viscosimeter	0.0026	1.72
Redwood Admiralty viscosimeter	0.027	20.0
Engler viscosimeter	0.00147	3.74

Table A-22. Prandtl Numbers $c_p\mu/k$ for Gases and Vapor at 1 Atm and 212°F[a]

	$\dfrac{c_p\mu}{k}$	$\left(\dfrac{c_p\mu}{k}\right)^{0.3}$	$\left(\dfrac{c_p\mu}{k}\right)^{1/3}$	$\left(\dfrac{c_p\mu}{k}\right)^{0.4}$	$\left(\dfrac{c_p\mu}{k}\right)^{2/3}$
Air, hydrogen	0.69	0.894	0.884	0.866	0.781
Ammonia	0.86	0.956	0.951	0.941	0.904
Argon	0.66	0.883	0.871	0.847	0.759
Carbon dioxide, methane	0.75	0.917	0.909	0.891	0.826
Carbon monoxide	0.72	0.906	0.896	0.877	0.803
Helium	0.71	0.902	0.892	0.872	0.796
Nitric oxide, nitrous oxide	0.72	0.906	0.896	0.877	0.803
Nitrogen, oxygen	0.70	0.899	0.888	0.867	0.789
Steam (low pressure)	1.06	1.018	1.020	1.024	1.040

[a] From F. G. Keyes, *Tech. Rept.* 37, Project Squid (Apr. 1, 1952).

TABLE A-23. NORMAL TOTAL EMISSIVITY OF VARIOUS SURFACES
(Compiled by H. C. Hottel)

Surface	t, deg F[a]	Emissivity	Reference number
A. Metals and Their Oxides			
Aluminum:			
Highly polished plate, 98.3%, pure	440–1070	0.039–0.057	26
Polished	212	0.095	1
Rough polish	212	0.18	1
Rough plate	100	0.055–0.07	25
Commercial sheet	212	0.09	1
Oxidized at 1110°F	390–1110	0.11–0.19	23
Heavily oxidized	200–940	0.20–0.31	2
Aluminum oxide	530–930	0.63–0.42	21
Aluminum oxide	930–1520	0.42–0.26	21
Al-surfaced roofing	100	0.216	15
Aluminum alloys[b]			
Alloy 75 ST; A, B_1, C	75	0.11, 0.10, 0.08	36
Alloy 75 ST; A^c	450–900	0.22–0.16	36
Alloy 75 ST; B_1^c	450–800	0.20–0.18	36
Alloy 75 ST; C^c	450–930	0.22–0.15	36
Alloy 24 ST; A, B_1, C	75	0.09	36
Alloy 24 ST; A^c	450–910	0.17–0.15	36
Alloy 24 ST; B_1^c	450–940	0.20–0.16	36
Alloy 24 ST; C^c	450–860	0.16–0.13	36
Calorized surfaces, heated at 1110°F			
Copper	390–1110	0.18–0.19	23
Steel	390–1110	0.52–0.57	23
Brass:			
Highly polished			
73.2 Cu, 26.7 Zn	476–674	0.028–0.031	26
62.4 Cu, 36.8 Zn, 0.4 Pb, 0.3 Al	494–710	0.033–0.037	26
82.9 Cu, 17.0 Zn	530	0.030	26
Hard-rolled, polished, but direction of polishing visible	70	0.038	25
Hard-rolled, polished, but somewhat attacked	73	0.043	25
Hard-rolled, polished, but traces of stearin from polish left on	75	0.053	25
Polished	212	0.06	1
Polished	100–600	0.10	15
Rolled plate, natural surface	72	0.06	25
Rolled plate, rubbed with coarse emery	72	0.20	25
Dull plate	120–660	0.22	32
Oxidized by heating at 1110°F	390–1110	0.61–0.59	23
Chromium (see nickel alloys for Ni-Cr steels):			
Polished	100–2000	0.08–0.36	7–17
Polished	212	0.075	1

TABLE A-23. NORMAL TOTAL EMISSIVITY OF VARIOUS SURFACES.—(*Continued*)

Surface	t, deg F[a]	Emissivity	Reference number
Copper:			
Carefully polished electrolytic copper	176	0.018	16
Polished	242	0.023	34
Polished	212	0.052	1
Commercial emeried, polished, but pits remaining	66	0.030	25
Commercial, scraped shiny, but not mirror-like	72	0.072	25
Plate, heated long time, covered with thick oxide layer	77	0.78	25
Plate heated at 1110°F	390–1110	0.57	23
Cuprous oxide	1470–2010	0.66–0.54	4
Molten copper	1970–2330	0.16–0.13	4
Dow metal:[b]			
A; B_1; C	75	0.15, 0.15, 0.12	36
A^c	450–750	0.24–0.20	36
$B_1{}^c$	450–800	0.16	36
C^c	450–760	0.21–0.18	36
Gold:			
Pure, highly polished	440–1160	0.018–0.035	26
Inconel:[b]			
Types X and B; surface A, B_2, C	75	0.19–0.21	36
Type X; surface A^c	450–1620	0.55–0.78	36
Type X; surface $B_2{}^c$	450–1575	0.60–0.75	36
Type X; surface C^c	450–1650	0.62–0.73	36
Type B; surface A^c	450–1620	0.35–0.55	36
Type B; surface $B_2{}^c$	450–1740	0.32–0.51	36
Type B; surface C^c	450–1830	0.35–0.40	36
Iron and steel (not including stainless):			
Metallic surfaces (or very thin oxide layer)			
Electrolytic iron, highly polished	350–440	0.052–0.064	26
Steel, polished	212	0.066	1
Iron, polished	800–1880	0.14–0.38	27
Iron, roughly polished	212	0.17	1
Iron, freshly emeried	68	0.24	25
Cast iron, polished	392	0.21	23
Cast iron, newly turned	72	0.44	25
Cast iron, turned and heated	1620–1810	0.60–0.70	22
Wrought iron, highly polished	100–480	0.28	32
Polished steel casting	1420–1900	0.52–0.56	22
Ground sheet steel	1720–2010	0.55–0.61	22
Smooth sheet iron	1650–1900	0.55–0.60	22
Mild steel;[b] A, B_2, C	75	0.12, 0.15, 0.10	36
Mild steel;[b] A^c	450–1950	0.20–0.32	36
Mild steel;[b] $B_2{}^c$	450–1920	0.34–0.35	36
Mild steel;[b] C^c	450–1950	0.27–0.31	36

TABLE A-23. NORMAL TOTAL EMISSIVITY OF VARIOUS SURFACES.—(*Continued*)

Surface	t, deg F[a]	Emissivity	Reference number
Oxidized surfaces			
Iron plate, pickled, then rusted red	68	0.61	25
Iron plate, completely rusted	67	0.69	25
Iron, dark gray surface	212	0.31	1
Rolled sheet steel	70	0.66	25
Oxidized iron	212	0.74	28
Cast iron, oxidized at 1100°F	390–1110	0.64–0.78	23
Steel, oxidized at 1100°F	390–1110	0.79	23
Smooth oxidized electrolytic iron	260–980	0.78–0.82	26
Iron oxide	930–2190	0.85–0.89	6
Rough ingot iron	1700–2040	0.87–0.95	22
Sheet steel			
Strong, rough oxide layer	75	0.80	25
Dense, shiny oxide layer	75	0.82	25
Cast plate, smooth	73	0.80	25
Cast plate, rough	73	0.82	25
Cast iron, rough, strongly oxidized	100–480	0.95	32
Wrought iron, dull oxidized	70–680	0.94	32
Steel plate, rough	100–700	0.94–0.97	15
Molten surfaces			
Cast iron	2370–2550	0.29	31
Mild steel	2910–3270	0.28	31
Steel, several different kinds with 0.25–1.2% C (slightly oxidized surface)	2840–3110	0.27–0.39	3
Steel	2730–3000	0.42–0.53	14
Steel	2770–3000	0.43–0.40	18
Pure iron	2760–3220	0.42–0.45	8
Armco iron	2770–3070	0.40–0.41	18
Lead:			
Pure (99.96%), unoxidized	260–440	0.057–0.075	26
Gray oxidized	75	0.28	25
Oxidized at 300°F	390	0.63	23
Magnesium:			
Magnesium oxide	530–1520	0.55–0.20	21
Magnesium oxide	1650–3100	0.20	10
Mercury	32–212	0.09–0.12	11
Molybdenum:			
Filament	1340–4700	0.096–0.202	37
Massive, polished	212	0.071	1
Monel metal:[b]			
Oxidized at 1110°F	390–1110	0.41–0.46	23
K Monel 5700; A, B_2, C	75	0.23, 0.17, 0.14	36
K Monel 5700; A^c	450–1610	0.46–0.65	36
K Monel 5700; $B_2{}^c$	450–1750	0.54–0.77	36
K Monel 5700; C^c	450–1785	0.35–0.53	36

TABLE A-23. NORMAL TOTAL EMISSIVITY OF VARIOUS SURFACES.—(*Continued*)

Surface	t, deg F[a]	Emissivity	Reference number
Nickel:			
Electroplated, polished	74	0.045	25
Technically pure (98.9% Ni, +Mn), polished	440–710	0.07–0.087	26
Polished	212	0.072	1
Electroplated, not polished	68	0.11	25
Wire	368–1844	0.096–0.186	29
Plate, oxidized by heating at 1110°F	390–1110	0.37–0.48	23
Nickel oxide	1200–2290	0.59–0.86	5
Nickel alloys:			
Chromnickel	125–1894	0.64–0.76	29
Copper-nickel, polished	212	0.059	1
Nichrome wire, bright	120–1830	0.65–0.79	30
Nichrome wire, oxidized	120–930	0.95–0.98	30
Nickel-silver, polished	212	0.135	1
Nickelin (18–32 Ni; 55–68 Cu; 20 Zn), gray oxidized	70	0.262	25
Type ACI-HW (60 Ni; 12 Cr)			
Smooth, black, firm adhesive oxide coat from service	520–1045	0.89–0.82	24
Platinum:			
Pure, polished plate	440–1160	0.054–0.104	26
Strip	1700–2960	0.12–0.17	11
Filament	80–2240	0.036–0.192	9
Wire	440–2510	0.073–0.182	13
Silver:			
Polished, pure	440–1160	0.020–0.032	26
Polished	100–700	0.022–0.031	15
Polished	212	0.052	1
Stainless steels:[b]			
Polished	212	0.074	1
Type 301; A, B_2, C	75	0.21, 0.27, 0.16	36
Type 301; A^c	450–1740	0.57–0.55	36
Type 301; $B_2{}^c$	450–1725	0.54–0.63	36
Type 301; C^c	450–1650	0.51–0.70	36
Type 316; A, B_2, C	75	0.28, 0.28, 0.17	36
Type 316; A^c	450–1600	0.57–0.66	36
Type 316; $B_2{}^c$	450–1920	0.52–0.50	36
Type 316; C^c	450–1920	0.26–0.31	36
Type 347; A, B_2, C	75	0.39, 0.35, 0.17	36
Type 347; A^c	450–1650	0.52–0.65	36
Type 347; $B_2{}^c$	450–1610	0.51–0.65	36
Type 347; C^c	450–1650	0.49–0.64	36
Type 304 (8 Cr; 18 Ni)			
Light silvery, rough, brown, after heating	420–914	0.44–0.36	24
After 42 hr heating at 980°F	420–980	0.62–0.73	24

TABLE A-23. NORMAL TOTAL EMISSIVITY OF VARIOUS SURFACES.—(*Continued*)

Surface	t, deg F[a]	Emissivity	Reference number
Type 310 (25 Cr; 20 Ni)			
Brown, splotched, oxidized from furnace service	420–980	0.90–0.97	24
Allegheny metal No. 4, polished	212	0.13	1
Allegheny alloy No. 66, polished	212	0.11	1
Tantalum filament	2420–5430	0.19–0.31	37
Thorium oxide	530–930	0.58–0.36	21
Thorium oxide	930–1520	0.36–0.21	21
Tin:			
Bright tinned iron	76	0.043 and 0.064	25
Bright	122	0.06	30
Commercial tin-plated sheet iron	212	0.07, 0.08	1
Tungsten:			
Filament, aged	80–6000	0.032–0.35	12
Filament	6000	0.39	38
Polished coat	212	0.066	1
Zinc:			
Commercial 99.1% pure, polished	440–620	0.045–0.053	26
Oxidized by heating at 750°F	750	0.11	23
Galvanized sheet iron, fairly bright	82	0.23	25
Galvanized sheet iron, gray oxidized	75	0.28	25
Zinc, galvanized sheet	212	0.21	1

B. Refractories, Building Materials, Paints, and Miscellaneous

Surface	t, deg F[a]	Emissivity	Reference number
Alumina (99.5–85 Al_2O_3; 0–12 SiO_2; 0–1 Fe_2O_3). Effect of mean grain size, microns (μ)	1850–2850		20
10 μ		0.30–0.18	
50 μ		0.39–0.28	
100 μ		0.50–0.40	
Alumina-silica (showing effect of Fe)	1850–2850		20
80–58 Al_2O_3; 16–38 SiO_2; 0.4 Fe_2O_3		0.61–0.43	
36–26 Al_2O_3; 50–60 SiO_2; 1.7 Fe_2O_3		0.73–0.62	
61 Al_2O_3; 35 SiO_2; 2.9 Fe_2O_3		0.78–0.68	
Asbestos:			
Board	74	0.96	25
Paper	100–700	0.93–0.94	15
Brick[d]			
Red, rough, but no gross irregularities	70	0.93	25
Grog brick, glazed	2012	0.75	22
Building	1832	0.45	30
Fireclay	1832	0.75	30
Carbon:			
T-carbon (Gebrüder Siemens) 0.9% ash. This started with emissivity at 260°F of 0.72, but on heating changed to values given	260–1160	0.81–0.79	26

TABLE A-23. NORMAL TOTAL EMISSIVITY OF VARIOUS SURFACES.—(Continued)

Surface	t, deg F^a	Emissivity	Reference number
Filament	1900–2560	0.526	19
Rough plate	212–608	0.77	1
Rough plate	608–932	0.77–0.72	1
Graphitized	212–608	0.76–0.75	1
Graphitized	608–932	0.75–0.71	1
Candle soot	206–520	0.952	33
Lampblack-waterglass coating	209–440	0.96–0.95	16, 26
Thin layer of same on iron plate	69	0.927	25
Thick coat of same	68	0.967	25
Lampblack, 0.003 in. or thicker	100–700	0.945	15
Lampblack, rough deposit	212–932	0.84–0.78	1
Lampblack, other blacks	122–1832	0.96	30
Graphite, pressed, filed surface	480–950	0.98	21
Carborundum (87 SiC; density 2.3)	1850–2550	0.92–0.82	20
Concrete tiles	1832	0.63	30
Enamel, white fused, on iron	66	0.90	25
Glass:			
Smooth	72	0.94	25
Pyrex, lead, and soda	500–1000	Ca 0.95–0.85	21
Gypsum, 0.02 in. thick on smooth or blackened plate	70	0.903	25
Magnesite refractory brick	1832	0.38	30
Marble, light gray, polished	72	0.93	25
Oak, planed	70	0.90	25
Oil layers on polished nickel (lubricating oil)	68	....	25
Polished surface, alone		0.045	
+0.001, 0.002, 0.005 in oil		0.27, 0.46, 0.72	
Thick oil layer		0.82	
Oil layers on aluminum foil (linseed oil)		....	28
Aluminum foil	212	0.087	
+1, 2 coats oil	212	0.561, 0.574	
Paints, lacquers, varnishes:			
Snow-white enamel varnish on rough iron plate	73	0.906	25
Black shiny lacquer, sprayed on iron	76	0.875	25
Black shiny shellac on tinned iron sheet	70	0.821	25
Black matte shellac	170–295	0.91	35
Black or white lacquer	100–200	0.80–0.95	15
Flat black lacquer	100–200	0.96–0.98	15
Oil paints, 16 different, all colors	212	0.92–0.96	28
Aluminum paints and lacquers:			
10% Al, 22% lacquer body, on rough or smooth surface	212	0.52	28
Other Al paints, varying age and Al content	212	0.27–0.67	28
Al lacquer, varnish binder, on rough plate	70	0.39	25

TABLE A-23. NORMAL TOTAL EMISSIVITY OF VARIOUS SURFACES.—(Continued)

Surface	t, deg F[a]	Emissivity	Reference number
Al paint, after heating to 620°F	300–600	0.35	26
Radiator paint; white, cream, bleach	212	0.79, 0.77, 0.84	1
Radiator paint, bronze	212	0.51	1
Lacquer coatings, 0.001–0.015 in. thick on aluminum alloys[c]	100–300	0.87 to 0.97	36
Clear silicone vehicle coatings, 0.001–0.015 in. thick:[c]			
On mild steel	500	0.66	36
On stainless steels, 316, 301, 347	500	0.68, 0.75, 0.75	36
On Dow metal	500	0.74	36
On Al alloys 24 ST, 75 ST	500	0.77, 0.82	36
Aluminum paint with silicone vehicle, two coats on Inconel[c]	500	0.29	36
Paper, thin, pasted on tinned or blackened plate	66	0.92, 0.94	25
Plaster, rough lime	50–190	0.91	32
Porcelain, glazed	72	0.92	25
Quartz:			
Rough, fused	70	0.93	25
Glass, 1.98 mm thick	540–1540	0.90–0.41	21
Glass, 6.88 mm. thick	540–1540	0.93–0.47	21
Opaque	570–1540	0.92–0.68	21
Roofing paper	69	0.91	25
Rubber:			
Hard, glossy plate	74	0.94	25
Soft, gray, rough (reclaimed)	76	0.86	25
Serpentine, polished	74	0.90	25
Silica (98 SiO_2; Fe-free), effect of grain size, microns (μ)	1850–2850		20
10 μ		0.42–0.33	
70–600 μ		0.62–0.46	
(See also Alumina-silica and quartz)			
Water	32–212	0.95–0.963	
Zirconium silicate	460–930	0.92–0.80	21
Zirconium silicate	930–1530	0.80–0.52	21

[a] When temperatures and emissivities appear in pairs separated by dashes, they correspond; and linear interpolation is permissible.

[b] Identification of surface treatment: surface A, cleaned with toluene, then methanol; B_1, cleaned with soap and water, toluene, and methanol in succession; B_2, cleaned with abrasive soap and water, toluene, and methanol; C, polished on buffing wheel to mirror surface, cleaned with soap and water.

[c] Results after repeated heating and cooling.

[d] See also under material type.

[e] Calculated from spectral data.

TABLE A-23. NORMAL TOTAL EMISSIVITY OF VARIOUS SURFACES.—(*Continued*)

References:
1. B. T. Barnes, W. E. Forsythe, and E. Q. Adams, *J. Opt. Soc. Am.*, **37** (10), pp. 804–807 (1947).
2. E. R. Binkley, private communication, 1933.
3. J. E. Bacon and J. W. James, "Proceedings of the General Discussion on Heat Transfer," pp. 117–121, Institution of Mechanical Engineers, London, and American Society of Mechanical Engineers, New York, 1952.
4. G. K. Burgess, *Natl. Bur. Standards, Bull.* 6, *Sci. Paper* 121, 111 (1909).
5. G. K. Burgess and P. D. Foote, *Natl. Bur. Standards, Bull.* 11, *Sci. Paper* 224, 41–64 (1914).
6. G. K. Burgess and P. D. Foote, *Natl. Bur. Standards, Bull.* 12, *Sci. Paper* 249, 83–89 (1915).
7. W. H. Coblentz, *Natl. Bur. Standards, Bull.* 7, 197 (1911).
8. M. N. Dastar and N. A. Gokcen, *J. Metals*, **1** (10), trans. 665–667 (1949).
9. C. Davisson and J. R. Weeks, Jr., *J. Opt. Soc. Am.*, **8**, 581–606 (1924).
10. C. Féry, *Ann. phys. chim.*, **27**, 433 (1902).
11. P. D. Foote, *Natl. Bur. Standards, Bull.* 11, *Sci. Paper* 243, 607 (1914); *J. Wash. Acad. Sci.*, **5**, 1 (1914).
12. W. E. Forsythe and A. G. Worthing, *Astrophys. J.*, **61**, 146–185 (1925).
13. W. Geiss, *Physica*, **5**, 203 (1925).
14. G. N. Goller, *Trans. Am. Soc. Metals*, **32**, 239 (1944).
15. R. H. Heilman, *Trans. ASME*, **FSP51**, 287–304 (1929).
16. K. Hoffmann, *Z. Physik*, **14**, 310 (1923).
17. E. O. Hulbert, *Astrophys. J.*, **42**, 205 (1915).
18. D. Knowles and R. J. Sarjant, *J. Iron Steel Inst. (London)*, **155**, 577 (1947).
19. O. Lummer, *Elektrotech. Z.*, **34**, 1428 (1913).
20. M. Michaud, Sc.D. Thesis, University of Paris, 1951.
21. M. Pirani, *J. Sci. Instr.*, **16**, (12) (1939).
22. V. Polak, *Z. tech. Physik*, **8**, 307 (1927).
23. C. F. Randolph and M. J. Overholtzer, *Phys. Rev.*, **2**, 144 (1913).
24. H. S. Rice, Chemical Engineering Thesis, Massachusetts Institute of Technology, 1931.
25. E. Schmidt, *Gesundh.-Ing.*, Beiheft 20, Reihe 1, 1–23 (1927).
26. H. Schmidt and E. Furthmann, *Mitt. Kaiser-Wilhelm-Inst. Eisenforsch. Düsseldorf, Abhandl.*, **109**, 225 (1928).
27. F. D. Snell, *Ind. Eng. Chem.*, **29**, 89–91 (1937).
28. Standard Oil Development Company, personal communication, 1928.
29. V. A. Suydam, *Phys. Rev.*, (2)**5**, 497–509 (1915).
30. M. W. Thring, "The Science of Flames & Furnaces," Chapman & Hall, London, 1952.
31. C. B. Thwing, *Phys. Rev.*, **26**, 190 (1908).
32. F. Wamsler, *Z. Ver. deut. Ing.*, **55**, 599–605 (1911); *Mitt. Forsch.*, **98**, 1–45 (1911).
33. M. Wenzl and F. Morawe, *Stahl u. Eisen*, **47**, 867–871 (1927).
34. W. Westphal, *Verhandl. deut. physik. Ges.*, **10**, 987–1012 (1912)
35. W. Westphal, *Verhandl. deut. physik. Ges.*, **11**, 897–902 (1913).
36. G. B. Wilkes, Final Report on Contract No. W33-038-20486, Air Materiel Command; Wright Field, Dayton, Ohio, DIC Report, Massachusetts Institute of Technology (1950).
37. A. G. Worthing, *Phys. Rev.*, **28**, 190 (1926).
38. C. Zwikker, *Arch. néerland. sci.*, **9** (Pt. IIIA), 207 (1925).

HEAT TRANSMISSION

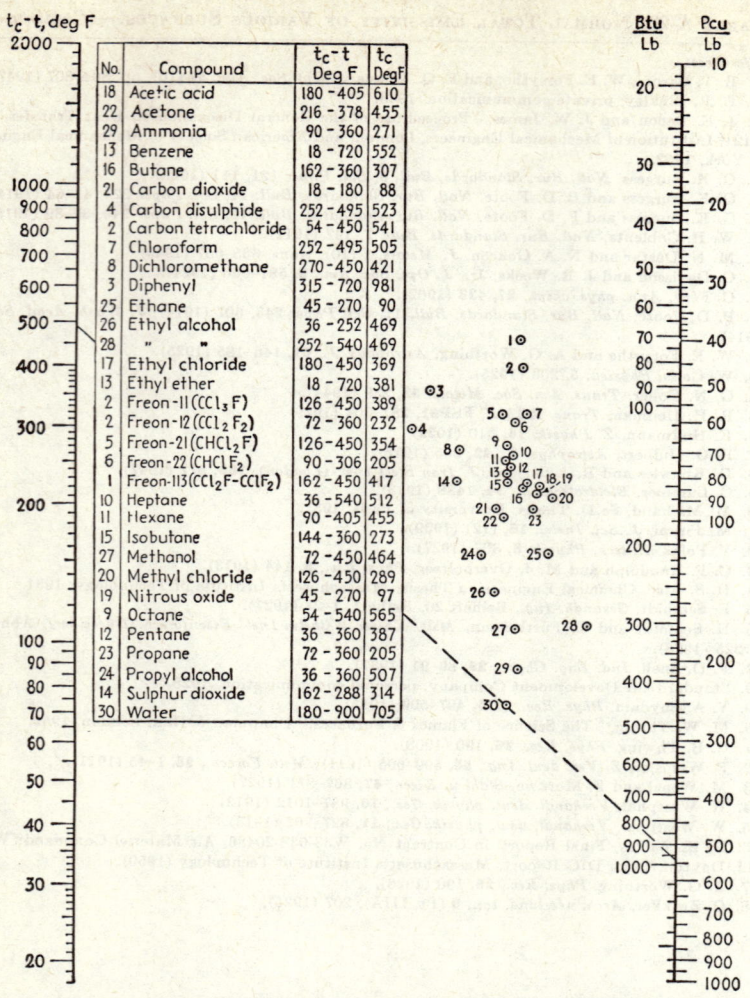

No.	Compound	t_c-t Deg F	t_c Deg F
18	Acetic acid	180-405	610
22	Acetone	216-378	455
29	Ammonia	90-360	271
13	Benzene	18-720	552
16	Butane	162-360	307
21	Carbon dioxide	18-180	88
4	Carbon disulphide	252-495	523
2	Carbon tetrachloride	54-450	541
7	Chloroform	252-495	505
8	Dichloromethane	270-450	421
3	Diphenyl	315-720	981
25	Ethane	45-270	90
26	Ethyl alcohol	36-252	469
28	"	252-540	469
17	Ethyl chloride	180-450	369
13	Ethyl ether	18-720	381
2	Freon-11 (CCl_3F)	126-450	389
2	Freon-12 (CCl_2F_2)	72-450	232
5	Freon-21 ($CHCl_2F$)	126-450	354
6	Freon-22 ($CHClF_2$)	90-306	205
1	Freon-113 ($CCl_2F-CClF_2$)	162-450	417
10	Heptane	36-540	512
11	Hexane	90-405	455
15	Isobutane	144-360	273
27	Methanol	72-450	464
20	Methyl chloride	126-450	289
19	Nitrous oxide	45-270	97
9	Octane	54-540	565
12	Pentane	36-360	387
23	Propane	72-360	205
24	Propyl alcohol	36-360	507
14	Sulphur dioxide	162-288	314
30	Water	180-900	705

FIG. A-8. Latent heats of vaporization. For water at 212°F, $t_c - t = 705 - 212 = 493$, and the latent heat of vaporization is 970 Btu/lb. (*Chilton, Colburn, and Vernon, personal communication, based mainly on data from "International Critical Tables."*)

A proposed general correlation of latent heats given by V. A. Klein, *Chem. Eng. Progr.*, **45**, 675 (1949), is as follows:

$$\Delta H = \frac{1.987 T \ln P_R}{T_R - 1} \left(1 - \frac{P_R}{T_R^3}\right)^{1/2}$$

where ΔH is the latent heat of vaporization in Btu per lb mole, T is the absolute temperature in degrees Rankine, and T_R and P_R are the dimensionless reduced temperatures and pressure, respectively.

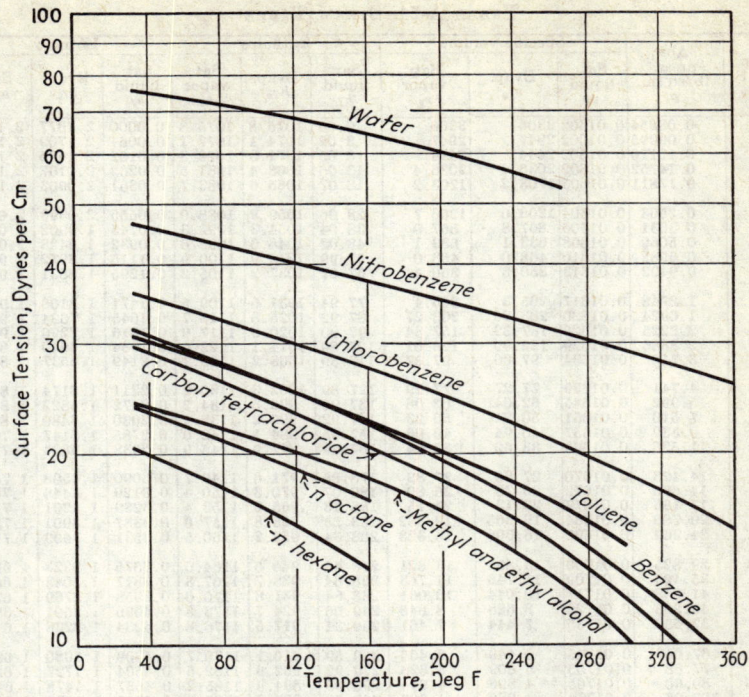

FIG. A-9. Surface tension of liquids vs. temperature. (*Based mainly on data from "International Critical Tables."*)

TABLE A-24. STEAM TABLE[a]

Temp., deg F t	Abs press. lb/sq in. p	Specific volume Sat. liquid v_f	Specific volume Evap. v_{fg}	Specific volume Sat. vapor v_g	Enthalpy Sat. liquid h_f	Enthalpy Evap. h_{fg}	Enthalpy Sat. vapor h_g	Entropy Sat. liquid s_f	Entropy Evap. s_{fg}	Entropy Sat. vapor
32	0.08854	0.01602	3306	3306	0.00	1075.8	1075.8	0.0000	2.1877	2.1877
35	0.09995	0.01602	2947	2947	3.02	1074.1	1077.1	0.0061	2.1709	2.1770
40	0.12170	0.01602	2444	2444	8.05	1071.3	1079.3	0.0162	2.1435	2.1597
45	0.14752	0.01602	2036.4	2036.4	13.06	1068.4	1081.5	0.0262	2.1167	2.1429
50	0.17811	0.01603	1703.2	1703.2	18.07	1065.6	1083.7	0.0361	2.0903	2.1264
60	0.2563	0.01604	1206.6	1206.7	28.06	1059.9	1088.0	0.0555	2.0393	2.0948
70	0.3631	0.01606	867.8	867.9	38.04	1054.3	1092.3	0.0745	1.9902	2.0647
80	0.5069	0.01608	633.1	633.1	48.02	1048.6	1096.6	0.0932	1.9428	2.0360
90	0.6982	0.01610	468.0	468.0	57.99	1042.9	1100.9	0.1115	1.8972	2.0087
100	0.9492	0.01613	350.3	350.4	67.97	1037.2	1105.2	0.1295	1.8531	1.9826
110	1.2748	0.01617	265.3	265.4	77.94	1031.6	1109.5	0.1471	1.8106	1.9577
120	1.6924	0.01620	203.25	203.27	87.92	1025.8	1113.7	0.1645	1.7694	1.9339
130	2.2225	0.01625	157.32	157.34	97.90	1020.0	1117.9	0.1816	1.7296	1.9112
140	2.8886	0.01629	122.99	123.01	107.89	1014.1	1122.0	0.1984	1.6910	1.8894
150	3.718	0.01634	97.06	97.07	117.89	1008.2	1126.1	0.2149	1.6537	1.8685
160	4.741	0.01639	77.27	77.29	127.89	1002.3	1130.2	0.2311	1.6174	1.8485
170	5.992	0.01645	62.04	62.06	137.90	996.3	1134.2	0.2472	1.5822	1.8293
180	7.510	0.01651	50.21	50.23	147.92	990.2	1138.1	0.2630	1.5480	1.8109
190	9.339	0.01657	40.94	40.96	157.95	984.1	1142.0	0.2785	1.5147	1.7932
200	11.526	0.01663	33.62	33.64	167.99	977.9	1145.9	0.2938	1.4824	1.7762
210	14.123	0.01670	27.80	27.82	178.05	971.6	1149.7	0.3090	1.4508	1.7598
212	14.696	0.01672	26.78	26.80	180.07	970.3	1150.4	0.3120	1.4446	1.7566
220	17.186	0.01677	23.13	23.15	188.13	965.2	1153.4	0.3239	1.4201	1.7440
230	20.780	0.01684	19.365	19.382	198.23	958.8	1157.0	0.3387	1.3901	1.7288
240	24.969	0.01692	16.306	16.323	208.34	952.2	1160.5	0.3531	1.3609	1.7140
250	29.825	0.01700	13.804	13.821	218.48	945.5	1164.0	0.3675	1.3323	1.6998
260	35.429	0.01709	11.746	11.763	228.64	938.7	1167.3	0.3817	1.3043	1.6860
270	41.858	0.01717	10.044	10.061	238.84	931.8	1170.6	0.3958	1.2769	1.6727
280	49.203	0.01726	8.628	8.645	249.06	924.7	1173.8	0.4096	1.2501	1.6597
290	57.556	0.01735	7.444	7.461	259.31	917.5	1176.8	0.4234	1.2238	1.6472
300	67.013	0.01745	6.449	6.466	269.59	910.1	1179.7	0.4369	1.1980	1.6350
310	77.68	0.01755	5.609	5.626	279.92	902.6	1182.5	0.4504	1.1727	1.6231
320	89.66	0.01765	4.896	4.914	290.28	894.9	1185.2	0.4637	1.1478	1.6115
330	103.06	0.01776	4.289	4.307	300.68	887.0	1187.7	0.4769	1.1233	1.6002
340	118.01	0.01787	3.770	3.788	311.13	879.0	1190.1	0.4900	1.0992	1.5891
350	134.63	0.01799	3.324	3.342	321.63	870.7	1192.3	0.5029	1.0754	1.5783
360	153.04	0.01811	2.939	2.957	332.18	862.2	1194.4	0.5158	1.0519	1.5677
370	173.37	0.01823	2.606	2.625	342.79	853.5	1196.3	0.5286	1.0287	1.5573
380	195.77	0.01836	2.317	2.335	353.45	844.6	1198.1	0.5413	1.0059	1.5471
390	220.37	0.01850	2.0651	2.0836	364.17	835.4	1199.6	0.5539	0.9832	1.5371
400	247.31	0.01864	1.8447	1.8633	374.97	826.0	1201.0	0.5664	0.9608	1.5272
410	276.75	0.01878	1.6512	1.6700	385.83	816.3	1202.1	0.5788	0.9386	1.5174
420	308.83	0.01894	1.4811	1.5000	396.77	806.3	1203.1	0.5912	0.9166	1.5078
430	343.72	0.01910	1.3308	1.3499	407.79	796.0	1203.8	0.6035	0.8947	1.4982
440	381.59	0.01926	1.1979	1.2171	418.90	785.4	1204.3	0.6158	0.8730	1.4887
450	422.6	0.0194	1.0799	1.0993	430.1	774.5	1204.6	0.6280	0.8513	1.4793
460	466.9	0.0196	0.9748	0.9944	441.4	763.2	1204.6	0.6402	0.8298	1.4700
470	514.7	0.0198	0.8811	0.9009	452.8	751.5	1204.3	0.6523	0.8083	1.4606
480	566.1	0.0200	0.7972	0.8172	464.4	739.4	1203.7	0.6645	0.7868	1.4513
490	621.4	0.0202	0.7221	0.7423	476.0	726.8	1202.8	0.6766	0.7653	1.4419
500	680.8	0.0204	0.6545	0.6749	487.8	713.9	1201.7	0.6887	0.7438	1.4325
520	812.4	0.0209	0.5385	0.5594	511.9	686.4	1198.2	0.7130	0.7006	1.4136
540	962.5	0.0215	0.4434	0.4649	536.6	656.6	1193.2	0.7374	0.6568	1.3942
560	1133.1	0.0221	0.3647	0.3868	562.2	624.2	1186.4	0.7621	0.6121	1.3742
580	1325.8	0.0228	0.2989	0.3217	588.9	588.4	1177.3	0.7872	0.5659	1.3532
600	1542.9	0.0236	0.2432	0.2668	617.0	548.5	1165.5	0.8131	0.5176	1.3307
620	1786.6	0.0247	0.1955	0.2201	646.7	503.6	1150.3	0.8398	0.4664	1.3062
640	2059.7	0.0260	0.1538	0.1798	678.6	452.0	1130.5	0.8679	0.4110	1.2789
660	2365.4	0.0278	0.1165	0.1442	714.2	390.2	1104.4	0.8987	0.3485	1.2472
680	2708.1	0.0305	0.0810	0.1115	757.3	309.9	1067.2	0.9351	0.2719	1.2071
700	3093.7	0.0369	0.0392	0.0761	823.3	172.1	995.4	0.9905	0.1484	1.1389
705.4	3206.2	0.0503	0	0.0503	902.7	0	902.7	1.0580	0	1.0580

[a] Reprinted from abridged edition of "Thermodynamic Properties of Steam," by Joseph H. Keenan and Frederick G. Keyes, John Wiley & Sons, Inc., New York, 1937, with the permission of the authors and publisher.

TABLE A-25. SOME PHYSICAL PROPERTIES OF AIR[a]

Temperature, deg F	c_p, Btu/(lb)(deg F)	μ, lb/(ft)(hr)	k, Btu/(hr)(sq ft)(deg F per ft)	N_{Pr}	$Y = \dfrac{\rho^2 \beta g c_p}{\mu k}$, $\dfrac{1}{(\text{cu ft})(\text{deg F})}$
−300	0.2393	0.0153	0.00478	0.77	514×10^6
−200	0.2393	0.0242	0.00763	0.76	47.3×10^6
−100	0.2393	0.0322	0.0104	0.74	9.82×10^6
−50	0.2394	0.0360	0.0118	0.73	5.25×10^6
32	0.2396	0.0417	0.0140	0.72	2.21×10^6
100	0.2401	0.0460	0.0157	0.71	1.21×10^6
200	0.2412	0.0521	0.0181	0.69	0.570×10^6
300	0.2421	0.0578	0.0204	0.69	0.300×10^6
400	0.2446	0.0630	0.0225	0.68	0.174×10^6
500	0.2474	0.0679	0.0246	0.68	0.107×10^6
600	0.2504	0.0726	0.0266	0.68	6.98×10^4
700	0.2535	0.0770	0.0285	0.68	4.74×10^4
800	0.2566	0.0813	0.0303	0.69	3.34×10^4
900	0.2598	0.0854	0.0320	0.69	2.42×10^4
1000	0.2630	0.0892	0.0337	0.69	1.80×10^4
1200	0.2687	0.0967	0.0369	0.70	1.05×10^4

[a] From F. G. Keyes, *Tech. Rept.* 37, Project Squid (Apr. 1, 1952).

TABLE A-26. ENTHALPY OF SATURATED AIR[a]

Basis: Dry air at 0°F, 29.921 in. Hg. Water as liquid, 32°F.
(Expressed in Btu per lb. of dry air content)

Deg F	0.0	2.0	4.0	6.0	8.0
0	0.835	1.408	1.991	2.583	3.188
10	3.803	4.432	5.076	5.735	6.412
20	7.106	7.820	8.557	9.317	10.103
30	10.915	11.758	12.585	13.438	14.319
40	15.230	16.172	17.149	18.161	19.211
50	20.301	21.436	22.615	23.84	25.12
60	26.46	27.85	29.31	30.83	32.42
70	34.09	35.83	37.66	39.57	41.58
80	43.69	45.90	48.22	50.66	53.23
90	55.93	58.78	61.77	64.92	68.23
100	71.73	75.42	79.31	83.42	87.76
110	92.34	97.18	102.31	107.73	113.46
120	119.54	125.98	132.8	140.1	147.8
130	155.9	164.7	174.0	183.9	194.4
140	205.7	217.7	230.6	244.4	259.3

[a] From data compiled by John A. Goff and S. Gratch in "Heating Ventilating Air Conditioning Guide," 31st ed., 1953. Reprinted by permission.

TABLE A-27. SOME PROPERTIES OF SATURATED LIQUID WATER[a]

Deg F	μ, lb/(hr)(ft)	k, Btu/(hr)(ft)(deg F)	v, cu ft/lb	$\phi = \left(\dfrac{k^3\rho^2 g}{\mu^2}\right)^{1/3}$
32	4.34	0.317	0.01602	1400
40	3.73	0.324	0.01602	1590
50	3.17	0.331	0.01603	1800
60	2.71	0.338	0.01604	2040
70	2.36	0.345	0.01606	2290
80	2.08	0.351	0.01608	2520
90	1.85	0.357	0.01610	2780
100	1.66	0.363	0.01613	3040
120	1.36	0.372	0.01620	3530
140	1.14	0.379	0.01629	4040
160	0.970	0.385	0.01639	4550
180	0.840	0.390	0.01651	5050
200	0.738	0.393	0.01663	5520
220	0.654	0.395	0.01677	5970
240	0.585	0.396	0.01692	6420
260	0.528	0.396	0.01709	6820
280	0.482	0.396	0.01726	7210
300	0.448	0.395	0.01745	7490
350	0.38	0.390	0.01799	8090
400	0.33	0.380	0.01864	8450
450	0.29	0.365	0.0194	8630
500	0.26	0.348	0.0204	8550
550	0.23	0.324	0.0218	8260
600	0.21	0.291	0.0236	7480

[a] Values of μ up to 300°F are from the "International Critical Tables," McGraw-Hill, New York, 1929; values of μ over 300°F are based on the equation of G. A. Hawkins, W. L. Sibbitt, and H. L. Solberg, *Trans. ASME*, **70**, 19 (1948).

Values of k are based on data of D. L. Timrot and N. Vargaftik, *J. Tech. Phys.*, *USSR*, **10**, 1063 (1940).

Values of v are from J. H. Keenan and F. G. Keyes, "Thermodynamic Properties of Steam," Wiley, 22d impression, New York, 1950.

In equations for film-type condensation, and in other cases, the term $\phi = (k^3\rho^2 g/\mu^2)^{1/3}$ appears. The above table gives values of t, μ, v, k, and ϕ expressed in units involving feet, pounds, hours, degrees Fahrenheit, and Btu; ϕ is expressed in Btu/(hr)(sq ft)(deg F).

TABLE A-28. PROPERTIES OF MOLTEN METALS[a]

Metal and melting point	Temperature, deg F	k, $\dfrac{\text{Btu}}{(\text{hr})(\text{ft})(\text{deg F})}$	ρ, $\dfrac{\text{lb}}{\text{cu ft}}$	c_p, $\dfrac{\text{Btu}}{(\text{lb})(\text{deg F})}$	μ, $\dfrac{\text{lb}}{(\text{ft})(\text{hr})}$
Bismuth............	600	9.5	625	0.0345	3.92
(520°F)	1000	9.0	608	0.0369	2.66
	1400	9.0	591	0.0393	1.91
Lead...............	700	10.5	658	0.038	5.80
(621°F)	900	11.4	650	0.037	4.65
	1300	.....	633	.....	3.31
Mercury............	50	4.7	847	0.033	3.85
(−38°F)	300	6.7	826	0.033	2.66
	600	8.1	802	0.032	2.09
Potassium..........	300	26.0	50.4	0.19	0.90
(147°F)	800	22.8	46.3	0.18	0.43
	1300	19.1	42.1	0.18	0.31
Sodium.............	200	49.8	58.0	0.33	1.69
(208°F)	700	41.8	53.7	0.31	0.68
	1300	34.5	48.6	0.30	0.43
Na, 56 wt %........	200	14.8	55.4	0.270	1.40
K, 44 wt %	700	15.9	51.3	0.252	0.570
(66.2°F)	1300	16.7	46.2	0.249	0.389
Na, 22 wt %........	200	14.1	53.0	0.226	1.19
K, 78 wt %	750	15.4	48.4	0.210	0.500
(12°F)	1400	.....	43.1	0.211	0.353
Pb, 44.5 wt %......	300	5.23	657	0.035	
Bi, 55.5 wt %	700	6.85	639	0.035	3.71
(257°F)	1200	.....	614	.....	2.78

[a] Based largely on the "Liquid-Metals Handbook," 2d ed., United States Government Printing Office, Washington, D.C., 1952.

Table A-29. Standard Atmosphere[a]

Altitude, ft	Temperature, deg F	Pressure ratio,[a] $(p/p_0) \times 10^5$	Density ratio[a] $(\rho/\rho_0) \times 10^5$
0	59.0	100,000	100,000
5,000	41.2	83,200	86,160
10,000	23.3	68,760	73,840
15,000	5.5	56,420	62,910
20,000	−12.3	45,940	53,270
25,000	−30.2	37,090	44,800
30,000	−48.0	29,680	37,400
35,000	−65.8	23,520	30,980
40,000	−67.0	18,520	24,470
50,000	−67.0	11,490	15,170
60,000	−67.0	7,125	9,413
70,000	−67.0	4,419	5,838
80,000	−67.0	2,741	3,621
90,000	−67.0	1,699	2,245
100,000	−67.0	1,054	1,392
140,000	73.9	199.1	193.5
180,000	170.6	56.98	46.89
220,000	93.0	16.56	15.54
260,000	−27.4	3.507	4.208
300,000	28.0	0.7228	0.6865
340,000	108.1	0.2153	0.1626
380,000	188.2	0.07751	0.05130

[a] Prepared from values in L. S. Marks, "Mechanical Engineers' Handbook," McGraw-Hill, 5th ed., 1951. NOTE: p_0 = 14.696 lb/sq in., ρ_0 = 0.07651 lb/cu ft.

TABLE A-30. STANDARD CONDENSER-TUBE DATA[a]

Outside diameter, in.	Size number, BWG	Wt per ft, lb[b]	Thickness, in.	Inside diameter, in.	Surface, sq ft per ft of length Outside	Surface, sq ft per ft of length Inside	Inside sectional area, sq in.	Velocity, ft/sec for 1 U.S. gal/min	Capacity at 1 ft/sec velocity U.S. gal/min	Capacity at 1 ft/sec velocity Lb water/hr
1/2	12	0.493	0.109	0.282	0.1309	0.0748	0.0624	5.142	0.1945	97.25
	14	0.403	0.083	0.334	0.1309	0.0874	0.0876	3.662	0.2730	136.5
	16	0.329	0.065	0.370	0.1309	0.0969	0.1076	2.981	0.3352	167.5
	18	0.258	0.049	0.402	0.1309	0.1052	0.1269	2.530	0.3952	197.6
	20	0.190	0.035	0.430	0.1309	0.1125	0.1452	2.209	0.4528	226.4
5/8	12	0.656	0.109	0.407	0.1636	0.1066	0.1301	2.468	0.4053	202.7
	14	0.526	0.083	0.459	0.1636	0.1202	0.1655	1.939	0.5157	258.9
	16	0.424	0.065	0.495	0.1636	0.1296	0.1925	1.667	0.5999	300.0
	18	0.329	0.049	0.527	0.1636	0.1380	0.2181	1.472	0.6793	339.7
	20	0.241	0.035	0.555	0.1636	0.1453	0.2420	1.326	0.7542	377.1
3/4	10	0.962	0.134	0.482	0.1963	0.1262	0.1825	1.758	0.5688	284.4
	12	0.812	0.109	0.532	0.1963	0.1393	0.2223	1.442	0.6935	346.8
	14	0.644	0.083	0.584	0.1963	0.1528	0.2678	1.198	0.8347	417.4
	16	0.518	0.065	0.620	0.1963	0.1623	0.3019	1.063	0.9407	470.4
	18	0.400	0.049	0.652	0.1963	0.1706	0.3339	0.9611	1.041	520.5
7/8	10	1.16	0.134	0.607	0.2291	0.1589	0.2893	1.108	0.9025	451.3
	12	0.992	0.109	0.657	0.2291	0.1720	0.3390	0.9465	1.057	528.5
	14	0.769	0.083	0.709	0.2291	0.1856	0.3949	0.8126	1.230	615.0
	16	0.613	0.065	0.745	0.2291	0.1951	0.4360	0.7360	1.358	679.0
	18	0.472	0.049	0.777	0.2291	0.2034	0.4740	0.6770	1.477	738.5
1	10	1.35	0.134	0.732	0.2618	0.1916	0.4208	0.7626	1.311	655.5
	12	1.14	0.109	0.782	0.2618	0.2048	0.4803	0.6681	1.497	748.5
	14	0.887	0.083	0.834	0.2618	0.2183	0.5463	0.5874	1.702	851.0
	16	0.708	0.065	0.870	0.2618	0.2277	0.5945	0.5398	1.852	926.0
	18	0.535	0.049	0.902	0.2618	0.2361	0.6390	0.5022	1.991	995.5
1 1/4	10	1.74	0.134	0.982	0.3271	0.2572	0.7575	0.4236	2.362	1181
	12	1.45	0.109	1.032	0.3271	0.2701	0.8369	0.3834	2.608	1304
	14	1.13	0.083	1.084	0.3271	0.2839	0.9229	0.3477	2.877	1439
	16	0.898	0.065	1.120	0.3271	0.2932	0.9852	0.3257	3.070	1535
	18	0.675	0.049	1.152	0.3271	0.3015	1.043	0.3075	3.253	1627
1 1/2	10	2.12	0.134	1.232	0.3925	0.3227	1.193	0.2688	3.720	1860
	12	1.76	0.109	1.282	0.3925	0.3355	1.292	0.2482	4.030	2015
	14	1.36	0.083	1.334	0.3925	0.3491	1.398	0.2292	4.362	2181
	16	1.09	0.065	1.370	0.3925	0.3585	1.473	0.2180	4.587	2294
2	10	2.94	0.134	1.732	0.5233	0.4534	2.355	0.1362	7.342	3671
	12	2.40	0.109	1.782	0.5233	0.4665	2.494	0.1287	7.770	3885
	14	1.85	0.083	1.834	0.5233	0.4803	2.643	0.1213	8.244	4122
	16	1.47	0.065	1.870	0.5233	0.4896	2.747	0.1168	8.562	4281

[a] Prepared by T. B. Drew.
[b] In brass, specific gravity = 8.56; specific gravity of steel = 7.8.

TABLE A-31. STANDARD DIMENSIONS FOR STANDARD-WEIGHT WROUGHT-IRON PIPE (Crane Company)

Nominal size, in.	Nominal size, mm	Actual diameters External, in.	Approximate internal, in.	Nominal thickness, in.	Circumference External, in.	Circumference Internal, in.	Transverse areas External, sq in.	Transverse areas Internal, sq in.	Transverse areas Metal	Length of pipe per sq ft External surface, ft	Length of pipe per sq ft Internal surface, ft	Length of pipe containing 1 cu ft, ft	Nominal weight, lb/ft Plain ends	Nominal weight, lb/ft Threaded and coupled	Number of threads per in. of screw
⅛	3	0.405	0.269	0.068	1.272	0.845	0.129	0.057	0.072	9.431	14.199	2533.775	0.244	0.245	27
¼	6	0.540	0.364	0.088	1.696	1.144	0.229	0.104	0.125	7.073	10.493	1383.789	0.424	0.425	18
⅜	10	0.675	0.493	0.091	2.121	1.549	0.358	0.191	0.167	5.658	7.747	754.360	0.567	0.568	18
½	13	0.840	0.622	0.109	2.639	1.954	0.554	0.304	0.250	4.547	6.141	473.906	0.850	0.852	14
¾	19	1.050	0.824	0.113	3.299	2.589	0.866	0.533	0.333	3.637	4.635	270.034	1.130	1.134	14
1	25	1.315	1.049	0.133	4.131	3.296	1.358	0.864	0.494	2.904	3.641	166.618	1.678	1.684	11½
1¼	32	1.660	1.380	0.140	5.215	4.335	2.164	1.495	0.669	2.301	2.767	96.275	2.272	2.281	11½
1½	38	1.900	1.610	0.145	5.969	5.058	2.835	2.036	0.799	2.010	2.372	70.733	2.717	2.731	11½
2	50	2.375	2.067	0.154	7.461	6.494	4.430	3.355	1.075	1.608	1.847	42.913	3.652	3.678	11½
2½	64	2.875	2.469	0.203	9.032	7.757	6.492	4.788	1.704	1.328	1.547	30.077	5.793	5.819	8
3	76	3.500	3.068	0.216	10.996	9.638	9.621	7.393	2.228	1.091	1.245	19.479	7.575	7.616	8
3½	90	4.000	3.548	0.226	12.566	11.146	12.566	9.886	2.680	0.954	1.076	14.565	9.109	9.202	8
4	100	4.500	4.026	0.237	14.137	12.648	15.904	12.730	3.174	0.848	0.948	11.312	10.790	10.889	8
4½	113	5.000	4.506	0.247	15.708	14.156	19.635	15.947	3.688	0.763	0.847	9.030	12.538	12.642	8
5	125	5.563	5.047	0.258	17.477	15.856	24.306	20.006	4.300	0.686	0.756	7.198	14.617	14.810	8
6	150	6.625	6.065	0.280	20.813	19.054	34.472	28.891	5.581	0.576	0.629	4.984	18.974	19.185	8
7	175	7.625	7.023	0.301	23.955	22.063	45.664	38.738	6.926	0.500	0.543	3.717	23.544	23.769	8
8	200	8.625	8.071	0.277	27.096	25.356	58.426	51.161	7.265	0.442	0.473	2.815	24.696	25.000	8
8	200	8.625	7.891	0.322	27.096	25.073	58.426	50.027	8.399	0.442	0.478	2.878	28.554	28.809	8
9	225	9.625	8.941	0.342	30.238	28.089	72.760	62.786	9.974	0.396	0.427	2.294	33.907	34.188	8
10	250	10.750	10.192	0.279	33.772	32.019	90.763	81.585	9.178	0.355	0.374	1.765	31.201	32.000	8
10	250	10.750	10.136	0.307	33.772	31.843	90.763	80.691	10.072	0.355	0.376	1.785	34.240	35.000	8
10	250	10.750	10.020	0.365	33.772	31.479	90.763	78.855	11.908	0.355	0.381	1.826	40.483	41.132	8
11	275	11.750	11.000	0.375	36.914	34.558	108.434	95.033	13.401	0.325	0.347	1.515	45.557	46.247	8
12	300	12.750	12.090	0.330	40.055	37.982	127.676	114.800	12.876	0.299	0.315	1.254	43.773	45.000	8
12	300	12.750	12.000	0.375	40.055	37.699	127.676	113.097	14.579	0.299	0.318	1.273	49.562	50.706	8

APPENDIX 489

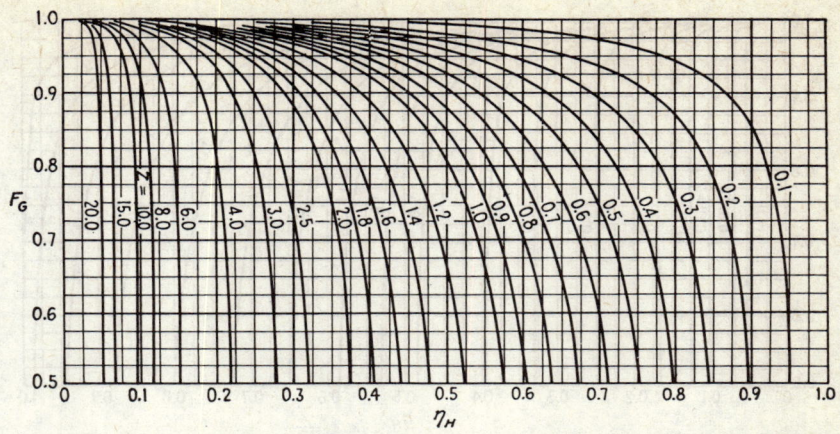

FIG. A-10. Mean temperature difference with 1 shell pass and 2, 4, 6, etc., tube passes. See Fig. 8-6A. (*Courtesy of Tubular Exchanger Manufacturers Association.*)

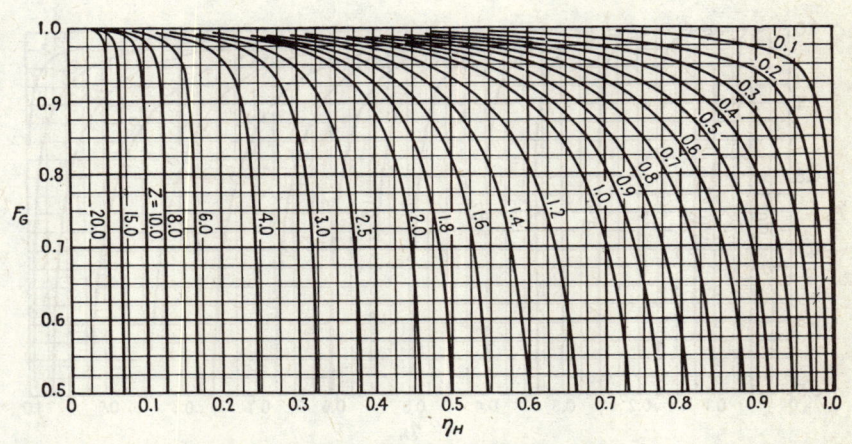

FIG. A-11. Mean temperature difference with 2 shell passes and 4, 8, 12, etc., tube passes. See Fig. 8-6B. (*Courtesy of Tubular Exchanger Manufacturers Association.*)

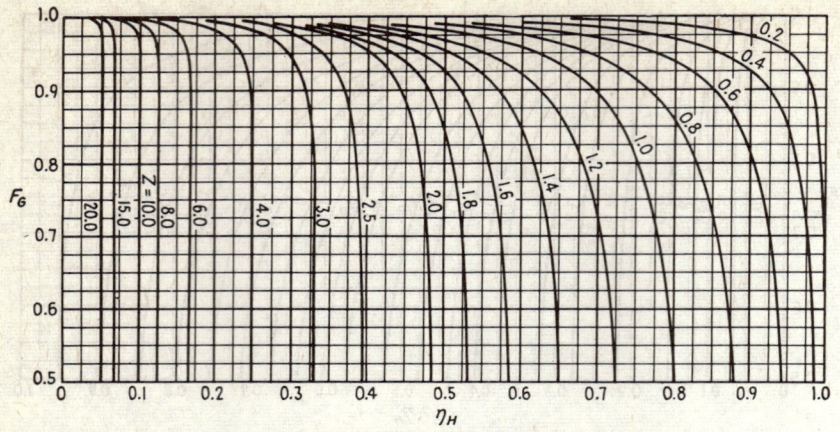

FIG. A-12. Mean temperature difference with 3 shell passes and 6, 12, 18, etc., tube passes. See Fig. 8-6C. (*Courtesy of Tubular Exchanger Manufacturers Association.*)

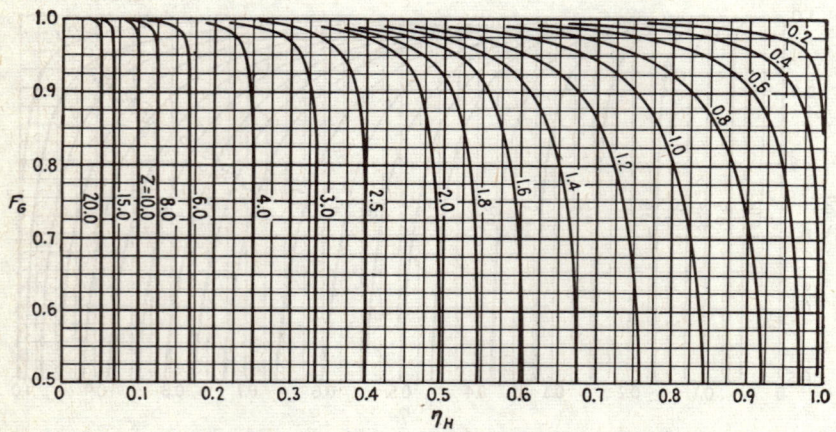

FIG. A-13. Mean temperature difference with 4 shell passes and 8, 16, 24, etc., tube passes. See Fig. 8-6D. (*Courtesy of Tubular Exchanger Manufacturers Association.*)

BIBLIOGRAPHY AND AUTHOR INDEX

In the references to journal articles, the volume number appears in **boldface** type, followed by the range of page numbers and year of publication. The numerals in the right-hand column are page numbers in this book; where such numbers appear in parentheses the literature reference pertains to the subject matter of the page given although it is not cited specifically.

CHAPTER 1. INTRODUCTION TO HEAT TRANSMISSION

PAGE

1. Boltzmann, L., *Wied. Ann.*, **22**, 291 (1884)... 4
2. Fourier, J. B. J., "Théorie analytique de la chaleur," Gauthier-Villars, 1822; English translation by Freeman, Cambridge, 1878.................................. 3
3. Hottel, H. C., "Notes on Radiant Heat Transmission among Surfaces Separated by Non-absorbing Media," Technology Store, Cambridge, Mass, 1951 ... 5
4. Newton, I., *Phil Trans Roy. Soc.* (*London*), **22**, 824 (1701)..................... 5
5. Stefan, J., *Sitzber. Kais. Akad. Wiss. Wien, Math.-Naturw. Kl.*, **79**, 391 (1879).. 4

CHAPTER 2. STEADY CONDUCTION

1. Awbery, J., and F. Schofield, *Proc. Intern. Congr. Refrig.*, 5th Congr., **3**, 591–610 (1929)... 19
2. Barratt, T., *Proc. Phys. Soc.* (*London*), **28**, 14–20 (1915)....................... 17
3. Bates, O. K., G. Hazzard, and G. Palmer, *Ind. Eng. Chem., Anal. Ed.*, **10**, 314–318 (1938); see also Bates, *Ind. Eng. Chem.*, **25**, 431 (1933)............ 27
3a. Boelter, L. M. K., H. P. Poppendiek, and J. T. Gier, *Natl. Advisory Comm. Aeronaut., Wartime Repts.* W-24 (1944); Boelter, L. M. K., H. F. Poppendiek, R. V. Dunkle, and J. T. Gier, *ibid.*, W-30 (1944)................................ 18
4. Bridgman, P. W., *Proc. Am. Acad. Arts Sci.*, **59**, 141 (1923)..................... 27
5. Brunot, A. W., and F. F. Buckland, *Trans. ASME*, **71**, 253–257 (1949)... 18
6. Cetinkale, T. N., and M. Fishenden, "Proceedings of the General Discussion on Heat Transfer," pp. 271–275, 309, Institution of Mechanical Engineers, London, American Society of Mechanical Engineers, New York, 1951..... 18
7. Dusinberre, G. M., "Numerical Analysis of Heat Flow," McGraw-Hill, New York, 1949.. 21, 22
8. Emmons, H. W., *Trans. ASME*, **65**, 607–612 (1943); *Quart. Appl. Math.*, **2**, 173–193 (1944)... 21, 22
9. Eucken, A., *Physik. Z.*, **12**, 1101–1107 (1911), **14**, 324 (1913)............... 28
10. Fourier, J. B. J., "Théorie analytique de la chaleur," Gauthier-Villars, Paris, 1822; English translation by Freeman, Cambridge, 1878............................ 7
11. Griffiths, E., *Food Investigation Board, Spec. Rept.* 5, Department of Scientific and Industrial Research, H. M. Stationery Office, London, 1921...... 10
12. Jakob, M., "Heat Transfer," 1st ed., Vol. 1, Wiley, New York, 1949..... 9, 10, 19, 22

15. Jeans, J. H., "Dynamical Theory of Gases," 4th ed., p. 291, Cambridge, London, 1925.. 28
14. Kardos, A., *Forsch. Gebiete Ingenieurw.*, **5**, 14 (1934)............... 28
15. Kennard, E. H., "Kinetic Theory of Gases," McGraw-Hill, New York, 1938 29
16. Langmuir, I., E. Q. Adams, and G. S. Meikle, *Trans. Am. Electrochem. Soc.*, **24**, 53 (1913)... 25
17. Lehmann, Th., *Elektrotech. Z.*, **30**, 995, 1019 (1909)............... 19
18. "Liquid-Metals Handbook," NAVEXOS P-733, 2d ed., Superintendent of Documents, Government Printing Office, Washington, D.C., 1952..... 27, 28
18a. London, A. L., personal communication, Jan. 2, 1952................. 25
19. Lorenz, L., *Ann. Physik*, **13**, 422 (1882)........................... 9
20. MacLean, J. D., *Trans. Am. Soc. Heating Ventilating Engrs.*, **47**, 323, 354 (1941)... 10
20a. McAdams, W. H., "Some Recent Developments in Heat Transfer," *Purdue Univ., Eng. Expt. Sta., Bull.* 104, **32** (2) (March, 1948)............. 21
21. Maxwell, J. C., "Collected Works," Vol. II, p. 1, Cambridge, London, 1890. 10, 28
21a. Moore, A. D., *J. Appl. Phys.*, **20**, 790–804 (1949)................. 19
22. Nichols, P., *J. Am. Soc. Heating Ventilating Engrs.*, **30**, 35 (1924); *Ind. Eng. Chem.*, **16**, 490 (1924).. (18)
23. Nickerson, T. S., and G. M. Dusinberre, *Trans. ASME*, **70**, 903–906 (1948) 14
24. Nukiyama, S., and H. J. Yosikata, *Trans. Soc. Mech. Engrs. (Japan)*, **33** (1) (1930)... 26
25. Powell, R. W., "Proceedings of the General Discussion on Heat Transfer," pp. 290–295, Institution of Mechanical Engineers, London, and American Society of Mechanical Engineers, New York, 1951..................... 10
26. Powell, R. W., *J. Iron Steel Inst.*, **162**, 315–324 (July, 1949)........ 28
26a. Riedel, L., inaugural dissertation, Technische Hochschule Fridericiana, Karlsruhe, C. F. Muller Verlag, Karlsruhe, 1948..................... 28
27. Smith, J. F. D., *Ind. Eng. Chem.*, **22**, 1246 (1930)................... 27
28. Smith, J. F. D., *Trans. ASME*, **58**, 719 (1936)....................... 28
29. Southwell, R. V., "Relaxation Methods in Engineering Science," Oxford, New York, 1940... 21
30. Taylor, T. S., *Trans. ASME*, **41**, 605 (1919); *Refrig. Eng.*, **10**, 179 (1923).. 17
31. Trayer, G. W., and H. W. March, *Natl. Advisory Comm. Aeronaut., Rept.* 334 (1930).. 19
32. Van Dusen, M. S., and J. L. Finch, *J. Research Natl. Bur. Standards* 6, Paper 291, pp. 493–522 (1931).. 18
33. Veron, M., "Champs thermiques et flux calorifiques," *Bull. tech. soc. franc. construct., Babcock & Wilcox* 23, 24 (1950–1951)..................... 20
34. Walker, W. H., W. K. Lewis, and W. H. McAdams, "Principles of Chemical Engineering," 2d ed., McGraw-Hill, New York, 1927................. 12
35. Weills, N. D., and E. A. Ryder, *Trans. ASME*, **71**, 259–267 (1949)..... 18
36. Wilkes, G. B., "Heat Insulation," 1st ed., Wiley, New York, 1950.. 10, 11, 18
37. Wright, L. T., Jr., *Cornell Univ., Eng. Expt. Sta., Bull.* 31 (1943)........ 21

CHAPTER 3. TRANSIENT CONDUCTION

1. Anthony, M. L., "Proceedings of the General Discussion on Heat Transfer," pp. 236–262, Institution of Mechanical Engineers, London, and American Society of Mechanical Engineers, New York, 1951.................... 36
2. Beuken, Cl. L., *Econ. Tech. Tidskr.*, **19**, 43 (1939)................... 51
3. Binder, L., dissertation, Munich, 1911............................. 43, 46

BIBLIOGRAPHY AND AUTHOR INDEX 493

PAGE

4. Byerly, W. E., "Elementary Treatise on Fourier Series," Ginn, Boston, 1928 35
5. Carslaw, H. S., "Mathematical Theory of Heat," Macmillan, New York, 1921, "Fourier's Series and Integrals," Macmillan, New York, 1930; see also especially H. S. Carslaw, and J. C. Jaeger, "Conduction of Heat in Solids," Oxford, New York, 1947.................................... 35, 52
6. Coyle, M. B., "Proceedings of the General Discussion on Heat Transfer," pp. 265–267, 309, Institution of Mechanical Engineers, London, and American Society of Mechanical Engineers, New York, 1951.................. 51
7. Dusinberre, G. M., *Trans. ASME*, **67**, 703–712 (1945)................ 44, 51
8. Dusinberre, G. M., "Numerical Analysis of Heat Flow," McGraw-Hill, New York, 1949... 44, 45, 46, 47
9. Elmer, L., *Refrig. Eng.*, **24**, 17 (1932)......................... 52
10. Emmons, H. W., *Trans. ASME*, **65**, 607–612 (1943); *Quart. Appl. Mech.*, **2**, 173–193 (1944).. 50, 51
11. Fell, E. W., "Proceedings of the General Discussion on Heat Transfer," pp. 276–278, 310, Institution of Mechanical Engineers, London, and American Society of Mechanical Engineers, New York, 1951.................. 52
12. Fischer, J., *Ingenieur-Arch.*, **10**, 95–112 (1939)...................... 35
13. Fishenden, M., and O. A. Saunders, "An Introduction to Heat Transfer," Oxford, New York, 1950.. 35
14. Fourier, J. B., "Théorie analytique de la chaleur," Gauthier-Villars, Paris, 1822; German translation by Weinstein, Springer, Berlin, 1884; *Ann. chim. et phys.*, **37** (2), 291 (1828); *Pogg. Ann.*, **13**, 327 (1828)................ 35
15. Gemant, A., *J. Appl. Phys.*, **17**, 1076 (1946)........................ 40, 42
16. Goldschmidt, H., and E. P. Partridge, "Industrial Heat Transfer," Wiley, New York, 1933.. 35, 52
17. Grober, H., "Warmeübertragung," Springer, Berlin, 1926; Grober, H., and S. Erk, "Die Grundgesetze der Warmeübertragung," Springer, Berlin, 1933.. 35, 52
18. Gurney, H. P., Heating and Cooling of Solid Shapes, unpublished monograph, Massachusetts Institute of Technology Library.................. 35
19. Gurney, H. P., and J. Lurie, *Ind. Eng. Chem.*, **15**, 1170–1172 (1923).35, 40, 41
20. Heisler, M. P., *Trans. ASME*, **68**, 493 (1946)........................ 51
21. Heisler, M. P., *Trans. ASME*, **69**, 227–236 (1947)................... 35
22. Hooper, F. C., and S. C. Chang, *Heating, Piping Air Conditioning*, **24**, 125–129 (October, 1952).. 40
23. Hottel, H. C., personal communication, 1938................... 36, 37, 38
24. Houghten, F. C., J. L. Blackshaw, E. M. Pugh, and D. McDermott, *Heating, Piping Air Conditioning*, **4**, 288 (1932)............................... 52
25. Ingersoll, L. R., O. J. Zobel, and A. C. Ingersoll, "Heat Conduction, with Engineering and Geological Applications," McGraw-Hill, New York, 1948.. 35, 52
26. Jakob, M., "Heat Transfer," Vol. 1, Wiley, New York, 1949............. 52
27. Joslyn, M. A., and G. L. Marsh, *Ind. Eng. Chem.*, **22**, 1192–1197 (1930)... 52
28. MacLean, J. D., *Proc. Am. Wood-Preservers' Assoc.*, 1930, p. 197....... 31, 51
29. Mersman, W. A., W. P. Berggren, and L. M. K. Boelter, *Univ. Calif., Publs. Eng.*, **5**, 1–22 (1942).. 49, 51, 52
30. Moore, A. D., *Ind. Eng. Chem.*, **28**, 704–708 (1936)..................... 51
31. Nessi, A., *Bull. soc. encour. ind. nat.*, **131**, 289 (1932)................... 51
32. Nessi, A., and L. Nisolle, "Méthodes graphiques pour l'étude des installations de chauffage," pp. 46*ff.*, Dunod, Paris, 1929....................... 51
33. Newman, A. B., *Trans. Am. Inst. Chem. Engrs.*, **24**, 44 (1930)...... 35, 39, 41

34. Olson, F. C. W., and O. T. Schultz, *Ind. Eng. Chem.*, **34,** 874–877 (1942). 38, 40
35. Paschkis, V., ASHVE Section, *Heating, Piping Air Conditioning*, **14,** 133 (1942) .... (51)
36. Paschkis, V., and H. D. Baker, *Trans. ASME*, **64,** 105 (1942) 51
37. Pekeris, C. L., and L. B. Slichter, *Appl. Phys.*, **10,** 135–137 (1939)........ 52
38. Perks, A. A., and R. W. Griffiths, *India-Rubber J.*, **67,** 24 (1924)......... 51
39. Planck, R., *Z. ges. Kälte-Ind.*, **39,** 56 (1932)................................ 52
40. Rosenthal, D., *Trans. ASME*, **68,** 849 (1946)...................... 43
41. Schack, A., *Stahl u. Eisen*, **50,** 1289 (1930)........................ 35, 52
42. Schmidt, E., "Foppls Festschrift," pp. 179–198, Springer, Berlin, 1924; abstracted by Sherwood and Reed............................... 43, 44, 45, 46
43. Schmidt, E., *Forsch. Gebiete Ingenieurw.*, **13,** 177–184 (1942).......... 48, 49
44. Shephard, J. R., and W. B. Wiegand, *Ind. Eng. Chem.*, **20,** 953–959 (1928). 51
45. Sherwood, T. K., *Ind. Eng. Chem.*, **20,** 1181 (1928)...................... 51
46. Sherwood, T. K., and C. E. Reed, "Applied Mathematics in Chemical Engineering," p. 211, McGraw-Hill, New York, 1939.................... 45
47. Van Ardsel, W. B., *Trans. Am. Inst. Chem. Engrs.*, **43,** 13–24 (1947)...... 51
48. Véron, M., "Champs thermiques et flux calorifiques," *Bull. tech. soc. franc. construct.*, Babcock & Wilcox 23, 24 (1950–1951)...................... 49
49. Williamson, E. D., and L. H. Adams, *Phys. Rev.*, **14,** 99–114 (1919).... 35, 43
50. Wirka, J., *Proc. Am. Wood-Preservers' Assoc.*, 1924, p. 285, 1925, p. 271... 51

CHAPTER 4. RADIANT-HEAT TRANSMISSION

1. Barret, P., *Recherche aéronaut.*, (4) (1948)............................ 104
2. Bearden, J. A., and H. M. Watts, *Phys. Rev.*, **81,** 73–81 (1951).......... 59
3. Binkley, E. R., "Heat Transfer," pp. 40–46, American Society of Mechanical Engineers, 1933–1934.. 61
4. Broeze, J. J., G. Ribaud, and O. A. Saunders, *J. Inst. Fuel*, **24,** S1 (1951). 100
5. Brooks, F. A., Observations of Atmospheric Radiation, *Papers Phys. Oceanog. Meteorol.*, *Mass. Inst. Technol. and Woods Hole Oceanog. Inst.*, **8** (2), 1941... 84
6. Brunet, M., *Journée études flammes*, June, 1953....................... 100
7. Chang, T. Y., Sc.D. Thesis in Chemical Engineering, Massachusetts Institute of Technology, 1941..................................... 99
8. Coblentz, W. H., "Investigation of Infra-red Spectra," Carnegie Institution, Washington, D.C., 1905... 95
9. Cohen, E. S., S.M. Thesis in Chemical Engineering, Massachusetts Institute of Technology, 1951.. 91, 92
10. Cohen, E. S., Progress Report on Sc.D. Thesis, Massachusetts Institute of Technology, 1953... 92
11. Comerford, F. M., *J. Inst. Fuel*, **24,** S4 (1951)...................... (100)
12. Cooper, M. A., Sc.D. Thesis in Physics, University of the Witwatersrand, Union of South Africa, 1932...................................... 104
12a. Crout, P. D., *Trans. AIEE*, **60,** 1235 (1941)........................ 115
13. Daws, L. F., and H. Herne, *J. Inst. Fuel*, **25,** S17 (1952)............. 100
14. Daws, L. F., and M. W. Thring, *J. Inst. Fuel*, **25,** S28 (1952)....... 100, 104
15. Daws, L. F., and M. W. Thring, "An Attempt to Obtain Evidence from the First Combustion Mechanism Trials as to the Factors Governing Local Flame Emissivity," *Flame Rad. Research Doc.* D2/b/6 (1952)............ 100
16. de Graaf, J. E., and M. W. Thring, *J. Inst. Fuel*, **25,** S31 (1952)........ 100
17. Eberhardt, J. E., and H. C. Hottel, *Trans. ASME*, **58,** 185–193 (1936); *Heat Treating and Forging*, **22,** 144–149, 193–198 (1936)................. 119

BIBLIOGRAPHY AND AUTHOR INDEX

		PAGE
18.	Eckert, E. R. G., *Forschungsheft*, **387**, 1–20 (1937)........................	84
19.	Elsasser, W. M., Heat Transfer by Infrared Radiation in the Atmosphere, *Harvard Meteorological Studies* 6, Harvard University, 1942..............	84
20.	Fishenden, M., *Engineering*, **138**, 478 (1934)............................	(82)
21.	Foote, P. D., *J. Wash. Acad. Sci.*, **5**, 1 (1915)............................	61
22.	Gerald, C., Sc.D. Thesis in Chemical Engineering, Massachusetts Institute of Technology, 1941..	99
23.	Guerrieri, S. A., Report on Research at Massachusetts Institute of Technology, 1933..	95
23a.	Hamilton, D. C., and Morgan, W. R., NACA TN 2836, December, 1952..	69
23b.	Haslam, R. T., and M. W. Boyer, *Ind. Eng. Chem.*, **19**, 4–6 (1927)........	99
24.	Haslam, R. T., and H. C. Hottel., *Trans. ASME*, **FSP50**, 9 (1928)........	104
24a.	Heiligenstädt, W., *Arch. Eisenhüttenw.*, **7**, 25, 103 (1933)................	119
24b.	Heiligenstädt, W., "Wärmetechnische Rechnungen für Industrieöfen," 3d ed., Verlag Stahleisen m.b. H., Düsseldorf, 1951......................	119
25.	Hooper, F. C., and I. S. Juhasz, Paper 52-F-19, Fall Meeting, American Society of Mechanical Engineers (September, 1952)....................	69
26.	Hottel, H. C., *Trans. Am. Inst. Chem. Engrs.*, **19**, 173 (1927); *Ind. Eng. Chem.*, **19**, 888 (1927)..	87, 96
27.	Hottel, H. C., *Trans. World Power Conf.*, 2d Conf., **18**, Sec. 32 (243) (1930)	69
28.	Hottel, H. C., *Mech. Eng.*, **52**, 699–704 (1930)............................	69
29.	Hottel, H. C., *Trans. ASME*, **FSP53**, 265–273 (1931)....................	69
30.	Hottel, H. C., Notes on Radiant Heat Transmission, Chemical Engineering Department, Massachusetts Institute of Technology (a) 1938, (b) 1951, (c) 1953...	110, 119
31.	Hottel, H. C., and F. P. Broughton, *Ind. Eng. Chem., Anal. Ed.*, **4**, 166–175 (1932)..	100, 102
32.	Hottel, H. C., and R. B. Egbert, *Trans. Am. Inst. Chem. Engrs.*, **38**, 531–565 (1942)..	82, 84
33.	Hottel, H. C., and J. D. Keller, *Trans. ASME*, Iron and Steel, **55–56**, 39–49 (1933)..	70
34.	Hottel, H. C., and H. G. Mangelsdorf, *Trans. Am. Inst. Chem. Engrs.*, **31**, 517–549 (1935)...	82, 84
35.	Hottel, H. C., and V. C. Smith, *Trans. ASME*, **57**, 463–470 (1935).....	82, 84
36.	Husson, G., *Journée études flammes* (June, 1953)........................	100
37.	Kurlbaum, F., *Physik. Z.*, **3**, 187 (1902)................................	104
38.	Lent, H., *Wärme*, **49**, 145 (1926)..	99
39.	Levêque, M., C. Levy, G. Louvet, and M. Michaud, *Journée études flammes* (June, 1953)..	99
40.	Levêque, M., "Examen au microscope électronique du carbone contenu dans une flamme de Mazout," *Flame Radiation Research Doc.* D2/f/2 (1952)...	99
41.	Lindmark, T., and H. Edenholm, *Ingeniors Vetenskaps Akad. Handl.* 66 (1927)..	104
42.	Lindmark, T., and L. Kignell, *Ing. Vetenskaps Akad. Handl.* 91, 1–36 (1929), 109, 1–20 (1931)..	104
43.	Lobo, W. E., and J. E. Evans, *Trans. Am. Inst. Chem. Engrs.*, **35**, 743–778 (1939), **36**, 173–175 (1940)..	120
43a.	McAdams, W. H., "Heat Transmission," 2d ed., Chap. III, McGraw-Hill, New York, 1942...	110
44.	Malcor, H., *Journée études flammes* (June, 1953)........................	100
45.	Mayorcas, R., *Iron Steel Inst.*, (London), *Spec. Rept.* 37, 129 (1946).......	99
46.	Mayorcas, R., *J. Inst. Fuel*, **24**, S15 (1951)...............................	100

47. Mayorcas, B., and M. Rivière, "Description of Trials and Results," *Flame Rad. Research Doc.* D3/b/8 (1952).................................... 100
48. Mekler, L. A., *Natl. Petroleum News*, **30** (30), R355–368, 398–400, 402, 405 (1938).. 120
49. Michaud, M., Sc.D. Thesis, University of Paris, 1951.................. 61
49a. Michaud, M., *Journée études flammes* (June, 1953).................. 100
50. Mie, G., *Ann. phys.*, **25**, 377 (1908)................................ 100
51. Mullikin, H. F., *Trans. ASME*, **57**, 517–530 (1935)................ 120
52. Naeser, G., and W. Pepperhoff, *Arch. Eisenhüttenw.*, 9 (January–February, 1951).. 99
53. Parker, W. G., and H. G. Wolfhard, *J. Chem. Soc.*, 2038 (1950)....... 99
54. Penner, S. S., *J. Appl. Phys.*, **21**, 685–695 (1950)................ 96
55. Penner, S. S., *J. Appl. Mechanics*, **18**, 33–38 (1951)............. 96
56. Port, F. J., Sc.D. Thesis in Chemical Engineering, Massachusetts Institute of Technology, 1940....................................... 88, 89, 96, 97
57. Rivière, M., *Journée études flammes* (June, 1953)................... 100
58. Schack, A., *Z. tech. Physik*, **5**, 266 (1924)...................... 96
59. Schack, A., "Der Industrielle Wärmeübergang," Verlag Stahleisen m.b. H., Düsseldorf, 1929.. 96
60. Schack, A., *Arch. Eisenhüttenw.*, **19**, D11 (1948)................. 96
61. Schmidt, E., *Forsch. Gebiete Ingenieurw.*, **3**, 57 (1932).......... 84
62. Schmidt, E., and E. Eckert, *Forsch. Gebiete Ingenieurw.*, **8**, 87 (1937)...... 87
63. Schmidt, H., *Ann. Physik*, **29**, 998 (1909)........................ 103
64. Schmidt, H., and L. Furthmann, *Mitt. Kaiser-Wilhelm-Inst. Eisenforsch. Düsseldorf*, Abhandl., **109**, 225 (1928)........................... 61
65. Seibert, O., *Wärme*, **54**, 737–739 (1931).......................... 69
66. Senftleben, H., and E. Benedict, *Ann. Physik.*, **60**, 297 (1919)... 100
67. Sherman, R. A., *Trans. ASME*, **56**, 177–185 (1934)................. 99
68. Sherman, R. A., *Trans. ASME*, **56**, 401–410 (1934)................. 104
68a. Sieber, W., *Z. tech. Phys.*, 130–135 (1941)......................... 63
69. Simpson, H., Progress Report on Sc.D. Thesis Research, Massachusetts Institute of Technology, 1953...................................... 99
69a. Smith, D., Ph.D. Thesis, Department of Fuel Technology, University of Sheffield, England, 1952... 119
70. Smith, D., Chemical Engineering Department Progress Report, Massachusetts Institute of Technology, 1953........................... 122
70a. Trinks, W., "Industrial Furnaces," 4th ed., Vol. 1, Wiley, New York, 1952 119
71. Trinks, W., and J. D. Keller, *Trans. ASME*, **58**, 203 (1936)....... 99
72. Ullrich, W., Sc.D. Thesis in Chemical Engineering, Massachusetts Institute of Technology, 1935.. 95, 96
73. Wilson, D. W., W. E. Lobo, and H. C. Hottel, *Ind. Eng. Chem.*, **24**, 486 (1932).. 120
74. Wohlenberg, W. J., and E. L. Lindseth, *Trans. ASME*, **48**, 849–937 (1926) 104
75. Wohlenberg, W. J., and D. G. Morrow, *Trans. ASME*, **47**, 127–176 (1925)... 104, 121
76. Wohlenberg, W. J., and H. F. Mullikin, *Trans. ASME*, **57**, 531–540 (1935) 121
77. Wolfhard, H. G., and W. G. Parker, *Proc. Phys. Soc. (London)*, **B62**, 523 (1949).. 101
78. Yagi, S., *J. Soc. Chem. Ind., Japan*, **40**, 50B, (1937)............ 100

CHAPTER 5. DIMENSIONAL ANALYSIS

1. Bridgman, P. W., "Dimensional Analysis," Yale University Press, New Haven, 1931.. 126

2. Comings, E. W., *Ind. Eng. Chem.*, **32**, 984–987 (1940).................... 129
3. Gibson, A. H., *Engineering*, **117**, 325–357, 391, 422 (1924)................ 136
4. Hunsaker, J. C., and B. G. Rightmire, "Engineering Applications of Fluid Mechanics," Wiley, New York, 1951................................ 130, 131
5. Hyman, S. C., C. F. Bonilla, and S. W. Ehrlich, "Preprints of Heat Transfer Symposium," pp. 55–76, American Institute of Chemical Engineers, December, 1951.. 133
6. Klinkenberg, A., and H. H. Mooy, *Chem. Eng. Progr.*, **44**, 17–36 (1948).... 135
7. Langhaar, H. L., "Dimensional Analysis and the Theory of Models," Wiley, New York, 1951.. 126, 130
8. Nusselt, W., *Z. Ver. deut. Ing.*, **53**, 1750, 1808 (1909); *Mitt. Forsch.*, **89**, 1 (1910)... 133
9. Nusselt, W., *Gesundh.-Ing.*, **38**, 477, 490 (1915)....................... 133
10. Rayleigh, Lord, *Nature*, **95**, 66 (1915); *Phil. Mag.*, **34**, 59 (1892).......... 129
10a. Rushton, J. H., *Chem. Eng. Progr.*, **48**, 33–38, 95–102 (1952)............. 136
11. Stokes, G. G., *Trans. Cambridge Phil. Soc.*, **8**, (1845); reprinted in "Mathematical and Physical Papers," pp. 95 and 104, Cambridge, London, 1905.. 133
12. Tribus, M., "Elementary Heat Transfer," *Univ. Calif.*, Syllabus Ser. 317, University of California Press, 1950................................... 134

CHAPTER 6. FLOW OF FLUIDS

1. Alves, G. E., D. F. Boucher, and R. L. Pigford, *Chem. Eng. Progr.*, **48**, 385–393 (1952)... 148
1a. Andreas, J. M., S.M. Thesis in Chemical Engineering, Massachusetts Institute of Technology, 1927....................................... 162
2. Atherton, D. H., *Trans. ASME*, **48**, 145–175 (1926).................... 158
3. Bakhmeteff, B. A., "The Mechanics of Turbulent Flow," Princeton University Press, Princeton, N. J., 1936................................... 154
4. Bergelin, O. P., G. A. Brown, and S. C. Doberstein, *Trans. ASME*, **74**, 953–960 (1952)... 163
5. Boelter, L. M. K., R. C. Martinelli, and F. Jonassen, *Trans. ASME*, **63**, 447–455 (1941).. 154
6. Bond, W. N., *Proc. Phys. Soc. (London)*, **43**, Pt. I, 46–52 (1931).......... 144
6a. Boucher, D. F., and C. E. Lapple, *Chem. Eng. Progr.*, **44**, 117–134 (1948).. 163
7. Boussinesq, J., "Théorie analytique de la chaleur," Vol. III, Gauthier-Villars, Paris, 1903; *J. math. pures appl.*, **1**, 285 (1905)....................... 149
8. Chilton, T. H., and R. P. Genereaux, *Trans. Am. Inst. Chem. Engrs.*, **29**, 161–173 (1933)... 162, 163
9. Clapp, M. H., and O. FitzSimons, S.M. Thesis in Chemical Engineering, Massachusetts Institute of Technology, 1928........................... 149
10. Cornish, R. J., *Proc. Roy. Soc. (London)*, **A120**, 69 (1928)............ 150, 158
11. Couch, W. H., and C. E. Herrstrom, Thesis in Chemical Engineering, Massachusetts Institute of Technology, 1924........................... 152
12. Davies, S. J., and C. M. White, *Proc. Roy. Soc. (London)*, **A119**, 92–107 (1928); *Engineering*, **128**, 69, 98, 131 (1929)........................ 150
13. Dönch, F., *Mitt. Forsch.*, **282**, 1–58 (1926)............................ 160
14. Deissler, R. G., *Natl. Advisory Comm. Aeronaut., Tech. Note* 2138 (July, 1950).. 154
15. Drew, T. B., personal communication, 1931............................ 151
16. Drew, T. B., personal communication, 1932...................... 150, 158
17. Drew, T. B., and R. P. Genereaux, *Trans. Am. Inst. Chem. Engrs.*, **32**, 17–19 (1936)... 155, 157

		PAGE
18.	Drew, T. B., E. C. Koo, and W. H. McAdams, *Trans. Am. Inst. Chem. Engrs.*, **28**, 56–72 (1932)	155, 157
19.	Dryden, H. L., and A. M. Keuthe, *Natl. Advisory Comm. Aeronaut.*, *Rept.* 320 (1929)	152
20.	Elias, F., *Abhandl. aerodyn. Inst., Tech. Hochschule Aachen*, **9**, 10 (1930); *Z. angew. Math. u. Mech.*, **9**, 434 (1929), **10**, 1 (1930); translated in *Natl. Advisory Comm. Aeronaut., Tech. Mem.* 614 (April, 1931)	152
21.	Ergun, S., *Chem. Eng. Progr.*, **48**, 89–94 (1952)	162
22.	Eustice, J., *Proc. Roy. Soc. (London)*, **A84**, 107–118 (1910), **A85**, 119–131 (1911)	150
23.	Fage, A., and H. C. H. Townsend, *Proc. Roy. Soc. (London)*, **A135**, 656–677 (1932)	144
24.	Froessel, W., *Forsch. Gebiete Ingenieurw.*, **7**, 75–84 (1936)	157
25.	Fromm, K., *Z. angew. Math. u. Mech.*, **3**, 339 (1923). Data in *Abhandl. aerodyn. Inst., Tech. Hochschule Aachen*, 1923	158
26.	Gibson, A. H., "Hydraulics and Its Applications," 4th ed., Van Nostrand, New York, 1930	160
27.	Graetz, L., *Z. Math. Physik*, **25**, 316, 375 (1880)	149
28.	Greenhill, A. G., *Proc. London Math. Soc.*, **13**, 43 (1881)	149
29.	Grimison, E. D., *Trans. ASME*, **59**, 583–594 (1937), **60**, 381–392 (1938)	162, 163
30.	Hansen, M., *Abhandl. aerodyn. Inst., Tech. Hochschule Aachen*, **8** (1928); *Z. angew. Math. u. Mech.*, **8**, 185 (1928); translated in *Natl. Advisory Comm. Aeronaut., Tech. Mem.* 585 (October, 1930)	152
31.	Hersey, M. D., and H. S. Snyder, *J. Rheol.*, **3**, 298 (1932)	149
31a.	Huebscher, R. G., *Heating, Piping Air Conditioning*, **19**, 127–135 (December, 1947)	158
32.	Hughes, H. J., and A. T. Safford, "Hydraulics," p. 298, Macmillan, New York, 1911	159
33.	Humble, L. V., W. H. Lowdermilk, and L. G. Desmon, *Natl. Advisory Comm. Aeronaut., Rept.* 1020 (1951)	157
33a.	Isakoff, S. E., and T. B. Drew, "Proceedings of the General Discussion on Heat Transfer," pp. 479–480, Institution of Mechanical Engineers, London, and American Society of Mechanical Engineers, New York, 1951.	147
34.	Jakob, M., *Trans. ASME*, **60**, 384–386 (1938)	162
35.	Jürges, W., *Gesundh.-Ing.*, **19** (1), 1 (1924)	152
36.	von Kármán, T., *J. Aeronaut. Sci.*, **1**, 1–20 (1934); *Engineering*, **148**, 210–213 (1939); *Trans. ASME*, **61**, 705–710 (1939)	154
36a.	Kays, W. M., "An Investigation of Losses of Flow Stream Mechanical Energy at Abrupt Changes in Flow Cross Section," *Tech. Rept.* 1, Navy Contract N6-ONR-251, Task Order U1 (NR-035-104), Stanford Department of Mechanical Engineering, Stanford University, Calif., 1948	159
37.	Keenan, J. H., and E. P. Neumann, *Natl. Advisory Comm. Aeronaut., Tech. Note* 963 (1945)	157, 158
38.	Keevil, C. S., Sc.D. Thesis in Chemical Engineering, Massachusetts Institute of Technology, 1930	149
39.	Keevil, C. S., and W. H. McAdams, *Chem. Met. Eng.*, **36**, 464 (1929)	149
40.	Kemler, E., *Trans.*, HYD 55, 7–32 (1933)	157
41.	Koo, E. C., Sc.D. Thesis in Chemical Engineering, Massachusetts Institute of Technology, 1932. See also ref. 18	155
42.	Kratz, A. P., H. J. Macintire, and R. E. Gould, *Univ. Ill., Eng. Expt. Sta., Bull.* 222 (1931)	158

BIBLIOGRAPHY AND AUTHOR INDEX

	PAGE
43. Kröner, R., *Mitt. Forsch.*, **222**, 1–85 (1920)	160
44. Lamb, H., "Hydrodynamics," 5th ed., pp. 555, 556, Cambridge, London, 1924	149
45. Lea, F., *Phil. Mag.*, **11**, 1235 (1931)	150, 158
46. McAdams, W. H., "Heat Transmission," 1st ed., McGraw-Hill, New York, 1933	149, 157
46a. Martinelli, R. C., C. J. Southwell, H. L. Craig, E. B. Weinberg, N. F. Lansing, and L. M. K. Boelter, *Trans. Am. Inst. Chem. Engrs.*, **38**, 493–530 (1942)	149
47. Mikrjukov, V., *Tech. Phys.* (USSR), **4**, 961–977 (1937)	158
48. Miller, B., *Trans. ASME*, **71**, 357 (1949)	154
48a. Moody, L. F., *Trans. ASME*, **66**, 671–684 (1944); *Mech. Eng.*, **69**, 1005–1006 (1947)	157
49. Nikuradse, J., *Mitt. Forsch.*, 1929, p. 289	160
50. Nikuradse, J., *Ing.-Arch.*, **1**, 306 (1930)	150, 158
51. Nikuradse, J., *Forschungsheft*, **361**, 1–22 (1933); *Petroleum Engr.*, **11** (6), 164–166 (1940), **11** (8), 75–82 (1940), **11** (9), 124–130 (1940), **11** (11), 38–42 (1940), **11** (12), 83 (1940)	153, 154
52. Perry J. H., "Chemical Engineer's Handbook," 3d ed., McGraw-Hill, New York, 1950	150, 159, 161
53. Pigott, R. J. S., *Mech. Eng.*, 55, 497–501 (1933)	157
54. Prandtl, L., *Z. Ver. deut. Ing.*, **77**, 105–113 (1933)	153
55. Reynolds, O., "Scientific Papers of Osborne Reynolds," Vol. I, pp. 81–85, Cambridge, London, 1901	140
56. Schiller, L., *Physik. Z.*, **26**, 566–594 (1925)	144
57. Schutt, H. C., *Trans. ASME*, HYD, **51**, 83 (1929)	159
58. Sieder, E. N., and G. E. Tate, *Ind. Eng. Chem.*, **28**, 1429–1436 (1936)	157
59. Stanton, T. E., *Proc. Roy. Soc.* (*London*), **A85**, 366 (1911); reprinted in *Collected Researches, Natl. Phys. Lab.* (*Teddington, England*), **8**, 75 (1912), **9**, 1 (1913); abstracted in *Trans. Inst. Naval Arch.*, **54**, 48 (1912)	153
60. Stanton, T. E., and J. R. Pannell, *Trans. Roy. Soc.* (*London*), **A214**, 199–224 (1914); reprinted in *Collected Researches, Natl. Phys. Lab.* (*Teddington, England*), **11**, 294 (1914)	155
61. Taylor, G. I., *Proc. Roy. Soc.* (*London*), **A124**, 243 (1929)	150
62. Trahey, J. C., and W. S. Smith, S.M. Thesis in Chemical Engineering, Massachusetts Institute of Technology, 1930	150
63. Van der Hegge Zijnen, B. G., "Measurements of the Velocity Distribution in the Boundary Layer along a Plane Surface," Thesis, Delft, 1924; printed by J. Waltman, Jr., Delft; *Proc. Acad. Sci. Amsterdam*, **31**, 500 (1928)	152, 153
64. White, C. M., *Proc. Roy. Soc.* (*London*), **A123**, 645 (1929)	150
65. White, J. B., S.M. Thesis in Chemical Engineering, Massachusetts Institute of Technology, 1928; data are given in ref. 2, Chap. 9	149
66. Wilson, R. E., W. H. McAdams, and M. Seltzer, *Ind. Eng. Chem.*, **14**, 105–119 (1922); abstracted in *Eng. News-Record*, **89**, 690 (1922)	148

CHAPTER 7. NATURAL CONVECTION

1. Ackermann, G., *Forsch. Gebiete Ingenieurw.*, **3**, 42–50 (1932)	177
2. Bailey, A., and N. C. Lyell, *Engineering*, **147**, 60–62 (1939)	179
2a. Beckmann, W., *Forsch. Gebiete Ingenieurw.*, **2**, 165–178, 213–219 (1931)	182
3. Boussinesq, J., *J. phys.*, **4**, 1, 65 (1902)	171
4. Chilton, T. H., A. P. Colburn, R. P. Genereaux, and H. C. Vernon, *Trans. ASME, Petroleum Mech. Eng.*, **55**, 7–14 (1933); *Petroleum Mech. Eng.*, **55**, 7–14, (1933)	177, 180

5. Colburn, A. P., and O. A. Hougen, [*Ind. Eng. Chem.*, **22**, 522 (1930);] *Univ. Wis., Eng. Expt. Sta., Bull.* 70, p. 29 (1930).................................. 172
6. Davis, A. H., *Phil. Mag.*, **44**, 920 (1922)................................... 177
7. Eberle, C., *Z. Ver. deut. Ing.*, **52**, 481–487, 539–547, 569–574, 626–632, 663–668 (1908); *Mitt. Forschungsarb.*, **78**, 1 (1909)........................... 177
8. Eckert, E. R. G., and E. Soehnghen, *USAF Tech. Rept.* 5747 (1939), Wright Patterson Air Force Base, Dayton, Ohio.................................... 168, 171
9. Eckert, E. R. G., and E. Soehnghen, "Proceedings of the General Discussion on Heat Transfer," pp. 321–323, 381, 387–388, Institution of Mechanical Engineers, London, and American Society of Mechanical Engineers, New York, 1951... 168, 170
10. Fishenden, M., and O. A. Saunders, "An Introduction to Heat Transfer," Oxford, New York, 1950... 180, 182
11. Griffiths, E., and A. H. Davis, *Food Investigation Board, Spec. Rept.* 9, Department of Scientific and Industrial Research, H.M. Stationery Office, London, 1922. See also rev. ed., 1931.. 167
12. Heilman, R. H., *Ind. Eng. Chem.*, **16**, 445–452 (1924).................... 179
13. Hermann, R., *Forschungsheft*, 7 (319), 1–24 (1936)........................... 175
14. Hyman, S. C., C. F. Bonilla, and S. W. Ehrlich, "Preprints of Heat Transfer Symposium," pp. 55–76g, American Institute of Chemical Engineers, 1951. 180
15. Jakob, M., "Heat Transfer," 1st ed., Vol. 1, pp. 534–542, Wiley, New York, 1949.. 181, 182
16. Jakob, M., and W. Linke, *Forsch. Gebiete Ingenieurw.*, **4**, 75–78 (1933)..... 172
17. Jodlbauer, K., *Forsch. Gebiete Ingenieurw.*, **4**, 157–172 (1933)............ 175
18. Keenan, J. H., and J. Kaye, "Gas Tables," Wiley, New York, 1948......... 173
19. Kennard, R. B., *Natl. Bur. Standards J. Research*, **8**, 787 (1932); "Temperature, Its Measurement and Control in Science and Industry," pp. 685–706, Reinhold, New York, 1941... 167, 168, 176
20. King, W. J., III. Free Convection, *Mech. Eng.*, **54**, 347 (1932).......... 172
21. Koch, W., *Gesundh.-Ing.*, **22** (1), 1–27 (1927)........................... 167, 177
22. Kraussold, H., *Forsch. Gebiete Ingenieurw.*, **5**, 186 (1934)................ 182
23. Lorenz, L., *Wied. Ann.*, **13**, 582 (1881)................................... 170
24. McAdams, W. H., "Heat Transmission," 2d ed., McGraw-Hill, New York, 1942.. 176
25. McMillan, L. B., *Trans. ASME*, **37**, 961 (1915)........................... 179
26. Madden, A. J., Jr., and E. L. Piret, "Proceedings of the General Discussion on Heat Transfer," pp. 328–333, 382, 388, Institution of Mechanical Engineers, London, and American Society of Mechanical Engineers, New York, 1951... 180
27. Martinelli, R. C., and L. M. K. Boelter, *Proc. Intern. Congr. Appl. Mech.*, 5th Congr., Massachusetts Institute of Technology, 1938, pp. 578–584..... 177
28. Mull, W., and H. Reiher, *Gesundh.-Ing.*, **28** (1), 1–26 (1930)........... 181, 182
29. Nusselt, W., and W. Jurges, *Z. Ver. deut. Ing.*, **72**, 597–603 (1928)....... 167
30. Ostrach, S., *Natl. Advisory Comm. Aeronaut., Tech. Note* 2635 (February, 1952)... 171
31. Rice, C. W., *Trans. AIEE*, **42**, 653 (1923); abstracted in *J. AIEE*, **42**, 1288 (1923), revised in *Trans. AIEE*, **43**, 131 (1924); abstracted in *Ind. Eng. Chem.*, **16**, 460 (1924); revised in "International Critical Tables," Vol. V, p. 234, McGraw-Hill, New York, 1929..................................... 177
32. Saunders, O. A., *Proc. Roy. Soc. (London)*, **A157**, 278–291 (1936)...... 171, 172
33. Schmidt, E., *Z. ges Kälte-Ind.*, **35**, 213 (1928) 167

BIBLIOGRAPHY AND AUTHOR INDEX

34. Schmidt, E., and W. Beckmann, *Tech. Mech. u. Thermodynam.*, **1**, 341, 391 (1930)... 167, 171, 172
35. Squire, see S. Goldstein, "Modern Developments in Fluid Dynamics," Oxford, New York, 1938.. 172
36. Touloukian, Y. S., G. A. Hawkins, and M. Jakob, *Trans. ASME*, **70**, 13 (1948)... 175
37. Tibus, M., "Elementary Heat Transfer," *Univ. Calif., Syllabus Ser.* 317 (1950).. 170
38. Wamsler, F., *Mitt. Forsch.*, **98, 99**, 1 (1911)................... 177
39. Weise, R., *Forsch. Gebiete Ingenieurw.*, **6**, 281–292 (1935)...... 172

CHAPTER 8. INTRODUCTION TO FORCED CONVECTION

1. Bailey, N. P., *Mech. Eng.*, **53**, 797–804 (1931), **54**, 553 (1932)........... 199
2. Baker, E. M., and A. C. Mueller, *Trans. Am. Inst. Chem. Engrs.*, **33**, 531–558 (1937); *Ind. Eng. Chem.*, **29**, 1067–1072 (1937)...................... 198
3. Baker, E. M., and U. Tsao, *Trans. Am. Inst. Chem. Engrs.*, **36**, 517–539, 783 (1940); *Ind. Eng. Chem.*, **32**, 1115–1121 (1940)..................... 198
4. Biskamp, H., *Z. tech. Physik*, **12**, 30–33 (1931)...................... 188
5. Boelter, L. M. K., and R. W. Lockhart, *Natl. Advisory Comm. Aeronaut., Tech. Note* 2427 (1951)... 199
6. Bowman, R. A., A. C. Mueller, and W. M. Nagle, *Trans. ASME*, **62**, 283–294 (1940).. 194, 195
7. Callendar, H. L., and J. T. Nicolson, *Engineering*, **64**, 481–482 (1897)..... 198
8. Colburn, A. P., *Ind. Eng. Chem.*, **25**, 873–877 (1933)............. 191, 192
9. Colburn, A. P., and O. A. Hougen, *Ind. Eng. Chem.*, **22**, 522 (1930)... 198, 199
10. Elias, F., *Abhandl. aerodyn. Inst., Tech. Hochschule Aachen*, **9**, 10 (1930); *Z. angew. Math. u. Mech.*, **9**, 434 (1929), **10**, 1 (1930); translated in *Natl. Advisory Comm. Aeronaut., Tech. Mem.* 614 (1931)..................... 199
11. Foote, P. D., C. O. Fairchild, and T. R. Harrison, *Natl. Bur. Standards (U.S.), Tech. Paper* 170 (Feb. 16, 1921)................................. 198
12. Gardner, K. A., *Ind. Eng. Chem.*, **33**, 1215–1223 (1941)............ 195
13. Gardner, K. A., *Ind. Eng. Chem.*, **33**, 1495–1500 (1941)............ 196
14. Gardner, K. A., *Ind. Eng. Chem.*, **33**, 1083–1087 (1942)............ 198
15. Hebbard, G. M., and W. L. Badger, *Ind. Eng. Chem., Anal. Ed.*, **5**, 359–362 (1933).. 199
16. Hebbard, G. M., and W. L. Badger, *Trans. Am. Inst. Chem. Engrs.*, **30**, 194–216 (1933–1934); *Ind. Eng. Chem.*, **26**, 420–424 (1934)............ 199
17. Hurd, N. L., *Ind. Eng. Chem.*, **38**, 1266–1271 (1946)................ 196
18. Insinger, T. H., Jr., and H. Bliss, *Trans. Am. Inst. Chem. Engrs.*, **36**, 491–516 (1940).. 199
19. Kays, W. M., A. L. London, and D. W. Johnson, "Gas Turbine Plant Heat Exchangers," pp. 24–25, American Society of Mechanical Engineers, New York, April, 1951... 197
20. Knoblauch, O., and K. Hencky, "Introduction to Accurate Technical Temperature Measurements," p. 59, R. Oldenbourg, Munich and Berlin, 1926. 199
21. Morgan, F. H., and W. E. Danforth, *J. Appl. Phys.*, **21**, 112–113 (1950).. 198
22. Othmer, D. F., and H. B. Coates, *Ind. Eng. Chem.*, **20**, 124–128 (1928).... 198
23. Partridge, E. P., *Univ. Mich., Eng. Research Bull.* 15 (1930)............ 188
24. Patton, E. L., and R. A. Feagan, Jr., *Ind. Eng. Chem.*, **33**, 1237–1239 (1941). 198, 199
25. Roeser, W. F., *J. Research Natl. Bur. Standards*, **7**, 485–494 (1931)....... 199

26. Roeser, W. F., and E. Mueller, *J. Research Natl. Bur. Standards*, **5**, 793 (1930).. 199
27. Smith, I. B., *Trans. AIEE*, **42**, 349 (1923)................................ 199
28. Smith, J. F. D., *Ind. Eng. Chem.*, **22**, 1246 (1930)...................... 198
29. Spear, E. B., and J. B. Purdy, *Ind. Eng. Chem.*, **15**, 842–845 (1923)...... 199
30. Stanton, T. E., *Trans. Roy. Soc.* (*London*), **A190**, 67 (1897)............. 198
31. Sucksmith, W., *Phil. Mag.*, (6) **43**, 223–240 (1922)....................... 198
32. "Temperature, Its Measurement and Control in Science and Industry," Reinhold, New York, 1941.. 198
33. Ten Broeck, H., *Ind. Eng. Chem.*, **30**, 1041–1042 (1938)................. 197
34. Tubular Exchanger Manufacturers Association, "Standards TEMA," 3d ed., New York, 1952.. 189
35. Underwood, A. J. V., *J. Inst. Petroleum Technol.*, **20**, 145–158 (1934)..... 193

CHAPTER 9. HEATING AND COOLING INSIDE TUBES

1. Adams, F. W., G. Broughton, and A. L. Conn, *Ind. Eng. Chem.*, **28**, 537–541 (1936)... 246
2. Bailey, A., and W. F. Cope, *Aeronaut. Research Comm.* (*Great Britain*), *Tech. Rept.* 43, p. 199 (1933).. 248
3. Bays, G. S., Sc.D. Thesis in Chemical Engineering, Massachusetts Institute of Technology, 1936.. 244
4. Bays, G. S., and W. H. McAdams, *Ind. Eng. Chem.*, **29**, 1240–1246 (1937). 245, 246, 247
5. Bernardo, E., and C. S. Eian, *Natl. Advisory Comm. Aeronaut., Wartime Rept.* E-136, originally issued as ARRE5FO7 (August, 1945)............. 222
6. Boelter, L. M. K., R. C. Martinelli, and F. Jonassen, *Trans. ASME*, **63**, 447–455 (1941).. 210
7. Boelter, L. M. K., G. Young, and H. W. Iversen, *Natl. Advisory Comm. Aeronaut., Tech. Note* 1451 (1948)........................ 225, 226, 239
8. Bonilla, C. F., A. Cervi, J. J. Colven, and S. J. Wang, "Preprints of the Heat Transfer Symposium," American Institute of Chemical Engineers, December, 1951.. 223
9. Brinn, M. S., S. J. Friedman, F. A. Gluckert, and R. L. Pigford, *Ind. Eng. Chem.*, **40**, 1050–1061 (1948)... 232
10. Carpenter, F. G., A. P. Colburn, E. M. Schoenburn, and A. Wurster, *Trans. Am. Inst. Chem. Engrs.*, **42**, 165–187 (1946)........................... 243
11. Chapman, D. R., and M. W. Rubesin, *J. Aeronaut. Sci.*, **16**, 547–561 (1949). 224
12. Chen, C. Y., G. A. Hawkins, and H. L. Solberg, *Trans. ASME*, **68**, 99 (1940).. 243
12a. Chilton, T. H., A. P. Colburn, R. P. Genereaux, and H. C. Vernon, *Trans. ASME, Petroleum Mech. Eng.*, **55**, 7–14 (1933)......................... 227
13. Cholette, A., *Chem. Eng. Progr.*, **44**, 81–88 (1948)............... 238, 239
14. Colburn, A. P., *Trans. Am. Inst. Chem. Engrs.*, **29**, 174–210 (1933)....... 219, 220, 232, 249
15. Colburn, A. P., and O. A. Hougen, *Ind. Eng. Chem.*, **22**, 522 (1930); *Univ. Wis., Eng. Expt. Sta., Bull.* 70, p. 29 (1930)............................ 228
16. Cope, W. F., "Proceedings of the General Discussion on Heat Transfer," pp. 453–458, Institution of Mechanical Engineers, London, and American Society of Mechanical Engineers, New York, 1951....................... 218
17. Cope, W. F., *Proc. Inst. Mech. Engrs.* (*London*), **145**, 99–105 (1941)..... 223
18. Davis, E. S., *Trans. ASME*, **65**, 755–759 (1943)................. 242, 243

BIBLIOGRAPHY AND AUTHOR INDEX

19. Deissler, R. G., *Natl. Advisory Comm. Aeronaut., Tech. Note* 2138 (July, 1950).. 210, 217
20. Deissler, R. G., *Natl. Advisory Comm. Aeronaut., Tech. Note* 2242 (December, 1950).. 216, 217
21. Deissler, R. G., *Natl. Advisory Comm. Aeronaut., Tech. Note* 2410 (July, 1951).. (217)
22. Deissler, R. G., and C. S. Eian, *Natl. Aavisory Comm. Aeronaut., Tech. Note* 2629 (February, 1952).. 216, 217, 218
23. Dittus, F. W., and L. M. K. Boelter, *Univ. Calif., Pubs. Eng.*, **2**, 443 (1930) 219
24. Desmon, L. G., and E. W. Sams, *Natl. Advisory Comm. Aeronaut., Research Mem.* E 50 H23 (November, 1950).. 221, 222
25. Drake, R. M., Jr., *J. Appl. Mech.*, **16**, 1–8 (March, 1949).............. 250
26. Drew, T. B., *Trans. Am. Inst. Chem. Engrs.*, **26**, 26 (1931)...... 230, 232, 233
27. Drew, T. B., personal communication, 1938........................... 245
28. Drew, T. B., J. J. Hogan, and W. H. McAdams, *Ind. Eng. Chem.*, **23**, 936–945 (1931); *Trans. Am. Inst. Chem. Engrs.*, **26**, 81 (1931).............. 235
29. Drew, T. B., H. C. Hottel, and W. H. McAdams, *Trans. Am. Inst. Chem. Engrs.*, **32**, 271–305 (1936); *Chem. Eng. Congr. World Power Congr., London*, **3**, 713–745 (1936).. 232
29a. Drexel, R. E., and W. H. McAdams, *Natl. Advisory Comm. Aeronaut., Wartime Rept.* 108 (February, 1945).. 226
30. Eckert, E. R. G., and G. M. Low, *Natl. Advisory Comm. Aeronaut., Tech. Note* 2401 (July, 1951).. (248)
31. Eubank, O. C., and W. S. Proctor, S.M. Thesis in Chemical Engineering, Massachusetts Institute of Technology, 1951........................... 235
32. Fage, A., and V. M. Falkner, *Brit. Advisory Comm. Aeronaut., Rept. Mem.* 1408 (1931).. 249
33. Fitzpatrick, J. P., S. Baum, and W. H. McAdams, *Trans. Am. Inst. Chem. Engrs.*, **35**, 97–107 (1939).. 246
34. Foust, A. S., and G. A. Christian, *Trans. Am. Inst. Chem. Engrs.*, **36**, 541–554 (1940).. 242
35. Frank, A., *Gesundh.-Ing.*, **52**, 541 (1929).. 250
36. Friedman, S. J., and A. C. Mueller, "Proceedings of the General Discussion on Heat Transfer," pp. 138–142, Institution of Mechanical Engineers, London, and American Society of Mechanical Engineers, New York, 1951. 250
37. Graetz, L., *Ann. Physik*, **18**, 79–94 (1883).. 230
38. Graetz, L., *Ann. Physik*, **25**, 337 (1885).. 230
39. Harrison, W. B., and J. R. Menke, *Trans. ASME*, **71**, 797–803 (1949).... 215
40. Haucke, E., *Arch. Wärmewirtsch.*, **116**, 53 (1930).. 250
40a. Hausen, H., *Z. Ver. deut. Ing.*, Beiheft Verfahrenstechnik (4), 91–98 (1943). 241
40b. Hoffmann, E., *Forsch. Gebiete Ingenieurw.*, **11**, 159–169 (1940)............ 210
41. Holden, P. B., Thesis in Chemical Engineering, Massachusetts Institute of Technology, 1923. Data are published in ref. 28....................... 235
42. Humble, L. V., W. H. Lowdermilk, and L. G. Desmon, *Natl. Advisory Comm. Aeronaut., Rept.* 1020 (1951).. 222
43. Isakoff, S. E., and T. B. Drew, "Proceedings of the General Discussion on Heat Transfer," pp. 405–409, 479–480, Institution of Mechanical Engineers, London, and American Society of Mechanical Engineers, New York, 1951.. 208, 214, 215, 216
43a. Jakob, M., "Heat Transfer," Vol. I, p. 551, Wiley, New York, 1949...... 243
44. Jakob, M., and K. A. Rees, *Trans. Am. Inst. Chem. Engrs.*, **37**, 619–648 (1941).. 244

45. Jenkins, R., "Heat Transfer and Fluid Mechanics Institute, Preprints of Papers," pp. 147–148, Stanford University Press, Stanford, Calif., 1951... 215
46. Jeschke, D., *Z. Ver. deut. Ing.*, **69**, 1526 (1925); *Z. Ver. deut. Ing. Erganzungsheft*, **24**, 1 (1925)... 228
47. Jurges, W., *Gesundh.-Ing.*, **19** (1), 1 (1924)........................... 249
48. von Kármán, T., *Proc. Intern. Congr. Appl. Mech.*, 4th Congr., 1934, pp. 54–91; *Engineering*, **148**, 210–213 (1939); *Trans. ASME*, **61**, 705–710 (1939). 210
49. Karim, B., and H. C. Travers, S.M. Thesis in Chemical Engineering Practice, Massachusetts Institute of Technology, 1949. See ref. 55.. 238, 239
50. Kays, W. M., *Tech. Rept.* 14, NR-035-104, Department of Mechanical Engineering, Stanford University, Calif. (June 15, 1951)................ 248
51. Keenan, J. H., and J. Kaye, "Gas Tables," 1948 ed., 3d printing, Wiley, New York, September, 1950... 240
52. Kern, D. Q., and D. F. Othmer, *Trans. Am. Inst. Chem. Engrs.*, **39**, 517–555 (1943).. 235, 237
52a. Keyes, F. G., and D. J. Sandell, Jr., *Trans. ASME*, **72**, 767 (1950)...... 243
53. Kirst, W. E., W. M. Nagle, and J. B. Kastner, *Trans. Am. Inst. Chem. Engrs.*, **36**, 371–394 (1940).. 222
54. Koo, E. C., Sc.D. Thesis in Chemical Engineering, Massachusetts Institute of Technology, 1932... 208
55. Kroll, C. L., Sc.D. Thesis in Chemical Engineering, Massachusetts Institute of Technology, 1951. This reference includes the thesis data of Margolin and of Karim and Travers.. 238, 240
56. Latzko, H., *Z. angew. Math. u. Mech.*, **1**, 268–290 (1921); English translation, *Natl. Advisory Comm. Aeronaut., Tech. Mem.* 1068 (1944).............. 225
57. Lévêque, J., *Ann. mines*, (12), **13**, 201, 305, 381 (1928).................. 231
58. Liepmann, H. W., and G. H. Fila, *Natl. Advisory Comm. Aeronaut., Rept.* 890 (1947).. (224)
59. Lowdermilk, W. H., and M. D. Grele, *Natl. Advisory Comm. Aeronaut., Research Mem.* E50E23 (August, 1950); *Research Mem.* E8L09 (March, 1949)... 222
60. Lubarsky, B., *Natl. Advisory Comm. Aeronaut., Research Mem.* E51G02 (September, 1951).. 215
61. Lyon, R. N., *Chem. Eng. Progr.*, **47**, 75–79 (1951)............. 213, 215, 216
62. McAdams, W. H., "Heat Transmission," 2d ed., McGraw-Hill, New York, 1942... 219, 220
63. McAdams, W. H., T. B. Drew, and G. S. Bays, Jr., *Trans. ASME*, **62**, 627–631 (1940)... 245, 246, 247
64. McAdams, W. H., W. E. Kennel, and J. N. Addoms, *Trans. ASME*, **72** (Fig. 10), 428 (1950), based on data of Keyes and Sandell, Jr., for physical properties of steam (ref. 52a)....................................... 243
65. Margolin, S. V., S.M. Thesis in Chemical Engineering, Massachusetts Institute of Technology, 1950. Data are given by Kroll, ref. 55...... 238, 239
66. Martinelli, R. C., *Trans. ASME*, **69**, 947–959 (1947).................... 210, 211, 212, 213, 214, 216
66a. Martinelli, R. C., and L. M. K. Boelter, *Univ. Calif. (Berkeley), Publs. Eng.*, **5** (2), 23–58 (1942)..................................... 233, 234, 235
67. Martinelli, R. C., C. J. Southwell, G. Alves, H. L. Craig, E. B. Weinberg, N. F. Lansing, and L. M. K. Boelter, *Trans. Am. Inst. Chem. Engrs.*, **38**, 493–530 (1942)... 233
68. Miller, B., "Heat Transfer Lectures," NEPA 804-IER-10, pp. 93–120, AEC, Oak Ridge, Tenn., 11-30-48-800-A13800; *Trans. ASME*, **71**, 357 (1949).. 210

BIBLIOGRAPHY AND AUTHOR INDEX

69. Mizushina, Tokuro, "Proceedings of the General Discussion on Heat Transfer," pp. 191–192, 221, Institution of Mechanical Engineers, London, and American Society of Mechanical Engineers, New York, 1951 ... 243
70. Monrad, C. C., and J. F. Pelton, American Institute of Chemical Engineers, Boston meeting, May, 1942 ... 242
71. Montillon, G. H., K. L. Rohrbach, and W. L. Badger, *Ind. Eng. Chem.*, **23**, 763–769 (1931) ... 237
71a. Mueller, A. C., *Trans. Am. Inst. Chem. Engrs.*, **38**, 613–629 (1942) ... 242
72. Nikuradse, J., *Proc. Intern. Congr. Appl. Mech.*, 3d Congr., Stockholm, 1930, **1**, 239 (1931); *Forschungsheft*, **361**, 1–22 (1933); *Petroleum Engr.*, **11** (6), 164–166 (1940), **11** (8), 75–82 (1940), **11** (9), 124–130 (1940), **11** (11), 38–42 (1940), **11** (12), 83 (1940) ... 210
73. Norris, R. H., and D. D. Streid, *Trans. ASME*, **62**, 525–533 (1940). 231, 239, 248
74. Nusselt, W., *Gesundh.-Ing.*, **49**, 97 (1922) ... 250
75. Nusselt, W., *Z. Ver. deut. Ing.*, **67**, 206 (1923) ... 245, 248
76. Pannell, J. R., *Brit. Aeronaut. Research Comm., Rept. Mem.* 243 (June, 1916) ... 206
77. Parmelee, G. V., and R. G. Huebscher, *Trans. Am. Soc. Heating Ventilating Engrs.*, **53**, 245–284 (1947) ... 250
77a. Pigford, R. L., personal communication; presented at St. Louis meeting of American Institute of Chemical Engineers, December, 1953 ... 235, 236
78. Pohlhausen, E., *Z. angew. Math. u. Mech.*, **1**, 115 (1921) ... 224
79. Poisson, S. D., "Théorie mathématique de la chaleur," Bachelier, Paris, 1335 ... 230
80. Poppendiek, H. F., and L. D. Palmer, *Oak Ridge Natl. Lab.*, ORNL-914 (May, 1952), Oak Ridge, Tenn ... (216), (225)
81. Prandtl, L., *Z. Physik*, **11**, 1072 (1910); *Physik. Z.*, **29**, 487–489 (1928) ... 210
82. Reynolds, O., *Proc. Manchester Lit. Phil. Soc.*, **8** (1874); reprinted in "Scientific Papers of Osborne Reynolds," Vol. II, Cambridge, London, 1901 ... 210
83. Rowley, F. B., A. B. Algren, and J. L. Blackshaw, *Heating, Piping Air Conditioning*, **2**, 501–508 (1930) ... 250
84. Sams, E. W., and L. G. Desmon, *Natl. Advisory Comm. Aeronaut., Research Mem.* E9D12 (June, 1949) ... 222
85. Schlinger, W. G., V. J. Berry, J. L. Mason, and B. H. Sage, "Proceedings of the General Discussion on Heat Transfer," pp. 150–153, 215–217, 221–222, Institution of Mechanical Engineers, London, and American Society of Mechanical Engineers, New York, 1951 ... 215
86. Seban, R. A., and T. T. Shimazaki, *Trans. ASME*, **73**, 803–809 (1951) ... 208, 217, 218
87. Sherwood, T. K., D. D. Kiley, and G. E. Mangsen, *Ind. Eng. Chem.*, **24**, 273–277 (1932) ... 235
88. Sieder, E. N., and G. E. Tate, *Ind. Eng. Chem.*, **28**, 1429–1436 (1936) ... 219, 232, 235, 237, 241
89. Taylor, G. I., *Brit. Advisory Comm. Aeronaut., Rept. Mem.* 272, 31, pp. 423–429 (1916) ... 210
90. Taylor, C. F., and A. Rehbock, *Natl. Advisory Comm. Aeronaut., Tech. Mem., Notes* 331 (1930) ... 250
91. Thomson, A. K. G., *J. Soc. Chem. Ind.*, **56**, 380T–384T (1937) ... 246
92. Trefethen, L. M., "Proceedings of the General Discussion on Heat Transfer," pp. 436–438, Institution of Mechanical Engineers, London, and American Society of Mechanical Engineers, New York, 1951; U.S. Atomic Energy Commission, *Rept.* NP-1788 (1950), Technical Information Service, Oak Ridge, Tenn ... 216, 244

93. Tribus, M., *Trans. ASME*, **73**, 808 (1951).................... 224
94. Wagener, G., *Gesundh.-Ing.*, Beiheft, **1** (24), 5 (1929).................. 250
95. Watsinger, A., and D. G. Johnson, *Forsch. Gebiete Ingenieurw.*, **10**, 182–196 (1939)... 234
96. West, F. B., and A. T. Taylor, *Chem. Eng. Progr.*, **48**, 39–43 (1952)...... 223
97. White, J. B., S.M. Thesis in Chemical Engineering, Massachusetts Institute of Technology, 1928. Data are given in ref. 28........................ 235
98. Wiegand, J. H., *Trans. Am. Inst. Chem. Engrs.*, **41**, 147–153 (1945).. 242, 243
99. Woolfenden, L. B., Thesis in Chemical Engineering, Massachusetts Institute of Technology, 1927... 206, 207
100. Zerban, A. H., Ph.D. Thesis, University of Michigan, 1940. Data are cited by J. H. Wiegand and E. M. Baker, *Trans. Amer. Inst. Chem. Engrs.*, **38**, 569–611 (1942).. 242

CHAPTER 10. HEATING AND COOLING OUTSIDE TUBES

1. Benke, R., *Arch. Wärmewirt.*, **19**, 287–291 (1938)....................... 258
1a. Bergelin, O. P., G. A. Brown, and S. C. Doberstein, *Trans. ASME*, **74**, 953–960 (1952).. 277
1b. Bergelin, O. P., A. P. Colburn, and H. L. Hull, *Eng. Expt. Sta., Bull.* 2 *Univ. Del.*, (June, 1950)... 277
2. Boelter, L. M. K., and R. W. Lockhart, *Natl. Advisory Comm. Aeronaut., Tech. Note* 2427 (July, 1951)... 264
3. Boelter, L. M. K., F. E. Romie, A. G. Guibert, and M. A. Miller, *Natl. Advisory Comm. Aeronaut., Tech. Note* 1452 (1948)....................... 264
4. Bryant, L. W., E. Ower, A. S. Halliday, and V. M. Falkner, *Brit. Aeronaut. Research Comm., Rept. Mem.* 1163 (May, 1928)............................ 265
5. Carrier, W. H., and F. L. Busey, *Trans. ASME*, **33**, 1055 (1911)......... 273
6. Chappel, E. L., and W. H. McAdams, *Trans. ASME*, **48**, 1201–1231 (1926) 273
6a. Chilton, T. H., T. B. Drew, and R. H. Jebeus, *Ind. Eng. Chem.*, **36**, 510–515 (1944).. (266)
7. Cichelli, M. T., *Ind. Eng. Chem.*, **40**, 1032–1039 (1948)................ (262)
8. Colburn, A. P., *Trans. Am. Inst. Chem. Engrs.*, **29**, 174–210 (1933)...... 272
9. Comings, E. W., J. T. Clapp, and J. F. Taylor, *Ind. Eng. Chem.*, **40**, 1076–82 (1948).. 261
10. Compan, P., *Ann. chim. et phys.*, **26** (7), 488–573 (1902)................ 265
11. Dahl, A. I., and E. F. Fiock, *Trans. ASME*, **71**, 153 (1949)............. (261)
11a. Davis, A. H., *Phil. Mag.*, **47**, 972, 1057 (1924); reprinted in *Collected Researches, Natl. Phys. Lab. (Teddington, England)*, **19**, 243 (1926)...... 266, 268
12. Drew, T. B., and W. P. Ryan, *Ind. Eng. Chem.*, **23**, 945–953 (1931); *Trans. Am. Inst. Chem. Engrs.*, **26**, 118–147 (1931)...................... 255, 257
13. Fage, A., *Brit. Aeronaut. Research Comm., Rept. Mem.* 1179 (1928)....... 257
14. Fage, A., and V. M. Falkner, *Brit. Aeronaut. Research Comm., Rept. Mem.* 1369 (1931)... 257
15. Gardner, K. A., *Trans. ASME*, **67**, 621–31 (1945).............. 269, 270, 271
16. Gibson, A. H., *Phil. Mag.*, (6) **47**, 324 (1924)........................ 258
17. Giedt, H. W., M.S. Thesis, University of California, Berkeley............ 257
18. Griffiths, E., and J. H. Awbery, *Engineering*, **136**, 692–694 (1933); *Proc. Inst. Mech. Engrs. (London)*, **125**, 319–382 (1933); *Ice and Cold Storage*, **37**, 14–16 (1934)... 273
19. Grimison, E. D., *Trans. ASME*, **59**, 583–594 (1937), **60**, 381–392 (1938).273, 274
20. Harris, R. G., L. E. Caygill, and R. A. Fairthorne, *Brit. Aeronaut. Research Comm., Rept. Mem.* 1326 (June, 1930)..................................... 265

BIBLIOGRAPHY AND AUTHOR INDEX

21. Hartmann, I., *J. Franklin Inst.*, **218**, 593–612 (1934).................... 265
22. Hauser, E. A., and D. R. Dewey, *J. Phys. Chem.*, **46**, 212 (1942)..... 255, 256
23. Hilpert, R., *Forsch. Gebiete Ingenieurw.*, **4**, 215–224 (1933).... 257, 258, 260, 265
24. Hixson, A. W., and S. J. Baum, *Ind. Eng. Chem.*, **33**, 1433 (1941)....... (266)
25. Huge, E. C., *Trans. ASME*, **59**, 573–581 (1937)........................ 273
26. Hughes, J. A., *Phil. Mag.*, (6) **31**, 118 (1916)...................... 258, 265
26a. Ingebo, R. D., *Natl. Advisory Comm. Aeronaut., Tech. Note* 2368 (July, 1951).. 265
27. Jakob, M., "Heat Transfer," Vol. I, Wiley, New York, 1949......... 265, 269
27a. Kays, W. M., and R. K. Lo, *Tech. Rept.* 15, NR-035-104, Department of Mechanical Engineering, Stanford University, Calif., Aug. 15, 1952.. 274, 275
28. Kennelly, A. E., and H. S. Sanborn, *Proc. Am. Phil. Soc.*, **53**, 55–77 (1914) 258
29. Kennelly, A. E., C. A. Wright, and J. S. van Bylevelt, *Trans. AIEE*, **28**, 363–393 (1909).. 258
30. Kerr, E. W., *Trans. ASME*, **38**, 67 (1916)............................. 258
31. King, W. J., *Trans. ASME*, **65**, (1943).............................. 263
32. King, L. V., *Trans. Roy. Soc. (London)*, **A214**, 373 (1914)............... 258
33. Kramers, H., *Physica*, **12**, 61–80 (1946)................... 258, 265, 267, 268
34. Kreisinger, H., and J. F. Barkley, *U. S. Bur. Mines, Bull.* 145 (1918)..... 262
35. Lautman, L., and W. C. Droege, Air Material Command, AIRL A6118, 50-15-3 (August, 1950).. 266
36. Lindmark, T., *Tek. Tidskr.*, **56**, 125–131 (1926) 273
37. Lohrisch, W., "Handbuch der Experimental Physik," 1929 ed., Vol. 9, Pt. 1, pp. 299, 302, 303; *Mitt. Forsch.*, **322**, 1 (1929)........ 253, 255, 256, 272
38. McAdams, W. H., "Heat Transmission," 2d ed., McGraw-Hill, New York, 1942... 260, 271, 275
38a. McAdams, W. H., and S. D. Turner, *J. Am. Soc. Heating Ventilating Engrs.*, **34**, 385 (1928); discusses Sage, *ibid.*, **33**, 707 (1927)............... 269
38b. Maisel, D. S., and T. K. Sherwood, *Chem. Eng. Progr.*, **46**, 131–138 (1950) 260
39. Maisel, D. S., and T. K. Sherwood, *Chem. Eng. Progr.*, **46**, 172–175 (1950) 261
40. Moffatt, E. M., "Errors in High Temperature Probes for Gases," *ASME, Paper* 48-A-52 (December, 1948).. 262
41. Mullikin, H. F., "Temperature, Its Measurement and Control," pp. 775–804, Reinhold, New York, 1941.. 262
42. Mullikin, H. F., and W. J. Osborn, "Temperature, Its Measurement and Control," pp. 805–829, Reinhold, New York, 1941...................... 262
43. Norris, R. H., and W. A. Spofford, *ASME, Advance Paper*, New York (December, 1941)... 269, 270
44. Nusselt, W., *Z. Math. Mech.*, **10**, 105 (1930)........................... 255
45. Paltz, W. J., and C. E. Starr, Thesis in Chemical Engineering, Massachusetts Institute of Technology, 1931. Data are published in reference..... 256
46. Pierson, O. L., *Trans. ASME*, **59**, 563–572 (1937)................... 273, 274
47. Piret, E. L., W. James, and M. Stacy, *Ind. Eng. Chem.*, **39**, 1088–1103 (1947).. 267, 268
48. Praminik, S. C., *Proc. Indian Assoc. Cultivation Sci.*, **7**, 115–123 (1922)... 252
49. Prandtl, L., *Engineering*, **123**, 627 (1927); *Natl. Advisory Comm. Aeronaut., Tech. Mem.* 452 (1928).. 253
50. Ray, B. B., *Proc. Indian Assoc. Cultivation Sci.*, **5**, 95 (1920)........ 252, 253
51. Reiher, H., *Mitt. Forsch.*, **269**, 1–85 (1925)..... 253, 260, 261, 265, 271, 273, 274
52. Rietschel, H., *Mitt. Prufungsanstalt f. Heizungs u. Luftungseinrichtungen, Konigl. Tech. Hochschule Berlin*, **3** (September, 1910)................ 273, 274
53. Rizika, J. W., and W. M. Rohsenow, *Ind. Eng. Chem.*, **44**, 1168–1171 (1952). 265

	PAGE
54. Rubach, H. L., *Mitt. Forsch.*, **185**, 1–35 (1916)	253
55. Schmidt, E., and Wenner, K. W., *Natl. Advisory Comm. Aeronaut., Tech. Mem.* 1050 (1943)	257
56. Sherwood, T. K., and C. E. Reed, "Applied Mathematics in Chemical Engineering," McGraw-Hill, New York, 1939	270
56a. Tubular Exchanger Manufacturers Association, "Standards TEMA," 3d ed., New York, 1952	276
57. Stewart, R. W., *Trans. Roy. Soc. (London)*, **A184**, 569–590 (1893)	269
58. Theodorsen, T., and W. C. Clay, *Natl. Advisory Comm. Aeronaut., Rept.* 403 (1932)	265
59. Thoma, H., "Hochleistungskessel," Springer, Berlin, 1921	255
60. Thomson, A. S. T., A. W. Scott, A. McK. Laird, and H. S. Holden, "Proceedings of the General Discussions on Heat Transfer," pp. 177–180, Institution of Mechanical Engineers, London, and American Society of Mechanical Engineers, New York, 1951	260
61. Tinker, T., "Proceedings of the General Discussion on Heat Transfer," pp. 89–116, Institution of Mechanical Engineers, London, and American Society of Mechanical Engineers, New York, 1951	276, 278, 279, 280
62. Tucker, W. B., S.M. Thesis in Chemical Engineering, Massachusetts Institute of Technology, 1936	272, 273
62a. Ulsamer, J., *Forsch. Gebiete Ingenieurw.*, **3**, 94 (1932)	267
63. University of Delaware, Chemical Engineering Department, Newark, Del.	279
64. Vyroubov, V., *J. Tech. Phys. (USSR)*, **9**, 1923–1931 (1939)	265, 266
65. Wells, W. C., "Essay on Dew," Constable, London, 1818	262
66. Winding, C. C., *Ind. Eng. Chem.*, **30**, 942–947 (1938)	274, 275
67. Winding, C. C., and Cheney, A. J., Jr., *Ind. Eng. Chem.*, **40**, 1087–1094 (1948)	256, 257
68. Williams, G. C., Sc.D. Thesis in Chemical Engineering, Massachusetts Institute of Technology, 1942	265, 266

CHAPTER 11. COMPACT EXCHANGERS, PACKED AND FLUIDIZED SYSTEMS

1. Ambrosio, A., C. D. Coulbert, R. P. Lipkis, P. F. O'Brien, and F. E. Romie, *ASME, Paper* 51-SA-34	298
2. Ackermann, G., *Z. angew. Mech.*, **11**, 192 (1931)	296
3. Aronson, D., *ASME, Paper* 51-A-107	290
4. Aronson, D., *Trans. ASME*, **72**, 967–978 (1950)	290
5. Baerg, A., J. Klassen, and P. E. Gishler, *Can. J. Research*, **F28**, 287–307 (1950)	301, 303, 305
5a. Brinn, M. S., S. J. Friedman, F. A. Gluckert, and R. L. Pigford, *Ind. Eng. Chem.*, **40**, 1050–1061 (1948)	299
6. Bunnell, D. G., H. B. Irvin, R. W. Olson, and J. M. Smith, *Ind. Eng. Chem.*, **41**, 1977–1981 (1949)	291
7. Campbell, J. R., and F. Rumford, *J. Soc. Chem. Ind.*, **69**, 373–377 (1950)	306
8. Colburn, A. P., *Purdue Univ., Eng. Bull.*, **26**, 47–50 (January, 1942)	288
9. Coppage, J. E., and A. L. London, *ASME, Paper* 52-A-93	162, 298
10. Damköhler, in A. Eucken and M. Jakob (eds.), "Der Chemie Ingenieur," Vol. III, Pt. 1, pp. 359–485; Akademische Verlagsgesellschaft, m.b. H., Leipzig, 1937	290
11. Deissler, R. G. and C. S. Eian, *Natl. Advisory Comm. Aeronaut., Research Mem.* E52C05 (June, 1952)	291

BIBLIOGRAPHY AND AUTHOR INDEX

	PAGE
12. Dow, W. M., and M. Jakob, *Chem. Eng. Progr.*, **47**, 637–648 (1951)	304
13. Fairbanks, D. F., Sc.D. Thesis in Chemical Engineering, Massachusetts Institute of Technology, 1953	306
14. Felix, J. R., and W. K. Neill, "Preprints of Heat Transfer Symposium," pp. 123–170, American Institute of Chemical Engineers, December, 1951	292
15. Fong, J. T., *ASME*, Paper 51-A-98	290
16. Gamson, B. W., G. Thodos, and O. A. Hougen, *Trans. Am. Inst. Chem. Engrs.*, **39**, 1–35 (1943)	295
17. Gilliland, E. R., *Chem. in Can.*, 119–127 (1950)	299, 300
18. Gilliland, E. R., and E. A. Mason, *Ind. Eng. Chem.*, **41**, 1191–1196 (1949), **44**, 218–224 (1952)	299
19. Girouard, H. D., personal communication	303
20. Hausen, H., *Tech. Mech. u. Thermodynam.*, **1**, 219 (1930); *Z. angew. Math. Mech.*, **11**, 105 (1931)	296, 297
21. Hausen, H., *Z. angew. Math. Mech.*, **9**, 173–200 (1929)	296
22. Heilingenstadt, W., "Regeneratoren, Rekuperatoren, Winderhitzer," p. 96, Spamer, Leipzig, 1931	296
23. Hottel, H. C., personal communication	296, 297
24. Hougen, O. A., and E. L. Piret, *Chem. Eng. Progr.*, **47**, 295–303 (1951)	292
25. Hryniszak, W., "Proceedings of the General Discussion on Heat Transfer," pp. 460–464, 475, 477–479, Institution of Mechanical Engineers, London, and American Society of Mechanical Engineers, New York, 1951	298
26. *Ind. Eng. Chem.*, **41**, 1249–1250 (1949)	299
27. Kays, W. M., *Tech. Rept.* 14, Navy Contract N-ONR-251 T.O. 6, Stanford University Press, Stanford, Calif., 1951	288, 289
28. Kays, W. M., and A. L. London, *Trans. ASME*, **72**, 1075–1085, 1086–1097 (1950)	288
29. Kays, W. M., and A. L. London, "Proceedings of the General Discussion on Heat Transfer," pp. 127–132, Institution of Mechanical Engineers, London, and American Society of Mechanical Engineers, New York, 1951	288
30. Kays, W. M., A. L. London, and D. W. Johnson, "Gas Turbine Plant Heat Exchangers," American Society of Mechanical Engineers, 1951	287, 288, 289
31. Latinen, G. A., and R. H. Wilhelm, Paper presented at Annual Christmas Symposium, Division of Industrial and Engineering Chemistry, American Chemical Society, New Haven, December, 1952	294
32. Leva, M., *Ind. Eng. Chem.*, **39**, 857–862 (1947)	293
33. Leva, M., *Ind. Eng. Chem.*, **42**, 2498–2501 (1950)	293
34. Leva, M., "Proceedings of the General Discussion on Heat Transfer," pp. 421–425, 480, Institution of Mechanical Engineers, London, and American Society of Mechanical Engineers, New York, 1951	304
35. Leva, M., and M. Grummer, *Ind. Eng. Chem.*, **46**, 415–419 (1948)	293
36. Leva, M., and M. Grummer, *Chem. Eng. Progr.*, **48**, 307–313 (1952)	304
37. Leva, M., M. Weintraub, and M. Grummer, *Chem. Eng. Progr.*, **45**, 563–572 (1949)	302, 304
38. Leva, M., M. Weintraub, M. Grummer, M. Pollchik, and H. H. Storch, *U.S. Bur. Mines, Bull.* 504 (1951)	301, 302
39. Levenspiel, O., and J. S. Walton, "Proceedings of Heat Transfer and Fluid Mechanics Institute," pp. 139–146, Berkeley, Calif., 1949	302
40. Lewis, W. K., E. R. Gilliland, and W. C. Bauer, *Ind. Eng. Chem.*, **41**, 1104–1117 (1949)	302
41. London, A. L., and W. M. Kays, *Trans. ASME*, **73**, 529–542 (1951)	290

	PAGE
42. London, A. L., and C. K. Ferguson, *Trans. ASME*, **71**, 17–26 (1949)	288
43. London, A. L., and W. M. Kays, *Trans. ASME*, **72**, 611–621 (1950)	283
44. Lubojatzky, S., *Metall u. Erz*, **28**, 205–214 (1931)	296
45. McCune, L. H., and R. H. Wilhelm, *Ind. Eng. Chem.*, **41**, 1124–1134 (1949)	295
46. Mickley, H. S., and C. A. Trilling, *Ind. Eng. Chem.*, **41**, 1135–1147 (1949)	303, 306
47. Miller, C. O., and A. K. Logwinuk, *Ind. Eng. Chem.*, **43**, 1220–1226 (1951)	301, 306
48. Munro, W. D., and N. R. Amundson, *Ind. Eng. Chem.*, **42**, 1481–1488 (1950)	299
49. Norris, R. H., and W. A. Spofford, *Trans. ASME*, **64**, 489–496 (1942)	288
50. Nusselt, W., *Ver. deut. Ing.*, **71**, 85 (1927), **72**, 1052 (1928)	296
51. Polack, J. A., Sc.D. Thesis in Chemical Engineering, Massachusetts Institute of Technology, 1948	290
52. Ruhl, C., *Wärme*, **55**, 189 (1932)	298
53. Rummel, K., *J. Inst. Fuel*, **4**, 160–173 (1931)	296, 298
54. Rummel, K., *Arch. Eisenhüttenw.*, **4**, 367 (1931)	298
55. Rummel, K., and A. Schack, *Stahl u. Eisen*, **49**, 1300 (1929)	296
56. Saunders, O. A., and S. Smoleniec, *Appl. Mechanics Revs.*, November, 1950, *Proc. 7th Intern. Congr. Appl. Mech.*, **3**, 91–105 (1948)	298
57. Schack, A., *Arch. Eisenhüttenw.*, **2**, 223, 481 (1929); *Stahl u. Eisen*, **51**, 163–164 (1931)	296
58. Schack, A., "Der Industrielle Wärmeübergang," 3d ed., pp. 234–288, Verlag Stahlheisen m.b. H., Düsseldorf, 1948	296
59. Schultz, B. H., "Proceedings of the General Discussion on Heat Transfer," pp. 440–443, Institution of Mechanical Engineers, London, and American Society of Mechanical Engineers, New York, 1951	296
60. Schumann, T. E. W., and V. Voss, *Fuel*, **13**, 249 (1934)	290
61. Singer, E., and R. H. Wilhelm, *Chem. Eng. Progr.*, **46**, 343–357, 1950	293, 294
62. Trinks, W., "Industrial Furnaces," Vol. I, p. 152, Wiley, New York, 1926	296
63. van Heerden, C., P. Nobel, and D. W. van Krevelen, "Proceedings of the General Discussion on Heat Transfer," pp. 358–360, 389, Institution of Mechanical Engineers, London, and American Society of Mechanical Engineers, New York, 1952	305
64. Verschoor, H., and G. C. A. Schuit, *Appl. Sci. Research*, **A2**, 97–119 (1950)	293
65. Vreedenberg, H. A., "Proceedings of the General Discussion on Heat Transfer," pp. 373–375, 389, 487, Institution of Mechanical Engineers, London, and American Society of Mechanical Engineers, New York, 1951	306
66. Widell, T. A., and S. I. Juhasz, *Trans. Roy. Inst. Technol.*, Stockholm, (54), 1–50 (1952)	296
67. Wilhelm, R. H., W. C. Johnson, R. Wynkoop, and D. W. Collier, *Chem. Eng. Progr.* **44**, 105–116 (1948)	291
68. Wilke, C. R., and O. A. Hougen, *Trans. Am. Inst. Chem. Engrs.*, **41**, 445–451 (1945)	295

CHAPTER 12. HIGH VELOCITY FLOW; RAREFIED GASES

1. Alancraig, C. R., and L. A. Bromley, *Chem. Eng. Progr.*, **48**, 357–361 (1952)	324
2. Amdur, I., *J. Chem. Phys.*, **14**, 339–342 (1946)	320
3. Brown, G. P., A. DiNardo, G. K. Cheng, and T. K. Sherwood, *J. Appl. Phys.*, **17**, 802–813 (1946)	323, 324

BIBLIOGRAPHY AND AUTHOR INDEX

PAGE

4. Brown, W. B., and P. L. Donoughe, *Natl. Advisory Comm. Aeronaut., Tech. Note* 2479 (1951)... 315, 322
5. Drake, R. M., Jr., and E. D. Kane, "Proceedings of the General Discussion on Heat Transfer," pp. 117–121, Institution of Mechanical Engineers, London, and American Society of Mechanical Engineers, New York, 1951... 322
6. Dushman, S., "Scientific Foundations of Vacuum," Wiley, New York, 1949
   320, 322
7. Eckert, E., and W. Weise, *Forsch. Gebiete Ingenieurw.*, **13**, 246–254 (1942). 312
8. Fischer, W. W., and R. H. Norris, *Trans. ASME*, **71**, 457–469 (1949).... 315
9. Garbett, C. R., *Dept. Mech. Eng.*, Stanford Univ., Tech. Rept. 1 + S − 4 (Oct. 1, 1951).. 315
10. Higgins, R. W., and C. C. Pappas, *Natl. Advisory Comm. Aeronaut., Tech. Note* 2351 (1951)... 312
11. Hottel, H. C., and A. Kalitinsky, *J. Appl. Mechanics*, **12**, A–25–32 (1945)
    312, 313
12. Hunsaker, J. C., and B. G. Rightmire, "Engineering Applications of Fluid Mechanics," McGraw-Hill, New York, 1947........................... 316
13. Johnson, H. A., and M. W. Rubesin, *Trans. ASME*, **71**, 447–456 (1949)
    311, 315
14. Kaye, J., J. H. Keenan, K. K. Klingensmith, G. M. Ketchum, and T. Y. Toong, *ASME Paper* 51–A–29(a)..................................... 312
15. Kaye, J., J. H. Keenan, and R. H. Shoulberg, "Proceedings of the General Discussion on Heat Transfer," pp. 133–137, 221, Institution of Mechanical Engineers, London, and American Society of Mechanical Engineers, New York, 1951... 312
16. Kaye, J., T. Y. Toong, and R. H. Shoulberg, *ASME, Paper* 51–A–29(b).. 312
17. Keenan, J. H., "Thermodynamics," Wiley, New York, 1941............. 316
18. Kennard, E. H., "Kinetic Theory of Gases," McGraw-Hill, New York, 1938.
    319, 322
19. Knudsen, M., *Ann. Physik*, **34**, 593–656 (1911)....................... 320
20. Kovasznay, L. S. G., *J. Aeronaut. Sci.*, **17**, 565–572 (1950).............. 315
21. Kundt, A., and E. Warburg, *Pogg. Ann.*, **155**, 337, 525 (1875).......... 323
22. McAdams, W. H., L. A. Nicolai, and J. H. Keenan, *Trans. Am. Inst. Chem. Engrs.*, **42**, 907–925 (1946); also *Natl. Advisory Comm. Aeronaut., Tech. Note* 985 (1945).. 313, 314
23. Madden, A. J., and E. L. Piret, "Proceedings of the General Discussion on Heat Transfer," pp. 328–333, 388, Institution of Mechanical Engineers, London, and American Society of Mechanical Engineers, New York, 1951. 321
24. Maxwell, J. C., "The Scientific Papers of James Clerk Maxwell," Vol. 2, p. 708, University Press, Cambridge, 1890.............................. 323
25. Nielsen, J. N., *Natl. Advisory Comm. Aeronaut.*, ARR L4C16 (October, 1944); reissued as *Wartime Rept.* L–179............................ 316, 318
26. Oppenheim, A. K., *J. Aeronaut. Sci.*, **20**, 49–58 (1953)................ 322
27. Rubesin, M. W., and H. A. Johnson, *Trans. ASME*, **71**, 383–388 (1949).. 311
28. Scadron, M. D., and I. Warshawsky, *Natl. Advisory Comm. Aeronaut., Tech. Note* 2599 (January, 1952).. 315
29. Scherrer, R., *Natl. Advisory Comm. Aeronaut., Rept.* 1055 (1951)..... 313, 315
30. Seban, R. A., Ph.D. Thesis, University of California, Berkeley, 1948..... 312
31. Shapiro, A. H., and W. R. Hawthorne, *Trans. ASME, J. Appl. Mechanics*, **14**, A317–A336 (1947)... 316, 319
32. Sibulkin, M., and W. K. Koffel, *Natl. Advisory Comm. Aeronaut., Tech. Note* 2067 (March, 1950).. 319

33. Slack, E. G., *Natl. Advisory Comm. Aeronaut., Tech. Note* 2686 (April, 1952).... 321
34. Stalder, J. R., G. Goodwin, and M. O. Creager, *Natl. Advisory Comm. Aeronaut., Tech. Note* 2244 (December, 1950); also see *Natl. Advisory Comm. Aeronaut., Rept.* 1032 (1951)................................. 321
35. Stalder, J. R., G. Goodwin, and M. O. Creager, *Natl. Advisory Comm. Aeronaut., Tech. Note* 2438 (August, 1951)............................. 321
36. Stalder, J. R., G. Goodwin, and M. O. Creager, "Proceedings of the General Discussion on Heat Transfer," pp. 143–149, 223, Institution of Mechanical Engineers, London, and American Society of Mechanical Engineers, New York, 1951... 320, 322
37. Stine, H. A., and R. Scherrer, *Natl. Advisory Comm. Aeronaut., Tech. Note* 2664 (March, 1952).. 313
38. Tsien, H. S., *J. Aeronaut. Sci.*, **13**, 653–664 (1946).................. 320
39. Valerino, M. F., *Natl. Advisory Comm. Aeronaut., Research Mem.* E8G23 (October, 1948).. 319
40. Valerino, M. F., and R. B. Doyle, *Natl. Advisory Comm. Aeronaut., Tech. Note* 2328 (April, 1951).. 318
41. Wiedman, M. L., and P. R. Trumpler, *Trans. ASME*, **68**, 57–64 (1946)..... 320

CHAPTER 13. CONDENSING VAPORS

1. Anne, J. W., *Power Plant Eng.*, **35**, 1018–1019 (1931)................... 347
2. Badger, W. L., *Trans. Am. Inst. Chem. Engrs.*, **33**, 441–446 (1937); *Ind. Eng. Chem.*, **29**, 910–912 (1937)................................... 333, 334
3. Badger, W. L., C. C. Monrad, and H. W. Diamond, *Ind. Eng. Chem.*, **22**, 700 (1930)... 333, 334
4. Baker, T., *Ind. Eng. Chem.*, **27**, 977 (1935).......................... 360
5. Baker, E. M., E. W. Kazmark, and G. W. Stroebe, *Trans. Am. Inst. Chem. Engrs.*, **35**, 127–134 (1939); *Ind. Eng. Chem.*, **31**, 214–218 (1939)......... 333
6. Baker, E. M., and A. C. Mueller, *Trans. Am. Inst. Chem. Engrs.*, **33**, 531–558 (1937); *Ind. Eng. Chem.*, **29**, 1067–1072 (1937)........ 339, 340, 352, 357
7. Baker, E. M., and U. Tsao, *Trans. Am. Inst. Chem. Engrs.*, **36**, 517–539, 783 (1940); *Ind. Eng. Chem.*, **32**, 1115–1121 (1940)................ 340, 352, 357
8. Baum, S., S.M. Thesis in Chemical Engineering, Massachusetts Institute of Technology, 1936.. 348
9. Bays, G. S., and L. M. Blenderman, S.M. Thesis in Chemical Engineering, Massachusetts Institute of Technology, 1935......................... 348
10. Beatty, K. O., Jr., E. B. Finch, and E. H. Schoenborn, *Agr. Eng. Univ. N. Carolina, Dept. Eng. Research, Carolina State Coll., Reprint Bull.* 24 (August, 1951); "Proceedings of the General Discussion on Heat Transfer," pp. 32–37, Institution of Mechanical Engineers, London, and American Society of Mechanical Engineers, New York, 1951...................... 364
11. Billings, R. T., and W. F. Wadt, S.M. Thesis, Massachusetts Institute of Technology, 1933.. 347
12. Boelter, L. M. K., *Heating, Piping Air Conditioning*, **11**, 639–647 (1939); "Heating, Ventilating, and Air Conditioning Guide," p. 251, American Society of Heating and Ventilating Engineers, New York, 1941; London, A. L., W. E. Mason, and L. M. K. Boelter, *Trans. ASME*, **62**, 41 (1941).. 352
13. Bosnjakovic, F., *Forsch. Gebiete Ingenieurw.*, **3**, 135 (1932)............ 351
14. Braunlich, R. H., S.B. Thesis in Chemical Engineering, Massachusetts Institute of Technology, 1940....................................... 336

15. Bromley, L. A., *Ind. Eng. Chem.*, 44, 2966 (1952)..................... 331
16. Carpenter, E. F., and A. P. Colburn, "Proceedings of General Discussion on Heat Transfer," pp. 20–26, Institution of Mechanical Engineers, London, and American Society of Mechanical Engineers, New York, 1951......... 336
17. Chilton, T. H., and A. P. Colburn, *Ind. Eng. Chem.*, 27, 255–260 (1935).. 359
18. Chilton, T. H., A. P. Colburn, R. P. Genereaux, and H. C. Vernon, *Trans. ASME, Petroleum Mech. Eng.*, 55, 7–14 (1933)........................ 337
19. Chu, J. C., R. K. Flitcraft, M. R. Holeman, *Ind. Eng. Chem.*, 41, 1789–1794 (1949).. 346
20. Coffey, B. H., and G. S. Dauphinee, *Am. Soc. Refrig. Engrs. J.*, 8, 177–202 (1921), 9, 74 (1922).. 360
21. Coffey, B. H., and G. A. Horne, *Am. Soc. Refrig. Engrs. J.*, 7, 173–201 (1920). 360
22. Cogan, C. A., S.M. Thesis in Chemical Engineering, Massachusetts Institute of Technology, 1934... 354
23. Colburn, A. P., personal communications, 1932, 1940................... 346
24. Colburn, A. P., *Trans. Am. Inst. Chem. Engrs.*, 30, 187–193 (1933–34); *Ind. Eng. Chem.*, 26, 432–434 (1934)..................................... 335
25. Colborn, C. E., *Power*, 58, 803 (1923)............................... 347
26. Colburn, A. P., "Proceedings of the General Discussion on Heat Transfer," pp. 1–11, Institution of Mechanical Engineers, London, and American Society of Mechanical Engineers, New York, 1951..................... (330)
27. Colburn, A. P., and T. B. Drew, *Trans. Am. Inst. Chem. Engrs.*, 33, 197–212 (1937).. 352
28. Colburn, A. P., and O. A. Hougen, *Univ. Wis., Eng. Expt. Sta., Bull.* 70, p. 29 (1930); *Ind. Eng. Chem.*, 26, 1178–1182 (1934)................ 329, 355, 356
29. Colburn, A. P., L. L. Millar, and J. W. Westwater, *Trans. Am. Inst. Chem. Engrs.*, 38, 447–468 (1942).. 338
30. Cooper, C. M., T. B. Drew, and W. H. McAdams, *Trans. Am. Inst. Chem. Engrs.*, 30, 158–169 (1933–1934); *Ind. Eng. Chem.*, 26, 428–431 (1934).... 331
31. Drew, T. B., personal communication, 1935............................ 350
32. Drew, T. B., personal communication, 1938...................... 330, 332
33. Drew, T. B., W. M. Nagle, and W. Q. Smith, *Trans. Am. Inst. Chem. Engrs.*, 31, 605–621 (1935).. 347
34. Emmons, H., *Trans. Am. Inst. Chem. Engrs.*, 35, 109–122 (1939)..... 347, 348
35. Fatica, N., and D. L. Katz, *Chem. Eng. Progr.*, 45, 661–674 (1949)........ 347
36. Fitzpatrick, J. P., S.M. Thesis in Chemical Engineering, Massachusetts Institute of Technology, 1936; also research reports, February and September, 1937; data are published in ref. 37............................. 348
37. Fitzpatrick, J. P., S. Baum, and W. H. McAdams, *Trans. Am. Inst. Chem. Engrs.*, 35, 97–107, (1939).................................... 348, 349, 350
38. Friedman, S. J., and C. O. Miller, *Ind. Eng. Chem.*, 33, 885–891 (1941)... 331
39. *Gas J.*, 196, 135 (1931); *Z. Ver. deut. Ing.*, 75, 1193 (1931).............. 347
40. Geibel, C., *Mitt. Forsch.*, 242, 1–98 (1921)...................... 360, 361
41. Ginabat, A., *Wärme*, 47, 573–578, 588–592 (1924)..................... 332
42. Goodman, W., *Heating, Piping Air Conditioning*, 10, 697–701, 707, 777–781 (1938), 11, 13–19, 83–86, 157–160, 233–236, 305–308 (1939)............. 358
43. Grigull, U., *Forsch. Gebiete Ingenieurw.*, 13, 49–57 (1942)................ 335
44. Grober, H., "Wärmeübertragung," Springer, Berlin, 1926................ 329
45. Grosvenor, W. M., *Trans. Am. Inst. Chem. Engrs.*, 1, 184–202 (1908)..... 357
46. Hagenbuch, W. H., S.M. Thesis in Chemical Engineering, Massachusetts Institute of Technology, 1941................................... 333, 346

47. Hampson, H., "Proceedings of the General Discussion on Heat Transfer," pp. 58–61, 84, Institution of Mechanical Engineers, London, and American Society of Mechanical Engineers, New York, 1951.................... 333, 334
47a. Hardy, J. K., K. C. Hales, and G. Mann, *Food Investigation Board, Spec. Rept.* 54, Department of Scientific and Industrial Research. H.M. Stationery Office, London, 1951.. 364
48. Haselden, G. G., and S. Prosad, *Inst. Chem. Engrs., Low Temperature Group Phys. Soc. (London), Advance Copy* (Nov. 8, 1949)........................ 353
49. Hayes, V. R., and J. A. Bartol, M.S. Thesis in Naval Architecture, Massachusetts Institute of Technology, 1944................................. 349
50. Hebbard, G. M., and W. L. Badger, *Ind. Eng. Chem., Anal. Ed.*, **5**, 359–362 (1933)... 333
51. Hensel, S. L., Jr., and R. E. Treybal, *Chem. Eng. Progr.*, **48**, 362–370 (1952) 363
52. Holloway, F. A., personal communication, 1940....................... 332
53. Horne, G., *Refrig. Eng.*, **9**, 143 (1922)................................ 344
54. Horne, G., and F. Ophuls, *Refrig. Eng.*, **11**, 1–13 (1924)................ 344
55. Jakob, M., *Mech. Eng.*, **58**, 643–660, 729–739 (1936); reprinted in *Univ. Ill., Bull.* 34, No. 37, pp. 1–75 (1937)................................. 329
56. Jakob, M., and S. Erk, *Mitt. Forsch.*, **310**, 1 (1928); *Z. Ver. deut. Ing.*, **73**, 176, 761 (1929)... 351
57. Jakob, M., S. Erk, and H. Eck, *Forsch. Gebiete Ingenieurw.*, **3**, 161 (1932); *Physik. Z.*, **36**, 73 (1935); Jakob, M., *Mech. Eng.*, **58**, 729 (1936)......... 336
58. Jeffrey, J. O., and J. R. Moynihan, *Mech. Eng.*, **55**, 751–754 (1933)... 347, 351
59. Johnstone, H. F., and A. D. Singh, *Ind. Eng. Chem.*, **29**, 286–298 (1937).. 360
60. Kaiser, F., *Z. bayer. Revisions-Ver.*, **33**, 167 (1929); *Arch. Wärmewirtsch.*, **11**, 247–250 (1930)... 351
61. Kamei, S., T. Mizushina, S. Kifune, and T. Koto, *Chem. Eng. (Japan)*, **14**, 53–60 (1950)... 364
62. Katz, D. L., and J. M. Geist, *Trans. ASME*, **70**, 907–914 (1948).......... 346
63. Kirkbride, C. G., *Ind. Eng. Chem.*, **25**, 1324–1331 (1933)......... 340, 352, 354
64. Kirkbride, C. G., *Trans. Am. Inst. Chem. Engrs.*, **30**, 170–186 (1933–1934); *Ind. Eng. Chem.*, **26**, 425–428 (1934)............................. 334, 335
65. Kirschbaum, E., *Arch. Wärmewirtsch.*, **12**, 265 (1931)................... 351
66. Kratz, A. P., H. J. Macintire, and R. E. Gould, *Univ. Ill., Eng. Expt. Sta., Bull.* 171 (1927), 186 (1928), 209 (1930)............................... 343
67. Krujilin, G., *Tech. Phys. (USSR)*, **5**, 59–66 (1938), **5**, 289–297 (1938)..... 329
68. Lewis, W. K., *Trans. Am. Inst. Chem. Engrs.*, **20**, 9 (1927); *Chem. Met. Eng.*, **34**, 735 (1927).. 355
69. Lockhart, C., personal communication, 1929; data abstracted in ref. 71... 355
70. London, A. L., W. E. Mason, and L. M. K. Boelter, *Trans. ASME*, **62**, 41 (1940).. 360, 361
71. McAdams, W. H., "Heat Transmission," 1st ed., McGraw-Hill, New York, 1933.. (359)
72. McAdams, W. H., "Heat Transmission," 2d ed., McGraw-Hill, New York, 1942.. 351, 360
73. McAdams, W. H., and T. H. Frost, *Ind. Eng. Chem.*, **14**, 13–18 (1922).... 340
74. McAdams, W. H., J. B. Pohlenz, and R. C. St. John, *Chem. Eng. Progr.*, **45**, 241–252 (1949)... 361
75. McAdams, W. H., T. K. Sherwood, and R. L. Turner, *Trans. ASME*, **48**, 1233 (1926)... 346
76. Meisenburg, S. J., R. M. Boarts, and W. L. Badger, *Trans. Am. Inst. Chem. Engrs.*, **31**, 622–637 (1935), **32**, 100–104 (1936)...................... 333, 334

BIBLIOGRAPHY AND AUTHOR INDEX

PAGE

77. Merkel, F., "Die Grundlagen der Wärmeübertragung," T. Steinkopf, Leipzig, 1927... 329, 351
78. Merkel, F., *Z. ges. Kälte-Ind.*, **34**, 117 (1927)............................. 358
79. Mickley, H. S., *Chem. Eng. Progr.*, **45**, 739–745 (1949)............. 363. 364
80. Monrad, C. C., and W. L. Badger, *Trans. Am. Inst. Chem. Engrs.*, **24**, 84–116 (1930); *Ind. Eng. Chem.*, **22**, 1103 (1930)........................ 329, 334
81. Montillon, G. H., K. L. Rohrbach, and W. L. Badger, *Ind. Eng. Chem.*, **23**, 763–769 (1931)... 340
81a. Moore, W. T., *Mech. Eng.*, **55**, 748–750 (1933)....................... (347)
82. Nagle, W. M., Sc.D. Thesis in Chemical Engineering, Massachusetts Institute of Technology, 1934... 347
83. Nagle, W. M., G. S. Bays, Jr., L. M. Blenderman, and T. B. Drew, *Trans. Am. Inst. Chem. Engrs.*, **31**, 593–604 (1935)........................... 348
84. Neiderman *et al.*, *Heating, Piping Air Conditioning*, **13**, 591–597 (1941). 360, 361
85. Nusselt, W., *Z. Ver. deut. Ing.*, **60**, 541, 569 (1916)................... 329, 338
86. Orrok, G. A., *Trans. ASME*, **32**, 1773 (1910); **32**, 1139 (1910)........... 345
87. Othmer, D. F., *Ind. Eng. Chem.*, **21**, 576 (1929)....... 340, 342, 343, 353, 355
88. Othmer, D. F., and R. E. White, *Trans. Am. Inst. Chem. Engrs.*, **37**, 135–156 (1941)... 340
89. Parr, S. W., *Engineer*, **113**, 559 (1921)................................... 338
90. Patton, E. L., and R. A. Feagan, Jr., *Ind. Eng. Chem.*, **33**, 1237–1239 (1941) 340
91. Perry, J. H., "Chemical Engineers' Handbook," 3d ed., McGraw-Hill, New York, 1950... 360
92. "Report Prime Movers Committee," *NELA Publ.* 069 (July, 1930)... 346, 347
92a. Robinson, C. S., *Refrig. Eng.*, **9**, 169–173 (1922); *Mech. Eng.*, **45**, 99–102 (1923)... 361
93. Schmidt, E., W. Schurig, and W. Sellschopp, *Tech. Mech. u. Thermodynam.*, **1**, 53 (1930).. 325, 329, 347
94. Schulman, H. L., and J. J. DeGouff, Jr., *Ind. Eng. Chem.*, **44**, 1915–1922 (1952).. 363
95. Shea, F. L., Jr., and N. W. Krase, *Trans. Am. Inst. Chem. Engrs.*, **36**, 463–489 (1940)...................................... 332, 333, 334, 339, 348
96. Sherwood, T. K., *Refrig. Eng.*, **13**, 253 (1923).......................... 347
97. Sherwood, T. K., *Ind. Eng. Chem.*, **33**, 424–429 (1941)............. 358, 359
98. Short, B. E., and H. E. Brown, "Proceedings of the General Discussion on Heat Transfer," pp. 27–31, Institution of Mechanical Engineers, London, and American Society of Mechanical Engineers, New York, 1951........ 341
99. Simpson, W. M., and T. K. Sherwood, *J. Am. Soc. Refrig. Eng.*, **52**, 535 (1946)... 359, 360
100. Spoelstra, H. J. *Arch. Suikerind.*, **3**, (23), 903 (1931)................... 347
101. Stender, W., *Z. Ver. deut. Ing.*, **69**, 905 (1925)...................... 329, 351
102. Taecker, R. G., and O. A. Hougen, *Chem. Eng. Progr.*, **45**, 188–193 (1949). 363
103. Ten Bosch, M., "Die Wärmeübertragung," Springer, Berlin, 1922........ 329
104. Tsao, U., Ph.D. Dissertation, University of Michigan, 1941............. 340
105. Ulloch, D. S., and W. L. Badger, *Trans. Am. Inst. Chem. Engrs.*, **33**, 417–440 (1937); *Ind. Eng. Chem.*, **29**, 905–910 (1937).......................... 334
106. Verschoor, H., *Trans. Inst. Chem. Engrs. (London)*, **16**, 66–76 (1938)...... 336
107. Walker, W. H., W. K. Lewis, W. H. McAdams, and E. R. Gilliland, "Principles of Chemical Engineering," 3d ed., p. 136, McGraw-Hill, New York, 1937.. 332, 358
108. Wallace, J. L., and A. W. Davison, *Ind. Eng. Chem.*, **30**, 948–953 (1938) 340, 353
109. Wilson, E. E., *Trans. ASME*, **37**, 47 (1915)........................... 343, 344

110. Young, F. L., and W. J. Wohlenberg, *Trans. ASME*, **64**, 787 (1942)...... 339
111. Zumbro, F. R., *Refrig. Eng.*, **13**, 49–63 (1926)........................... 344

CHAPTER 14. BOILING LIQUIDS

1. Abbott, M. D., and W. D. Comley, S.M. Thesis in Chemical Engineering, Massachusetts Institute of Technology, 1938......................... 379
2. Addoms, J. N., Sc.D. Thesis in Chemical Engineering, Massachusetts Institute of Technology, 1948................................. 382, 384, 385
3. Akin, G. A., and W. H. McAdams, *Trans. Am. Inst. Chem. Engrs.*, **35**, 137–155 (1939); *Ind. Eng. Chem.*, **31**, 487–491 (1939)................... 386
4. Badger, W. L., *Trans. Am. Inst. Chem. Engrs.*, **13** (Pt. II), 139 (1920).... 407
5. Badger, W. L., "Heat Transfer and Evaporation," Chemical Catalog Company, Inc., New York, 1926................................... 394
6. Badger, W. L., and W. L. McCabe, "Elements of Chemical Engineering," 2d ed., McGraw-Hill, New York, 1936............................ 394
7. Badger, W. L., C. C. Monrad, and H. W. Diamond, *Ind. Eng. Chem.*, **22**, 700 (1930)... 404
8. Badger, W. L., and P. W. Shepard, *Trans. Am. Inst. Chem. Engrs.*, **13** (Pt. I), 101–137 (1920).. 407
9. Beecher, Norman, S.M. Thesis in Chemical Engineering, Massachusetts Institute of Technology, 1948...................................... 378
10. Bergelin, O. P., and Carl Gazley, Jr., "Proceedings of Heat Transfer and Fluid Mechanics Institute," pp. 5–18, Berkeley, Calif., 1949............ 396
11. Bergelin, O. P., P. K. Kegel, F. G. Carpenter, and Carl Gazley, Jr., "Proceedings of Heat Transfer and Fluid Mechanics Institute," pp. 19–28, Berkeley, Calif., 1949... 395
12. Bernath, L., *Ind. Eng. Chem.*, **44**, 1310–1313 (1952)..................... 378
13. Boarts, R. M., W. L. Badger, and S. J. Meisenburg, *Trans. Am. Inst. Chem. Engrs.*, **33**, 363–389 (1937); *Ind. Eng. Chem.*, **29**, 912–918 (1937)..... 400, 406
14. Bonilla, C. F., and A. A. Eisenberg, *Ind. Eng. Chem.*, **40**, 1113–1122 (1948). 383
15. Bonilla, C. F., and C. H. Perry, *Trans. Am. Inst. Chem. Engrs.*, **37**, 685–705 (1941).. 375, 381, 383
16. Bonnet, W: E., and J. A. Gerster, *Chem. Eng. Progr.*, **47**, 151–158 (1951).. 404
17. Braunlich, R. H., S.M. Thesis in Chemical Engineering, Massachusetts Institute of Technology, 1941................................. 381, 384, 386
18. Bromley, L. A., *Chem. Eng. Progr.*, **46**, 221–227 (1950)............. 387, 388
19. Brooks, C. H., and W. L. Badger, *Trans. Am. Inst. Chem. Engrs.*, **33**, 392–413 (1937); *Ind. Eng. Chem.*, **29**, 918–923 (1937)...................... 406
20. Buchberg, H., F. Romie, R. Lipkis, and M. Greenfield, "Proceedings of Heat Transfer and Fluid Mechanics Institute," pp. 177–192, Stanford, Calif., 1951... 393
21. Bulkley, W. L., S.M. Thesis in Chemical Engineering, Massachusetts Institute of Technology, 1940.. 377
22. Castles, J. T., S.M. Thesis in Chemical Engineering, Massachusetts Institute of Technology, 1947....................................... 371, 387
23. Cichelli, M. T., and C. F. Bonilla, *Trans. Am. Inst. Chem. Engrs.*, **41**, 755–787 (1945), 411–412 (1946); "Heat Transfer Lectures," Vol. II, pp. 150–186, NEPA-979-IER-13, AEC, Oak Ridge, Tenn., 6-27-49-900, A14480 .... 381, 382, 383, 384
24. Claassen, H. *Mitt. Forsch.*, **4**, 49 (1902)............................. 382
25. Cleve, K., *Mitt. Forsch.*, **322**, 1 (1929).......................... 404, 405

	PAGE
26. Coates, J., and W. L. Badger, *Trans. Am. Inst. Chem. Engrs.*, **32**, 49–61 (1936)	404
27. Coffey, J. F., S.B. Thesis in Chemical Engineering, Massachusetts Institute of Technology, 1939	377
28. Colburn, A. P., E. M. Schoenborn, and C. S. Sutton, *Natl. Advisory Comm. Aeronaut., Tech. Note* 1498 (1948)	390
29. Cooper, H. B. H., S.M. Thesis in Chemical Engineering, Massachusetts Institute of Technology, 1938	375
30. Cryder, D. S., and A. C. Finalborgo, *Trans. Am. Inst. Chem. Engrs.*, **33**, 346–361 (1937)	381, 383
31. Dengler, C. E., Sc.D. Thesis in Chemical Engineering, Massachusetts Institute of Technology, 1952	394, 400, 401, 402, 403
32. Deutsch, R. K., and J. C. Rhode, B.S. Thesis in Chemical Engineering, Massachusetts Institute of Technology, 1940	375
33. Dew, Jess E., S.M. Thesis in Chemical Engineering, Massachusetts Institute of Technology, 1948	391
34. Dreselly, R. A., S.M. Thesis in Chemical Engineering, Massachusetts Institute of Technology, 1937	379
35. Drew, T. B., and A. C. Mueller, *Trans. Am. Inst. Chem. Engrs.*, **33**, 449–471 (1937)	370, 379, 386
36. Farber, E. A., and R. L. Scorah, *Trans. ASME*, **70**, 369–383 (1948)	387
37. Foust, A. S., E. M. Baker, and W. L. Badger, *Trans. Am. Inst. Chem. Engrs.*, **35**, 45–71 (1939); *Ind. Eng. Chem.*, **31**, 206–214 (1939)	405
38. Gunderson, L. O., and W. L. Demnan, *Ind. Eng. Chem.*, **40**, 1363 (1948)	376
39. Gunther, F. C., *Trans. ASME*, **73**, 115–123 (1951)	393
40. Gunther, F. C., and F. Kreith, "Proceedings of Heat Transfer and Fluid Mechanics Institute," pp. 113–126, American Society of Mechanical Engineers, 1949	391
41. Harvey, B. F., and A. S. Foust, "Reprints of Heat Transfer Symposium," American Institute of Chemical Engineers, pp. 289–329, December, 1951	402
42. Haywood, R. W., "Proceedings of the General Discussion on Heat Transfer," pp. 63–65, Institution of Mechanical Engineers, London, and American Society of Mechanical Engineers, New York, 1951	404
43. Hildebrandt, F. M., and K. H. Warren, *Ind. Eng. Chem.*, **41**, 754–760 (1949)	394
44. Insinger, T. H., Jr., and H. Bliss, *Trans. Am. Inst. Chem. Engrs.*, **36**, 491–516 (1940)	375, 376, 381
45. Jacoby, A. L., and L. C. Bischmann, *Ind. Eng. Chem.*, **40**, 1360–1363 (1948)	376
46. Jakob, M., and W. Fritz, *Forsch. Gebiete Ingenieurw.*, **2**, 434–447 (1931)	372, 373, 374
47. Jakob, M., and W. Linke, *Forsch. Gebiete Ingenieurw.*, **4**, 75–78 (1933)	376, 380
48. Jens, W. H., and P. A. Lottes, *Argonne Natl. Lab., Rept.* ANL-4627 (May 1, 1951)	393
49. Johnson, H. A., and A. H. Abou-Sabe, *ASME, Paper* 51-A-111 (1951)	395, 396, 397
50. Kane, D. E., Jr., S.M. Thesis in Chemical Engineering, Massachusetts Institute of Technology, 1951	388
51. Kaulakis, A. F., and L. M. Sherman, S.B. Thesis in Chemical Engineering, Massachusetts Institute of Technology, 1938	379, 381, 386
52. Kazakova, E. A., *Engr's. Digest*, **12**, 81–85 (1951)	384, 385
53. Kirschbaum, E., B. Kranz, and D. Starck, *VDI-Forschungheft* 375 (1935)	405, 406
54. Knowles, J. W., *Can. J. Research*, **26**, 268–270 (1948)	390

PAGE

55. Kreith, F., and M. Summerfield, *Trans. ASME*, **71**, 805–815 (1949), **72**, 869 (1950) 393
56. Lang, C., *Trans. Inst. Engrs. Shipbuilders, Scot.*, **32**, 279–295 (1888) 370
57. Larson, R. F., *Ind. Eng. Chem.*, **37**, 1004–1009, 1010–1018 (1945) 378
58. Ledinegg, M., *Engr's. Digest*, **10**, 85–89 (1949) 404
59. Leidenfrost, J. G., "De aquae communis nonnullis qualitatibus tractatus," Duisburg, 1756 .... 370, 371
60. Linden, C. M., and G. H. Montillon, *Trans. Am. Inst. Chem. Engrs.*, **24**, 120 (1930) 406, 407
61. Lockhart, R. W., and R. C. Martinelli, *Chem. Eng. Progr.*, **45**, 39–48 (1948); presented in October, 1947 .... 395, 396, 399
62. Lukomskii, S. M., *Chem. Ind. (USSR)*, **6**, 8–14 (1944) 383
63. McAdams, W. H., J. N. Addoms, P. M. Rinaldo, and R. S. Day, *Chem. Eng. Progr.* **44**, 639–646 (1948) 379, 388
64. McAdams, W. H., W. E. Kennel, C. S. Minden, R. Carl, P. M. Picornell, and J. E. Dew, *Ind. Eng. Chem.*, **41**, 1945–1953 (1949) 390, 391
65. McAdams, W. H., W. K. Woods, and R. L. Bryan, *Trans. ASME*, **63**, 545–552 (1941) 389, 398
66. McAdams, W. H., W. K. Woods, and L. C. Heroman, Jr., *Trans. ASME*, **64**, 193–200 (1942) 399
67. Martinelli, R. C., and D. B. Nelson, *Trans. ASME*, **70**, 695–702 (1948); presented in December, 1947 399
68. Mead, B. R., F. E. Romie, and A. G. Guibert, "Proceedings of Heat Transfer and Fluid Mechanics Institute," pp. 209–216, Stanford, Calif., 1951 ... (372)
69. Moore, T. V., and H. D. Wilde, Jr., *AIMME, Petroleum Development Technol.*, **92**, 296–319 (1931) 396
70. Morgan, A. I., L. A. Bromley, and C. R. Wilke, *Ind. Eng. Chem.*, **41**, 2767–2769 (1949) 376
71. Moscicki, I., and J. Broder, *Roczniki Chem.*, **6**, 319–354 (1926) 389
72. Myers, J. E., and D. L. Katz, "Reprints of Heat Transfer Symposium," pp. 330–350, American Institute of Chemical Engineers, December, 1951 .. 380
73. Nukiyama, S., *J. Soc. Mech. Engrs. (Japan)*, **37**, 367–374, S53–54 (1934) 370, 379, 389
74. Oliver, E., S.M. Thesis in Chemical Engineering, Massachusetts Institute of Technology, 1939 .... 400, 401
75. Perry, J. H., "Chemical Engineers' Handbook," 3d ed., McGraw-Hill, New York, 1950 394
76. Pridgeon, L. A., and W. L. Badger, *Ind. Eng. Chem.*, **16**, 474–478 (1924) .. 376
77. Rhodes, F. H., and C. H. Bridges, *Trans. Am. Inst. Chem. Engrs.*, **35**, 73–93 (1939); *Ind. Eng. Chem.*, **30**, 1401–1406 (1938) 376
78. Rosenblad, C., *Pulp & Paper Mag. Can.*, **51** (6), 85–94 (1950) 394
79. Rohsenow, W. M., and J. A. Clark, *Trans. ASME*, **73**, 609–620 (1951) ... 391
80. Rohsenow, W. M., and J. A. Clark, "Proceedings of Heat Transfer and Fluid Mechanics Institute," pp. 193–208, Berkeley, Calif., 1951 393
81. Sauer, E. T., S.M. Thesis in Chemical Engineering, Massachusetts Institute of Technology, 1935 .... 372, 375
82. Sauer, E. T., H. B. H. Cooper, G. A. Akin, and W. H. McAdams, *Mech. Eng.*, **60**, 669–675 (1938) 375, 386
83. Schweppe, J. L., and Alan S. Foust, "Reprints of Heat Transfer Symposium," pp. 242–291, American Institute of Chemical Engineers, December, 1951 492

BIBLIOGRAPHY AND AUTHOR INDEX

84. Stroebe, G. W., E. M. Baker, and W. L. Badger, *Trans. Am. Inst. Chem. Engrs.*, **35**, 17–41 (1939); *Ind. Eng. Chem.*, **31**, 200–206 (1939) 406
85. Taylor, J. J., personal communication, 1943 390
86. Tibbetts, E. F., and J. B. Cohen, personal communication to A. R. Kaufmann, Massachusetts Institute of Technology 389
87. Van Marle, D. J., *Ind. Eng. Chem.*, **16**, 458–459 (1924) 406
88. Vener, R. E., and A. P. Thompson, *Ind. Eng. Chem.*, **42**, 464–467 (1950) 377, 378
89. Verschoor, H., and S. Stemerding, "Proceedings of the General Discussion on Heat Transfer," pp. 201–204, Institution of Mechanical Engineers, London, and American Society of Mechanical Engineers, New York, 1951 396, 397
90. Walker, W. H., W. K. Lewis, W. H. McAdams, and E. R. Gilliland, "Principles of Chemical Engineering," 3d ed., p. 136, McGraw-Hill, New York, 1937 394
91. Webre, A. L., and C. S. Robinson, "Evaporation," Chemical Catalog Company, Inc., New York, 1926 .... 377, 378
92. Woods, W. K., Sc.D. Thesis in Chemical Engineering, Massachusetts Institute of Technology, 1940 .... 394, 398, 399, 407
93. Yoder, R. J., and B. F. Dodge, "Proceedings of the General Discussion on Heat Transfer," pp. 15–19, Institution of Mechanical Engineers, London, and American Society of Mechanical Engineers, New York, 1951 404

CHAPTER 15. APPLICATIONS TO DESIGN

1. Bagley, G. D., *Trans. ASME*, **40**, 667 (1918) 414
2. Carberry, J. J., *Chem. Eng.*, **61**, 225–227 (1953) 439
3. Cardwell, F. D., *ASME*, Paper 50–S–6 .... 432, 437
4. Chilton, C. H., *Chem. Eng.*, **56** (6), 97–106 (1949) 422
5. Colburn, A. P., *Purdue Univ.*, *Eng. Bull.* 26 (1) (1942) 435, 439
6. Colburn, A. P., personal communication, 1942 431
7. Day, V. S., *Univ. Ill., Eng. Expt. Sta., Bull.* 117 (1920) 415
8. Drew, T. B., H. C. Hottel, and W. H. McAdams, *Trans. Am. Inst. Chem. Engrs.*, **32**, 271–305 (1936); *Chem. Eng. Congr., World Power Congr., London*, **3**, 713–745 (1936) 431, 435
9. Dufton, A. F., *Heating, Piping Air Conditioning*, **4**, 355 (1932) 414
10. Fabry, C., *Wärme*, **55**, 163 (1932) 414
11. Fong, J. T., *ASME*, Paper 51–A–98 436
12. Gilliland, E. R., D. Bareis, and G. Feick, "The Science of Nuclear Power," Vol. II, p. 120, Addison-Wesley, Cambridge, Mass., 1949 439
13. Lewis, W. K., J. T. Ward, and E. Voss, *Ind. Eng. Chem.*, **16**, 467 (1924) . 431, 435
14. McMillan, L. B., *Trans. ASME*, **37**, 961 (1951) 414
15. McMillan, L. B., *Trans. ASME*, **48**, 1269–1317 (1926) 414
16. McMillan, L. B., *Mech. Eng.*, **51**, 349 (1929) 414
17. Nichols, P., *Refrig. Eng.*, **9**, 152 (1922), 254 (1923) 414
18. Parsons, P. W., and B. J. Gaffney, *Trans. Am. Inst. Chem. Engrs.*, **40**, 655–673 (1944) 439
19. Patton, T. C., *Heating, Piping Air Conditioning*, **4**, 6–11 (1932) 414
20. Porter, A. W., and T. R. Martin, *Phil. Mag.*, **20**, 511 (1910) 414
21. Sanbern, E. N., *J. Am. Soc. Heating Ventilating Engrs.*, **34**, 197 (1928) 414
22. Sieder, E. N., *Chem. Met. Eng.*, **46**, 322–325 (1939) 416, 423
23. Stone, J. H., *Refrig. Eng.*, **10**, 137 (1923) 414

24. Tubular Exchanger Manufacturers Association, "Standards TEMA," 3d ed., New York, 1952............................. 416, 419, 420, 421
25. Weidlein, E. R., *Trans. Am. Inst. Chem. Engrs.*, **13** (Pt. II), 25–42 (1920–1921)... 414
26. Wood, A. J., and P. X. Rice, *Trans. ASME*, **45**, 497 (1923); *Refrig. Eng.*, **10**, 357 (1924).. 414
27. Wunderlich, M. S., *J. Am. Soc. Heating Ventilating Engrs.*, **34**, 189 (1928)... 414
28. Zeiner, E. F., *Power*, **73**, 957 (1931)................................ 414

SUBJECT INDEX

A

Absolute temperature, definition, 55
Absorption strength of luminous flames, 102, 105
Absorptivity of gases, 82, 89, 90, 106, 111
  of surfaces, 58, 60, 62
Accommodation coefficient, 320
Acoustic velocity, 8, 27, 157, 311, 316
Aerocat particles, 304, 306
Aerofoil, 265
Air, over finned tubes, 268–271
  friction (see Friction)
  gaps, 17
  natural convection, 165–183, 233, 235
  normal to banks, 271–280
  normal to single cylinder, 252–265
  in pipes, recommended equation for, 219–229
  over plates, 311
  specific heat of, 464, 465
  in steam, effect of, 355
  (See also Gases)
Air conditioning, 355–365
Algae, prevention of growth of, 347
Alloys, surface emissivities of, 472
  thermal conductivities of, 445–447
Alumina packing, 291
Aluminum foil, 11
Ammonia, condensation of, 343
  radiation from, 96
Analogies, between conduction of electricity and heat, 51
  between heat transfer and friction, 206–218
  (See also specific analogies)
Analysis (see Dimensional analysis)
Annealing furnace, 125
Annular spaces, friction in, 157
  heat transfer in, 182, 241–244
  hydraulic radius of, 182, 241
  radiation in, 241–242
Areas, arithmetic-mean, 13
  average, for hollow cylinder, 12, 414

Areas, average, for hollow parallelepiped, 25
  for hollow sphere, 14
  for radiation, 90–94
  geometric-mean, 15
  of heating surface, 186–187
  logarithmic-mean, 12, 414
  for pipes, tubes, 487, 488
Asphalt heating, 237
Average temperature, of fluids, 184–186
  of gas, 158
  for thermal conductivity, 11
Average temperature difference, 190–198, 231, 332, 424, 489

B

Baffles, 276–280, 419–421
Bar fins, 269
Base pipes, heat loss from, 179, 261, 411
Beam length, 87
Billet reheating furnace, 119
Biot number, 135
Bituminous coal, 99, 104
Black body, 59
Blast-furnace stoves, 295
Boiler furnace, 105
Boilers, gases in (see Gases)
Boiling liquids, 368–409
  addition agents, 376
  annuli, boiling in, 390
  film boiling, 387
  forced circulation, 397
  in horizontal tubes, 398
  Leiden frost point, 371
  maximum flux, 386
  mechanism, 394
  mixtures, 404
  natural circulation, 404
  nature of liquid, 383
  nucleate boiling, 372
  peak-heat flux, 383
  pressure, effect of, 380
  rectangular ducts, 393

Boiling liquids, regimes, 370
  results, reproducibility of, 374
  saturated liquids, 368
  scale deposits, 376
  subcooled liquids, 389
  submerged surfaces, 368
  surface, nature of, 374
  temperature difference, critical, 383
    effect of, 377
  tube arrangement, 379
  tube size, 379
  in tubes, surface boiling, 393, 394
  velocity, critical, 402
    effect of, 378
  in vertical tubes, 400
Boltzmann, 59
Boundary layer, 151–155, 206–218
Brick checkerwork, 298
Buffer layer, 151, 206
Building materials, thermal conductivities of, 445–454

C

Carbon dioxide, 82, 90
Carbon monoxide, 95
Carborundum muffle, 80
Cauchy number, 135
Celite packing, 292
Centipoise, 444
Chlorination of cooling water, 347
Choking, 316
Chromium alloys, thermal conductivities of, 447
Cleaning of heating surface, 347, 377, 417, 420
Clouds of particles, radiation from, 99–105
Coal, powdered, 99, 104
Coefficients, 425
  contact, 17
  effective, 313, 315
  individual, definition, 186, 187
  over-all, 187
    in condensers, 343
    local, 6, 184
  in packed cylinders, 291
  radiation, 78
  of rarefied gas, 321
Coils, 150, 228
Colloidal particles, 144, 255
Color band, 144, 150, 152

Color screens in optical pyrometry, 102
Combustion chamber, radiation in, 117–121
Compact exchanger, 282, 289
  comparison of, 287
  design data, 288
  heat transfer, 286
  liquid-coupled, 290
  optimization of, 287
  pressure drop, 286
Compactness, 288
Compressible flow, 309
  calculations for, 316
  Nielsen plot, 318
Condensation, 325–367
  direct-contact, 356
  dropwise, 347
    data for, 348
    vs. film-type, 325
    mechanism, 347
    promoters for, 347–351
  film-type, alignment chart for, 337
    of ammonia, 343
    average coefficients, 333
    data for, 332
    finned tubes, 346
    graphical analysis, 343
    horizontal tubes, 338–340
    laminar flow of condensate, 329
    local coefficients, 333
    noncondensable gas, 334, 355
    over-all coefficients, 343
    recommended relations, for horizontal tubes, 342
      for vertical tubes, 337
    simplified relations, 331
    subcooling of condensate, 338
    tube-wall temperature, 332
    turbulence, effect of, 334
    vapor velocity, effect of, 336
    vertical vs. horizontal tubes, 341
    Wilson method, 343
  frost formation, 364
    mechanism, 325
  noncondensable gases, calculations for, 355
    effect of, 334, 355
    mechanism, 355
  number, 135
  vapors, mixtures, 351
    single-phase condensate, 353
    superheated, 351
    two-phase condensate, 354

SUBJECT INDEX

Condenser tubing, standard dimensions of, 487
Condensers, cleaning of, 347
  scale coefficient in, 189
  surface, 325–365
Condensing vapors, 325–367
Conditioning of air, 355–365
Conductance, definition, 15
Conduction, 7–30
  buried heat source, 25
  definition, 1
  Fourier equation for, 7
  internal-heat generators, 18
  periodic, 43, 52
  steady, definition, 7
    through flat walls, 12
    in fluids, 26–29
    graphical method for, 16
    through hollow cylinder, 12, 14
    nomenclature for, 8
    through nonhomogeneous material, 9
    through pipe covering, 12
    through porous materials, 9
    problems in, 29
    resistance concept of, 15
    in series, 15
    in solids, 9–26, 31–52, 186
    temperature gradient for, 8–26
    three-dimensional, 25
      relation method, 21
    two-dimensional, mapping, 19
  transient, 31–52
    analogue methods for, 51
    approximate method for, 43–50
    in brick-shaped solid, 39
    in cube, 43
    in cylinder, 40
    definition, 7
    dimensionless ratios in, 31–43
    graphical methods for, 43–50
    midplane, definition, 35
    nomenclature table for, 32
    problems in, 52–54
    Schmidt method, 44
    in semi-infinite solid, 39
    in slab, 35–38
    in sphere, 14, 40
    in square bar, 43
    surface resistance, 41
    theory of, 33–35
Conductivity, thermal (see Thermal conductivity)
Conservation, of energy, 145

Conservation, of matter, 141
Consistent units, 8, 32, 56, 128, 142, 166, 185, 204, 254, 284, 309, 369, 412
Contact coefficients, 17
Continuity equation, 141
Continuum flow, 319
Contraction loss, 159
Convection, definition, 1
  emissivity due to, 79
  forced (see Forced convection)
  free (see Natural convection)
Conversion factors, for coefficients of heat transfer, 444
  for Reynolds number, 137
  for thermal conductivity, 444
  for viscosity, 444
Cooling towers, 356–365
Core resistance, 153
Cosine law, 64, 71
Costs, 415, 422, 430–441
Countercurrent, 190
Critical Δt, 383
Critical radius, 414
Critical velocity, 144
Cross-flow heat exchanger, 195
Crout method for evaluating determinants, 115
Curved pipe, friction in, 150
  heat transfer in, 228
Curvilinear square, 20

D

Dehumidification, 355–365
Deposits, scale, 376
Design, of exchangers, 410–441
  of heat-transfer equipment, 415, 441
  of insulation, 410
Determinants in radiation calculations, 74, 115
Diameter, equivalent, 88, 144, 148, 157, 182, 245, 265, 275
Diameters of pipes and tubes, 487, 488
Diathermanous substances, 11
Diffuse reflection, 71, 113
Diffusion of vapor through gas, 351–365
Diffusivity, thermal, 34, 51
Dimensional analysis, 126–239
  in heat transfer, 126–139
  limitations of, 135
  nomenclature for, 127, 128, 135
  pi theorem in, 129, 130
  problems in, 138
  theory of models, 135

Dimensional constant, 127
Dimensionless groups, ratios, 135
Direct contact, coolers, 361
   of liquid and gas, 356–364
Direct plus reradiation factor, 69
Dirt deposits, 188, 189, 223, 343, 347, 356
Double-pipe condenser, 344
Drip coolers, 244
Drip-type condenser, 339
Dropwise condensation (see Condensation)
Ducts, friction in, 157
   heat transfer in, 241
Dynamics of fluids, 140–146

E

Economical frequency of cleaning condensers and evaporators, 347
Economical temperature, temperature difference in exchangers, 440
Economical thickness of insulation, 411
Economical velocity in exchangers, 427–438
Eddy viscosity, 209
Effective heat-transfer coefficient, 313
   for across cylinders, 315
   for flat plate, 315
   for wedges and cones, 315
Effective temperature difference, 313
Efficiency of extended surfaces, 286
Emission of radiation, 87, 89
Emissivity, of gases, ammonia, 96
     carbon dioxide, 82, 90
     carbon monoxide, 95
     definition, 82
     sulfur dioxide, 94
     water vapor, 84, 90
   of surfaces, definition, 60
     effective, of tube row, 117
     metals vs. nonmetals, 61
   tables of, 472–479
   total hemispherical, 61
Energy balance, 145
Energy conservation, 145
Energy distribution in spectrum, 60
Enlargement loss, 159–162
Enthalpy, balance, 423
   definition, 145
   table for steam and water, 482
Enthalpy potential, 357–365
Entropy of steam and water, 482
Equipment (see Design)

Equivalent diameter (see Diameter, equivalent)
Euler number, 135
Evaporation (see Boiling liquids)
Evaporators, inclined-tube, 406
   liquid level, effect of, 407
   long-tube, 405
   short-tube, 404
Extended heating surface, 268–271, 282, 283

F

Factors, fouling, 188, 189, 223, 343, 347, 356
   friction, 156, 426
   geometrical, 12–15
     in radiation, 63, 69, 72
   of safety, 145, 156
   shape, for steady conduction, 12
Fanning equation, 145, 156
Fanning friction factor, 135, 156, 426
Film resistance, 210
Film temperature, 212, 219, 255, 327
Film-type coolers, 244
Finned surfaces, 268
Finned-tube exchanger, 283
Fittings, resistance of, 161
Fixed bed, 290
Flames (see Luminous flames)
Flow, compressible, 309, 316, 318
   of fluids, 140–164
     dimensional analysis applied to, 134
     end effects, 159
        sudden contraction, 159
        sudden enlargement, 159
     equivalent length of fittings, 161
     Fanning equation, 145
     force balance, 141
     Hagen-Poiseuille law, 148, 322
     hydraulic radius, 150
     material balances, 141
     mechanism of, 140–157
        streamline, 140, 148–151
        turbulent, 140, 151–159
   in packed tubes, 162
   Poiseuille law, 148, 322
   Reynolds number, critical, 144
     effect of curvature on, 151
   streamline, 148–151
     in annular spaces, 149
     effect of curvature on, 150
     in miscellaneous shapes, 151
     in noncircular ducts, 149

Flow, of fluids, streamline, nonisothermal, 149
    normal to tubes, 163
    tubes and pipes, 148
    turbulent, 151–159
      in annular spaces, 157
      buffer layer in, 153
      friction-factor plot, 156
      friction velocity, 153
      laminar sublayer, 153
      inside pipes, 153, 155
      along plate, 152
      in rectangular ducts, 157
      roughness, effect of, 153, 157
      turbulent core, 153
      velocity, 153
    two-phase, 395
      heat transfer in, 396
      liquid holdup in, 396
      pressure drop in, 395
Flue gas, radiation from, 98
Fluidization, type of, 299
    aggregative, 300
    critical velocity, 301
    dense-phase bed, 300
    disperse, 300
    incipient, 301
    lean-phase bed, 306
    minimum, 300
    particulate, 301
    quiescent bed, 300
    slugging bed, 300
Fluidized bed, 299
Fluidized solids, 299
Fluidized systems, 299
  apparent thermal conductivity, 303
  external coefficient, 304, 306
  fraction voids, 301
  heat-transfer properties, 302
  internal coefficient, 305
  pressure drop, 301
Fluorescence, 58
Flux density in radiation, 73
Forced convection, 184–201, 233, 235
  definition, 1
  heat-transfer coefficients, individual, 186, 187
    local, 184
  mean-temperature differences, 190–195
  over-all coefficients, 184, 187
  resistance concept, 188
  temperature gradient in, 186
  (*See also* Liquids)

Fouling factors, 188, 189, 223, 343, 347, 356
Fourier equation, 3
Fourier number, 135
Fourier series, 34
Free convection (*see* Natural convection)
Free-molecule flow, 319
  pressure drop in, 322
Freezing, of fruit juices, 52
  by radiation, 125
Friction, analogy to heat transfer, 206–218
  in annular sections, 149, 157
  contraction loss, 159
  in curved pipes, 150
  definition, 145
  enlargement loss, 159
  for fluid flow, 145, 156
  in packed tubes, 162
  in rectangular sections, 149, 151
  in straight pipes, 156
  across tube banks, 162–164
Friction factors, 156, 426
Friction power, 287
Fromm waffle plate, 153
Froude number, 135
Furnace design, 100, 105, 106, 109, 117–121
Fused salts, 222

G

Gases, absorptivity of, 82, 89, 90, 106, 111
  in annular spaces, 182, 241
  effect of temperature on Prandtl number, 471
  kinetic theory of, 319
  liquefied, specific heat of, 465
  natural convection, 165–183
  in packed tubes, 290–299
  over plane surfaces, 249
  radiation from, 82–84
    effect of shape on, 86–89
  rarefied, 319–321
  real, radiation from, 111
  outside single tubes, forced-convection data, 258–265
    dimensional analysis, 258
    mechanism, 252
    nomenclature, 254
    simultaneous radiation, 199, 261
  specific heat of, 464
  temperature of, errors due to radiation, 199, 261

Gases, thermal conductivity of, 28–29, 457–459
  transmittance of, for radiation, 92, 106, 111
  outside tube banks, data for, 271
    forced convection, 271–275
    radiation from, 98
  in tubes, heating and cooling of, 221, 226–228
    approximate relations for (charts), 226, 227
    calculations for, 227
    in coils, 228
    condensation of (see Condensation)
    effect on, of dust, 224
      of length, 224
    vs. Reynolds equation, 210
    simplified equation, for diatomic, 226
      for superheated steam, 226
  viscosity vs. temperature of, 469
  (See also Compressible flow)
Geometric mean, area, 15
  temperature difference in radiation, 119
  view factor, 119
Geometrical factors, 12–15
Geometry of gas radiation, 86
Glass beads, 293, 305
Graetz number, 135
Grashof number, 135
Gray gas, radiation from, 106–110
Gray surfaces, definition, 63
  effect of, on gas radiation, 121
  enclosure of, 72
Gurney-Lurie charts, 35–41

H

h, individual coefficient of heat transfer, definition, 187
Hagen-Poiseuille law, 148, 322
Heat, mechanical equivalent of, 142, 147
  total (see Enthalpy)
Heat balance, 423
Heat capacity, 461–465
Heat economizers, 189
Heat exchangers, compact (see Compact exchanger)
  cross-flow, 195
  design of, 410–441
    optimum, 429
  economical velocity in, 427–438
  finned-tube, 283

Heat exchangers, multipass, 193–198, 489
  mean temperature difference in, 193, 489–490
  pin-fin, 283
  plate-fin, 283
  shell-and-tube, 194, 271, 276, 278, 416
Heat loss, base-pipe, coefficient for, 179, 261, 411
Heat meter, 19
Heat regenerators, 290–299
  dimensionless period, 296
  dimensionless size, 296
  nonsymmetrical cycle, 297
  symmetrical cycle, 296
  temperature efficiency, 296
  temperature fluctuation, 298
Heat transfer, analogy to fluid friction 206–218
  coefficient of (see Coefficients)
  over-all, definition, 187
  range of, 5
  in two-phase flow, 395–396
Heat-transfer fluid, selection of, 438
Heating and cooling equations, 202, 252
Hemispherical intensity, definition, 58
Hot-wire anemometer, 152, 176
Humidification, 355–365
Hydraulic radius (see Diameter, equivalent)
Hydrocal, 51

I

Ice formation, 52, 364
  by radiation, 125
Index of refraction, effect on radiation laws, 64
Insulating materials, thermal conductivity of, 448–454
Insulation, critical radius, 414
  heat loss decreased by, 14
  heat loss increased by, 415
  optimum thickness, 411
Intensity of radiation, definition, 64
  total hemispherical, 58
Interference fringes, 168
Investment charges, 287
Iron-pipe dimensions, 488

K

Kármán analogy, 210
Kinematic viscosity, 144

SUBJECT INDEX

Kinetic energy, discussion of, 146
Kinetic theory of gases, 319
Kirchhoff's law, 58
Knudsen number, 135, 319

L

Laminar flow film, 101, 210
Laminar motion, 101, 229
Latent heat of vaporization, of liquids, 480
  of water, 482
Lead balls, 293, 294
Liquids, heating and cooling of, forced convection, 202–281
  in annular spaces, 182, 241–244
  in baffled exchangers, 276, 280, 419
  drip coolers, 244
  normal to single tubes, 266
  plates, 268
  in tubes, streamline flow, 229
    turbulent flow, 205
  natural convection, 165–181
    latent heat of vaporization of, 480
    specific heat of, 462–465
    thermal conductivity of, 28, 37–38, 455–456
    viscosity vs. temperature of, 466
Logarithmic-mean area, 13
Logarithmic-mean-temperature difference, 191, 424
Louvered plate fins, 289
Luminous flames, absorption strength of, 102, 105
  measurements on, 101
  problems on, 105
  radiation from, 99–105
Lungstrom preheater, 298

M

Mach number, 135, 311
Martinelli analogy, 210
Mass-transfer coefficient, 357–364
Mass velocity, definition, 141, 204, 254, 272
Material balance, 141
Maximum velocity, 141, 206, 286
Mean free path, 319
Mean hydraulic radius (see Diameter, equivalent)
Mean mixing length, 215
Mean path length, gas radiation, 88

Mean temperature difference, 190–198, 231, 332, 424, 489
Mechanical equivalent of heat, 142, 147
Metals, thermal conductivity of, 445–447
Meter, heat, 19
Midplane, definition, 35
Minimum hydraulic radius, 286
Models, theory of, 136
Molecular speed ratio, 319
Monochromatic emissive power, 59
Motion, streamline, vs. turbulent, 101
Moving bed, 290, 298
Muffle furnace, example, 80, 124
Multipass heat exchanger, 193–198, 489

N

Natural convection, 165–183, 233, 235
  in annular spaces, 182
  definition, 1
  in enclosed spaces, 181
  from horizontal cylinder, 175
  inside horizontal pipes, 181, 235
  from horizontal plates, 180
  in liquid metals, 180
  at low Prandtl numbers, 171
  in molten salts, 180
  from planes, 173, 180, 181
  simplified relations for, 173, 177, 180
  simultaneous radiation loss in, 179
  single cylinders, 176
  temperature distribution in, 167, 168
  inside tubes, 233
  velocity distribution in, 167
  from vertical cylinders, 172, 233
  from wires, 176
  (*See also* Liquids)
Newton's law, of cooling, 5, 184
  of motion, 127
Nielsen plot, 318
No-flux surface, definition, 70
Nomenclature, boiling liquids, 369
  compact exchangers, 284
  condensing vapors, 326
  design, 413
  flow of fluids, 142
  fluidized systems, 284
  forced convection, 185
  heating and cooling, inside tubes, 204
    outside tubes, 254
  high-velocity flow, 310
  natural convection, 166
  packed systems, 284

Nomenclature, radiant-heat transmission, 56
　steady conduction, 32
　transient conduction, 8
Nongray enclosures, effect on gas radiation, 111
　radiation in, 77
Nonluminous gases, radiation due to, 82–98
Nonuniformity, in temperature distribution, 206
　in velocity distribution, 206
Nusselt number, 135

O

Oil, mineral, thermal conductivity of, 455
Oil furnace, 97
Oils (see Liquids)
Optimum amount of coolant, 439
Optimum frequency of cleaning condenser and evaporator, 347, 377
Optimum temperature difference in exchangers, 190, 435
Optimum thickness of insulation, 14, 411
Optimum velocities in exchangers, 189, 430–437
Over-all coefficient, definition, 6, 187
　graphical analysis of, 343
Oxide films, 17

P

Packed bed, 290
　heat-transfer coefficient in, 294
　mass-transfer coefficient in, 295
　radiation within, 290
　thermal conductivity of, 290
Packed cylinder, 291
　apparent conductivity, 291, 292
　heat-transfer coefficient in, 291
Packed solids, 299
Packed tubes, friction in, 162
　gases in, 290–299
　heat transfer in, 290
Parallel plates, radiation between, 69, 79
Peclet number, 135
　modified, 294
Perfect-gas law, 147
Perfect radiator, 58
Periodic conduction, 43, 52
Petroleum heaters, 119
Petroleum oils (see Liquids)

Phosphorescence, 58
Photographs, of boiling, 371, 372, 390
　of condensation, 329
　of finned tubes, 268
　of flow, over banks of tubes, 272
　　over plates, 168–170
　　across tubes, 253, 256, 272
　of heat-transfer equipment, 417
Pi theorem, 129
Pin-fin exchanger, 283
Pipes, banks of, 271
　curved, 150, 228
　dimensions of, 487, 488
　friction in, 156
　photographs of, 253, 256
　radiation in banks of, 98
Planck equation, 59
　Wien's modification of, 101
Plate-fin exchangers, 283
Poiseuille law, 148, 322
Powdered coal, radiation from, 104
Power costs, 287
Prandtl analogy, 210
Prandtl number, 28, 135
Pressure, effect of, on gas radiation, 82, 84
　on natural convection, 176
　on viscosity, 468
Primary surface, 283, 287
Pulverized-coal flame, radiation from, 99, 104
Pumping power, 287

Q

Quartz window, transmittance for radiation, 79

R

Radiation of heat, 55–125
　absorption strength, 102
　absorptivity (see Absorptivity)
　from ammonia, 96
　in annular spaces, 241–242
　areas for, average, 90–94
　in banks of pipes, 98
　black-body, 59
　clouds of particles, 99–105
　coefficient of transfer due to, 78
　in combustion chamber, 117–121
　constants, 59, 60
　vs. convection, 55
　cosine law, 64, 71

SUBJECT INDEX

Radiation of heat, emissivity (*see* Emissivity)
  in enclosure, 105–124
  errors in pyrometry due to, 199, 261
  from flue gas, 98
  freezing by, 125
  from gases, 82–84
    geometry of, 86
  geometrical factors, 63, 69, 72
  from gray gas, 106–110
  intensity of, definition, 64
  from luminous flames, 99–105
  nature of, 55–63
  in nongray enclosures, 77
  from nonluminous gases, 82–98
  within packed bed, 290
  inside pipes, 97
  Planck equation, 59
  plus convection, coefficients for pipes, 179
  problems, 124
  pulverized-coal flame, 99, 104
  from real gas, 111
  reflectivity for, 73
  Stefan-Boltzmann equation for, 59
  from sulfur dioxide, 94–95
  between surfaces, 63–82
    adjacent rectangles, 68
    disks, 69
    parallel rectangles, 69, 79
    planes and tubes, 69, 81
  temperature in, of no-flux surfaces, 110
    space variation in, 76
  thermal, definition, 58
  total emissive power, 58
  total hemispherical intensity, 58
  volume-emission of, 87, 89
  to wall tubes, 81
  from water vapor, 84, 90
  zones in, number of, 75
Rarefied gas, 319
  heat-transfer coefficients, 321
  natural convection, 321
  parallel flat plates, 322
  pressure drop, 322
Raschig rings, 361
Real gas, radiation from, 111
Recovery of waste heat, 189, 295
Recovery factor, 311
  for cones, 313
  for flat plates, 311
  for probes, 312
  for tubes, 312

Reflectivity of radiation in enclosures, 73
Refractory surfaces, allowance for radiation in, 69
  effect of grayness on gas radiation, 121
Refrigerants, 343, 380
Regenerators, 295
Reheating furnace, 119
Resistance, thermal, contact, 17
  graphical analyses of, 343
  individual, definition, 16, 187
  over-all, definition, 187
  total, definition, 16
  of valves, 161
Resistances, example, 17
  in parallel, 15
  in series, 15
Reynolds analogy, 209
Reynolds number, 127, 134, 135
  critical value, 144, 157, 239
Ripples, 331
Roughness, 136, 157, 223
Rubber, heating of, 37, 51
Ruffled fins, 293

S

Safety factor, 228, 425
Saturation pressure of water, 482
Scale deposits, 188, 189, 223, 343, 347, 356, 376
Schmidt method, 44
Schmidt number, 135
Shape factors for steady conduction, 12
Shell-and-tube condenser, 344
Shell-and-tube exchangers, data, 271, 276, 278, 416
  mean temperature difference, 194
Similarity principle (*see* Dimensional analysis)
Slip flow, 319
  pressure drop in, 322
Solids, specific heat of, 461
  temperature distribution in, 9–26, 31–52, 186
  thermal conductivity of, 9–10
Sonic velocity, 27, 309, 311, 316
Soot, radiation from, in flames, 99
Source-sink surfaces, definition, 69
Specific heat, 147, 461–465
  of gases, 464, 465
  of liquefied gases, 465
  of liquids, 462
  of solids, 461

Spheres, forced convection with, 265
  radiation from gas, 88, 105
Spray ponds, 361
Stagnation temperature, 309
Stanton number, 135
Steady state, definition, 4, 7
Steam, condensation of, 325, 343
  superheated, 351
Steam-boiler furnaces, 120, 122
Steam table, 482
Steaming of wood, 51
Steel balls, 293, 294
Steel-pipe dimensions, 488
Stefan-Boltzman equation, 59
Stethoscope, 144
Stratification, 206
Strip-fin exchanger, 283
Strips, parallel, radiation between, 79
Streamline flow (see Flow of fluids, streamline)
Streamline shapes, 265
Subcooling, 329, 338
Submerged conductors, 25
Sulfur dioxide, radiation due to, 94–95
Superheat in vapors, removal of, 351
Superheated steam, condensing of, 351
Supersonic flow, 309
Surface condensers, 325–365
Surface temperature, measurement of, 198, 322
Surface tension, 128

T

Tables, of beam lengths, 87
  of conversion factors, for h and U, 444
    for Reynolds number, 127
    for thermal conductivity, 444
    for viscosity, 444
  of emissivities of surfaces, 472–480
  of enthalpy of steam, 483
  of latent heat of vaporization, 480
  nomenclature (see Nomenclature)
  of pipe dimensions, 487, 488
  of Prandtl number for gases, 471
  of specific heats, 461–465
  of surface tensions, 481
  of thermal conductivities, 445–460
  of tube dimensions, 487, 488
  of vapor pressure of water, 482
  of viscosities, 466–471
Temperature, absolute, definition, 55
  average (see Average temperature)
  effect on interchange factor, 79

Temperature, mean value in gas radiation, 90–92
  measurement, 198, 261, 322
  of moving stream, 309
  of no-flux surfaces in radiation, 110
  saturated, of water vapor, 482
  space variation in, in radiation, 76
  stagnation, 309
  total, 309
Temperature difference, average, 190–198, 231, 332, 424, 489
  critical, 383
  economical, in exchangers, 440
  mean, 190–194
    in countercurrent flow, 190
    in multipass exchangers, 193, 489–490
    in parallel flow, 190
  optimum, 440
  in exchangers, 190, 435
Temperature distribution, in moving fluids, 206, 253
  in solids, 9–26, 31–52, 186
Temperature efficiency, 286
Thawing of fruit juices, 52
Theory of models, 136
Thermal conductivity, 7–53, 445
  of alloys, 445–447
  average temperature for, 11
  of building and insulating materials, 445–454
  conversion of units for, 444
  effect on, of density, 9–10
    of porosity, 10
    of temperature, 9
  of gases, 38–39, 457–459
  of liquids, 37–38, 455
    empirical equation for, 28
  of solids, 9–10
  temperature coefficient of, 9
  units, 9
  of water, 455
Thermal diffusivity, 34
  of wood, 51
Thermal radiation, definition, 58
Thermocouples, 198–199, 261
  conduction correction, 264
  errors, due to radiation, 261
    reduction of, 262
  traveling, 199
Total emissive power, 58
Total heat of steam and water, 482
Total hemispherical emissivity, 61
Total hemispherical intensity, 58

SUBJECT INDEX

Towers, heat transfer to gases in, 296
  packed, friction in, 162
Transition region, 239
Transmittance of gases for radiation, 92, 106, 111
Tubes, dimensions of, 487, 488
  heat transfer in, 202–251
    annular spaces, 241
      laminar flow in, 242
      rodlike flow in, 243
      turbulent flow in, 242
    flow in layer form, 244
      falling-film heaters, 246
      horizontal tubes, 245, 247
      vertical tubes, 244, 247
    streamline flow, 229–239
      forced convection, 233
      gases, data for, 238
      natural convection, 233
      parabolic distribution, 230
      radial variations, 239
      rodlike flow, 232, 239
    transition region, 239
      gases, 239
      liquids, 240
    turbulent flow, 205–229
      analogies with momentum, 208–219
      analogy for uniform wall temperature, 217
      correlation of data, 219–229
      curvature, effect of, 228
      design charts for, 227
      eddy diffusivities, 214
      gases, data for, 221
      inlet region, 225
      length to diameter, effect of, 224
      liquid metals, data for, 215
      liquids, recent data, 222
      Martinelli analogy, 210
      mechanism, 205
      mercury, data for, 215
      pulsations, effect of, 223
      radial variation, 216
      simplified equations for, 226, 228
      slurries, data for, 223
      surface condition, 223
      vorticity theory, 218
  heat transfer outside, 252–281
    banks of tubes, cross-baffled exchangers, 276
      gases, 271
      liquid, 275

Tubes, heat transfer outside, banks of tubes, recommended procedure for air, 275
  single cylinders, 252–268
    air, data for, 258
    analogies, 256
    average coefficient, 257
    effect of turbulence, 261
    gas, true temperature, 261
    liquids, data for, 266
    local coefficients, 255
    mechanism, 252
    simplified equation, 261
  radiant interchange between, 80
Turbulent motion, 140, 205
Two-phase flow, 395
  liquid holdup in, 396
  pressure drop in, 395

U

U, over-all coefficient of heat transfer, definition, 187
Units, consistent (*see* Consistent units)
  conversion of, for coefficient of heat transfer, 444
    for thermal conductivity, 444
    for viscosity, 444
Unsteady conduction (*see* Conduction, transient)
Unsteady state, definition, 3

V

Vacuum, evaporation under, 380
Valves, resistance of, 161
Vapor-diffusion theory, 351–365
Vapor pressure of water, 482
Vaporization, due to boiling (*see* Boiling liquids)
  latent heat of, 480, 482
Vapors, condensation of mixed, 351
  condensing, 325–367
  desuperheating of, with condensation, 351
  without condensation, 351
Venting, 421
Velocity, acoustic, 8, 27, 157, 311, 316
  effect on heat transfer, 189, 202–281
  mass, weight, definition, 141, 204, 254, 272
  maximum, 141, 206, 286
  vs. average, 141

Velocity, optimum, 189, 430
Velocity distribution, isothermal, over rough surfaces, 141, 153, 157
   in smooth pipes, 141, 153
View-factor, 63–69
Viscosity, conversion factor for, 444
   definition, 149
   eddy, 209
   effect of pressure on, 468
   of gases vs. temperature (chart), 469
   of liquids vs. temperature (chart), 466
   units, 444
   of water, 466, 471
Viscous motion (see Flow of fluids, streamline)
Void fraction, 290, 293, 301
Volume-emission of radiation, 87, 89
von Kármán analogy, 210

W

Waste-heat recovery, 189, 440
Water, chlorination of, 347
   enthalpy of, 482
   entropy of, 482
   evaporation of, 368–408
      into air, 356–365
   heating and cooling of, baffled exchangers, 276, 419

Water, heating and cooling of, drip coolers, 244
   forced convection in annular spaces, 241
   normal to single tubes, 266
   plates, 249
   latent heat of vaporization, 480
   optimum amount of, 439
   in pipes, cleaning of tubes, 379
      heating and cooling of, 222, 240
      use of chlorine in, 347
   specific volume, of liquid, 482
      of vapor, 482
   thermal conductivity of, 455
   vapor pressure of, 482
   viscosity vs. temperature, 466
Water vapor, radiation due to, 84, 90
   saturated temperature of, 482
White surface, 113
Wien's displacement law, 59
Wien's modification of Planck equation, 101
Wire gauze, 298
Wood, steaming of, 51
Wrought-iron pipe, table of dimensions, 488

Z

Zones, choice of number in radiation, 75